Lecture Notes in Computer Science 1300

Edited by G. Goos, J. Hartmanis and J. van Leeuwen

Advisory Board: W. Brauer   D. Gries   J. Stoer

Springer-Verlag Berlin Heidelberg GmbH

Christian Lengauer   Martin Griebl
Sergei Gorlatch (Eds.)

Euro-Par'97 Parallel Processing

Third International Euro-Par Conference
Passau, Germany, August 26-29, 1997
Proceedings

Springer

Series Editors

Gerhard Goos, Karlsruhe University, Germany

Juris Hartmanis, Cornell University, NY, USA

Jan van Leeuwen, Utrecht University, The Netherlands

Volume Editors

Christian Lengauer
Martin Griebl
Sergei Gorlatch
Universität Passau, Fakultät für Mathematik und Informatik
Innstr. 33, D-94030 Passau, Germany
E-mail: (lengauer/griebl/gorlatch)@fmi.uni-passau.de

Cataloging-in-Publication data applied for

Die Deutsche Bibliothek - CIP-Einheitsaufnahme

Parallel processing : proceedings / Euro-Par '97, Third International
Euro-Par Conference, Passau, Germany, August 26 - 29, 1997.
Christian Lengauer ... (ed.).
(Lecture notes in computer science ; Vol. 1300)
ISBN 978-3-540-63440-9   ISBN 978-3-540-69549-3 (eBook)
DOI 10.1007/978-3-540-69549-3

CR Subject Classification (1991): C.1-4, D.1-4, F.1-2, G.1-2, E.1, H.2

ISSN 0302-9743
ISBN 978-3-540-63440-9

© Springer-Verlag Berlin Heidelberg 1997
Originally published by Springer-Verlag Berlin Heidelberg New York in 1997

Typesetting: Camera-ready by author
SPIN 10546317   06/3142 – 5 4 3 2 1 0   Printed on acid-free paper

Preface

Euro-Par

Euro-Par is the annual European Conference on Parallel Processing. Its purpose is to gather people interested in any aspects of parallel computing and parallel architectures. Euro-Par emerged in 1995 from the former CONPAR-VAPP and PARLE conference series. Euro-Par'95 was held in Stockholm and Euro-Par'96, in the new workshops format, in Lyon.

Euro-Par'97 consists of a large panel of workshops on all aspects of parallel processing, from theory to practice and from academia to industry. These workshops are expected to present the latest advances in their respective domains and are chaired by leading researchers in the field.

The workshop format of Euro-Par was incepted in 1996. As explained in the proceedings of Euro-Par'96, the idea behind it is to attract more tightly knit interest groups, so that the attendees of this large, general-purpose conference can be sure to meet people with similar technical background and interest.

Each workshop is headed by a small programme committee (usually of four people). One member of this committee, the global chair, is reponsible for fleshing out the workshop topic in the call for papers, and for attracting high-quality submissions. Another member, the local chair, is the interface to the local organization team and, possibly, to related workshops of Euro-Par. Typically, the local chair represents the workshop at the programme committee meeting. The other members help in the collection of reviews and in the advertising of the workshop. Thus, each submission to a Euro-Par conference is (or, at least, should be) judged by four different peers.

The workshop papers are divided into three categories. The category determines the length of the paper and of the presentation; a few extra pages (but no extra time) can be purchased:

category	page limit	extra pages	presentation time
distinguished	12	3	30 min.
regular	8	2	30 min.
short	4	1	15 min.

The conference commences with a day of three-hour high-level tutorials whose aim is to expose the audience to practical topics of wide interest in parallel processing. The workshops follow, framed by one-hour invited talks which should summarize recent developments in research.

Euro-Par'97

Twenty workshops called for papers for Euro-Par'97. New this year are the topics *Object-Oriented Programming, Programming Models and Methods* and *Real-Time Systems and Constraints*. We added also a separate workshop on

ESPRIT projects. After several merges, 16 workshops remained, which were distributed across 6 parallel tracks over four half days, the other two half days remaining for the 6 centrally presented invited talks.

As in the year before, submissions were almost exclusively electronic (except for 3 papers); reviewing was exclusively electronic. We received 297 submissions for the 20 regular workshops, and we collected 1101 reviews. That is an average of 3.7 reviews per submission. Accepted and appearing in this proceedings are 154 papers in all: 9 papers in the distinguished category (5.9%), 94 papers in the regular category (61%), and 51 papers in the short category (33.1%). Distinguished papers are indicated in the table of contents with a superscribed asterisk behind the title, short papers with a superscribed dagger.

It is Euro-Par policy that members of the programme committee should exercise restraint in submitting papers. We received 10 submissions (3.4%), of which 6 were accepted as regular and 1 as short.

Four papers ended up in a different workshop than the one they were submitted to.

Submissions were received from 43 different countries; 28 countries are represented in the conference. The three leading countries with almost equally many accepted papers are Germany (24), the UK (23) and France (22). The U.S. have a similar count.[1] The submission count of all other countries is in the single digits.

Acknowledgements

A conference composed of 20 workshops depends on the help and good will of a lot of people.

First, we would like to thank our helpers at the University of Passau: the Euro-Par'97 secretary, Ulrike Peiker, and the secretary of the Chair for Programming, Johanna Bucur, who handled the correspondence and some of the local arrangements. Much of the programming and electronic data management was carried out by student assistants: Sven Anders designed our Web pages and implemented them together with Andreas Dischinger and Oliver Nyderle. Michael Erl was our "grip", and filled this rôle with energy and reliability. A special *thank you* goes to our colleagues Ulrike Lechner for preventing a catastrophe by a successful restore of accidentally deleted data, and Nils Ellmenreich for helping instantly every time it was necessary, not only in system management. Klaus Schießl and Detlef Menzel, the technicians of our department, gave advice and assistance in the set-up and maintenance of projection devices and computer equipment during the conference.

Of essential help were the experience and connections of Hildegard Buchhart and Joan Maria Brown from the Events Office of the University of Passau (HA5). Without this office, we would not have bid for Euro-Par'97.

---

[1] This count is harder to determine, since the Euro-Par statistics software sorts by the email suffix of the submitter, and there is no suffix for the U.S.

Next, we would like to thank the Euro-Par'97 programme committee, composed of 86 members, especially the local and global chairs who identified with the idea of their workshops and of Euro-Par and who kept on soliciting submissions and pushing for reviews, or doing them themselves—and the 415 reviewers, who are not members of the programme committee and who helped us attain an exceptional average number of reviews per paper.

Of the Euro-Par steering committee, Ron Perrott was tireless in helping to make policy decisions as the planning of Euro-Par'97 progressed. He also helped in representing Euro-Par'97 at the EC. The team of Euro-Par'96 at the LIP of ENS Lyon, Luc Bougé, Pierre Fraigniaud, Anne Mignotte, and Yves Robert, gave us detailed advice, based on their experience. Also, crucial to the success of our management was the Euro-Par software for handling submissions and reviews, which originated at KTH in 1995 and was rewritten and extended to the workshop format by Luc Bougé.

We would also like to thank the sponsors. The Deutsche Forschungsgemeinschaft (DFG) and the European Commission (EC) granted us a generous amount. Springer-Verlag gave us a good deal on the proceedings. The University of Passau supported us with money and advice. Commercial sponsors include NAG, the Parsys Group, Visual Numerics, Oberösterreichisches Bankhaus Passau, Pustet University Bookstore, Sparkasse Passau, and Sun Microsystems.

Finally, we would like to welcome the attendants of the meeting on "HPCN in Developing Countries", whose papers are not part of this proceedings, to Euro-Par'97.

Passau, June 1997   Christian Lengauer, Martin Griebl and Sergei Gorlatch

Euro-Par Steering Committee

Ron Perrott, Chair (Queen's University Belfast, UK), r.perrott@qub.ac.uk
Emilio Zapata, Vice Chair (University of Malaga, Spain),
    ezapata@atc.ctima.uma.es
Luc Bougé (ENS Lyon, France), bouge@lip.ens-lyon.fr
Paul Feautrier (University of Versailles, France), paul.feautrier@prism.uvsq.fr
Lucio Grandinetti (University of Calabria, Italy), lugran@ccusc1.unical.it
Seif Haridi (SICS, Sweden), seif@sics.se
Peter Kacsuk (KFKI, Hungary), kacsuk@sunserv.kfki.hu
Christian Lengauer (University of Passau, Germany),
    lengauer@fmi.uni-passau.de
Karl Dieter Reinartz (University of Erlangen, Germany),
    reinartz@informatik.uni-erlangen.de
Paul Spirakis (CTI, Greece), spirakis@cti.gr
Marian Vajtersic (Slovak Academy, Slovakia), marian@ifi.savba.sk
Richard Wait (MidSweden University, Sweden), richard@midgard.nts.mh.se
Makoto Amamiya (Kyushu University, Japan), amamiya@is.kyushu-u.ac.jp
Agnes Bradier (EC, Belgium), agnes.bradier@dg3.cec.be
Ian Foster (Argonne National Lab, USA), foster@mcs.anl.gov

Euro-Par'97 Programme Committee

Workshop 01: Support Tools and Environments

Global Chair: Bernard Tourancheau (ENS Lyon, France),
    Bernard.Tourancheau@lip.ens-lyon.fr
Local Chair: Thomas Ludwig (TU Munich, Germany),
    ludwig@informatik.tu-muenchen.de
Vice-Chairs:
- Helmar Burkhart (University of Basel, Switzerland), burkhart@ifi.unibas.ch
- Allen Malony (University of Oregon, USA), malony@cs.uoregon.edu

Workshop 02: Routing and Communication
in Interconnection Networks

Global Chair: Chita R. Das (Pennsylvania State University, USA),
    das@cse.psu.edu
Local Chair: Ernst W. Mayr (TU Munich, Germany),
    mayr@informatik.tu-muenchen.de
Vice-Chairs:
- Pierre Fraigniaud (ENS Lyon, France), Pierre.Fraigniaud@ens-lyon.fr
- Abhiram Ranade (I.I.T. Bombay, India), ranade@cse.iitb.ernet.in

Workshop 03: Automatic Parallelization and High-Performance Compilers

Global Chair: Yves Robert (ENS Lyon, France), yrobert@lip.ens-lyon.fr
Local Chair: Jean-François Collard (University of Versailles, France),
    Jean-Francois.Collard@prism.uvsq.fr
Vice-Chairs:
- Bill Pugh (University of Maryland, USA), pugh@cs.umd.edu
- Jingling Xue (University of New England, Australia),
    xue@zermelo.une.edu.au

Workshop 04+08+13: Parallel and Distributed Algorithms

Global Chairs:
- Clyde Kruskal (University of Maryland, USA), kruskal@cs.umd.edu
- Keith Marzullo (UC San Diego, USA), marzullo@cs.ucsd.edu
- Cynthia Phillips (Sandia National Laboratories, USA), caphill@cs.sandia.gov
Local Chairs:
- Frans Kaashoek (Massachusetts Institute of Technology, USA,),
    kaashoek@pdos.lcs.mit.edu
- Michael Kaufmann (University of Tübingen, Germany),
    mk@informatik.uni-tuebingen.de
- Klaus-Jörn Lange (University of Tübingen, Germany),
    lange@informatik.uni-tuebingen.de
- Friedemann Mattern (TU Darmstadt, Germany),
    mattern@informatik.th-darmstadt.de
Vice-Chairs:
- Özalp Babaoğlu (University of Bologna, Italy), ozalp@cs.unibo.it
- Fran Berman (UC San Diego, USA), berman@cs.ucsd.edu
- Friedhelm Meyer auf der Heide (University of Paderborn, Germany),
    fmadh@uni-paderborn.de
- Bill McColl (Oxford University, UK), mccoll@comlab.ox.ac.uk
- Paul Spirakis (Computer Technology Institute, Greece), spirakis@cti.gr
- Shang-Hua Teng (University of Minnesota, USA), steng@cs.umn.edu

Workshop 05+06: Programming Languages and Concurrent Object-Oriented Programming

Global Chairs:
- Ron Perrott (Queen's University, Belfast, UK), R.Perrott@qub.ac.uk
- Gul Agha (University of Illinois, USA), Agha@cs.uiuc.edu
Local Chairs:
- Luc Bougé (ENS Lyon, France), bouge@lip.ens-lyon.fr
- Martin Wirsing (University of Munich, Germany),
    wirsing@informatik.uni-muenchen.de

Vice-Chairs:
- José Fiadeiro (University of Lisbon, Portugal), llf@di.fc.ul.pt
- Ian Foster (Argonne National Lab, USA), foster@mcs.anl.gov
- Seif Haridi (SICS, Sweden), seif@sics.se
- Benjamin Pierce (University of Indiana, USA), pierce@cs.indiana.edu

Workshop 07: Programming Models and Methods

Global Chair: David Skillicorn (Queen's University, Canada),
    skill@qucis.queensu.ca
Local Chair: Helmuth Partsch (University of Ulm, Germany),
    partsch@informatik.uni-ulm.de
Vice-Chairs:
- Daniel Le Métayer (IRISA, Rennes, France), lemetayer@irisa.fr
- Jan Prins (UNC Chapel Hill, USA), prins@cs.unc.edu

Workshop 09: Parallel Numerical Algorithms

Global Chair: Ulrich Langer (University of Linz, Austria),
    ulanger@numa.uni-linz.ac.at
Local Chair: Hans-Joachim Bungartz (TU Munich, Germany),
    bungartz@informatik.tu-muenchen.de
Vice-Chairs:
- David E. Keyes (ICASE and Old Dominion University, USA),
    keyes@cs.odu.edu
- Marian Vajtersic (Slovak Academy, Slovakia), marian@par.univie.ac.at

Workshop 10+11+14: Parallel Computer Architecture and Image Processing

Global Chairs:
- Patrice Quinton (IRISA-CNRS, Rennes, France), Patrice.Quinton@irisa.fr
- Per Stenström (Chalmers University, Sweden), pers@ce.chalmers.se
- Lothar Thiele (ETH Zurich, Switzerland), thiele@tik.ee.ethz.ch
Local Chairs:
- Peter Marwedel (University of Dortmund, Germany),
    marwedel@ls12s.informatik.uni-dortmund.de
- Karl Dieter Reinartz (University of Erlangen, Germany),
    reinartz@informatik.uni-erlangen.de
- Hartmut Schmeck (University of Karlsruhe, Germany),
    schmeck@aifb.uni-karlsruhe.de

Vice-Chairs:
- Ed F. Deprettere (Delft University of Technology, The Netherlands),
  ed@cas.et.tudelft.nl
- Nikil Dutt (UC Irvine, USA), dutt@ics.uci.edu
- Edward A. Lee (UC Berkeley, USA), eal@eecs.berkeley.edu
- André Seznec (IRISA, Rennes, France), seznec@irisa.fr
- David Snelling (Fujitsu European Centre for Information Technology),
  snelling@fecit.co.uk

Workshop 12: Applications of High-Performance Computing

Global Chair: John Murphy (British Aerospace, UK),
  John.Murphy@src.bae.co.uk
Local Chair: Wolfgang Gentzsch (GENIAS Software, Germany),
  gentzsch@genias.de
Vice-Chairs:
- Frédéric Desprez (ENS Lyon & INRIA Rhone-Alpes, France),
  desprez@lip.ens-lyon.fr
- Walter Tichy (University of Karlsruhe, Germany), tichy@info.uni-karlsruhe.de

Workshop 15: Scheduling and Load Balancing

Global Chair: Vipin Kumar (University of Minnesota, USA), kumar@cs.umn.edu
Local Chair: Reinhard Lüling (University of Paderborn, Germany),
  rl@uni-paderborn.de
Vice-Chairs:
- Takashi Chikayama (University of Tokyo, Japan),
  chikayama@logos.t.u-tokyo.ac.jp
- Catherine Roucairol (University of Versailles, France),
  Catherine.Roucairol@prism.uvsq.fr

Workshop 16: Performance Evaluation and Prediction

Global Chair: Jack Dongarra (University of Tennessee, USA),
  dongarra@cs.utk.edu
Local Chair: Arndt Bode (TU Munich, Germany),
  bode@informatik.tu-muenchen.de
Vice-Chairs:
- Ulrich Herzog (University of Erlangen, Germany),
  herzog@immd7.informatik.uni-erlangen.de
- A. van der Steen (University of Utrecht, The Netherlands),
  A.vanderSteen@fys.ruu.nl

Workshop 17: Instruction-Level Parallelism

Global Chair: Chris Jesshope (Massey University, New Zealand),
  C.R.Jesshope@massey.ac.nz
Local Chair: Damal K. Arvind (University of Edinburgh, UK), dka@dcs.ed.ac.uk

Vice-Chairs:
- Kemal Ebcioglu (IBM Watson, USA), kemal@watson.ibm.com
- Michael Schlansker (Hewlett-Packard, USA), schlansk@hplabs.hpl.hp.com
- Michael Smith (Harvard University, USA), smith@eecs.harvard.edu

Workshop 18: Parallel and Distributed Database Systems

Global Chair: Andreas Reuter (University of Stuttgart, Germany),
    Andreas.Reuter@informatik.uni-stuttgart.de
Local Chair: Burkhard Freitag (University of Passau, Germany),
    freitag@fmi.uni-passau.de
Vice-Chairs:
- Martin Kersten (CWI, Amsterdam, The Netherlands), Martin.Kersten@cwi.nl
- M. Tamer Özsu (University of Alberta, Canada), ozsu@cs.ualberta.ca

Workshop 19: Symbolic Computation

Global Chair: Manuel Hermenegildo (TU Madrid (UPM), Spain),
    herme@clip.dia.fi.upm.es
Local Chair: Hoon Hong (RISC Linz, Austria), Hoon.Hong@risc.uni-linz.ac.at
Vice-Chairs:
- Kevin Hammond (University of St. Andrews, UK), kh@dcs.st-and.ac.uk
- Wolfgang Küchlin (University of Tübingen, Germany),
    kuechlin@informatik.uni-tuebingen.de

Workshop 20: Real-Time Systems and Constraints

Global Chair: Gérard Berry (Ecole des Mines, France), berry@cma.cma.fr
Local Chair: Hans-Jürgen Siegert (TU Munich, Germany),
    siegert@informatik.tu-muenchen.de
Vice-Chairs:
- Rajeev Alur (AT&T Bell Laboratories, USA), alur@research.att.com
- Günter Hommel (TU Berlin, Germany), hommel@cs.tu-berlin.de

Esprit Workshop

Global Chair: Sergei Gorlatch (University of Passau, Germany),
    gorlatch@fmi.uni- passau.de
Local Chair: Christian Lengauer (University of Passau, Germany),
    lengauer@fmi.uni-passau.de
Vice-Chairs:
- Ron Perrott (Queen's University Belfast, UK), r.perrott@qub.ac.uk
- Richard Wait (MidSweden University, Sweden), richard@midgard.nts.mh.se

Euro-Par'97 Referees
(excluding members of the programme committee)

Aavermiddig, Alfons
Achatz, Klaus
Adams, Andrew
Adve, Vikram
Alouini, Ilyes
Anik, Sadun
Arbab, Farhad
Astley, Mark
Attardi, Giuseppe
Axford, Tom
Aylward, Stephen R.
Azevedo, Ana
Bailey, David
Baldoni, Roberto
Banerjee, A.
Barker, Ken
Barklund, Jonas
Barth, Dominique
Barthou, Denis
Bäumker, Armin
Beckman, Pete
Beguelin, Adam
Bellosa, Frank
Berenbrink, Petra
Bermond, Jean-Claude
Berthomé, Pascal
Berthomieu, Bernard
Bierens, Laurens
Bik, Aart
Bischof, Stefan
Blayo, Eric
Blum, Joachim
Blumofe, Bobby
Bodin, François
Boiten, Eerke
Bonacina, Maria Paola
Bono, Viviana
Boulet, Pierre
Boura, Younes
Brandes, Thomas
Bräunl, Thomas
Brownhill, Carrie

Brunie, Lionel
Buffat, Marc
Bündgen, Reinhard
Cabeza, Daniel
Cai, Wentong
Calder, Brad
Calland, Pierre-Yves
Caromel, Denis
Carriero, Nicholas
Carro, Manuel
Casanova, Henri
Catthoor, Francky
Cesari, Giovanni
Chakravarty, Manuel T.
Chang, Chung-yen
Charot, François
Chaumette, Serge
Chen, Gang
Cheriyan, Joseph
Chow, Peter
Ciancarini, Paolo
Clérot, Fabrice
Clint, Maurice
Coelho, Fabien
Cohen, Albert
Cole, Murray
Constantinou, Chris
Cosnard, Michel
Cung, Van-Dat
Dahlgren, Fredrik
Danelutto, Marco
Darte, Alain
Davy, John
De Lyon, Bernard
Debbabi, Mourad
Decker, Thomas
Denzinger, Jörg
Dershowitz, Nachum
D'Hollander, Erik H.
Diekmann, Ralf
Dinechin, Florent de
Dittrich, Wolfgang

Domas, Stéphane
Dominique, Lavenier
Dosch, Walter
Drum, Philipp
Dunigan, Tom
Eckert, Zulah
Eisenbeis, Christine
Erlebach, Thomas
Etiemble, Daniel
Evans, Brian L.
Fang, Niandong
Färber, Georg
Fay, Don
Feeley, Michael
Feldmann, Rainer
Fernau, Henning
Figueira, Silvia
Fischer, Clemens
Fitzpatrick, Stephen
Fleury, Eric
Fox, Geoffrey
Fradet, Pascal
Freytag, Johann-Christoph
Friedetzky, Tom
Fruchtl, Herbert
Galicia, Geroncio
Gavoille, Cyril
Geerling, Max
Gehring, Jörn
Geib, Jean-Marc
Geigher-Hilk, Ralph
Gengler, Marc
Gerbessiotis, Alexandros
German, Reinhard
Geser, Alfons
Giavitto, Jean-Louis
Gibbons, Jeremy
Gibbons, Phil
Glauert, John
Grahn, Håkan
Gupta, Sandeep
Gupta, Shail Aditya
Gupta, Vineet
Gutzmann, Michael M.
Hächler, Guido

Hackstadt, Steven
Haenssgen, Stefan
Hains, Gaetan
Halang, Wolfgang
Han, Eui-Hong
Hansen, Lars
Hardwick, Jonathan
Harmer, Terence
Harrison, Peter
Harrop, Christopher
Hart, William
Hatcher, Phil
Heindl, Armin
Hellwagner, Hermann
Helm, B. Robert
Herley, Kieran
Heun, Volker
Heusdens, Richard
Heydemann, Marie-Claude
Hill, Jonathan
Ho, C.-T. Howard
Hoeflinger, Jay
Hoffmann, Rolf
Hollingsworth, Jeff
Hromkovic, Juraj
Hsu, Tsan-sheng
Hübsch, Volker
Huckle, Thomas
Hummel, Joe
Hwang, Yuan-Shin
Irigoin, François
Jagadeesan, Lalita
Jamali, Nadeem
Jay, Barry
Jegou, Yvon
Johnson, Richard
Jonker, Pieter
Joshi, Mahesh
Ju, Roy
Juurlink, Ben
Kale, L.V.
Karabatis, George
Karanjkar, Sushrut
Karlsson, J.S.
Karlsson, Magnus

Karlsson, Roland
Kearns, Phil
Keller, Gabriele
Kelly, Paul H.J.
Kelly, Wayne
Kemper, Alfons
Kennaway, Richard
Kenyon, Claire
Kienhuis, Bart
Kilpatrick, Peter
Kim, WooYoung
Kirchner, Claude
Klasing, Ralf
Kohn, Markus
Konig, Jean-Claude
Kotsis, Gabriele
Kranzlmueller, Dieter
Krizanc, Danny
Kshemkalyani, Ajay
Ktari, Bechir
Kucera, Ludek
Kuchen, Herbert
Kuhlmann, Thomas
Kuhn, Walter
Kunde, Manfred
Lagendijk, Reginald
Lanfear, Timothy
Lano, Kevin
Lavenier, Dominique
Leberecht, Markus
Lederer, Edgar
Lefèvre, Laurent
Lefurgy, Charles
Leinberger, William
Leopold, Claudia
Leung, Ho-fung
Li, Kei Chun
Lindenstrauss, Naomi
Linton, Steve
Lisper, Björn
Lo, Virginia
Loewenstein, Paul
Loi, Michel
Loidl, Hans-Wolfgang
López-García, Pedro

Lusk, Ewing
Mäder, Roman
Maggs, Bruce
Mahlke, Scott A.
Maier, Ursula
Mainwaring, Alan
Malhotra, Dalvinder S.
Malumbres, M.P.
Mariño, Julio
Marro, Jean-Louis
Marzetta, Ambros
McCanny, John
McParland, Patrick J.
Meister, Gerd
Mémin, Étienne
Merker, Renate
Middendorf, Martin
Mills, Peter
Mitschang, Bernhard
Mohapatra, Prasant
Mohr, Bernd
Möller, Bernhard
Moore, Reagan
Moreno-Navarro, Juan José
Morse, Bryan
Mudge, Trevor
Mueller, Fritz
Mullins, Robert
Murao, Hirokazu
Murthy, Praveen
Mutka, Matt
Nagar, Shailabh
Nagle, David
Naim, Oscar
Namyst, Raymond
Nett, Edgar
Netzer, Robert
Neubacher, Andreas
Niedermeier, Rolf
Nitsche, Thomas
Nyland, Lars
O'Donnell, John
Oosterlee, Kees
Opatrny, Jarda
Orlando, Salvatore

Paap, Hans-Georg
Palermo, Daniel
Panda, Preeti
Pande, Santosh
Papadopoulos, Philip
Pasquale, Joseph
Pazat, Jean-Louis
Pelagatti, Susanna
Pelc, Andrzej
Pellegrini, François
Penttonen, Martti
Pérez, Christian
Perraudeau, Laurent
Peters, Randal
Petiton, Serge
Peyton Jones, Simon
Pfaffinger, Alexander
Philippsen, Michael
Plank, James
Prechelt, Lutz
Preis, Robert
Pretot, Gerald
Priol, Thierry
Prylli, Loic
Rajopadhye, Sanjay
Ramme, Friedhelm
Randriamaro, Cyril
Rangaswami, Roopa
Ratschan, Stefan
Rau, Bob
Ravada, Sivakumar
Raynal, Michel
Reed, Dan
Reekie, John
Reinhardt, Klaus
Ren, Shangping
Rendl, Franz
Reymann, Olivier
Richter, Harald
Riely, James
Rieping, Ingo
Rinaldo, Roberto
Rinard, Martin
Risset, Tanguy
Rivière, Michel

Roantree, Donal
Roever, Willem-Paul de
Rogers, Owen
Roman, Jean
Rönngren, Robert
Rover, Diane
Rowse, David
Rüb, Werner
Sabry, Amr
Sagnol, David
Sahlin, Dan
Saini, Subhash
Sanders, Peter
Sands, David
Scharff, Christelle
Schlansker, Michael
Schloegel, Kirk
Schmitt, Klaus
Schnieder, Eckehard
Schnoebelen, Philippe
Schramm, Andreas
Schreiber, Robert
Schreiner, Wolfgang
Schröder, Heiko
Schrott, Gerhard
Schulte, Wolfram
Schwabe, Eric
Shende, Sameer
Sheth, Amit
Sibeyn, Jop
Silva, António Rito
Simons, Martin
Sinclair, Robert
Sivasubramaniam, Anand
Slowik, Adrian
Sodan, Angela
Sotelo-Salazar, Salvador
Sotteau, Dominique
Spezialetti, Madalene
Spring, Neil
Stadtherr, Hans
Stellner, Georg
Ster, Mircea Chis
Stewart, Alan
Strohmaier, Erich

Südholt, Mario
Suel, Torsten
Sulzmann, Martin
Sunderam, Vaidy
Sussman, Alan
Sussman, Jeremy
Sutcliffe, Geoff
Talia, Domenico
Tang, Peiyi
Tanyi, Benedict
Taveniku, Mikael
Tel, Gerard
Theel, Oliver
Theunis, Rik
Thia-Kime, Gerard
Thiemann, Peter
Thierauf, Thomas
To, Hing Wing
Torres-Rojas, Francisco
Trinitis, Joerg
Trystram, Denis
Tsai, Jenn-Yuan
Ubeda, Stephane
Uht, Augustus
Ujaldon, Manuel
Unger, Walter
Ungerer, Theo
Unrau, Ronald C.
Vaidya, Aniruddha
Varavithya, Vara
Vasconcelos, Vasco

Vasekin, Vladimir
Vassiliadis, Stamatis
Vigouroux, Xavier
Virot, Bernard
Vivien, Frédéric
Voruganti, Kaladhar
Vullinghs, Ton
Waldby, James
Walker, David
Warschko, Thomas
Waveren, Matthijs van
Weikum, Gerhard
Whittle, John
Wilhelm, Reinhard
Wilkerson, Daniel
Wilson, Robert
Wismueller, Roland
Wolf, Odile
Wong, Kam-Fai
Worley, Patrick
Wray, Paul
Yahmadi, Imed
Yang, Tao
Young, James
Zehendner, Eberhard
Zijal, Robert
Zimmer, Stefan
Zimmermann, Armin
Zimmermann, Wolf

Contents

Workshop 04+08+13:
Parallel and Distributed Algorithms   375

Workshop 05+06:
Programming Languages and Concurrent Object-Oriented
Programming   505

Workshop 07:
Programming Models and Methods   609

Workshop 09:
Parallel Numerical Algorithms 683

Workshop 10+11+14:
Parallel Computer Architecture and Image Processing 761

Workshop 12:
Applications of High-Performance Computing   825

Workshop 15:
Scheduling and Load Balancing 877

Workshop 18:
Parallel and Distributed Database Systems   1113

Workshop 19:
Symbolic Computation   1165

Workshop 20:
Real-Time Systems and Constraints   1227

Workshop 09

Parallel Numerical
Algorithms

Workshop 09: Parallel Numerical Algorithms

Hans-Joachim Bungartz

During the last years, new developments in both high-performance computers and networks and in numerical software have led to a considerable gain in
importance of numerical simulation in science and engineering. Today, in many
scientific fields, new discoveries, insight, or knowledge are based on both theoretical and experimental results and on numerical simulations. The proceedings
of Workshop 12 (Applications of High-Performance Computing) in this volume
illustrate the increasing importance and industrial relevance of numerical simulation techniques.

For most of the computational work arising in high-performance scientific
computing, robust and efficient parallel algorithms for the basic problems of
numerical mathematics like the solution of large sparse linear systems are crucial.
Therefore, it seems to be more than adequate for a conference like Euro-Par'97
to pick out these numerical kernel tasks as a central theme – by an invited
presentation on iterative algorithms on high-performance architectures given by
Ulrich Rüde as well as by a special workshop on parallel numerical algorithms.
Thus, Workshop 09 has been conceived to be a forum for the presentation and
discussion of new developments in this area and to cover all aspects from the
algorithmic idea and the software prototyping to the efficient implementation on
modern parallel architectures and, finally, to the performance analysis.

Obviously, aspects of parallelization have led to a new valuation of numerical
algorithms. Concerning the classical iterative schemes for systems of linear equations, e.g., the Jacobi iteration has gained in attractiveness in comparison with
the Gauß-Seidel scheme due to its advantages with respect to parallelization. On
the other hand, good parallelization properties can certainly not be the sole reason for a revival of methods that have already been sorted out in the sequential
world. Consequently, parallelization efforts should primarily focus on sequential
algorithms that represent the state of the art. Furthermore, although today's research is still concentrated on the development and optimization of parallelized
implementations of existing algorithms and code, the design of numerical algorithms that are developed in a parallel context from the very beginning and
which, thus, are not the result of an explicit parallelization process, will surely
become a crucial issue in the near future.

Though the development of parallel numerical software, today, seems to belong more to the applied mathematics or scientific computing community than
to be a central topic of computer science, the response to our announcement
was very lively. Out of all contributions being submitted to this workshop, nine
have been accepted for the conference and the proceedings: one distinguished, six
regular, and two short papers. The contributions cover both direct and iterative
methods, both kernel tasks from numerical linear algebra (like the computation
of eigenvalues) and applications (like computational fluid dynamics), and the

topics discussed range from more theoretical issues up to implementations on different parallel computers. In order to fit into Euro-Par's schedule, our workshop has been organized in two sessions, one on numerical linear algebra and the other one on partial differential equations and optimization.

In the first session, which deals with standard tasks from numerical linear algebra, the distinguished paper of T. Rauber, G. Rünger, and C. Scholtes studies the degree of parallelism, the scalability, and the scheduling overhead for different sparse Cholesky factorization approaches on a shared memory machine with up to 4096 processors. Direct solvers are also considered in the paper of E. Santos, where tridiagonal linear systems are in the centre of interest. Here, asymptotic bounds on the execution time on the LogP model are given for two algorithms, the odd-even cyclic reduction and the prefix summing. M. Szularz, J. Weston, and M. Clint deal with the efficient parallel computation of closely clustered eigenvalues. Two single-vector Lanczos algorithms based on an explicit restarting strategy are analysed. Numerical results on a Connection Machine are given for several test matrices from the Harwell-Boeing collection. The short paper of A. Kiper is probably more of a theoretical interest. Here, it is shown how a standard LAL^T matrix transformation with $n \times n$-matrices A (symmetric) and L (lower triangular) can be done in $O(\log(n))$ parallel time on a shared memory machine, if $O(\frac{n^3+n^2}{2})$ intermediate storage and $O(\frac{n^3+n^2}{2})$ processors are provided. Finally, the second short paper of this session by T. Rossi and J. Toivanen presents a parallel fast direct solver for systems of linear equations. It is based on a divide-and-conquer approach for systems resulting from finite element discretizations of the two-dimensional Poisson equation. Numerical experiments on a Cray T3E show the efficiency of this approach.

The second session contains four papers with more application-oriented topics. The first three contributions deal with the parallelization of solvers for partial differential equations, and the fourth one studies parallel least squares algorithms. The paper of G. Haase taxonomizes two data-parallel preconditioners for elliptic partial differential equations with a classification scheme based on different possibilities of how to store the interface data on distributed memory machines: in a redundant way (full information on each subdomain or processor, resp.) or partially (only a subdomain's contribution is stored on this subdomain). The consequences of these different data partition schemes for an efficient parallel matrix-vector multiplication – the preconditioners' algorithmic kernel – are studied in detail. W. Huber compares the parallelization properties of the combination technique with standard domain decomposition approaches for problems from computational fluid dynamics. By combining linearly discrete solutions calculated on different standard (coarse) grids, a sparse grid solution of a full (fine) grid approximation quality can be obtained with a significant reduction of the number of necessary degrees of freedom. Thus, via its efficient sparse grid discretization scheme and its easily parallelizable algorithmic structure, the combination technique opens the way to tackle problems like the direct numerical simulation of turbulence. The paper of A. Frommer, T. Lippert, and K. Schilling deals with a parallelization scheme based on the so-called Eisenstat-trick for

SSOR-preconditioned Krylov subspace iterative solvers. The different algorithmic variants presented are applied to problems resulting from lattice gauge theory computations in quantum chromodynamics and are compared with respect to the number of iterations necessary. Finally, T. Steihaug and Y. Yalçinkaya report on the application of asynchronous relaxations to least squares minimization problems. The effect of several parameters in these group iterative methods is studied for sparse matrices from the Harwell-Boeing collection. Especially, it turns out that some kind of synchronization is necessary to avoid deteriorating convergence rates, if the number of processors (and, thus, the influence of old information) is increased. For the numerical experiments of this contribution, an Intel Paragon has been used.

The various interesting and promising results in the above-mentioned contributions to our workshop show that the field of parallel numerical mathematics is a very active one, and we are already looking forward to the submissions concerning this topic to next year's Euro-Par conference.

Scalability of Parallel Sparse Cholesky Factorization

Thomas Rauber[1], Gudula Rünger[2], Carsten Scholtes[2]

[1] Institut für Informatik,Universität Halle-Wittenberg, Kurt-Mothes-Str.1, 06120 Halle (Saale), Germany, rauber@informatik.uni–halle.de
[2] Fachbereich Informatik, Universität des Saarlandes, PF 151150, 66041 Saarbrücken, Germany, ruenger@cs.uni–sb.de, scholtes@cs.uni–sb.de

Abstract. A variety of different algorithms has been proposed for sparse Cholesky factorization. This article investigates shared-memory implementations for several variants in a task-oriented execution model with dynamic scheduling. We concentrate on the degree of parallelism, the scalability, and the scheduling overhead of the different algorithms for relatively large numbers of processors. As execution platform, we use the SB-PRAM, a shared-memory machine with up to 4096 processors. The article can be considered as a case study in which we try to answer the question which performance we can hope to get for a typical irregular application on an ideal machine on which the locality of memory accesses can be ignored but for which the overhead for the management of data structures still takes effect.

1 Introduction

Large sparse systems of linear equations arise in many applications from science and engineering. If the system is described by a symmetric positive definite matrix, Cholesky factorization can be used to solve the system. Algorithms for the Cholesky factorization of sparse matrices are irregular algorithms due to data access patterns that depend on the sparsity structure of the input matrix and that cannot be analyzed at compile time. Investigations for an efficient execution of sparse Cholesky factorization on different execution platforms include simple memory systems [5], hierarchical-memory systems [11], vector supercomputers [2], and parallel systems with shared-memory and distributed-memory organization [10, 8]. Most investigations for parallel systems consider machines with only a small or medium number of processors. But investigations for larger numbers of processors are especially important as there is considerable effort to build such machines. For hundreds or thousands of processors, it is essential to increase the exploitable degree of parallelism which represents an upper bound for the achievable speedup. Moreover, sequential portions of the parallel implementations, the granularity of the computations, and load balancing issues have a much larger influence on the efficiency than for smaller numbers of processors.

In this article, we consider the efficiency of different algorithms for sparse Cholesky factorization on parallel machines with a large number of processors. These algorithms include left-looking, right-looking, and supernodal variants. In particular, we investigate the inherent degree of parallelism which determines the number of independent subcomputations that can be executed in parallel. The investigations are executed with a simulator of the SB-PRAM [1]. The machine provides a global shared memory with uniform access time, i.e., from a virtual processor's point of view, an access to the global memory takes the same time as an arithmetic operation, independently of the memory location that is addressed. Because of this memory organization, locality properties can be ignored and the investigations can concentrate on the exploitation of the maximum degree of parallelism and scalability properties. This provides an upper bound on the attainable speedup, which also takes into consideration the overhead for shared data structures and the task management overhead.

The investigations show that the dependence of the efficiency on the number of executing processors is quite different for different parallel algorithms for Cholesky factorization and that for different numbers of processors, a different parallel implementation shows the best efficiency. For a small or medium number of processors, the supernodal algorithm leads to the smallest execution times because of having the smallest execution overhead. For a large number of processors, the column-based left-looking and right-looking algorithms are better than the supernodal algorithm because of a better exploitation of the available degree of parallelism better.

The rest of the article is organized as follows. Section 2 introduces sequential algorithms for sparse Cholesky factorization and outlines their degree of parallelism. Section 3 gives an overview of the programming model and the execution platform. Section 4 describes the parallel implementations. Section 5 presents experiments and Section 6 concludes.

2   Sequential Sparse Cholesky Factorization

For a symmetric and positive definite $n \times n$-matrix $A \in I\!R^{n \times n}$ there exists a unique triangular factorization $A = LL^T$ where L is a lower triangular matrix with positive diagonal elements and L^T denotes the transposed matrix of L [13]. Using this factorization, the solution x of a system of equations $Ax = b$ with $b \in I\!R^n$ is determined in two steps by solving the triangular systems $Ly = b$ and $L^T x = y$ one after another. A Cholesky factorization yields the factorization $A = LL^T$ for a given matrix A [5]. Here we consider the Cholesky algorithm that computes $L = (l_{ij})_{i=0,..n-1, j=i,..n-1}$ from $A = (a_{ij})_{i,j=0,...,n-1}$ column by column from left to right. For dense matrices A, computing the Cholesky factorization requires $O(n^2)$ storage space and $O(n^3/6)$ arithmetic operations [13]. For sparse matrices, drastic reductions in storage and execution time can be achieved by storing and computing only the nonzero entries of A.

According to [8], we denote the sparsity structure of column j and row i of L (excluding diagonal entries) by $Struct(L_{*j}) = \{k > j : l_{kj} \neq 0\}$ and $Struct(L_{i*}) = \{k < i : l_{ik} \neq 0\}$. $Struct(L_{*j})$ contains the row indices of all nonzeros of column j of matrix A; $Struct(L_{i*})$ contains the column indices of all nonzeros of row i. The *left-looking* algorithm for the Cholesky factorization of A computes the columns one after another from left to right in the following way:

(I)
```
left_cholesky =
    for j = 0, ..., n − 1 do {
        for each k ∈ Struct(L_{j*}) do: cmod(j, k);
        cdiv(j); }
```

The entries of column j are modified after all columns *to the left* of j have completely been computed. Procedure cmod(j, k) subtracts a multiple of column k from column j; procedure cdiv(j) divides column j by the square root of its diagonal entry [8, 11].

```
cmod(j,k) = for each i ∈ Struct(L*k) with i ≥ j do: aij = aij − ljklik ;
cdiv(j) = ljj = √ajj ;
           for each i ∈ Struct(L*j) do: lij = aij/ljj ;
```

The same destination column j is used for a number of consecutive cmod() operations. An alternative way is to use the entries of column j after the complete computation of j to modify all columns k *to the right* of j, that depend on column j, i.e., $k \in Struct(L_{*j})$. This results in the *right-looking* algorithm which uses the same source column j for a number of consecutive cmod() operations.

(II)

```
right_cholesky =
       for j = 0, ..., n − 1 do {
           cdiv(j);
           for each k ∈ Struct(L*j) do : cmod(k, j); }
```

An important concept in sparse Cholesky factorization is that of supernodes [8, 11]. A supernode is a set $I(p) = \{p, p+1, \ldots, p+q-1\}$ of contiguous columns in L with the following property for i with $p \leq i \leq p + q - 2$:

$$Struct(L_{*i}) = Struct(L_{*(p+q-1)}) \cup \{i+1, \ldots, p+q-1\},$$

A supernode has a dense triangular block above (and including) row $p+q-1$ and an identical sparsity structure for each column below row $p+q-1$, see Figure 1 for an example. An important property of a supernode is that each member column modifies the same set of destination columns outside its supernode [11]. Thus, the factorization can be expressed in terms of supernodes modifying columns. The procedure smod(j,J) modifies column j with all columns of supernode J:

```
smod(j, J) = r = min{j − 1, last(J)};
             for k = first(J), ..., r do: cmod(j, k);
```

where $first(J) = p$ and $last(J) = p+q-1$ for a supernode $J = \{p, p+1, \ldots, p+q-1\}$. For column j belonging to the supernode J only the columns of J to the left of j are used to modify j. This leads to the following right-looking supernodal computation scheme:

(III)

```
supernode_cholesky =
       for each supernode J do from left to right {
           cdiv(first(J));
           for j = first(J) + 1, ..., last(J) do {
               smod(j, J); cdiv(j); }
           for k ∈ Struct(L*(last(J))) do: smod(k, J);
       }
```

For the computation of the first column of a supernode J only a cdiv() operation is necessary; the modification with all columns to the left are already performed. The columns of J are computed in a left-looking way, i.e., column j is computed by first modifying it with all columns of J to the left of j and then performing a cdiv() operation. After the computation of all columns of J, all columns to the right of J are modified with each column in J. An alternative would be a right-looking computation of the columns of J. An advantage of the supernodal algorithm lies in an increased locality of memory accesses.

Sources of Parallelism: There are several sources of parallelism in the algorithms $(I), (II), (III)$. The sparsity structure of L may lead to an additional source of parallelism when different columns (and the columns having effect on them) have disjoint sparsity structures. The data dependencies that still exist

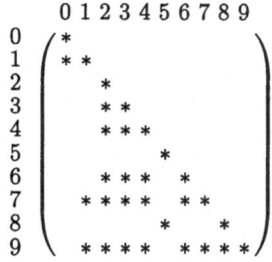
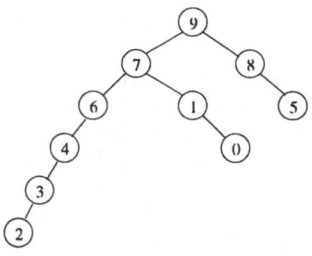

Fig. 1. Matrix L with supernodes $I(0) = \{0\}$, $I(1) = \{1\}$, $I(2) = \{2,3,4\}$, $I(5) = \{5\}$, $I(6) = \{6,7\}$, $I(8) = \{8,9\}$ and corresponding elimination tree.

can be described by *elimination trees* [8, 7]. According to [8], for each column j, $0 \le j < n$, we define: $parent(j) = min\{i \mid i \in Struct(L_{*j})\}$ if $Struct(L_{*j}) \ne \emptyset$, i.e., $parent(j)$ is the row index of the first off-diagonal nonzero in column j. If $Struct(L_{*j}) = \emptyset$, we define $parent(j) = j$. Moreover we define for $0 \le j < n$: $children(i) = \{j < i \mid parent(j) = i\}$ i.e., $children(i)$ contains all columns j that have their first off-diagonal nonzero in row i.

The elimination forest of L is a directed graph $T = (V, E)$ with one node for each column, i.e., $V = \{0, \ldots, n-1\}$. There is an edge $(i, j) \in E$ iff $i = parent(j)$ and $i \ne j$. T is a forest because i is uniquely defined for each j. T is a tree, if matrix A is *irreducible*[3] which we assume without loss of generality. Figure 1 shows a matrix and its corresponding elimination tree. In the following, we denote the subtree with root j by $T[j]$. The elimination tree T for a matrix A specifies the order in which the columns must be evaluated: the definition of *parent* implies that column i must be evaluated before column j, if $j = parent(i)$. Column j does not depend on any column that is not in $T[j]$ and it can be computed in parallel with columns i if $T[i]$ and $T[j]$ are disjoint subtrees.

For a single `cmod(j,k)` or `cdiv(j)` operation modifying column j, the manipulations of the entries of column j are independent from each other and provide fine-grained parallelism. In the left-looking scheme (I), the computation of column j requires the modification of j with all columns $k \in Struct(L_{j*})$. This can be done in parallel only if the target machine allows the concurrent manipulation of a shared variable by several processors. Otherwise, column j must be protected by a locking mechanism as described below. In the right-looking scheme (II), the modification of the columns in $Struct(L_{*j})$ by column j are data independent. But the target columns in $Struct(L_{*j})$ must be locked to prevent a concurrent manipulation of the same column by different processors performing manipulations by other columns than j. In the supernodal scheme (III), the modifications described by one `smod(k,J)` operation all modify the same column k and therefore require a concurrent manipulation mechanism or a locking mechanism for a parallel execution. The columns of a supernode J have to be computed sequentially from left to right because the computation of column $j \in J$ requires the computation of all columns of J to the left of j. When performing the modification of the columns of supernode J in a right-looking way, the modification of the columns to the right of j can be done in parallel.

---

[3] Matrix A is called reducible, if A can be permuted such that it is block-diagonal. For A reducible, the blocks can be factorized independently.

3 Programming Model and Execution Platform

A task-oriented programming model is used for the implementation. Simultaneous accesses to shared data structures are usually protected by a locking mechanism which ensures that at any time only one processor manipulates the data structure. To access a data structure that can also be accessed by other tasks at the same time, a task executes a LOCK operation. Only the task initiating the lock operation is allowed to manipulate the corresponding data structure. All other tasks that try to access the data structure have to wait until the active task executes an UNLOCK operation. If different tasks simultaneously execute LOCK operations, the access to the critical data structure is granted only to one task T and all other tasks have to wait until T executes an UNLOCK operation.

As execution platform, we use the SB-PRAM [1] on which the memory is accessed as a linear shared memory. Each physical processor simulates a fixed number v of *virtual* processors in a pipelined way, i.e., after having executed a machine instruction of a virtual processor i, a physical processor performs a context switch and executes the next machine instruction of virtual processor $(i+1) \bmod v$. The number of virtual processors per physical processor is chosen such that, with respect to the cycle time of a virtual processor, the latency of the network is hidden with one delay cycle for a load and no delay cycle for a store operation. The simulation of virtual processors by physical processors is hidden from the programmer and the SB-PRAM appears as a parallel machine with $p \cdot v$ (virtual) processors and a global shared memory with a uniform memory access time. Currently, a prototype of the SB-PRAM with 128 physical processors each simulating 32 virtual processors is under construction.

Besides the usual load and store operations to access memory cells, the SB-PRAM also offers *multiprefix* instructions which enable several processors to perform simple operations on a memory cell in parallel which we illustrate briefly with a multiprefix addition MPADD. Let $p_1, \ldots, p_n$ be the executing processors where processor p_i contributes a local value o_i. Let s be a shared memory cell with value o. If $p_1, \ldots, p_n$ execute the MPADD operation synchronously, i.e., processor p_i executes $\mathrm{MPADD}(s, o_i)$, then after the operation, processor p_j holds value $o + \sum_{i=1}^{j-1} o_i$, and s contains the value $o + \sum_{i=1}^{n} o_i$. A multiprefix operation is performed in two time units, independently of the number of participating processors. Often the locking mechanism for a concurrent access to shared data structures can be avoided by using appropriate multiprefix operations.

4 Parallel Implementation

A sparse lower triangular matrix L is stored in a compressed storage scheme of size $O(n + nz)$ where n is the number of rows (or columns) in L and nz is the number of nonzeros. We adopt the storage scheme of the SPLASH implementation which, according to [6], stores a sparse matrix in a compressed manner similar to [4]. This storage scheme allows a fast access to columns and sparsity structures but results in irregular data access pattern, see [9] for details.

4.1 Parallel left-looking algorithm

The parallel implementation of the left-looking algorithm (I) is based on *column tasks* $Tcol(j)$, $0 \le j < n$, comprising the execution of $\mathrm{cmod}(j, k)$ for all

```
parallel_left_cholesky =
    c = 0;
    while ((j = MPADD(c, 1)) < n) {
        for i = 0, ..., |Struct(L_{j*})| - 1 do {
            while (S_j empty) wait();
            k = pop(S_j);
            cmod(j, k); }
        cdiv(j);
        for (i ∈ Struct(L_{*j})) do: push(j, S_i);
    }
```

Fig. 2. Parallel left-looking algorithm according to variant L2.

$k \in Struct(L_{j*})$ and the execution of cdiv(j). The parallel implementation maintains a central pool of column tasks. The assignment of tasks to processors is dynamic, i.e., when a processor is idle, it takes a task from the task pool.

We consider three implementation variants. Variant L1 inserts column task $Tcol(j)$ into the task pool not before all column tasks $Tcol(k)$, $k \in Struct(L_{j*})$, have been finished. Hence, a processor that has accessed task $Tcol(j)$ can execute the task without having to wait for other tasks to be finished. Variant L2 allows the execution of $Tcol(j)$ to start without requiring that it can be executed to completion immediately. The task pool is initialized with all column tasks available. Column tasks are accessed by the processors dynamically from left to right, i.e., an idle processor accesses the next column that has not yet been assigned to a processor. On the SB-PRAM, this scheduling is implemented by an array containing the column tasks. To take a column out of the task pool, a processor accesses the array via a shared counter c by executing an MPADD(c,1) operation. The counter is initialized with 0. To control the execution of the column tasks, each column j is assigned a parallel stack S_j containing all columns $k \in Struct(L_{j*})$ for which cmod(j, k) can already be executed. When a processor finishes the execution of the column task $Tcol(k)$ (by executing cdiv(k)), it pushes k onto the stack S_j for each $j \in Struct(L_{*k})$. Because different processors might try to access the same stack at the same time, a *parallel* stack must be used. The processor executing $Tcol(j)$ pops columns k from S_j and executes cmod(j, k). If S_j is empty, the processor waits for another processor to insert new columns. Task $Tcol(j)$ is completed by a cdiv(j) operation, when all $|Struct(L_{j*})|$ cmod() operations have been executed. Figure 2 shows the corresponding implementation.

Variant L3 is a variation of L2 that takes the structure of the elimination tree into consideration. The columns are not assigned strictly from left to right to the processors, but according to their height in the elimination tree, i.e., the children of a column j in the elimination tree are assigned to processors before their father j. This variant tries to complete the column tasks in the order in which the columns are needed for the completion of the other columns.

4.2 Parallel right-looking algorithm

The parallel implementation of the right-looking algorithm (II) is based on column tasks $Tcol(j)$, $0 \le j < n$, comprising the execution of cdiv(j) and cmod(k,j) for all $k \in Struct(L_{*j})$. A task $Tcol(j)$ is inserted into a parallel task pool as soon as the operations cmod(j,k) for all $k \in Struct(L_{j*})$ are executed.

```
parallel_right_cholesky =
        c = 0;
        while ((j = MPADD(c, 1)) < n) {
            while (task pool empty) wait();
            j = get_column(j);
            cdiv(j);
            for (k ∈ Struct(L_{*j})) {
                LOCK(k); cmod(k, j); UNLOCK(k);
                if (MPADD(c_k, 1) + 1 == |Struct(L_{k*})|) add_column(k); }
        }
```

Fig. 3. Parallel right-looking algorithm.

At the beginning, all tasks $Tcol(j)$ that correspond to leaves in the elimination tree are inserted into the task pool. Because the number of tasks inserted to and removed from the task pool is fixed, the task pool can also be implemented as an array T of size n with a global read counter R and a global write counter W.

Task scheduling is implemented by maintaining a counter c_j for each column j. The counter is initialized with 0 and is incremented after the execution of each $\texttt{cmod}(j, *)$ operation by an $MPADD(c_j, 1)$ operation. For the execution of a $\texttt{cmod}(k, j)$ operation by a task $Tcol(j)$, column k must be locked to prevent other tasks from modifying the same column at the same time. Figure 3 sketches the corresponding implementation.

The difference between the right-looking implementation and the left-looking variant L2 lies in the execution order of the $\texttt{cmod}()$ operations and in the executing processor: In the L2 variant, the operation $\texttt{cmod}(j, k)$ is initiated by the processor computing column k by pushing it onto stack S_j, but the operation is executed by the processor computing column j. This execution needs not to be performed immediately after the initiation of the operation. In the right-looking variant, the operation $\texttt{cmod}(j, k)$ is not only initiated but also executed by the processor that computes column k.

4.3 Parallel supernodal algorithm

The parallel implementation of the supernode algorithm (III) is based on supernode tasks where task $Tsup(J)$ for $0 \le J < N$ comprises the execution of $\texttt{smod}(j, J)$ and $\texttt{cdiv}(j)$ for each $j \in J$ from left to right and the execution of $\texttt{smod}(k, J)$ for all $k \in Struct(L_{*(last(J))})$. Tasks $Tsup(J)$ ready for execution are held in a central task pool that is accessed by idle processors. The pool is initialized with supernodes that are ready for completion, i.e., by those supernodes whose first column is a leaf in the elimination tree. If the execution of $\texttt{smod}(k, J)$ operations by a task $Tsup(J)$ finishes the modification of another supernode $K > J$, the task $Tsup(K)$ is inserted into the task pool. Because the number of tasks is fixed, the task pool can be implemented by an array.

The scheduling of supernode tasks is realized by maintaining a counter c_j for each column j of a supernode. Each counter is initialized with 0 and is incremented for each modification that is executed for column j. Ignoring the modifications with columns inside a supernode, a supernode task $Tsup(J)$ is ready for execution, if the counters of the columns $j \in J$ reach the value $|Struct(L_{j*})|$. The column counters c_j are manipulated by an $MPADD(c_j, 1)$ operation to coordinate concurrent accesses by different processors. For the manipulation of a

```
parallel_supernode_cholesky =
    c = 0;
    while ((J = MPADD(c, 1)) < N) {
        while (task pool empty) wait();
        J = get_supernode();
        cdiv(first(J));
        for j = first(J) + 1, ..., last(J) do {
            smod(j, J); cdiv(j); }
        for (k ∈ Struct(L_*(last(J)))) do: {
            LOCK(k); smod(k, J); UNLOCK(k);
            MPADD(c_k, 1);
            if (supernode K to k ready for execution) add_supernode(K); }
    }
```

Fig. 4. Parallel supernodal algorithm.

column $k \notin J$ by a smod(k, J) operation, column k is locked to avoid concurrent manipulation. Figure 4 shows the corresponding implementation.

4.4 Parallel single-modify queue algorithm

The single-modify queue algorithm is based on the right-looking algorithm, but uses separate tasks for each cdiv() or cmod() operation. A task for a cmod(j, k) operation is inserted into the task pool as soon as the source column k has been completely computed. The task pool is initialized with tasks cdiv(j) for all columns j that are leaves in the elimination tree. For the execution of a cmod(j, k) operation, the destination column j must be locked to prevent other processors from modifying column j at the same time. Similar as for the right-looking variant, a counter c_j counts the number of cmod$(j, *)$ operations executed for column j. The counter is incremented by an $MPADD(c_j, 1)$ operation after the completion of each cmod() operation. The processor executing the last modification of a column j is also executes the cdiv(j) operation to complete column j. After the completion of column j, all tasks cmod(k, j) for $k \in Struct(L_{*j})$ are inserted into the task pool. This results in the implementation in Figure 5. Because the number of tasks is fixed (n cdiv() tasks and nz cmod() tasks), the task pool is again implemented as an array.

```
parallel_single_mod_cholesky =
    c = 0;
    while ((j = MPADD(c, 1)) < nz) {
        while (task pool empty) wait();
        get_cmod(j, k);
        LOCK(j); cmod(j, k); UNLOCK(j);
        if (MPADD(c_j, 1) + 1 == |Struct(L_{j*})|) {
            cdiv(j);
            for (k ∈ Struct(L_{*j})) do: add_cmod(k, j); }
    }
```

Fig. 5. Parallel right-looking algorithm with separate cmod() tasks.

5   Experiments and Results

In this section, we consider the application of the parallel algorithms from the last section to sparse matrices of different size, see Table 1. The implementations result from modifications of the supernodal algorithm from the SPLASH application program suite [12] which uses the p4 library for task coordination. The test matrices are taken from the Harwell-Boeing collection [3] and describe problems from structural engineering. The value n ist the number of rows or columns in the matrix, $nz(A)$ is the number of nonzeros below the diagonal in the original (symmetric) matrix A, $nz(L)$ is the number of nonzeros below the diagonal in the resulting matrix L, N is the number of supernodes, $\bar{n}z$ is the average number of nonzeros per column, Δnz is the standard deviation from this value.

matrix	n	$nz(A)$	$nz(L)$	N	$\bar{n}z$	Δnz
BCSSTK14	1806	30824	110461	503	35	9.6
BCSSTK15	3948	56934	647274	1295	30	7.6
BCSSTK29	13992	302748	1680804	3231	44	16.0

Table 1. Characteristics of test problems.

Figure 6 compares the runtimes for different implementations of the algorithms for the matrix BCSSTK14. The implementations include the left-looking variants L1, L2, L3, the right-looking variant RL, the single modify-queue variant MQ and and three versions of the supernodal algorithms S0, S1, S2. S0 represents the supernodal algorithm with a central, lock-protected task queue whereas S1 represents an algorithm that uses the array implementation of a task queue as described in Subsection 4.2. S1 uses a fixed number of lock variables, S2 uses a lock variable for each column.

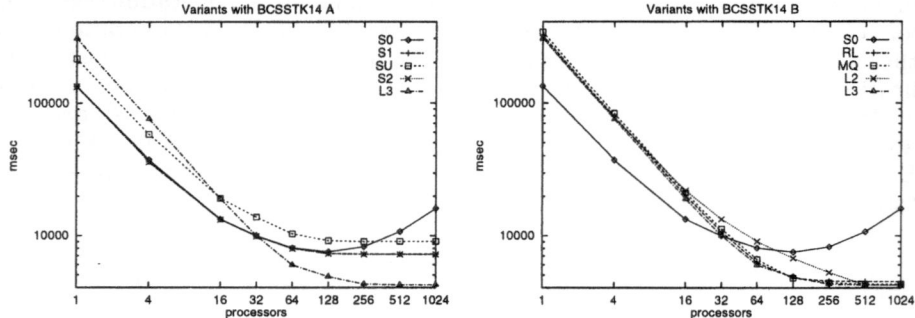

Fig. 6. Comparison of runtimes of the different algorithms for BCSSTK14.

Figures 7, 8, and 9 compare the runtimes and speedup values of the most promising variants applied to the different matrices. The speedup values are obtained by a comparison with a sequential version of the supernodal algorithm which is the fastest sequential algorithm among the investigated variants. Table 2 shows the speedup values obtained for the largest number of processors (1024). The absolute speedup values are not very large, because the sequential version of the supernodal algorithm is up to a factor of 2.5 faster than the other algorithms.

The left-looking variant L1 only reaches limited speedup values of about 2. The speedup values do not increase with the number of processors. Moreover, the sequential variant of L1 requires a larger execution time than the original supernodal algorithm because of a larger task management overhead (L1 uses n tasks whereas the supernodal implementation uses $N < n$ tasks). The limited speedup values indicate that L1 does not fully exploit the available degree of parallelism. This might caused by sequentializations that result from inserting only those tasks $Tcol(j)$ into the task queue which can be executed to completion. Variant L1 will not be considered further in this section. The left-looking variant L2 can be considered as an overlapped version of L1 which does not delay the execution of the cmod() operations if the destination column has already been assigned to a processor. The overlapping has a dramatic effect on the attained speedup, because the modification of a column can be started as soon as the first column it depends on is available. The sequential execution time of L2 is larger than the sequential execution time of L1 because L2 requires maintaining the column stacks for controlling the overlapped execution. The sequential version of L2 is about four times slower than the sequential version of the supernodal algorithm. L2 may still cause sequentializations because a processor computing column j may be idle waiting for another processor to finish its computation. Variant L3 tries to minimize the overhead of these sequentializations by assigning the columns according to their height in the elimination tree to the processors. This leads to smaller execution times for a medium number of processors (between 16 and 256) because of less waiting time. For larger numbers of processors, the positive effect of the ordering vanishes, because L2 starts the execution of enough columns such that most of the leaves of the elimination tree are among them.

Also the right-looking algorithm RL has a larger sequential execution time than the supernodal algorithm because of a larger task management overhead. The execution time is about the same as for the L3 variant. For larger numbers of processors, the attainable speedup is limited because the probability that different processors try to modify the same destination column at the same time increases with the number of processors. Moreover, the processors may have to wait for new tasks to be created and sequentializations occur.

The supernodal algorithm has the smallest sequential execution time, because of a small overhead for the task management. For the parallel version, a parallel task queue that allows concurrent accesses by different processors has only a small advantage over a central, lock-protected task queue. The reason is that the tasks are large and that the processors do not have to access the task queue very often. So, only a few sequentializations occur because of locking. Since the destination column k of a smod(k, J) operation must be locked, there should be a lock variable available for each column. Providing only a fixed number of lock variables leads to performance reduction for larger numbers of processors because processors may have to wait for a free lock variable. For smaller numbers of processors (up to 32 for BCSST14), the supernodal algorithm has the smallest

Speedups	S0	S2	RL	MQ	L3
BCSSTK14	8.2944	18.6977	29.9380	31.4240	31.8718
BCSSTK15	32.6159	41.1327	66.5425	78.8768	80.9700
BCSSTK29	45.2442	52.7331	86.0342	98.4819	100.9220

Table 2. Speedup values of the different algorithms for 1024 processors.

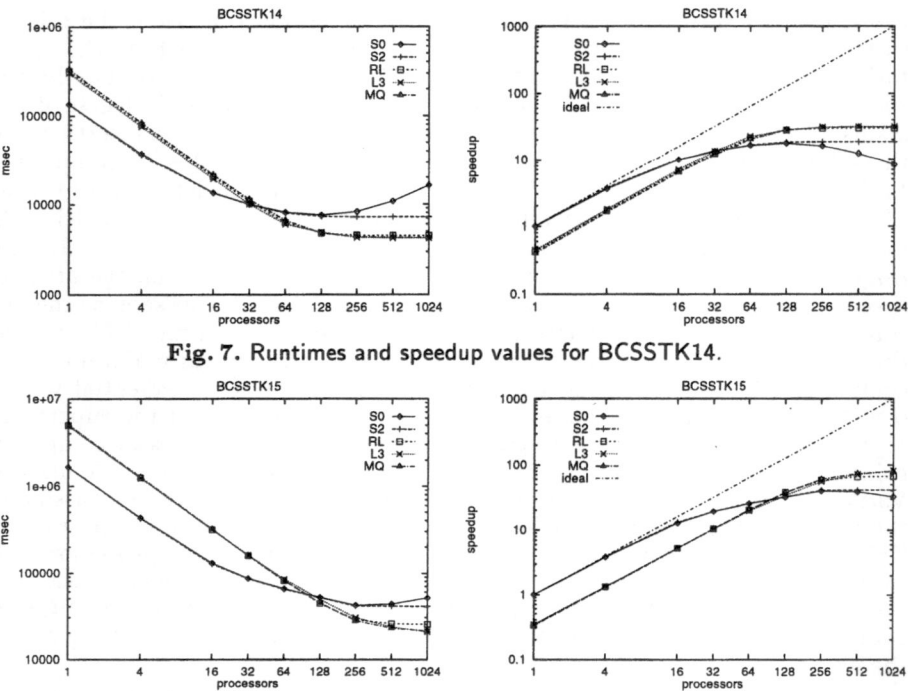

Fig. 7. Runtimes and speedup values for BCSSTK14.

Fig. 8. Runtimes and speedup values for BCSSTK15.

parallel execution time; for larger numbers of processors (more than 128 for BCSST14), the left-looking and right-looking algorithms are faster because of the larger potential of parallelism due to smaller tasks.

The modify-queue (MQ) algorithm which uses single cmod() operations as tasks has about the same sequential execution time as the L3 variant, because both require the execution of a queue or stack operation for each cmod() operation. The parallel implementation of the MQ algorithm has about the same execution time as the L3 or the right-looking algorithm. Similar to the L3 variant, the algorithm requires $O(n + nz)$ task or stack manipulations. Similar to the right-looking algorithm, the attainable speedup is limited for larger numbers of processors because the probability that different processors try to modify the same destination column at the same time is increasing.

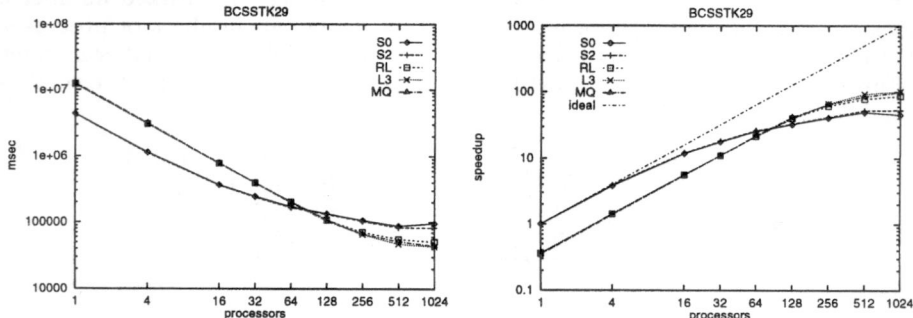

Fig. 9. Runtimes and speedup values for BCSSTK29.

6 Conclusion

The left-looking, right-looking and the supernodal algorithms are variants of sparse Cholesky factorization. In this paper, we have described a variety of different parallel implementations of those algorithms which mainly differ in the task management and task execution order. All parallel variants have been implemented using the p4 library for task coordination and have been applied to test matrices of different sizes. The runtime tests have been performed on a simulator of the SB-PRAM, a shared memory machine with uniform memory access time. The investigations have shown that for larger numbers of processors it is difficult to get large absolute speedup values because fast sequential algorithms do not scale up as well as algorithms that have a considerably larger sequential execution time. The exploitation of the available degree of parallelism for larger numbers of processors with dynamic scheduling requires an expensive task management even for a parallel machine that behaves much like an ideal PRAM machine. Nevertheless, these algorithms are useful because they are - depending on the input matrix - up to a factor of 2 faster than the parallel version of better sequential algorithms and because they at least result in moderate speedup values. The inherent degree of parallelism in this irregular application is large enough for an execution on a large number of processors.

References

1. F. Abolhassan, J. Keller, and W.J. Paul. On the Cost–Effectiveness of PRAMs. In *Proceeding of the 3rd IEEE Symposium on Parallel and Distributed Processing*, pages 2–9, 1991.
2. C. Ashcraft, R.G. Grimes, J.G. Lewis, B.W. Peyton, and H.D. Simon. Recent Progress in Sparse Matrix Methods for Large Linear Systems. *Int. J. Supercomputer Applications*, 1(4):10–30, 1987.
3. I. Duff, R. Grimes, and J. Lewis. User's Guide for the Harwell-Boeing Sparse Matrix Collection (Release I). Technical Report TR/PA/92/86, Boeing Computer Services, 1992.
4. A. George, J. Liu, and E. Ng. User's Guide for SPARSPAK: Waterloo Sparse Linear Equations Package. Technical Report Research Report CS-78-30, Department of Computer Science, University of Waterloo, 1980.
5. A. George and J. W-H Liu. *Computer Solution of Large Sparse Positive Definite Systems*. Prentice-Hall, 1981.
6. D.E. Lenoski and W. Weber. *Scalable Shared-Memory Multiprocessing*. Morgan Kaufmann Publishers, Inc, 1995.
7. Joseph W. H. Liu. The Role of Elimination Trees in Sparse Factorization. *SIAM J. Matrix Anal. Appl.*, 11:134–172, 1990.
8. E.G. Ng and B.W. Peyton. A Supernodal Cholesky Factorization Algorithm for Shared-Memory Multiprocessors. Mathematical Sciences Section, P.O. Box 2009, Bldg. 9207-A, Oak Ridge, TN 37831-8083, 1991.
9. T. Rauber, G. Rünger, and C. Scholtes. Scalability of parallel sparse cholesky factorization. Technical Report 02/97, University Saarbrücken, 1997.
10. E. Rothberg. Alternatives for Solving Sparse Triangular Systems on Distributed-Memory Multiprocessors. *Parallel Computing*, 21:1121–1136, 1995.
11. E. Rothberg and A. Gupta. An Evaluation of Left-Looking, Right-Looking and Multifrontal Approaches to Sparse Cholesky Factorization on Hierarchical-Memory Machines. *Int. J. High Speed Computing*, 5(4):537–593, 1993.
12. J.P. Singh, W.D. Weber, and A. Gupta. SPLASH: Stanford Parallel Applications for Shared-Memory. *Computer Architecture News*, 20(1):5–44, 1992.
13. J. Stoer and R. Bulirsch. *Introduction to Numerical Analysis*. Springer, New York, 1990.

Optimal Parallel Algorithms for Solving Tridiagonal Linear Systems

Eunice E. Santos[1]

Department of Electrical Engineering and Computer Science,
Lehigh University, Bethlehem, PA 18015, USA.
Research partially supported by an NSF CAREER Grant.

Abstract. We consider the problem of solving tridiagonal linear systems on parallel distributed-memory machines. We present tight asymptotic bounds for solving these systems on the LogP model using two very common direct methods : odd-even cyclic reduction and prefix summing. For each method, we begin by presenting lower bounds on execution time for solving tridiagonal linear systems. Specifically, we present lower bounds in which it is assumed that the number of data items per processor is bounded, a general lower bound, and lower bounds for specific data layouts commonly used in designing parallel algorithms to solve tridiagonal linear systems. Moreover, algorithms are provided which have running times within a constant factor of the lower bounds provided. Lastly, the bounds for odd-even cyclic reduction and prefix summing are compared.

1 Introduction

Solving tridiagonal linear systems is a basic problem in scientific computing. This problem is especially useful in solving ordinary or partial differential equations which occur naturally in many applications such as fluid dynamics, plasma physics, etc. In this paper, we consider the problem of solving tridiagonal linear systems using direct methods on distributed-memory machines. Although much research has been spent exploring this problem, most deal with designing and analyzing algorithms that solve these systems on specific types of interconnection networks [1, 7, 9, 10] such as hypercube or butterfly. Although some of these algorithms are believed to be efficient, so far no formal proofs are available to substantiate them. One of the first attempts at establishing a lower bound for solving tridiagonal linear systems using odd-even cyclic reduction on distributed-memory machines was advanced by Johnsson [5] and subsequently by Lakshmivarahan and Dhall [8]. However, their results are applicable only to algorithms designed under severe constraints such as a very specific partitioning of tasks onto processors.

The main objective of this paper is to present asymptotically tight bounds on the running time for solving tridiagonal systems which utilize common direct methods; in particular, odd-even cyclic reduction and prefix summing. The results obtained in this paper will provide not only a means for measuring efficiency of existing algorithms but also provide a means of determining what kinds of data layouts and communication patterns are needed to achieve optimal running times.

In order to make our results applicable to a wide spectrum of distributed-memory machines, we will work within LogP [2], a recently proposed model for parallel distributed-memory machines. Important characteristics of parallel machines can be represented using the parameters in the model. An important feature of LogP is that the interconnection network of the machine is modeled by its performance as viewed by the user, rather than its detailed interconnection structure. Algorithms designed on this model are portable from one distributed-memory machine to another and the running times of these algorithms will vary from machine to machine according to the parameter values associated with these machines.

We shall show, among other things, that using common data layouts and straightforward communication patterns do not result in significantly higher complexities than assuming that all processors have access to all data items regardless of communication pattern. In fact, in most cases, common data layouts and straightforward communication patterns can be used to obtain optimal running times. We shall also show that the communication parameters of a network have a significant effect on the complexity of this problem.

The paper is divided as follows. Section 2 contains a description of the LogP model. In Section 3 we consider the odd-even cyclic reduction method for solving tridiagonal systems. We begin by deriving lower bounds on execution time independent of the data layout for this method. Next, we derive lower bounds for data layouts in which the number of data items per processor is bounded. In particular we will show that the skewness of the distribution of data has no significant effect on complexity. We then derive lower bounds for specific data layouts commonly used in designing parallel algorithms for solving tridiagonal linear systems. Lastly, running times for algorithms which are within a constant factor of the lower bounds are provided. Section 4 considers the prefix summing method and derives similar upper and lower bounds on running time. For brevity, proofs and algorithms are not provided in this paper. Many of the proofs and algorithms can be found in [11]. Section 5 gives the conclusion and summary of results.

2 The LogP Model

LogP [2] is a parallel distributed-memory model that specifies the performance characteristics of an interconnection network without describing the structure of the network. Communication between processors is assumed to be point-to-point. The following description uses terminology specific to the problem on hand. The main parameters of the model are:

P: the number of processor/memory modules.

L: an upper bound on the *latency*, or delay, incurred in communicating a message containing a numerical value from its source module to its target module.

o: the *overhead*, defined as the length of time that a processor is engaged in the transmission or reception of each message; during this time, the processor cannot perform other operations.

g: the *gap*, defined as the minimum time interval between consecutive message transmissions or consecutive message receptions at a processor.[1]

The parameters L, o and g are measured as multiples of a processor cycle. A processor cycle is the (unit) time a processor takes to execute an arithmetic operation not requiring communication. Also, at any time step, at most $\lceil L/g \rceil$ messages can be in transit to or from any particular processor. If an attempt to exceed this limit is made by a processor by transmitting a message, the processor stalls until the message can be sent without exceeding the capacity limit. All algorithms discussed in this paper satisfy this capacity constraint, and we do not mention it henceforth.

3 Odd-Even Cyclic Reduction Method

The Problem: Given $M\mathbf{x} = \mathbf{b}$ solve for $\mathbf{x}$, where M is a tridiagonal $N \times N$ matrix, $\mathbf{b} = (b_j)$ is a vector of size N, and $\mathbf{x} = (x_j)$ a vector of size N.

We assume for both the discussion of odd-even cyclic reduction and prefix summing that $1 < P \leq N$. An algorithm is simply a set of arithmetic operations such that each processor is assigned a sequential list of these operations. An initial assignment of data to the processors is called a data layout. A list of message transmissions and receptions between processors is called a communication pattern. These three components (algorithm, data layout, communication pattern) are needed in order to determine running time.

Odd-even cyclic reduction [3, 4, 9] is a recursive method for solving tridiagonal systems of size $N = 2^n - 1$. This method is divided into two parts: reduction and back substitution.

The first step of reduction is to remove each odd-indexed x_i and create a tridiagonal system of size $2^{n-1} - 1$. We then do the same to this new system and continue on in the same manner until we are left with a system of size 1. This requires n phases. We refer to the tridiagonal matrix of phase j as M^j and the vector as $\mathbf{b}^j$. The original M and $\mathbf{b}$ are denoted M^0 and $\mathbf{b}^0$. The three non-zero items of each row i in M^j are denoted l_i^j, m_i^j, r_i^j (left, middle, right). Below are the list of operations needed to determine the items of row i in matrix M^j.

$$e_i^j = -\frac{l_i^{j-1}}{m_{i-2^{j-1}}^{j-1}}, \qquad f_i^j = -\frac{r_i^{j-1}}{m_{i+2^{j-1}}^{j-1}}, \qquad l_i^j = e_i^j l_{i-2^{j-1}}^{j-1}, \qquad r_i^j = f_i^j r_{i+2^{j-1}}^{j-1},$$

$$m_i^j = m_i^{j-1} + e_i^j r_{i-2^{j-1}}^{j-1} + f_i^j l_{i+2^{j-1}}^{j-1}, \qquad b_i^j = b_i^{j-1} + e_i^j b_{i-2^{j-1}}^{j-1} + f_i^j b_{i+2^{j-1}}^{j-1}$$

Clearly each system is dependent on the previous systems. In this paper, we assume that if an algorithm employs odd-even cyclic reduction, we assume that a processor computed items of whole rows of a matrix (i.e. the three non-zero data items) and the appropriate item in the vector.

---

[1] We assume $o \leq g$.

The back substitution phase is initiated after the system of one equation has been determined. We recursively determine the values of the x_i's. The first operation is $x_{2^n-1} = \frac{b_{2^n-1}^{n-1}}{m_{2^n-1}^{n-1}}$. For the remaining variables x_i, let j denote the last phase in the reduction step before x_i is removed then $x_i = \frac{b_i^{j-1} - l_i^{j-1} x_{i-2^{j-1}} - r_i^{j-1} x_{i+2^{j-1}}}{m_i^{j-1}}$.

The serial complexity of this method is $S(N) = 17N - 12\log(N+1) + 11$.

In Section 3.1, we present lower bounds on running time for odd-even cyclic reduction algorithms. Specifically, this section contains the general lower bounds on running time independent of data layout which are applicable to all algorithms utilizing odd-even cyclic reduction, lower bounds on running time for data layouts in which the number of data items initially assigned to a processor is bounded, and lower bounds on running time for common data layouts for this problem. In Section 3.2 we present optimal algorithm running times.

3.1 Lower bounds for Odd-Even Cyclic Reduction

We begin by listing some definitions necessary for the discussion of lower bounds.

Definition 1. The class of all communication patterns is denoted by $\mathcal{C}$. The class of all data layouts is denoted by $\mathcal{D}$. A data layout D is said to be a *single-item layout* if each non-zero matrix-item is initially assigned to a unique processor.

Definition 2. Let A be an odd-even cyclic reduction algorithm. For $i = 1, 2, \cdots n$ and $j = 1, 2, \cdots 2^{n-i+1} - 1$, define $T_{A,D,C}(i,j)$ to be the minimum time at which the items of row j of level i are computed using algorithm A and communication pattern C and assuming data layout D, and define $T_{A,D,C}(i) = min_{1 \leq j \leq 2^{n-i+1}-1} T_{A,D,C}(i,j)$.

It follows that for all odd-even cyclic algorithms A, data layout D, communication pattern C, and any $i < n$, $T_{A,D,C}(i+1) > T_{A,D,C}(i)$.

Definition 3. Let A be an odd-even cyclic reduction algorithm and let $\mathcal{O}$ denote the class consisting of all odd-even cyclic reduction algorithms. For all $i \leq n$, $T_{A,D}(i) = min_{C \in \mathcal{C}} T_{A,D,C}(i)$, $T_A(i) = min_{D \in \mathcal{D}} T_{A,D}(i)$ and $T_{\mathcal{O}}(i) = min_{A \in \mathcal{O}} T_A(i)$.

In the following sections we shall provide lower bounds on $T_{A,D}(n)$ for algorithms $A \in \mathcal{O}$ and certain types of data layouts D. The lower bounds hold regardless of the choice of communication pattern. For simplicity, we state our results in the special case $L = g$ of the LogP model. Although the proofs of the lower bounds are based on the assumption that $o = 0$, they are clearly applicable to arbitrary o.

3.1.1 A General Lower Bound for Odd-Even Cyclic Reduction

In this subsection we assume the data layout is the one in which each processor has a copy of every non-zero entry of M^0 and b^0. We denote this layout by $\bar{D}$. Since $\bar{D}$ is the most favorable data layout, $T_{\mathcal{O}}(i) = min_{A \in \mathcal{O}} T_{A,\bar{D}}(i)$.

Theorem 4. *Let A be an odd-even cyclic reduction algorithm. The following is a lower bound for A regardless of data layout and communication pattern:*

$$\begin{cases} 14(N-n) = \Omega(N) & \text{if } g \geq 14(N-n) \\ max(g(n-n'), \frac{S(N)}{P}, n) = \Omega(g\log(\frac{N}{g}) + \frac{N}{P} + \log N) & \text{otherwise} \end{cases}$$

where n' is the smallest integer i such that $g < 14(2^i + i - 1)$.

In Section 3.2 we will provide optimal algorithms, i.e. the running times are within a constant factor of the lower bounds.

3.1.2 Lower Bounds for $\mathcal{O}$ on $\frac{c}{P}$-data layouts

Many algorithms designed for solving tridiagonal linear systems assume that the data layout is single-item and that each processor is assigned roughly $\frac{1}{P}^{th}$ of the rows of M^0 where P is the number of processors available. In this section, we consider single-item data layouts in which each processor is assigned at most a fraction $\frac{c}{P}$ of the rows of M where $1 \leq c \leq \frac{P}{2}$.

Definition 5. Consider c where $1 \leq c \leq \frac{P}{2}$. A data layout D on P processors is said to be a $\frac{c}{P}$-*data layout* if D is single-item, no processor is assigned more than a fraction $\frac{c}{P}$ of the rows of M^0, and each processor is assigned at least one row of M^0. Denote the class of $\frac{c}{P}$ data layouts by $\mathcal{D}(\frac{c}{P})$.

Theorem 6. *If $D \in \mathcal{D}(\frac{c}{P})$, then for any $A \in \mathcal{O}$,*

$$T_{A,D}(n) \geq max(g\lceil\log(P-1)\rceil, g\frac{\lfloor\log\frac{N(P-c)}{P(P-1)}\rfloor - 6}{4}, \frac{S(N)}{P}, n) = \Omega(g\log N + \frac{N}{P}).$$

Analyzing the result given in the above theorem, we see that for any odd-even cyclic algorithm and any $\frac{c}{P}$-data layout, the running time (regardless of communication pattern) is $\Omega(ng + \frac{N}{P})$. Comparing results against the general lower bound, we see that for sufficiently large N, the lower bounds are within constant factors of one another. In Section 3.2, we present algorithms that are within constant factors of the bounds.

Since $c \leq \frac{P}{2}$, every $\frac{c}{P}$-data layout is a $\frac{1}{2}$-data layout. This leads to the following corollary:

Corollary 7. *If $D \in \mathcal{D}(\frac{1}{2})$, then for any $A \in \mathcal{O}$,*

$$T_{A,D}(n) \geq max(g\lceil\log(P-1)\rceil, g\frac{\lfloor\log\frac{N}{P-1}\rfloor - 7}{4}, \frac{S(N)}{P}, n).$$

The complexity of any algorithm $A \in \mathcal{O}$ using a $\frac{c}{P}$-data layout is $\Omega(g\log N + \frac{N}{P})$. We see that the "communication part" of the bound grows only logarithmically in problem size N and is independent of $P(> 1)$, whereas the "computation part" grows linear in N and is dependent on P. In addition, although the number of rows assigned to different processors may vary from one to $\frac{1}{2}$ of the total number of initial rows, the above results and the algorithms in Section 3.2 show that the skewness of the distribution of data has no significant effect on complexity.

3.1.3 Lower Bounds for $\mathcal{O}$ on standard data layouts

In this section we present lower bounds on the running time for odd-even cyclic reduction algorithms using specific data layouts commonly used by algorithm designers, namely cyclic and blocked data layouts. Definitions are given below.

Definition 8. A single-item data layout on $(1 \leq)P(\leq N)$ processors $p_1, \cdots p_P$ is *cyclic* if for all $i \leq N$, $r_i^0, m_i^0, l_i^0, b_i^0$ are assigned to processor p_j where $j+1 \equiv i$ mod P. We denote this layout by D_C.

Definition 9. A single-item data layout on $(1 \leq)P(\leq N)$ processors $p_1, \cdots p_P$ is *blocked* if for all $i \leq P$, p_i is assigned the nonzero items in rows $(i-1)\frac{N}{P}+1$ to $i\frac{N}{P}$ of M^0 and $\mathbf{b}^0$. We denote this layout by D_B.

The following definitions are needed for the discussion of lower bounds in this section.

Definition 10. A row j of level i is said to be an *original* row of some processor p if the items of rows $j - 2^{i-1} + 1$ to $j + 2^{i-1} - 1$ of level 0 are originally assigned to p. If a row is not an original of p it is said to be a *non-original* row of p.

Definition 11. Two rows j and $j+1$ of level i are referred to as *neighbors*.

Theorem 12. *Let A be an algorithm in $\mathcal{O}$, i.e. A is an odd-even cyclic reduction algorithm. The following are lower bounds assuming the appropriate data layout:*

$$max(g\lceil \tfrac{n}{2} \rceil, \tfrac{S(N)}{P}, n) \qquad \text{Blocked Data Layout}$$
$$max(g\lceil \tfrac{n}{2} \rceil, g\tfrac{N}{2P}, \tfrac{S(N)}{P}, n) \quad \text{Cyclic Data Layout}.$$

Analyzing these lower bounds, we see that blocked and cyclic data layouts have lower bounds $\Omega(g \log N + \frac{N}{P})$ and $\Omega(g(\log N + \frac{N}{P}))$ respectively. In Section 3.2 we provide an algorithm using blocked data layout whose running time is $O(g \log N + \frac{N}{P})$. This clearly shows that the complexity for blocked data layout is as good as and in most cases better than that of cyclic data layout. Therefore, we are able to formally confirm, for the first time, that the widely held belief that cyclic layout is no better than block layout is indeed true for a wide range of parallel machines.

Comparing the complexity of blocked layout with the lower bounds for $\frac{c}{P}$-data layouts, we see that the complexity of blocked layout is within a constant factor of the lower bounds for $\frac{c}{P}$-data layouts. Furthermore, comparing the complexity of blocked layout with the general lower bound (which we show is achievable up to a constant factor in Section 3.2), we see that for sufficiently large n the complexity of blocked layout is within a constant factor of the general lower bound. Therefore we can use the much more realistic blocked data layout rather than $\bar{D}$ and still achieve the lower bounds (up to a constant factor).

3.2 Algorithms and Communication Patterns

We have designed algorithms and communication patterns where when used with the appropriate data layouts have running times matching the lower bounds presented, i.e. the running times differ from the lower bounds by at most a small constant factor. Below is a table of running times. For brevity, algorithms and communication schedules have been omitted. Full algorithm and communication schedules can be found in [11].

Running Time:	Data Layout
$10n(g + 2o) + \frac{S(N)}{P} = O(g \log N + \frac{N}{P})$	Blocked Layout
$10(n - n')(g + 2o) + max(\frac{S(N)}{P}, 19n)$	$\bar{D}$ – Best Layout
$= O(g \log \frac{N}{g} + \frac{N}{P} + \log N)$	
$S(N) = O(N)$	Serial Algorithm

4 Prefix Summing Method

The Prefix Summing Problem: Given N items $s_1, s_2, \cdots, s_N$ and an operator $\otimes$, determine $S_1, S_2, \cdots S_N$ where $S_i = x_1 \otimes x_2 \otimes x_3 \otimes \cdots \otimes x_i$.

We describe below how we can solve a tridiagonal system of equations by transforming it into a non-commutative prefix summing problem [10]. We assume that $1 < P \leq N$.

We begin by reformulating every equation in the system as a matrix-vector product. Specifically, consider the ith equation $l_i x_{i-1} + m_i x_i + r_i x_{i+1} = b_i$ where $1 \leq i < N$ and $l_1 = 0$. The corresponding matrix-vector product is the following:

$$\begin{pmatrix} x_{i+1} \\ x_i \\ 1 \end{pmatrix} = G_i \begin{pmatrix} x_i \\ x_{i-1} \\ 1 \end{pmatrix} \text{ where } G_i = \begin{pmatrix} -\frac{m_i}{r_i} & -\frac{l_i}{r_i} & \frac{b_i}{r_i} \\ 1 & 0 & 0 \\ 0 & 0 & 1 \end{pmatrix} \text{ (assume } l_1 = 0\text{)}$$

From repeated substitution, we see that

$$\begin{pmatrix} x_{i+1} \\ x_i \\ 1 \end{pmatrix} = H_i \begin{pmatrix} x_1 \\ 0 \\ 1 \end{pmatrix} \text{ where } H_i = G_i \cdots G_2 G_1.$$

To solve the tridiagonal system, after H_{N-1} is computed, we solve the following system:

$$\begin{pmatrix} x_N \\ x_{N-1} \\ 1 \end{pmatrix} = H_{N-1} \begin{pmatrix} x_1 \\ 0 \\ 1 \end{pmatrix}, l_N x_{N-1} + m_N x_N = b_N.$$

Therefore, once H_{N-1} is computed we can compute the value of x_1 in 9 steps. We then determine the remaining values of x_i in a similar fashion.

The values of H_i can be computed by prefix summing where $s_i = G_i$ and $\otimes$ is 3×3 matrix multiplication (an associative, non-commutative operator).

The serial complexity of this method is $\Theta(N)$.

In Section 4.1, we present a general lower bound (independent of data layout) on running time for prefix summing by considering the complexity of summing. Also, a lower bound for single-item layouts (and therefore a lower bound for $\frac{c}{P}$-data layouts) is provided. Furthermore, a lower bound for blocked data layout is provided. In Section 4.2 we present running times for optimal or near optimal algorithms which use specific types of data layouts for prefix summing. One of the data layouts considered is blocked data layout.

4.1 Lower bounds for Prefix Summing

We present lower bounds on running time for prefix summing algorithms. These lower bounds are based on the lower bounds for summation of N items and for broadcasting an item to P processors. Lower bounds for summation, broadcast and prefix summing are presented in [6, 12]. For simplicity, we state our results in the special case $L = g$ of the LogP model.

Theorem 13. *Any algorithm which solves a tridiagonal linear system of size N using prefix summing requires at least $T_{PS}(N, P)$ where*

$$T_{PS}(N, P) = \begin{cases} N - 1, & \text{if } N \leq g \\ max(\frac{g}{2}\log(\frac{N}{g}), \frac{N-1}{P}) & \text{if } N \geq g. \end{cases}$$

Theorem 14. *Any algorithm which solves a tridiagonal linear system of size N using prefix summing and using a single-item data layout requires at least*

$$max(g\log(P-1), \frac{N-1}{P}, \frac{g}{2}\log\frac{N}{g}) = \Omega(g\log P(\frac{N}{g})^{\frac{1}{2}} + \frac{N}{P}).$$

Clearly the lower bound for single-item layouts is the lower bound for $\frac{c}{P}$-data layouts

Theorem 15. *Any algorithm which solves a tridiagonal linear system of size N using prefix summing and using a blocked data layout requires at least $T_{PS,D_B}(N, P)$ where*

$$T_{PS,D_B}(N, P) = max(g\log(P-1), \frac{N-1}{P}, \frac{g}{2}\log\frac{N}{g}) = \Omega(g\log P(\frac{N}{g})^{\frac{1}{2}} + \frac{N}{P}).$$

We see that the lower bound for single-item data layouts (and for $\frac{c}{P}$-data layouts) is within a constant factor of the lower bound for blocked layout. Moreover, when N is sufficiently large $T_{PS,D_B}(N, P) = max((g + 1)\log(P - 1), \frac{N-1}{P}, \frac{g}{2}\log\frac{N}{g}) = \Omega(g\log N + \frac{N}{P})$. Since algorithms are presented in Section 4.2 whose running times are within a constant factor of the lower bounds derived in this section, clearly (1) the complexity for blocked data layout is within a constant factor of the lower bound for $\frac{c}{P}$-data layouts, and (2) for sufficiently large N, the complexity for blocked data layout is within a constant factor of the general lower bound. (3) although the number of rows assigned to different processors may vary from 1 to $N - P + 1$ of the total number of initial rows, the above results and the algorithms presented in the following section show that the skewness of the distribution of data has no significant effect on complexity (in fact, the skewness of distribution that we have proven for prefix summing is greater than the one we've proven for odd-even cyclic reduction).

4.2 Algorithms and Communication Patterns

In this section, we provide a table of running times for algorithms for prefix summing which are within a constant factor of the lower bounds derived in Section 4.1. All of these algorithms are based on the optimal summing, prefix summing and broadcast algorithms presented in [6, 12]. Below is a table of running times. For brevity, algorithms and communication schedules have been omitted. Algorithms and communication schedules can be found in [11].

Running Time:	Data Layout
$O(T_{PS,D_B}(N,P))$	Blocked Layout
$O(T_{PS}(N,P))$	$\bar{D}$ – Best Layout
$N-1$	Serial Algorithms

5 Conclusion

We considered the problem of solving tridiagonal linear systems using two very common direct methods: odd-even cyclic reduction and prefix summing. We were able to derive tight asymptotic bounds on the execution time for these problems and provided algorithms which achieve these bounds.

Specifically, we proved that the complexity for solving tridiagonal linear systems for both methods regardless of data layout is $\Theta(g \log \frac{N}{g} + \frac{N}{P})$ for $N = \Omega(g)$ and $\Theta(N)$ for $N = O(g)$. When we added the realistic assumption that the data layouts are single-item and the number of data items assigned to a processor is bounded, we derived the following bounds for odd-even cyclic reduction and prefix summing respectively, $\Theta(g \log N + \frac{N}{P})$ and $\Theta(g \log P(\frac{N}{g})^{\frac{1}{2}} + \frac{N}{P})$. Analyzing these bounds we see that for sufficiently large N the complexity for solving these systems using odd-even cyclic reduction and prefix summing are both $\Theta(g \log N + \frac{N}{P})$. Next comparing these bounds to one another suggests that except for extreme values of P the decision to choose one method over the other may be based on other factors such as numerical stability. We also see that the skewness of data distribution does not significantly affect the complexity of the problem. Specifically, for odd-even cyclic reduction this result is true for single-item data layouts in which no processor is assigned more than $\frac{1}{2}$ of the rows. For prefix summing, the result holds for data layouts which only need to be single-item. In fact, comparing these bounds with the general lower bounds, we see that restricting the proportion of data items assigned to a processor to $\frac{N}{P}$ does not result in a significantly higher complexity than assuming all processors have all the data items for sufficiently large N.

We also derived bounds for blocked and cyclic data layouts. Although it is widely believed that cyclic data layout is no better than blocked data layout, using our results we were able to formally confirm, for the first time, that the "belief" is indeed valid for a wide range of parallel machines.

Lastly, we show that there are algorithms, data layouts, and communication patterns whose running times are within a constant factor of the lower bounds provided. This provides us with the Θ-bounds stated above. To achieve the general lower bound, i.e. the complexity for these methods regardless of data layout, we use $\bar{D}$ the best data layout, i.e. the data layout in which every processor is assigned all the data items. For the $\frac{1}{2}$-data layout lower bound for odd-even cyclic reduction and for the single-item layout lower bound for prefix summing, we use blocked data layout. Clearly blocked data layouts is more realistic than D and is easy to assign across processors. Also, since for sufficiently large N these two lower bounds are asymptotic to their respective general lower bounds, this makes the algorithms and communication patterns provided for blocked data layout practical and efficient for these methods.

Since all of the optimal algorithms discussed were variants of standard algorithms using straightforward communication patterns, this shows that it is futile to search for sophisticated techniques, and complicated communication patterns to significantly improve the running times of algorithms which solve tridiagonal systems using odd-even cyclic reduction or prefix summing.

Many of the lower bounds obtained in this paper hold for an extended model with multi-broadcast capability, i.e. the lower bounds hold even under the assumption that any value computed by a processor p is available to that processor immediately and to all other processors $L + 2o$ steps later.

References

1. C. Amodio and N. Mastronardi. A parallel version of the cyclic reduction algorithm on a hypercube. *Parallel Computing*, 19, 1993.
2. D. E. Culler, R. M. Karp, D. A. Patterson, A. Sahay, E. Santos, K. E. Schauser, R. Subramonian, and T. von Eicken. LogP: A Practical Model of Parallel Computation. *Communications of the ACM*, May 1996.
3. D. Heller. A survey of parallel algorithms in numerical linear algebra. *SIAM J. Numer. Anal.*, 29(4), 1987.
4. A. W. Hockney and C. R. Jesshope. *Parallel Computers*. Adam-Hilger, 1981.
5. S. L. Johnsson. Solving tridiagonal systems on ensemble architectures. *SIAM J. Sci. Stat. Comput.*, 8, 1987.
6. R. M. Karp, A. Sahay, E. E. Santos, and K.E. Schauser Optimal Broadcast and Summation on the LogP Model. In *Proceedings of the Fifth Annual ACM Symposium on Parallel Algorithms and Architectures*, 1993.
7. S. P. Kumar. Solving tridiagonal systems on the butterfly parallel computer. *International J. Supercomputer Applications*, 3, 1989.
8. S. Lakshmivarahan and S. D. Dhall. A Lower Bound on the Communication Complexity for Solving Linear Tridiagonal Systems on Cube Architectures. In *Hypercubes 1987*, 1987.
9. S. Lakshmivarahan and S. D. Dhall. *Analysis and Design of Parallel Algorithms : Arithmetic and Matrix Problems*. McGraw-Hill, 1990.
10. F. T. Leighton. *Introduction to Parallel Algorithms and Architectures: Arrays-Trees-Hypercubes*. Morgan Kaufmann, 1992.
11. E. E. Santos. Direct methods for solving tridiagonal linear systems in parallel. Technical Report TR-95-029, International Computer Science Institute, 1995.
12. E. E. Santos. Optimal and efficient parallel algorithms for summing and prefix summing. In *Proceedings of the Eighth Annual IEEE Symposium on Parallel and Distributed Processing*, 1996.

Robust Parallel Lanczos Methods for Clustered Eigenvalues*

M. Szularz[1], J. Weston[1], and M. Clint[2]

[1] School of Information & Software Engineering, University of Ulster, Coleraine
BT52 1SA, Northern Ireland
[2] Department of Computer Science, The Queen's University of Belfast, Belfast BT7
1NN, Northern Ireland

Abstract. In this paper two recently proposed single-vector Lanczos methods based on a simple restarting strategy are analysed and their suitability for the computation of closely clustered eigenvalues is evaluated. Both algorithms adopt an approach which yields a fixed k-step restarting scheme in which one eigenpair at a time is computed using a deflation technique in which each Lanczos vector generated is orthogonalized against all previously converged eigenvectors. In the first algorithm each newly generated Lanczos vector is also orthogonalised with respect to all of its predecessors; in the second, a selective orthogonalisation strategy permits re-orthogonalization between the Lanczos vectors to be almost completely eliminated. 'Reverse communication' implementations of the algorithms on an MPP Connection Machine CM-200 with 8K processors are discussed. Advantages of the algorithms include the ease with which they cope with genuinely multiple eigenvalues, their guaranteed convergence and their fixed storage requirements.

Key words : Lanczos, restart, deflation, orthogonalization, MPP.

1 The Lanczos Algorithm

Essentially, the Lanczos algorithm [2], [5] generates a sequence of tridiagonal matrices T_j, each with values $(\alpha_1, \ldots, \alpha_j)$ on its main main diagonal and values $(\beta_1, \ldots, \beta_{j-1})$ on its main sub- and super-diagonals, together with a sequence of orthonormal matrices of Lanczos vectors $Q_j = [q_1, \ldots, q_j]$, where $j \leq n$, such that $Q_j^t A Q = T_j$. It can be shown that Q_j is an orthonormal basis for $\mathcal{K}(A, q_1, j)$, the Krylov subspace of order j generated by A and q_1. Let $X^T A X = \mathrm{diag}(\lambda_1, \ldots, \lambda_n)$ where $\lambda_1 \geq \lambda_2 \geq \cdots \geq \lambda_n$, and where $X = [x_1, \ldots, x_n]$ is orthonormal, be the spectral decomposition of A. Similarly, let $S_j^T T_j S_j = \mathrm{diag}(\theta_1(T_j), \ldots, \theta_j(T_j))$ where $\theta_1 > \theta_2 > \cdots > \theta_j$, and $S_j = (s_{pq})$, be the spectral decomposition of T_j. Then $y_i \in \Re^n$ in $Q_j S_j = Y_j = [y_1, \ldots, y_j]$ is known as the i-th Ritz vector of A for the subspace $range(Q_j)$, and $\theta_i(T_j)$

* This work was supported by the Engineering and Physical Sciences Research Council under grants GR/J41857 and GR/J41864 and was carried out using the facilities of the University of Edinburgh Parallel Computing Centre

is known as the corresponding Ritz value. It can be shown that, for surprisingly small j, the Ritz pair (θ_i, y_i) closely approximates the i-th eigenpair of A, (λ_i, x_i), provided that the ratio $|\lambda_i|/\|A\|_2$ is reasonably close to unity.

1.1 Estimating the Largest Eigenvalue

Let $q_1 \in \Re^n$ be such that $\|q_1\|_2 = 1$. Then the Lanczos algorithm for the computation of the largest eigenvalue *only* may be expressed as follows:

Algorithm 1 function$[(\theta_1, y_1)] = \text{Lanczos}(A, q_1, k)$
    $\beta_0 \leftarrow 1, q_0 \leftarrow 0$
    for $j = 1, 2, \ldots, k$
        $\alpha_j \leftarrow q_j^T A q_j$
        $q_{j+1} \leftarrow (A - \alpha_j I)q_j - \beta_{j-1}q_{j-1}$
        $\beta_j \leftarrow \|q_{j+1}\|_2$
        if $\beta_j = 0$ **then** STOP (eigenvalues of T_j are the j largest eigenvalues of A)
        $q_{j+1} \leftarrow q_{j+1}/\beta_j$
        (A) orthonormalize q_{j+1} against $q_1, \ldots, q_j$
    end_for
    compute (θ_1, y_1).

It follows that the Ritz pair (θ_1, y_1) returned by Algorithm 1 is an approximation to the eigenpair (λ_1, x_1) of A. The accuracy of this approximation may be estimated before y_1 is computed on the basis of the error bounds $|\beta_j||s_{j1}|$ [2], since

$$|\beta_j||s_{j1}| = \|Ay_1 - \theta_1 y_1\|_2 \tag{1}$$

Observe that the orthogonality of the Lanczos vectors has been guaranteed throughout by the inclusion of a complete re-orthogonalization of the Lanczos vectors in each iteration (**Step A**).

A major problem associated with the above algorithm is that k must be chosen to be sufficiently large to guarantee the required accuracy in the solution. Moreover, the value of k required is not known in advance. However, two explicit restarting schemes for the computation of the $p \ll n$ largest eigenvalues of a symmetric matrix A have been developed [8], each of which incorporates a *fixed* k-step variant of Algorithm 1 where k is assumed to be *small*. Brief outlines of these schemes are now given.

2 Lanczos with Explicit Restart: EXPRES1

Let approximated quantities be decorated with the $\hat{\ }$ symbol. Thus, at the j-th Lanczos step, the Ritz pair (θ_i, y_i) is synonymous with $(\hat{\lambda}_i, \hat{x}_i)$. Suppose that the approximated eigenpairs $(\hat{\lambda}_1, \hat{x}_1), \ldots, (\hat{\lambda}_i, \hat{x}_i)$, where $i < p$, are given. Let $\hat{\mathcal{X}}_i = \text{span}\{\hat{x}_1, \ldots, \hat{x}_i\}$ and let $\hat{\mathcal{X}}_i^\perp$ be its orthogonal complement in $\Re^n$. Observe that, if the Lanczos vectors $q_1, \ldots q_k$ are constrained to stay in the subspace $\hat{\mathcal{X}}_i^\perp$, the Lanczos algorithm will converge to the Ritz values of A in the subspace $\hat{\mathcal{X}}_i^\perp$, viz, the desired approximations $\hat{\lambda}_{i+1}, \hat{\lambda}_{i+2},\ldots$. This observation provides the basis for the two fixed k-step restarting schemes described below. The first may be expressed as the following algorithm:

Algorithm 2 function$[(\hat{\lambda}_1, \hat{x}_1),\ldots,(\hat{\lambda}_p, \hat{x}_p)] = \mathrm{EXPRES1}(A, p, k, tol)$
    $\hat{\mathcal{X}}_0 = 0$
    (1) for $i = 1 : p$
        (1) choose $q_1 \notin \hat{\mathcal{X}}_{i-1}$
        (2) $y_1 \leftarrow q_1$; $\theta_1 \leftarrow y_1^t A y_1$
        (3) while ($\frac{\|Ay_1 - \theta_1 y_1\|_2}{|\theta_1|} > tol$)
            (1) $(\theta_1, y_1) \longleftarrow \mathrm{Lanczos}(A, q_1, k)$
            (2) $q_1 \leftarrow y_1$
        end_while
        (4) $(\hat{\lambda}_i, \hat{x}_i) \leftarrow (\theta_1, y_1)$
        (5) if $i < p$ **then** $\hat{\mathcal{X}}_i \leftarrow \hat{\mathcal{X}}_{i-1} \oplus \mathrm{span}\{\hat{x}_i\}$
    end_for

tol is the user supplied tolerance (normally set to **u**, the relative machine accuracy). It is assumed that Algorithm 1, as used in **Step (1.3.1)** above, is modified to include a mechanism for projecting each newly generated Lanczos vector (including q_1) into $\hat{\mathcal{X}}_i^\perp \neq 0$. This can obviously be achieved by the explicit orthogonalization of each q_j against $\hat{x}_1,\ldots,\hat{x}_i$. Algorithm 2 as described above is henceforth referred to as EXPRES1. Observe that, since k is completely independent of p, it can always be chosen to be small.

Clearly, since the i-th eigenvalue of A is sought in the subspace $\hat{\mathcal{X}}_{i-1}^\perp$, Algorithm 2 is theoretically ideal for coping with closely clustered and genuinely multiple eigenvalues of A. This paper provides numerical evidence that this is indeed the case in practice. Further, it is demonstrated that, in many cases where closely clustered and genuinely multiple eigenvalues occur, Algorithm 2 is significantly more efficient than Sorensen's state-of-the-art routine when implemented in a massively parallel SIMD environment.

3 Lanczos with Explicit Restart: EXPRES2

A finely tuned version of Algorithm 1 has also been developed for use with Algorithm 2 in which the *complete* re-orthogonalization of the Lanczos vectors (**Step A**) is replaced by a *selective* re-orthogonalization strategy [5], [6]. Thus, as soon as the error bound associated with a Ritz vector satisfies

$$| \beta_j | \, | \, s_{ji} | \le \sqrt{\mathbf{u}} \, \| A \|_2 \tag{2}$$

the Lanczos process is immediately restarted with the current value of y_1, even when $j < k$. Further, if y_1 triggered the restart, all subsequent q_{j+1} computed are not only projected into $\hat{\mathcal{X}}_i^\perp \neq 0$ but are also orthogonalized against the recently computed, 'converging' Ritz vector y_1. This orthogonalization strategy *purges* all unwanted, 'converging' Ritz vectors, $y_i : i \neq 1$, from the system and considerably reduces the computational overhead associated with the orthogonalisation process. An outline of this version of Algorithm 1 is given below:

Algorithm 3 function$[(\theta_1, y_1)] = $ Lanczos$(A, q_1, k, converging)$

$\quad \beta_0 \leftarrow 1, q_0 \leftarrow 0$

$\quad$ **(1) for** $j = 1, 2, \ldots, k$

$\quad\quad$ **(1) if** $i \neq 0$ **then** orthogonalize q_j against $\hat{x}_1, \ldots \hat{x}_i$

$\quad\quad$ **(2)** $\alpha_j \leftarrow q_j^T A q_j$

$\quad\quad$ **(3)** $q_{j+1} \leftarrow (A - \alpha_j I)q_j - \beta_{j-1}q_{j-1}$

$\quad\quad$ **(4)** $\beta_j \leftarrow \| q_{j+1} \|_2$

$\quad\quad$ **(5) if** $\beta_j = 0$ **then** STOP (eigenvalues of T_j are the j largest eigenvalues of A)

$\quad\quad$ **(6)** $q_{j+1} \leftarrow q_{j+1}/\beta_j$

$\quad\quad$ **(7) if** $converging = false$ **then**

$\quad\quad\quad$ **(1)** compute $s_{j1}, \ldots, s_{jj}$

$\quad\quad\quad$ **(2) if** $\min\{| \beta_j \| s_{j1} |, \ldots, | \beta_j \| s_{jj} |\} = | \beta_j \| s_{jr} | \leq \sqrt{u} \| A \|_2$

$\quad\quad\quad$ **then**

$\quad\quad\quad\quad$ **(1)** compute (θ_1, y_1)

$\quad\quad\quad\quad$ **(2) if** $r = 1$ **then** $converging \leftarrow true$

$\quad\quad\quad\quad$ **(3) exit**

$\quad\quad$ **else**

$\quad\quad\quad$ **(3)** orthogonalize q_{j+1} against q_1

$\quad\quad$ **end_if**

$\quad$ **end_for**

$\quad$ **(2)** compute (θ_1, y_1).

The variant of Algorithm 2 which incorporates the finely tuned version of Algorithm 1 above is referred to as EXPRES2. Note also that this variant requires $converging \leftarrow false$ to be added to Step **(1.2)** in Algorithm 2 and Step **(1.3.1)** to be replaced by $(\theta_1, y_1) \longleftarrow $ Lanczos$(A, q_1, k, converging)$. Further, it can be established formally that each of the restart schemes, EXPRES1 and EXPRES2, guarantees convergence to the p required eigenvalues of A.

4 Implementation on the Connection Machine CM-200

The algorithms have been implemented using a *Reverse Communication* strategy [7], [3] in which the user is responsible for supplying the code for all matrix-vector products of the form Aq_j. Such an approach enables the user to take advantage of the architecture of the target machine and of any available highly optimised code when implementing these computationally expensive operations. Thus, all vectors of length n are declared and stored as distributed CM arrays, whereas all other quantities are confined to the 'front-end' machine. It follows that all massive *saxpy* type operations involving the n-vectors are *fine-grain* parallel operations and, consequently, they are performed on the CM itself using the highly optimised matrix-vector product routines which are available in the *Connection*

Machine Scientific Support Library. These operations include, for example, the dense matrix-vector products in the reorthogonalization steps **(1.1)** and **(1.7.3)** of Algorithm 3 and they occur also in the computation of the Ritz vectors in steps **(2)** and **(1.7.2.1)** of this algorithm. In contrast, the remaining computations, viz. those involving the matrices T_j, are computationally inexpensive and are performed on the 'front-end' machine, thereby enabling heavy use to be made of functions from the BLAS library.

5 Numerical Experience

The performances of the restarting algorithms, EXPRES1 and EXPRES2, have been compared with the performance of the appropriate driver program calling the symmetric Arnoldi (Lanczos) routine 'SSAUPD' from the ARPACK library [4]. Single precision CM-200 versions of the three methods, SSAUPD, EXPRES1 and EXPRES2, have been constructed and numerical experiments using a variety of matrices with closely clustered and/or multiple eigenvalues taken from real world applications have been conducted. Test results for four sparse matrices

	PLAT1919			NOS7		
	$n = 1,919$ $(nz = 17159)$ $\lambda_1 = 2.9216371$ $\lambda_{64} = 1.3470327$			$n = 729$ $(nz = 2,673)$ $\lambda_1 = 9.86403 \times 10^6$ $\lambda_{64} = 9.62953 \times 10^2$		
p	SSAUPD $k = \max(16, 2p)$	EXPRES1 $k = 32$	EXPRES2 $k = 32$	SSAUPD $k = \max(16, 2p)$	EXPRES1 $k = 32$	EXPRES2 $k = 32$
1	1.5 (18%)	1.1 (63%)	0.4 (93%)	0.7 (21%)	0.4 (41%)	0.4 (53%)
2	3.3 (22%)	2.2 (62%)	0.9 (92%)	1.4 (18%)	8.1 (42%)	2.5 (83%)
4	3.9 (18%)	3.7 (59%)	2.2 (90%)	4.2 (11%)	24.0 (33%)	3.2 (82%)
8	8.9 (11%)	8.1 (55%)	5.1 (88%)	5.1 (8%)	30.3 (40%)	4.8 (79%)
16	28.1 (5%)	21.9 (50%)	11.9 (81%)	8.5 (4%)	78.9 (34%)	24.1 (65%)
32	123.3 (2%)	53.2 (43%)	29.0 (71%)	18.9 (3%)	107.5 (31%)	48.3 (56%)
64	568.8 ($< 1\%$)	162.4 (33%)	90.7 (62%)	192.8 (1%)	208.2 (24%)	85.0 (79%)
128				1816.4 ($\approx 0\%$)	585.9 (15%)	274.5 (28%)
	6lu	11u	14u	$11\sqrt{u}$	$4\sqrt{u}$	$17\sqrt{u}$

Table 1. Time (in seconds) for matrices PLAT1919 and NOS7 from the PLATZ and LANPRO collections, respectively (figures in brackets show the percentage of total time taken for the computation of matrix-vector products).

selected from the Harwell-Boeing Sparse Matrix Collection [1] where p ranges from 1 to 128 are presented in Tables 1 and 2. In Table 1 the matrix PLAT1919 provides a well known difficult sparse symmetric eigenproblem whose eigenvalues occur in pairs (except for an isolated singleton at zero); the condition number of matrix NOS7 in the same table has the value 1.8×10^9. The matrix BCSSTK19

	BCSSTK19			BCSSTK25		
	$n = 817$ $(nz = 3835)$ $\lambda_1 = 1.92216 \times 10^{15}$ $\lambda_{64} = 6.12629 \times 10^{14}$			$n = 15,439$ $(nz = 133840)$ $\lambda_1 = 1.06002 \times 10^{16}$ $\lambda_{32} = 4.03627 \times 10^{13}$		
p	SSAUPD $k = \max(16, 2p)$	EXPRES1 $k = 32$	EXPRES2 $k = 32$	SSAUPD $k = 32$	EXPRES1 $k = 32$	EXPRES2 $k = 32$
1	1.8 (32)	0.4 (32)	0.2 (34)	18.9 (32)	3.1 (32)	3.0 (34)
2	1.7 (32)	1.1 (96)	0.6 (100)	18.8 (32)	6.2 (64)	11.6 (132)
4	4.1 (58)	2.2 (192)	1.2 (200)	18.8 (32)	12.5 (128)	23.4 (264)
8	6.6 (73)	4.3 (352)	2.7 (432)	42.8 (49)	28.3 (288)	38.2 (432)
16	15.6 (99)	10.0 (736)	6.4 (896)	40.9 (40)	54.4 (544)	71.5 (800)
32	64.2 (148)	28.8 (1696)	21.2 (2,368)		121.5 (1,152)	163.0 (1,760)
64	185.9 (192)	90.9 (3,968)	51.1 (4,480)			
	32u	14u	12u	29u	269u	1,205u

Table 2. Time (in seconds) for matrices BCSSTK19 and BCSSTK25 from the BC-SSTRUC3 collection (figures in brackets show the number of matrix-vector products).

in Table 2 is very poorly conditioned and its computed eigenvalues occurred in clusters. The computed eigenvalues of the matrix BCSSTK25 also occur in clusters. Further, storage restrictions prevented the SSAUPD routine from obtaining partial eigensolutions for this matrix for those cases where $p = 32$ and 64. Table 3 presents the results for a tridiagonal matrix A, all of whose diagonal and off-diagonal elements are set to 2 and 1, respectively. This matrix (henceforth referred to as the (1,2,1) matrix) arises in a variety of discretization problems in numerical mathematics and is known to be fairly pathological and slow to converge. In the header of each Table nz denotes the number of nonzero elements in the upper triangle of the matrix. Also in the case of the SSAUPD routine k

	(1,2,1) Matrix		
	$n = 10,000$ $(nz = 19,999)$ $\lambda_1 = 0.39999959 \times 10^1$, $\lambda_{16} = 0.39999380 \times 10^1$		
p	SSAUPD $k = 32$	EXPRES1 $k = 32$	EXPRES2 $k = 32$
1	32.8 (64)	1.4 (96)	0.7 (100)
2	427.9 (1310)	23.2 (1,538)	9.0 (1,190)
4	474.8 (1,242)	82.0 (5,248)	30.3 (3,882)
8	1114 (2,094)	227.2 (13,632)	85.6 (10,226)
16	2735 (2,763)	584.3 (31,360)	220.8 (23,554)
	706u	1006u	1006u

Table 3. Time (in seconds) for the (1,2,1) matrix(figures in brackets show the number of matrix-vector products).

denotes the number of Lanczos steps after which an 'implicit' restart is made; the chosen mode is 'exact shifts' [4], [7]. The final row of each Table gives the accuracy of the computed solution as $c\mathbf{u}$ (or $c\sqrt{\mathbf{u}}$), where c is a natural number. This accuracy has been computed 'independently' as

$$\max(\frac{\| A\hat{x}_i - \hat{\lambda}_i \hat{x}_i \|_2}{| \hat{\lambda}_i |}) \; ; \; i = 1, \ldots, \max(p)$$

In all cases the initial value of q_1 is chosen to be $\frac{1}{\sqrt{n}} \times [1, 1, \ldots, 1]$ and the products Aq_j are computed using the Connection Machine Scientific Support Library routine 'sparse_matvec_mult'. With the exception of the (1,2,1) matrix, the requested tolerance has been set in all cases to $\mathbf{u}$.

6   Conclusions

The results presented in Tables 1 and 2 show clearly that Sorensen's implicit Arnoldi routine and the two explicit restart routines discussed in this paper are capable of computing very accurate partial eigensolutions of real world large sparse symmetric matrices whose eigenvalue distribution contains closely clustered and/or multiple eigenvalues. Thus, in particular, EXPRES1 and EXPRES2 both computed partial eigensolutions of the matrix PLAT1919 of order up to 64 with accuracy better than 14$\mathbf{u}$. Observe that SSAUPD computed the same solutions with accuracy better than 61$\mathbf{u}$. However, in the case of the (1,2,1) matrix, it was not possible to obtain partial eigensolutions of similar accuracy using any of the algorithms as is indicated in Table 3. Nevertheless, considering the pathological nature of this matrix, the single precision solutions obtained were good.

Table 3 shows that, for the pathological (1,2,1) matrix, the explicit restart routines developed by Szularz et al [8] perform signicantly better than Sorensen's routine [4]. Further, it seems to be the case that, in general, the explicit restart routine EXPRES2 is much more efficient than Sorensen's routine.

It is clear from Tables 2 and 3 that, for any given example, the restart strategies discussed in this paper require significantly more matrix vector products of the form Aq_j than does the SSAUPD routine. Further, Table 1 suggests that a greater proportion of EXPRES2 consists of these products than is the case for the other two algorithms. Considering the efficiency with which these products can be implemented in the highly parallel Connection Machine environment, it is this property of the algorithm which enables it consistently to outperform the others on the CM-200.

The results show also that, with the exception of the matrix BCSSTK25, EXPRES2 is considerably more efficient than EXPRES1. The reduction in the amount of re-orthogonalization required appears to be the significant factor although another important factor is that, if reorthogonalization can be avoided, a 'purer' Krylov subspace is constructed and, consequently, convergence is faster.

The reason why EXPRES1 is more efficient than EXPRES2 in the case of BC-SSTK25 is currently unknown. However, it is our opinion that, for those matrices where orthogonalization is an issue, the use of a solver which incorporates full re-orthogonalization of the Lanczos vectors is to be preferred to one which incorporates only a selective re-orthogonalizaton. The matrix BCSSTK25 may belong to this category.

In conclusion, it has been shown that the two algorithms discussed in this paper (together with Sorensen's routine taken from the ARPACK library) are well suited to the computation of partial eigensolutions of large sparse symmetric matrices containing closely clustered or multiple eigenvalues. It has also been demonstrated that one of the algorithms (EXPRES2) is highly competitive with Sorensen's algorithm for problems of this kind when implemented in a highly parallel SIMD environment.

The algorithms are currently being implemented on the Cray T3D and an analysis of their performances on this highly parallel shared distributed memory machine will form the basis of a future paper.

References

1. Duff, I.S., Grimes, R.G., and Lewis, J.G., (1992), 'User's Guide for the Harwell-Boeing Sparse Matrix Collection' (release I), available online ftp orion.cerfacs.fr.
2. Golub, G., and Van Loan, C.F., (1989), 'Matrix Computations', John Hopkins University Press, London.
3. O.A. Marques (1995), 'BLZPACK: Description and User's Guide', CERFACS Report TR/PA/95/30, Toulouse, France, October 1995.
4. Lehoucq, R., Sorensen, D.C., and Vu, P.A., (1994), 'SSAUPD : Fortran subroutines for solving large scale eigenvalue problems', Release 2.1, available from netlib@ornl.gov in the scalapack directory.
5. Parlett, B.N., and Scott, D.S., (1979), 'The Lanczos algorithm with selective orthogonalization', *Math. Comput.*, **33**, 217-238.
6. Simon, H.D., (1984), 'Analysis of the Symmetric Lanczos Algorithm with Reorthogonalization Methods', *Linear Algebra Appl.*, **61**, 101-131.
7. Sorensen, D.C., (1992), 'Implicit Application of Polynomial Filters in a k-step Arnoldi Method', *SIAM J. Matrix Anal. Appl.*, **13**, 357-385.
8. Szularz, M., Weston, J., Clint, M., and Murphy, K., (1996), 'A Highly Parallel Explicitly Restarted Lanczos Algorithm', in *Applied Parallel Computing Industrial Computation and Optimization*, J. Wasniewski, J. Dongarra, K. Madsen and D. Olesen (Eds.), LNCS 1184, Springer-Verlag, 651-660.

A Fully Parallel Symmetric Matrix Transformation

Ayse Kiper

Department of Computer Engineering, Middle East Technical University
Ankara-Turkey

Abstract. A parallel algorithm for the transformation LAL^T of an nxn symmetric A through a nonsingular triangular L is presented. Speedup and efficiency successively are of $O((n^3 + n^2)/(1 + \lceil \log n \rceil))$ and $O(1/(1 + \lceil \log n \rceil))$ with $O(1 + \lceil \log n \rceil)$ parallel time.

1 Introduction

Similarity and orthogonal transformations have great significance in eigenvalue problems since the efficient numerical methods are based on these transformations. In this work we describe a parallel algorithm to generate the lower triangular part of the transformation

$$S = LAL^T \tag{1}$$

where A is an $n \times n$ symmetric and L is a non-singular lower triangular matrices. Method is heavily based on elementwise multiplication and addition of auxiliary matrices. Parallel assignment operations are also used but the time consumed for these is negligible compared to that required for the complete algorithm (hereafter referred to as PST). PST is particularly suitable for shared memory systems.

2 Parallel Symmetric Transformation Algorithm (PST)

The matrices A, L and S in (1) can be defined as

$$A = \left[a_{ij} \right], \quad i = 1, 2, \ldots, n \; ; \; j = 1, 2, \ldots, n$$

$$L = \left[l_{ij} \right], \quad j = 1, 2, \ldots, i \; ; \; i = 1, 2, \ldots, n \; \text{ and } \; l_{ij} = 0 \quad \text{ for } \; j \rangle i$$

$$S = \left[s_{ij} \right], \quad i = 1, 2, \ldots, n \; ; \; j = 1, 2, \ldots, n.$$

Let the lower triangular part of S is defined by

$$S' = \left[\tilde{s}_1, \tilde{s}_2, \ldots, \tilde{s}_n \right] \tag{2}$$

where $\tilde{s}_k$ denotes the kth column of S' and can be expressed as

$$\tilde{s}_k = (\tilde{o}^T, \tilde{v}_k^T)^T, \quad k = 1, 2, \ldots, n$$

where $\tilde{o}$ is a null vector of size $(k-1)$ and $\tilde{v}_k$ is a $(n-k+1)$ vector whose ith element is denoted by v_{ki} and defined as

$$v_{ki} = \sum_{r=1}^{k} l_{kr} p_{ir}, \quad i = k, k+1, k+2, \ldots, n \tag{3}$$

where p_{ir} represents the sum of products

$$p_{ir} = \sum_{s=1}^{i} l_{is} a_{sr}, \quad i = r, r+1, r+2, \ldots, n. \tag{4}$$

PST algorithm can be described in the following six main steps:

Step 1.*(Matrix construction by parallel assignment)* Auxiliary matrices C and D of size $(n(n+1)/2) \times n$ are constructed by parallel assignment as

$$C = \begin{bmatrix} C_1^T & C_2^T & \cdots & C_{n-1}^T & C_n^T \end{bmatrix}^T$$

$$D = \begin{bmatrix} D_1^T & D_2^T & \cdots & D_{n-1}^T & D_n^T \end{bmatrix}^T$$

where C_i and D_i are submatrices of size $(n-i+1) \times n$ and obtained from the elements of L and A successively as

$$C_i = \begin{bmatrix} l_{i1} & \cdots & & l_{ii} & & & 0 \\ l_{i+1,1} & \cdots & & l_{i+1,i} & l_{i+1,i+1} & & \\ \vdots & & & \vdots & & \ddots & \\ \vdots & & & \vdots & & & \ddots \\ l_{n1} & \cdots & & l_{ni} & l_{n,i+1} & \cdots & l_{nn} \end{bmatrix}$$

$$D_i = \begin{bmatrix} a_{i1} & a_{i2} & \cdots & a_{ii} & & & 0 \\ a_{i1} & a_{i2} & \cdots & a_{ii} & a_{i,i+1} & & \\ \vdots & \vdots & & \vdots & & \ddots & \\ \vdots & \vdots & & \vdots & & & \ddots \\ a_{i1} & a_{i2} & \cdots & a_{ii} & a_{i,i+1} & \cdots & a_{in} \end{bmatrix}$$

Step 2.(*Elementwise matrix-matrix multiplication*) Termwise multiplication of C and D yields

$$M = C * D$$

whose elements are the product terms $l_{is}a_{sr}$ $(i = r, r+1, ..., n \;\; ; \;\; r = 1, 2, ..., n)$ in (4).

Step 3.(*Eleentwise vector-vector additions by fan-in*) Columns of

$$M = \left[\tilde{m}_1, \tilde{m}_2, ..., \tilde{m}_n \right]$$

are added in parallel by fan in to yield

$$\tilde{p} = \sum_{i=1}^{n} \tilde{m}_i$$

in $\lceil \log n \rceil$ steps. Elements of $\tilde{p}$ are the sum of product terms p_{ir} given in (4). $\tilde{p}$ can be partitioned as

$$\tilde{p} = (\tilde{p}_1^T, \tilde{p}_2^T, ..., \tilde{p}_n^T)^T$$

with

$$\tilde{p}_r = (p_{ir}), \quad i = r, r+1, ..., n \;\; ; \;\; r = 1, 2, ..., n$$

Step 4.(*Matrix construction by parallel assignment*) E and F are of size $n \times (n(n+1)/2)$ and are constructed by assignment. Columns of

$$E = \left[\tilde{e}_1, \tilde{e}_2, ..., \tilde{e}_{n(n+1)/2} \right]$$

and

$$F = \left[\tilde{f}_1, \tilde{f}_2, ..., \tilde{f}_{n(n+1)/2} \right]$$

are obtained respectively by

$$\tilde{e}_{(i(i-1)/2)+r} \leftarrow (\tilde{o}_i^T, \tilde{p}_r^T)^T, \quad i = r, r+1, ..., n \;\; ; \;\; r = 1, 2, ..., n \tag{5}$$

and

$$\tilde{f}_{(i(i-1)/2)+r} \leftarrow (\tilde{o}_i^T, \tilde{l}_{ir}^T)^T, \quad r = 1, 2, ..., i \;\; ; \;\; i = 1, 2, ..., n. \tag{6}$$

In (5) and (6) $\tilde{o}$ denotes a null vector of size $(i-1)$ and in (6) $\tilde{l}_{ir}$ denotes a $(n-i+1)$ vector of equal elements defined as $\tilde{l}_{ir} = (l_{ir})$.

Step 5.(Elementwise matrix-matrix multiplication) Elementwise multiplication of E and F yields

$$R = E * F$$

whose elements are the sum of product terms p_{ir} when multiplied by appropriate constants l_{kr} as described in (3) and (4).

Step 6.(Elementwise vector-vector additions by fan-in) The appropriate columns of

$$R = \left[\tilde{r}_1, \tilde{r}_2, \ldots, \tilde{r}_{n(n+1)/2} \right]$$

are added by fan-in method using the formula

$$\tilde{s}_k = \sum_{i=1}^{k} \tilde{r}_{(k(k-1)/2)+i}, \quad k = 1, 2, \ldots, n$$

to yield the matrix S'. $\tilde{s}_n$ is obtained in $\lceil \log n \rceil$ steps which determines the parallel time for Step 6.

3 Performance of PST

PST algorithm requires 2 multiplication steps which are independent of problem size and $2\lceil \log n \rceil$ addition steps. Memory or data alignment costs are negligible. It can easily be adapted to multiprocessor systems having limited number of processors with a minor effort by splitting the objects into subobjects. PST is equally suitable and adaptable for vector processors. These make the algorithm general.

Performance parameters speedup (T_1 / T_p) and efficiency (S_p / p_{max}) are

$$S_p = \frac{n(3n^2 + 2n + 1)}{6(1 + \lceil \log n \rceil)}$$

and

$$E_p = \frac{3n^2 + 2n + 1}{3n(n+1)(1 + \lceil \log n \rceil)},$$

while the maximum number of processors used is

$$p_{max} = n^2(n+1)/2.$$

PST is an addition dominant algorithm particularly for large size of problems.

Numerical Experiments with a Parallel Fast Direct Elliptic Solver on Cray T3E

Tuomo Rossi and Jari Toivanen

University of Jyväskylä, Department of Mathematics, Laboratory of Scientific Computing, P.O. Box 35, FIN-40351 Jyväskylä, Finland.

Abstract. A parallel fast direct $\mathcal{O}(N \log N)$ solver is shortly described for linear systems with separable block tridiagonal matrices. A good parallel scalability of the proposed method is demonstrated on a Cray T3E parallel computer using MPI in communication. Also, the sequential performance is compared with the well–known BLKTRI–implementation of the generalized cyclic reduction method using a single processor of Cray T3E.

1 Introduction

We study a parallel fast direct solution algorithm for linear systems with block tridiagonal separable matrices. In [4], this algorithm was named as the Divide & Conquer method (DC–method). It was originally introduced in [11] and further considered in [2,4]. The stability of the DC–method based on the partial solution technique [1,3] and the close relation to the cyclic odd–even reduction is analyzed in [5]. For more information about fast direct solvers and their parallelization see, for example, [7,9,10], and references therein.

In the next section, we give a very brief description of the algorithm and parallel implementation using MPI. For details, we refer to [6]. In the numerical experiments, the parallel scalability and efficiency is demonstrated in a Cray T3E distributed memory parallel computer.

Our model problem is the Dirichlet boundary value problem for the two–dimensional Poisson equation in a rectangle. By using a rectangular, possibly nonuniform triangulated finite element mesh with $n_1 \times n_2$ interior nodes and the piecewise linear finite elements, we obtain the linear system $Au = f$. When the mesh nodes are numbered column–wise, the separable coefficient matrix has the form

$$A = A_1 \otimes M_2 + M_1 \otimes A_2, \quad A_i, M_i \in \mathbb{R}^{n_i \times n_i}, \ i = 1, 2,$$

where $\otimes$ is the matrix tensor (Kronecker) product ($A \in \mathbb{R}^{k \times k}$, $B \in \mathbb{R}^{s \times s}$, $A \otimes B = \{A_{i,j}B\}_{i,j=1}^{k} \in \mathbb{R}^{ks \times ks}$), and A_1, A_2 and M_1, M_2 are tridiagonal and diagonal matrices, respectively.

2 The parallel Divide & Conquer algorithm

For simplicity, we only consider the case when $n_1 = 4^k - 1$, where k is a fixed integer satisfying $k \geq 1$. The generalization for arbitrary n_1 is described in

[6]. Let us first define a hierarchical sequence of integer sets J_i as follows:

$$J_i = \bigcup_{j=1}^{4^{k-i}-1} \{j \cdot 4^i\}, \quad i = 0, \ldots, k-1, \quad J_k = \emptyset.$$

Next, we define the projection matrices P_i, $i = 0, \ldots, k$, by setting

$$P_i = \tilde{P}_i \otimes I_{n_2}, \quad \tilde{P}_i = \text{diag}\{p_{i,1}, \ldots, p_{i,n_1}\}, \quad p_{i,j} = \begin{cases} 0, & \text{if } j \in J_i, \\ 1, & \text{otherwise}, \end{cases}$$

where the matrix I_{n_2} is the identity matrix of order n_2.

DC–algorithm: Set $f_1 = f$ and $u_{k+1} = 0$.
First stage: For $i = 1, \ldots, k-1$, compute

$$f_{i+1} = P_i f_i + (I - P_i)(f_i - A(P_i A P_i)^+(I - P_{i-1})f_i).$$

Second stage: For $i = k, \ldots, 1$, compute

$$u_i = (I - P_{i-1})(P_i A P_i)^+((I - P_{i-1})f_i - A(I - P_i)u_{i+1}) + (I - P_i + P_{i-1})u_{i+1}.$$

Then, $u = u_1 = A^{-1}f$.

By analyzing the previous algorithm, it turns out that in both stages the vectors multiplied by the pseudo–inverse $(P_i A P_i)^+$ have rather few nonzero components. Also, only sparse sets of the components of resulting vectors are required. These problems under consideration can be efficiently solved by the partial solution technique using $\mathcal{O}(n_1 n_2)$ flops. Since $\mathcal{O}(\log_4 n_1)$ steps are required in both stages, the total cost of DC–algorithm is $\mathcal{O}(n_1 n_2 \log_4 n_1)$ flops and for our specific implementation, it is $\mathcal{O}((44 \log_4(n_1 + 1) - \frac{163}{4})n_1 n_2)$ flops [6].

The number of processors p is assumed to be a power of two in our implementation. The communication is performed using MPI message passing. This leads to a portable implementation for distributed memory parallel computers. The right–hand side vector f is split to almost equally sized parts, one for each process. The solution u will be split in the same way, since it is computed on top of the vector f. In the first stage and second stage, only the steps $i > k - \lceil \log_4 p \rceil$ require interprocess communication. For example, when $p = 8$ communication is needed only in the steps $k - 1$ and k. The resulting communication pattern for this case is shown in Fig. 1. A detailed description of the parallelization can be found in [6].

3 Numerical experiments

The parallel DC–algorithm has been implemented as a FORTRAN subroutine PDC2D, which can handle arbitrary matrix dimensions. The numerical

724

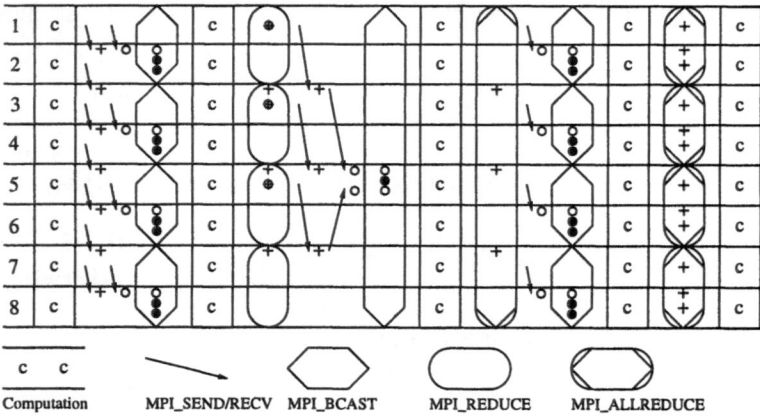

Fig. 1. A sample communication pattern in the case of eight processes. The plus sign, balls and circles all denote communication involving a vector of length n_2.

experiments are performed in a Cray T3E. Fig. 2 illustrates the parallel efficiency of PDC2D when $n_1 = n_2 = n$. For example, with 64 processors, the speedups are more than 32 already when $n = 447$. Starting from $n = 1151$, the speedups are higher than 48. The solution of a problem with 1023^2 unknowns takes 0.082 seconds of wall clock time using 64 processors. The jumps in the efficiency curves, leading even to speedups greater than the number of processors, are caused by the effects of the cache memory of the computer.

In Table 1, the single processor efficiency of PDC2D is compared with the well–known BLKTRI–subroutine [8]. The initialization of PDC2D, essentially the solution of the arising generalized one–dimensional eigenvalue problems using the EISPACK routines, is slower than in the BLKTRI subroutine, but the actual solution is several times faster. The initialization times are, however, negligible when several problems are solved with the same coefficient matrix, since the subroutines have to be initialized only once in this case.

Acknowledgments. The authors are grateful to Prof. Yuri A. Kuznetsov for valuable discussions about the DC–method.

Table 1. Comparison of the CPU–time usage (in seconds) between the PDC2D– and BLKTRI–subroutines in single processor runs.

$n_1 = n_2$	63	127	255	511	1023	2047
PDC2D init.	0.03	0.11	0.45	1.81	7.59	38.52
BLKTRI init.	0.01	0.02	0.07	0.26	1.08	4.12
PDC2D solve	0.01	0.03	0.14	0.75	3.95	31.35
BLKTRI solve	0.02	0.09	0.85	2.96	29.44	173.89

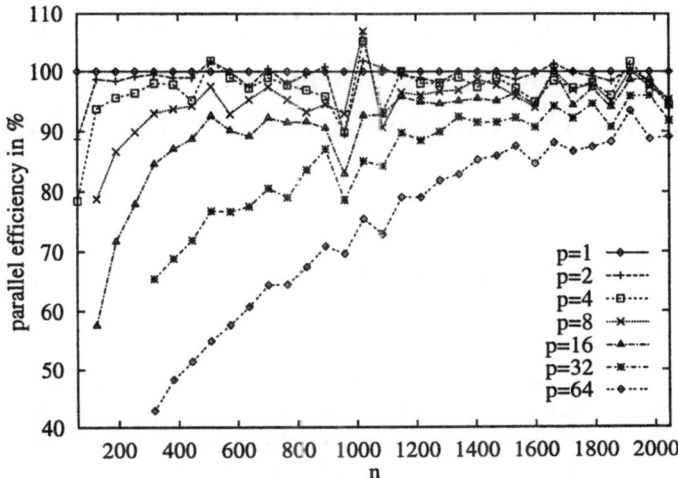

Fig. 2. The parallel efficiency of the subroutine PDC2D in a Cray T3E computer.

References

1. Banegas, A.: Fast Poisson solvers for problems with sparsity. Math. Comp. **37** (1978) 441–446
2. Kuznetsov, Yu. A.: Numerical methods in subspaces. in Vychislitel'-nye Processy i Sistemy II, Marchuk, G. I., ed., Nauka, Moscow, 1985, 265–350 In Russian
3. Kuznetsov, Yu. A., Matsokin, A. M.: On partial solution of systems of linear algebraic equations. Sov. J. Numer. Anal. Math. Modelling **4** (1989) 453–468
4. Kuznetsov, Yu. A., Rossi, T.: Fast direct method for solving algebraic systems with separable symmetric band matrices. East–West J. Numer. Math. **4** (1996) 53–68
5. Rossi, T., Toivanen, J.: New variants of Divide & Conquer method arising from block cyclic reduction type formulation. Tech. Rep. 20, Laboratory of Scientific Computing, University of Jyväskylä, 1996 Submitted
6. Rossi, T., Toivanen, J.: A parallel fast direct solver for block tridiagonal systems with separable matrices of arbitrary dimension. Tech. Rep. 21, Laboratory of Scientific Computing, University of Jyväskylä, 1996 Submitted
7. Swarztrauber, P. N.: Fast Poisson solvers. in Studies in Numerical Analysis, Golub, G. H., ed., vol. 24 of MAA Studies in Mathematics, Mathematical Association of America, 1985, 319–370
8. Swarztrauber, P. N., Sweet, R. A.: Efficient FORTRAN subprograms for the solution of separable elliptic partial differential equations. ACM Trans. Math. Software **5** (1979) 352–364
9. Swarztrauber, P. N., Sweet, R. A.: Vector and parallel methods for the direct solution of Poisson's equation. J. Comput. Appl. Math. **27** (1989) 241–263
10. Vajtersic, M.: Algorithms for Elliptic Problems: Efficient Sequential and Parallel Solvers. Kluwer Academic Publisher, Dordrecht, 1993.
11. Vassilevski, P. S.: Fast algorithm for solving a linear algebraic problem with separable variables. C. r. Acad. Bulg. Sci. **37** (1984) 305–308

New Matrix-by-Vector Multiplications Based on a Nonoverlapping Domain Decomposition Data Distribution

Gundolf Haase

University Linz, Inst. of Math., Altenbergerstr. 69, A–4040 Linz, Austria

Abstract. The nonoverlapping domain decomposition (DD) method allows to store vectors as accumulated or as distributed (each process stores a part of that value) vectors. The case of distributed stored matrices is fully investigated, whereas the use of accumulated matrices is not customary in the DD community. This paper is concerned with the accumulated matrices and investigates conditions under which matrix-by-vector multiplications can be realized. Especially, we derive restrictions on the matrix shape and the finite element (FE) mesh. As a new result, the special structure of these admissible accumulated matrices leads directly to global incomplete factorizations as preconditioners in CG-like methods. Also, the well-known DD preconditioners fit into the general framework of the matrix-by-vector operations presented.

Keywords : Parallel Iterative Solvers, Incomplete Factorization, Preconditioning, Domain Decomposition, Finite Element Method.

1 Introduction

Let us consider an abstract elliptic and bounded variational problem arising from the weak formulation of a scalar second–order, uniformly elliptic boundary value problem (BVP) given in a plane bounded domain $\Omega \subset \mathbb{R}^2$ with a piecewise smooth boundary $\Gamma = \partial\Omega$. Defining the usual linear FE nodal basis the FE isomorphism results in a large-scale sparse system

$$K \underline{u} = \underline{f} \tag{1}$$

of equations with the positive definite stiffness matrix K. For simplicity, we assume triangular elements in 2d and tetrahedrons in 3d. We focus our interest on the 2d case; the 3d case is similar.

The nonoverlapping domain decomposition (DD) data distribution following the ideas in [7,8] allows to store vectors as accumulated (each process stores the full value) or as distributed (each process stores a part of that value) vectors. The same storage classes exist for matrices. The case of distributed stored matrices is fully investigated (see [7,8]). So, we focus our interest on the question why the matrix-by-vector multiplication is not allowed for accumulated matrices. The general answer was given in Groh [2] but without any further investigations. A detailed answer, given in Sect. 2.2, includes requirements on mesh and matrix shape to allow the proper multiplications. In Sect. 3, we investigate in general the preconditioners of the preconditioning step in CG-like methods and distinguish between distributed and accumulated type. It turns out that the DD preconditioners [1,5] are of accumulated type. Now, due to the assumptions on the mesh also global incomplete factorization preconditioners are feasible which are examined in detail in Sect. 4. Some numerical experiences are presented in Sect. 5.

2 The Matrix-by-Vector Multiplication in the DD

2.1 The DD Data Distribution

First we subdivide our 2d-domain Ω into p subdomains $\overline{\Omega}_i$ $(i = 1, 2, \ldots, p)$ and discretize the subdomains with linear triangular elements. The triangulation has to be conform in the whole domain Ω. Figure 1 illustrates this for an unit square subdivided into 4 congruent squares.

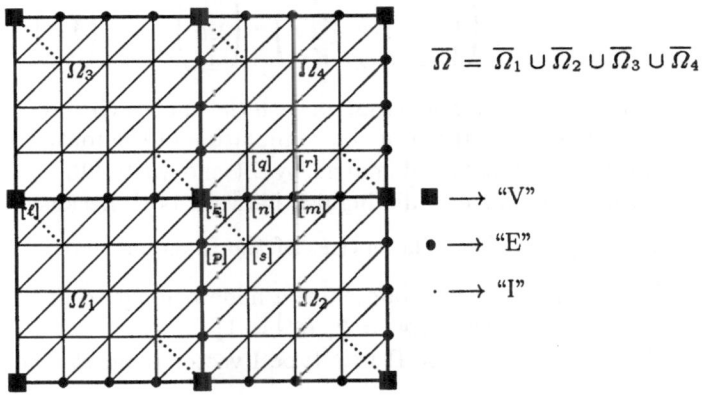

$$\overline{\Omega} = \overline{\Omega}_1 \cup \overline{\Omega}_2 \cup \overline{\Omega}_3 \cup \overline{\Omega}_4$$

$\blacksquare \longrightarrow$ "V"

$\bullet \longrightarrow$ "E"

$\cdot \longrightarrow$ "I"

Fig. 1. Domain decomposition, initial (/) and modified ($\cdots$) triangulation.

Let us introduce 3 classes of nodes: the inner nodes further denoted by the subscript "I", the nodes on the interior of a coupling edge (edge nodes) denoted by "E" and the vertex nodes (cross points) "V". If no distinction is necessary then the latter are called coupling nodes indicated by "C". The unknowns are ordered such that we number first the vertices, then the edge nodes (subsequently on each edge) and last the inner nodes in the subdomains $\Omega_1 \ldots, \Omega_p$.

According to the p subdomains $\overline{\Omega}_i$ $(i = 1, 2, \ldots, p)$ we distribute all matrices and vectors to the p processors P_i $(i = 1, 2, \ldots, p)$ of the parallel machine and define A_i as the connectivity matrix of $\overline{\Omega}_i$, i.e., the matrix A_i is the boolean matrix of dimension $N_i \times N$ mapping a global vector $\underline{g} \in \mathbb{R}^N$ into a local vector $\underline{g}_i \in \mathbb{R}^{N_i}$.

Now we have the opportunity to define two types of vectors, the accumulated (type I) and the distributed (type II) vector :

I : $\underline{u}$ and $\underline{w}$ are stored in processor P_i ($\hat{=} \overline{\Omega}_i$) as $\underline{u}_i = A_i \underline{u}$ and $\underline{w}_i = A_i \underline{w}$, i.e., in each subdomain every interface node possesses the full value of its vector components.

II : $\underline{r}, \underline{f}$ are stored in P_i as $\underline{r}_i, \underline{f}_i$, so that $\underline{r} = \sum_{i=1}^{p} A_i^T \underline{r}_i$ etc. is valid, i.e., the nodes on the interfaces store just a part of the actual value which will be recovered by accumulation.

The matrix K (1) will be stored in a distributive sense similar to a type-II vector. We call it a type-II matrix $\mathsf{K} = \sum_{i=1}^{p} A_i^T \mathsf{K}_i A_i$, where K_i is the stiffness matrix belonging to the subdomain $\overline{\Omega}_i$. Looking at $\overline{\Omega}_i$ as a macro finite element its element stiffness matrix K_i is identical to the proper local part of the distributed stored matrix K. The ordering of the nodes given above and the types of the vectors lead to the following presentation of equation (1) :

$$
\begin{pmatrix} \mathsf{K}_V & \mathsf{K}_{VE} & \mathsf{K}_{VI} \\ \mathsf{K}_{EV} & \mathsf{K}_E & \mathsf{K}_{EI} \\ \mathsf{K}_{IV} & \mathsf{K}_{IE} & \mathsf{K}_I \end{pmatrix} \cdot \begin{pmatrix} \underline{u}_V \\ \underline{u}_E \\ \underline{u}_I \end{pmatrix} = \begin{pmatrix} \underline{f}_V \\ \underline{f}_E \\ \underline{f}_I \end{pmatrix} . \tag{2}
$$

Here K_I is a block diagonal matrix with the entries $\mathsf{K}_{I,i}$. $\mathsf{K}_{IC}, \mathsf{K}_{CI}, \mathsf{K}_{IV}, \mathsf{K}_{VI}$ are block matrices, too. If a global accumulation of the matrix K is performed then we will denote that type-I matrix by $\mathfrak{M}$ and write $\mathfrak{M}_i = A_i \mathfrak{M} A_i^T$. Although $\mathsf{K} \equiv \mathfrak{M}$ holds we have to distinguish between both because of the different local storing, i.e., $\mathsf{K}_i \neq \mathfrak{M}_i$. The entries of the diagonal matrix $R = \sum_{i=1}^{p} A_i^T A_i$ are equal to the number of subdomains a node belongs to (e.g., according to Fig. 1 $R^{[k]} = 4$, $R^{[n]} = R^{[m]} = R^{[p]} = 2$, $R^{[q]} = 1$).

The change from type II to a type I vector requires communication :

$$
\underline{w}_i = A_i \sum_{i=1}^{p} A_i^T \underline{r}_i . \tag{3}
$$

Changing a type I vector into a type II is not unique. One opportunity is to divide locally the value by the number of neighbours, i.e., $\underline{r}_i = R^{-1} \underline{w}_i$ with R defined in the paragraph above.

2.2 Matrix-by-Vector Operations

In the following we investigate briefly the matrix-by-vector multiplication with respect to accumulated matrices and different vector types. Sub- and superscripts are used in the following way : $v_{C,i}^{[n]}$ denotes the n-th component (local or global ordering) of a vector $\underline{v}$ stored on processor i. The subscript "C" indicates that this part of the vector belongs to the coupling nodes. A similar notation will be used for the matrices.

1. Type-I matrix × type-I vector cannot be realized with general type-I matrices $\mathfrak{M}$. Let us have a look at the node n in Fig. 1 and the operation $\mathfrak{M} \cdot \mathfrak{w}$. The local multiplications $\underline{u}_i = \mathfrak{M}_i \cdot \mathfrak{w}_i$ $(i = 2, 4)$ for node n result in

$$u_2^{[n]} = \mathfrak{M}_2^{[n,n]}\mathfrak{w}_2^{[n]} + \mathfrak{M}_2^{[n,m]}\mathfrak{w}_2^{[m]} + \mathfrak{M}_2^{[n,k]}\mathfrak{w}_2^{[k]} + \mathfrak{M}_2^{[n,p]}\mathfrak{w}_2^{[p]} + \mathfrak{M}_2^{[n,s]}\mathfrak{w}_2^{[s]}$$
$$u_4^{[n]} = \mathfrak{M}_4^{[n,n]}\mathfrak{w}_4^{[n]} + \mathfrak{M}_4^{[n,m]}\mathfrak{w}_4^{[m]} + \mathfrak{M}_4^{[r,k]}\mathfrak{w}_4^{[k]} + \mathfrak{M}_4^{[n,q]}\mathfrak{w}_4^{[q]} + \mathfrak{M}_4^{[n,r]}\mathfrak{w}_4^{[r]} \ .$$

In the equations above, the terms 4 and 5 of the right-hand sides differ. The processors 2 and 4 achieve different results, instead of the unique one, since processor 2 uses data not available to processor 4 and vice versa. Denoting the set of processors a node j belongs to by ω_j we can ensure identical results in the equations above only if we use nodes j fulfilling $\omega_n \subseteq \omega_j$ $(j \in \{n, k, m\})$ in the multiplication. If the matrix is represented by a directed graph then certain edges of that graph are not allowed. For example, in Fig. 1 the entries $q \rightarrow k$, $q \rightarrow n$, $s \rightarrow n$, $s \rightarrow m$, $s \rightarrow p$, $p \rightarrow k$, $n \rightarrow k$, $p \rightarrow n$, $n \rightarrow p$, $\ell \rightarrow k$, $k \rightarrow \ell$ are not admissible. So, only a matrix of the shape

$$\mathfrak{M} = \begin{pmatrix} \mathfrak{M}_V & 0 & 0 \\ \mathfrak{M}_{EV} & \mathfrak{M}_E & 0 \\ \mathfrak{M}_{IV} & \mathfrak{M}_{IE} & \mathfrak{M}_I \end{pmatrix} \quad \Longrightarrow \quad \underline{u} = \mathfrak{M} \cdot \mathfrak{w} \tag{4}$$

with block-diagonal matrices $\mathfrak{M}_I$, $\mathfrak{M}_E$, $\mathfrak{M}_V$ ensures coherent data stored as a type-I vector. While the block-diagonality of $\mathfrak{M}_I$ is guaranteed by the data decomposition, for the remaining two matrices the mesh has to fulfill the requirements
(a) No connection between vertices belonging to different sets of subdomains.
(b) No connection between edges belonging to different sets of subdomains.
Requirement 1a can be easily fulfilled if at least one node is located on all edges between two vertices. A modification of the given mesh generator can also guarantee requirement 1b, e.g., if the edge between nodes p and n is omitted. The dotted lines in Figure 1 represent the necessary changes in the mesh.

2. As in the previous point, Type-I matrix × type-II vector cannot be performed by general type-I matrices $\mathfrak{M}$, even if the necessary requirements 1a and 1b are fulfilled by a modified mesh (see Fig. 2). Calculating $M_{IC} \cdot \underline{r}_C$ locally on processor 4 results in

$$f_4^{[r]} = \mathfrak{M}_{IC,4}^{[r,n]} \cdot r_4^{[n]} + \mathfrak{M}_{IC,4}^{[r,m]} \cdot r_4^{[m]} \ .$$

The interior node r should always store the full value, but from a global view the entries $\mathfrak{M}_{IC,2}^{[r,n]} \cdot r_2^{[n]}$ and $\mathfrak{M}_{IC,2}^{[r,m]} \cdot r_2^{[m]}$ are missing. So, this operation is not valid. Now, using the sets ω_j from the previous point we have to ensure $\omega_r \supseteq \omega_j$ $(j \notin \{n, m\})$ to perform the multiplication under consideration, e.g., the entries $\mathfrak{M}^{[k,n]}$ and $\mathfrak{M}^{[k,q]}$ are allowed but not the transposed ones. So, just a matrix of the shape

$$\mathfrak{M} = \begin{pmatrix} \mathfrak{M}_V & \mathfrak{M}_{VE} & \mathfrak{M}_{VI} \\ 0 & \mathfrak{M}_E & \mathfrak{M}_{EI} \\ 0 & 0 & \mathfrak{M}_I \end{pmatrix} \quad \Longrightarrow \quad \underline{f} = \sum_{i=1}^{P} A_i^T \underline{f}_i = \sum_{i=1}^{P} A_i^T \left(\mathfrak{M}_i \underline{r}_i \right) \tag{5}$$

with block-diagonal matrices $\mathfrak{M}_I$, $\mathfrak{M}_E$, $\mathfrak{M}_V$ can be used in this type of matrix-by-vector multiplication resulting in a type-II vector.

Remark : The factorization of a matrix $\mathfrak{M}$ into a lower and an upper triangular matrix $\mathfrak{L}^{-1}$ and $\mathfrak{U}^{-1}$ can be realized in two ways. Applying it in the preconditioning step $\underline{w} = C^{-1}\underline{r}$ of CG-like methods results in :

$$\underline{w} = \mathfrak{L}^{-1}\mathfrak{U}^{-1} \cdot \underline{r} := \mathfrak{L}^{-1} \sum_{i=1}^{P} A_i^T \mathfrak{U}_i^{-1} \cdot \underline{r}_i \tag{6a}$$

$$\underline{w} = \mathfrak{U}^{-1}\mathfrak{L}^{-1} \cdot \underline{r} := \sum_{i=1}^{P} A_i^T \mathfrak{U}_i^{-1} R_i^{-1} A_i \mathfrak{L}^{-1} \cdot \left(\sum_{j=1}^{P} A_j^T \underline{r}_j\right) \tag{6b}$$

The remaining equations can be easily derived by type conversion.

3 Preconditioners for the Iteration Form

3.1 Type-I Preconditioners

We are in the position to choose a factorized type-I preconditioner from the equations (6a) and (6b). To minimize the required communication, the first one will be investigated in the remaining sections.

Remark : In the symmetric case, the well known ASM-DD-preconditioner [5] fits exactly in the scheme presented by equation (6a). Therein, Schur complement preconditioners proposed in [1,10] are used in which the vertex nodes do not form a block-diagonal matrix. This is no contradiction to the requirements on accumulated matrices because of the usual serial and global handling of the related system of equations.

3.2 Type-II Preconditioners

Applying a type-II preconditioning matrix $C^{-1} = \sum_{i=1}^{P} A_i^T C_i^{-1} A_i$ we are in need to convert the vectors twice so that the preconditioning step looks like

$$\underline{w} = C^{-1} \sum_{j=1}^{P} A_j^T \underline{r}_j := \sum_{i=1}^{P} A_i^T \left(C_i^{-1} \cdot A_i \sum_{j=1}^{P} A_j^T \underline{r}_j\right) . \tag{7}$$

The matrix K is not accumulated and so the submatrices K_i represent a 2nd order PDE in Ω_i with homogeneous Neumann B.C. on the inner boundaries $\partial\Omega_i \backslash \partial\Omega$. If $\partial\Omega_i \cap \Gamma_D = \emptyset$ the matrix K_i may become singular. Therefore we cannot choose C_i in the same way. One opportunity is to assemble the stiffness Matrix K and set $C_i = \mathfrak{K}_i$ so that the local preconditioner represents a 2nd order PDE in $\widetilde{\Omega}_i$ with homogeneous Dirichlet B.C. on $\partial\widetilde{\Omega}_i$, where $\widetilde{\Omega}_i$ is the domain Ω_i expanded by the next layer of nodes (elements) over the inner boundary. So we end up with an additive Schwarz overlapping preconditioner which is out of scope of this work; for further investigations see [9].

4 An Incomplete Factorization Preconditioner

The choice $\mathfrak{C}^{-1} = \mathfrak{L}^{-1} \cdot \mathfrak{U}^{-1}$ (6a) indicates a factorization $\mathfrak{U} \cdot \mathfrak{L}$ of the given accumulated stiffness matrix $\mathfrak{K}$, but the fill-in does not fit into our DD data concept and would scale down the problem size handled in parallel. So, we are concerned with the classic ILU factorization [3] preserving the given pattern of the matrix and take advantage of the block structure of matrix K (2).
Note that the matrix has to be accumulated at first during the factorization.

Start	K	
Determine		Why parallel ?
U_I, L_I	$K_I = U_I \cdot L_I$	DD $\to$ blocks matrices
L_{IE}	$K_{IE} = U_I \cdot L_{IE}$	DD $\to$ blocks matrices
L_{IV}	$K_{IV} = U_I \cdot L_{IV}$	DD $\to$ blocks matrices
U_{EI}	$K_{EI} = U_{EI} \cdot L_I$	DD $\to$ blocks matrices
U_{VI}	$K_{VI} = U_{VI} \cdot L_I$	DD $\to$ blocks matrices
Modify		
	$\mathsf{K}_E := \mathsf{K}_E - U_{EI} \cdot L_{IE}$	same matrix type
	$\mathsf{K}_{EV} := \mathsf{K}_{EV} - U_{EI} \cdot L_{IV}$	same matrix type
	$\mathsf{K}_{VE} := \mathsf{K}_{VE} - U_{VI} \cdot L_{IE}$	same matrix type
	$\mathsf{K}_V := \mathsf{K}_V - U_{VI} \cdot L_{IV}$	same matrix type
Accumulate $\mathfrak{K}_E, \mathfrak{K}_{EV}, \mathfrak{K}_{VE}$		
e.g.	$\mathfrak{K}_{EV} := \sum\limits_{i=1}^{P} A_{E,i}^T \mathsf{K}_{EV,i} A_{V,i}$	—
Determine		
$\mathfrak{U}_E, \mathfrak{L}_E$	$\mathfrak{K}_E = \mathfrak{U}_E \cdot \mathfrak{L}_E$	DD, mesh $\to$ blocks matrices
$\mathfrak{U}_{VE}$	$\mathfrak{K}_{VE} = \mathfrak{U}_{VE} \cdot \mathfrak{L}_E$	DD, mesh $\to$ blocks matrices
$\mathfrak{L}_{EV}$	$\mathfrak{K}_{EV} = \mathfrak{U}_E \cdot \mathfrak{L}_{EV}$	DD, mesh $\to$ blocks matrices
Modify		
	$\mathsf{K}_V := \mathsf{K}_V - \mathfrak{U}_{VE} \cdot R_E^{-1} \cdot \mathfrak{L}_{EV}$	same matrix type (6b)
Accumulate $\mathfrak{K}_V$		—
Determine		
$\mathfrak{U}_V, \mathfrak{L}_V$	$\mathfrak{K}_V = \mathfrak{U}_V \cdot \mathfrak{L}_V$	mesh $\to$ diagonal matrix

Fig. 2. Parallelized IUL-factorization

Similar to (6a) the application of $\mathfrak{C}^{-1}$ requires communication within the vector type conversion. Figure 3 presents the proper parallel algorithm.

I) $\underline{u}_I := U_I^{-1} \underline{r}_I$
$\underline{u}_E := \mathfrak{U}_E^{-1}(\underline{r}_E - U_{EI}\underline{u}_I)$
$\underline{u}_V := \mathfrak{U}_V^{-1}(\underline{r}_V - \mathfrak{U}_{VE}\underline{u}_E - U_{VI}\underline{u}_I)$

II) $\underline{u} := \sum\limits_{i=1}^{P} A_i^T \underline{u}_i$

III) $\underline{w}_V := \mathfrak{L}_V^{-1}\underline{u}_V$
$\underline{w}_E := \mathfrak{L}_E^{-1}(\underline{u}_E - \mathfrak{L}_{EV}\underline{w}_V)$
$\underline{w}_I := L_I^{-1}(\underline{u}_I - L_{IE}\underline{w}_E - L_{IV}\underline{w}_V)$

Fig. 3. IUL Preconditioning $\mathfrak{U} \cdot \mathfrak{L} \cdot \underline{w} = \underline{r}$

5   Numerical Results

Due to the restriction of the given code to symmetric problems a parallelized CG-method with an incomplete Cholesky preconditioning, similar to (3), was used in the numerical experiments. The CG iteration stops at a relative accuracy of 10^{-6} measured in the $KC^{-1}K$-norm of the error. The code runs on a 16 processor XPLORER-machine by PARSYTEC with 32 MB per processor. For simplicity we solve the Laplace equation $-\Delta u(x, y) = 1$ with homogeneous Dirichlet B.C. in the unit square as a test example.

Table 1. Constant global problem size : 37.121 degrees of freedom

# processors	# iterations	factorization	CG	one CG step
1	253	0.31 sec	64.4 sec	0.254 sec
4	245	0.08 sec	17.1 sec	0.070 sec
16	202	0.03 sec	4.8 sec	0.024 sec

In Table 1, the classical speed up for 4 and 16 processors reaches 3.8 and 14.6, respectively. It is quite clear that this speed up will decrease with growing number of processors due to Amdahl's law. Moreover, the relation between communication and arithmetic work becomes rather poor due to the fact that in the 16 processor case less then 6% of available memory is used. With respect to the requirements on the mesh, the automatic mesh generator produced different global meshes for different numbers of subdomains. This results in different iteration counts.

Table 2. Constant local problem size

P	#unknowns	# iterations	factorization	CG	one CG step
1	37 121	253	0.31 sec	64.4 sec	0.254 sec
4	147 969	498	0.32 sec	131.4 sec	0.264 sec
16	590 849	845	0.34 sec	238.8 sec	0.277 sec

In principle, the scaled speed up should be much better than the classical speed up. Examining the 3 time columns in Table 2 the scaled speed up from 1 to 16 processors achieves 14.6 for the factorization and one CG step. This ensures us that the parallelization for itself works perfect. In coincidence with Gustafsson [3] the number of CG iterations grows like the square root of the number of unknowns $(= P \cdot N_{local})$ so that the scaled speed up for the whole algorithm has to be less then $\sqrt{P}$. Actually we have got a value of 3.7 .

6 Conclusions

The IUL-preconditioners proposed in Sect. 4 is one opportunity to code a parallel version of a given solver and works also for multiple degrees of freedom per node. The only restrictions are the requirements 1a and 1b on the mesh which have to be extended in a similar way in the 3D case.

The classical speedup is rather good whereas the scale up is poor due to the growing of the condition number with the number of unknowns. So, the parallelized factorizations are also well suited for use as a smoother in a parallelized global multi-grid method [6]. The overall solution times are approximately 4-10 times longer than for the fastest parallel solvers based on a CG with DD-ASM (Additive Schwarz Method) preconditioners [4] or the parallelized multi-grid method [6]. But in contrast to those multi-level methods we need only one mesh.

In the nonsymmetric case we have to distinguish between strong and weak nonsymmetry. In case of a convection dominated problem the numbering proposed in Sect. 2.1 does not represent the necessary numbering along the stream lines. A decrease in the numerical efficiency has to be expected caused by cutting the stream lines on domain interfaces. On the other hand if the problem is not convection dominated but nonsymmetric then the preconditioners under consideration can be used in the parallelized GMRES. The proper tests will be done in the future.

References

1. J. H. Bramble, J. E. Pasciak, and A. H. Schatz. The construction of preconditioners for elliptic problems by substructuring I – IV. *Mathematics of Computation*, 1986, 1987, 1988, 1989. 47, 103–134, 49, 1–16, 51, 415–430, 53, 1–24.
2. U. Groh. Local realization of vector operations on parallel computers. Preprint SPC 94-2, TU Chemnitz, 1994. in german.
3. I. Gustafsson. A class of first order factorization methods. *BIT*, 18:142–156, 1978.
4. G. Haase. Hierarchical extension operators plus smoothing in domain decomposition preconditioners. *Applied Numerical Mathematics*, 23(3):327–346, May 1997.
5. G. Haase, U. Langer, and A. Meyer. The approximate Dirichlet decomposition method. part I,II. *Computing*, 47:137–167, 1991.
6. M. Jung. On the parallelization of multi-grid methods using a non-overlapping domain decomposition data structure. *Applied Numerical Mathematics*, 23(1), 1997.
7. K. H. Law. A parallel finite element solution method. *Computer and Structures*, 23(6):845–858, 1989.
8. A. Meyer. A parallel preconditioned conjugate gradient method using domain decomposition and inexact solvers on each subdomain. *Computing*, 45:217–234, 1990.
9. B. Smith, P. B. rstad, and W. Gropp. *Domain Decomposition : Parallel Multilevel Methods for Elliptic Partial Differential Equations*. Cambridge University Press, 1996.
10. C. H. Tong, T. F. Chan, and C. J. Kuo. Multilevel filtering preconditioners: Extensions to more general elliptic problems. *SIAM J. Sci. Stat. Comput.*, 13:227–242, 1992.

A Comparison Between Different Parallelization Methods on Workstation Clusters to Solve CFD-Problems

W. Huber

Institut für Informatik, Technische Universität München,
Arcisstrasse 21, D-80290 München, Germany

Abstract. The efficient parallel solution of flow problems on parallel computers requires highly efficient numerical methods as well as highly efficient parallelization methods. Parallelization is mostly done using domain decomposition methods. With the introduction of the combination method a big jump in computational speed is possible. Here, a numerical solution is computed on a sparse grid. The algorithmic concept is based on the independent solution of many problems with reduced size and their linear combination. So, this method is perfectly suitable for parallelization. In this paper we compare both very efficient methods. We focus on the parallel solution of typical problems arising in the field of flow simulations. A comparison is done using the overall runtimes of both methods as well as the parallel speedup. The implementations are done on a workstation cluster of 16 HP machines.

1 Introduction

Domain decomposition methods are well known and have been applied to a wide variety of problems esp. to the parallelized numerical computing of partial differential equations (PDEs). These techniques are very flexible for a large range of problems. In the context of parallel computing, domain decomposition refers to the decomposition of data structures that can be computed independently. So, quite often *data decomposition* is a synonym term on it.

In this paper, we like to introduce a new discretization method as well as a new parallelization strategy. The so-called *sparse grid* technique is promising for the solution of linear and nonlinear PDEs. There, for two-dimensional problems, only $O(h_m^{-1}\mathrm{ld}(h_m^{-1}))$ grid points are needed instead of $O(h_m^{-2})$ grid points as in the conventional full grid case. Here, $h_m = 2^{-m}$ denotes the mesh size in the two-dimensional region. However, the accuracy of the sparse grid solution is of the order $O(h_m^2 \mathrm{ld}(h_m^{-1}))$ (with respect to the L_2- and L_∞-norm). This is only slightly worse than the order $O(h_m^2)$ obtained for the usual full grid solution. For further details on sparse grids, see [1, 7].

For the solution of problems arising from the sparse grid discretization approach, we like to use the so-called *combination method*. There, the solution is obtained on a sparse grid by a certain linear combination of discrete solutions on different full, but smaller grids.

2 Domain Decomposition Method

The numerical simulation of fluid flow problems involves the solution of the Navier-Stokes equations. To reduce the execution time parallelization by using domain decomposition methods have been applied.

The first step in computing fluid flow problems is to discretize the given set of PDEs. The usual approach is to discretize the problem by a FE, FV, or FD method.

Furthermore appropriate boundary conditions have to be used. The discretization leads to a linear or linearized system of equations $L_{h_m,h_m} u_{h_m,h_m} = f_{h_m,h_m}$ on grid Ω_{h_m,h_m} with mesh sizes $h_m = 2^{-m}$ in the different directions. To get a solution u_{h_m,h_m} iterative methods are used (for example algebraic multigrid-methods –AMG or SOR-methods, see [2, 6]).

Reducing the overall computing time parallelization is used. Normally the grid Ω_{h_m,h_m} is decomposed into several subdomains $\Omega_{h_m,h_m}^{(1)}, \Omega_{h_m,h_m}^{(2)}, ..., \Omega_{h_m,h_m}^{(P)}$ (i.e. data decomposition). Here, P denotes the number of used workstations. Several methods exists to make an efficient decomposition. In our discussion we like to decompose the full grid in overlapping stripes. The overlapping area is one stripe.[1]

3 Sparse Grids and the Combination Method

Now, we like to discuss the main ideas of the combination method. For reasons of simplicity, we focus on the two dimensional region $\Omega =]0, 1[\times]0, 1[\subset \mathbf{R}^2$ in the x, y coordinate system. The principles are independent of the used discretization method.

Extending this standard approach, we now study linear combinations of discrete solutions of the problem on different regular grids. To this end, let Ω_{h_i,h_j} denote a grid on the two dimensional region with mesh sizes $h_i = 2^{-i}$ in the x-direction, $h_j = 2^{-j}$ in the y-direction. In [4], the *combination method* was introduced. We recall the definition of the combined solution u_{h_m,h_m}^c:

$$u_{h_m,h_m}^c := \sum_{i+j=m+1} u_{h_i,h_j} - \sum_{i+j=m} u_{h_i,h_j}. \tag{1}$$

Here i, j range from 1 to m, where m is defined by the mesh size $h_m = 2^{-m}$ of the associated full grid. The different solutions u_{h_i,h_j} represent solutions of the same fluid flow. Thus, we have to solve m problems arising from the discretization of the Navier-Stokes equations (i.e. $i+j = m+1$) each with about 2^m unknowns and $m-1$ problems (i.e. $i+j = m$) each with about 2^{m-1} unknowns. Then, we combine their interpolated solutions. Interpolations can be done by using constant or linear functions. This leads to a solution defined on the sparse grid Ω_{h_m,h_m}^s in Cartesian coordinates. The sparse grid Ω_{h_m,h_m}^s is a subset of the associated full grid Ω_{h_m,h_m} (see Fig. 1). Its grid points

Fig.1 Sparse grid Ω_{h_m,h_m}^s and associated full grid Ω_{h_m,h_m} with mesh width $h_m = 1/16$.

are defined to be the union of the grid points of the grids Ω_{h_i,h_j}, where $i+j = m+1$

---

[1] Communication takes place after each iteration of the iterative solver. So, the number of iterations of the parallel implementation is the same as in our sequential implementation (numerical efficiency of both implementations is the same).

and i, j range from 1 to m. Altogether, the combination method involves $O(h_m^{-1}\text{ld}(h_m^{-1}))$ unknowns, in contrast to $O(h_m^{-2})$ unknowns for the conventional full grid approach. Additionally, the accuracy of the combination solution u_{h_m,h_m}^c is of the order $O(h_m^2\text{ld}(h_m^{-1}))$. This is only slightly worse than for the associated full grid, where the error is of the order $O(h_m^2)$. For details on sparse grids and the combination method, see [1, 4, 7].

To apply the combination method also in the case where *different* amounts of grid points for the different coordinate directions are used, we have to generalize Eqn. (1). To this end, we have to change the notation. Let L, M denote the number of volume cells in the x- and y-directions, respectively, and let $\Omega_{(L,M)}$ denote the corresponding grid. Without loss of generality, let L, M be even and positive. Now, we seek the *factorization* of L and M, where

$$L =: 2^l \cdot \kappa_l, \quad M =: 2^m \cdot \kappa_m, \tag{2}$$

with $\kappa_l, \kappa_m, m, n \in \mathbb{N}, \kappa_l, \kappa_m \geq 2$ and l, m maximal. The value p, which will be needed later in Eqn. (5), is determined by

$$p := min(l, m) + 1. \tag{3}$$

Now, we define the mesh sizes h_i, h_j of a grid Ω_{h_i,h_j} to be

$$h_i = k_L^{-1} \cdot 2^{-i}, \quad h_j = k_M^{-1} \cdot 2^{-j}, \tag{4}$$

with $k_L, k_M \in \mathbb{R}$. The mesh sizes h_i and h_j then take the values k_L and k_M into account, which are defined via $L =: 2^p \cdot k_L, M =: 2^p \cdot k_M$. The corresponding solutions are described by u_{h_i,h_j}. In this way, we are able to give a generalized combination formula for the case of different amounts of grid points/cells for the different coordinate directions by

$$u_{(L,M)}^c := \sum_{i+j=p+1} u_{h_i,h_j} - \sum_{i+j=p} u_{h_i,h_j}. \tag{5}$$

Now i, j range from 1 to p, where p is given by Eqn. (3). $u_{(L,M)}^c$ provides a solution on the *generalized* sparse grid $\Omega_{(L,M)}^s$ which is associated with $\Omega_{(L,M)}$ (by Eqns. (4) and (5)). Its grid points are now defined to be the union of the points of the grids Ω_{h_i,h_j} according to Eqns. (4) and (2), where $i + j = p + 1$ and each index i and j ranges from 1 to p.

The use of extremely distorted grids with a very high aspect ratio of the mesh sizes might cause certain problems (sensitive discretization scheme, slowing down of the solver, no resolution of physical features). Therefore, we must be able to limit the range of solutions involved and to use only a subset of all possible grids. We achieve this by introducing a new parameter $q \in \mathbb{N}, 1 \leq q \leq p$, into the combination method. We define

$$u_{(L,M)}^{c,q} := \sum_{i+j=q+1} u_{h_i,h_j} - \sum_{i+j=q} u_{h_i,h_j} \tag{6}$$

where each index i, j now ranges only from 1 to q. The mesh sizes h_i and h_j for grid Ω_{h_i,h_j} are now defined by $h_i = k_L^{-1} \cdot 2^{-i}$ and $h_j = k_M^{-1} \cdot 2^{-j}$, where k_L and k_M are defined by

$$k_L = L/2^q, k_M = M/2^q. \tag{7}$$

For the case $q = p$, we obtain the earlier Eqn. (5), but for values $q < p$ we limit the range of the combination method and thus exclude solutions u_{h_i,h_j} on extremely distorted grids. In the case $q = 1$, we just obtain the full grid solution $u_{(L,M)}$ and no combination takes place. The corresponding generalized sparse grid $\Omega^{s,q}_{(L,M)}$ consists of the union of the points of the grids Ω_{h_i,h_j} according to Eqns. (4) and (7), where $i + j = q + 1$ and i, j range from 1 to q. Now, in Eqn. (6), we have to solve q problems (i.e. $i + j = q + 1$) and $q - 1$ problems (i.e. $i + j = q$) arising from the discretization of the Navier-Stokes equations at various mesh refinements.

Quite often, engineers are interested in typical coefficients, like the lift coefficient. In this case, it is possible to combine these typical values only. In a formal way, we get once again a further modification of the combination formula In Eqn. (8), F denotes a typical coefficient.

$$F(u^{c,q}_{(L,M)}) := \sum_{i+j=q+1} F(u_{h_i,h_j}) - \sum_{i+j=q} F(u_{h_i,h_j}). \tag{8}$$

4 Parallelization of the Combination method

The parallelization of our combination algorithm can be achieved straightforwardly on a relative coarse grain level (that means code-partitioning). The computations for the different subproblems that are associated with the grids Ω_{h_i,h_j} (i.e. $i + j = q + 1$ and $i + j = q$, compare Eqn. (6) and (8)) can be performed *fully in parallel*.[2] Here, we restrict ourselves on the coarse grain code-partitioning. Using this relative coarse grain parallelization method, the number of usable workstations is restricted by the number of problems, see Table 1.

However, the *explicit* storage and assemblage of the sparse grid solution are not really necessary. In [3] it is described how the explicit combination to the sparse grid can be avoided.[3] On different values of q, Table 1 shows the fraction of the amount of memory necessary to store the data belonging to *one*, i.e. the largest grid arising in Eqn. (6), versus the amount of memory necessary to store the data belonging to the full grid $\Omega_{(L,M)}$. In brackets we note the reduction rate in percent. This resembles the case that a sufficient amount of workstations is available.

Table 1 Requirements for storing.

q	Number of grids involved in the combination method	Shared storage in comparison with the full grid	Distributed storage in comparison with the full grid
1	1	1 (100.00%)	1 (100.00%)
2	3	5/4 (125.00%)	1/2 (50.00%)
3	5	8/8 (100.00%)	1/4 (25.00%)
4	7	11/16 (68.75%)	1/8 (12.50%)
5	9	14/32 (43.75%)	1/16 (6.25%)
6	11	17/64 (26.56%)	1/32 (3.13%)

---

[2] Of course, the execution times of our parallel implementation can be reduced by using a data partitioning on the arising subproblems or by a vectorization. At the end, we get a two-level parallelization by using the sparse grid combination method.

[3] This is achieved by *directly* exchanging relevant data between the workstations that store the data associated with the grids Ω_{h_i,h_j} ($i + j = q + 1$ and $i + j = q$) and by updating the data on these grids.

5 Numerical Results

For the discretization of the incompressible Navier-Stokes equations a second order finite differencing method in space on a staggered grid is used. For time discretization we use a first order explicit discretization method. Solving the arising algebraic system of equations a AMG and a SOR are implemented. So, we are able to cope with complex geometries. On more details of the Navier-Stokes solver NaSt2D, see [2].

First of all, we focus on the parallel solution of a heat driven fluid flow, see Fig. 2. Here, we have to solve the Navier-Stokes equations and a PDE to describe the temperature. The Prandtl-Number[4] is $Pr = 7.0$, the Reynolds-number is $Re = 11063$, and the Rayleigh-number[5] is $Ra = 2.0 \cdot 10^5$. Normally, engineers are interested only in specific parameters like the Nußelt-number[6]. So, we have to combine this value only (see also Eqn. (8)). The computation of the solution u_{h_i,h_j} on the different regular grids can be done fully in parallel and no communication takes place.

Table 2 shows the numerical results on the associated full grid, whereas Table 3 shows the computed results using the combination method. Our results using the combination method are in a good agreement with the computations on the associated full grid. Increasing the number of combined grids (i.e. increasing q) the numerical results are getting worser in comparison to the numerical results on the associated full grid. That's not very surprising, because the sparse grids $\Omega_{h_n,h_n}^{s,q}$ consists less and less grid points and so characteristic structures of the velocity field are not be able to compute.

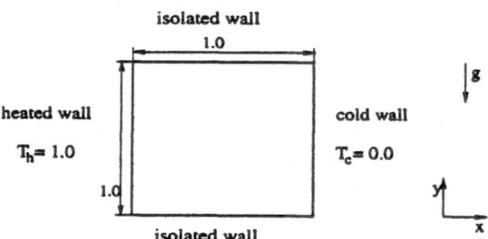

Fig.2 *Configuration for a heat driven fluid flow.*

full grids	Nußelt-number
$\Omega_{16,16}$	6.75050
$\Omega_{32,32}$	6.08627
$\Omega_{64,64}$	5.87789
$\Omega_{128,128}$	5.82428

Table 2 Numerical results on various full grids for various mesh refinements.

sparse grids, q	Nußelt-number
$\Omega_{(64,64)}^{s,2}$	5.87727
$\Omega_{(64,64)}^{s,3}$	5.87517
$\Omega_{(64,64)}^{s,4}$	5.88193
$\Omega_{(64,64)}^{s,5}$	5.91919
$\Omega_{(64,64)}^{s,6}$	6.01629

Table 3 Numerical results using the combination method based on the associated full grid $\Omega_{(64,64)}$.

---

[4] defined as $Pr = \frac{\nu}{\alpha}$. Here, ν denotes the kinematic viscosity.
[5] The Rayleigh-number Ra is defined as $Ra = Pr \cdot Gr$ and Gr the Grashof-number,
[6] The Nußelt-number is a non dimensional value computed by the quotient of convection and diffusion.

The second test case are steady calculations of a fluid flow around a few obstacles at a Reynolds-number of $Re = 100$ (see Fig. 3). Some definitions have to be introduced to specify the fluid flow. The obstacles are located at D1 $((0.5, 0.625) - (0.8125, 0.8125))$, D2$((0.5, 0.3125) - (0.6875, 0.4375))$ and D3 $((1.0, 0.375) - (1.125, 0.625))$. The Reynolds-number is defined as $Re = \frac{\overline{U} \cdot H}{\nu}$ with the mean velocity $\overline{U}(t) = 1.0$. Here, H denotes the channel height $(H = 1m)$ and ν the kinematic viscosity $(\nu = 10^{-2}m^2/s)$. The inflow condition is $U(0, y) = 4 \cdot U_m \cdot y(H - y)/H^2$ with $U_m = 1.5m/s$, yielding the Reynolds-number of $Re = 100$. Neumann outflow boundary conditions are used. Comparing different numerical simulations, we compute the lift coefficient on all obstacles

$$c_{li} = \frac{2F_a}{\rho \cdot \overline{U}^2 H_i} \quad i = 1, 2, 3$$

with H_i the height of the different obstacles, ρ denotes the fluid density $(\rho = 1.0kg/m^3)$. Results of mesh-refinement studies are shown in Table 4. Using the combination method on the associated full grid $\Omega_{512,128}$ only $q = 2, 3$ and on grid $\Omega_{1024,256}$ only $q = 2, 3, 4$ are possible. It depends on the less number of grid points in the different directions. Numerical results are listed in Table 5. Once again a quite good agreement between both simulations on the different grid representations (i.e. full and sparse grids) is established. Similar to the first test case, the accuracy using the combination method drops down by increasing the number of used grids. It depends on the reduced number of grid points on the sparse grid.

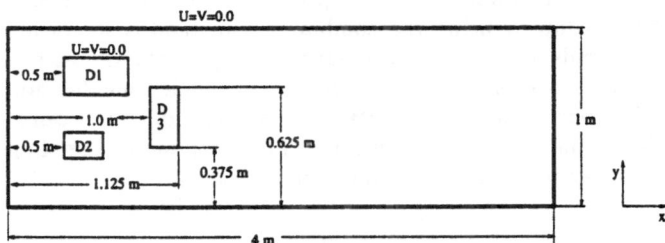

Fig.3 *Configuration and boundary conditions for a fluid flow around a few obstacles (D1 - D3).*

full grids	c_{l1} at D1	c_{l2} at D2	c_{l3} at D3
$\Omega_{256,64}$	2.188	1.374	1.300
$\Omega_{512,128}$	2.222	1.382	1.298
$\Omega_{1024,256}$	2.238	1.388	1.302

Table 4   Numerical results on various full grids.

sparse grids	c_{l1} at D1	c_{l2} at D2	c_{l3} at D3
$\Omega^{c,2}_{(512,128)}$	2.212	1.372	1.291
$\Omega^{c,3}_{(512,128)}$	2.214	1.351	1.277
$\Omega^{c,2}_{(1024,256)}$	2.229	1.376	1.293
$\Omega^{c,3}_{(1024,256)}$	2.204	1.340	1.266
$\Omega^{c,4}_{(1024,256)}$	2.166	1.282	1.219

Table 5   Numerical results using the combination method.

6 Parallelization Results

We implemented both parallelization methods (i.e. combination method and domain decomposition method) on a cluster of 16 HP 9000/720 workstations. The computers are organized like a binary tree. Communication is done by using PVM 3.3.6.

To compare the different sequential and the different parallel implementations of our Navier-Stokes solver a few metrics are used. First of all we focus on the parallel speedup $S_{par}(P)$ on P workstations to evaluate the performance of our implementations.[7]

On the heat driven flow we get the results seen in Table 6. The reduction in computing time is similar to the reduction in storage requirement. The different values of the parallel speedup for various numbers of combined grids depend on the non optimal number of used workstations.[8]

$P \backslash q$	2	3	4	5
1	1.00	1.00	1.00	1.00
2	1.67	2.00	1.83	2.00
4	-	4.00	3.67	3.50
8	-	-	-	7.00

Table 6 Parallel speedup using the combination method for various numbers of used workstations P on the first test case.

Now let us come to the test case with the three obstacles. In Table 7 we can see the parallel speedup using a domain decomposition method. The reason for the poor speedup-values is the small number of grid points on the different subdomains.

The parallel speedup by using the combination method is shown in Table 8. The different values depends on the different made time step for the various numbers of grids Ω_{h_i,h_j} of the Navier-Stokes solver arising from the explicit time discretization and on the non optimal number of used workstations (compare Table 1). It can be seen clearly that the parallelization properties using the combination method are in general better in comparison to the used domain decomposition method.

P	$\Omega_{512,128}$	$\Omega_{1024,256}$
1	1.00	1.00
2	1.80	1.88
4	2.73	3.37
8	3.85	6.21
16	4.92	9.68

Table 7 Parallel speedup using a domain decomposition method on various numbers of used workstations P on the second test case.

$P \backslash q$	$\Omega^{c,2}_{(512,128)}$	$\Omega^{c,3}_{(512,128)}$	$\Omega^{c,2}_{(1024,256)}$	$\Omega^{c,3}_{(1024,256)}$	$\Omega^{c,4}_{(1024,256)}$
1	1.00	1.00	1.00	1.00	1.00
2	1.72	1.97	1.71	1.92	1.92
3	2.12	2.78	1.71	2.61	2.81
4	-	-	-	-	3.59

Table 8 Parallel speedup using the combination method on various numbers of used workstations P on the second test case.

The different time steps on the various numbers of grids arising in the combination formula as well as the various numbers of grid point on the full and the sparse grids

---

[7] Dealing on the parallel speedup, the same code is used on different numbers of workstations.

[8] The number of used workstations is limited by the number of independent problems arising in the combination formula (compare Table 1), so a optimal load balancing is not always possible.

are reasons why the parallel speedup is not sufficient to compare both parallelization methods. So, we compare the total execution times of both parallelization strategies also, see Table 9. Here, we compute

$$S(P) = \frac{\text{execution times of full grid version on } P \text{ workstations}}{\text{execution times of the combination method on } P \text{ workstations}}.$$

$P\backslash q$	$\Omega^{c,2}_{(512,128)}$	$\Omega^{c,3}_{(512,128)}$	$\Omega^{c,2}_{(1024,256)}$	$\Omega^{c,3}_{(1024,256)}$	$\Omega^{c,4}_{(1024,256)}$
1	1.57	2.79	1.37	3.68	8.26
2	1.66	3.01	1.70	3.75	8.44
3	1.36	2.36	1.20	3.59	8.70
4	-	-	-	-	8.80

Table 9 Reduction on computing time using the combination method in comparison to the full grid version using a various number of workstations on the second test case.

The advantage of using the combination method increases by increasing the number of used grids. Using domain decomposition methods it is possible to realize an optimal load balancing in comparison to the combination method. Here, load balancing is done by the different numbers of problems only. So, load imbalancing is possible and as a result the run times in comparison to the full grid drops down in a different way. It mentioned before, that the used time step on the different grids has great influence on the run times. So, using an implicit time discretization the situation will be different.

7 Conclusion

In this paper, we compared different parallelization strategies to simulate incompressible laminar fluid flows. The parallel efficiency of the combination method is higher in comparison to standard domain decomposition methods. By increasing the number of grid points the overall run times using domain decomposition methods will increase much faster in comparison to the combination method. Implementations were done on a cluster of 16 HP workstations. On further numerical results, compare [5].

References

1. H. Bungartz. Dünne Gitter und deren Anwendung bei der adaptiven Lösung der dreidimensionalen Poisson-Gleichung. Dissertation, Institut für Informatik, TU München, 1992.
2. M. Griebel, Th. Dornseifer, and T. Neunhoeffer. Numerische Simulation in der Strömungsmechanik. Vieweg Verlag, 1996.
3. M. Griebel, W. Huber, T. Störtkuhl, and C. Zenger. On the parallel solution of 3D PDEs on a network of workstations and on vector computers. In A. Bode and M. Dal Cin, editor, Lecture Notes in Computer Science,732, pp. 276-291, Computer Architecture: Theory, Hardware, Software, Applications. Springer Verlag, Berlin, Heidelberg, New York, 1993.
4. M. Griebel, M. Schneider, and C. Zenger. A combination technique for the solution of sparse grid problems. In P. de Groen and R. Beauwens, editor, Proceedings of the IMACS International Symposium on Iterative Methods in Linear Algebra, pp. 263-281. Elsevier, Amsterdam, 1992.
5. W. Huber. Numerical turbulence simulation on different parallel computers using the sparse grid combination method. In Lecture Notes in Computer Science, volume 1124, pages 62-65. Springer-Verlag, 1996.
6. K. Stüben. Algebraic Multigrid (AMG), Experience and Comparisons. Appl. Math. Comp., 13:419-452, 1983.
7. Chr. Zenger. Sparse grids. In W. Hackbusch. editor, Parallel Algorithms for Partial Differential Equations, Proceedings of the Sixth GAMM-Seminar, Kiel, 1990, Notes on Numerical Fluid Mechanics, 31. Vieweg, Braunschweig, 1991.

Scalable Parallel SSOR Preconditioning for Lattice Computations in Gauge Theories

A. Frommer[1], Th. Lippert[2]* and K. Schilling[2]

[1]Department of Mathematics, University of Wuppertal, 42097 Wuppertal, Germany
[2]HLRZ, c/o KFA-Jülich, D-52425 Jülich, Germany and DESY, Hamburg, Germany

Abstract. We discuss a parallelization scheme for SSOR preconditioning of Krylov subspace solvers as applied in lattice gauge theory computations. Our preconditioner is based on a locally lexicographic ordering of the lattice points leading to a parallelism adapted to the parallel system's size. By exploitation of the 'Eisenstat-trick' within the bi-conjugate gradient stabilized iterative solver, we achieve a gain factor of about 2 in the number of iterations compared to conventional state-of-the-art odd-even preconditioning. We describe the implementation of the scheme on the APE100/Quadrics SIMD parallel computer in the realistic setting of a large scale lattice quantum chromodynamics simulation.

1 Introduction

Lattice gauge theory (LGT) deals with the controlled numerical evaluation of gauge theories like quantum chromodynamics (QCD) on a 4-dimensional space-time-grid. QCD [1] is considered as the fundamental theory of the strong forces that bind quarks with gluons to form the known hadrons like the proton or neutron. In the low energy regime, QCD cannot be solved by non-perturbative analytical methods. Therefore, numerical simulations become more and more important to provide theoretical input for current and future accelerator experiments that attempt to observe new physics beyond the *Standard Model* of elementary particle physics [2].

The heavy computational demands in LGT are due to repeated solution of a huge system of linear equations,

$$M\psi = (1 - \kappa D)\psi = \phi, \tag{1}$$

with M being the so-called fermion (quark) matrix (that can be considered as analogous to a discretised Laplace equation) of dimension $r = 3 \times 4 \times V$. V is the volume of the underlying 4-dimensional space-time lattice, D contains non-diagonal elements only. The solution ψ of (1), a Green's function, describes the time behavior of the quarks [3]. This Green's function is related to both the simulation of QCD with respect to dynamical quark-gluon interaction and the extraction of physical observables like hadron masses. The size of the solution vector is of order $O(10^7)$ elements in today's state-of-the-art simulations.

---

* Presented by Th. Lippert

Recently, the BiCGstab method [5] has been established as nearly optimal Krylov subspace method in lattice QCD applications [6], requring only slightly more less iterations than the full GMRES method. This quasi-optimality suggests to turn attention on multigrid methods and/or preconditioning in order to achieve further speed up in numerical methods for solving (1). The application of multigrid techniques, up to now, is impractical, however, due to the gauge noise of the gluonic background field entering the fermion matrix in form of matrix coefficients of the discrete differential operator. This fields represents the gluons in lattice QCD (particles analogous to photons in electrodynamics). Thus preconditioning techniques, *i. e.* methods to decrease the condition of M appear to be the only promising path to further accelerate Krylov subspace solvers like BiCGstab.

A widely used parallelizable preconditioning approach in lattice gauge computations rests upon an odd-even decomposition of the matrix M [7]. It yields an efficiency gain by a factor of 2 when solving (1). In the mid-eighties, Oyanagi [8] proposed to use a certain incomplete LU (ILU) factorization of M as a preconditioner. As it stands, Oyanagi's method works satisfactorily on vector machines. However, on local memory or grid-oriented parallel computers, this preconditioner can hardly be implemented efficiently.

In the present paper we will discuss the parallel aspects of a new SSOR preconditioner for Lattice QCD. We call it the *locally lexicographic SSOR preconditioner* (LL-SSOR). As opposed to multicolor preconditioners (like the odd-even preconditioner) which lead to a decoupling of variables on a very fine grain level, the LL-SSOR approach reduces the decoupling to the minimum which is necessary to achieve a given parallelism. As for any SSOR preconditioner, the Eisenstat Trick [9] is crucial to its efficient implementation.

Our numerical experiments show that LL-SSOR leads to the fastest known solution method on current parallel computers [10], if M represents the widely used standard Wilson fermion matrix. The SSOR preconditioner is applied on the Italian supercomputer APE100/Quadrics and on the Cray T3E within the large scale simulation project SESAM [11, 12].

2 Preconditioning

2.1 Symmetric Gauss-Seidel method

Generally, the preconditioning of (1) proceeds by application of two non-singular matrices V_1 and V_2. They play the role of a left and a right preconditioner, respectively:

$$V_1^{-1} M V_2^{-1} \tilde{\psi} = \tilde{\phi}, \quad \text{where } \tilde{\phi} = V_1^{-1} \phi, \ \tilde{\psi} = V_2 \psi. \tag{2}$$

Any Krylov subspace method could now be applied directly to (2), replacing each occurrence of M and ϕ by $V_1^{-1} M V_2^{-1}$ and $\tilde{\phi}$, respectively.

However, this would yield only the preconditioned iterates $\tilde{\psi}^k$ and preconditioned residuals. Therefore, one usually reformulates the algorithms incorporating an implicit transformation back to the unpreconditioned quantities. For

BiCGstab the resulting algorithm then requires two additional systems with matrix $V = V_1 V_2$ and two systems with matrix V_1 to be solved in each iterative step (see Ref. [5]).

We consider symmetric Gauß-Seidel (SSOR) preconditioning. Assuming that M is scaled to unit diagonal and denoting $M = I - L - U$ with L strictly lower and U strictly upper triangular, the SSOR preconditioner is specified by $V_1 = I - L$, $V_2 = I - U$.

2.2 Eisenstat-Trick

For the SSOR preconditioner the simple identity $V_1 + V_2 - M = I$ holds. This important relation allows to apply the so-called 'Eisenstat-trick' [9]: We can write $V_1^{-1} M V_2^{-1} = V_2^{-1} + V_1^{-1}(I - V_2^{-1})$. Thus the matrix vector product $w = V_1^{-1} M V_2^{-1} r$ amounts to a 2-step solve

$$v = V_2^{-1} r, \quad u = V_1^{-1}(r - v), \quad w = v + u. \tag{3}$$

Since the matrices $V_1 = I - L$ and $V_2 = I - U$ are triangular, the solves can be done directly via forward or backward substitution, respectively. In terms of computational cost, a forward followed by a backward solve is approximately as expensive as a multiplication with M (required in the unpreconditioned method).

As a result, the SSOR-preconditioned BiCGstab-method (as well as other Krylov subspace methods) can be implemented with basically the same amount of work per iteration as in the unpreconditioned method. Any occurence of a matrix vector multiply in the unpreconditioned method is replaced by a forward and a backward solve in the preconditioned method, see Algorithm 1.

2.3 Ordering

The generic form of the Wilson fermion matrix M can be seen from the relation

$$(M\psi)_x = \psi_x - \kappa \left(\sum_{\mu=1}^{4} m_{x,x-\mu}^{+} \psi_{x-\mu} m_{x,x+\mu}^{-} \psi_{x+\mu} \right), \tag{4}$$

where x is an index for the (4-dimensional) grid coordinate, and $x \pm \mu$, $\mu = 1, \ldots, 4$ stands for the nearest neighbors in dimension μ on the grid (periodic at the boundaries). At each grid point x we have 12 variables, i.e. $\psi_x \in \mathbf{C}^{12}$, and the coupling matrices $m_{x,x-\mu}^{-}$ and $m_{x,x+\mu}^{+}$ are of the form

$$m_{x,x-\mu}^{-} = (I + \gamma_\mu) \otimes U_\mu^H(x + \mu), \quad m_{x,x-\mu}^{+} = (I - \gamma_\mu) \otimes U_\mu(x).$$

Here, the $U_\mu(x)$ are 3×3 matrices from SU(3) which represent the gluonic degrees of freedom on the lattice. The matrices γ_μ are 4×4 Dirac matrices.

We have the freedom to choose any ordering scheme for the lattice points x. Different orderings yield different matrices M, permutationally similar to each other. The efficiency of the SSOR preconditioner depends on the ordering scheme chosen. If we assume an arbitrary numbering (ordering) of the lattice

{ initialization }
choose ψ_0, set $r_0 = \phi - M\psi^0$
solve $(I - L)\tilde{r}_0 = r_0$ to get $\tilde{r}_0$ { forward solve }
$\tilde{\tilde{r}}_0 = \tilde{r}_0$
set $\rho_0 = \rho_1 = \alpha_0 = \omega_0 = 1$
set $\tilde{v}_0 = \tilde{p}_0 = 0$
{ iteration }
for $i = 1, 2, \ldots$
    $\rho_i = \tilde{\tilde{r}}_0^\dagger \tilde{r}_{i-1}$
    $\gamma_i = (\rho_i/\rho_{i-1})(\alpha_{i-1}/\omega_{i-1})$
    $\tilde{p}_i = \tilde{r}_{i-1} + \gamma_i(\tilde{p}_{i-1} - \omega_{i-1}\tilde{v}_{i-1})$
    solve $(I - U)z_i = \tilde{p}_i$ to get z_i { backward solve }
    solve $(I - L)\tilde{w}_i = \tilde{p}_i - z_i$ to get $\tilde{w}_i$ { forward solve }
    $\tilde{v}_i = z_i + \tilde{w}_i$
    $\alpha_i = \rho_i/\tilde{\tilde{r}}_0^\dagger \tilde{v}_i$
    $\tilde{s}_i = \tilde{r}_{i-1} - \alpha_i\tilde{v}_i$
    solve $(I - U)y_i = \tilde{s}_i$ to get y_i { backward solve }
    solve $(I - L)\tilde{u}_i = s_i - y_i$ to get $\tilde{u}_i$ { forward solve }
    $\tilde{t}_i = y_i + \tilde{u}_i$
    $\omega_i = \tilde{t}_i^\dagger \tilde{s}_i/\tilde{t}_i^\dagger \tilde{t}_i$
    $\psi_i = \psi_{i-1} + \omega_i y_i + \alpha_i z_i$
    $\tilde{r}_i = \tilde{s}_i - \omega_i\tilde{t}_i$
end for

Algorithm 1 SSOR preconditioned BiCGstab. Due to the Eisenstat trick, only backward and forward solves are present, no matrix vector multiplications

points, then, considering a given grid point x, we find that the corresponding row in the matrix L or U contains exactly the coupling coefficients of those nearest neighbors of x which have been numbered before or after x, respectively.

Therefore, a generic formulation of the forward solve for this ordering is given by Algorithm 2. The backward solves are done similarly, now running through the grid points in *reverse* order and taking those grid points $x \pm \mu$ which were numbered *after* (instead of *before*) x.

2.4 Odd-even preconditioning

A particular ordering for M is the odd-even scheme where lattice sites are collected in two groups according to their color in a checkerboard-like coloring. The SSOR preconditioner for this scheme hitherto was considered as the only successful preconditioner for lattice QCD in a parallel computing environment. For this particular ordering the inverses of $I - L$ and $I - U$ can be determined

for all grid points x in the given order
  { update y_x }
  $y_x = p_x$
  for $\mu = 1, \ldots, 4$
    if $x - \mu$ was numbered before x then
      $y_x = y_x + \kappa \cdot m_{x, x-\mu}^{+} y_{x-\mu}$
  for $\mu = 1, \ldots, 4$
    if $x + \mu$ was numbered before x then
      $y_x = y_x + \kappa \cdot m_{x, x+\mu}^{-} y_{x+\mu}$

for all colors in lexicographic order
  for all processors
    $x :=$ site of that color
    { update y_x }
    $y_x = p_x +$
    $\kappa \left(\sum_{\mu, \, x-\mu \leq_{ll} x} m_{x, x-\mu}^{+} y_{x-\mu} \right.$
    $\left. + \sum_{\mu, \, x+\mu \leq_{ll} x} m_{x, x+\mu}^{-} y_{x+\mu} \right)$

Algorithm 2 Generic forward solve and ll-forward solve

directly. With the odd-even ordering, the matrix M has the form

$$M = \begin{pmatrix} I & -\kappa D_{oe} \\ -\kappa D_{eo} & I \end{pmatrix} \tag{5}$$

so that

$$I - L = \begin{pmatrix} I & 0 \\ -\kappa D_{eo} & I \end{pmatrix}, \quad \text{therefore } (I - L)^{-1} = \begin{pmatrix} I & 0 \\ \kappa D_{eo} & I \end{pmatrix}$$

and

$$I - U = \begin{pmatrix} I & -\kappa D_{oe} \\ 0 & I \end{pmatrix}, \quad \text{therefore } (I - U)^{-1} = \begin{pmatrix} I & \kappa D_{oe} \\ 0 & I \end{pmatrix}.$$

Hence

$$(I - L)^{-1} M (I - U)^{-1} = \begin{pmatrix} I & 0 \\ 0 & I - \kappa^2 D_{eo} D_{oe} \end{pmatrix},$$

where $I - \kappa^2 D_{eo} D_{oe}$ is called the matrix of the odd-even reduced system.

2.5 Lexicographic ordering

The Oyanagi preconditioner considers M to be given with respect to the natural (lexicographic) ordering of the lattice points. This means that grid point $x = (i_1, i_2, i_3, i_4)$ is numbered before $x' = (i_1', i_2', i_3', i_4')$ if and only if ($i_4 < i_4'$) or ($i_4 = i_4'$ and $i_3 < i_3'$) or ($i_4 = i_4', i_3 = i_3'$ and $i_2 < i_2'$) or ($i_4 = i_4', i_3 = i_3', i_2 = i_2'$ and $i_1 < i_1'$). Oyanagi [8] showed that SSOR preconditioning for the lexicographic ordering yields a further improvement over odd-even preconditioning as far as the number of iterations is concerned[2]. Unfortunately, the parallel implementation of Oyanagi's method on local memory machines is very difficult and not efficient.

---

[2] Actually, he employed ILU preconditioning. This turns out to be identical to SSOR in our case, however.

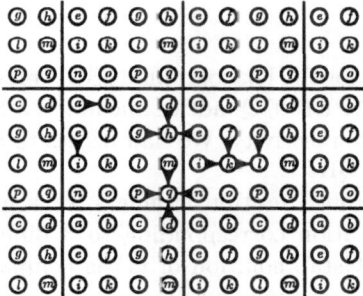

Fig. 1. *Locally lexicographic ordering* and forward solve in 2 dimensions

3 Parallel SSOR Preconditioning

The novel type of ordering we propose is adapted to the parallel computer used to solve equation (1). We assume that the processor connectivity is as a $p_1 \times p_2 \times p_3 \times p_4$ 4-dimensional grid (allowing $p_i = 1$ for certain i). The space-time lattice can be matched to the processor grid in an obvious natural manner, producing a local lattice of size $n_1^{loc} \times n_2^{loc} \times n_3^{loc} \times n_4^{loc}$ with $n_i^{loc} = n_i/p_i$ on each processor. The whole lattice is divided into $n^{loc} = n_1^{loc} n_2^{loc} n_3^{loc} n_4^{loc}$ groups. Each group corresponds to a fixed position of a site in the local grid, and a different color is associated with each of the groups, see Figure 1.

Let us consider the alphabetic ordering of the colors $a - q$ in Figure 1. Such an ordering is termed *locally lexicographic*. All nearest neighbors of a given grid point have colors different from that point. This implies that when performing the forward and backward solves in Algorithm 1, grid points having the same color can be worked upon in parallel, thus yielding an optimal parallelism of p, the number of processors.

A formulation of the *ll*-forward solve is given as Algorithm 2. Here, we use '$\leq_{ll}$' as a symbol for '*ll*-less than'. For grid points lying in the 'interior' of each local grid, we have $x - \mu \leq_{ll} x \leq_{ll} x + \mu$ for $\mu = 1, \ldots, 4$. The update in the forward solve $(I - L)y = p$ thus becomes

$$y_x = p_x + \kappa \left(\sum_{\mu=1}^{4} m_{x,x-\mu}^+ y_{x-\mu} \right),$$

whereas on the 'local boundaries' we will have between 0 (for the *ll*-first point) and 8 (for the *ll*-last point) summands to add to p_x.

The *ll*-forward and *ll*-backward solves can be carried out in parallel synchronously for the members of each color. The parallelism achieved is p and thus less than with the odd-even ordering but it is optimal since we have p processors. If we change the number of processors, the *ll*-ordering, and consequently the properties of the corresponding LL-SSOR preconditioner will change, too.

4 Results

Our numerical tests of the locally lexicographic SSOR preconditioner (LL-SSOR) have been performed on APE100/Quadrics machines, a SIMD parallel architecture with next-neighbor connectivity. We had access to a 32-node Quadrics Q4 and a 512-node Quadrics QH4.

In Fig. 2a, three different iteration numbers for a given thermalised QCD background field configuration corresponding to 3 different quark masses are given. The standard BiCGstab solution of (1) is compared with the odd-even preconditioned method and the new locally lexicographic ordering scheme. The measurements are carried out on a lattice of size $8^3 \times 16$, hence the computational problem is of granularity $g = N/p = 256$ on the Quadrics Q4 with 32 nodes. The gain in iterations is more than a factor of 2 for the κ-parameter values cho-

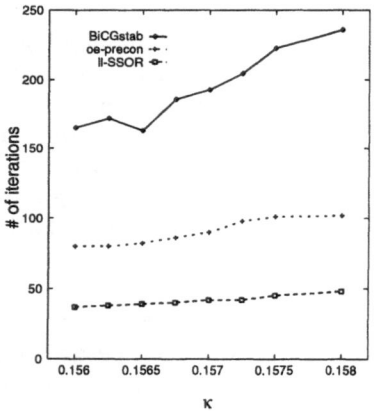 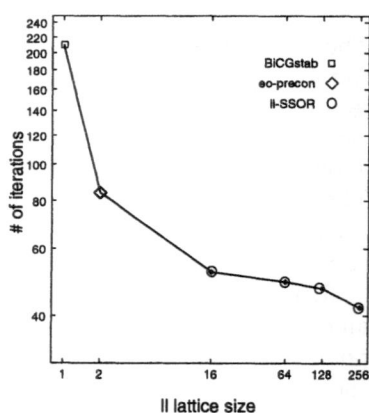

Fig. 2. Iteration numbers of unpreconditioned BiCGstab •, the odd-even- + and the *ll*-preconditioned □ version for a series of condition numbers on a $8^3 \times 16$ lattice. The second plot shows the dependence of iteration numbers on the local lattice size.

sen from a realistic setting and this translates into slightly smaller but similar gains in run time on the Q4. Compared to the unpreconditioned BiCGstab the gain is even a factor of 4. The convergence stopping criterion was to test for $||r||_2/||\phi||_2 < \epsilon$, where r is the *preconditioned* residual in the case of the preconditionmed methods. In either method we took $\epsilon = 10^{-8}$, for the preconditioned methods we also verified that the unpreconditioned residuals $Mx - \phi$, calculated explicitly from the solution x, also satisfied the stopping criterion.

The dependence of the iteration numbers on the local lattice size is depicted in Fig. 2b. Here the hopping parameter (controlling the condition number and the quark mass) of the matrix M is $\kappa = 0.1575$, closer to the critical value for which M becomes singular. The smallest local lattice size we can investigate within LL-SSOR is a 2^4 lattice. The unpreconditioned case can be interpreted as having local lattice size 1 and the odd-even preconditioning having local lattice size 2. As expected, smaller local lattices lead to a less efficient preconditioning. Eventually, for local= global lattice size, Oyanagi's results can be recovered.

5 Outlook

The parallel preconditioning scheme presented here has been proven very efficient in a large scale lattice QCD simulation, the SESAM project. Currently, we are investigating the application of LL-SSOR preconditioning to so-called improved actions, *i. e.*, new discretisation schemes that are able to improve lattice discretisations on the quantum level.

Acknowledgments. The work of Thomas Lippert is supported by the Deutsche Forschungsgemeinschaft DFG under grant No. Schi 257/5-1.

References

1. P. M. Zerwas and H. A. Kastrup (edts.), *QCD 20 years later* (World Scientific, Singapore, 1992).
2. Particle data group, Phys. Rev **D54** (1996) 1.
3. Th. Lippert, K. Schilling, and N. Petkov, 'Quark Propagator on the Connection Machine', Parallel Computing **18** (1992) 1291.
4. A. Frommer, V. Hannemann, Th. Lippert, B. Nöckel, and K. Schilling, 'Accelerating Wilson Fermion Matrix Inversions by Means of the Stabilized Biconjugate Gradient Algorithm' Int. J. Mod. Phys. **C5** (1994) 1073.
5. van der Vorst, H.: BiCGstab: A Fast and Smoothly Converging Variant of Bi-CG for the Solution of Non-symmetric Linear Systems, SIAM J. Sc. Stat. Comp. **13** (1992) 631–644.
6. Frommer, A., Hannemann, V., Nöckel, B., Lippert, Th. and Schilling, K.: Accelerating Wilson Fermion Matrix Inversions by Means of the Stabilized Bi-conjugate Gradient Algorithm, Int. J. of Mod. Phys. **C5** (1994) 1073–1088.
7. T. DeGrand and P. Rossi, Comp. Phys. Comm. **60** (1990) 211.
8. Oyanagi, Y.: An Incomplete LDU Decomposition of Lattice Fermions and its Application to Conjugate Residual Methods, Comp. Phys. Comm. **42** (1986) 333.
9. S. Eisenstat, 'Efficient Implementation of a Class of Preconditioned Conjugate Gradient Methods', SIAM J. Sci. Stat. Comput. **2** (1981) 1.
10. S. Fischer, A. Frommer, U. Glässner, S. Güsken, H. Hoeber, Th. Lippert, G. Ritzenhöfer, K. Schilling, G. Siegert, A. Spitz, 'A Parallel SSOR Preconditioner for Lattice QCD', in C. Bernard, M. Golterman, M. Ogilvie, and Jean Potvin, (edts.): Proceedings of *Lattice '96*, Nucl. Phys. B (Proc. Suppl.), **53** (1997) 990-992.
11. N. Attig, S. Güsken, P. Lacock, Th. Lippert, K. Schilling, P. Ueberholz and J. Viehoff, 'Highly Optimized Code for Lattice Quantum Chromodynamics on Cray T3E', accepted for Parco 97.
12. SESAM-Collaboration: U. Glässner, S. Güsken, H. Hoeber, Th. Lippert, X. Luo, G. Ritzenhöfer K. Schilling and G. Siegert: 'QCD with Dynamical Wilson Fermions – First Results from SESAM', in T. D Kieu, B. H. J. McKellar, and A. J. Guttmann, (edts.): Proceedings of *Lattice '95*, Nucl. Phys. B (Proc. Suppl.) **47** (1996) 386-393.

Deteriorating Convergence for Asynchronous Methods on Linear Least Squares Problems*

Trond Steihaug and Yasemin Yalçınkaya

University of Bergen,
Department of Informatics,
Bergen, Norway

Abstract. A block iterative method is used for solving linear least squares problems. The subproblems are solved asynchronously on a distributed memory multiprocessor. It is observed that an increased number of processors results in deteriorating rate of convergence. This deteriorating convergence is illustrated by numerical experiments. The deterioration of the convergence can be explained by contamination of the residual. Our purpose is to show that the residual is contaminated by *old information*. The issues investigated here are the effect of the number of processors, the role of *essential neighbors*, and synchronization. The characterization of old information remains an open problem.

1 Introduction

In this paper, a block iterative method is used for solving sparse linear least squares problems. A general framework for this method is introduced by Dennis and Steihaug [4], and the preliminary tests in [4] indicate that this method leads quickly to cheap solutions of limited accuracy.

Due to the rapid development and increasing usage of parallel computers and distributed computing, it is now important to adapt the methods to the new architectures. Baudet's [1] experimental results on systems of linear equations show a considerable advantage for iterative methods on parallel computers with no synchronization. This leads to experimentation with *totally asynchronous* [2] block iterative methods for the solution of linear least squares problems.

The block iterative method [4] partitions the columns of the coefficient matrix into disjoint blocks of columns and then projects the updated residual into each column subspace. This algorithm, which is called *column oriented successive subspace correction* (CSSC) in [8], is, in fact, Gauss-Seidel iteration on the normal equations. Each subproblem is substantially smaller than the original problem and hence is solved directly using QR factorization [7] and semi-normal equations (SNE) [3] on one processor. Each processor computes a correction of the solution vector restricted to the variables associated with the blocks of columns. The computation requires the residual, which is the global data. Each processor uses the residual available at the start of the computations without waiting for

---
* This research is supported by The Research Council of Norway.

the newest data. This way, the disadvantages resulting from the execution of synchronization primitives are avoided.

We do not address the issues of timing and speedup in this paper. Some timing and speedup results can be found in [11].

In the following, we first give the framework of the block iterative method for the linear least squares problem and state the sequential algorithm. The subproblems in this algorithm are to be solved using a direct method. A discussion of suitable factorization techniques for the subproblems is found in [10]. Then, we introduce the totally asynchronous algorithmic model and the algorithm for the asynchronous implementation of the method. In Section 4, the results of our experiments are presented. In the asynchronous implementation it is observed that increased number of processors results in contaminated residual and hence deteriorating convergence. This is due to the existence of *old information* in the system.

2 The Linear Least Squares Problem

Let A be an m by n real matrix, $m \geq n, b \in \mathbb{R}^m$. Let M be an m by m positive definite matrix. The weighted linear least squares problem is:

$$\min_{x \in \mathbb{R}^n} \|Ax - b\|_M, \tag{1}$$

where $\|y\|_M^2 = y^T M y$.

The columns of A are divided into g blocks $A_1, A_2, \ldots, A_g$, where $A_i \in \mathbb{R}^{m \times n_i}$. Assume, without loss of generality, that each A_i has full rank. The least squares problem (1) is equivalent to

$$\min\{\|A_1 x_1 + A_2 x_2 + \cdots + A_g x_g - b\|_M : x_i \in \mathbb{R}^{n_i}, i = 1, 2, \ldots, g\} . \tag{2}$$

Suppose that x^k is an approximation to a solution x^* to (1), and x^k is divided into $x_1^k, x_2^k, \ldots, x_g^k$ as above. Then from (2), the following successive replacements iteration can be obtained:

for $i = 1, 2, \ldots, g$ **do**
  Solve for $x_i^{k+1} \in \mathbb{R}^{n_i}$:
    $\min \| \sum_{j=1}^{i-1} A_j x_j^{k+1} + A_i x_i^{k+1} + \sum_{j=i+1}^{g} A_j x_j^k - b \|_M$.

This is equivalent [4, 12] to:

for $i = 1, 2, \ldots, g$ **do**
  $\hat{r} = r^{k+(i-1)/g}$.
  Solve for s_i^k : $\min\{\|A_i s_i + \hat{r}\|_M : s_i \in \mathbb{R}^{n_i}\}$. (3)
  Update the residual: $r^{k+i/g} = r^{k+(i-1)/g} + A_i s_i^k$.
  Update the solution: $x^{k+i/g} = x^{k+(i-1)/g} + \bar{s}_i^k$.

For $s_i \in \mathbb{R}^{n_i}$, introduce the vector $\bar{s}_i \in \mathbb{R}^n$, which is obtained by starting with a zero vector and placing the nonzero entries of s_i in the right positions.

This is block Gauss-Seidel iteration on the normal equations for (1). The residual $\hat{r} = r^k$, gives block Jacobi iteration on the normal equations. The intermediate residual is a combined Jacobi and Gauss-Seidel method. With the introduction of a relaxation parameter successive over-relaxation (SOR) method is obtained.

Algorithm 1.

    Subdivide A into g blocks.
    Choose $0 < \omega_i < 2, i = 1, 2, \ldots, g$.
    Choose $x_i^0, i = 1, 2, \ldots, g$, $x^0 = \sum_{i=1}^{g} \bar{x}_i^0$.
    Compute $r^0 = Ax^0 - b$.
    for $k = 0$ **step** 1 **until** *convergence* **do**
        for $i = 1, 2, \ldots, g$ **do**
            $\hat{r} = r^{k+(i-1)/g}$.
            Solve for s_i^k : $\min\{\|A_i s_i^k + \hat{r}\|_M\}$.
            $r^{k+i/g} = r^{k+(i-1)/g} + \omega_i A_i s_i^k$.
            $x^{k+i/g} = x^{k+(i-1)/g} + \omega_i \bar{s}_i^k$.
    Check for *convergence*.

The series of approximations $\{x^k\}$ from Algorithm 1 converge to x^*, a solution of the least squares problem (1), and $\|r^k\|_M$ is strictly monotonically decreasing [4].

3 Parallelization

In this section, we consider the parallel implementation of Algorithm 1, and formulate the main algorithm used in the experiments. First, we will see how we can get the parallel version of the sequential algorithm at hand, and then we will point out some of the advantages and disadvantages of asynchronous computation over the synchronous mode.

Jacobi type of iterations are straightforward to implement in parallel. In Algorithm 1, if use $\hat{r} = r^k$, we get Jacobi method. The main computation in the inner loop is the solution of (3). This system can be solved concurrently for each block i on multiple processors provided that the submatrices A_i are available on the processors. The processors, after computing their corrections on block components of x have to synchronize at the end of the loop before starting with the next loop. Now that we are able to compute the corrections from each block in one step, we have gained a considerable advantage over the sequential algorithm in terms of parallelization. In synchronized algorithms, the faster processors waiting for the slower ones to complete their computations causes an overhead. To get higher utilization of the available CPU power we can remove synchronization and the restriction on the order of the updates. By removing synchronization from the synchronous Jacobi algorithm and letting $\hat{r}$

get the latest available value of the residual in the system, we obtain a totally asynchronous algorithm.

Asynchronous algorithms reduce the synchronization penalty caused by fast processors waiting for slow processors to complete the computations, and for slow communication channels to deliver messages. The reason is that processors can execute more updates when they are not constrained to wait for the results of the computation on other processors. The only requirement on the computation of the updates is that, eventually, the values of an early update cannot be used any more in further evaluations. This condition is met as long as no processor falls out of the system. However, removing synchronization brings out the danger that the updates are performed on the basis of outdated (old) information.

An important disadvantage of asynchronism is that it can impede convergence properties that the algorithm may possess when executed synchronously or sequentially. In some cases, it is necessary to place limitations on the size of communication delays to guarantee convergence. Necessary conditions for the convergence of linear problems is given in Bertsekas and Tsitsiklis [2].

Before giving the asynchronous implementation of Algorithm 1, we will introduce the totally asynchronous algorithmic model.

3.1 The Totally Asynchronous Algorithmic Model

Let $\mathcal{T} = \{1, 2, \ldots\}$ be a set of times at which one block x_i of x is updated by some processor, and $\mathcal{T}^i$ = set of times at which x_i is updated.

The processor computing s_i may not have access to the most recent values of x_j at the time of the update. For $t \in \mathcal{T}^i$, $s_i(t)$ is computed using a residual $\hat{r} = \sum_{j=1}^{g} A_j x_j(\tau_j^i) - b$, where $\tau_j^i(t)$ are times satisfying

$$0 \leq \tau_j^i(t) \leq t - 1 .$$

At all times $t \notin \mathcal{T}^i$, x_i is left unchanged and

$$x(t) = x(t-1) + \bar{s}_i(t), \ t \in \mathcal{T}^i .$$

3.2 Asynchronous Implementation

The next algorithm is the asynchronous implementation of Algorithm 1 on an MPMD machine with $p = g + 1$ processors. In the algorithm, processor p_0 is used as the master and processors p_i, $i = 1, 2, \ldots, g$ act as slaves. **Send** and **Receive** are communications with the master. **Broadcast** is done by the master processor.

Algorithm 2.
  Subdivide A into g blocks.
  Choose $0 < \omega_i < 2, i = 1, 2, \ldots, g$.
  Choose $x_i(0), i = 1, 2, \ldots, g$, $x(0) = \sum_{i=1}^{g} \bar{x}_i(0)$.
  Compute $r(0) = Ax(0) - b$.

```
Initiate each processor i = 1, 2, ..., g :
    Receive(A_i, p_0).
    Receive(x̄_i(0), p_0).
    Receive(ω_i, p_0).
    Preprocess.
Broadcast(r(0)).
                        {Let t be a global counter of corrections and}
                        {let t₁ⁱ and t₂ⁱ be two consecutive elements in 𝒯ⁱ}
t = 0.
while not termination do
    if slave then
        Solve for s_i :
            min{‖A_i s_i + r(t₁ⁱ)‖_M : s_i ∈ ℝ^{n_i}}.
        Update x_i : x_i = x_i + ω_i s_i.
        Send(A_i s_i, p_0).
        Receive(r(t₂ⁱ), p_0).
    else                                        {master}
        Receive(A_i s_i, p_i).          {s_i computed using r at t₁ⁱ}
        r(t + 1) = r(t) + ω_i A_i s_i.
        Check for termination.
        if not termination then
            t = t + 1;
            Send(r(t), p_i).                    {t₂ⁱ = t}
```

4 Experiments

Test problems used in the experiments are taken from Harwell-Boeing sparse matrix test collection [5]. All the graphs refer to problem ASH958. Blocks are formed by taking blocks of consecutive columns. For the specified problem, the number of blocks g is 30. Each subproblem is solved using QR factorization and SNE. Both parallel and sequential implementations are done on Intel Paragon. Static assignment of blocks to processors is chosen to avoid the overhead of assignments during the computation phase. The number of processors p reported is the number of slave processors.

Algorithm 1 is a block Gauss-Seidel iteration which can be converted to Jacobi type iteration by taking $\hat{r} = r^k$, and the intermediate residual is a combined Jacobi and Gauss-Seidel iteration. To verify this, the residual is "fixed" for p-updates, i.e. $\hat{r} = r^{k+c/g}$ in (3), where $c = p\lfloor i/p \rfloor$. The quantity $\lfloor \cdot \rfloor$ is the largest integer not greater than its argument. Assume for simplicity that $g \bmod p = 0$. This routine implements Gauss-Seidel updates with a Jacobi iteration on groups of g/p block components. In this implementation $p = g$ gives the block Gauss-Seidel method, and $p = 1$ gives the block Jacobi method. We see in Fig. 1 that the resulting convergence is between sequential Gauss-Seidel and Jacobi methods acknowledging our statement. Increased number of processors results in increased deterioration of the convergence of the residual.

In the second experiment, a "time-lag" of p is used, where $\hat{r} = r^{k+(i-p)/g}$ in (3). The same block assignment as in the former case is used. Again, an increase in the number of processors results in increased deterioration as seen in Fig. 2. To state this result in a more formal way, let

$$\rho(p) = \sup_{x^0} \limsup_{k \to \infty} \|x^k - x^*\|^{1/k}$$

be the average rate of convergence using p processors. For a special class of linear systems Elsner, Neumann and Vemmer [6] prove $\rho(p+1) \geq \rho(p)$.

In the next experiment, a totally asynchronous iteration on different number of processors p, $p \leq g$, and $g \bmod p = 0$, is implemented. We see in Fig. 3 that for a small number of processors the rate of convergence lies in the neighborhood of sequential Gauss-Seidel method. If we increase the number of processors further to make better use of the available CPU power, the rate of convergence is degraded. In an asynchronous implementation, the order of the updates may change, and also the the total number of updates. In five different runs of the totally asynchronous implementation on $p = g$ processors, the number of updates before convergence varies between 958 and 1046.

An asynchronous implementation on homogeneous processors with negligible communication delays will, after some time, be almost cyclic in the updates. Let t_1 and t_2 be two consecutive updates of block i, $t_1, t_2 \in \mathcal{T}^i$. At t_2 the correction $s_i(t_1)$ is the solution of:

$$\min\{\|A_i s_i + r(t_1))\|_M\} \ .$$

The updated solution and residual are:

$$x(t_2) = x(t_2 - 1) + \omega_i \bar{s}_i(t_1), \quad r(t_2) = r(t_2 - 1) + \omega_i A_i s_i(t_1),$$

and for a cyclic implementation with cycle length g with p processors, an update will be computed with a residual which is p time units old. Hence, $t_1 = t_2 - p$, and we have a sequential implementation with time-lag of p.

We need to consider the effect of dependence between blocks on the convergence rate in the asynchronous implementation. Let $E_i = \{j \mid$ block i and block j have nonzero elements on the same row positions$\}$ be called the *essential neighbors* [9] of block i. Let $t_1, t_2 \in \mathcal{T}^i$ be consecutive times of update from block i. When we update the residual and approximate the solution at time t_2, $A_i^T r(t_2) = 0$ and $s_i(t_2) = 0$, unless any block $j, j \in E_i$ has sent an update between t_1 and t_2. In the next experiment, to avoid zero corrections, block i is forced to wait until an update from block $j, j \in E_i$ arrives at the master. We use $p = g$ processors in a heterogeneous environment with each processor's speed varying with a factor between 1 and 7. In Fig. 4, the curve marked WFE illustrates the implementation where the processors wait for their essential neighbors. The time between two nonzero updates of block i will be shorter if the processor waits for an essential neighbor of block i. Hence, we expect that the time-lag on the average will be reduced. In the numerical experiments, we see a decrease in the deteriorations, and as a result, an improvement in the convergence.

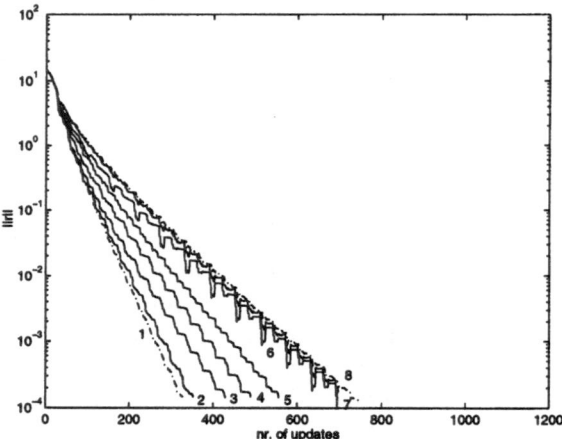

Fig. 1. Residual "fixed" for p updates. (1): Gauss-Seidel ($p = 1$)
(2): $p = 2$ (3): $p = 3$ (4): $p = 6$ (5): $p = 10$ (6): $p = 15$ (7): $p = 30$
(8): Jacobi

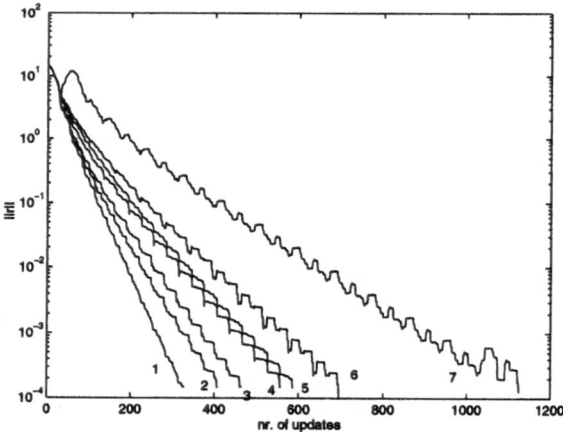

Fig. 2. "Time-lag" of p.   (1): Gauss-Seidel ($p = 1$)   (2): $p = 2$ (3): $p = 3$ (4): $p = 6$
(5): $p = 10$ (6): $p = 15$ (7): $p = 30$

The numerical testing depicted in Fig. 2 shows that increasing the number of processors means that older information is used to calculate any new iterate. This reduces the rate of convergence. To avoid that too old information is used to update the residual in the asynchronous implementation, we introduce a limit (Np) on the magnitude of time-lag, i.e., $t - Np \leq \tau_j^i(t) \leq t - 1$ for all i and j, and all $t \geq 0, t \in \mathcal{T}^i$. In [2], this is called a *partially asynchronous* iterative method. We use $p = g$ processors in a heterogeneous environment as in the former

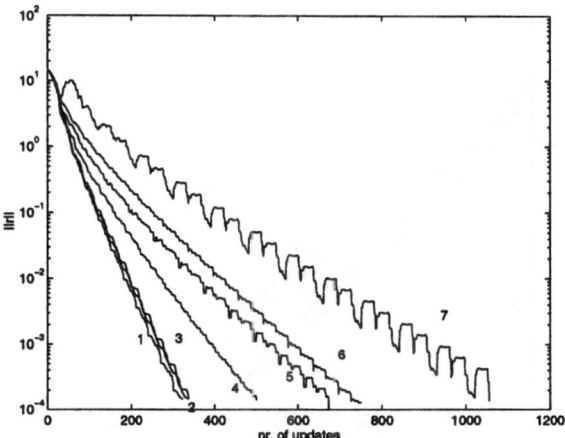

Fig. 3. Totally asynchronous on p processors. (1): $p = 6$ (2): $p = 3$
(3): Gauss-Seidel (4): $p = 10$ (5): $p = 15$ (6): Jacobi (7): $p = 30$

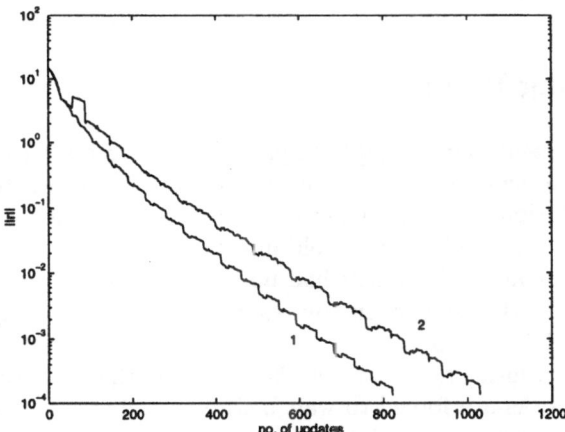

Fig. 4. Heterogeneous environment with $p = g$.
(1): "Wait-for-essential-neighbor"(WFE) (2): Totally asynchronous

experiment. When each processor has updated the residual a fixed number of
times (N), we flush the queue at the master and broadcast the new residual.
Letting $N = 1, 2, 3$, we observe that $N = 1$ gives (approximately) Jacobi's
method and $N \geq 3$ gives (approximately) totally asynchronous method (Fig. 5).

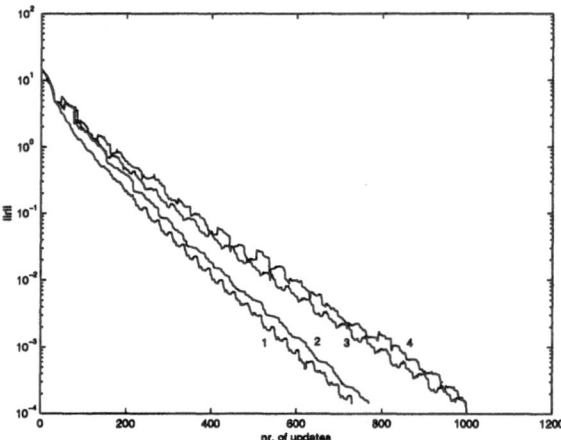

Fig. 5. Synchronization after N updates in a heterogeneous environment with $p = g$.
(1): $N = 1$ ($\approx$ Jacobi)   (2): $N = 2$ (3): Totally asynchronous (4): $N = 3$

5   Concluding Remarks

Elsner, Neumann and Vemmer [6] have proved, under certain assumptions, that increasing the number of processors decreases the convergence rate. This is shown in the implementation of asynchronous iterations on linear least squares problems and is explained as a result of using old information.

The heterogeneous environment had no significant effect on the rate of convergence. However, the changes in the residual is damped (compare curve 2 in Fig. 4 and curve 7 in Fig. 3).

The role of essential neighbors is also a factor that has to be taken into consideration. Blocks are forced to wait before receiving a new residual until a correction from their essential neighbors is received, and in the long run the time-lag is decreased. The result is an improvement in convergence and degradation in the deteriorations.

It is shown [4] that $\|r(t)\|_M$ is monotonically decreasing for the sequential case. In [12], the relaxation parameter ω_i is chosen such that $\|r(t)\|_M$ is minimized for every update. This reduces the effect of old information.

Another attempt to decrease deteriorations in the residual is the introduction of synchronization into the system. It is seen that synchronization after an *a priori* chosen number of corrections on the solution vector lessens the effect of old information and improves the convergence rate. However, the effect of the synchronization decreases rapidly with the "age" of the updates.

To our knowledge, the results of using asynchronous iterations on linear least squares problems have not been discussed in literature. There is no theory to characterize old information, only heuristics. A relaxation parameter can be used to reduce the deteriorations. Synchronization is needed in many cases, but there is no theory to support *when* to synchronize. Synchronization after only one or two updates from each block reduces the effect of old information, but later synchronizations do not cure the deteriorations.

In [10, 12] and in this paper several issues were studied to reduce the deterioration in convergence. However, the single most influential factor based on the numerical testing is the existence of old information in the computation of the updates.

References

1. Baudet, G. M.: Asynchronous Iterative Methods for Multiprocessors. Journal of the ACM. **25** (1978) 226–244
2. Bertsekas, D. P., Tsitsiklis, J. N.: Parallel and Distributed Computation, Numerical Methods. Prentice-Hall Inc., Englewood Cliffs, N. J. (1989)
3. Björck, Å.: Numerical Methods for Least Squares Problems. SIAM, Philadelphia, PA (1996)
4. Dennis, J. E. Jr., Steihaug, T.: On the Successive Projections Approach to Least Squares Problems. SIAM J. Numer. Anal. **23** (1986) 717–733
5. Duff, I., Grimes, R. G., Lewis, J. G.: User's Guide for the Harwell-Boeing Sparse Matrix Collection. Technical Report TR/PA/92/86, CERFACS (1992)
6. Elsner, L., Neumann, M., Vemmer, B.: The Effect of the Number of Processors on the Convergence of the Parallel Block Jacobi Method. Lin. Alg. Appl. **154–156** (1991) 311–330
7. George, J. A., Heath, M. T.: Solution of Sparse Linear Least Squares Problems Using Givens Rotations. Lin. Alg. Appl. **34** (1980) 69–83
8. Kolm, P., Arbenz, P., Gander, W.: Generalized Subspace Correction Methods for Parallel Solution of Linear Systems. Technical Report TRITA-NA-9509, C2M2, Nada, KTH, Sweden (1995)
9. Savari, S. A., Bertsekas, D. P.: Finite Termination of Asynchronous Iterative Algorithms. *Parallel Computing.* **22** (1996) 39–56
10. Steihaug, T., Yalçınkaya, Y.: Asynchronous Methods and Least Squares: An Example of Deteriorating Convergence. Technical Report No. 131. Department of Informatics, University of Bergen, Bergen, Norway (1997)
11. Yalçınkaya, Y.: Asynchronous Solution of Linear Least Squares Problems Using Generalized Group Iterative Methods. Master's thesis. University of Bergen, Norway (1995)
12. Yalçınkaya, Y., Steihaug, T.: Asynchronous Methods and Least Squares: An Example of Deteriorating Convergence. Proceedings of the 15th IMACS World Congress on Scientific Computation, Modelling and Applied Mathematics, August 24-29, 1997, Berlin, Germany (to appear) [part of [10]]

Workshop 10+11+14

Parallel Computer Architecture and Image Processing

Workshops 10+11+14: Parallel Computer Architecture and Image Processing

Per Stenström and Patrice Quinton

Introduction

This workshop consists of two sessions: one on parallel computer architecture and one on image processing. A common denominator for both sessions is to exploit parallelism at the algorithm and architectural level to reach a high performance. The first session is mostly concerned with this issue for general-purpose high-performance computers whereas the second session is concerned with the image processing application domain.

Parallel Computer Architecture

Parallelism has played a key role as a means to design computer architectures for high-performance applications over several decades – both regarding the design of individual processors as well as the design of large-scale parallel systems. The quest for even higher performance levels, however, keeps fueling the research in computer architecture. Important areas of research encompass further development of techniques to exploit parallelism at the instruction level as well as at the thread level. More parallelism naturally puts a heavier burden on the memory and message transfer subsystems which has led to close attention to techniques that aim at reducing or hiding latencies in the data transport. The papers in this session advance state of the art with new approaches to exploit parallelism at the processor architectural level as well as at the system level.

The first group of papers in this session is concerned with new approaches to design high-performance microprocessors. The RISC and MIPS projects at Berkeley and Stanford in the early 80s constituted a starting point for a new methodology to design high-performance microprocessors. Important facets of this methodology are to support common-case operations efficiently in hardware and to use pipelining in conjunction with static (compiler) analysis to exploit inherent instruction-level parallelism.

The first paper by Glossner and Vassiliadis is a good exercise along these lines applied to efficient execution of the Java language. The platform independence requirement of Java is to some extent contradictory to this goal. Their approach is to do on-the-fly translation of Java code to native code and in this process extract instruction-level as well as thread-level parallelism in a programmer-transparent way.

The second paper in this session by Arvind and Sotelo-Salazar reports on a new approach to reach a high instruction-execution rate when instruction latencies are not known. They approach this problem by using asynchronous functional units and present a new instruction scheduling method for asynchronous instruction pipelines.

Two prevailing programming models for multiprocessor systems are message passing and shared-memory. While process cooperation and coordination under these models are carried out through explicit sending and receiving of messages and implicitly through loads and stores to a single address space, respectively, they both introduce a fundamental architectural performance issue – how to reduce or hide latency. The second group of papers in this session is concerned with this problem.

In the third paper by Tomiyasu and his associates, the problem is to support fine-grain communication in an efficient way by hiding the latency of message transfers. They attack this problem using multithreading and presents the design of a co-processor that facilitates fine-grained message handling and communication.

Address translation is on the critical path between the processor and the memory system. Virtually-addressed caches can shorten this path but introduce costly operations to maintain consistency, especially in multiprocessors with multiple caches. The idea in the fourth paper by Kim and Lee is to split the cache system into a virtually-addressed partition that is used for private data and into a physically-addressed partition that is used for shared data.

Finally, the steadily increasing chip density provides an important opportunity for innovations in processor design. The last paper by Kisuki and his associates, considers on-chip multiprocessors – one interesting option – and how the on-chip memory subsystem should be designed to reduce memory access latencies.

Image Processing

Computation-intensive applications have always been at the source of research on parallel architectures. Among other application domains, image processing plays a key role in these research, due to the variety of algorithms involved – from low-level signal processing routines to high-level symbolic functions – and due to the huge quantity of computations needed for processing an image. This rôle is not likely to diminish in the near future, as images are at the heart of multimedia applications, whose importance is currently expanding, to say the least.

As for many applications of parallel processing, a central question is to match algorithms and architectures in such a way that good performance is reached. It is often the case that a good sequential algorithm behaves poorly once implemented on a parallel architecture, unless a significant effort is devoted to its optimization.

The first paper of this session, presented by Schmidt, Schimmler and Schröder, answers this question by modifying a classical algorithm in order to improve its performance, while fitting the characteristics of a flexible parallel architecture.

The authors present a new algorithm for line detection, combining the morphological approach to the Hough Transform. This algorithm is designed to take advantage of the performance of the *Systola1024* architecture, a systolic array with instruction flexibility. This illustrates the advantage of designing directly a parallel algorithm, instead of modifying a sequential one.

The same question is addressed in a similar way by the second paper of this session, authored by Lam and Furness. Algorithms for edge detection with built-in noise filtration and good localisation properties have an inherent iterative structure which is not well suited to parallel implementation. The paper introduces a novel edge detection method, based on the cascaded precursor technique, which enjoys a better adequation to parallel computing. This new algorithm is analyzed, and compared to other approaches.

Designing a new algorithm is not the only approach to solving the algorithm–architecture adequation problem. Instead, one may explore how to distribute a classical algorithm, in order to reach the best possible performance. This is the direction taken by the paper of Fleury, Downton, and Clark which studies the implementation of the Karhünen-Loève Transform on a network of processors running a real-time operating system. The exploration of the design trade-offs for this particular algorithm, on large- and fine-grained hardware, leads to conclusions which can be extended to other methods of the same class.

The last paper of this session, by Muchnick, Shafarenko, and Willink, tackles another source of inefficiency when implementing digital signal processing applications. They observe that the use of high-level languages to express algorithms, as required by a reasonable design methodology, leads to low quality code, and therefore, to poor performance on special-purpose architectures such as Digital Signal Processors. The current solution to this situation is to rewrite the algorithm in assembler, most often by hand, in order to exploit the characteristics of the target architecture. The paper addresses this problem by defining a high-level intermediate language, named F-code, developed for data parallelism. The authors explain how such a language allows the high-level phases of a compiler to exploit the peculiarities of the architecture in order to reach good performance code.

The DELFT-JAVA Engine: An Introduction

C. John Glossner[1,2] and Stamatis Vassiliadis[2]

[1] Lucent / Bell Labs, Allentown, Pa.
[2] Delft University of Technology, Department of Electrical Engineering
Delft, The Netherlands
(glossner,stamatis)@einstein.et.tudelft.nl

Abstract. In this paper we introduce the DELFT-JAVA multithreaded processor architecture and organization. The proposed architecture provides direct translation capability from the JAVA Virtual Machine instruction set into the DELFT-JAVA instruction set. The instruction set is a 32-bit RISC instruction set architecture with support for multiple concurrent threads and JAVA specific constructs. The parallelism is extracted transparently to the programmer. Except for kernel programs, programmers need only be concerned with the semantics of the JAVA programming language. In addition, programmers who desire to take greater advantage of parallelism can execute privileged instructions which provide additional capabilities for Multimedia and DSP processing.

1 Introduction

The JAVA language provides processor architects with opportunities for exploiting Instruction Level Parallelism (ILP). Rather than requiring the processor to extract all ILP from a single executing thread, the JAVA language intrinsically supports programmer specification of parallelism through threads. Our goal in the DELFT-JAVA architecture is to extract maximal parallelism as defined by the JAVA language without burdening the programmer to specify any additional parallelism that is not inherent in the language constructs. At the highest level, a programmer of a DELFT-JAVA processor views it as a JAVA Virtual Machine (JVM).

Using RISC architectural concepts, we present a concurrent multithreaded architecture and organization that fully utilizes the JAVA programming paradigm. Furthermore, the architecture allows mechanisms for increasing single thread performance by allowing a single thread to issue multiple instructions per cycle. The architecture is scalable in the number of concurrent threads that can be supported and in the number of execution units that can be implemented. In addition to JVM execution, the DELFT-JAVA architecture provides general support for C compilers and other operations that are required in general purpose processors. Architectural support for Multimedia SIMD and DSP instructions is also incorporated into the architecture.

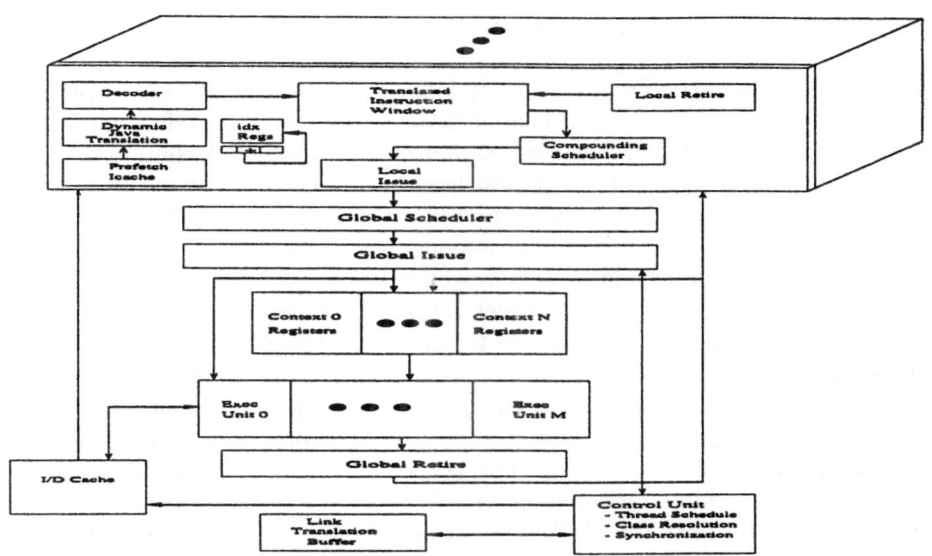

Fig. 1. Concurrent Multithreaded Organization.

2 DELFT-JAVA Architecture

The DELFT-JAVA architecture has two logical views: 1) a JVM Instruction Set Architecture (ISA) and 2) a RISC-based ISA. The JVM view is stack-based with support for standard datatypes, synchronization, object-oriented method invocation, arrays, and object allocation[1]. An important property of JAVA bytecodes is that statically determinable type state enables simple on-the-fly translation of bytecodes into efficient machine code[2]. We utilize this property to dynamically translate JAVA bytecodes into DELFT-JAVA instructions. Programmers who wish to take advantage of other languages which exploit the full capabilities of the DELFT-JAVA processor may do so but require a specific compiler. Some additional architectural features in the DELFT-JAVA processor which are not directly accessible from JAVA code include pointer manipulation, Multimedia SIMD instructions, unsigned datatypes, and rounding/saturation modes for DSP algorithms. The remaining sections of this paper refer to the DELFT-JAVA architecture and not to the JAVA architectural view.

Instructions: In DELFT-JAVA, nearly all instructions are executed as 32-bit fixed width instructions with an 8-bit opcode. A typical encoding useful for JAVA translation is `add.ind.w32 idx[0], ix, iy-1, it-1`. This instruction specifies a 32-bit, 2's complement integer addition. Normally, the register file is accessed with a direct reference (e.g. `add rx, ry, rt`). The DELFT-JAVA processor facilitates JVM translation by allowing indirect access through 3 indices into the register file which create a circular address. Using this mechanism, it is possible to provide both LIFO stack and FIFO vector operations.

3 Concurrent Multithreaded Organization

An organization of the DELFT-JAVA architecture which supports multiple concurrent execution of threads and shared global execution units is shown in Figure 1. We define a *context* as a hardware supported thread unit. A context does not include shared resources such as a first level (L1) cache, execution units, a register file, global instruction schedulers, nor global issue units. The term *thread* is generally used to refer to the programmer's view of a thread - a possibly concurrent stream of independent executing instructions[3]. In this paper, the term context denotes the hardware on which a thread may run.

Operation: All instructions are fetched from global shared memory and placed into a global L1 on-chip instruction cache. Each context also assumes a (logical) zero level (L0) instruction cache to provide concurrent per context *instruction fetch* capacity. During normal user-level operation, all instructions are fetched as JAVA instructions. The *control unit* is responsible for synchronization, cache locking, dynamic linking, I/O, loading instructions, and general system functions. Since the JVM does not provide all the functionality generally required by a full operating system, many of these functions have been grouped into a special control unit. A control unit is analogous to a context except that it contains additional resources that are not necessarily required within a JAVA context. After being fetched, most JAVA instructions are *dynamically translated* into the DELFT-JAVA instruction set. Because the instructions are stored in cache memory as JAVA instructions, branching and method invocation code produced by JAVA compilers will execute properly on the DELFT-JAVA architecture. After translation, the instructions are decoded and placed in a *local instruction window* which keeps track of issued and pending instructions. The *local instruction scheduler* takes translated instructions in a RISC form and schedules them for execution. The *local issue unit* determines if the instructions that have been locally scheduled can be issued to the global instruction scheduler.

All instructions which require access to shared resources must be forwarded to the *global instruction scheduler*. This unit schedules the aggregated instructions destined for execution units. The JAVA language specifies that in the absence of explicit synchronization, a JAVA implementation is free to update the main memory in any order[4]. This relaxed memory consistency model allows the scheduler to reorder the instructions from individual contexts to optimize the utilization of the shared execution units. The *global issue unit* ensures that global resources are available prior to issuing instructions. Instructions may be issued in one of two forms: single independent instructions and compound parcels[5].

After execution, all results are forwarded to the *global retire unit*. This unit removes the requirement for a general interconnection unit between all contexts and execution units. If instructions were not executed speculatively, the global retire unit writes the results to the register file. Otherwise, the result is maintained in the retire unit until the conditional outcome is known. The *local retire unit* removes the instruction from the window and commits state in the context. Each context may retire multiple instructions per cycle.

Translation: Most JVM bytecodes are translated. However, some more com-

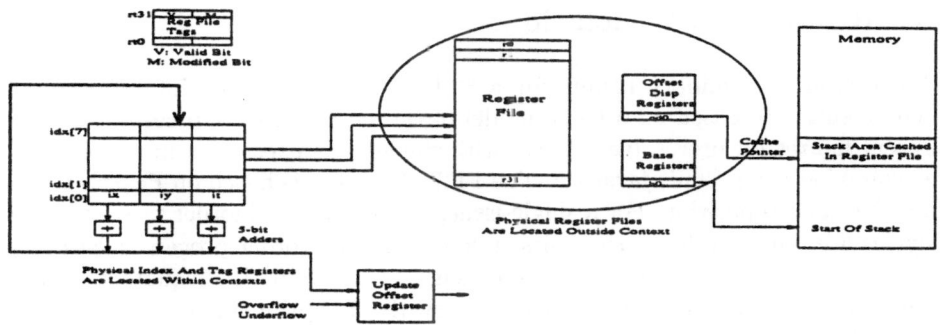

Fig. 2. Concurrent Multithreaded Registers.

plex instructions are directly incorporated into the DELFT-JAVA ISA. In particular, the following categories of instructions are not translated: a) synchronization, b) array management c) object management, d) method invocation, e) exception handling, and f) complex branching. Indirect access to the register file plays the largest role in the translation of JAVA bytecodes to DELFT-JAVA instructions. As shown in Figure 2, when executing JAVA instructions, the register file index registers create a circular buffer that is mapped to the stack in memory. A set of valid and modified bits are associated with each register. These bits are maintained logically within the local context. A JAVA instruction such as `iadd` goes through two intermediate translation phases. The first phase translates the instruction into a valid DELFT-JAVA instruction. In this case, an `add.ind.w32 idx[0] it, it-1, it-1` is generated by the translation logic. As an example, assume the top of stack referenced by `idx[0].rt` points to r5. The second phase decodes the indirect register reference and places the decoded instruction into the instruction window as `add.w32 r5, r4, r4`. When the instruction is placed in the window, the `idx[0].rt` is modified by the destination decrement. Functionally, this performs r5 + r4 → r4. In the DELFT-JAVA processor, the stack grows upward in both memory and in register file references. These registers automatically prefetch and spill as the stack size changes.

Link Translation Buffer: An important consideration in accelerating JAVA's dynamic linking and polymorphic method invocation is a *Link Translation Buffer* (LTB). The LTB acts as a global repository for dynamically resolved names. During dynamic invocation, the name to be resolved is contained in the constant pool. After a process called resolution[4], the name contained within the constant pool can be associated with a physical location in memory for each object. This association is placed in the Link Translation Buffer. If the control unit finds the constant pool address of the requesting object in the LTB and the requesting class has access permissions to the data, then the LTB efficiently returns the resolved address. A programmer may also completely disable the LTB or more judiciously issue `flushLTB` instructions.

4   Results and Conclusions

We present preliminary results for a 32-bit full complex 4-point FFT kernel. The results are based on a C++ model of the DELFT-JAVA processor and represent figures for preresolved classes with single-cycle execution units. The FFT is compiled using Sun's javac -O. The FFT algorithm is based on Pease's tensor product decomposition. For a single-issue, single context, inorder processor, 226 cycles are required. For a single-issue, four context, inorder processor, 84 cycles are required when amortized over 4 concurrent FFT's with adequate execution units. Because the javac compiler is conservative in optimizing loads and stores, a number of instructions are generated that could be further optimized. Because we are accelerating JAVA programs produced directly from a JAVA compiler, we do not use any of the multimedia datatypes which would enhance the FFT performance. We anticipate with better optimizations and multiple issue per thread, the FFT performance of the DELFT-JAVA processor will improve by 10x based on a similar algorithm used in [6].

In this paper we have introduced the DELFT-JAVA processor architecture and organization. The goal of the processor is to exploit the parallelism inherent in JAVA multithreaded programs without requiring the programmer to specify any additional information. We have accomplished this by designing a concurrent multithreaded organization using modern RISC techniques with multiple instruction issue capability per context. Compared to current techniques, our processor efficiently exploits the JAVA programming language to provide a very high performance JAVA processor architecture.

References

1. Tim Lindholm and Frank Yellin. *The Java Virtual Machine Specification*. The Java Series. Addison-Wesley, Reading, MA, USA, 1997.
2. James Gosling. Java Intermediate Bytecodes. In *ACM SIGPLAN Notices*, pages 111–118, New York, NY, January 1995. Association for Computing Machinery. ACM SIGPLAN Workshop on Intermediate Representations (IR95).
3. Bil Lewis and Daniel J. Berg. *Threads Primer: A Guide to Multithreaded Programming*. SunSoft Press - A Prentice Hall Title, Mountain View, California, 1996.
4. James Gosling, Bill Joy, and Guy Steele, editors. *The Java Language Specification*. The Java Series. Addison-Wesley, Reading, MA, USA, 1996.
5. S. Vassiliadis, B. Blaner, and R. J. Eickemeyer. SCISM: A Scalable Compound Instruction Set Machine. *IBM Journal of Research and Development*, 38(1):59–78, January 1994.
6. C. J. Glossner, G. G. Pechanek, S. Vassiliadis, and J. Landon. High-Performance Parallel FFT Algorithms on M.f.a.s.t. Using Tensor Algebra. In *Proceedings of the Signal Processing Applications Conference at DSPx'96*, pages 529–536, San Jose Convention Center, San Jose, Ca., March 11-14 1996.

Scheduling Instructions with Uncertain Latencies in Asynchronous Architectures

D. K. Arvind and S. Sotelo-Salazar

Department of Computer Science, The University of Edinburgh,
Mayfield Road, Edinburgh EH9 3JZ, Scotland.

Abstract. This paper addresses the problem of scheduling instructions in micronet-based asynchronous processors (MAP), in which the latencies of the instructions are not precisely known. A PTD scheduler is proposed which minimises true dependencies, and results are compared with two list schedulers - the Gibbons and Muchnick scheduler, and a variation of the Balanced scheduler. The PTD scheduler has a lower time complexity and produces better quality schedules than the other two when applied twenty-three loop- and control-intensive benchmark programs.

1 Introduction

There has been a revival of interest in the use of asynchrony, albeit in a restricted form known as self-timing, in the design of processor architectures. Asynchronous circuits offer some distinct advantages. Their power consumption is generally much lower compared to their synchronous equivalent. This is because at any time only parts of the asynchronous system are active as required, with the rest remaining in a quiescent state. Self-timed systems allow a modular approach to processor design whereby parts can be added and deleted with little impact on the rest of the system. These systems are also robust to environmental changes.

The feature which is of most interest to our work and which was first recognised in the Micronet model [1] is that asynchrony offers scope for fine-grain concurrency in the processor architecture. The micronet model exposes this feature naturally, and asynchronous architectures based on this model are better able to exploit instruction-level parallelism.

A micronet-based architecture is viewed as a network of typed functional units. These units operate concurrently and communicate asynchronously with the rest of the architecture. The functional units themselves can be described at different levels of abstraction. In this paper the architecture is composed of the following functional unit types: one or more Arithmetic Unit (AU), a Logic Unit (LU), a Memory Unit (MU) and a Branch Unit (BU).

The issue and execution of an instruction consist of a sequence of micro-operations involving the Issue Unit (IU), the Register Bank, and the appropriate functional unit. An instruction is issued when both its operands are available. Once the instruction has been issued, it runs to completion unless it is stalled due to contention for resources in the trajectory of the instruction at any one of these points: the read ports, the functional unit, the write-back port. The micronet model enables concurrent execution of the micro-operations of the different instructions in flight, and minimises the costs of instruction stalls due to resource contentions. The latency of the instruction depends on a number of factors: its type, the data on which it operates, and the contention for resources which depends on the mix of instructions.

This paper proposes a relatively inexpensive method for scheduling instructions within the basic block. The objective of the scheduler is to ensure the rapid issue of independent instructions, thereby minimising the number of stalls of the issue unit, and in reducing the contention for the functional units by enabling instructions of different types to be in flight at the same time. This is achieved by assigning penalties to data dependencies and successive instructions of the same type, and transforming the schedule by moving instructions to reduce the penalties. This results in a schedule in which dependent instructions are separated, and independent instructions of different types are issued in succession.

The next section describes the traditional list scheduling algorithms such as Gibbons and Muchnick and the Balanced schedulers.

2 Traditional scheduling heuristics

2.1 The Gibbons and Muchnick (GM) scheduler

This is a well-known example of a list scheduling algorithm proposed originally for scheduling instructions in pipelined architectures [2]. The algorithm selects the instructions to be scheduled from a directed acyclic graph, beginning at the roots. The instructions are selected for scheduling if all their immediate predecessors have been scheduled. These *ready* instructions are prioritised on the following basis: if possible, an instruction is scheduled that will not interlock with the one just scheduled; given a choice, an instruction will be scheduled which is most likely to cause interlocks with instructions after it. The complexity in the absence of any lookahead in the instructions is $\theta(n^2)$, where n is the number of instructions in a basic block.

2.2 The Balanced scheduler

The Balanced scheduler [3] was devised to take account of unpredictable memory access latencies. The idea is to compute weights for load instructions based on the number of available independent instructions. The instructions are scheduled as in a traditional list scheduler with independent instructions being distributed behind loads to buffer for unpredictable memory accesses. This idea is extended beyond the load instruction to all the instructions in the MAP architecture. The priority for ready instructions is based on a weighted sum of values derived from MAP tailored heuristics - whether the instruction uses the same resources as the previous scheduled one; the number of immediate successors of the instruction;

the length of the longest path from the instruction to the leaves of the DAG; and the number of source registers which are freed should the instruction be scheduled which effectively takes account of the register pressure.

3   The "Penalise True Dependencies" (PTD) scheduler

The essence of this heuristic is to identify true data and resource dependencies and re-order, where possible, the instructions such that their detrimental effect is reduced. The schedule is allocated a penalty measure based on the number and type of these dependencies. A true consecutive data dependency is penalised by one which is treated as the base case. If the dependency is with a branch or load instruction then it is penalised more severely. The actual value depends on the relative latencies of the functional units as shown in Table 1.

Instructions with resource dependencies are treated in a similar manner. If there are say p functional units of Type A, q units of Type B and r units of Type C, then a sequence containing more than p consecutive instructions of Type A, or q of Type B, or r of Type C will incur penalties. This assumes that the latencies of the three types of FUs are approximately the same; the run-length of the instructions can be suitably amended to take account of different latencies. The algorithm to derive this measure has a complexity of $\theta(n)$.

Cases of dependencies	Consecutive instructions	Separated by one inst.
True dependency with a load inst.	3	1
True dependency with a branch inst.	2	0
Resource dependency within mem. inst.	1	0
Normal true dependencies	1	0

Table 1. Table of penalties for true data dependencies.

We next demonstrate the correlation between the penalty measure considering only the true data dependencies and the makespans of the schedules for the program in Figure 1. The target asynchronous architecture has three types of functional units: an arithmetic unit (AU), logic unit (LU) and the memory unit (MU). The latency values for the units ranged over an interval, as shown in Table 2, with a Gaussian distribution. The results from a stochastic simulator which exhaustively simulated all the schedules (24,192) and averaged the results over 20 runs are shown in Figure 2. This result is representative of simulations of other programs with different spread of latencies. We can observe the trend that the penalty measure increases in step with the makespans of the schedules. This should ideally be a strict monotonic function, but the overlaps between the schedules of neighbouring penalties are tolerable for the heuristic approach. A scheduler based on minimising the penalty measure is introduced in the next section.

```
                              L4.main:
                                  muli    $13,$9,4
                                  la      $14,$29,0
  main() {                        addu    $15,$14,$13
                                  muli    $24,$9,4
                                  la      $25,$29,0
    int i, j, n = 10;             addu    $11,$25,$24
      int x[10];                  lw      $12,$11,0
                                  muli    $13,$10,4
    for (i = 0; i < n; i++)       la      $14,$29,0
      for (j = 0; j < n; j++)     addu    $24,$14,$13
        x[i] = x[i] * x[j];       lw      $25,$24,0
                                  mul     $11,$12,$25
                                  sw      $11,$15,0
  }                               addui   $10,$10,1
                                  slt     $12,$10,$8
                                  bt      $12,L4.main
```

Figure 1. C and MAP assembly code from our example.

Component type	Minimum latency	Maximum latency
Issue Unit (IU)	1.00 ns	2.00 ns
Input buses	2.00 ns	4.00 ns
Output buses	2.00 ns	4.00 ns
Arithmetic Unit (AU)	4.00 ns	8.50 ns
Logical Unit (LU)	2.00 ns	7.00 ns
Memory Unit (MU)	10.00 ns	20.00 ns

Table 2. Latencies values for the target architecture.

3.1   The PTD scheduler

The PTD scheduler works in two phases: in the first phase the contention for resources is minimised, and in the second phase consecutive data dependent instructions are separated.

In the first phase, the types of consecutive instructions are compared and instructions are moved, where possible, so that the overall penalty measure is reduced, such that the number of consecutive instructions of the same type is no greater than the number of functional units of that type.

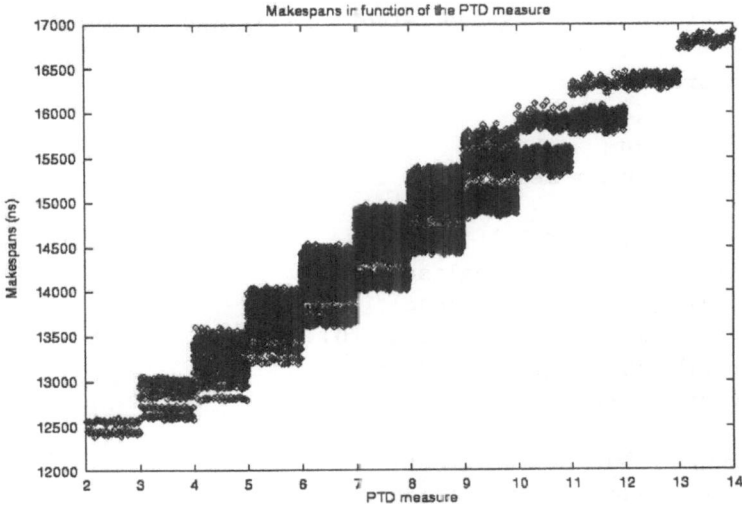

Figure 2. Execution distribution for the example.

In the second phase, the schedule is again scanned from start to finish, to identify consecutive data dependencies, and independent instructions are sandwiched in between them so that the overall penalty measure is reduced to zero or cannot be reduced any further due to the lack of suitable instructions. The details of the PTD scheduler are shown in Figure 3. The functions PTD_arrange_left() and PTD_arrange_right() traverse the schedule in both directions in search of independent instructions for insertion immediately after the penalised one. Two transformations are employed: a *swap* operation and a *move_ahead* operation and their use is illustrated in the following example.

Let l, m, n and o represent consecutive instructions in a schedule with a data dependency between n and o. This is represented by $n \rightarrow o$. The conditions for performing a $swap(m, n)$ transformation which eliminates (or reduces) the penalty to o, are the following:

- $m \parallel n$ (m is independent of n),
- $m \,!\!\rightarrow o$ (meaning not producing a penalty) and
- $l \,!\!\rightarrow n$

If the penalties go beyond consecutive instructions then in order to ensure that the penalty measure will be reduced after the *swap*, the necessary condition is that the sum of penalties before the movement is greater than the measure after the transformation is made.

The conditions for performing a $move\_ahead(x, n)$ (moves x ahead of n) to eliminate (or reduce) the penalty to o, are the following:

- $x \parallel a, ..., x \parallel l, x \parallel m, x \parallel n$,
- $x \,!\!\rightarrow o$ and
- $x_{-1} \,!\!\rightarrow x_{+1}$ where x_{-1} and x_{+1} are the instructions previous and following x, respectively.

```
void PTD_second_phase(dagnodes *root) {
  measure = PTD_measure(root, second_phase);
  if (measure > 0)
    do {
      node = root;
      last_measure = measure;

      while (node != NULL) {
        if (node -> PTD.penalised > 0)
          PTD_arrange_left (node);
        if (node -> PTD.penalised > 0)
          PTD_arrange_right(node);
        node = node -> next;
      }
      measure = PTD_measure(root, second_phase);
    } while (measure < last_measure && measure > 0);
}
```

Figure 3. The PTD scheduling algorithm - Phase 2.

Again to generalise the rules to allow a *move_ahead*, the sum of penalties before the insertion must be greater than the total number of penalties after the instruction has moved.

The conditions just outlined apply for the PTD_arrange_left() function which examines the left-hand side of the penalised instruction. The analogous conditions apply for the PTD_arrange_right() function but have been omitted for the sake of brevity. These conditions are sufficient to preserve the semantics of the program and reduce the PTD measure.

There will be cases where the only way to decrease the PTD measure of a schedule would be to replace a high penalty, i.e. load from memory, with a less expensive one, such as a "move register" instruction. So in terms of the penalty, one of 3 is reduced to 2 by moving an offending instruction, but the goal of reducing the overall measure is still accomplished.

The complexity of the PTD scheduler is $\theta(n\,e)$ where e is the number of penalties in the schedule. The worst case is one in which the schedule has at most $n-1$ consecutive dependencies (a pure sequential code) giving a complexity of $\theta(n^2)$ and the best case is $\theta(n)$. The linear-time complexity for the PTD scheduler is better than the $\theta(n^2)$ for the list scheduler [2] and $\theta(n^2\,\alpha\,n)$ [1] for the balanced scheduler [3].

4 Results

We next compare the quality of schedules produced by the Balanced, Gibbons and Muchnick (GM) and the PTD schedulers for a range of benchmarks which

---

[1] α is the inverse of the Ackerman function.

represent both loop-intensive (Livermore loops) and control-intensive categories of programs. These were compiled on the SUIF Compiler for the MAP target, but without any MAP-specific optimisations, and provided the same base schedule for the three schedulers under comparison.

The schedules were simulated on a discrete-event model of the MAP architecture. An architecture file describes the functionality and interconnection, and the spread of latencies as shown in Table 2. The distribution of latencies were chosen to best reflect the behaviour of the functional unit. The bimodal distribution for the Memory Unit captures the behaviour due to cache hits and misses. The distribution of the latencies for the Arithmetic Unit is based on the graph in Figure 4 in [4], and the distribution is uniform for the Logic Unit.

The simulation results presented in Figure 4, represent the average of five simulation runs for each program. They represent the percentage improvement with respect to the base case, i.e. the SUIF compiler output. The PTD scheduler outperforms the other two schedulers on both the control-intensive and loop-intensive programs.

When the number of AUs is increased from one to two (Fig. 5), we see a marked improvement in the schedules, but this tapers off when the AUs are increased further. This could be improved upon by scheduling instructions beyond the basic blocks. The favourable run-time complexity of the PTD algorithm makes this a practical proposition.

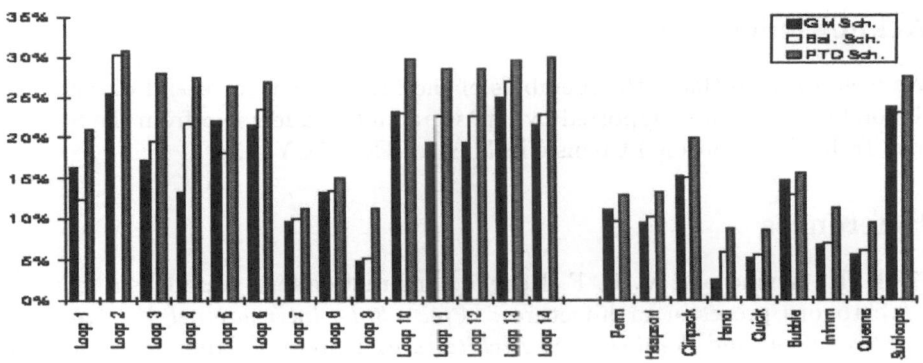

Figure 4. Average improvement for the whole set of benchmarks.

5 Conclusions

The PTD scheduler provides a simple yet effective method for scheduling instructions within basic blocks for programs running on MAP architectures. It has a better time complexity than the other two well-known list schedulers, and

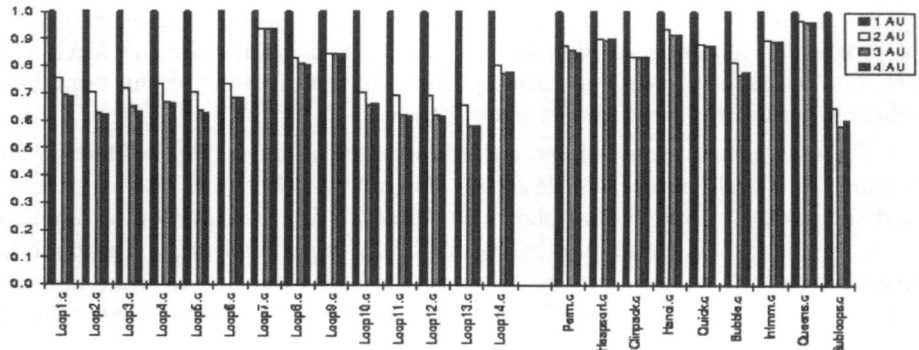

Figure 5. Ratio between the 1 AU and the other configurations.

the quality of the PTD schedules are better for a range of control- and loop-intensive benchmarks. The method reduces the stalls of the Issue Unit due to true data dependencies between instructions and enables better utilisation of the functional units by reducing the resource contention between instructions. The performance of the scheduler was investigated when the number of Arithmetic Units was scaled from 1 to 4. Future work will investigate the scheduling of instructions beyond the basic block boundaries for better utilisation as the functional units are scaled.

Acknowledgements

We would like to thank the members of the MAP group for useful discussions. S. Sotelo-Salazar was supported by a postgraduate studentship from the Science and Technology National Counsel in Mexico (CONACYT).

References

1. D. K. Arvind and V. E. F. Rebello. Instruction-level parallelism in asynchronous processor architectures. *Proc. 3rd. International Workshop on Algorithms and Parallel VLSI Architectures*, Leuven, Belgium, August 1994, pp. 203-215.
2. P. B. Gibbons and S. S. Muchnick. Efficient instruction scheduling for a pipelined architecture. *Proc. SIGPLAN 1986 Symposium on Compiler Construction*, SIGPLAN Notices, 21(7), July 1986, pp. 11-16.
3. D. R. Kerns and S. J. Eggers. Balanced scheduling: Instruction scheduling when memory latency is uncertain. *In ACM SIGPLAN 1993 Conference on Programming Language Design and Implementation*, SIGPLAN Notices, 28(6), June 1993, pp. 278-289.
4. D. J. Kinniment. An evaluation of asynchronous addition. *IEEE Transactions on Very Large Scale Integration (VLSI) systems*, March 1996, pp. 137-140.

Co-processor System Design for Fine-Grain Message Handling in KUMP/D

Hiroshi Tomiyasu[1], Shigeru Kusakabe[1],
Tetsuo Kawano[1]*, and Makoto Amamiya[1]

Department of Intelligent Systems Graduate School of Information Science and
Electrical Engineering Kyushu University Kasuga, Fukuoka 816 Japan

Abstract. In parallel processing, fine-grain parallel processing is quite
effective solution for latency problem caused by remote memory accesses
and remote procedure calls. We have proposed a processor architecture, called Datarol-II, that promotes efficient fine-grain multi-thread
execution by performing fast context switching among fine-grain concurrent processes. We are now building a prototype multi-media machine
KUMP/D (Kyushu University Multi-media Processor on Datarol-II) on
the basis of the fine-grain multi-threading architecture. In the design of
the KUMP/D, we used the commercial microprocessor for its processing
element, and designed a co-processor, called FMP(Fine-grain Message
Processor), for fine-grain message handling and communication control.
In this paper, we show the KUMP/D processor design and its performance evaluation.

1 Introduction

In parallel processing, one of the most critical issues is the latency problem
caused by remote memory accesses and remote procedure calls. To solve this
problem, fine-grain parallel processing is quite effective. We have proposed a
processor architecture, called Datarol-II[7][2]. The Datarol-II architecture is an
evolution of the Datarol architecture[1] which surmounts drawbacks of dataflow
computation scheme, such as unnecessary data copy caused by-value data access
mechanism, increasing memory access by operand matching. The main idea of
the Datarol architecture is to extract a multi-thread control program, called
Datarol, from a dataflow graph, to eliminate redundant flow controls in the
graph by introducing a by-reference data access concept.

The Datarol-II processor architecture is an optimized version of original
Datarol architecture. The Datarol-II executes fine-grain threads by means of
a program-counter-based execution pipeline with high-speed registers.

We are now building a prototype of multi-thread machine KUMP/D (Kyushu
University Multi-media Processor on Datarol-II)[2]. In the design of the KUMP/D,
we used a commercial high speed microprocessor Pentium rather than a completely new designed processor for its processing element.

By using a commercial high-end microprocessor, we can build a new machine
in a short design cycle that follows the mainstream of processor design technology. Current commercially available microprocessors are designed as a single processor or, at most, low-scale parallel processor, and they are tuned to offer very

---

* Currently, he is at NTT Software Laboratories.

high performance in internal sequential execution. However, their functions lack in the high speed communication for supporting the higher-scale multi-thread parallel processing. The key issue in designing a cost effective multi-threading parallel machine is to develop a very simple high speed message handling and communication control hardware device, or co-processor, which complements the high speed internal execution by commercial microprocessor.

We design a new simple mechanism, FMD (Fine-grain Message Driven mechanism). A co-processor, FMP (Fine-grain Message Processor), is an external hardware to realize efficient fine-grain message handling based on the FMD. The FMP incorporates fine-grain data and synchronization messages into the main computation minimally interrupting the core processor.

In the paper, we introduce the KUMP/D machine and its processing element construction with Pentium and FMP, and discuss the feasibility of the design from the viewpoint of cost/performance.

2 Overview of the KUMP/D

Fig. 1 shows an overview of the KUMP/D. Each processor element(PE) is based on the Datarol-II architecture [7], which is especially designed for fine-grain parallel processing. The PE construction is described in subsequent sections.

For the scalability, as the inter-PE network of the KUMP/D, we have chosen a 2-D torus network[3], which is also suitable for mapping images on PEs. This interconnection network is specialized for fine-grain message communication, which includes parameter passing and result value returning. This network has a deadlock-free structure based on its hierarchical packet buffers[4][12].

Since the interconnection network is not suitable for long I/O packet transmissions, KUMP/D has a specialized I/O network, which consists of serial links connecting the PEs in a ring structure, for communication of long packets such as video I/O packets and disk I/O packets[8]. One important feature of the I/O network is that it supports a mechanism of the synchronization of each video frame and process execution[12].

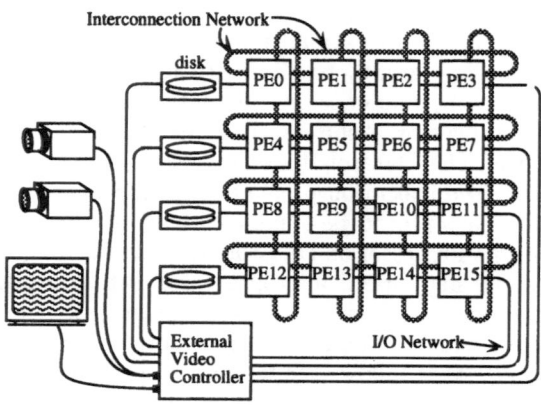

Fig. 1. Overview of the KUMP/D

3 FMD(Fine-grain Message Driven mechanism)

The FMD (Fine-grain Message Driven mechanism) is a revised version of the Datarol-II execution mechanism. The FMD is a message driven execution mechanism, and an FMD message is simpler and more general than that of the Datarol-II. Basic FMD message is an explicitly addressed remote memory write message which also contains a continuation thread after its write operation. The FMD uses a remote memory write message and a remote thread activation message.

Basic runtime model of the FMD is similar to that of the Datarol. A number of function instances are created during program executions. A function instance has a shared program code and private execution environment, which is called an instance frame. A program code is split into threads. A thread is a code block that is executed without any interrupts until its termination point. A context switch may occur at the end of a thread. A context is realized as an instance frame on memory.

By assuming the order of message sent to the same destination instance from an instance is preserved, the FMD simplifies link-receive parameter passing of the Datarol-II mechanism.

In the FMD, a caller instance sends messages which write argument data into the callee instance frame, and the caller also sends a message which activates the destination thread[2] as shown in Fig. 2.

In a function application, at first, a caller instance issues a message to get a new instance frame. Second, the caller sends a message to the callee that has newly obtained instance to activate initial thread. Then, the caller sends a set of parameter data by link message to the callee. After the caller sends all the data needed to an entry thread, the caller activate the entry thread by start instruction.

Synchronizations of threads in an instance are realized by means of messages. A thread that needs multiple inputs is activated after all the data arrived. In the FMD, threads are repeatedly activated by messages along with the dependency, and PE repeatedly executes activated thread during a program execution.

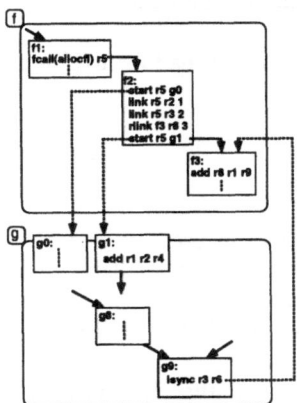

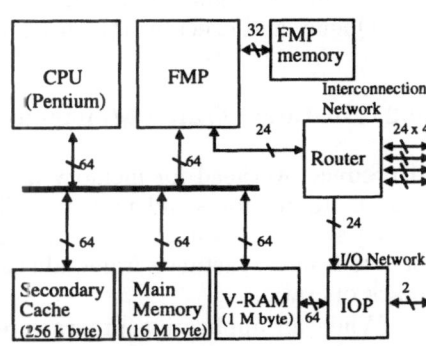

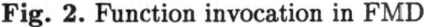

Fig. 2. Function invocation in FMD   **Fig. 3.** Processor Element(PE)

[2] We call such threads that are formally activated by receive messages in Datarol-II "entry threads" in the FMD

4 FMP(Fine-grain Message Processor)

The FMP is an implementation of the FMD mechanism, which assists fine-grain message passing, thread synchronization, remote memory access, and instance frame management.

The PE construction is shown in Fig. 3. When issuing a message, the CPU writes a set of data to the FMP. Messages from the FMP to the CPU is written to secondary cache, and the FMP starts a thread execution. In this thread, the CPU gets these messages by explicitly inserted load instructions. To reduce bass traffic, FMP has its own high speed memory called "FMP memory." In this section, we describe main functions of the FMP.

4.1 Thread Synchronization Mechanism

In KUMP/D, we use a structure data called "SyncCell" for thread synchronization. Each SyncCell has an instance frame address (frame), an instruction pointer to starting thread address (IP), and a synchronization counter (count). In function application, an initial thread gets SyncCells and initializes these values.

Message passing among threads and thread synchronizations are operated following way:

- A CPU sends arguments for the destination thread, and issues a SYNC FMP instruction. The SYNC FMP instruction specifies the SyncCell number and gives the FMP a synchronization point of a thread.
- When an FMP receives the SYNC message, the FMP decreases and checks synchronization counter. If a synchronization has proved successful, then the thread is activated.
- The FMP releases the SyncCell, and puts the instance frame address and the starting thread address of active thread into the queue called "Thread Queue."
- At the end point of each thread, a CPU gets a new active thread from Thread Queue, and executes the thread.

4.2 Instance Frame Management

To reduce overhead for memory management, the KUMP/D uses fixed size instance frames for small instances. The FMP has several frame stacks[3] These frame stacks consist of pointers which point to the start addresses of free instance frames. Instance frame allocation and release are implemented by these stack operation.

When an application program requires a larger size instance frame, an interruption is occurred, and a memory management routine supplies a required frame.

---

[3] Currently, the prototype of the KUMP/D supports four frame stacks.

5 FMP instructions

FMP instructions to issue a message are implemented as several kinds of memory access instructions in CPU. In this section, we explain the FMP instructions and FMP registers accessible from user programs, where FMP registers are internal registers in FMP and are provided as an interface to CPU. (For more details, see [2])

Table 1 shows a list of FMP instructions. In this table, Rs indicates a register name of the CPU, FRa and FRs are register names of the FMP. When a CPU issues an FMP instruction, the CPU sets a CPU register "Rs" and FMP registers beforehand. Registers in square brackets mean value of the register.

Instruction	Operand	Function and Message Format <OP,ADDR,DATA>	Notes
FMP Register Operation			
SETFRA	Rs	Set an FRa register	
READFRA	Rs	Read an FRa register	
SETFRS	Rs	Set an FRs register	
READFRS	Rs	Read an Frs register	
SyncCell Operation			
ALLOCSCELL	Rs, count	Get a SyncCell	$[Rs]$=Frame
SETSCELLIP	Rs	Set an Instruction Pointer(IP) into the SyncCell	$[Rs]$=IP
Message Issue			
LINK	Rs, offset	<SET,[FRa]+offset,[Rs]>	$[Rs]$=data [FRa]=Instance Frame
RLINK	Rs, offset	<SET,[FRa]+offset,([FRs],[Rs])>	$[Rs]$=offset [FRa]=Instance Frame [FRs]=SyncCell
START	Rs	<START,[FRa],[Rs]>	$[Rs]$=IP [FRa]=Instance Frame
SYNC	Rs	<SYNC,[Rs],->	$[Rs]$=SyncCell
LSYNC	Rs	<LSYNC,[FRa],[Rs]>	$[Rs]$=data [FRa]=SyncCell
Othres			
ALLOCFL	Rs	Get an Instance Frame	$[Rs]$=size
FREEFL	Rs	Release an Instance Frame	$[Rs]$=size

Table 1. FMP instructions (for user program)

6 Performance Analysis

In this section, we evaluate the performance of the KUMP/D. To put it concretely, we estimate the performance in sending/receiving messages for both the KUMP/D and the Datarol-II. Because the message passing is the operations very freaquently executed in fine-grain multi processing, evaluation of message handling performance is very important. The Datarol-II processor is suitable for performance comparison, since it has high speed hardware mechanism for message handling[7].

6.1 Preparatory Analysis

In a PE of the KUMP/D, since the CPU bus (memory bus) is used by FMP instructions to issue a message as well as usual memory accesses, the memory bus will easily be a bottleneck. In a PE of the Datarol-II, the potential bottleneck is also access to Register Buffer(RB), which is used as cache memory in Datarol-II[7]. Accordingly, we examine the capacity in handling messages (packets) for three typical operations by focusing on the frequency of memory bus occupations.

(1) **Function invocation (frame allocation and activation of initial thread):**
The call instruction of the Datarol-II PE corresponds to the following five KUMP/D instructions which perform a frame allocation and an activation of initial thread.

allocscell
setscellip } set SyncCell

allocfl } frame allocation

setfra
start } initial thread activation

(2) **Parameter passing:** In KUMP/D, a parameter passing uses following three instructions, while two instructions (link and receive) are used in Datarol-II.

setfra
link } message for argument data

start } entry thread activation

(3) **Receiving result value:** In KUMP/D, receiving a return value uses six instructions, while 3 instructions (rlink, return and receive) are used in Datarol-II.

allocscell
setscellip } set SyncCell

setfra
rlink } rlink message

setfra
lsync } return message

Table 2 shows the number of memory bus cycles in the Datarol-II PE. In the Datarol-II PE, a memory write (W_D) occurs in FU to set continuation data. When receiving a packet in CU, a memory read (R_D) for continuation data and a write for packet data (W_D) occur. Table 3 shows the number of memory bus cycles in the KUMP/D PE. In KUMP/D PE, CPU uses memory bus to issue an FMP instruction (I_K). FMP writes the contents of message to the memory when received a message(W_K).

	FU	CU	total
(1)	$1W_D$	$1R_D+1W_D$	$1R_D+2W_D$
(2)	$1W_D$	$1R_D+1W_D$	$1R_D+2W_D$
(3)	$2W_D$	$2R_D+2W_D$	$2R_D+4W_D$

Table 2. number of bus cycles (Datarol-II)

	FU	CU	total
(1)	$5I_K$	$1W_K$	$5I_K+1W_K$
(2)	$3I_K$	$1W_K$	$3I_K+1W_K$
(3)	$6I_K$	$2W_K$	$6I_K+2W_K$

Table 3. number of bus cycles (KUMP/D)

Where, R_D is the number of cycles used in read access from RB, and W_D in write access into RB. I_K is the number of memory bus cycles used when CPU issues an FMP instruction, and W_K in write access into memory.

To compare KUMP/D with Datarol-II, we estimate the memory bus occupation time by using a clock rate of a KUMP/D PE (66 M Hz)[4] as a standard. A PE of the KUMP/D executes with this shortest clock cycle, since FMP interface to receive instructions from CPU can be built with simple logic.

We assume the external cache is an S-RAM with $15ns$ access time, and memory access time of the KUMP/D (R_K and W_K) is 2-clock. On the other hand, we assume the clock speed of the Datarol-II will be an half of Pentium, since Datarol-II PE needs custom design. In Datarol-II PE, since one RB access will be finished within a basic cycle, we assume $R_D = W_D = 2$ clocks.

Table 4 lists the number of clocks for three operations discussed above by using this assumption. From this table, we can calculate that Datarol-II needs 32 clocks and the KUMP/D 46 clocks for a 2-arity function call with one result. The KUMP/D PE uses the memory bus about 1.4 times more than Datarol-II PE. If the access time of the memory bus dominates the execution time in PE, the capacity for message handling of the KUMP/D is about 70% compared with Datarol-II. Moreover, the number of instructions in KUMP/D is 17, while 8 in Datarol-II.

	Datarol-II	KUMP/D
(1)	6	12
(2)	6	8
(3)	12	16

Table 4. Total memory bus cycles. (All accesses are assumed as two clocks.)

As discussed above, regarding message handling, the performance of the KUMP/D PE is assumed to be about 70% compared to Datarol-II processor which runs at about half clock rate of the KUMP/D.

Since the increase of the clock speed-up of recent microprocessors pis so rapid, the gap of clock rate between such microprocessors and custom-processors will extend [5]. Judging from this expectation, we can conclude that message handling performance of the KUMP/D is almost the same as Datarol-II from a practical point of view. In addition to that, KUMP/D PE has much higher peak performance with fast clock rate and superscaler facility than that of the Datarol-II.

6.2 Performance of Message Handling

We use the behavior level simulations of the Datarol-II and KUMP/D for some sample programs (see Fig. 4 and Fig. 5) to estimate the performance of PE. In

---

[4] Current high-end microprocessors work at higher clock rate than 200 M Hz. However, these processors do not achieve drastic speed up at bus cycles, therefore we can apply these analysis of bus cycles for current microprocessors. If a new circuit technology produces higher bus cycle, we will be able to achieve better performance for message handling by designing the FMP for higher clock rate.

[5] Even though we estimate that Datarol-II PE will work at half of the KUMP/D PE (33MHz), it is hard to accomplish such a clock rate with custom design.

this simulation, we use hand compiled code. For **Fibonacci(n)** and **Queen(n)**, which are very fine-grain programs, elapsed time in Datarol-II is dominated by the frequency of memory bus usage. When executing such programs on KUMP/D, the memory bus will also be a bottleneck, since the memory bus usage in KUMP/D is 1.4 times more than Datarol-II. These results agree well with above estimates.

In larger problem size, i.e. n, on **Fibonacci(n)**, the reverse of performance is caused by the difference of cache line filling mechanism. In order to reduce the latency by cache miss hit, the Datarol-II processor has implicit loading mechanism[7], which predicts next instance frame from the Thread Queue and loads all cells of the frame. However, since the size of variables in **Fibonacci(n)** is too small, this mechanism has overhead.

In **Queen(n)**, which has a little longer threads and more variables, the difference of performances is smaller, there is no reverse of performance.

Since the frequency of message issues is smaller in practically used applications than these programs, the difference of performances will be smaller. Especially, in compute intensive applications, this approach of the KUMP/D will be a reasonable solution.

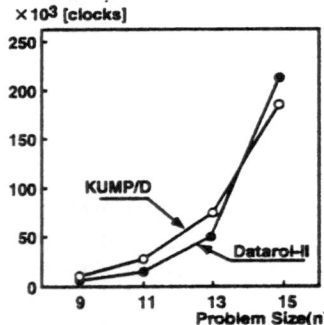 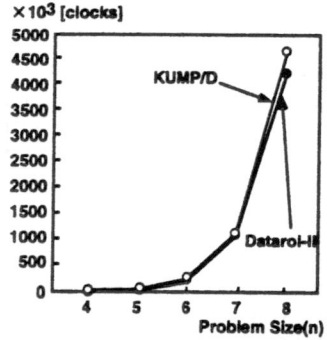

Fig. 4. Execution Time of Fibonacci(n)

Fig. 5. Execution Time of Queen(n)

6.3 Performance of the Prototype

In the prototype of the KUMP/D, we use a Pentium(66 M Hz) processor, and the number of processor is 16.

Since the speed of an I/O serial link is about 33M bytes/sec, and there are four serial lines, occupation rate by video data will be around 0.41. p

In this condition, Table 5 shows our performance predictions of the KUMP/D for image processing by using hand compiled codes.

A PE of the KUMP/D will have enough performance for real time image processing, except for extremely heavy application such as Ray-tracing.

In addition, to evaluate the balance of processor performance and inter connection network, we predict a FFT program on the 16 processor system. Since image mapping to each PE is completely free by using External Video Controller, we can perform FFT by nearest neighbor communications. In this configuration, the prototype of the KUMP/D will perform 512 × 512 point 2D FFT in 5 video frames(NTSC).

Even in the prototype of the KUMP/D, this system will have enough performance to perform real time image processing in NTSC video image. We will achieve higher performance easily by increasing the number of PEs and performance improvement of core processor.

Problem	Performance
Ray-tracing	1 M [polygon/sec]
	10 to 20 [objects/frame]
Polygon rendering	60 to 80 k [polygon/sec]
Canny filter	3.7 k [pixel/frame]
Region Segmentation	21.6 k [pixel/frame]

Table 5. Performance of PE

7 Related Work

In the design of an architecture, cost and performance trade-off should always be taken into account considering the current and future trends of the VLSI and microprocessor technology. One direction of multi-threading architecture research is to find a way to implement a cost effective machine still preserving the dataflow concept. Our research is on this direction, and several machine architectures were designed[7][2].

However, these architectures required special hardware designs, and could not get the benefit of the commercially available microprocessor technology.

Other projects like at MIT[10], ETL[11], and others[5] took the similar approach, and they shared the same problem. Our FMD approach is an attempt to use a off-the-shelf microprocessor for developing the fine-grain multi-threading architecture. In addition to using the off-the-shelf microprocessor, we considered two important points for the design of a massively parallel machine. One point is that the architecture should be tuned to efficient data/message communication, since communication is dominant in massively parallel computations. Another point is that the dataflow concept, whether it is in fine level or in coarse level, is quite natural as a parallel processing model.

Current commercially available microprocessors are designed as a single processor or, at most, low-scale parallel processor, and they are tuned to offer very high performance in internal sequential execution. However, their functions lack in the high speed communication for supporting the higher-scale multi-thread parallel processing. The key issue in desgining a cost effective multi-threading parallel machine is to develop a very simple high speed message handling and communication control hardware device, or co-processor, which complements the high speed internal execution done by commercial microprocessor.

*T[9] at MIT and EARTH[6] is on the similar approach. Different point of *T and EARTH from our approach is that these architectures use commercial microprocessors for both the internal execution and the communication and synchronization control. Since these architectures use the general purpose microprocessor for communication and synchronization control, message handling and synchronization functions have to be implemented in software. This causes an imbalance between the internal execution and the external communication control, and declines the performance in fine-grain multi-threading.

8 Conclusion

In this paper, we gave a co-processor design of the KUMP/D , and showed performance evaluation of the KUMP/D PE. We designed this architecture using a commercial high-end microprocessor and a external co-processor FMP which is a very simple high speed message handling and communication control hardware device. The KUMP/D PE has the same performance of fine-grain parallel processing as a custom designed Datarol-II PE, which works at half the rate of the KUMP/D clock speed. Since KUMP/D CPU itself has quadruple performance compared with Datarol-II PE, KUMP/D PE will achieve better performance than Datarol-II PE in practically used applications.

We can build a high-performance fine-grain multi-thread machine in a short term that follows the mainstream of processor design technology, because this approach solves a problem in special hardware design such as expensive development cost.

The KUMP/D efficiently handles asynchronous fine-grain parallel processes and has high throughput and high flexibility necessary for the next generation multi-media applications, which require complex computer vision and computer graphics algorithms.

References

1. M. Amamiya, and R. Taniguchi, "Datarol: A Massively Parallel Architecture for Functional Language," Proc. SPDP, pp. 726-735 (1990).
2. M. Amamiya, T. Kawano, H. Tomiyasu and S. Kusakabe,"A Practical Processor Design For Multithreading," Proc. of the Sixth Symposium on Frontiers of Massively Parallel Computing, pp.23-32, 1996.
3. W. J. Dally "Performance Analysis of k-ary n-cube Interconnection Networks," IEEE Transactions on Computer, Vol. 39, No. 6, pp. 775-785 (1990).
4. I. S. Gopal, "Prevention of Store-and-Forward Deadlock in Computer Networks, "IEEE Transactions on Communications, Vol. COM-33, No. 12, pp. 1258-1264 (1985).
5. V. G. Grafe and J. E. Hoch, "The Epsilon-2 Multiprocessor System," *Journal of Parallel and Distributed Computing*, Vol.10, No.4, pp.309-318, 1990.
6. H. H. J. Hum, G. R. Gao, et.al., "A Design Study of EARTH Multiprocessors," *Proc. 8th IEEE International Conference on Parallel Architecture and Compilation Techniques (PACT'95)*, pp.59-68, 1995.
7. T. Kawano, S. Kusakabe, R. Taniguchi, and M. Amamiya, "Fine-grain Multithread Processor Architecture for Massively Parallel Processin," Proc. of HPCA'95(First IEEE Symposium on High-Performance Computer Architecture), pp.308-317, 1995.
8. A. L. Narasimha, Reddy and James C. Wyllie, "I/O Issues in a Multimedia System, " IEEE COMPUTER, pp. 69-74 (1994).
9. R. S. Nikhil, G. M. Papadopoulos and Arvind, "*T: A Multithread Massively Parallel Architecture," Proc. 19th ISCA, pp.156-167, 1992.
10. G. M. Papadopoulos and D. E. Culler, "Monsoon: an Explicit Token-Store Architecture," Proc. 17th , pp.82-91, 1990.
11. S. Sakai, Y. Yamaguchi, K. Hiraki, and T.Yuba, "An Architecture of a Dataflow Single Chip Processor," Proc. 16th ISCA, pp.46-53, 1989.
12. H. Tomiyasu, T. Kawano, R. Taniguchi and M. Amamiya, "KUMP/D: the Kyushu University Multi-medea Processor," Proc. Computer Architectures for Machine Perception '95, pp.367-374, (1995).

A Virtual-Physical On-Chip Cache for Shared Memory Multiprocessors

Dongwook Kim and Joonwon Lee

Computer Architecture Lab., Computer Science Department,
Korea Advanced Institute of Science and Technology,
373-1 Kusung-dong Yusung-ku Taejon 305-701, South Korea.

1   Introduction

Two-level caches are popular in high performance multiprocessors because the cache at the first-level provides data very fast while the one at the second-level enables high hit ratios[1, 2]. The cache at the first-level, L1, is small and close to the processor usually on-chip for the fast access time, while the cache at the second-level, L2, is large in its capacity. Since the virtual cache makes the cache access time fast, it is adequate for the design of the L1 cache. When a virtual cache is used as the L1 cache, the synonym problem can be handled properly if the inclusion property is enforced[2]. An inclusion property means that all the data in the lower level cache is always in the higher level cache. However, the cache coherence problem in shared memory multiprocessors imposes another difficulty in designing two-level cache. Since physical addresses are used in the L2 cache, it necessitates the use of pointers between two levels to keep track of the mappings between virtual cache and physical cache for the coherence[5]. This limitation makes a write policy of the virtual cache more complex than that of the physical cache. This complexity will reduce the potential of the virtual cache to match the timing requirements of fast processors.

In this paper, we propose a novel multi-level cache architecture, where the L1 cache is decomposed into on-chip two-level virtual-physical caches.

2   The Architecture of a Virtual-Physical On-chip Cache

The main idea of our scheme is to overcome the drawbacks of a virtual cache by classifying memory requests into instruction, private data and shared data, and then storing them in separate caches according to the classified state. A simple hardware block diagram of a virtual-physical on-chip two-level cache is shown in Fig. 1(a).

The L1 virtual caches are accessed via virtual addresses which are also forwarded to the TLB at the shared data cache, and thus address translation can proceed concurrently with the access to the virtual caches. If there is a valid hit in the virtual caches, the translation and the access to the physical cache is aborted. Otherwise, the L1 physical cache is accessed by the translated address. The L2 cache would be accessed when the L1 caches cannot service the memory requests from a processor. The private data cache exclusively loads private data,

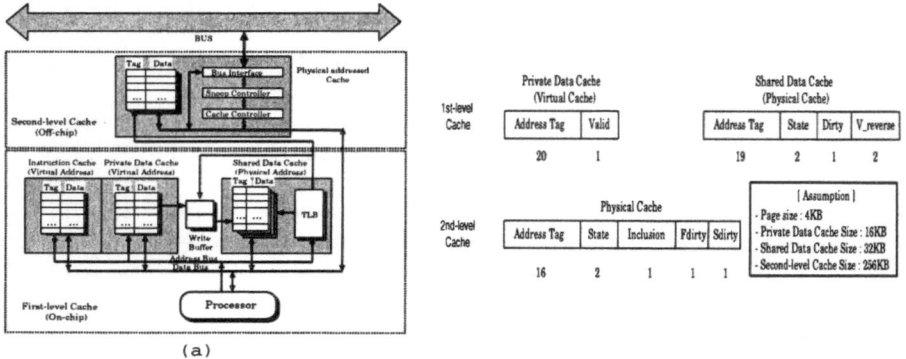

(a)

Fig. 1. Virtual-physical on-chip cache architecture

while the shared data cache can load both private and shared data. According to the state of the fetched cache line, it can be loaded in either cache. If the state is exclusive, the fetched cache line can be loaded in the private data cache and the shared data cache. Otherwise, it can be loaded in the shared data cache. The write buffer between the private data cache and the shared data cache can hide the memory latencies due to subsequent writes.

Fig. 1(b) shows the tags and control bits which are associated with each cache line. In the shared data cache, each tag entry contains a tag, state bits, a dirty bit and a v_reverse bits. The state bits indicate status of a line and it would be used for cache coherence control with other physical caches and classifying cache data into private or shared. A dirty bit indicates the cache line is updated. The v_reverse bits are reverse address pointer and are used to detect synonyms. The L2 cache is responsible for shielding cache coherence interference from other processors. The fdirty bit indicates if the corresponding cache line in the L1 cache is updated or not. It is used for the cache coherence protocol for multiprocessors. The sdirty bit indicates that the corresponding cache line in the L2 cache is updated. We assume that our scheme uses an invalidation cache coherence protocol for simplicity, although it will also work for other protocols as well.

3 Performance Evaluation

To study the performance implications, we use an event-driven simulator which runs on the MINT[4]. The test programs used in this study come from the SPLASH suite[3]. Those are FFT, Barnes, MP3D, Pthor and LU ,which are implemented for snoopy cache bus-based multiprocessors. We measure two performance metrics in the simulation : the cache coherence traffic and the average memory access time. An important advantage of the hybrid V-P cache is that the private data cache is not affected by irrelevant cache coherence interference. A processor does not stall to access the private data cache when a coherence

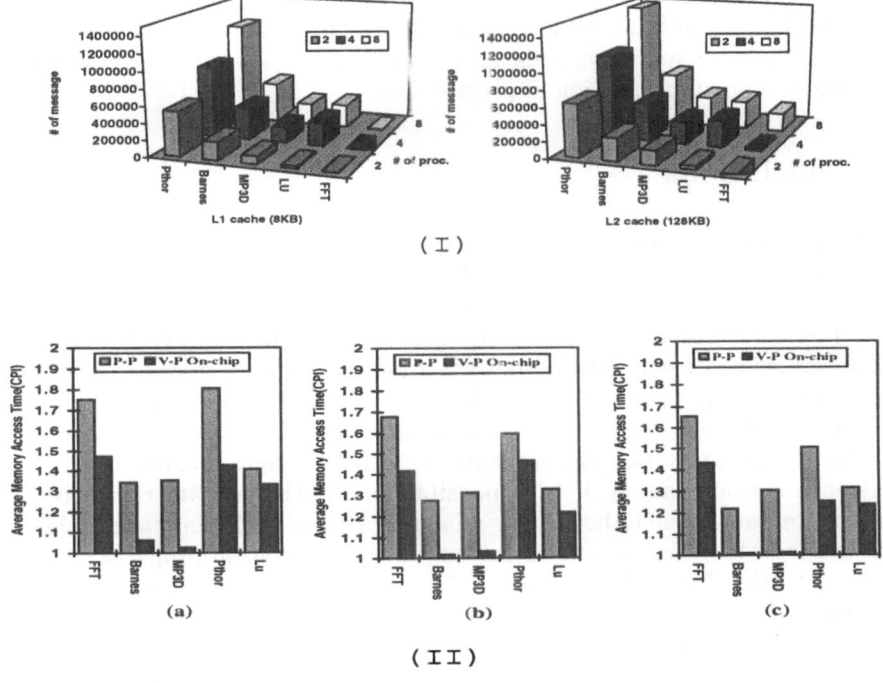

Fig. 2. (I) Number of coherence messages to the L1 cache(16K/128K) (II) Average memory access time(cycles) versus cache sizes, (a)16K/128K, (b)32K/256K, (c)64K/512K

message is issued from the bus to the L1 cache because the message affects only the shared data cache. Thus, a processor stall times can be significantly reduced by the private data cache. The results in Fig. 2(I) show that the inclusion property is essential for a two-level cache to reduce the number of coherence messages which are issued to the L1 cache. We believe that the performance gains by the private data cache will be more prominent as the number of processors increases. This is due to the fact that more bus coherence requests will be generated from a large number of processors, and the L1 cache will be disrupted more often and a processor stall time will be increased.

A higher hit ratio does not necessarily mean faster memory access time since cache access time for a cache hit is more important when the hit ratio is very high. The V-P on-chip cache provides fast cache access time for a cache hit at the private data cache. Therefore, average memory access time can be decreased in proportion to the hit ratios of the private data cache. Fig. 2(II)(a), Fig. 2(II)(b) and Fig. 2(II)(c) show average memory access times for various cache sizes. The V-P on-chip cache outperforms the P-P cache for all the tested applications.

There is a wide performance gap between the V-P on-chip cache and the P-P cache for the Barnes and Pthor, since a large proportion of data references of those applications are private data. Meanwhile, LU shows slight performance gap due to a small proportion of private data.

4 Conclusions

In this paper, we propose a new V-P on-chip cache scheme for multiprocessors. The key idea of the new scheme is that virtual caches and physical cache are incorporated at the same level as L1 caches, and the fetched data from a memory should be loaded in an exclusive cache according to whether it is shared data or private data. If a newly fetched data is a private data, it can be stored in the private data cache accessed by virtual addresses. Otherwise, it should be stored only in the physical cache called shared data cache. It is easy to optimize the circuits of a private data cache, since the logics for managing synonyms and cache coherence are not needed. It is responsible for the shared data cache to shield the private data cache either from synonyms or from irrelevant cache coherence interferences. In order to achieve this purpose, an inclusion property should be imposed to the shared data cache as like an L2 cache. Simulation results show that the V-P on-chip cache outperforms other cache structures.

Recent VLSI technology enables a very fast cycle time for a processor, and thus it becomes more difficult for the memory system to meet such a fast cycle time. In a shared memory multiprocessor, this problem is even more aggravated due to the cache coherence problem. We believe that our proposed cache scheme addresses such issues appropriately, and thus it will be a promising soultion for memory system of high performance multiprocessors.

References

1. C. Anderson. Improving performance of bus-based multiprocessors. *PhD thesis, University of Washington*, 1995.
2. J. L. Baer and W. H. Wang. On the inclusion property for multi-level cache hierarchies. *In Proceedings of the 15th Annual International Symposium on Computer Architecture*, pages 73–80, 1988.
3. J. P. Singh, W. Weber, and A. Gupta. SPLASH:Stanford Parallel Applications for Shared Memory. *Computer Architecture News*, 20(1):5–44, 1992.
4. J. E. Veenstra and R. J. Fowler. MINT tutorial and user manual. *TR452, Computer Science Department, Univ. of Rochester*, 1994.
5. W. H. Wang, J. L. Baer, and H. M. Levy. Organization and performance of a two level virtual real cache hierarchy. *In Proceedings of the 16th Annual International Symposium on Computer Architecture*, pages 140–148, 1989.

Shared vs. Snoop: Evaluation of Cache Structure for Single-Chip Multiprocessors

Toru Kisuki[†],Masaki Wakabayashi[†],Junji Yamamoto[†],Keisuke Inoue[†],
Hideharu Amano[†]

[†]Department of Computer Science, Keio University
3-14-1, Hiyoshi Yokohama 223 Japan
{kisuki,masaki,junji,keisuke,hunga}@aa.cs.keio.ac.jp

Abstract. The shared cache structures and snoop cache structures for single-chip multiprocessors are evaluated and compared using an instruction level simulator. Simulation results show that 1-port large shared cache achieves the best performance if there is no delay penalty for arbitration and accessing the bus. However, if 1-clock delay is assumed for accessing the shared cache, a snoop cache with internal wide bus and invalidate style NewKeio protocol overcomes shared caches.

1 Introduction

In order to make the best use of a large silicon area, single-chip microprocessor approach using simple microprocessors has been investigated[1][2]. The precise simulation results demonstrated that this approach is hopeful compared with complicated superscaler processors for parallel applications[1].

In these evaluations, processors are connected with a shared cache which is suitable for on-chip implementation. However, a private cache with snoop mechanism which is mainly used in board level implementation is also hopeful structure for on-chip implementation, since each cache can be connected with a high bandwidth on-chip bus. The snoop cache protocol can be optimized for on-chip implementation. In this paper, the structure of snoop cache for single-chip multiprocessor is discussed and the performance is compared with shared cache using an instruction level simulation.

2 Cache structures for Single-Chip Multiprocessors

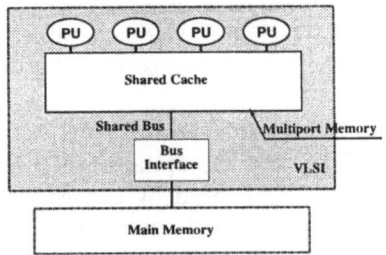

Fig. 1. Shared Cache Structure

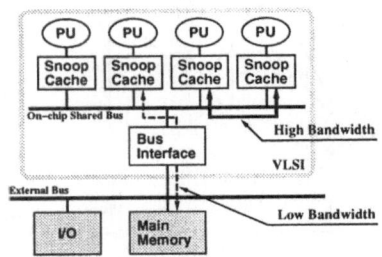

Fig. 2. Snoop Cache Structure

2.1 Shared cache structure In many possible connection architectures for single-chip multiprocessors, the most simple structure is connecting processors to shared cache directly. Fig.1 shows the most simple and quite realistic structure of the single-chip multiprocessor. In this structure, a large SRAM shared cache is also connected to main memory outside the chip. Due to the overhead caused by the conflict, multi-port memory which allows simultaneous access is sometimes introduced. But the access time is often stretched by a large fan-out required in cells of m-port memory.

2.2 Snoop cache structure Private cache with snoop mechanism is an alternative approach to efficient cache for single-chip multiprocessors. In this approach, each processor provides its private cache connected with each other by a shared bus. Each cache checks the address and data on the shared bus and maintains the consistency according to the cache consistency protocol. This structure is commonly used in the recent multiprocessor workstations. In the board level implementation, since the common 64 bits address/data multiplexed bus is used, the cost for maintaining cache consistency often dominates the performance.

However, in the single-chip multiprocessor, the cost of wire is much less than that in the board-level implementation, and a bus with a large bandwidth can be used. Here, we propose the snoop cache with a wide bus for single-chip multiprocessors. In this cache, the following bus is used: (1) the size of the data bus is the same as the cache line, and (2) 64-bit address bus is provided independent from the data bus. In such a snoop cache, a cache line can be transferred between caches only with a clock, and overhead for maintaining the consistency is drastically reduced.

2.3 NewKeio protocol In a snoop cache with a wide bus, the gap between inside and outside of the chip is an essential problem. In order to cope with this problem, we have proposed a coherent protocol which minimizes the data transfer between cache and main memory.

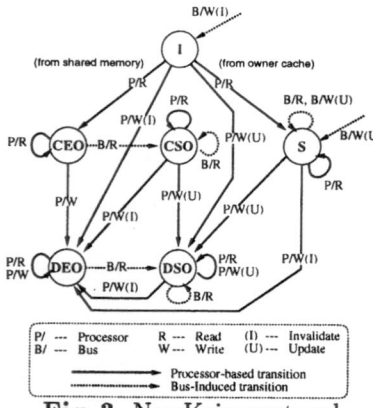

Fig. 3. NewKeio protocol

This protocol is called NewKeio Protocol[3]. As shown in Fig.3, the ownership is introduced even for a clean line, and the miss-hit cache line is transferred from cache as possible. Moreover, the attribute is attached to the page table to select whether the protocol uses invalidation or updating.

By using the invalidation protocol for instruction or local data, and the update protocol for shared data, the loss caused by consistency is minimized. In Fig.3, (U),(I) indicates Update and Invalidate type, respectively.

3 Simulation Environment

3.1 Target systems By the end of 1997, it is possible to make a processor chip which has $500mm^2$ die area in $0.25\mu m$ process technology. Since a sophisticated processors requires a large area, the possible number to be implement becomes small. Therefore, there are many combinations between the class and number of processors. Here, we select rather simple R3000 class RISC processor. In this case, from four to six processors and 256KB SRAM can be mounted. 500MHz system clock is assumed, and a latency accessing the outside main memory becomes 50 clocks.

Here, we evaluate the following cache structures:

- 4-Port Shared Cache
  In this cache, 4-port memory is used. Each processor has its own port and accesses cache memory without any conflicts. However, the total cache size is set to be a forth (64KB) of the 1-Port Shared Cache.
- 1-Port Shared Cache
  A single-port memory is used for this cache. Since each processor shares a port, a conflict occurs when multiple processors access the port simultaneously. However, its simple structure allows a large cache size (256KB).
- Snoop Cache(Illinois protocol)
  A common cache protocol (Illinois protocol[4]) is used as snoop cache protocol. Each processor has its own 64Kbyte cache, so total cache size is 256KB.
- Snoop Cache(NewKeio protocol)
  The same structure as the snoop cache with Illinois protocol, but NewKeio protocol is used.

3.2 Instruction-level simulator ISIS Since the performance of cache is depending on the detail behavior of hardware including pipeline structure, a precise simulation is required.

For this purpose, we developed an instruction level simulator called ISIS. ISIS simulates the behavior of the RISC processor pipeline stage for every instruction. Precise operation of each pipeline stage including delay slots can be simulated.

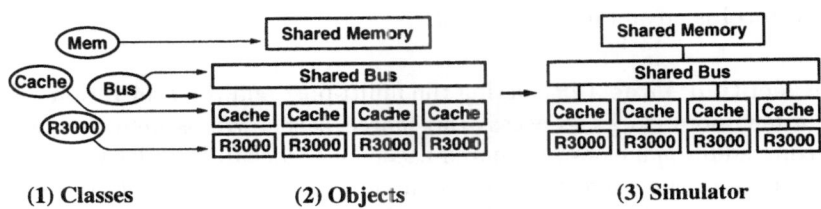

Fig. 4. Organization of ISIS

Fig.4 shows the organization of ISIS. All units such as processor, cache, bus and memory are implemented as classes. At first, each class is constructed, then each object is connected through an interface. In this way, various architectures can be simulated by replacing units.

Three parallel applications from SPLASH benchmark programs[5] are selected for evaluation.
- FFT(data points is set to be 4096)
- MP3D(100 steps with 8000 particles at test.geom)
- LU(128×128 matrix with a block size 16)

4 Simulation Results

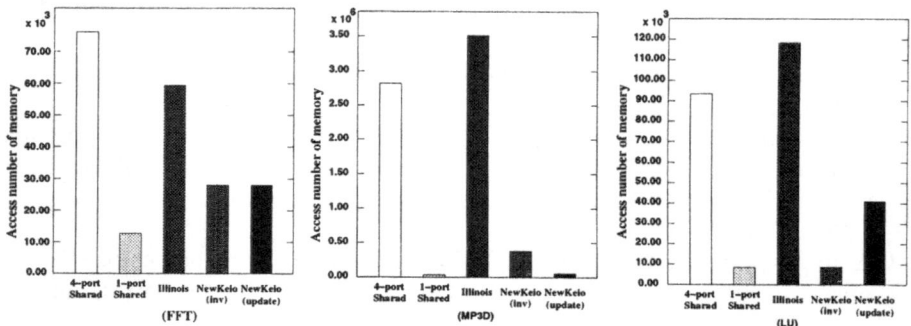

Fig. 5. Access Number of Memory

4.1 Access frequency of the outside memory Fig.5 shows the number of accessing outside memory. Since there is a large gap between the bandwidth of inside and outside chip, a large number of access to the main memory causes performance degradation.

In FFT, 4-port shared cache shows most frequent memory access by the capacity miss. Illinois protocol is better than 4-port shared cache, since cache to cache inside the chip is provided for access miss of a clean line. NewKeio protocol shows further better performance than Illinois protocol. Unlike Illinois protocol which requires write back when the dirty line is required from other processor, the write back occurs only at replacing in NewKeio protocol. This causes the difference of the access number, especially in MP3D which requires a large amount of interprocessor communication. Due to this reason, the update type protocol is advantageous in MP3D.

4.2 Execution Time Considering the multi-port and the arbitration delay, it takes a few more clocks to access the shared cache. We assume this delay by penalizing additional clocks to shared cache (1-3 clocks). In order to investigate the influence of the conflicts which occur in 1-port shared cache, the result of ideal execution time without conflicts is also shown in Fig.6.

Without delay, the performance of 1-port shared cache is the best in all applications. In 1-port shared cache, the influence of the conflict is not so large if no delay is assumed. As mentioned above, the performance of 4-port shared cache is degraded mainly by the communication with outside memory.

However, with the delay penalty, the performance of 1-port shared cache is severely degraded because of the conflict. Even with 1-clock delay, invalidate

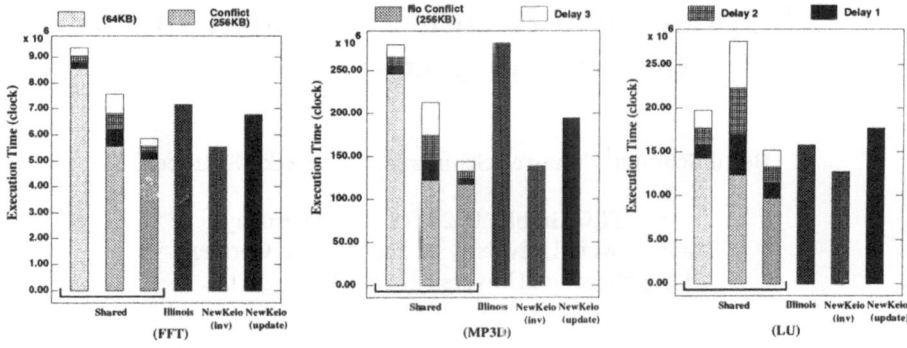

Fig. 6. Execution Time

type NewKeio protocol shows better performance than 1-port shared cache. Since there are a few shared line and interprocessor communication, each private cache can be used efficiently and reduce the overhead to maintain the consistency. From this reason, the performance of invalidate type NewKeio protocol is better than that of update type. Compared with Illinois protocol, NewKeio protocol improves the performance in 15%-20%.

5 Conclusion

The simulation results show that with the parameters used here, 1-port large shared cache achieves the best performance if there is no delay penalty for arbitration and accessing the bus. However, if 1-clock delay is assumed for accessing the shared cache, a snoop cache with internal wide bus and invalidate style NewKeio protocol shows better performance. Further simulations with various parameters and applications are required for investigating optimized cache structure for single-chip multiprocessors.

References

1. Basem A. Nayfeh, Lance Hammond, and Kunle Olukotun. Evaluation of Design Alternatives for a Multiprocessor Microprocessor. *ISCA*, 1995.
2. Marco Fillo, Stephen W. Keckler, William J. Dally, Nicholas P. Carter, Andrew Chang, Yevgeny Gurevish, and Whay S. Lee. The M-Machine Multicompuer. 1995.
3. T. Terasawa, S. Ogura, K. Inoue, and H. Amano. A Cache Coherence Protocol for Multiprocessor Chip. *In proc. of IEEE International Conference on Wafer Scale Integration*, pages 238–247, January 1995.
4. M.S.Papamarcos and J.H.Patel. A Low-overhead Coherence Solution for Multiprocessors with Private Cache Memoryes. *ISCSA84*, pages 348–354, 1992.
5. Steven Cameron Woo, Moriyoshi Ohara, Evan Torrie, Jaswinder Pal Singh, and Anoop Gupta. The SPLASH-2 Programs: Characterization and Methodological Considerations. *ISCA*, pages 24–36, June 1995.

Morphological Hough Transform on the Instruction Systolic Array

Bertil Schmidt[1], Manfred Schimmler[2] and Heiko Schröder[3]

[1] ISATEC GmbH, D-24118 Kiel, Germany
[2] Braunschweig University of Technology, Germany
[3] Loughborough University of Technology, UK

Abstract. Instruction systolic arrays have been developed in 1987 [La1] in order to combine the speed and the simplicity of systolic arrays with the flexibility of MIMD parallel computer systems. In this paper a new algorithm for line detection is presented which applies the morphological approach to the well known Hough transform. The quality of its results is significantly higher than that of the classical Hough transform. This new algorithm has been tailored towards the capabilities of the instruction systolic array. It has been implemented on the *Systola 1024*, the first parallel computer of this particular architecture on the market. Systola 1024 is an low cost add-on board for standard PC's with PCI slots.

1 Introduction

Instruction systolic arrays (ISAs) provide a programmable high performance hardware for specific computation intensive applications [DS1,Sch1]. Typically, such an array is connected to a sequential host, thus operating like a coprocessor which solves only the computationally intensive tasks within a global application. The ISA model is a mesh connected processor grid, where the processors are controlled by three streams of control information: instructions, row-selectors and column-selectors. The concept of instruction systolic arrays is explained in detail in the second section.

Hough transform is a standard technique for line detection in image processing. It is based on the accumulation of information about straight lines intersecting with points (pixels) of the image plane. The details of Hough transform are given in section 3. The classical Hough transform creates artefacts if there are structures in the image like dashed lines or small line segments. Therefore, modifications of Hough transform have been developed [BPS1,Lea1]. In this paper an approach based on mathematical morphology is taken to find an efficient implementation of Hough transform that minimises the number of artefacts in the transform space. This algorithm is explained in section 4.

The implementation of this new algorithm is presented in section 6. For this purpose the Systola 1024 is shortly explained in section 5. It is the instruction systolic computer architecture on which the algorithm has been implemented.

Section 7 discusses the performance in comparison to implementations on a sequential architecture. The results of the original Hough transform and those of the new algorithm are demonstrated in an example.

2   Principle of the ISA

The ISA is a quadratic array of identical processors, each connected to its four direct neighbours by data wires. The array is synchronised by a global clock. The processors are controlled by instructions, row- and column-selectors.

The instructions are input in the upper left corner of the processor array, and from there they move step by step in horizontal and vertical direction through the array. This guarantees that within each diagonal of the array the same instruction is active during a single clock cycle. Processor $(i+1,j)$ and $(i,j+1)$ execute in clock cycle $k+1$ an instruction that has been executed by processor (i,j) in clock cycle k.

The selectors also move systotically through the array: the row-selectors horizontally from left to right, the column-selectors vertically from top to bottom (Fig. 1). The selectors mask the execution of the instructions within the processors, i.e. an instruction is executed if and only if both selector bits, currently in that processor, are equal to one.

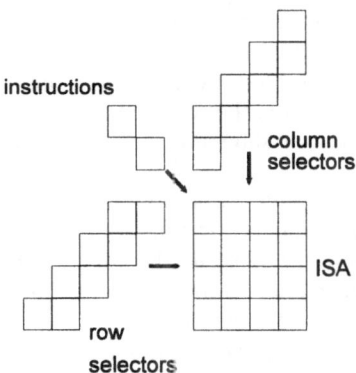

instructions

column selectors

ISA

row selectors

Fig. 1: Control flow in an ISA

Every processor has read and write access to its own memory. Beside that, it has a designated *communication register (C-register)* that can also be read by the four neighbour processors. Two adjacent processors can exchange data by writing on their own C-register and afterwards reading the C-register of the other in two subsequent clock phases. Within one clock phase the reading access is always performed before the writing access. This convention on the one hand avoids read/write-conflicts, on the other hand it creates the possibility to broadcast information across a whole processor-row or -column with one single instruction. This property can be exploited for an efficient calculation of row-sums and row-ringshifts which are the key-operations in the parallel Hough transform implementation described in section 6:

Row-sum. One important advantage of ISAs is the capability of performing aggregate functions within one (or a constant number of) instructions. Aggregate functions are operations where every processor needs information of all processors with smaller column index and the same row index (or vice versa). The simplest example is the computation of the row sums: Each processor computes the sum of the C-registers of its left neighbour and itself. Since the execution of this operation is pipelined along the row each processor accumulates the sum of all C-registers up to its own which is identical to the prefix sum.

Row-ringshift: The contents of the C-registers can be ringshifted along the processor rows by two instructions. Every two horizontally adjacent processors exchange data (using one read left and one read right operation). Because of the instruction flow from west to east this implements a ringshift. Of cause, a column-ringshift can be executed in the same way.

3 Hough Transform

The classical Hough transform (HT) [IK1] is a very powerful technique to detect straight lines in a binary image. It transforms the image into a parameter space by counting pixels on straight lines. For every possible straight line in the image a counter is introduced which accumulates the number of pixels which satisfy the line equation $y = mx + c$. In the parameter space we represent a straight line by these two parameters: the slope m and the intercept (the intersection point with the y-axis) c. This representation is rather simple. The set of straight lines intersecting in one point of the original image is represented by a straight line in the m-c-space. The disadvantage of this representation is the fact that vertical lines (with infinite slope) cannot be represented. However, as shown in section 6, this does not cause problems in our implementation.

The Hough transform algorithm proceeds now in the following way: An accumulator array $B[m_k, c_l]$ is introduced to represent the parameter space. For every white pixel in the image and every slope m_k the corresponding value c_l is computed from the line equation and the counters B for all straight lines intersecting in this pixel are incremented by one. By that at the end of the execution the value of each counter represents the fraction of the corresponding straight line in the original image. The local maxima of the B array indicate the presence of lines of certain slopes and intersects in the image.

The HT algorithm in the simplest form for a binary image $I(i,j)$ of size NxN and an accumulator array $B[m_k, c_l]$, where $k = 0, ..., M-1$, $m_k = k/M$, looks like this:

Alg. 1: Standard HT

for i:=0 to N-1 do
  for j:=0 to N-1 do
    if I(i,j)=1 then
      for k:=0 to M-1 do increment $B[m_k, round(j-i \cdot k/M)]$

One disadvantage of HT is the high requirement in computing power such that on-line computations are impossible on existing sequential computers. The work is of the order N^2M.

Therefore, a parallel implementation is useful and necessary for many applications. Another disadvantage is the fact that HT creates artefacts by counting structures in the image that finally turn out to be only very small fractions of straight lines (e.g. only one dot). In [BPS1] a new method to solve this problem is presented. Its basic idea and the new algorithm derived from it is presented in the next section.

4 Morphological Hough Transform

From the point of mathematical morphology the identification of a white pixel is the erosion operation with a structural element that consists of only one dot. Whenever the Hough transform algorithms finds a white pixel, it increments the counter for every straight line containing this pixel. By using a structural element which represents a straight line by more than one dot we can significantly reduce the number of counter increments. In particular, all those cases where the white pixel was isolated or consists of only a small white segment do not lead to an increment of any counter. This suppresses the artefacts in the m-c-space. Ideally, a straight line can be represented by two pixels (because there is exactly one straight line intersecting with these two pixels). By using the morphological approach with a structural element consisting of two pixels for every possible slope we increase a counter if and only if both pixels match in the original image.

Alg. 2 shows an implementation of this method (with $I(i,j)$, $B[m_k, c_l]$ like in Alg. 1 and the structuring element for each slope k has the distance (q_k, p_k) from the actual pixel-coordinates). In order to produce a parallel implementation on the ISA a much more efficient version of the morphological approach becomes possible if the loops in this program are swapped as in Alg. 3.

Alg. 2: MHT

```
for i:=0 to N-1 do
 for j:=0 to N-1 do
  if I(i,j)=1 then
   for k:=0 to M-1 do
    if I(i+qₖ,j+qₖ)=1 then
     increment B[mₖ,round(j-i·pₖ/qₖ)]
```

Alg. 3: MHT with swapped loops

```
for k:=0 to M-1 do
 for i:=0 to N-1 do
  for j:=0 to N-1 do
   if I(i,j)=1 and I(i+qₖ,j+pₖ)=1 then
    increment B[mₖ,round(j-i·pₖ/qₖ)]
```

Now we perform the accumulation operations for the complete image slope by slope. Every slope is represented by a structural element of two pixels, e.g. *(0,0)* and *(x,y)*. By shifting the image by *(-x,-y)* we get two copies of the image on top of each other. Now the morphological erosion can be performed by one AND-operation of the original image with the shifted image.

In reality, there may occur uncertainties due to rounding effects: Not every straight line has a slope which can be exactly represented by two pixels with a constant distance. The solution of [BPS1] selects a collection of structural elements that are almost equidistant. Skewing the image and choosing an appropriate structural element can avoid this problem. The details of this solution as well as the parallel implementation of the morphological erosion with the shifted image are given in the section 6.

5 Architecture of Systola 1024

The parallel computer Systola 1024 is an low cost add-on board for standard PCs [LMS1], providing a *32x32*-ISA. It consists of a *4x4* array of processor chips. Each chip contains *64* Processors, arranged as an *8x8* square.

In order to exploit the computation capabilities of this unit, it is necessary to provide data and control information at an extremely high speed. Therefore, a cascaded memory concept is implemented on board that forms a fast input and output environment for the parallel processing unit (Fig. 2).

For the fast data exchange with the processor array there are rows of intelligent memory units at the northern and western borders of the array. These units are called *interface processors*. Eight interface processors are integrated onto one chip. Each interface processor is connected to its adjacent array processor by a single wire for data transfer in each direction.

The interface processors have access to an on board memory by means of special fast data channels, those at the northern interface chips with the northern board RAM, and those of the western chips with the western board RAM. The board RAM can communicate bidirectionally with the PC memory. The data transfer between every two memory units within this hierarchy is controlled by an on board controller chip. In particular, it controls the channels between the interface processors and the board RAMs as well as between the board RAMs and the PCI bus of the PC.

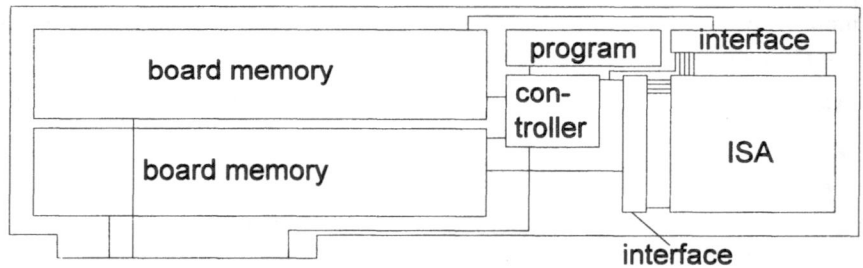

Fig. 2: Block diagram of the Systola1024-board

In addition, the controller supplies the interface processors and the array processors with instructions and selectors that are stored in an additional memory on board, the *ISA program memory*. The ISA program memory can contain a large variety of programs for the processor array, according to the different tasks that may be necessary to perform during one application. Every word clock cycle the controller transmits one instruction, one row-selector and one column-selector from the ISA program memory to the interface processors. The instruction consists of *32* bits as well as the row-selector and the column-selector. A triple of instruction, row-selector and column-selector is called a program diagonal. The ISA program memory can contain *16 K* program diagonals.

The controller receives its instructions either directly from the PC as so *board instructions*, or it can operate autonomously. In the second case it receives *controller instructions* from an *instruction queue*, which is located on the board, too. The instruction queue can be loaded from the PC and it consists of up to *256* controller instructions, each of them in *16* bit format.

Systola 1024 has a peak performance of *3200* MIPS. In practice the performance lies, dependent on the application, between *1000* and *3000* MIPS.

6 Parallel Implementation of Morphological Hough Transform

For the implementation of the MHT we take a NxN, $N=512$, binary image. Every processor of *Systola 1024* stores a *16x16* subimage in *16* internal registers. So each of this registers stores *16* adjacent pixels of the same row. The capabilities of ISAs to compute row (column)-sums and -ringshifts very efficient (see section 2) are exploited in the parallel implementation:

The erosion and accumulation for horizontal (vertical) lines can be done very fast: For a structuring element of two pixels with distance d, we can shift a copy of the image above the origin by d pixels. Now a counter has to be incremented if a pixel of both, the origin and the shifted image are set. With shearing of the image other line angles are transformed in horizontal position, such that the efficient horizontal operations are always applied.

Horizontal Accumulation. This operation means to sum up the pixels of each image-row. The implementation first builds the subrow-sum within each processor in its C-register. Afterwards the efficient processor-row-sum computes the whole image-row-sum. This operation is repeated *16* times for every subimage-row of each processor. At the end of the iteration the last processor-column stores the results. Because of the limited memory (*32* registers) of each processor the results are shifted one processor-column to west in every second iteration step. The *i.th* processor-column always collects accumulator-entries for the line-intercept i. When all processor-columns are covered the accumulator-array-results are transferred to the board-RAM.

Horizontal Erosion. Let the structuring element have the distance d from the image-origin. Then an image-row-ringshift for d positions and an AND-Operation computes the horizontal erosion. In Systola an image-subrow is ringshifted one processor. Afterwards a shifting for *(d mod 16)* positions within each processor is performed. Then a ringshift for a *(d div 16)*-processor-distance computes the image-row-ringshift. Finally one AND-Operation between the origin sub-row-register and the ringshifted sub-row-register in each processor calculates the erosion. After that the result is vertical accumulated.

Shearing. The shearing for line-angles between $0°$ and $45°$ is an image-column-ringshift, where each column has a different shifting-distance depending on the line-angle. To avoid complex operations this distances are precalculated.

For the implementation we take N slopes between $0°$ and $45°$. So the following line equations are considered: $y = (k/(N-1))x + c$; $k=0...N-1$; $c=0...N-1$. The different line-angles are handled in increasing order. Then the shear operation only needs to incorporate the relative shear between neighbouring angles. So the image-column-ringshift only has to consider the next column. The precalculated shearing-distances turns to a binary image-row-mask: Only the image-columns with an entry in the corresponding mask-position has to be ringshifted.

The angles between $0°$ and $-45°$ are computed by loading the input image with transposed rows on the ISA and applying the same program. The angles between $45°$ and $90°(-45°$ and $-90°)$ are calculated by storing the *16x16*-subimage column-wise in each processor and applying the reflected ISA-program (swapping row- and column-selectors, changing west and north, east and south) for $0°$ to $45°$ ($0°$ and $-45°$). The finally accumulator size is $4N$ (slopes) x N (intercepts).

7 Performance and Experimental Results

To compare the performance of the parallel algorithm in section 6, we compare the implementation on Systola 1024 with two sequential versions on a PC (Pentium Processor, 200 MHz, optimised code). The first version is the standard version of the HT-algorithm (referred to as *"P200 Standard"* in Table 1). The second version is a sequential simulation of the algorithm implemented on Systola 1024 (referred to as *"P200 Parallel"* in Table 1). For a comparison to other parallel architectures for the Hough transform see [Fer1].

For the implementations we use binary images of size NxN, $N=512$, and an *8-bit* accumulator array consisting of *4N* slopes and *N* intercepts. Systola needs *0,25 s* (*375* processor instructions per slope) for the standard HT and *0,31 s* (*471* instructions per slope) for the MHT. Since computing time dominates communication time in this application, data transfers can be almost totally hidden as it can be executed concurrently to the computation.

The standard HT (Alg. 1) depends on the number of white pixels in the input image, but has the disadvantage of a complex computation of the accumulator bins to be incremented. This means a computing time of *70,0 s* in the worst case (all pixels white). The used edge images from the pick-and-place process of multichip-modules (Fig. 7) have *10%* seeded pixels on an average, which corresponds to a computing time of *7,0 s*.

The standard MHT (Alg. 2) depends on the number of seeded pixels in the input image and in each eroded image. An average multichip-module edge image (Fig. 7) has *10%* seeded pixels in the input and *1%* in each eroded image, which leads to a computing time of *5,7 s* in the average case.

The sequential version of the parallel algorithm in section 6 is independent of the number of white pixels, but has the advantage of an easier accumulation. This leads to a constant computing time of *5,0 s* for the HT and *6,2 s* for the MHT for each image.

Table 1. Comparisons of the average performance of HT/MHT for a *512x512* image and *2048* different slopes on Systola 1024 (including data transfers) and Pentium 200

	Systola 1024	P200 Standard	P200 Parallel	Speedup
HT	0,25 s	7,0 s	5,0 s	28 / 20
MHT	0,31 s	5,7 s	6,2 s	18 / 20

In order to evaluate the qualitative performance of the MHT we have compared it with the HT on several different images.

Fig. 3 shows three lines close together. The relevant part of the accumulator array from the HT and MHT (with distance $d=50$ for the structuring element) is displayed as a height field in Fig. 4 and 5. It can be seen that both, HT and MHT produce artefacts (in the Figures they are behind the three peaks representing the lines). In case of MHT these artefacts are significantly reduced. The MHT also creates an enhanced accumulator contrast, which simplifies the higher order image analysis.

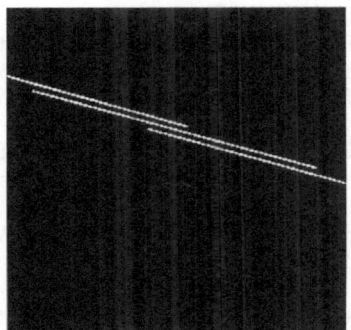

Fig. 3: Original Image

Fig. 4: Accumulator from HT   **Fig. 5**: Accumulator from MHT

Comparisons of the effectiveness of the HT and MHT methods are displayed in Fig. 6-9. Fig. 6 shows an image from the pick-and-place process of multichip-modules. Fig. 7 shows the result after an edge detector is performed on the image. Fig. 8 and 9 show the result after a HT and MHT ($d=10$), where accumulator entries greater than *90* and *80* are detected. The resulting lines are displayed after an AND-operation with the edge image. It can be readily seen that the HT picks up many more false lines than the MHT in the dotted areas.

Obviously there is still lots of scope for fine tuning the combination of HT and morphology to best fit a given application. For example, the structuring elements could be optimised to search for lines of a given thickness (using larger structuring elements), of a given length or which are dotted. We attempt to work with this method further in the area of automatic quality control and image classification [Ko1].

References

1. Beresford-Smith, B., Pham, B., Schröder, H.: A parallel morphological implementation of the Hough transform, Proc. 25th. HICSS, 111-119 (1992).
2. Dittrich, A., Schmeck, H.: Given's Rotation on the Instruction Systolic Array. In: Wolf, G., Legendi, T., Schendel, U. (eds.): PARCELLA '88, Mathematical Research, Bd. 48, Akademie Verlag, Berlin, 340-346 (1988)

3. Ferretti, M., Albanesi, M.G.: Architectures for the Hough transform, Computer Vision, Graphics and Image Processing 44 (1996) 542-551
4. Illingworth, J., Kittler, J.: A survey of the Hough transform. Computer Vision, Graphics and Image Processing, 44, 87-116 (1988)
5. Kolbe, W.: Online Qualitätskontrolle von Oberflächen mit den ISATEC Surface Quality Scannern SQS, Journal für Oberflächentechnologie, 3 (1997)
6. Lang, H.-W., The Instruction Systolic Array, a parallel architecture for VLSI, integration, the VLSI Journal 4, 65-74, (1986)
7. Lang, H.-W., Maaß, R., Schimmler, M.: The Instruction Systolic Array - implementation of a low cost parallel architecture as add-on board for Personal Computers, Proc. HPCN 94, Munich (1994)
8. Leavers, V.F.: Which Hough transform, Computer Vision, Graphics and Image Processing 58 (1993) 250-265
9. Schimmler, M.: Fast sorting on the Instruction Systolic Array, Technical Report No. 8709, Christian-Albrechts-University, Kiel, (1987)

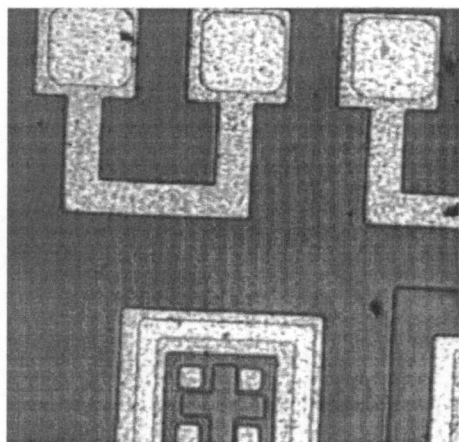

Fig. 6: Original image

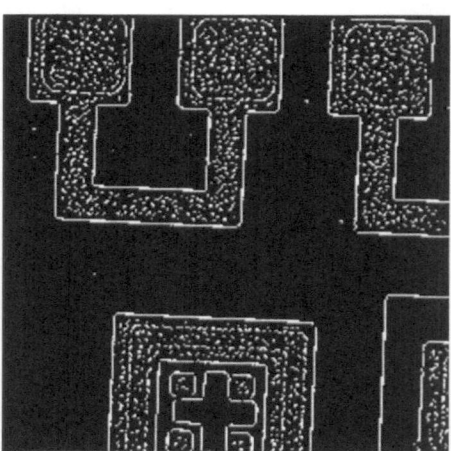

Fig. 7: Edge image

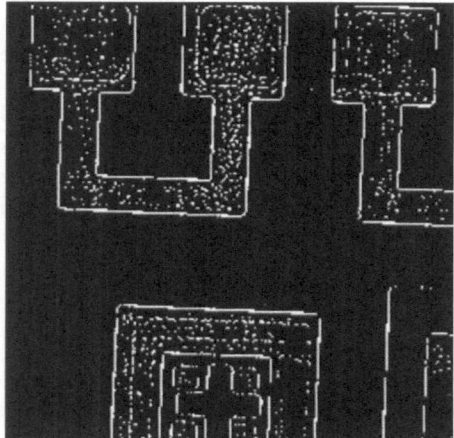

Fig. 8: Result from a HT

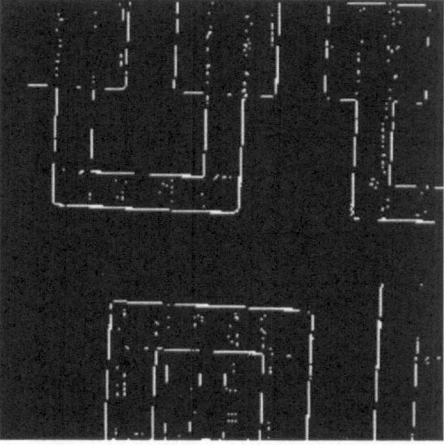

Fig. 9: Result from a MHT

An Analytical Design of High-Speed Pixel Transformation for Object Boundary Enhancement

KP Lam and A Furness

Automatic Identification Research Laboratory,
Centre for Electronic Engineering, University of Keele,
STAFFS ST5 5BG, UK

Abstract. Sophisticated algorithms for edge detection with built-in noise filtration and good localisation properties often require an inherently iterative and/or adaptive processing structure, which is difficult to parallelise for high speed operation. This paper describes a boundary enhancement technique which can be adopted as a cascaded precursor for edge detection. In addition to the demonstrable efficacy in enhancing edge features, the precursor offers the distinct advantage of exploiting both the image and operator parallelism. Further, unlike many other enhancement methods, its characteristics can be studied analytically. The performance of the proposed detector is examined and compared with established techniques.

1 Introduction

Improvements in speed and efficiency for edge detection continue to pose a challenge in 2-D real-time image processing. The gradient operators most commonly used as a measure of edge profiles fall into three general categories: (1) Operators such as Robert and Sobel which approximates image function derivatives using spatial differences with convolution masks. (2) More sophisticated algorithms with built-in noise filtering capabilities, as exemplified by Marr-Hildreth [12] and Canny [1] operators, based on the zero crossings of image function second derivatives. (3) Parametric operators, such as that of Hueckel [13] and Tan et al [16], which match and/or optimise an image function to a specific model of the edge profiles. Computationally many of these techniques apply an appropriate sequence of neighbourhood operations simultaneously on all pixels of an image, thus enabling them to exploit image parallelism when implemented on parallel computers [4]. In terms of algorithmic complexity, handling the first class of techniques is relatively simple, whilst the other two generally require significantly more iterative as well as adaptive processing [2]. In general, the approach to edge enhancement, followed by thresholding in a variety of forms, and characterising operators from the first class in particular, is largely heuristic in nature. Similarly the higher-order derivative and parametric approaches, although mathematically formulated, also suffer from this type of setback.

This paper presents a parallel, computationally efficient boundary enhancement technique which can be adopted as an effective precursor for edge detection.

Here the precursor significantly steepens the sheerness of image discontinuities, thus enabling edges with widely varying strengths and qualities to be determined using a single, non-interactive thresholding operation after the application of a gradient-based detector. In terms of computational efficiency, the technique takes advantage of the combined robustness of non-linear mathematical techniques and the simplicity of gradient operators. Adding to these the uniformity of design and and regularity of operations, the characteristic blueprints for efficient parallel algorithms for low-level image processing, makes the proposed precursor technique a practical approach to high-speed edge detection.

The non-linear technique of generalised nearest-neighbour (GNN) transformation which forms the mathematical basis of the enhancement method is discussed in Section 2. An analytical treatment which leads to the development of the proposed precursor is also presented. Sections 3 and 4 examine and discuss techniques to exploit the intrinsic parallelism of the detector method devised. Section 5 describes the analytical tests and results for the proposed detector. Finally, concluding remarks are included in Section 6.

2 Generalised Nearest-Neighbour Transformation

The enhancement method of low pass filtering to suppress noise and eliminate spurious edges have introduced two major problems: Firstly, edges that appear in close proximity are merged and thus become indistinguishable. Secondly, the widening of an edge due to smoothing leads to a much weakened gradient amplitude, which in turn restricts the practical range of thresholds at which edges can be detected. The boundary enhancement method presented here overcomes these limitations. On the one hand, it facilitates the minimisation of the local variance of pixel gray-level values, thus counteracting the undesirable effects of smoothing. On the other hand, it maximises the separation of boundaries between relatively homogeneous image areas, thus causing the useful ranges of detection thresholds to overlap.

2.1 Rank filters

Earlier work on the properties and applications of these filters for noise suppression and image enhancements can be found in [7,9-10]. The extremum operators, min_n, and max_n, which compute rank positions at the two extremes, have the important effect of reducing the gray-scale variance of an image. Here, for a square neighbourhood window W_P of $N(= L^2)$ pixels, $P = \{p_{ij}^k, k = 1..N\}$, centered upon pixel p_{ij}, we evaluate $min_P = R_1(P)$ and $max_P = R_N(P)$, where R_r gives the r-th element of the N ordered values such that $p_{ij}^1 \leq p_{ij}^2 \leq \ldots p_{ij}^r \leq \ldots p_{ij}^N$. Without loss of generality, the effect of max_P and min_P on the image gray-levels can be analysed with a 1-D sequence, say $\{x_i, i = 1..n\}$, using a moving window max_e of extent 3. This is depicted in Fig. 1 below. With $x_{k+2} = x_m' = max\{x_i, k-1 \leq i \leq k+5\}$, it is clear that max_e has the net effect of

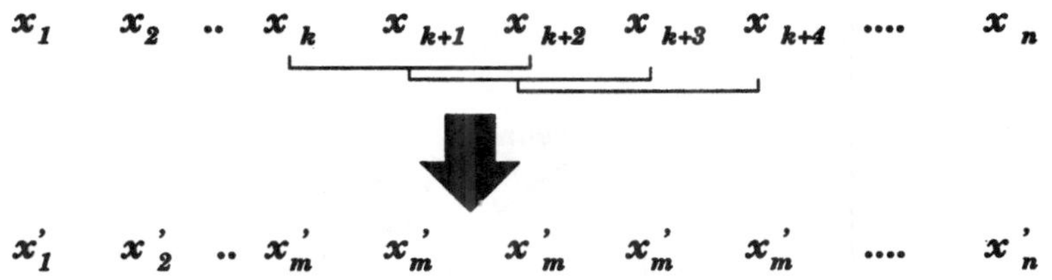

$$x_1 \quad x_2 \quad .. \quad x_k \quad x_{k+1} \quad x_{k+2} \quad x_{k+3} \quad x_{k+4} \quad .... \quad x_n$$

$$x'_1 \quad x'_2 \quad .. \quad x'_m \quad x'_m \quad x'_m \quad x'_m \quad x'_m \quad .... \quad x'_n$$

Fig. 1. The effect of max_e operating on $\{x\}$

reducing the statistical variance $Var(x)$ of $\{x\}$, since, $Var(x) = E(x^2) - E^2(x)$, is a decreasing function which attains its minimum ($=0$) when

$$x_1 = x_2 = x_3 = \ldots = x_n \tag{1}$$

It can be shown by induction proof that the equality of (1) is the only condition whereby $Var(x)$ is a minimum.

2.2 Integrating the max and min operators

The *min* and *max* operators can be integrated to take advantage of the characteristics illustrated in the preceding section. Here the principal objective is to maximise separation of the gray-scale boundaries between the relatively homogeneous image areas achieved by the min/max transformation. The Nearest-Neighbour (NN) approach offers a mathematically robust technique to achieve this. Analytically, let $d(x_i, x_j)$ be the distance between x_i and x_j in the 1-D Euclidean space, the operator $g(x)$, defined as, $g(x) = min\{d(x, x_1), \ldots .d(x, x_n)\}$, represents the nearest neighbour of x amongst $\{x_i\}$. The generalisation of NN provides a key to the unification of the two operators for the characterisation of individual image points. The essence of the principle is illustrated in Fig. 2, where pixels are categorised with reference to the *a priori* measurement of focal points within a predefined neighbourhood of pixels. Individual pixel gray-levels can then be determined according to their proximity to these focal points computed by the extremum operators.

The net effects of integrating the min/max operators using the NN method are two-fold; Firstly, variations in brightness values across different edges is significantly reduced. Secondly, characterisation of the boundaries of gray-scale discontinuities representing the separation of distinctively monotonic image areas is achieved. These results are also confirmed by the analytical experiments presented in Section 5. Combining the discussion above with that of the preceding section, the GNN-based precursor transformation can now be stated as follows. If S and D represents the source and destination image respectively, then;

$$D_{i,j} = \begin{cases} M, & \text{if } (S_{i,j} - m) > (M - S_{i,j}) \\ m, & \text{otherwise.} \end{cases} \tag{2}$$

where $M = max[S_{x,y}; (x,y) \in W_P]$ and $m = min[S_{x,y}; (x,y) \in W_P]$. Equation (2) gives a rounding down condition if $S_{i,j}$ is equidistant from m and M.

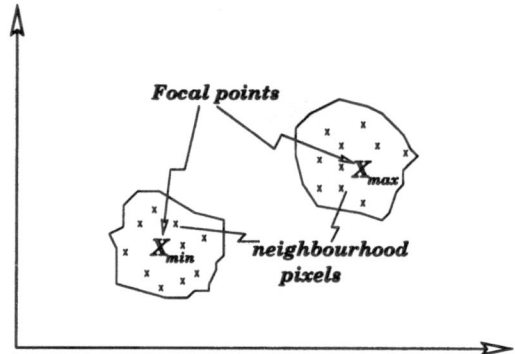

Fig. 2. Principle of the GNN method

3   Parallelism within the Cascaded Precursor

The boundary enhancement algorithm comprises of two stages of point-wise image operation; an edge-preserving smoothing and the Generalised Nearest-Neighbour transformation (GNN) as discussed earlier. The first processing stage facilitates a noise-smoothing filtration for the source image. As well as possessing both powerful noise reduction and good edge-preserving properties [3], the K-AVE operator selected and the succeeding GNN transformation are both derived from the same class of mathematical techniques and so they are highly compatible in terms of computational characteristics and requirements. The latter consideration is particularly desirable from a theoretical viewpoint, since the GNN is formulated to "sharpen" the spatial features of the K-AVE filtered image, thereby counteracting the effect of the much weakened edge gradients that critically limits the practicality of subsequent gradient calculations. To facilitate demonstrating the technique as an effective precursor to edge detection, the well-known Sobel operator [14] is chosen in this work primarily because of its relatively high accuracy and robustness [5]. More importantly, the convolution operation required is well suited to high-speed (parallel) implementation [11].

The detector described above is inherently parallel; the directly cascadable operation of the GNN precursor results in a computationally efficient processing structure which can exploit two levels of parallelism, namely the *image* and *operator* parallelism [4].

1. *Image Parallelism* is directly derivable from the K-AVE, GNN and Sobel operators and offers a fine-grained parallelism which is characteristic of many low-level image processing algorithms. Specifically individual processors can apply the appropriate sequence of computations specified by each operator to different (sets of) pixels concurrently.
2. *Operator Parallelism* is achievable by pipeline processing successive stages of the K-AVE, GNN and Sobel computation. This offers a coarse-grained parallelism in that individual sets of processors may be cascaded in the form of a pipeline whereby each set operates on its inputs independently.

As with many iconic image transformations, it is evident that the relative multiplicity of data arising from each of of the spatial operators in 1. above determines the maximum parallelism achievable, *ie.* the totality of data in a practical image offers maximum units of parallelism [15]. In this respect, the most CPU-intensive is the calculation of different ranks required by the K-AVE filter and the GNN transformation. Adopting the notations in Section 2, the former entails the selection of R_j, $j = 1..K$, from the defined window neighbourhood of W_P, whilst the latter requires the determination of R_1 and R_N.

4 Sort-And-Merge Technique

Previous work has demonstrated that the data-parallel model of computation can be used effectively in realising a general class of rank filters on massively parallel platforms [8]. For the GNN transformation, it can be shown that the two-phase divide-and-conquer algorithm described in [8] can easily be adapted to facilitate an efficient data-parallel implementation. Specifically the computation of the extrema in a LxL window is performed by first sorting L subsequences along one dimension, say the column, of the moving window, then merging them in the orthogonal direction. A simple sorting method is used as L is small. Similarly, values of R_1 and R_N are determined by comparing the smallest and largest elements of the appropriate groups of L sorted columns respectively. A representation of the method is depicted in Fig. 3. Since each of the L sorted

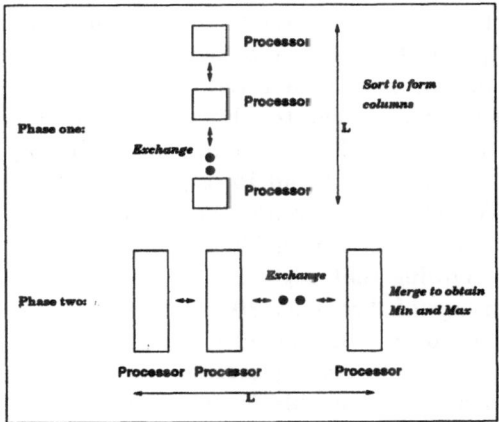

Fig. 3. The two-stage Sort-and-Merge selection method

columns can again be used for computing a further $L-1$ calculations of extrema in the neighbourhood, the method thus facilitates $O(k_1 L^2)$ comparisons to be executed on L processors with an $O(k_2 L)$ complexity, where k_1 and k_2 are small constants.

In performing the K-AVE filtration, it should be pointed out that the sorted subsequences obtained in the first phase have to be re-computed by their $L-1$ immediate neighbours, since the reference centre pixel $p_{i,j}$ from which the nearest distances are derived changes as the window W_P moves across the image. Further, it should also be noted that, for a relatively large K, $\frac{N}{2} << K < N$, it would be more advantageous to compute the $N-K$ largest elements in P, since $\sum_{i=1}^{K} R_i = \sum_{i=1}^{N} R_i - \sum_{i=K+1}^{N} R_i$. Consequently, the K-AVE operator can be computed as a spatial convolution with unity mask weights (*cf.* Average-filter), less the sum total of the individual ranks $R_{K+1}+..+R_N$. These rank values are determined using the Sort-and-Merge method depicted in Fig. 3.

5   Performance and Discussion

The approach of Pratt [14] is used to quantify the efficacy of the proposed GNN precursor. A 64x64 gray-scale (256 levels) image is constructed, which contains a steep ramp edge of height 30, separating two uniform regions of gray values 100 and 130 respectively in fixed steps of 15. The noise used is additive Gaussian, $N(0,\sigma)$, with signal-to-noise ratio $(SNR) = (h/\sigma)^2$, where $h(= 30)$ represents the edge height. The quality of edge detection, ψ, is defined as:

$$\psi = (\frac{100}{I_N}) \sum_{i=1}^{I_A} \frac{1}{1 + \alpha d_2} \tag{3}$$

where $I_N = max(I_I, I_A)$, I_I = number of ideal edge points, I_A = number of detected edge points, α = scale factor to penalise offset edge points, taken as 1/9 here and in [14], and d = distance of the detected edge point from the true edge. Of particular interest here is the variation of ψ with the threshold T of edge detection following the Sobel operator. Using a 3x3 window mask, *ie.* $L = 3$, with $K = 6$ for computational efficiency and good results [6], plots of ψ *vs.* T are presented in Fig. 4 for SNR = 7.5 and 15 respectively. These plots clearly show that whilst the K-AVE filtration smooths edges, the GNN precursor significantly widens the range of detectable thresholds as predicted. In particular, a plateau generally develops, showing that edges can be detected with over 90% quality from about a wide range of thresholds $30 \le T \le 120$.

For comparison purposes, the cascaded precursor approach is also considered with the Canny operator, using several real-world digitised images. The scene was captured using a portable camera unit whose frame buffer size is 754x502, with 0-255 gray-levels. Thus picture degradation from input circuitry provides an approximately Gaussian additive noise field. Computationally the GNN-based detector method is inexpensive; using the techniques described in Section 4 where applicable, the total processing time required to produce a good edge image ranges $1.26-1.99$ seconds on a P166 CPU, the maximum corresponds to a double application of K-AVE and GNN operators. The Canny algorithm typically requires $34 - 52$ seconds, depending upon the image and σ used. In all cases, the GNN process has kept close edges narrow, with weak edges steepened

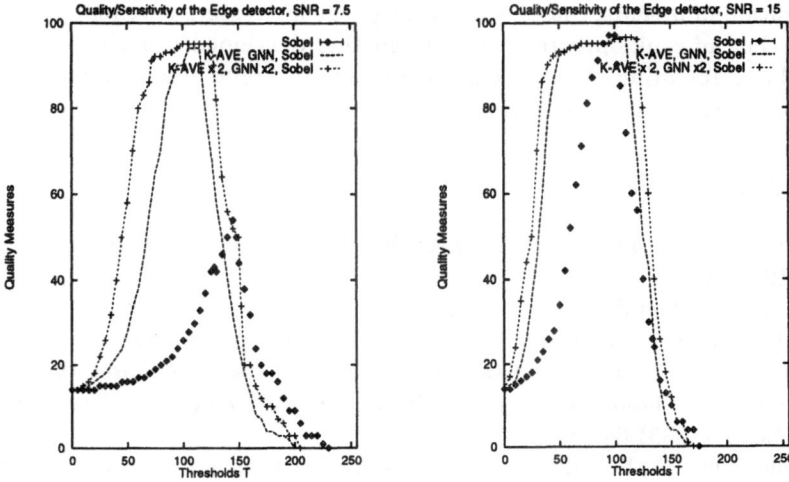

Fig. 4. SNR=7.5 and SNR=15 respectively

and the noise suppressed sufficiently so that they become distinguishable within the overlapped detection ranges for strong edges.

6   Summary and Conclusion

The detection of edges is simplified if the edge strengths are suitably enhanced. Many edge operators approximate local derivatives with spatial differencing techniques, which emphasise high frequency features including false edges due to noise points. If the noise is smoothed by low-pass filters, the strength of edges detected can be significantly weakened. Also such a non-selective smoothing causes close edges to merge, limiting the threshold range at which true edges may be detected. The proposed GNN precursor resolves these problems by adopting a non-linear mathematical technique to enhance spatial discontinuities of an image. When combined with the edge-preserving K-AVE filter, it has been shown to exhibit the desirable properties of sharpening edge features, thus enabling them to be determined irrespective of the widely varying strengths and qualities. These characteristics have been shown analytically, and the preliminary experimental results obtained confirm the analysis.

Computationally the characteristics of the proposed GNN match well with that of the K-AVE, both being derived from the same class of operators. By adopting the data-parallel technique developed in an earlier work, it has been shown that the uniformity of the K-AVE and GNN computations has the important advantage that they could both be implemented efficiently on highly parallel platforms. Further, given the the relative robustness and ease of parallelisation of the Sobel operator, the precursor-based edge detection method has considerable potential for real-time applications in in practical situations. Moreover, the

814

investigation of precursor method also raises the interesting possibilities that the general class of rank operators could be further developed to facilitate selective removal and retention of image edge features for optimal detection (*cf.* WMF in [8]). This warrants further research.

References

1. Canny JF.: A computational approach to edge detection. IEEE Trans. PAMI. **8** (1986) 1-14
2. Chen SS., Chou JS., Lin WC., Pelizzari CA.: Edge and surface searching in medical images. Medical Imaging **II** - SPIE (1988) 594-599
3. Chin RT., Yeh CL.: Quantitative evaluation of some edge-preserving noise-smoothing techniques, Computer Vision, Graphics and Image Processing Vol **23** No. 1 (1983) 64-91
4. Danielsson PE., Levialdi S.: Computer architectures for pictorial information systems, IEEE Computer, **14**(11) (1981) 53-67
5. Davies ER., Circularity - A new principle underlying the design of accurate edge orientation operators. Image and Vision Computing **2** (1984) 134-142
6. Davis LS., Rosenfeld A.: Noise cleaning by iterative local averaging. IEEE Trans. Syst. Man Cybern. SMC-**7** (1978) 256-261
7. Hodgeson RM.: Properties, implementations and applications of rank filters. Image and Vision Computing **3** (1985) 3-14
8. Lam KP., Horne E.: A data-parallel approach to the implementation of weighted medians technique on parallel/super-computers. Procs. of Fifth International Symposium on Parallel and Distributed Processing (1993) 734-738
9. Lee WH.: Signal processing techniques for CCD image sensors. Applied Optics **23** December (1984) 4280-4289
10. Levy-Mandel et al: Knowledge-based land-marking of cephalograms. Computers and Biomedical Research **19** (1986) 282-309
11. Manning LJ., Dew PM., Wang H.: Programming models for VLSI Array Processors with application to low-level vision processing. Procs. of Parallel Architecture and Computer Vision Workshop, UK (1987) UK1.5
12. Marr D. and Hildreth E.: Theory of edge detection. Procs of the Royal Society London B**207** (1980) 187-217
13. Nevatia R.: Evaluation of simplified Hueckel edge-line detector. Computer Graphics and Image Processing. **6**(6) (1977) 582-588
14. Pratt WK., Abdou IE.: Quantitative design and evaluation of enhancement/thresholding edge detectors. Procs IEEE Vol **67-5** May (1979)
15. Quinn MJ., Hatcher PJ.: Data-parallel programming on multi-computers. IEEE software **7** (1990) 69-76
16. Tan HK., Gelfand SB., Delp EJ.: A cost minimization approach to edge detection using simulated annealing, IEEE Trans of PAMI **14**(1) (1992) 42-56

Karhünen-Loève Transform: An Exercise in Simple Image-Processing Parallel Pipelines

Martin Fleury, Andy C. Downton and Adrian F. Clark

Department of Electronic Systems Engineering, University of Essex, Wivenhoe Park,
Colchester, CO4 3SQ, U.K
tel: +44 - 1206 - 872795
fax: +44 - 1206 - 872900
e-mail fleum@essex.ac.uk

Abstract. Practical parallelizations of multi-phased low-level image-processing algorithms may require working in batch mode. The features of a common processing model, employing a pipeline of processor farms, are described. A simple exemplar, the Karhünen-Loève transform, is prototyped on a network of processors running a real-time operating system. The design trade-offs for this and similar algorithms are indicated, when a general solution is sought. Eventual implementation on large- and fine- grained hardware is considered. The chosen exemplar is shown to have some features, such as strict sequencing and unbalanced processing phases, which militate against a comfortable parallelization.

1 Introduction

Many low-level image-processing (IP) algorithms, such as spatial filters, are completely localized in their data references. If adjacent image data are overlapped at boundaries then at a small additional cost a data-farming programming paradigm can be employed, in which the only communication is between worker process and data farmer. Using separability and/or linearity, it is also possible to decompose other algorithms, such as orthogonal transforms, rather than employ a global access pattern. If these latter algorithms are viewed as single-image library functions then a difficulty commonly arises because it is necessary to centralize between the data-farming phases. However, since IP is often in batch mode, it only requires a slight shift in perspective towards continuous data flows in order to realise that effective parallelizations may occur if a pipeline is the normal form of processing.

In this paper, this basic concept is applied to a Karhünen-Loève transform (KLT), which is generally engaged in batch mode. As a development environment, we have used a network of microprocessors running the VxWorks real-time operating system [1]. Compared to other distributed environments, the VxWorks-based system is attractive for algorithm prototyping: because it is an isolated environment, because the thread structure is not superimposed on top of heavy-weight processes, and because event response times are optimized. Since the VxWorks system is not the final target parallel system, we are interested in the computational complexity rather than the performance on VxWorks.

2 The VxWorks Real-Time Kernel

VxWorks is intended as a Unix-like single-user operating system (O.S.) for real-time development work. The KLT program modules were written in 'C' and cross-compiled on a PC running the NextStep O.S. Once compiled, the modules are loaded and linked on a set of 68030 boards, each hosted by a PC. The 68030 processors are connected by an Ethernet LAN segment. For inter-processor communication, VxWorks includes a source-compatible BSD socket API.[1] Each program module consists of one or more threads, each of which can be spawned as required. Remote spawning is accomplished by providing an iterative server, which responds to requests over the LAN from the central farmer module.

Priority-based scheduling of threads is employed on worker modules, giving priority to communication-handling threads. Messages are stored in circular buffers, with access by the communication threads regulated through semaphores.[2] A **message queue** primitive is available in VxWorks which allows the application thread to **rendezvous** with any communication thread. On the farmer module, the **select** socket call is employed in a way that de-multiplexes work requests in a fair manner. A number of features enable the software structure to be reused, e.g. tagged messages, the selection of a streamed communication mode and a strict interface to application functions.

3 The Karhünen-Loève Transform

The KLT [2] is most widely used in applications such as multi-spectral analysis of satellite-gathered images [3] through the resulting spectral signature of imaged regions. Significant data reductions are also achieved in the storage of satellite images if the multi-spectral set are transformed to KLT space. The KLT has recently been applied to the recognition of facial images [4]. Notice that the size of the image set is potentially much larger in the latter two applications.

Consider a sample set of real-valued images from an ensemble of images. Create vectors with the equivalent pixel taken from each of the images, i.e. if there are D images each of size $M \times N$ then form the column vectors $\mathbf{x}_k = (x_{ij}^0, x_{ij}^1, \ldots, x_{ij}^{D-1})^T$ for $k = 0, 1, \ldots, MN - 1$, $i = 0, 1, \ldots, M - 1$ and $j = 0, 1, \ldots, N - 1$ (with superscript T representing the transpose). Calculate the sample mean vector: $\mathbf{m}_x = \frac{1}{M} \sum_{k=0}^{MN-1} \mathbf{x}_k$. Use a computational formula to create the sample covariance matrix: $[\mathbf{C}_x] = \frac{1}{M} \left(\sum_{k=0}^{MN-1} \mathbf{x}_k \mathbf{x}_k^T \right) - \mathbf{m}_x \mathbf{m}_x^T$. Form the eigenvector set: $[\mathbf{C}_x]\mathbf{u}_k = \lambda_k \mathbf{u}_k$, $k = 0, 1, \ldots D - 1$, where $\{\mathbf{u}_k\}$ are the eigenvectors with associated eigenvalue set $\{\lambda_k\}$. The KLT kernel is a unitary matrix, $[\mathbf{V}]$, whose columns, $\{\mathbf{u}_k\}$ (arranged in descending order of eigenvalue amplitude), are used to transform each zero-meaned vector: $\mathbf{y}_k = [\mathbf{V}]^T(\mathbf{x}_k - \mathbf{m}_x)$.

---

[1] The VxWorks system is available in a tightly-coupled variant, by means of processors linked by a VME bus, but again sockets form the principal communication mode.

[2] The 68030 has two compare-and-swap instructions as well as support for cache management.

4 KLT Parallelization

The time complexity of the operations is analysed as follows, where no distinction is made between a multiplication and an add operation: form the mean vector with $O(MND)$ element-wise operations; calculate the set of outer products and sum, $\sum_{k=0}^{MN-1} \mathbf{x_k x_k^T}$, in $O(MND^2)$ time; form $\mathbf{m_x m_x^T}$; subtract matrices to find $[\mathbf{C_x}]$; and find the eigenvectors of $[\mathbf{C_x}]$ (the eigenvector calculation is $O(D^3)$); convert the $\{\mathbf{x_k}\}$ to zero-mean form in $O(MND)$; and form the $\{\mathbf{y_k}\}$ by $O(MND^2)$ operations. Since the covariance matrix is generally too small to justify parallelization, the total parallelizable complexity is $O(MND+MND^2)$, i.e. the eigenvectors are found sequentially.

Consider the KLT as applied to a single image in one-off mode. One parallelization method would be to send a cross-section through the images to each process, selecting the cross-section on the basis of image strips. In a first phase, the mean vector of each cross-section image strip is found and returned to a central farmer along with a partial vector sum, forming the strip matrix: $[\mathbf{T_i}] = \frac{1}{MN} \sum_{k=0}^{(MN-1)/n} \mathbf{x_k^i} (\mathbf{x_k^i})^T$, $i = 1, 2, \ldots n$, for n strips. In a second phase, the farmer can find $[\mathbf{V}]$ from $[\mathbf{C_x}]$, which is now broadcast so that for each strip the calculation of $\{\mathbf{y_k^i}\}$ can go ahead. However, the duplication of sub-image distribution (once for the partial sums and once to compute the transform) is inefficient even though the duplication is strictly necessary if demand-based data-farming is employed.

A possibility is to retain the data that are farmed out in the first phase at the worker processes. On a system with store-and-forward communication the first farming phase will have established an efficient distribution of the workload given the characteristics of the network. Therefore, the second phase will already have approximately the correct workload distribution. This is not a solution on a shared network of workstations as processor load and network load is time dependent. The solution is also not a general one since other two-phased low-level IP algorithms do not usually use the same data in both phases, though the time complexity can be similar. The method of finding a workload distribution by a demand-based method and then re-using the distribution for a static load-balance in subsequent independent runs may have general potential.

An alternative static load-balancing scheme is to exchange partial results amongst the worker processes so that the calculation of matrix $\mathbf{V}$ can be replicated on each worker process. A suitable exchange algorithm for large-grained machines is available if the processors can be organized in a uni-ring. A second method of performing a KLT is to consider each image as a single vector formed by stacking rows. In [5], this scheme, whatever its merits for particular applications, is shown to have the same time complexity but to be less flexible in regard to load-balancing.

5 Pipeline Decompositions

A simple pipeline can be formed by the sequence: covariance, eigenmatrix and image-transform modules. In a preliminary implementation of this pipeline on the VxWorks-based system, both farmers were placed on the same processor (Fig. 1) since the same data are needed for forming the covariance matrix and for transforming to eigenspace.[3] In principle, double buffering of image sets allows loading of one image set to proceed while the previous image set is transformed. However, for a target VLSI implementation this implies a total buffer size of 5 Mbytes and upwards would be needed for (say) 10 images of 512×512 size.

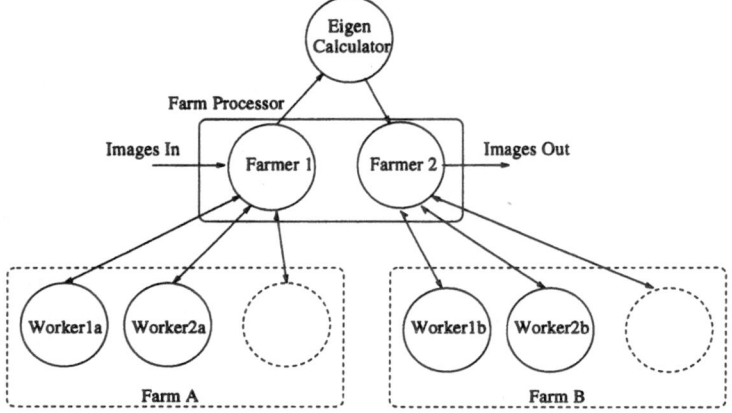

Fig. 1. KLT Pipeline Partitioning

The first pipeline stage can be further partitioned into calculation of the mean vector and calculation of the outer products, since the two calculations are independent and therefore can take place in parallel. Additionally, the second stage calculations can be split further between reducing the image set to zero-mean form and transforming the image set, though these calculations are not independent. However, the reduction to zero-mean form is independent of the eigenvector calculation and could take place in parallel with that calculation. These partitioning possibilities are shown in Fig. 2. Assume that the two farms in the first pipeline partition can be operated in parallel, by means of two farmers on the same processor feeding from a common buffer. Since the maximum time complexity of each stage of the new pipeline is reduced from $O(MND + MND^2)$ to $O(MND^2)$, then the number of processors on any one farm that will reduce pipeline throughput is reduced. However, the bandwidth requirements are increased. Since both the components of the second partition are dependent on the completion of all the calculations on the first partition, the pipeline traversal latency will not be reduced by decomposing the image into smaller components.

The pipeline of Fig. 2 is relevant as the basis of a VLSI scheme, possibly through a systolic array. For a large-grained parallelization, the arrangement of Fig. 1 but merging the eigenvector calculation into the work of the second farmer is practical. The scheduling regime on the processor hosting the two farmers is round-robin for fairness. Since the time complexity of both stages of the pipeline

---

[3] A worker module can also be placed on the same processor to soak up any spare processing capacity.

is the same it is now easily possible to scale the throughput in an incremental fashion.

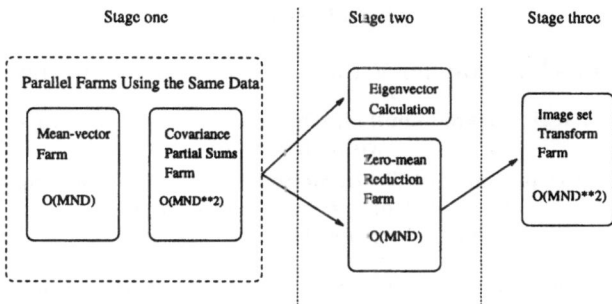

Fig. 2. Alternative Partitioning of the KLT Pipeline

6 Conclusion

The Karhünen-Loève Transform has been prototyped on a pipeline of parallel processor farms. A two-phased single farm arrangement is described, in one mode of which the initial workload distribution, arrived at by a demand-based method, is reused in a static load-balance. Two different pipeline decompositions are explored. To achieve a completely balanced pipeline for all but the largest of jobs will be prohibitive in practical terms. The simpler of the two pipelines is appropriate for large-grained applications, whereas a further decomposition may be relevant to fine-grained VLSI implementations. The strict sequencing in the KLT algorithm prevents attempts to improve the pipeline traversal latency.

Acknowledgement

This work is being carried out under EPSRC research contract GR/K40277 'Portable software tools for embedded signal processing applications' as part of the EPSRC Portable Software Tools for Parallel Architectures directed programme.

References

1. Wind River Systems, Inc., 1010, Atlantic Avenue, Almeda, CA. *VxWorks Programmer's Guide*, 1993. Version 5.1.
2. J. J. Gerbrands. On the relationship between SVD, KLT and PCA. *Pattern Recognition*, 14(1-6):375–381, 1981.
3. P. J. Ready and P. A. Wintz. Information extraction, SNR improvement and data compression in multispectral imagery. *IEEE Trans. on Comms.*, 31(10):1123–1130, 1973.
4. M. Kirby and L. Sirovich. Application of the Karhunen-Loève procedure for the characterization of human faces. *IEEE Trans. PAMI*, 12(1):103–108, 1990.
5. M. Fleury. *Efficient Parallel Image-Processing Software*. PhD thesis, University of Essex, 1996.

Use of F-Code as a Very High Level Intermediate Laguage for DSP

Edward D. Willink[1], Alexander V. Shafarenko[2] and Vyacheslav B. Muchnick[2]

[1] Racal Research Limited, Worton Drive, Reading, England
willink@rrl.co.uk

[2] Department of Electronic and Electrical Engineering,
University of Surrey, Guildford, England
{a.shafarenko,v.muchnick}@ee.surrey.ac.uk

Abstract: Block diagram languages provide an effective approach to developing Digital Signal Processing applications. The tools that support block diagram languages use existing compilation systems to produce code. The inefficiencies of the compilation systems are compounded with inefficiencies interfacing to them.

Generation of intermediate code direct from the block diagram bypasses these inefficiencies. We describe the direct generation of F-code, a very high level intermediate language developed for data parallel applications.

1 Introduction

Development of DSP applications is amenable to a graphical approach using a schematic editor to enter a block diagram, and to parameterise each block. Research tools such as Ptolemy [2] and commercial products such as COSSAP and SPW are based upon this approach.

Executable code is generated by pasting together template code fragments for each block. An appropriate order for the fragments is determined by dataflow analysis [1]. Independent fragments are required for each execution environment: simulation, target processor and ASIC. Moreover, each environment often requires multiple implementations of each fragment, VHDL or Verilog for ASIC, and C or assembler for a target processor. Each of these implementations may be further varied to suit the eccentricities of the subsequent compiler and target: Synopsys or Vantage VHDL, DSP32C or TMS320C30 C or DSP56000 assembler. Distinct library blocks are required for each data type: integer, real, complex, fixed point. For some tools each permutation of array rank and data type requires additional implementations. Full support of a new basic block imposes such a heavy development burden that few blocks are consistently available (and tested) for all uses. (Ptolemy contains 15 out of 20 expected adder implementations).

The efficiency with which fragments are pasted together varies between tools. The poor quality of DSP compilers encourages the tool implementors to provide as much assistance as possible by generating inline code rather a subroutine call per block. When inlining is performed at the assembler level, opportunities for inter-block optimisation are sacrificed.

In this paper we propose using the regularity and data parallelism of F-code intermediate language to capture a block definition. This definition can then be used to generate existing implementations. Alternatively, it may act as input to a subsequent compilation system. Generation of efficient DSP code from F-code is beyond the scope of this paper.

2 F-code

The F-code intermediate language is the result of research in data parallelism [3]. Only the briefest summary can be provided here. A formal treatment may be found in [4]. F-code uses a LISP style syntax and a high level program representation similar to a parse tree.

```
(assign (target y)                  (assign (target y)
    (if (ge (value x) (const 5))        (if (ge (value x) @11 (const 5))
        (max (value x) (value z))           (max (value x) (value z))
        (add (value x) (const 2))))          (add (value x) @11 (const 2)))))
```

The left hand example corresponds to the scalar C statement

```
y = x >= 5 ? max(x, z) : x + 2;
```

The right hand example is an element-wise generalisation to two dimensions, with the 11 binary mask projecting the scalar constants along each axis. Data parallelism is consistently supported for all arithmetic operations. Regular operations can be expressed compactly:

```
(add @100 (mul @110 (value x) @101 (value y)))
```

performs a matrix multiply: The 110 and 101 binary masks control projection of x and y from two to three dimensions, where an element-wise product is reduced by adding elements along the dimension identified by the 100 mask. Non-computational operations can be expressed using geometric operations, much more in the style of APL2 than HPF.

Some support for process parallelism is provided by OCCAM-style par and seq constructs. Communication and synchronisation is supported by defining a pipe of any type (except a pipe). A pipe is a FIFO and maybe bounded or unbounded.

The original F-code base types have been reviewed to more accurately reflect typical hardware, and to permit their idealised behaviour to be tailored to impose minimum or exact precisions constraints and to define any required overflow and rounding behaviours. Thus.

```
(type-def clip16 real (huge 1.0) (epsilon 3.36e-5)
    (maximum 32767) (minimum -32768) (clip) (convergent))
```

defines clip16 as a type whose fundamental behaviour is that of the real base type but whose implementation need only handle magnitudes up to 1.0 with integral mantissa representations in the range -32768 to +32767 inclusive. A code generator for a DSP target may recognise that this can be satisfied by a 16 bit representation. In addition, it is specified that overflows should be resolved by clipping out of range numbers and that truncation errors should be rounded to the nearest number, mid-point values being rounded to the even value.

Explicit precisions may be imposed on operations using a signature. Thus, with appropriate definition of frac16, frac32 and frac40 types,

```
(add (signature frac40 frac40 frac32) (value acc)
    (mul (signature frac32 frac16 frac16) (value x) (value y)))
```

imposes the minimum characteristics of 16 bit multiplication with a 32 bit product and 40 bit accumulation on acc + x * y, while allowing higher precision to be used where more efficient.

3 Example

The parallel F-code specification of a two input adder is applicable to any type X for which addition is defined. In F-code this comprises the natural,

integer, real and complex base types, types derived from them by precision
tailoring and types derived by formation of arrays.

```
(pipe signal X)                    -- signal is an unbounded FIFO of X
(function add2                     -- declare the add2 function
    (signature hole                -- no function return value
        (target signal out)        -- out to unbounded FIFO of X
        (value signal in1)         -- in1 from unbounded FIFO of X
        (value signal in2))        -- in2 from unbounded FIFO of X
    (put                           -- output to
        (target out)               --     out pipe
        (add                       --     the sum of
            (get (value in1))      --     input from in1 pipe
            (get (value in2))))))  --     and input from in2 pipe
```

The token flow, 1+1 in and 1 out, is directly deducible from the get and
put constructs.

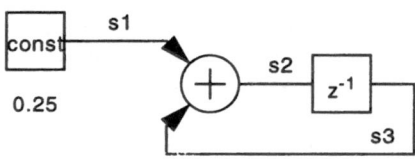

A simple integrator program (less initialisation and declarations) may be
captured by a par construct to define a set of child threads that execute in
parallel, an infinite loop construct and by a function call for each block.

```
(par
    (loop
        (call const (target s1) (const 0.25)))
    (loop
        (call latch (name latch_context) (target s3) (value s2)))
    (loop
        (call add2 (target s2) (value s3) (value s1))))
```

This form of specification is appropriate when dataflow scheduling
activities are implemented as an F-code to F-code translation. The
excessively parallel implementation of each block is replaced by a scheduled
sequential implementation. Fixed buffers replace pipes

```
(loop                                  -- program runs forever
    (seq                               -- each block is sequential
        (call const (target s1) (const 0.25)))
        (call add2 (target s2) (value s3) (value s1)))
        (call latch (name latch_context) (target s3) (value s2))))
```

The resulting sequential F-code implementation of the adder is

```
(function add2
    (signature hole (target X out) (value X in1) (value X in2))
    (assign (target out) (add (value in1) (value in2))))
```

Standard compiler optimisations should inline the function bodies, and
propagate the constant. Optimisation for an F-code program is a little easier
than for a C program: the availability of array operations simplifies the
dependence analysis prior to loop strength optimisations, and reduces the
number of pointers and consequent aliases.

4 F-code source

F-code is a high level intermediate language and although it has a textual
representation, F-code is not meant to be a programming language. We
require a high level representation that can be sensibly translated into
F-code. At present there are no such mainstream languages, because in many
respects F-code is at a higher level than high level languages.

DSP applications developed using block diagram tools use an ad hoc graphical language whose operations are defined by the available block library and whose syntax is constrained by the limitations of the support environment. Additional operations may be added by extending the library.

Conversion of the block diagram to F-code is not supported by any tool. However generation of C or VHDL is supported by all tools. Therefore a C or VHDL to F-code translator is required. Although usage of C is most prevalent, ANSI C is unsuitable for fixed point DSP applications because of the lack of support for fixed point arithmetic. VHDL has fewer limitations, supports user defined types and description of parallel processes. We briefly pursued a VHDL to F-code approach in order to exploit as much of the potential of F-code as possible without incurring undue dependence on the block diagram tool. However working around the limitations of type polymorphism and the distinction between the VHDL event and DSP token flow models required a degree of stylisation that was hard to justify.

Ptolemy is extensible and so can be configured to generate F-code directly, treating F-code as just another code generation domain. This enables all the scheduling facilities of Ptolemy to be used, without constraining compiler optimisation unduly. Each leaf library block (star) requires a sequential F-code implementation to be added. Implementations for other domains can be generated from the F-code implementation. In order for the full capability of F-code to be exploited, a parallel F-code implementation of each block should be used, with scheduling performed after conversion to F-code.

Direct generation of F-code is undesirable, but acceptable for simple blocks. A better textual programming language is required for large blocks.

5 Summary

We have indicated how F-code may be used to specify the behaviour of blocks in a DSP block diagram. Polymorphism and data parallelism enable a single implementation to replace many existing implementations. Process parallelism offers an opportunity to integrate dataflow scheduling with subsequent compilation activities. Use of an intermediate compiler representation that can express the characteristics of DSP computations avoids unnecessary problems associated with the use of ANSI C. Generation of good quality code from F-code for fixed point DSPs will be described in a future paper.

References

1. Bhattacharyya, S., Murthy, P., Lee, E.: *Software synthesis from dataflow graphs*. Kluwer academic publishers, January 1995.
2. Buck, J., Ha, S., Lee, E., Messerschmitt, D.: *Ptolemy: a framework for simulating and prototyping heterogeneous systems*. International Journal on Computer Simulation, vol. 4, 155-182, April 1994.
3. Muchnick, V., Shafarenko, A., Sutton, C.: *F-code and its implementation: a portable software platform for data parallelism*. The Computer Journal, vol. 36, no. 8, 712-722, 1993.
4. Muchnick, V., Shafarenko, A.: *Data parallel computing: the language dimension*. Thomson Computer Press, 1996.

Workshop 12

Applications of
High-Performance
Computing

Workshop 12:
Applications of High-Performance Computing

Wolfgang Gentzsch

In the past 20 years, high performance computing (HPC) has largely focused on issues of scientific simulations to speed the process of scientific and technological innovation, e.g., in the fields of physics, chemistry, and biology.

Today, HPC hardware and software systems are relatively mature compared to the parallel computing situation, e.g,. 10 years ago, not only by technology improvements, but also simply because governments and the CEC have put a lot of effort and funding into the development of tools and applications in HPC (e.g., PHAROS, EUROPORT). This led to some acceptance of these systems also in the R&D departments in industry, especially for large-scale, complex, three-dimensional, time-dependent simulations, such as, e.g., car crash, fluid flow, electromagnetics, molecular dynamics, seismic data analysis, decision support, economic modeling, and computer graphics. However, we still have a long way to go, particularly with respect to intelligent simulation and the compute resource needed to drive it.

This workshop on applications in high-performance computing presents a selection of papers dealing with applications in the areas of aerospace design, fluid dynamics, combustion, electrodynamics, radio wave propagation, semiconductor device simulation, industrial visual inspection, and polycrystal deformation. We want to demonstrate that HPC is now on the verge of being used more widely in large and especially industrial applications, thus complementing or even replacing more conservative experimental methods in, e.g., industrial design and development.

With the integration of this workshop within the Euro-Par conference, we intend to underline the strong interdisciplinary dependencies of industrial HPC and all kinds of computer science aspects like hardware technology, high-performance compilers, algorithms and software tools. Also, on the other hand, it is especially important for computer scientists and applied mathematicians to consider requirements of large-scale applications already very early in their research activities, to guarantee wide acceptance and efficient usage of HPC in research and industry in the future.

Experiments on Using WPVM
for Industrial Visual Inspection Problems

Barbosa, J.G.*[†]; Padilha, A. J.[†]; Madier, J.-P.[‡]; Neubert, T.[‡]

[†]FEUP-INEB (P) [‡]REALIX Technologies (F)

Abstract. This paper reports the main results obtained in some experiments conducted in the context of an European project to evaluate the performance and feasibility of using WPVM as a tool to support the parallelisation of application software for industrial visual inspection problems.

1 Introduction

In image analysis applications it is quite common that a specific sequence of operations has to be applied repeatedly over a stream of input data. The general application context addressed in this paper considers such a situation, with the additional assumption that the input image data consists of a flow of independent items (that we name fragments), each item being composed by a variable number of images (that we name views).

The following sections describe a number of experiments that were conducted in order to assess the value and performance benefits obtained for a specified set of assumptions, present details of a real application problem and the results obtained, and discuss the main issues regarding application performance. In the last section, the main conclusions are drawn.

2 Windows PVM

WPVM is an implementation of PVM for the MS Windows operating system, developed at University of Coimbra, Portugal [1, 2]. The software can be downloaded from http://dsg.dei.uc.pt/wpvm. WPVM offers the same set of functions as standard PVM and allows interaction between WPVM and PVM hosts.

3 Problem definition and basic experiments

The problem to be addressed can be briefly stated as one satisfying the following context:

A set of networked personal computers is used as the target virtual parallel machine to accommodate the image processing and analysis of a continuous

---

* e-mail: jbarbosa@tom.fe.up.pt

sequence of fragments. Each fragment is composed by a number of views, each one of these being one grey-level image of non-fixed size.

The experiments made aimed at measuring the system performance in order to specify the maximum speedup that can be obtained with the targeted virtual parallel computer. They are designed to measure the communication overheads for different image sizes. In order to eliminate problems related to an unbalanced distribution, every fragment was assumed to contain three images of the same size. The processing time was considered constant for each 10 kilobytes of image data. In Figure 1, for the processing time of 0.25s/10kb, a test case corresponding to a 50kb image results in a processing time of 1.25s. This simplified definition of the processing time is justified by the reasonable assumption that the image processing algorithms are dominantly local operators and, therefore, depend mostly on the image size.

Fragments are sent by an UNIX process that simulates the acquisition system.

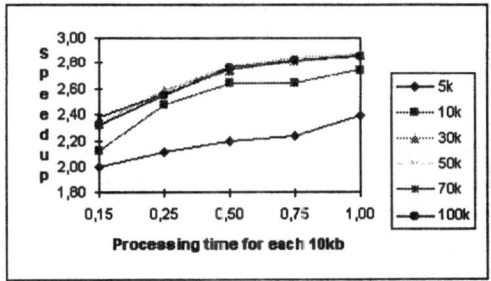

Fig. 1. Speedup with 3 slave processors

The speedup is defined as the ratio between the processing time when executing a single process on a single node, and the processing time achieved when using two or more processes (each executing on its own node), so that the speedup values reflect only the communication overheads.

Figure 1 represents the speedup obtained for three processes, and the speedup values range from 2.15 for 10kb images to 2.86 for 100kb images, corresponding to efficiencies of 72% and 95% respectively.

In the range of sizes of the images to be processed, 10kb to 100kb, it can be said that the virtual parallel computer achieves a good performance, having higher speedup values as the processing time per 10kb of image data increases.

4   Results for a real application

An application system has been designed for the automatic inspection of optical WORM disks. The system is composed by three parts, as shown in Figure 2: Acquisition/Detection Module, Master Process and Image Processing Elements.

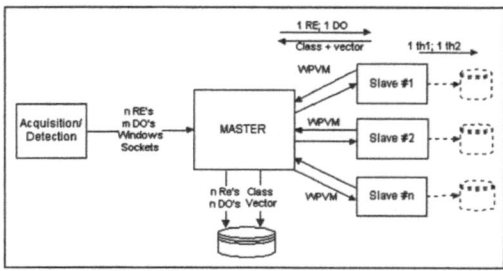

Fig. 2. System configuration

original without lines white obj black obj

Fig. 3. Results of pre-processing; original image, without lines result, white and black objects identification

The master process has the function of co-ordinating all the activity related to the disk analysis, disk classification, communication management and user interaction, acting therefore as a centralised scheduler of workload.

The image processing elements, or slave processes, perform the image processing and analysis algorithms in order to qualify and quantify the defect in each fragment. Several copies of this process can be concurrently executed in different machines to speedup the disk analysis. Figure 3 shows some results of the image processing algorithms.

The parallelisation is at the program level so that it is classified as coarse grain. This choice is adequate as high communication rates are not possible in a parallel virtual machine made of PC's connected by a shared Ethernet. Also, the porting of existing image processing serial code is greatly simplified.

A second experiment was run to assess the performance when using data from the real application, processed by the application algorithms. The sequences of images acquired from the surfaces of the disks are mostly composed by small fragments, with sizes between 10kb and 20kb. Only a few defect types originate fragments of size above 50kb, therefore sequences are often composed exclusively by small fragments.

Three sequences were built to test the system; these are composed by 10 fragments arranged in different orders, so that all sequences have the same amount of data to process. Table 1 displays the identities and sizes of the fragments.

The composition of the sequences and the results obtained are shown in Table 2. Considerably different processing times are obtained depending only on the order by which the fragments are processed. The speedup varies from 1.56 to 1.90, reflecting an unbalanced load distribution, since the slave processors have the same computation power and are fully available to process the fragments.

	1	2	3	4	5	6	7	8	9	10
Fragment	ar1	bd1	fr1	ps1	rd1	rvp1	rvp2	td2	rd2	fr2
Size (kb)	10	8	20	14	120	18	14	65	86	17

Table 1. Fragments used in the experiments

Sequence	Fragment Number										Processing Time(s)	Speedup
1	1	2	3	4	10	5	6	9	8	7	3.9	1.56
2	1	8	9	10	7	2	3	6	5	4	3.5	1.73
3	4	8	9	10	3	5	6	2	1	7	3.2	1.90

Table 2. Composition of the sequences

5 Conclusions and further work

The results obtained in the first experiment show that a network of PC's running WPVM for communication management can achieve good performance in the problem domain addressed.

As far as the defect identification is concerned, the image processing software performed well in the steps of pre-processing, segmentation and classification.

The general strategy delineated in the paper proved to be of value in a real image analysis problem where the mentioned constraints and assumptions hold.

This strategy will be further evaluated in various other industrial inspection problems, which are currently being addressed. The specific issue of load balancing, briefly mentioned at the end of the previous section, will be investigated in the broader context of the full range of inspection problems under study.

6 Acknowledgments

The work herein described was carried out in the context of a case study within the EU Esprit project no. 20059, Hipercosme - High Performance Computing for Small and Medium Size Enterprises. The authors wish to express their gratitude to the CEC and to the other colleagues in the project consortium.

References

1. A. Alves, L. Silva, J. Carreira and J. Silva, "WPVM: Parallel Computing for the People", Proceedings of HPCN'95, Italy, 1995
2. A. Alves and J. Silva, "Evaluating the Performance of WPVM", Byte, May 1996
3. Al Geist et al, "PVM 3 User's Guide and Reference Manual", Oak Ridge National Laboratory, 1994
4. M. J. Quinn, "Parallel Computing, Theory and Practice", McGraw-Hill, 1994

Object-Oriented Parallel Software for Radio Wave Propagation Simulation in Urban Environment

Frédéric Guidec, Patrice Calégari, Pierre Kuonen

Swiss Federal Institute of Technology Lausanne
Theoretical Computer Science Laboratory
CH-1015 Lausanne (Switzerland)
E-mail: {guidec,calegari,kuonen}@di.epfl.ch

1 Introduction

Because of the rapid growth of cellular phone networks, radio wave propagation simulation is of great interest to telecommunication operators, for it makes it possible to predict the shape of the cells of any potential future networks. The objective of the project STORMS[i] is to develop a software tool to be used for the design and the planning of the future UMTS (Universal Mobile Telecommunication System) network. This software tool, which is planned to be used mostly interactively, shall include radio wave propagation simulation algorithms for both urban and rural environments. Since a single cellular network may consist of thousands of cells, radio wave propagation simulation must be as fast as possible.

The ParFlow method permits the simulation of outdoor radio wave propagation in urban environment, describing the physical system in terms of the motion of fictitious microscopic particles over a lattice. It is appropriate for simulating radio wave propagation when fixed antennas are placed below rooftops, as is the case in urban networks composed of micro-cells.

This paper reports the design and the implementation of ParFlow++, an object-oriented, irregular implementation of the ParFlow method targeted at distributed memory parallel platforms. To date the use of object-oriented programming is not very common in parallel supercomputing. This is the reason why the parallel implementation of the ParFlow method using object-oriented techniques appeared to us as an appealing challenge.

2 The ParFlow Method

The ParFlow method was designed at the University of Geneva by Chopard, Luthi and Wagen [4]. It compares with the so-called Lattice Boltzman Model, that describes a physical system in terms of motion of fictitious microscopic particles over a lattice [1]. According to the Huygens principle, a wave front consists of a number of spherical wavelets emitted by secondary radiators. The

---

[i] Software Tools for the Optimization of Resources in Mobile Systems.

ParFlow method is based on a discrete formulation of this principle. Space and time are represented in terms of finite elementary units Δr and Δt, related by the velocity of light $C_0{}^{ii}$. Space is modeled by a grid with a mesh size of length Δr, and flow values are defined on the edges connecting neighbouring grid points. The flows entering a grid point at time t are scattered at time $t + \Delta t$ among the four neighbouring points. Because of the discretization of time, it is convenient to distinguish between the flows coming into a grid point, and those going out of this point. Outgoing flows at time t are a linear combination of incoming flows at that time, and an incoming flow at time t corresponds to the outgoing flow calculated on a neighbouring grid point at time $t - \Delta t$. These rules apply for all grid points but those modelling the source (the transmitter) and obstacles. The source point radiates a signal through its four outgoing flows, but it does not propagate incoming flows. Obstacles (typically, city buildings) are modelled by two kinds of grid points: wall points, and indoor points (see Figure 1). As, in the current ParFlow model, it is assumed that radio waves do not penetrate buildings, indoor points are not involved in the computation. Wall points are perfectly reflecting points that return any incident wave with opposite sign, and whose reflection coefficient and matrix elements can be modified in order to model different kinds of walls [5].

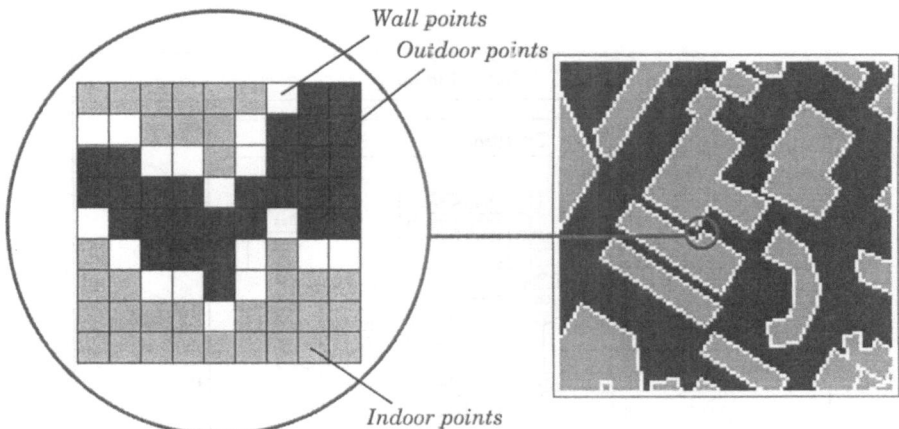

Fig. 1. The ParFlow method operates on a city district described as a bitmap that distinguishes between indoor points, outdoor points, and wall points.

---

^{ii} $\Delta t = \dfrac{\Delta r}{C_0 \sqrt{2}}.$

3   Design and Implementation

3.1   Object-Oriented Irregular Application

Since the ParFlow method does not simulate radio wave propagation through buildings, it can be most interesting not to model indoor points. Indeed, experience shows that when modelling an urban area, buildings can represent up to 30 % of the surface considered. Not modelling indoor points avoids a waste of memory space and a waste of computational power. Such an approach inevitably leads to the creation and the management of an irregular data structure. As they provide powerful features to describe and to manipulate complex, irregular data structures, object-oriented languages are perfectly suited to an irregular implementation of the ParFlow method.

Design and Object-Oriented Implementation ParFlow++ was developed in C++, according to the fundamental principles of software engineering. The preliminary analysis and design of the application were achieved using the Fusion method [2]. The main classes and the relationships that were identified during the analysis phase constitute an object model, that is partially reproduced in Figure 2.

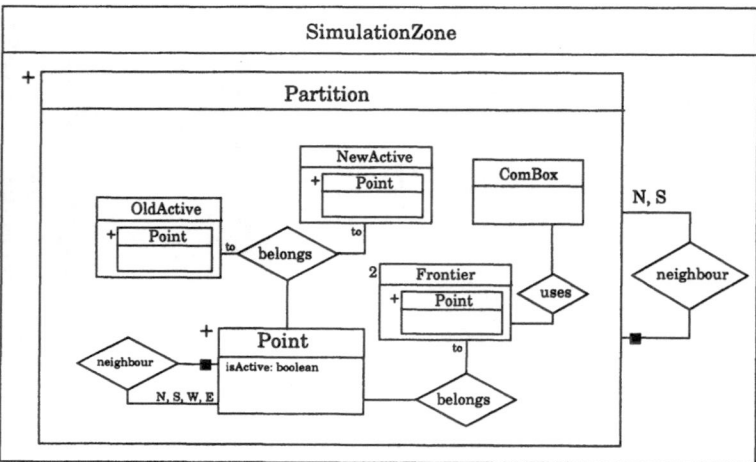

Fig. 2. Fusion object model of ParFlow++.

All kinds of non-indoor points are described in a hierarchy of C++ classes. Instances of these classes can be assembled at runtime so as to constitute an irregular structure that preserves the neighbourhood relationships.

In ParFlow the amount of computation required during a simulation step is not constant. Because of the discretization of time and space in the simulation process, a wave radiated by the source point propagates step by step throughout the simulation zone. ParFlow++ was designed in such a way that the amount of points implied during any computation step is always kept at a minimum. At any time, a point that has not been reached by the wave yet is said to be inactive (initially, all points are inactive but the source point). Points that have just been reached by the wave (during the current simulation step) are said to be *new* active points, whereas those that were reached during former simulation steps are referred to as *old* active points.

At each step of the simulation flow values need only be calculated for active points, and the propagation of outgoing flows is only required from active points to their neighbours. To achieve these goals ParFlow++ manages several structures internally. Every newly activated point is inserted in a list *newActive*, whose role is to permit an efficient propagation of the activity status after each computation step: the only points that are in a position to be activated are those that are neighbours to members of the *newActive* list, and that are still inactive. Iterating through members of the *newActive* list facilitates the identification of new active points. Once a member of *newActive* has activated its neighbours, it is transferred into another list *oldActive*. Hence, this point will never be considered again when looking for new active points, although it will still participate in the computation.

Figure 3 confirms the advantage of activating points dynamically. It shows how the workload per simulation step increases as the radio wave propagates and covers a growing surface, and how it reaches a ceiling when the simulation zone is completely covered.

Figure 3 also confirms the advantage of using a data structure that does not model indoor points. The speed at which the workload increases depends on the amount of obstacles met by the radio wave during its propagation, and the maximal workload depends on the amount of outdoor points in the simulation zone.

3.2 Data Parallel Implementation

A major advantage of the ParFlow method is that, although the calculations made during each iteration step are theoretically synchronous, updates of points require independent computation. The ParFlow algorithm is therefore a good candidate to parallel implementation.

The simulation zone is split in horizontal bands, which are then allocated to processors based on a round-robin policy, as shown in Figure 4.

In ParFlow++ a partition basically consists of a structured collection of points and of the two lists that are used to maintain references to 'new' and 'old' active points (see Section 3.1). Each partition also manages two 'frontiers' internally. A frontier can be perceived as a collection of references to points that are either on the northern edge or on the southern edge of a partition. The behaviour of frontier points differs slightly from that of other points, for they

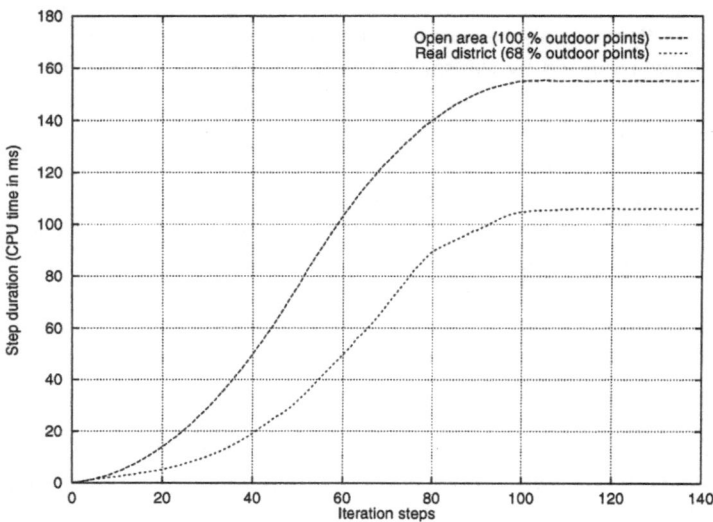

Fig. 3. Evolution of the workload per simulation step during a sequential simulation on a 100 × 100 points area. These results show the workload evolution observed when achieving a simulation on an open area (upper curve), and on a real district of the city of Geneva (lower curve). The CPU times reported were measured on one processor of a Cray T3D supercomputer.

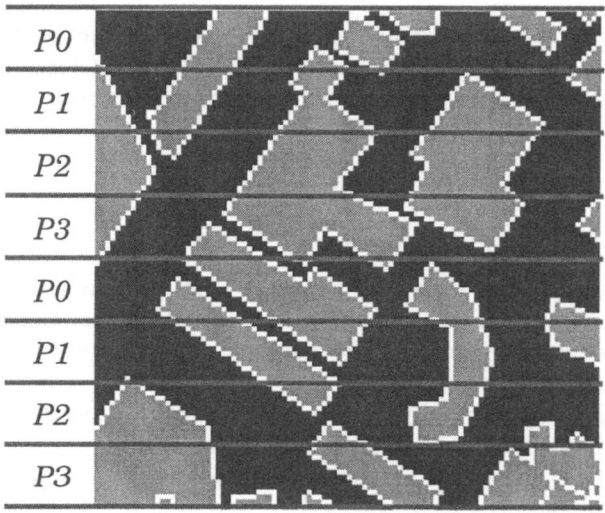

Fig. 4. Simulation zone partitioning and mapping.

have to interact with 'neighbouring' points that are actually stored on remote processors. A frontier thus ensures the propagation of flows between a partition and one of its neighbours. It also ensures the propagation of the activity status, since a newly active point located on a frontier may activate a point in the opposite frontier of a neighbouring partition.

4 Experimental Results

ParFlow++ was compiled on a Cray T3D. We ran a 800 step radio wave propagation simulation on a 500x500 point zone modelling a 1 km^2 district of the city of Geneva. Figure 5 shows the resulting path-loss map.

Fig. 5. Results of a radio wave propagation simulation on a district of the city of Geneva. The broken arrow shows the position of the emitter. The computation of such a figure takes about 18 minutes on a single processor of the Cray T3D.

On the Cray T3D we also measured the speedups observed for various simulation zone sizes. Figure 6 confirms the scalability of the parallel implementation. (For the 1024 × 1024 and 2048 × 2048 zones, the simulation was not possible on a single processor because of memory limitation, so we had to estimate the sequential reference times).

5 Related Work

The performances reported in the former section are quite satisfactory. Yet experiments have shown that partitions dimensioning is a crucial issue. In the future we would like to experiment with an heterogeneous partitioning policy. Since many bands are required near the source point in order to give better workload

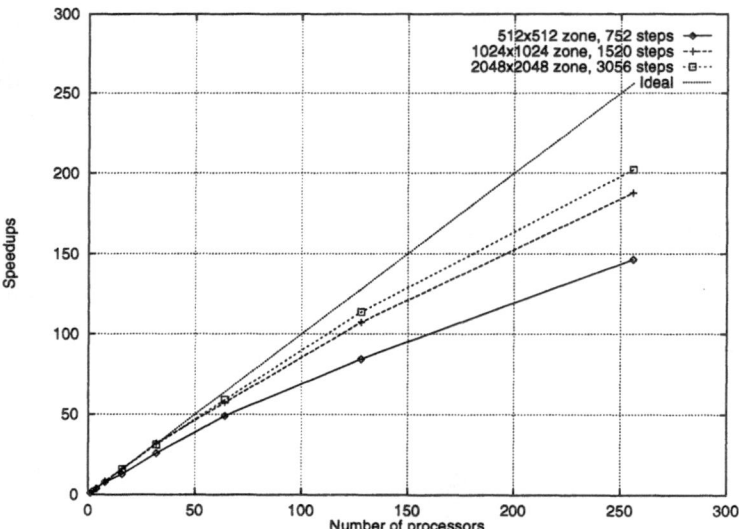

Fig. 6. Speedups observed on a Cray T3D, when achieving simulations for various district sizes.

balancing, and since too many bands lead to a degradation of the communication performances, a compromise would be to allocate thinner bands near the source. Some work has been achieved to build a mathematical model of the workload and of communications in ParFlow++[iii]. This model should help select the best partitioning policy for any simulation problem case and any target platform.

6   Conclusion

In this paper we have reported the development of ParFlow++, a new parallel object-oriented piece of software that permits the prediction of radio wave propagation in urban environments, based on a bidimensional simulation over a digital city map.

   The main originality of ParFlow++ with respect to former implementations of the ParFlow method [5] is that it was implemented as an irregular application. Since the method does not allow for radio waves to propagate through buildings, ParFlow++ does not model indoor points. Experience shows that this approach permits a significant reduction of computation times.

   ParFlow++ was designed in an object-oriented way, and implemented in C++ for MIMD-DM[iv] platforms. It was developed so as to be highly portable, and its current implementation relies on the PVM library [3]. Although this

---

[iii] *'Performance Analysis of a Parallel Program for Wave Propagation Simulation'*, paper also published in these proceedings.
[iv] Multi Instruction stream, Multi Data stream, Distributed Memory

implementation can still be improved in many ways, experiments achieved on the Cray T3D at the Swiss Federal Institute of Technology show the scalibility of the code. The results also demonstrate that an object-oriented irregular implementation does not necessarily lead to poor performances on MIMD-DM platforms.

The object-orientation of ParFlow++ brings in much versatility. The set of classes that constitute its source code could easily be extended, or reused in another context. For example, in the near future ParFlow++ will be modified so as to simulate tri-dimensional radio wave propagation. Its classes should also serve as basic building blocks to develop software tools capable of simulating other physical phenomena, such as fluid dynamics or reaction-diffusion processes.

Acknowledgments

The STORMS project is a European ACTS project, funded by the European Community and by the Swiss government (OFES grant). The Cray T3D experimentations were conducted on the machine of the Swiss Federal Institute of Technology in Lausanne.

References

1. R. Benzi, S. Succi, and M. Vergassola. The Lattice Boltzmann Equation: Theory and Applications. *Physics Reports*, 222(3):145–197, 1992.
2. D. Coleman and al. *Object-Oriented Development - The Fusion Method*. Prentice Hall Object-Oriented Series, Englewood Cliffs, NJ, 1995. ISBN0-13-338823-9.
3. A. Geist, A. Beguelin, J. Dongarra, W. Jiang, R. Manchek, and V. Sunderam. PVM: Parallel Virtual Machine. A User's Guide and Tutorial for Networked Parallel Computing. MIT Press, 1994.
4. P. O. Luthi and B. Chopard. Wave Propagation with Transmission Line Matrix. Technical report, University of Geneva and Swiss Telecom PTT, 1994.
5. P. O. Luthi, B. Chopard, and J.-F. Wagen. Wave Propagation in Urban Micro-cells: a Massively Parallel Approach using the TLM Method. In *Proceeding of PARA '95, Workshop on Applied Parallel Scientific Computing, Copenhagen*, aug 1995. Also in COST 231 TD(95) 33.

A Portable Parallel Implementation of a 3D Semiconductor Device Simulator

Ali Bouaricha and Stephan Mueller

Silvaco International, 4701 Patrick Henry Dr., Bldg. 3, Santa Clara, CA 95054

Abstract. We describe a parallel implementation of a 3D semiconductor device simulator based on the MPI message passing interface. In this article, we primarily focus on the effective implementation of the numerical algorithms employed in the 3D device simulator, in particular the Newton-Krylov-Schwarz method for solving the large, sparse nonlinear equations problems resulting from the semiconductor device modeling on unstructured meshes. Test results obtained on a shared-memory multiprocessor computer are presented and analyzed.

1 Introduction

Semiconductor device simulation ([8], [6]) is the modeling of the electrical characteristics of a fraction of a semiconductor chip based on a computer model of its geometry and impurity distribution. This modeling is done by solving numerically one or more of the following equations:

$$-\nabla \cdot (\varepsilon \nabla \kappa) = q \left(p - n + N_D^+ - N_A^-\right) \tag{1}$$
$$\nabla \cdot (\kappa \nabla T) = H(n, p, \psi) \tag{2}$$

$$\nabla \cdot J_n = q \left(\frac{\partial n}{\partial t} + R - G\right) \tag{3}$$

$$-\nabla \cdot J_p = q \left(\frac{\partial p}{\partial t} + R - G\right) \tag{4}$$

Equation (1), "Poisson equation", and equation (2), "Heat transport equation", can be discretized using finite element methods. The other two equations ("Electron and hole continuity equations") are commonly discretized by a finite volume discretization technique [2].

The numerical solution technique commonly used to solve these equations is a Newton-Raphson method which internally requires the repeated solution of large sparse linear systems. More details can be found in [1]. The discretization, which is typically done on a tetrahedral mesh, results in typical linear system sizes of the order of 100000 unknowns with approximately 40 nonzero entries per row. This makes it mandatory to use an iterative linear solving technique. The discretized form of the semiconductor device equations also shows an unusually unstable numerical system. Condition numbers larger than 10^{20} are not uncommon. For this reason extreme care has to be taken to use stable numerical

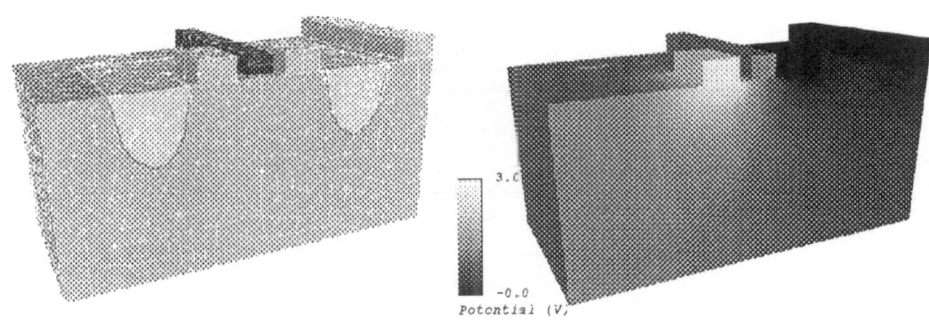

Fig. 1. Typical 3D device example. Left: Structure. Right: Potential distribution

methods. For example a simple diagonal preconditioner is entirely unsuitable for this application. Reference [7] gives a more detailed explanation of the numerical problems mentioned above.

Because of the very large sizes and the complexity of the above systems of linear equations, conventional uniprocessor architectures become overwhelmed by the computational demands. The use of scalable, high-performance parallel computers to satisfy such requirements has become important in the continued success of large-scale device simulation. In this work, we use a portable message passing interface, namely, MPI to implement the numerical algorithms required for the solution of the above system of equations.

The remainder of this article is organized as follows. In section 2 we describe the semiconductor device simulation application. In section 3 we describe the numerical algorithms implemented in the 3D simulator. Section 4 discusses the parallel implementation issues. Finally, in section 5 we present and analyze our experimental results on a shared-memory multiprocessor computer.

2 Semiconductor Device Simulation Application

A typical example of the 3D structures used for this study is shown in Figure 1. Figure 1 shows an NMOS transistor using a trench isolation technique. The first picture shows the materials and the second a sample potential distribution computed by the simulator.

Th implementation of the 3D semiconductor device simulator is done in a very object oriented manner in C++. This allows for a very flexible and modular implementation but turned out to be also beneficial for the parallelization. A schematic outline of the objects used in this code is shown in Figure 2. Figure 2 shows a nonlinear solver object that receives several nonlinear equations together with boundary conditions and interface conditions.

The parallelization of this device simulator benefited from this concept. One particular example for this is the linear solver object that provided an easy mechanism to combine various linear solver methods which will be explained in more detail in section 3.

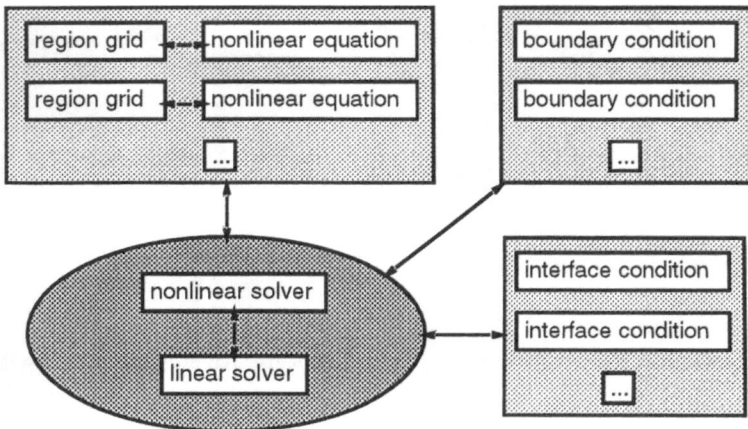

Fig. 2. Outline of the object oriented implementation of the device simulator.

Another set of objects that was useful for the parallelization are the region grid objects. (A region grid is a connected subsection of the tetrahedral mesh used to discretize the device.) The nonlinear solver class is able to handle arbitrary combinations of region grid/nonlinear equation pairs together with accompanying boundary conditions and interface conditions. This proves to be very helpful for the load balancing algorithm as will be explained in section 4

3 Numerical Algorithms

In this section we give a brief description of the numerical algorithms used in the implementation of the 3D device simulator, namely, Newton-Krylov-Schwarz methods for solving the large, sparse nonlinear equations resulting from the semiconductor device modeling, with an emphasis on domain decomposition, Krylov iterative methods, graph partitioning algorithms, and scaling.

3.1 Newton-Krylov-Schwarz Methods

A Newton-Krylov-Schwarz method combines a Newton method with a Krylov-Schwarz technique (domain decomposition) for solving the linearized system of equations that arise in each Newton iteration. Krylov-Schwarz methods employ an overlapping Schwarz domain decomposition to derive a preconditioner for the Krylov accelerator that depends primarily on local information, for data-parallel concurrency. Mathematically, if

$$Au = f \qquad (5)$$

is the linearized system of equations, Schwarz operates by:

1. Decomposing the variable space $u : \mathcal{U} = \sum_{k=1}^{p} \mathcal{U}_k$, where p is the number of subdomains;

2. Finding the restriction of A to each $\mathcal{U}_k : A_k = R_k A R_k{}^T$, for some restriction operators $R_k : \mathcal{U} \to \mathcal{U}_k$ and extension operators $R_k{}^T : \mathcal{U}_k \to \mathcal{U}$;
3. Forming M^{-1} from the $A_k{}^{-1}$, where the inverse of A_k is well defined within the k-th subspace.

A Schwarz preconditioner is

$$M^{-1} = \sum_{k=1}^{p} R_k{}^T \tilde{A}_k^{-1} R_k; \qquad (6)$$

where $\tilde{A}_k$ is an approximation to $A_k = R_k A R_k{}^T$. In our work, $\tilde{A}_k$ is an incomplete LU (ILU(0)) decomposition of A_k if A_k is nonsymmetric, and the improved incomplete Cholesky factorization (IIC) of Jones and Plassmann [5] if A_k is symmetric. For $k = 1, 2, \cdots$, the R_k and $R_k{}^T$ are simply gather and scatter operators, respectively. We refer the reader to [4] for details on domain decomposition algorithms.

The motivations for choosing the Schwarz domain decomposition as preconditioner are (1) ease of parallelization and good parallel performance and (2) superior convergence properties.

3.2 Iterative Solvers

The 3D semiconductor device simulator incorporates two iterative solvers that represent the current state-of-the-art for solving large sparse linear systems of equations: Conjugate Gradient (CG) and Biconjugate Gradient Stabilized (Bi-CGSTAB). Both of these Krylov solvers are used in conjunction with the Schwarz domain decomposition preconditioner. Parallel implementations of these solvers are not difficult to create; the bulk of the work is a sparse matrix vector multiply which is inherently parallel. (See section 4).

3.3 Graph Partitioning Algorithms

The partitioning of the sparse coefficient matrix in (5) in p submatrices A_k of roughly the same size, where p is the number of processors, requires the solution to a graph partitioning problem. This is the key to an efficient parallel implementation of the domain decomposition method on multiprocessor computers. To this end, we have developed a graph partitioning program, MGP++, based on multilevel graph partitioning techniques.

3.4 Scaling

The coefficient matrix A in (5) above is scaled so that its diagonal is the identity. That is,

$$A_m = D^{-1/2} A D^{-1/2},$$

where $D = diag(A)$. In practice, this simple and cheap scaling has proved to be remarkably effective as it considerably reduces the condition number of the matrix A.

4 Parallel Implementation

4.1 Master-Slave Implementation

The 3D semiconductor device simulator was implemented using a master-slave approach. That is, whenever a task is to be performed in parallel, the following sequence is executed:

- The master sends the appropriate execution command to the slaves;
- The slaves receive the command, interpret it, and waits for the master to send further data;
- The slaves perform the appropriate tasks then send the results back to the master.

4.2 Load Balancing

The major load balancing problem we encountered during the testing of the 3D device simulator was due to the large difference in sizes of the region grids objects (see section 2 above). The solve this problem, we have implemented a dynamic load balancing strategy which repeatedly selects the region grid that has the largest number of nodes and partitions it in half with MGP++. This process continues until either a given total number of partitions was reached or all the regions grids are approximately the same size.

4.3 Latency Hiding and Message Combining

The 3D device simulator code masks the latency in interprocessor communication as much as possible by using asynchronous communication and overlapping work with communication. Furthermore, processes in the 3D device simulator use the MPI_Pack facility whenever possible to combine several data (possibly of different types) into a single user buffer and communicate it at once.

4.4 Parallel Krylov-Schwarz methods

Krylov-Schwarz methods are known to offer excellent scalability. For more detail on this subject we refer the reader to [3]. The parallel implementation for these methods is given in Figure 3.

5 Performance

We tested our 3D device simulator on two problems (Problem 1 and 2) of the type described in section 1. The mesh sizes of Problem 1 and 2 are 78025 and 114002 points, respectively. Figures 4 and 5 show the speedups of the 3D device simulator on a Sun multiprocessor computer with nprocs=1,2,3,4,5,6 for Problem 1 and 2, respectively. We define speedup as follows:

$$speedup = E_1/E_p,$$

CG:

Compute $r^{(0)} = f - Au^{(0)}$ for some initial guess $u^{(0)}$ in **PARALLEL**
for i=1,2,$\cdots$
    solve $Mz^{(i-1)} = r^{(i-1)}$ in **PARALLEL** (see below)
    $\rho_{i-1} = r^{(i-1)T}z^{(i-1)}$
    if $i = 1$
        $p^{(1)} = z^{(0)}$
    else
        $\beta_{i-1} = \rho_{i-1}/\rho_{i-2}$
        $p^{(i)} = z^{i-1} + \beta_{i-1}p^{(i-1)}$
    endif
    $q^i = Ap^{(i)}$ in **PARALLEL**
    $\alpha_i = \rho_{i-1}/\rho^{(i)T}q^{(i)}$
    $u^{(i)} = u^{i-1} + \alpha_i p^{(i)}$
    $r^{(i)} = r^{i-1} - \alpha_i q^{(i)}$
    check convergence; continue if necessary
end

Schwarz domain decomposition preconditioner to solve $Mz = r$:

Step 0. Perform an IIC factorization on A_k: $A_k = \tilde{L}_k \tilde{L}_k^T = R_k^T A R_k$ in **PARALLEL**,
where R_k and R_k^T are defined in section 3.1
Step 1. Solve $\tilde{L}_k \tilde{L}_k^T v_k = R_k^T f$ in **PARALLEL**
Step 2. Assemble the solution $z = \sum_{k=1}^{p} R_k^T v_k$ in **PARALLEL**

Fig. 3. Parallel conjugate-gradient-Schwarz solver

where E_1 is the execution time of the serial code and E_p is the execution time of the parallel code using p processors.

On the basis of Figures 4 and 5, the following observations can be made:

- The speedups of the 3D device simulator appear to stall when $p > 5$ and $p > 4$ for Problem 1 and 2, respectively. This is due to the fact that the communication/computation ratio becomes relatively significant, thereby causing the speedups to stay constant for certain values of p, and to eventually deteriorate as p gets larger.
- Figure 4 shows superlinear speedups for $p \leq 3$. This can be explained by noting that the most dominant computational part of the Newton solver is the domain decomposition preconditioner, which in Problem 1 employs a direct solver at the subdomain level, that is computed in order to solve the linear system of equations (also called the Newton equations) that arise in each Newton iteration. Since the domain decomposition preconditioner partitions the Jacobian matrix into p submatrices of roughly the same size, the time required to factor those p Jacobian submatrices is significantly less than that required by the factorization of the whole Jacobian. Also, if we

consider that the number of linear iterations required to solve the Newton equations does not vary a lot as p increases from 1 to 3, then the total time spent in the Newton solver of the parallel code is considerably less than p/E_n, where E_n is the total time spent in the Newton solver of the serial code, which explains the superlinear speedups of Figure 4.

– Figure 5 shows inferior speedups to those shown in Figure 4. This remark can be justified by noting that the total time spent in the Newton solver is dominated by the linear iterative solver (i.e., CG), which require a considerable amount of global communication. This claim can be explained as follows. Due to the large memory requirements of the Jacobian matrix of Problem 2, we used an ILU factorization at the subdomain level for the domain decomposition solver. Since the ILU factorization is much cheaper to compute than a direct solver, it is less effective as a preconditioner. Thus, to solve the Newton equations, the domain decomposition solver takes many more linear iterations to converge than when a direct solver is used. Consequently, the dominant computational part is spent in computing the linear iterations. Furthermore, if we consider that the number of iterations required to solve the Newton equations increases considerably as p increases, it is clear that the obtained speedups are not competitive.

A serious limitation of our 3D MPI-based implementation on shared-memory computers is the fact that large amount of data such as the region grids is duplicated for each process that uses them, thereby making the total memory requirements prohibitive for sufficiently large problems. Moreover, the communication of large region grids between processes can result in a real bottleneck. It turns out that the remedy to the problems cited above is to implement a multithreaded version of the MPI library while keeping the same MPI interface. The advantage of this approach is that all the read-only data such as the region grids are not duplicated; instead only the addresses of these data are being communicated. In a future article, we plan to report the performance of the 3D device simulator using a multithreaded implementation of MPI and compare it to MPI implementation described in this article.

References

1. Bank, R. E., Rose, D. J.: Global approximate newton methods Numer. Math., **37** (1981) 279–295
2. Bank, R. E., Rose, D. J., Fichtner, W.: Numerical methods for semiconductor device simulation IEEE Trans., **ED-30** (1983) 1031–41
3. Cai, X. C., Keyes, D. E., Venkatakrishnan, V.: Newton-Krylov-Schwarz: An implicit solver for CFD John Wiley & Sons Ltd **v3.0**, 1995
4. Chan, T. F., Mathew, T. P.: Domain decomposition algorithms Preprint CAM 94-2, Department of Mathematics, University of California, Los Angeles, 1994
5. Jones, M. T., Plassmann, P. E.: An improved incomplete Cholesky factorization, Preprint MCS-P206-0191, Argonne National Laboratory, Argonne, Illinois, 1991
6. Markowich, P. A., Ringhofer, C. A., Schmeiser, C.: Semiconductor Equations Springer, 1990

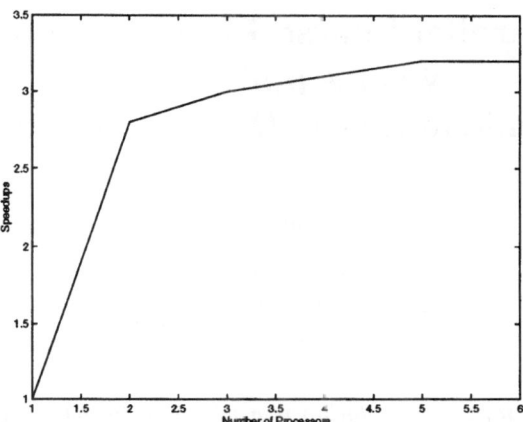

Fig. 4. Speedups for Problem 1

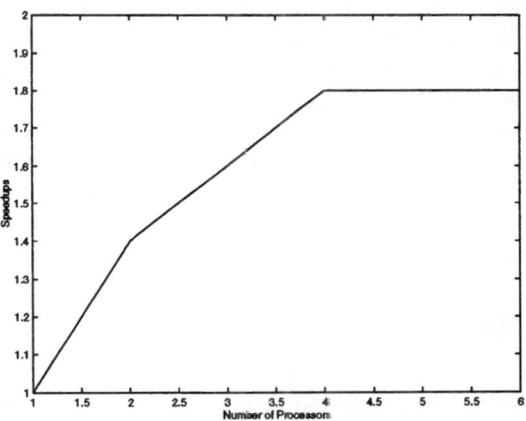

Fig. 5. Speedups for Problem 2

7. Pommerell, C.: Solution of Large Unsymmetric Systems of Linear Equations PhD thesis, Integrated Systems Laboratory, ETH Zurich, Switzerland, 1992 Published by: Hartung Gorre Verlag, Konstanz, Germany
8. Selberherr, S.: Analysis and Simulation of Semiconductor Devices Springer, 1984

A Parallel Sparse LU Decomposition with Application to Semiconductor Device Simulation

Mounir Hahad

SILVACO International
4701 Patrick Henry Drive, bldg # 3, Santa Clara, CA-95054
e-mail : mounir.hahad@silvaco.com

Abstract – *Computer simulation is a vital tool to today's advanced semiconductor device design and manufacturing. Unfortunately, precision requirements makes the shortest meaningful simulation run at least for hours on top of the line desktop workstations. To overcome this performance issue, we resort to parallel computers. In this paper, we describe the parallel linear solver based on LU decomposition that we introduce to solve the linear systems arising from the discretization of partial differential equations involved in semiconductor device simulation. This solver is now part of a software package called ATLAS that is widely distributed. Experiments are performed on the latest shared memory and distributed shared memory parallel computers, namely Sun Enterprise servers, Silicon Graphics Origin2000 and HP/Convex S-Class Exemplar.*

1 INTRODUCTION

The semiconductor industry is driven by novel technologies (new materials, manufacturing procedures, ...) beeing introduced at a very high pace. It is unrealistic to try to master these technologies by running conventional experiments on real wafers for two main reasons: first, it takes a long time to set up a production line to accomodate the desired experiment parameters. Second, it is very expensive to do so. Therefore, computer simulation is a must in every semiconductor research laboratory.

Surprisingly, very few simulation software development groups consider parallel processing as a viable alternative. As a proof of concept, we have successfully parallelized a commercially available and widely spread device simulation software on shared memory and distributed shared memory parallel computers. ATLAS [2] is the framework of this software package that predicts the electrical behavior of specified semiconductor structures. The core tool is a finite element discretization of the Boltzman charge transport equation in two of its simpler forms [8,2], the drift-diffusion model and the energy balance model. Including the Poisson solution for potential and a lattice heat equation, we end up with six coupled equations to be solved at each time step at each node of the discretization mesh. Since the system's matrix is unsymmetric, we use the LU decomposition.

2 Parallel LU Decomposition

Assuming that U is unit upper triangular, then $U_{i,i} = 1, \forall i$. The remaining entries in L and U are computed according to the following:

$$\begin{cases} L_{i,j} = A_{i,j} - \sum_{k=1}^{k=j-1} L_{i,k} U_{k,j}, & \forall i > j; \\ L_{i,i} = A_{i,i} - \sum_{k=1}^{k=i-1} L_{i,k} U_{k,i}; \\ U_{i,j} = (A_{i,j} - \sum_{k=1}^{k=i-1} L_{i,k} U_{k,j})/L_{i,i}, & \forall i < j; \end{cases}$$

To reduce the fill-in that occurs during the factorization, we experiment using the reverse Cuthill McKee (RCM), the minimum degree (MD) and the Nested Dissection (ND) heuristics [5,6]. Afterwards, the numerical factorization, which is by far, numerical factorization the most time consuming phase is performed. Only this phase will be considered for parallelization.

Previous work has been conducted on parallel LU factorization, but very little on general sparse matrices. In [9], Van Der Stappen et al. describe an implementation specifically tuned to transputer networks. In [1], Alaghband describes a parallel pivoting algorithm for unstable systems. In [4], Buoni et al. describe their approach to factorize upper Hessenberg matrices.

In the parallel algorithm we introduce, all the computations to evaluate one row are assigned to a single processor. This scheme has 2 main advantages: larger grain as compared to distributed memory approaches for similar factorization algorithms [3]; and each row is only modified by one single processor leading to very limited synchronization. It relies on the elimination tree [7,5]. There are several methods to walk the elimination tree and produce a valid scheduling of the rows.Each node in the tree is assigned to a specific *level* depending on the tree-walk method used. These levels exhibit the property that all rows within a single level are entirely independent, and therefore can be simultaneously scheduled for parallel execution. In the Bottom-Up tree-walk that we use, all the leaves of the elimination tree belong to the first level. Their parents belong to the second level, and so on. Once all the nodes are assigned to a level, we assign nodes to processors in a cyclic way, starting by the level one, then moving to the next level. The transition from one level to another should require a synchronization between the processors, but a careful examination of the dependencies shows that a global synchronization is not necessary. Instead, we introduce an *events-based* synchronization, such that:

- to the completion of each row computation is associated one event;
- the processor in charge of computing row i signals that the event i has occured to all the processors once it has finished computing row i;
- a row which associated event has not yet been signaled cannot be used to update another row.

The implementation of this synchronization mechanism requires two primitives: *signal(i)* and *wait(i)*. An outline of the parallel algorithm is as follows:

```
build Task(i) as a level numbering of the elimination tree
Do step=1,n
    i = Task[step]
    Do k in Struct(L_{i*})
        wait(k)
        update row i with row k
    EndDo
    signal(i)
EndDo
```

3 Experimental results

The parallel architectures that we consider are either shared memory or distributed shared memory parallel computers. The advantage of such architectures is the much easier parallelization of large existing scientific codes thanks to the programming model they support. We conducted our experiments on several platforms from the major supercomputer vendors, including SGI Origin2000 (O-2000), SGI PowerChallenge with both Mips R10000 processor and Mips R8000 processor, Convex Exemplar S-Class and Exemplar SPP1200, Sun Enterprise4000 (E-4000) and SparcCenter1000. All these platforms provide light weight processes to map user threads to processors. The maximum number of processors used is voluntarily limited to those configurations the final users of the software are expected to use. Most of the experiments have been run at the vendors' facilities. /the code itself has not been fine tuned to any specific platform.

The table 1 reports the execution times for 3 test problems with different sizes on 3 different platforms. The number of rows in these matrices are 3433, 8052 and 17656 for Problem1, Problem2 and Problem3 respectively. The corresponding number of non zero elements in the LU factors are 255847, 3856892 and 6756600.

	Problem1			Problem2			Problem3		
# CPUs	E-4000	O-2000	S-Class	E-4000	O-2000	S-Class	E-4000	O-2000	S-Class
1 CPU	0.97	0.36	0.33	50.85	16.61	12.81	84.94	28.98	26.02
2 CPUs	0.51	0.21	0.20	26.95	8.70	6.80	45.35	14.37	13.88
4 CPUs	0.30	0.12	0.12	15.06	4.71	3.65	24.37	8.09	7.92

Table 1. Execution times in seconds for test structures.

We have reordered the matrices with the minimum degree (MD) heuristic as well as reverse Cuthill McKee (RCM) and nested dissection (ND). As far as execution time is concerned, the best results for Problem2 were obtained with the RCM heuristic in spite of the larger number of fill-in as compared with MD. This is due to the better use of the cache when the matrix bandwidth is reduced. The two other problems perform far better with MD.

4 Conclusion

In this paper, a large grain algorithm for sparse LU decomposition using events-based synchronization is introduced, with focus on semiconductor device simulation application. The algorithm accesses the matrix exclusively by rows, thus requiring less storage overhead. The major benefit from a large grain task model is the limited need for synchronization, as we have shown in our algorithm. Furthermore, this type of synchronization is implemented at the user level, with a flag array of the size of the order of the matrix, avoiding expensive system calls for locks usually used in synchronizing shared memory algorithms.

We have shown a very good performance on a wide array of platforms with the sample problems. The influence of the reordering heuristic is still not yet fully caracterized, although most of the problems benefit more from the MD reordering than any other one. In general, Parallel ATLAS shows a speed up of up to 5.3 on an 8 cpu platform, depending on the problem caracteristics.

References

1. Gita Alaghband. A parallel pivoting algorithm on a shared memory multiprocessor with controlled fill-in. In *Proceedings of the 1988 International Conference on Parallel Processing*, volume III, Algorithms and Applications, pages 177–180, University Park, Penn, August 1988. Penn State. U. Colorado, Denver.
2. Y. Apanovich, P. Blakey, R. Cottle, E. Lyumkis, B. Polsky, A. Shur, and A. Tcherniaev. Numerical simulation of submicrometer devices including coupled nonlocal transport and nonisothermal effects. *IEEE Transactions on Electron Devices*, pages 890–898, 1995.
3. Cleve Ashcraft, Stanley Eisenstat, and Joseph Liu. A fan-in algorithm for distributed sparse numerical factorization. *Siam Journal of Scientific and Statistical Computations*, pages 593–599, 1990.
4. J. J. Buoni, P. A. Farrell, and A. Ruttan. Algorithms for LU decomposition on a shared memory multiprocessor. *Parallel Computing*, 19(8):925–937, August 1993.
5. J.W.H. Liu. Equivalent sparse matrix reordering by elimination tree rotations. *Siam journal of scientific and statistical computing*, pages 424–445, 1988.
6. J.W.H. Liu. Reordering sparse matrices for parallel elimination. *Parallel computing*, pages 73–91, 1989.
7. J.W.H. Liu. The role of elimination trees in sparse factorization. *Siam journal of matrix analysis applications*, pages 134–172, 1990.
8. S. Selberherr. *Analysis and Simulation of Semiconductor Devices*. Springer Verlag, 1981.
9. A. van der Stappen, R. H. Bisseling, and J. G. G. van der Vorst. Parallel sparse LU decomposition on a mesh network of transputers. *SIAM Journal on Matrix Analysis and Application*, 14:853–879, 1993.

A Parallel Simulation of a Quantitative Large-Strain Polycrystal Deformation

M. Juganaru, I.Sakho, C. Maurice, and F. Montheillet

Ecole Nationale Supérieure des Mines de St. Etienne
F-42023 St. Etienne cedex 2, France

1 Introduction

Predicting behaviour of materials when submitted to large strain is a fundamental problem of mechanics. In general, numerical simulation is one of the more widely used prediction methods. The problem related to numerical predictions is twofold. First, we have to define a simple reliable model which describes the major aspects of the evolution of material. Secondly, we have to face the need for computational power which conventional computers do not provide.

This paper deals with the problem of predicting the behaviour of polycrystals submitted to large strain. When polycrystalline metals are submitted to large strains associated with metal forming processes, such as rolling, extrusion, and forging, significant strain inhomogeneities occur at a grain scale. This is, first, due to the various crystallographic orientations of the grains, which induce a scatter of the flow stresses. However, two other factors can significantly affect the local behaviour of the aggregate: the aspect ratio of the grains (morphological texture) and the local distribution of the grains (topological texture). Such effects are not properly taken into account by the classical laws of mixture. In the present work, a new method was used to model the deformation of a polycrystal. It has been described elsewhere [4] and will be summarized in the next section.

To meet the computational requirements this model is simulated on a distributed memory parallel computer.

2 Outline of the model

The aggregate is represented by a two dimensional wraparound array of grains (see figure 1) generated in a plane (x_1, x_2) by a Voronoï tessellation [3].

The material is assumed to be incompressible and power law viscoplastic, i.e. $\sigma_i = k_i \dot{\varepsilon}_i^m$ where σ_i and $\dot{\varepsilon}_i$ are the flow stress and strain rate of grain i, respectively, k_i denotes the viscosity, and the strain rate sensitivity parameter $m = 0.2$ for all grains. The neighbourhood of each grain consists of the set of the n adjacent crystals (here, $n \approx 6$); it remains constant during straining.

The aggregate is submitted to plane strain compression along axis x_2, and elongates along x_1. For each overall strain step, all grains are considered in turn. The average viscosity of the neighbourhood of grain i is first calculated as:

$\overline{k_i} = \sum_{j=1}^{n} \alpha_i k_i$. The weighting coefficient α_i is proportional to the length of the boundary between grains i and j, and $\sum_{j=1}^{n} \alpha_j = 1$.

The strain rate of grain i is then derived as follows (it should be noted that the strain rate tensor reduces in the present case to a single component). The localization factor δ_i is : $\delta_i = \dfrac{(1 + aF)(1 - \tanh((b + cF)(\Sigma_i - 1)))}{1 + aF + (-1 + dF)\tanh((b + eF)(\Sigma_i - 1))}$. In this equation $F = \dfrac{2 * \lambda_i}{\lambda_i^2 + 1}$, where λ_i is the aspect ratio of the grain, which is defined here as the ratio of the average intercepts of the grain parallel to the x_2 and x_1 axes. Σ_i denotes the hardness ratio $\dfrac{k_i}{\overline{k_i}}$, and a, b, c, d, and e are constants which have been determined numerically. The strain rate can then be written in the form: $\dot{\varepsilon}_i = \dfrac{\delta_i}{< \delta_i >}\dot{\varepsilon}_\infty$ where $\dot{\varepsilon}_\infty$ is the remote prescribed strain rate, and the average $< \delta_i >$ is extended over all the grains of the aggregate (each δ_i is weighted by the surface of the associated grain). The strain increment of each grain is then obtained as $d\varepsilon_i = \dot{\varepsilon}_i\, dt$. Finally, the overall viscosity of the polycrystal is : $k = \dfrac{< k_i \delta_i^m >}{< \delta_i >^m}$. In this work, each k_i was assumed to be strain dependent according to the classical equation: $k_i = \left[k_{si}^2 - (k_{si}^2 - k_{0i})\exp(-r_i\varepsilon_i) \right]^{1/2}$ where k_{0i} and k_{si} are the viscosity values for incipient straining (elastic limit), and at large strains (steady state), respectively, and r_i is a dynamic recovery parameter.

3 Parallel simulation

In the previous section we presented the model of a polycrystal deformation. Now we deal with the parallelism of the model and the problems which arise when implementing it on a real parallel computer.

It is obvious that the model is inherently parallel and has a synchronous behavior; therefore it is well suited to be implemented on a parallel computer.

However, it presents some pitfalls. This problem is similar to the graph partitioning problem [2]. Indeed, the polycrystal aggregate can be viewed as a graph, the nodes corresponding to the grains and the edges corresponding to common edges of the grains. Optimal partitioning algorithms are known to be very expensive; thus in this project, because our final goal is the simulation of continuous dynamic recrystallization, we decided a balanced partitioning that fairly allocates the grains in the order their generation according to the Voronoï tessellation.

On the other hand, the computation of global variables requires gathering information disseminated throughout the network. In this work these global variables are calculated according to a master/slave approach. Each slave processor computes the partial values of these variables on the basis of the grains it owns, then sends the result to the master processor.

Now let us examine the processing of a grain. It consists in the computation of : k_i, $\overline{k_i}$, Σ_i, δ_i, $< \delta_i >$, $\dot{\varepsilon}_i$, the temporary coordinates of each vertex V_i^l, the final coordinates of all vertices, k and other global metrics of the aggregate. By gathering the information disseminated in the network, as described above, we obtain the following algorithms for each grain i and for the overall aggregate :

Deform_grain(i)
begin
    compute k_i
    get neighbours' information k_j
    compute $\overline{k_i}$, Σ_i, δ_i
    get global information $< \delta_i >$
    compute $\dot{\varepsilon}_i$, σ_i and the temporary coordinates of each vertex W_i^l
    get the neighbours' coordinates of the common vertices W_j^m
    compute by the barycenter method the final coordinates V_i^l of all vertices
    provide k_i, δ_i for the computation of k
end

Deform_aggregate
begin
    for all i in parallel do
        Deform_grain(i)
    end

4   Implementation and results

We implemented the simulation of our model on a $T805^{TM}$ transputer-based parallel machine. It consists of 16 transputers interconnected in a hypercube of dimension 4 (4×4 torus). This implementation uses the package D4414A of SGS - Thomson Microelectronics ANSI C Toolset [1] which incorporates a routing kernel. The performance of the simulation depends critically on the performance of this routing kernel.

The simulation algorithm described in section 3 could be implemented as described. All the grains located on the same processor are simulated as a single sequential process. In this case it is common that, when computing the status of some grains, the status of its neighbours is not yet available. This situation occurs when a grain has a neighbour located on a distant processor. Symmetrically, the neighbour has also the same view. Thus each pair of such grains must exchange their status. In order to avoid several exchanges of small messages, each processor sends one message to each of the other processors which hold neighbors of its grains; each message contains the status of the grains on the originating processor which have neighbours located on the destination processor. The overall algorithm thus consists of as many processes as processors, completely connected and exchanging through a small number of virtual channels at most two messages : the first for the values of viscosities and the second for the values of the temporary coordinates.

The initial mechanical characteristics of each grain are set randomly according to a uniform law (see figure 1). The overall rate strain $\dot{\varepsilon}$ is constant while the local one $\dot{\varepsilon}_i$ varies with the viscosity, the aspect ratio and the neighbourhood of the grain. Figure 2 shows the final aggregate after 1000 deformation steps with $\dot{\varepsilon} = 10^{-3}$ on the aggregate in figure 1.

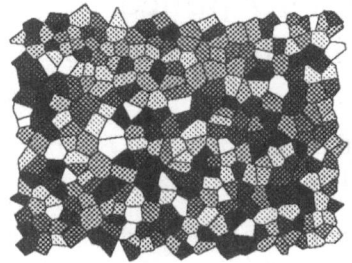

Fig. 1. An initial aggregate with 311 grains; the grey-level indicates the magnitude of viscosity

Fig. 2. The same aggregate after 1000 deformation steps, with $\dot{\varepsilon} = 0.001$; the grey-level indicates the magnitude of grain strain

Size of aggregate	60	311	2134	3847
Sequential time	0.143	0.743	5.100	12.242
Parallel time	0.0375	0.083	0.448	1.017
Speedup	3.82	8.89	11.36	12.04

Fig. 3. Time execution (in seconds) and speedup for a single step deformation

This simulation has permitted firstly to validate on larger-size aggregates of grains the material behaviour model described in section 2. Secondly, it allows us to obtain some new results about mechanical properties : only a very slight variation of the overall surface, has been obtained (0.02 %) and the three Hill macrohomogeneity conditions are fulfilled.

We can observe on figure 3 that the implementation strategy chosen is reasonably efficient. The speedup increases with the size of the aggregate.

5 Future work

This paper reported a parallel simulation of large-strain polycrystal deformation. A new model of deformation taking into account a large spectrum of deformation aspects (geometrical, topological and mechanical) was developed and validated.

The model is presently being extended to take into account the occurrence of continuous dynamic recrystallization at large strains. For that purpose, when a given grain i has reached some prescribed level of stored energy $w_i = \int k_i \dot{\varepsilon}_i^m \, d\varepsilon_i$, it is divided into new grains by a Voronoï tessellation, in the same manner as for the original structure.

The parallel simulation of continuous dynamic recrystallization requires dynamic load balancing. The related study is in progress.

References

1. *ANSI C Toolset User Guide.* SGS-Thomson Microelectronics, May 1995.
2. M. J. Berger and S. H. Bokhari. A partitioning startegy for nonuniform problems on multiprocessors. *IEEE Transactions on Computers*, C-36(5):570–580, May 1987.
3. J. Boissonnat and M. Yvinec. *Géométrie algorithmique*. Ediscience Intl., 1995.
4. F. Montheillet and P. Gilormini. Predicting the Mechanical Behavior of Two–Phase Materials with Cellular Automata. *Int. Journal of Plasticity*, 12(4):561–574, 1996.

Parallel Genetic Algorithms Applied to Optimum Shape Design in Aeronautics

Nathalie Marco, Stéphane Lanteri, Jean-Antoine Désidéri[1] and Bertrand Mantel, Jacques Périaux[2]

[1] INRIA, 2004 route des Lucioles, BP 93, 06902 Sophia Antipolis Cedex, France
[2] DASSAULT AVIATION, 78 Quai Dassault, 92214 Saint Cloud Cedex, France

Abstract. This paper presents a two-level strategy for the paralleliza-tion of a genetic algorithm coupled to a compressible flow solver designed on unstructured meshes. The resulting algorithm is used for the optimum shape design of aerodynamic configurations. Preliminary results are pre-sented for the optimization of two-dimensional airfoils.

1 Introduction

Genetic Algorithms (GAs) are efficient search methods based on principles of natural selection and population genetics. They lay on one of the most impor-tant principles of Darwin : *survival of the fittest.* Holland[4] in the 1970's was one of the first to incorporate in a computer algorithm such a technique in order to solve difficult problems through evolution. This technique needs a population of strings of binary digits (called *chromosomes*) which is submitted to many transformations called genetic operations (*selection, crossover* and *mutation* op-erators). The population evolves according to the fitness of the chromosomes during a number of generations, and, when a steady state is reached, the algo-rithm has converged to the best individual, solution of the given problem. The fitness function is directly related to the problem at hand e.g. the drag resulting from the calculation of a steady compressible flow around an airfoil geometry. More recently, Goldberg[3] brought GAs to non convex optimization theory and introduced a decisive thrust in the GAs research field.

In recent years, GAs have been introduced in aerodynamics shape design problems (see for example Quagliarella[9] and Périaux *et al.*[7]). The main rea-sons that make GAs more attractive among the aerodynamic design optimization methods are that they are much more robust than gradient based algorithms and they can approach a multi-objective optimization problem (i.e. when several cri-teria need to be taken into account simultaneously, see Poloni[8]). However, the main concern related to the use of GAs for aerodynamic design is the compu-tational effort needed for the accurate evaluation of a design configuration that, in the case of a crude application of the technique, might lead to unacceptable computer time if compared with more classical algorithms. In addition, hard problems need a bigger population and this translates directly into higher com-putational costs. GAs can be effectively parallelised and can therefore take full advantage of massively parallel computer architectures. The basic motivation behind many early studies of Parallel Genetic Algorithms (PGAs) was to reduce the processing time needed to reach an acceptable solution; later on, coarse grain PGAs based on sub-population models and migration operators were introduced and were shown to bring improvements on the convergence rate of classical GAs. We refer to Cantù-Paz[1] for a detailed review of various parallelization strategies applied to GAs.

In this paper we present a two-level strategy for the parallelization of a genetic algorithm coupled to a compressible flow solver. The resulting algorithm is used for the optimum shape design of aerodynamic configurations. The flow solver is based on a mixed finite volume/finite element MUSCL method designed on unstructured triangular (2D)/tetrahedral (3D) meshes. Preliminary results are presented for a direct optimization problem with the minimization of the drag-induced shock on a RAE2822 airfoil.

2   Shape parametrization for 2D airfoils

In the present study, the population individuals represent airfoil shapes. Following a strategy already adopted by several authors[7]-[8]-[9], the shape parametrization procedure is based on Bézier curves; A few control points are enough to represent the whole shape and, on the other hand, the smoothness properties of Bézier curves never imply the creation of non feasible shapes by the crossover operator. A Bézier curve of order n is defined by the *Bernstein polynoms* $B_{n,j}$:

$$B(t) = \sum_{i=0}^{n} B_{n,i} P_i \quad \text{with} \quad B_{n,i} = C_n^i t^i (1-t)^{n-i} \quad , \quad C_n^i = \frac{n!}{(n-i)!}$$

where $t \in [0,1]$, $P_i = (x_i, y_i)$ are the coordinates of the control points. In this work, a 8 order Bézier representation has been used :

$$x(t) = \sum_{i=0}^{8} C_8^i t^i (1-t)^{8-i} \, x_i \quad , \quad y(t) = \sum_{i=0}^{8} C_8^i t^i (1-t)^{8-i} \, y_i \qquad (1)$$

Two Bézier curves are defined respectively for the extrados and the intrados; the points $P_0 = (0,0)$ and $P_8 = (1,0)$ are fixed because they correspond to the leading and trailing edges of the airfoil. The values $x_i \in [0,1]$ are fixed and the only parameters that vary are the ordinates y_i. A chromosome is defined by a vector of $\mathbb{R}^{14}$ (see Fig. 1) :

$$chromosome = (\underbrace{y_1, \cdots, y_7}_{extrados}, \underbrace{y_8, \cdots, y_{14}}_{intrados})$$

The first contrainst we use is a geometrical constraint : the y_i's vary on intervals $[min_i, max_i]$; this has a direct implication on the search space. The parameters constituting the chromosomes are binary encoded.

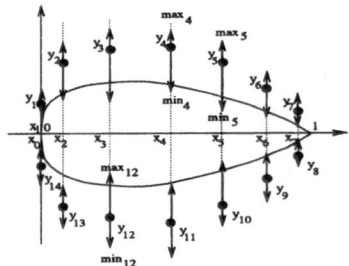

Fig. 1. Representation of an airfoil using Bezier curves

3 The Genetic Algorithm

The basic structure of a Genetic Algorithm consists of the following steps :
(1) initialize randomly a population of airfoils (the individuals), (2) evaluate
the individuals following their fitness (cost functional) value, (3) apply genetic
operators (crossover and mutation) to the population and goto (2) until the best
individual is reached. The genetic operators are the following :

- *Selection process (or reproduction)* : the selection process is based on a tournament approach. Globally, the tournament consists of picking randomly two
  individuals (for a tournament of size 2 or "binary tournament" which is the
  most current) from the population and selecting the best according to its
  fitness. The two individuals are then put back in the population and the
  process is restarted. For more details, see Oei *et al.*[6].
- *Crossover operator* : as reproduction does not create new individuals, the
  crossover operator is needed to increase diversity among the population.
  Crossover proceeds in two steps. First, strings of the newly reproduced population are mated at random ; this crossover operation is made with a probability p_c. Second, each pair of strings undergoes crossing-over as follows :
  an integer position k along the string is selected uniformly at random in
  $[1, l - 1]$, where l is the string length and two new strings are created by
  swapping all characters between positions $k + 1$ and l inclusively.
- *Mutation operator* : the mutation is needed because reproduction and crossover
  can occasionally loose some potentially useful genetic material (1's or 0's at
  particular locations). Mutation is then an insurance policy against premature
  loss of important notions. In the binary coding, this simply means changing
  a 1 to a 0 and vice versa, randomly, with a small probability p_m.

4 The flow solver

The underlying flow solver solves the 2D/3D Euler equations. The flow domain
Ω is assumed to be a polyhedral bounded region of $I\!R^3$. Let $\mathcal{T}_h$ be a standard
triangulation (2D)/tetrahedrization (3D) of Ω. At each mesh vertex S_i, a control

volume C_i is constructed as the union of local contributions obtained from the set of tetrahedra attached to S_i. The union of all these control volumes constitutes a dual discretization of domain Ω. The spatial approximation method adopted here makes use of a mixed finite element/finite volume formulation. Briefly speaking, for each control volume C_i, one has to solve for n *one-dimensional Riemann problems* at the control volume boundary (finite volume part), n being the number of neighbors of S_i. The spatial accuracy of the Riemann solver depends on the accuracy of the interpolation of the physical quantities at the control volume boundary. For first order accuracy, the values of the physical quantities at the control volume boundary are taken equal to the values in the control volume itself. Extension to second order accuracy can be performed via a "Monotonic Upwind Scheme for Conservative Laws" (MUSCL) technique of Van Leer[10] . It consists in giving the Riemann solver a better interpolated value, taking into account some gradient of the physical quantities (finite element part). Time integration relies on a linearised implicit formulation which is best suited to steady flow simulations[2]. We refer to [5] for a detailed description of the flow solver.

5 Parallelization strategy

5.1 Parallelization of the flow solver

The parallelization strategy adopted for the flow solver combines domain partitioning techniques and a message-passing programming model. The underlying mesh is assumed to be partitioned into several submeshes, each defining a subdomain. Basically the same "old" serial code is going to be executed within every subdomain. Applying this parallelization strategy to the previously described Euler flow solver results in modifications occuring in the main time-stepping loop in order to take into account one or several assembly phases of the subdomain results, depending on the order of the spatial approximation and on the nature of the time advancing procedure (explicit/implicit). This approach enforces data locality, and therefore is suitable for all parallel hardware architectures. For the partitioning of the unstructured mesh, two basic strategies can be considered. The first one is based on the introduction of an overlapping region at subdomain interfaces and is well suited to the mixed finite volume/element formulation considered herein. Mesh partitions with overlapping have a main drawback : they incur redundant floating-point operations. The second possible strategy is based on non-overlapping mesh partitions and incur no redundant floating-point operations. While updated nodal values are exchanged between the subdomains in overlapping mesh partitions, partially gathered quantities are exchanged between subdomains in non-overlapping ones. In the present study we will consider one triangle/tetrahedron wide overlapping mesh partitions. We refer to [5] for a detailed comparison between these two approaches.

5.2 Overall optimization loop

The coordination of subdomain calculations through information exchange at artificial boundaries is implemented using calls to functions of the MPI library.

Therefore, the paralellization described above aims at reducing the cost of the fitness function evaluation for a given individual. However, another level of parallelism can clearly be exploited here and is directly related to the binary tournament approach and the crossover operator. In practice, during each generation, individuals of the current population are treated pairwise; this applies to the selection, crossover, mutation and fitness function evaluation steps. Here, the main remark is that for this last step, the evaluation of the fitness functions associated to the two selected individuals, define independent operations. We have chosen to exploit this fact using the notion of process groups which is one of the main features of the MPI environment. Two groups are defined, each of them containing the same number of processes; this number is given by the number of subdomains in the partitioned mesh. Now, each group is responsible for the evaluation of the fitness function for a given individual. We note in passing that such an approach based on process groups will be also interesting in the context of sub-population based PGAs; this will be considered in a future work.

6 Numerical results

We consider now the external Euler flow around a RAE2822 airfoil ($2°$ degree incidence, freestream Mach number fixed to 0.73). The non-linear convergence tolerance has been fixed to 10^{-6}. The cost functional is defined by :

$$j(\gamma) = C_D + 10(C_L - C_L^{comput})^2$$

γ represents the airfoil shape, C_D is the drag coefficient, C_L is the lift coefficient, C_L^{comput} is the computed lift coefficient on the RAE2822. The aim is to reduce C_D while preserving C_L^{comput}. Here we present results for a coarse computational mesh consisting of 1892 vertices (76 vertices on the airfoil) and 3680 triangles. Concerning the GA parameters, the accuracy defining the binary coding of the parameters is equal to 10^{-6}. The probability of crossover is $p_c = 0.8$, the probability of mutation is $p_m = 0.03$ and the population consists of 30 individuals. sharing technique has been used in order to prevent premature convergence to a super-individual (see Goldberg[3] pp. 185-196 for more details). In addition to the geometrical ones, a constraint on the thickness of the airfoil has been taken into account : the thickness of the RAE2822 airfoil is preserved with a 5% maximum variation. After 50 generations, the shape has been modified and the shock has been notably reduced. The initial and final flows (iso-Mach values) are presented on Figure 2. The initial and final values of C_D and C_L are respectively given by $C_L^{comput} = 0.709$, $C_L^{opt} = 0.708$, $C_D^{init} = 0.0078$, and $C_D^{opt} = 0.0050$.

7 Parallel performance results

Preliminary results are presented for the above test case which is relatively challenging from the parallel performance viewpoint because of the way of addressing

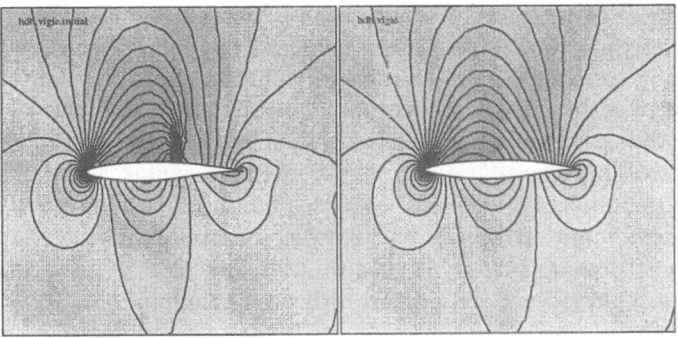

Fig. 2. Drag reduction : initial and optimised flows (steady iso-Mach lines)

the constraint on the airfoil thickness. In practice, when two individuals are selected, mated and evaluated, the Euler flows are effectively calculated only if
the corresponding airfoils do verify the constraint on the thickness. This means
that situations occur such that one of the two individuals is not subject to a
flow calculation while the other does; in other words, taking into account the
geometrical constraint in this simple way has a direct impact on the overall
computational load at a given generation.

7.1 Coarse mesh calculations

A first set of calculations have been performed on a SGI Power Challenge array
equipped with Mips R8000/90 Mhz processors. We have used the native MPI
library provided by SGI. Performance results are given for 64 bit arithmetic
computations. We compare timing measures for the overall optimization using
one and two process groups; the one group case corresponds to a sequential
execution mode (one group consisting of one process). For the two group case
we give results for configurations where each group consists of one, two and
four processes. Timings are given for 10 generations (optimization iterations)
in Tab. 1 where N_g and N_p respectively denote the number of process groups
and the total number of processes ($N_g = 2$ and $N_p = 4$ means 2 processes
for each of the two groups), "Total" is the total execution time and "Flow" is
the accumulated flow solver time; finally $S(N_p)$ is the calculated speed-up. For
the multiple process cases, the given timing measures always correspond to the
maximum value over the per process measures. The rapid degradation of the
speed-up observed here can be explained by the following arguments :

- the sequential part of the overall optimization loop remains quasi-constant.
  The operations accounted here are mainly the genetic operators (selection,
  crossover and mutation) that are executed on the same data (the chromosomes) by each process. The corresponding measure is given by the difference
  between the total execution time and the flow solver time (most of the communication time is spent in the flow solution and is actually accounted for

in the flow solver time). For the case $N_g = N_p = 2$ it represents 7% of the total execution time while for the last case this figure has raised to 12%;

- if we look in more details to the flow solution times for the last case , we observe that the faster process has spent 472 sec in this part of the computation; thus, there is a 9% load imbalance between the slower and the faster processes. This degradation of the parallel performances of the flow solver is even more important in the 3D case (see Lanteri[5]) and is mainly attributed to the redundant arithmetic calculations incurred by the overlapping in the mesh partitioning. In the 3D case we therefore plan to use a parallel solver based on a partitioning approach with no overlapping.

Table 1. Drag reduction : parallel perfomance results on a SGI **Power Challenge** array

N_g	N_p	Total	Flow	$S(N_p)$
1	1	2580 sec	2520 sec	1
2	2	1510 sec	1402 sec	1.7
2	4	877 sec	787 sec	3.0
2	8	587 sec	512 sec	4.4

7.2 Fine mesh calculations

We consider the same application using a mesh consisting of 8220 vertices (160 vertices on the airfoil) and 16000 triangles. We again give timings for 10 generations. Calculations are now performed on a cluster of PC P6/200 **Mhz** running the Linux operating system. The same code has been compiled usig the G77 gnu compiler with maximal optimization options (64 bit arithmetic computations). The cluster nodes are currently interconnected via a Fast Ethernet hub. Communications are handled using MPICH (version 1.1.0). Results are summarized in Tab. 2 where "Elapsed" denotes the total elapsed execution time and "CPU" denotes the total CPU time (taken as the maximum value over the local measures); the speedup is now measured based on the total elapsed execution times. We note that for the case $N_g = 2$ and $N_p = 8$ (i.e. 4 processes for each of the two groups) the total CPU time represents 82% of the total elapsed execution time which demonstrates that the concept of cluster computing is becoming an interesting alternative to true parallel computing on an integrated MIMD system. To illustrate this point, we conclude this section by giving in Tab. 3 a few timing results measured on a SGI **Origin** 2000 system equipped with **Mips R10000/195 Mhz** processors.

8 Future works

Future works concern the extension of the present methodology to the optimization of 3D geometries (wings). The parallel Euler flow solver is already existing[5]

Table 2. Drag reduction : parallel perfomance results on a Fast Ethernet PC cluster

N_g	N_p	Elapsed	CPU	Flow	$S(N_p)$
1	1	49980 sec	49691 sec	49596 sec	1
1	4	14760 sec	14015 sec	13886 sec	3.4
2	4	18720 sec	16296 sec	16010 sec	2.7
2	8	10920 sec	8956 sec	8817 sec	4.6

Table 3. Drag reduction : parallel perfomance results on a SGI Origin 2000 system

N_g	N_p	CPU	Flow
2	4	3117 sec	2781 sec
2	8	1379 sec	1298 sec

and the main problem will concern the construction of a shape parametrization procedure that will have to deal with an unstructured surface triangulation. In addition, future works will also address sub-population based PGAs[1].

References

1. Cantù-Paz, E: A summary of research on parallel genetic algorithms. University of Illinois at Urbana-Champaign, IlliGAL Report 95007 (1995)
2. Fezoui, L., Stoufflet, B.: A class of implicit upwind schemes for Euler simulations with unstructured meshes. J. of Comp. Phys. **84** (1989) 174-206
3. Goldberg, D.E.: Genetic algorithms in search, optimization and machine learning. Addison-Wesley Company Inc. (1989)
4. Holland, J.-H.: Adaptation in natural and artificial systems. MIT press/Bradford books, Cambridge, Massachussets (1992)
5. Lanteri, S.: Parallel solutions of compressible flows using overlapping and non-overlapping mesh partitioning strategies. Parallel Comput. **22** (1996) 943-968
6. Oei, C.K., Goldberg, D.E., Chang, S.J.: Tournament selection, niching and the preservation of diversity. University of Illinois at Urbana-Champaign, IlliGAL Report 91011 (1992)
7. Périaux, J., Sefrioui, M., Stoufflet, B., Mantel, B., Laporte, E.: Robust genetic algorithms for optimization problems in aerodynamic design. Genetic algorithms in engineering and computer science, G. Winter *et al* Eds., John Wiley & Sons (1995) 371-396
8. Poloni, C.: Hybrid GA for multi-objective aerodynamic shape optimization. Genetic algorithms in engineering and computer science, G. Winter *et al* Eds., John Wiley & Sons (1995) 397-415
9. Quagliarella, D.: Genetic algorithms applications in computational fluid dynamics. Genetic algorithms in engineering and computer science, G. Winter *et al* Eds., John Wiley & Sons (1995) 417-442
10. Van Leer, B.: Towards the ultimate conservative difference scheme V : a second-order sequel to Godunov's method. J. of Comp. Phys. **32** (1979) 361-370

Parallel Multidimensional Calculation of Steady-State and Time-Dependent Flows with Combustion

Samir Muzaferija[2], Volker Seidl[2], Hannes Fogt[1], and Aron Kneer[1]

[1] Battelle Ingenieurtechnik GmbH, Düsseldorfer Str. 9, D-65760 Eschborn
[2] Institut für Schiffbau, Lämmersieth 90, D-22305 Hamburg

Abstract. Calculations of flows with combustion in complex geometries impose high computational demands. This applies in particular to highly transitional flow, where accuracy requirements cause long computation time. Parallelization in space and time, based on domain decomposition techniques is applied and reduction on turn-around time is documented. On networked computers, large applications – beyond present single processor workstation capacity – can thus be solved. The approach presented has been implemented and benchmarked on different platforms. Using encapsulated message passing routines based on PVM and MPI libraries, it is possible to port the code on different hardware platforms ranging from workstation clusters up to massively parallel computers. For two examples, results are presented and performance issues are discussed.

1 Introduction

Fluid flows combined with combustion in complex geometries are evident in a multitude of industrial processes. Furnaces of steam generators and power plants, flow in reciprocating engines and gas turbines are typical examples.

A combustible gas can unintendedly be set free in buildings, which can be damaged or destroyed if the premixed gases are ignited. Experimental studies in medium and large scale multiroom buildings – filled with a mixture of air and hydrogen – have been carried out to investigate flame propagation and transient pressure loads. The experiments revealed that maximum pressure is basically influenced by local gas composition, position of the ignition point, and by the size and arrangement of rooms, vents and obstructions. Under adverse conditions, relatively high loads may be encountered even at low gas concentrations.

To investigate various accident scenarios and the effectiveness of counter-measures, numerical calculations of the transient combustion of premixed gas in buildings are employed. Combustion characteristics are strongly influenced by small scale geometry; computational grids must be sufficiently fine to take into account such details. Numerous control volumes and large systems of equations are the consequence, increasing CPU-time and memory requirements, in particular for three-dimensional applications. Rapid change of flow properties enforce very small time steps, which further increase computational time. Therefore, parallel computation was employed to enable the calculation of such problems using available computer resources.

In the present work, domain decomposition in both space [4] and time [6] is employed. Both workstation clusters and high performance platforms (Parsytec Xplorer with 16 processors and Cray T3D with 64 processors) have been addressed. Implementation of the *Single Process Multiple Data* (SPMD) [8] programming model with encapsulated message passing routines allows easy porting of the code and permits adaption to forthcoming message passing standards without changes of the main program.

2 Description of the CFD method

The integral formulation of conservation of mass, momentum, energy, and additional species (in this case hydrogen, oxygen, vapor as the reaction product, nitrogen, and – if necessary – inhibitor gas) constitute the mathematical model to be solved. Radiative heat transfer is neglected as short term solutions are sought. Turbulence effects are accounted for using standard k-ε–model and wall-functions. Different combustion models – of Arrhenius type for almost laminar flow during initial phase and the Magnussen model for fully turbulent flow – are applied to calculate local reaction rate.

The numerical program consists of two fundamental elements: The module COMET [1], [2] is a general purpose CFD solver. It is based on the finite-volume discretisation, using unstructured computational grids, which facilitate an efficient meshing of complicated geometries, keeping the number of grid points to the minimum. In addition, local grid refinement permits high numerical resolution in the vicinities of vents and obstructions without excessive increase of computational requirements. The solution procedure consists of a segregated algorithm of the SIMPLE type [9]. In this method inter-equation coupling and non-linearities are resolved iteratively in a predictor-corrector fashion within *outer iterations*. Linear equation systems are relaxed by Conjugate Gradient (CG) based solvers in *inner iterations* [7]. The program is equipped with a general user interface which permits individual modifications and attachments.

Depending on local flow conditions, the module COMBUST [3] calculates local fluid properties and chemical reaction rate and feeds them back to the fluid solver. Thus, formation and consumption of species and heat generation are taken into account while solving transport equations for mass and energy.

3 Parallelization Strategy

The concurrent algorithm is based on data parallelism in both space and time and has been implemented under the SPMD paradigm using message passing [5].

For space-parallelism the solution domain is decomposed into non-overlapping subdomains [4]. The assembly of the coefficient of the coefficient matrices and source terms in outer iterations can be performed in parallel as only variable values from the previous outer iteration are required. In inner iterations, the preconditioning of the CG Solver is performed locally in each domain, neglecting the coupling between subdomains. After the interface data between neighboring domains have been updated, the CG sweep itself is performed as in serial execution with global communication to calculate scalar products, thus preserving largely the convergence properties of the overall method.

For time parallelism, multiple time levels are considered simultaneously resulting in an enlarged system of equations which can be decomposed with respect to time: in an outer iteration the preliminary solutions from preceding time levels are used [6]. These temporary solutions are updated in subsequent time levels as soon as better solutions are available between outer iterations.

A combination of space and time parallelism is especially useful for problems of moderate size, where one strategy alone would be less efficient.

4 Applications

4.1 Example 1

Many production processes need large volumetric flow rates of hot gas, which is generated in combustion chambers. For proper operation, homogeneous temperature profile and velocities of the flue gas are required in subsequent plant sections. To inhibit unintended reaction of the hot gas with the other components involved in the process, the percentage of unburned gas is strictly limited. Lack of space frequently dictates to arrange the combustion chamber, the supply pipes and the exhaust channels in close vicinity, which gives rise to adverse operating conditions. Numerical calculations are applied to optimize the geometry and to improve the process. A simplified model of a thermal reactor is presented in Figure 1. The overall length is approximately 6 m and the diameter 1.5 m. Air is supplied almost in radial direction through the large inlet on the left side. Combustible gas is added through the central supply tube and expelled radially through orifices, where reaction with the air takes place. The flue gas is directed to the outlet, where it leaves the chamber in axial direction. The computational grid consists of nearly 400 000 Control-Volumes.

4.2 Example 2

Application of the program to safety analysis is demonstrated. Figure 2 illustrates a building which consists of three rooms. The length of the upper room is 22 m, the total volume of the building is approximately 4400 m^3. The rooms are connected by several horizontal and vertical vents. Obstacles and flow obstructions are present in both of the lower rooms. The numerical grid consist of roughly 100 000 Control-Volumes.

All rooms are initially filled with a mixture of 90 vol.% air and 10 vol.% hydrogen. Ignition is initiated close to one of the vents which connect the lower rooms. Short after ignition, the flame is driven by buoyancy and propagates at low velocity. Further grow of the flame causes increase of pressure and induces

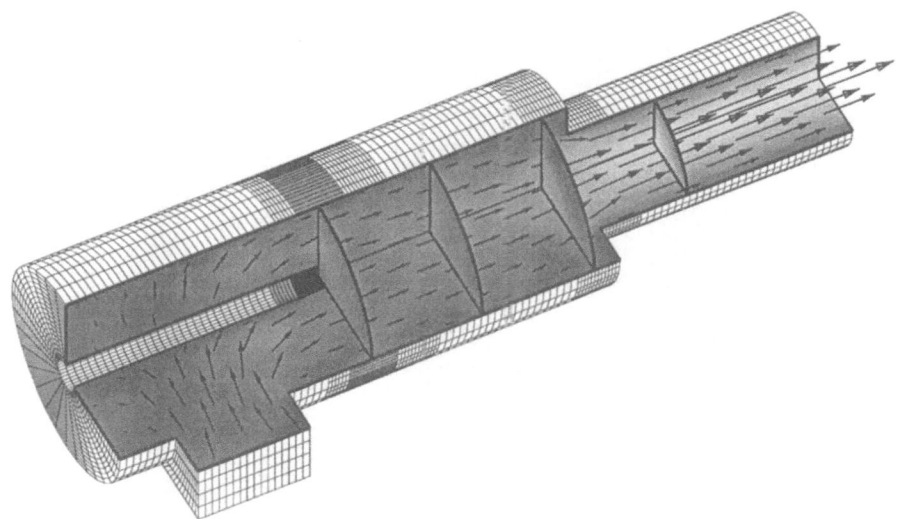

Fig. 1. Computational grid (domain decomposed) and fluid velocity in a combustion- and mixture chamber.

flow through the connecting vents. When the flame approaches the vent, local flow velocity increases significantly. The flame is sucked to the second room, where it meets a region of high turbulence. Forceful combustion, followed by a pressure rise in the second room are the consequences. Gas flow to the upper room and backflow to the ignition room occur. Reverse in flow direction leads to jet ignition in the first room with similar pressure rise. Calculated loads are in accordance with measurements, which have been carried out in large and medium size test facilities.

5   Parallel Performance

Benchmark calculations have been carried out to determine the performance of the parallelized program on different hardware, ranging form networked general purpose computers to high performance platforms:

- A cluster (four to eight workstations Unix HP Series 700) connected via PVM
- Parsytec Xplorer/16 with 8 and 16 processors
- Cray T3D with 64 processors

Comparison was made using identical programs. No specific adaption to hard- ware was made. Calculations started at relatively coarse meshes. For steady state

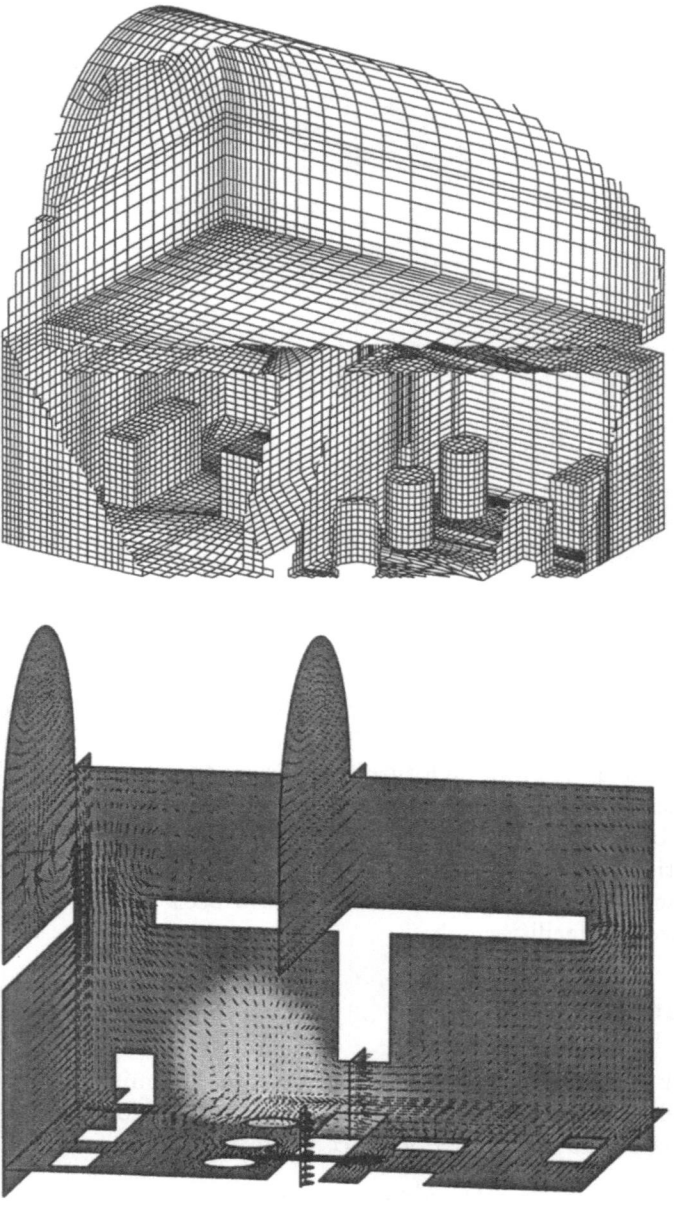

Fig. 2. Combustion induced flow in a building. Computational grid. Temperature and velocity vectors.

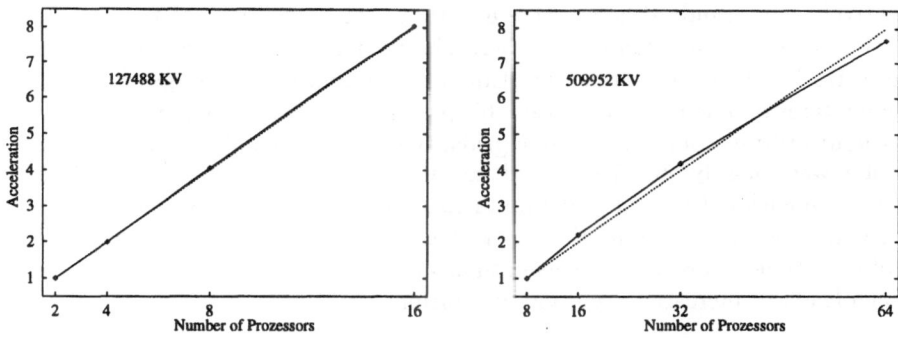

Fig. 3. Speed-up for two different grid sizes as a function of the number of processors.

calculations, the grids where gradually refined up to 432 000 Control-Volumes for the workstation cluster and up to a maximum node number of 510 000 for calculations on Cray T3D. Total efficiency for the parallelized program was obtained by comparing calculation time with the minimum number of processors necessary for that problem size. In general, CPU time can not be expected to decrease linearly with growing number of processors for the following major reasons:

- Partitioning and distribution of the load on several processors results in time-consuming data communication.
- Imperfect load balancing causes synchronization losses and decreased performance.
- Depending on the selected solution method, the numerical efficiency of the parallelized algorithm may be inferior to the serial version.

The measured speed-up on Cray T3D is, for two grid sizes and different number of processors, shown in Figure 3. On the smaller grid, the speed-up is almost linear up to 16 processors. The performance of the parallelized algorithm is thus almost ideal: increasing the number of processors results in a proportional decrease of CPU time. The efficiency on the larger grid was roughly 92% for 64 processors. The observed super-linear speed-up is due to a better memory management and performance of each individual processor for smaller problem size (so-called cache effects). Extrapolating this tendency to 128 processors, especially when the grid size is also increased, indicates good scalability even on massively parallel systems.

Identical measurements could not be carried out on networked computers, as the workstations where not isolated. Interference with other applications occurred, imposing relatively high uncertainties on performance data obtained. The tests where carried out on a cluster of heterogeneous computers, each of them with different speed and memory equipment. Without individual load balancing and specific optimization of block size and boundaries, additional synchronization losses occur. Processor to processor communication was based on relatively slow standard Ethernet connection, which was exposed to normal traffic. Large communication losses where observed for small grids, as in this case the ratio between communication time and computation is large. For realistic applications on fine grids, however, this performance loss turned out to be small. Without optimization, but for fine grids, the total efficiency on the workstation cluster was roughly 50 – 70%, resulting in a substantial decrease of CPU time. Three-dimensional unsteady computations of deflagration thus can be carried out within acceptable time. Equally important is the fact that now much larger problems (finer grids) can be solved than was possible on any single workstation, which leads to increased accuracy of solutions.

6 Conclusions

A parallel numerical method for the calculation of combustion-induced pressure loads in buildings has been adopted. Using finite-volume discretisation on unstructured and locally refined grids, accurate numerical predictions accounting for detailed flow features are possible with a minimum number of cells. Grid generation is facilitated by interfaces to popular grid generators. Combustion models – thoroughly verified in earlier two-dimensional applications – have been incorporated.

Efficiency of the parallel program based on domain decomposition was investigated on different hardware platforms. On massively parallel computers, an efficiency of approximately 90% was achieved. Performance measurements on networked computers are less documented. However appreciable speedups (between 50 to 70%) have been observed on systems exposed to typical load during office hours. Therefore, parallel technology opens up possibilities to solve large problems on networked computers, which are frequently available at industrial enterprises. These gains encourage sophisticated combustion modeling and offer new opportunities for three-dimensional calculations.

References

1. I. Demirdžić and S. Muzaferija. Numerical method for coupled fluid flow, heat transfer and stress analysis using unstructured moving meshes with cells of arbitrary topology. *Computer Methods in Applied Mechanics and Engineering*, 125:235 – 255, 1995.
2. I. Demirdžić, S. Muzaferija, and M. Perić. *Advances in Computation of Heat Transfer, Fluid Flow and Solid Body Deformation using Finite-Volume Approaches*, chapter 2. Advances in Numerical Heat Transfer. Taylor and Francis, 1996.
3. H. Fogt and A. Kneer. Entwicklung eines parallelen Berechnungsverfahrens für teilchenbeladene Strömungen mit Verbrennung. Technical report, Battelle Ingenieurtechnik GmbH, Dec 1996.
4. E. Schreck and M. Perić. Computation of fluid flow with a parallel multigrid solver. *International Journal for Numerical Methods in Fluids*, 16:303 – 327, 1993.
5. V. Seidl, M. Perić, S. Schmidt. Space- and Time-Parallel Navier-Stokes Solver for 3D Block-Adaptive Cartesian Grids. In A. Ecer, J. Periaux, N. Satufako, S. Taylor (eds), Parallel Computational Fluid Dynamics, 577-584, Elsevier, Amsterdam, 1995.

6. G. Horton. TIPSI – A Time-Parallel SIMPLE-Based Method for the Incompressible Navier-Stokes Equations. In K.G. Reinsch et al. (eds), Parallel Computational Fluid Dynamics, Elsevier, Amsterdam, 1991.
7. H.A. Van der Vorst and P. Sonneveld. CGSTAB: A more smoothly convergent variant of CG-S. Technical Report 90-50, Delft University of Technology, 1990.
8. M.J. Flynn. Some computer organizations and their effectiveness. IEEE Transactions on Computers, C-21(9), 948-960, 1972.
9. S.V. Patankar. Numerical Heat Transfer and Fluid Flow. McGraw-Hill, 1980.

A Two-Level Parallel Strategy for Rotorcraft Optimization and Design

Joseph W. Manke[1], Thomas M. Wicks[1],
Leo Dadone[2], Joel E. Hirsh[2], Byung Oh[2]

[1] Boeing Information and Support Services, Research and Technology Division
P.O. Box 3707, MS 7L-21, Seattle, WA 98124-2207
[2] Boeing Defense and Space Group, Helicopters Division,
P.O. Box 16858, MS P32-74, Philadelphia, PA 19142-0858

Abstract. We present results from our work to explore the viability of using parallel computers to solve multidisciplinary problems relevant to the definition of low noise, high-performance, low vibration rotorcraft. Our work has addressed issues related to the performance and scalability of parallel computers with respect to the rotor design process. We describe our recent work to couple parallel versions of a comprehensive rotor analysis code, a rotor aerodynamics code and a rotor optimization code to implement a two-level parallel rotor optimization strategy.

1 Introduction

Designing advanced rotorcraft has typically required performance, vibration and noise improvements which cannot be achieved without the accurate modeling of extremely complex rotor flow environments, which are also associated with complex blade motions and elastic deflections. Advanced features, like high performance, low vibration and low noise, require better modeling of the underlying physical phenomena. Designing for cost and safety calls for better interdisciplinary methods and integration with manufacturing processes and maintenance requirements. More accurate models of the physical phenomena, the expansion of analysis into multidisciplinary applications and, ultimately, optimization call for significantly larger computing resources and more robust procedures than commonly available today. In a three-year research project, we have explored the viability of using parallel computers to provide the needed computing resources.

2 Rotor Design and Analysis Tools

Figure 1. illustrates an ideal multidisciplinary design and optimization process combining comprehensive rotor analysis, rotor aerodynamics, and aeroacoustics. Boeing Helicopter's rotary wing analysis code TECH-01 [4] computes blade natural frequencies, rotor system performance, hub and control system loads (steady and vibratory), blade loads, motions and aerodynamic loading and associated parameters. The wake modeling includes a classical prescribed wake as well as a free wake [1]. The rotor aerodynamics code, FPR [6, 2] computes the aerodynamic

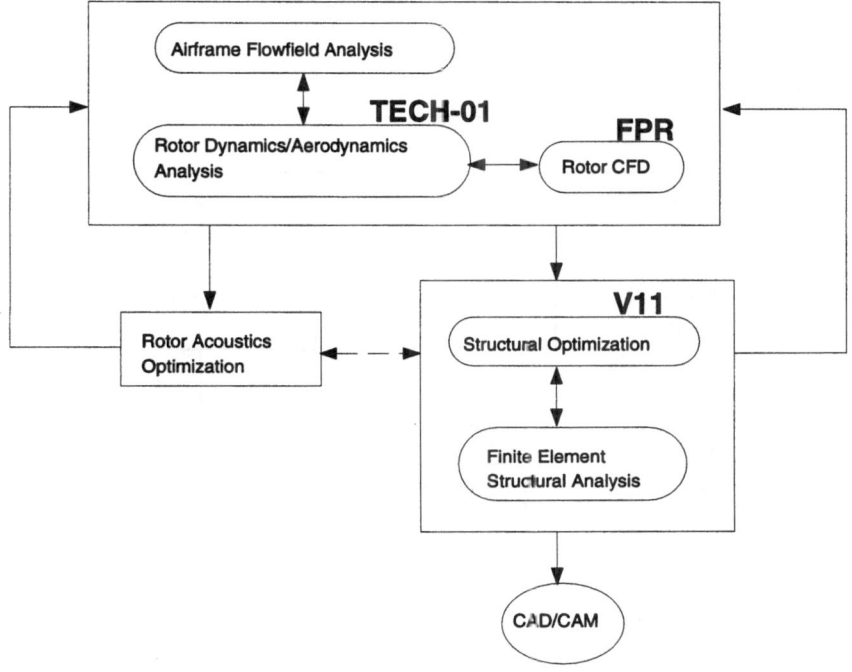

Fig. 1. Helicopter Multi-Disciplinary Design and Optimization Process

flow parameters on and in the immediate vicinity of a rotating wing. The blade is modeled as a spanwise series of body-fitted O-grids whose motion can be translational or rotational or combinations thereof making it ideally suited for rotary wings at transonic Mach numbers. The method is useful for blades of arbitrary planform at low to moderate angles of attack. The structural optimization code V11 couples TECH-01 with optimization procedures that allow the search for feasible blade structural properties which minimize blade response in the presence of vibratory airloads. The optimization method is a sequential quadratic programming method using the optimization code NPSOL [5]. We have implemented parallel versions of TECH-01, FPR and V11 and benchmarked them on the 160 processor IBM-SP2 at NASA Ames Research Center. We focused the parallelization effort on the most computationally intensive part of each code: the vortex method used for the free wake model in TECH-01, the approximate factorization (AF) solver used in FPR and the forward finite difference method used for the gradient calculation in V11.

Comprehensive rotor analysis codes, such as TECH-01, make use of tabulated sectional characteristics in the evaluation of the local blade airloads. While the determination of blade airloads from tabulated airfoil data is fast, it is not sufficiently accurate in the vicinity of the blade tip, and additional three-dimensional corrections, or "tip relief" effects, have to be accounted for by empirical means. The problem is that none of the existing empirical tip corrections are sufficiently

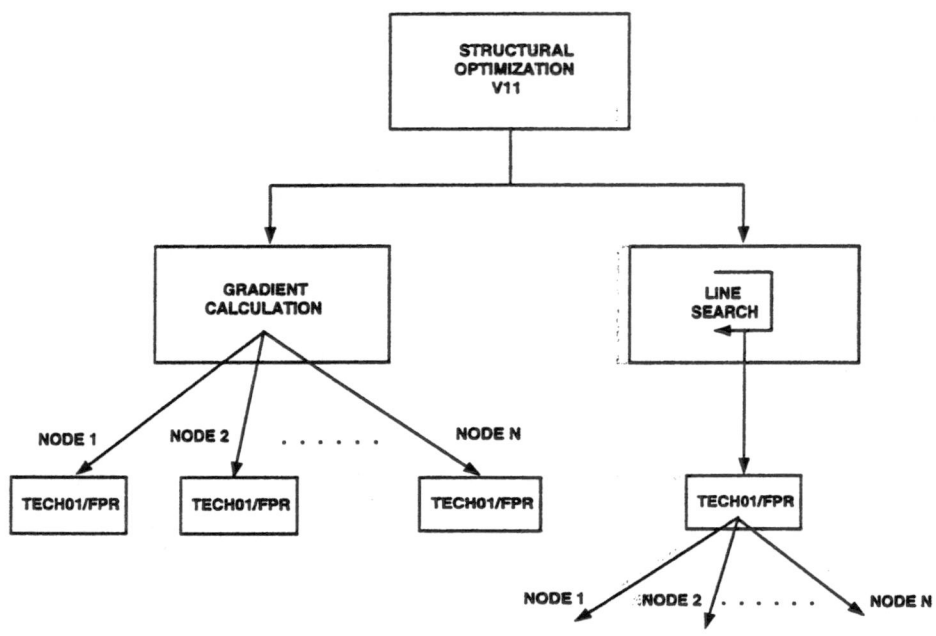

Fig. 2. Two-Level Parallel Strategy for Structural Optimization

accurate to support the design of advanced blade tips. We addressed this problem by coupling FPR to TECH-01 [3] and using FPR to model the aerodynamics of the blade tip. In effect, we replaced the empirical tip corrections by accurate tip corrections predicted by FPR. We also extended the coupling method to the parallel versions of TECH-01 and FPR.

3 Two-Level Parallel Rotor Optimization Strategy

The rotor optimization process illustrated in Figure 1. has two computational levels. The upper computational level is the optimization method performed by V11, which involves a gradient calculation followed by a line search in the direction of steepest decent. Both the gradient calculation and the line search require repeated evaluations of the objective function. The lower computational level is the evaluation of the objective function, which is performed by the coupled TECH-01/FPR code.

The gradient calculation, which uses a forward finite difference method, requires one evaluation of the objective function for each component of the gradient. The function evaluations are completely independent. Thus, we can parallelize the gradient calculation at the upper upper computational level by distributing the function evaluations. We perform the function evaluations in parallel using the parallel coupled TECH-01/FPR code on one (or possibly a few) processors. The line search requires a sequential evaluation of the objective function at selected points in the direction of steepest decent. Thus, we can parallelize the line search only at the lower computational level. We perform the sequential function evaluations using the parallel coupled TECH-01/FPR code on as many processors as possible. The resulting two-level parallel rotor structural optimization strategy is illustrated in Figure 2.

We have implemented this two-level parallel strategy on the 160 processor IBM-SP2 at NASA Ames Research Center. We used our benchmark data on the parallel versions of TECH-01, FPR and V11 to tune the two-level method to use the available processors efficiently for both the gradient calculation and the line search.

4 Conclusions

We have described our implementation of a two-level parallel rotor optimization strategy which couples parallel versions of a comprehensive rotor analysis code, a rotor aerodynamics code and a rotor optimization code. This work has carried us closer to our goal of realizing a robust structural optimization process for the design of low noise, high performance and low vibration rotorcraft.

References

1. D. BLISS, D. WACHSPRESS, AND T. QUAKENBUSH, *A New Methodology for Helicopter Free Wake Analysis*, in 39th Annual Forum of the American Helicopter Society, May 1983.
2. J. BRIDGEMAN, R. STRAWN, F. CARADONNA, AND C. S. CHEN, *Advanced Rotor Computations with a Corrected Potential Method*, in 45th Annual Forum of the American Helicopter Society, May 1989.
3. F. X. CARADONA, *Application of Transonic Flow Analysis to Helicopter Rotor Problems*, in Unsteady Transonic Aerodynamics, D. Nixon, ed., AIAA, 1989, pp. 263–285.
4. L. DADONE, R. C. DERHAM, B. K. OH, B. PANDA, L. A. SHULTZ, AND F. J. TARZANIN, *Interdisciplinary Analysis for Advanced Rotors - Approach, Capabilities and Status*, in American Helicopter Society Aeromechanics Specialists Conference, January 1994.
5. P. GILL, W. MURRAY, M. SAUNDERS, AND M. WRIGHT, *User's Guide for NPSOL (Version 4.0): A Fortran Package for Nonlinear Programming*, Tech. Rep. Stanford University Technical Report SOL 86-2, Department of Operations Research, Stanford University, 1986.
6. R. STRAWN AND F. CARADONA, *Conservative Full Potential Model for Unsteady Tran-sonic Rotor Flows*, AIAA Journal, 25 (1987), pp. 193–198.

Workshop 15

Scheduling and Load Balancing

Workshop 15: Scheduling and Load Balancing

Vipin Kumar, Reinhard Lüling and Catherine Roucairol

Mapping a parallel computation onto a parallel computer system is one of the most important questions in the design of efficient parallel algorithms. Especially for irregular data structures the problem of distributing the workload evenly onto a parallel system can become very complex. A large number of mapping, load balancing and scheduling problems have been investigated in the past.

To understand the different mapping, load balancing and scheduling problems, we will give some definition of the different problems and classify the papers presented within the workshop according to this framework. To do this, we view a parallel system with P processors as a graph $H = (U, F)$ with nodes $U = \{0, \ldots, P - 1\}$ and edges $F \subseteq U \times U$. We consider only homogeneous systems. Thus, we omit node and edge weights which would be needed to model heterogeneous machines.

A parallel application is modeled as a graph $G = (V, E)$ with nodes $V = \{0, \ldots, N - 1\}$, edges $E \subseteq V \times V$, node weights and edge weights. The meaning of nodes and edges can differ from application to application. For example, nodes can describe processes or data items, edges can stand for data dependencies or communication demands.

Using this kind of model, load balancing can be viewed as a graph embedding problem. The task is to find a mapping $\pi : G \rightarrow H$ of the application graph to the processor graph minimizing certain cost criteria. Depending on the application, the graph G can be *static* or *dynamic*, i.e., the computational load of the application nodes may change during run time, or not. The processor graph is usually considered to be static.

Static applications have to be mapped only once onto the processor network and do not change during run time. Thus, the load balancing problem can be reduced to a classical mapping problem where one graph has to be embedded into another. Most applications from the area of scientific computing behave statically in this context. Examples are all kinds of non-adaptive FEM-simulations.

Dynamic load balancing problems occur if the application graph changes during run time. The graph may grow or shrink, i.e., nodes and edges might be inserted or deleted, or the node and edge weights may vary. Load balancing problems can be classified according to the weight functions of the application graph. Especially the *granularity* of an application is of importance for the choice of the right balancing algorithm. Fine-granular applications consist of a large number of light-weighted nodes representing small processes or data items. Coarse-granular tasks are those with a small number of heavy processes.

The distinction often determines whether an application node can be migrated or not. Heavy processes are usually difficult to move from one processor to another during run time, while data items can be transfered without problems. Accordingly, we classify dynamic load balancing problems depending on the communication structure of the application and on the possibility to migrate processes.

The *Dynamic Mapping Problem* usually appears in the context of client-server applications. During run time, processes are generated dynamically and have to be placed onto processors. There are almost no dependencies between them (thus, only little communication), but they also cannot be migrated. Once placed, a process has to stay on "its" processor until termination. Load balancing algorithms have to take into account the tradeoff between increasing overhead (e.g., to gather information about the system state) and increasing balancing quality.

If there are strong communication dependencies between the processes and they additionally cannot be moved, the placement problem becomes quite difficult. This so-called *Dynamic Embedding* problem occurs in dynamic tree-search applications like configuration systems or game tree search.

The *Dynamic Load Balancing Problem* is *the* classical load balancing problem. For applications where migratable jobs without communication dependencies are generated dynamically, a large number of efficient load balancing algorithms are known.

Problems of the fourth class, the *Re-Embedding Problems*, usually occur in scientific computing applications. The application graph changes during run time, communication dependencies have to be considered, but nodes can be migrated between processors. Examples of such applications are adaptive finite element simulations where the mesh is refined depending on the solution.

Within the workshop, a number of problems described above are considered.

The paper presented by J. Galtier focuses on the graph partitioning of 3D meshes. These meshes arise in a large number of applications solved by numerical methods. The approach presented here is very attractive for practical applications as it considers the interfaces between subdomains. An evaluation using industrial benchmarks is presented in the paper. Applications coming from the area of numerical analysis are also targeted in the paper by F. d'Amore et. al. Here a number of techniques are presented to embed a 2-dimensional FEM graph into a grid. Some of the results are based on upper bounds, others are heuristics that are investigated using benchmark situations.

The Re-Embedding Problem, as defined above, is considered in the paper of A. Gerasoulis et. al. In this paper a number of scientific computing applications in the area of Computational Fluid Dynamics are studied. To remap these applications after refinement steps or after generating new tasks onto the processor architecture, two new schemes are presented and validated using experiments.

Dynamic Load Balancing Problems are investigated in the papers of A. Corradi et. al. and S. K. Das. Corradi et. al. present a number of different load balancing methods that are all based on the Diffusion Scheme. This scheme has already been found of interest in previous work. The paper presents a detailed investigation and comparison of some variations of the scheme. The paper of d'Amore et. al. investigates load balancing methods on symmetric broadcast networks. This network is mapped statically over the processor network and provides a topology-independent communication platform for the load balancing algorithm. Three load balancing algorithms based on this model are presented and evaluated using benchmark scenarios.

A number of other papers presented in the workshop study the scheduling of parallel tasks onto a processor network. X. Du et. al. study this problem for a network of workstations. Taking the special characteristic of these types of parallel computers into account, a new scheduling algorithm is presented and validated using the NAS benchmark suite. W. Löwe et. al. study the problem of scheduling of task graphs under the LogP model.

Performance Comparison of Load Balancing Policies based on a Diffusion Scheme

Antonio Corradi
DEIS - Università di Bologna
2, Viale Risorgimento - I-40136 Bologna
acorradi@deis.unibo.it

Letizia Leonardi, Franco Zambonelli
DSI - Università di Modena
213/b, Via Campi - I-41100 Modena
{leonardi, zambonelli}@dsi.unimo.it

Abstract

In the area of load balancing policies for massively parallel architectures, several load balancing policies can be inspired from the idea of diffusion: the paper defines a few basic strategies with different scope of locality and evaluates them depending on the properties of the system load.

1. Introduction

In dynamic applications, **load balancing** decisions must occur at execution time: these decisions must take into account the current system state and the application needs [3]. The main goal is prompt decisions that can be on-line employed to steer the system. The distributed nature of massively parallel architectures suggests decentralised control that avoid bottlenecks; distributed protocols, however, pay the cost of coordination toward a common goal. After ruling out a global perspective [1], solution that lacks of scalability, the **local approach** restricts each decision centre to a local view of the system to limit the coordination efforts [4].

Therefore, the load balancing (LB) algorithms presented in this paper base their decisions on limited load information: the achievement of the global load balancing goal derives from the composition of independent and asynchronous local actions. This is similar to the phenomenon of **diffusion**, that balances non homogeneous distributions, on the basis of the local state, by locally moving elementary items on the direction of a local balance. Similarly, a distributed LB policy can balance the load on the basis of local load information and by locally moving entities along the directions of local load imbalance. In particular, an LB policy is diffusive when [2]:

- it is based on **replicated decision** components all with the same behaviour;
- the scope of the actions for each component is bound to a local area of the system (**local domain**);
- the **LB goal is locally driven**; replicated components try to balance the load in their local domain only on the basis of load information limited to their domain;
- the whole system is covered by the union of these domain (**seamless coverage**).

The paper presents different load balancing algorithms all modelled after the diffusion, with different coordination degrees in their neighbourhood, i.e., different size of the local domain. The paper evaluates these algorithms and analyses their performance in balancing different load situations with different load patterns.

2. Basic Diffusive Algorithms

This section presents several diffusive LB policies characterised by different definitions of local **domain**. In the presentation of the policies, we refer to one policy handler, called Allocation Manager (AM for short), replicated in each node of the massively parallel system. Each AM is in charge of implementing the allocation policy by coordinating itself with other AMs.

2.1 The Direct Neighbour (DN) Policy

The direct neighbour policy (DN for short) minimises the domain size: each domain consists of only two nodes directly connected by a physical link (figure 1a). The presence of nodes interconnected to several other ones, thus belonging to more domains, achieves the seamless coverage. A balancing action within one local domain strives to equalise the load of the two component nodes.

Whenever a node receives updated load information from one of the neighbour nodes, the AM compares it with the current local load. If the local load exceeds the load of the neighbour by more than a threshold (i.e., a percentage), the AM starts a migration action.

The migration of entities takes place only if it does not reverse the roles of the involved nodes. In other words, when two neighbour nodes exchange load, the overloaded node cannot become underloaded in its turn and vice versa. This condition guarantees that balancing actions go always in the direction of balancing the load in one domain and avoid instability, even with entities of coarse granularity.

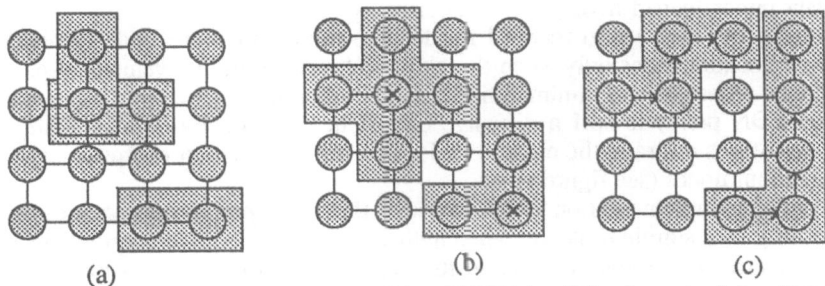

(a)   (b)   (c)

Figure 1. Domains in the DN (a), AN (b) and EDN (c) policies for a mesh topology

2.2 The Average Neighbourhood (AN) Policy

In the AN policy, one domain is constituted by one node and all its direct neighbours (figure 1b). If N is the number of the direct neighbours of a given node, that node belongs to at most to $N+1$ different domains: in one domain it acts as a controller (we call it the *master* of this domain), being a direct neighbour of all other nodes of the domain; in the others it plays as a peripheral node (*slave* node). The goal of the AN policy is to balancing the load of all the nodes within a domain.

The AM in one node is in charge of balancing the domain D it is the master, on the basis of the average domain load. This solution is natural, because the master is the only node in the domain D able to directly control each node of D.

The AM of the master node holds load information about every node of D, and computes the average domain load and two thresholds, T_{sender} and $T_{receiver}$, centred around it. A node is classified as sender in D if its load exceeds T_{sender}, receiver if its load is under $T_{receiver}$, neuter otherwise. Any time the master node receives load information, the AM calculates the new average domain load, updates T_{sender} and $T_{receiver}$ and evaluates the load situation of D. Then, the AM of the master node tries to push (and/or pull) the load of all the nodes in its domain (and not only the load of the master node itself) as close as possible to the average domain load. As in the DN policy, a migration must not reverse the role of nodes in the domain to grant stability.

2.3 The Extended AN Policy

The AN strategy defines domains in a direct neighbourhood relation with a master node: one can think to enlarge the size of domains. A domain can be defined as the set of nodes whose distance from a centre node is less than n. When n is 1, the strategy is the AN, when n is 2 the strategy can be defined as average neighbourhood of second level, AN-2 for short, and so on (AN-n strategy). Though still diffusive, the implementation of an AN-n strategy introduces higher implementation problems because it requires communication between non-neighbour nodes and because the enlarged domain size can it make difficult for the master to control the allocation for a larger number of nodes. For these reasons, we limit our considerations to a AN-2 policy.

2.4 The Extended DN Policy

The extended direct neighbour (EDN) algorithm exploits additional information to locate the destination of migrating entities. In the DN policy, once a sender-receiver couple is established, the moved load can be allocated only on the receiver node; however, the receiver node can see, in a different domain to which it belongs, an even under loaded node. The EDN strategy takes advantage of this situation by permitting the receiver to forward the load to more and more underloaded nodes. Load reallocation stops only when there are no longer useful movements, i.e., a node is reached whose load is minimal in its neighbourhood.

The EDN policy is still a diffusive one, with dynamically defined domains: as soon as entities migrate, the crossed nodes identify one domain composed of a chain of neighbour nodes (see figure 1c).

From the implementation point of view, the EDN policy implements the load forwarding by a simple protocol: when a direct-neighbour couple is established and a node i has to receive load, it looks in its neighbourhood for one less loaded node. If this node exists, the node i acts as a forwarder. The same protocol is repeated by any receiver node up to the final receiver, that is always one node with a local minimum load in its neighbourhood.

3. Performance Comparison

To evaluate the effectiveness of the presented policies, we have used as target a 100 nodes transputer-based architecture, by creating artificial execution load that permitted us to evaluate the policies on a wide rage of load situations. There are a few properties that influence the policy behaviour [2]. We consider here:

- *unbalancing* (σ), i.e., the deviation of the current load situation from the perfectly balanced one, measured by the global standard deviation σ of the load of all the system nodes, normalised to the average system load;
- *dispersion* (δ), i.e., the presence in the system of an uneven distribution of load, indicated by the presence of more or less concentrated regions of overloaded (or underloaded) nodes. Dispersion can be measured by the standard deviation of the load computed as if each node had a load equal to the average of its load and of its neighbourhood. A small dispersion parameter indicates an even distribution of the load. Higher dispersion (up to σ) indicates concentrated regions of imbalance.

All diffusion policies are stable: in any execution for all different load situations they always reach stability; there were no cases of entities engaged in endless migration.

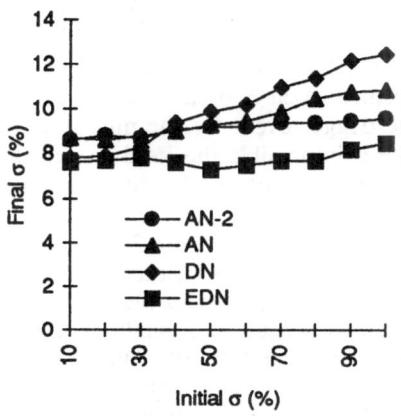

Figure 2. Influence of initial unbalancing on LB quality (δ = 50%)

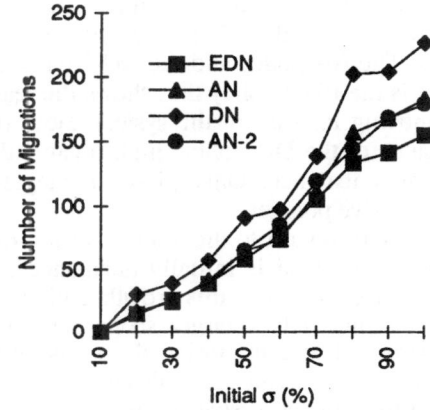

Figure 3. Influence of initial unbalancing on migration effort (δ = 50%)

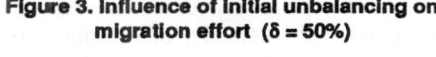

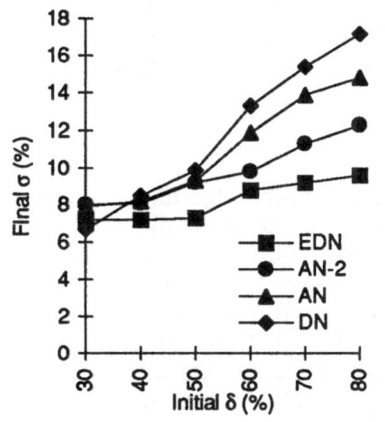

Figure 4. Influence of dispersion on LB quality (σ = 50%)

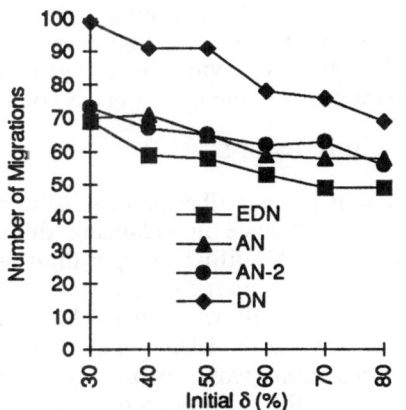

Figure 5. Influence of dispersion on migration effort (σ = 50%)

We analyse the capacity of achieving a good LB, i.e., a limited σ, connected with the related effort. While all policies are able to greatly diminish the initial unbalance, they do not achieve the same LB quality. The DN policy exhibits the worst behaviour: it is not able to achieve the same quality of the other policies and, moreover, it employs the highest number of migrations. The AN policy does not improve the quality (over the DN), but substantially reduces the number of migrations. The AN-2 policy achieves a little further improvement, both in terms of final balancing quality and number of migrations, due to the enlargement of the dimension of the domains. The EDN policy behaves even better, by obtaining the best balancing quality with the lowest migration effort.

To better understand these results, figure 2 shows that the dependence on the initial σ of the achieved LB is low for all policies; figure 4 shows that the significant load parameter is δ. When δ is low, i.e., the unbalance is evenly distributed in the system, all policies achieve good results. When δ increases and becomes comparable to σ, the local scope can make sometimes impossible to detect the global unbalance. In this case, the DN policy with its restricted locality often stops with high residual σ. The AN policy and the AN-2 one enlarge their locality and are less sensitive to δ. It is the EDN policy that shows the least dependence on δ, because of its capacity of moving load far in the system. However, the trigger event for the migration is the same of the DN policy: high δ can still make it impossible the local recognition of one unbalance. Only global strategies can fully overcome the limited scope of diffusive policies.

With regard to the number of migrations to achieve LB, figure 3 shows that the more the load is initially unbalanced, the more load has to be moved in order to balance. However, this growth is higher for the DN policy than for the AN and AN-2 ones: not only a larger scope permits to achieve a better LB quality, but it also permits better migration decisions and diminishes their number. The EDN policy achieves the balance with an even more limited number of migrations by moving entities, in successive "hops", between non neighbour nodes (a movement composed of several hops is considered as a single migration). With regard to δ, figure 5 shows that even if the capacity of balancing of all policies substantially diminish with the increase of δ, the migration effort does not always proportionally diminish: the number of migrations is high even if the quality of load balancing is poor. Only the EDN policy overcomes this problem because movements are no longer strictly local and they are effective even in case of high δ.

4. Conclusions

The paper describes and evaluates a few basic diffusion algorithms and analyses their performance in balancing different load situations with different kinds of unbalances. All diffusion algorithms are stable and achieve efficient load balancing. However, the enlargement of the locality scope of a policy produces significantly better results only when this enlargment is based on a dynamic identification of the domains, as in the EDN policy. Further experiments [2] show that policies with an enlarged scope works poorer in highly dynamic situations. On this base, we are working in the design of a tool that decides automatically the most suitable 'locality' by evaluating the dynamicity of the application.

References

[1] A. Corradi, L. Leonardi, F. Zambonelli, "Load Balancing Strategies for Massively Parallel Architectures", Parallel Processing Letters, Vol. 2, No. 2&3, Sept. 1992.
[2] A. Corradi, L. Leonardi, F. Zambonelli, "Diffusive Algorithm for Dynamic Load Balancing in Massively Parallel Architectures", DEIS Technical Report No. DEIS-LIA-96-001, University of Bologna, April 1996. Available at http://www-lia.deis.unibo.it/Research/TechReport.html
[3] N. G. Shivaratri, P. Krueger, M. Singhal, "Load Distributing for Locally Distributed System", IEEE Computer, Vol. 25, No. 12, Dec. 1992.
[4] M. Xu, B. Monien, R. Luling, F. C. M. Lau, "Nearest Neighbour Algorithms for Load Balancing in Parallel Computers", Concurrency: Practice and Experience, Vol. 7, No. 7, Oct. 1995.

Effectively Scheduling Parallel Tasks and Communications on Networks of Workstations *

Xing Du[1], Yingfei Dong[2], and Xiaodong Zhang[1]

[1] Department of Computer Science, College of William and Mary
Williamsburg, VA 23187, USA
[2] Department of Computer Science, University of Minnesota
Minneapolis, MN 55455, USA

Abstract. Coordinating parallel tasks and minimizing communication delays are two important issues for the design of scheduling policies on networks of workstations. We address the two issues by presenting a scheduling scheme, called *self-coordinated local scheduling*. It consists of two parts: computation scheduling and communication scheduling. The computation scheduler on each workstation schedules its parallel task and local user jobs independently based on a static power preservation in that workstation so that parallel tasks on different workstations are executed at the same pace to achieve a global coordination. To minimize the communication latency, each local scheduler uses a non-preemptive strategy to try to make communication activities complete as soon as possible. The scheduling scheme is evaluated by simulating the execution of four NAS benchmark programs, and comparing it with a Unix based scheduling policy and the co-scheduling policy. Our experiments show that the power preservation in each workstation guarantees the performance of both parallel and local jobs. Furthermore, the communication delay is reduced significantly by the communication scheduling compared with a spinning based scheme and a Unix based scheme.

1 Introduction

Networks of Workstations (NOWs) are cost-effective platforms for parallel computations. Effectively scheduling parallel jobs on NOWs is very important. However, most existing scheduling policies on multiprocessor/multicomputer systems (e.g. [3] [7]) are not applicable because of the heterogeneous and non-dedicated features of NOWs [1]. In order to effectively manage the interaction between parallel jobs and user jobs and provide good performance to both kinds of jobs, two issues must be well addressed: how to coordinate the execution of tasks of a parallel job on different workstations (inter-processor coordination), and how

* This work is supported in part by the National Science Foundation under grants CCR-9102854 and CCR-9400719, by the Air Force Office of Scientific Research under grant AFOSR-95-1-0215, and by the Office of Naval Research under grant ONR-95-1-1239. Xing Du is on leave from Computer Science Department, Nanjing University, Nanjing, P.R.China.

to manage the interaction between a parallel job and local user jobs on a work-station (intra-processor coordination). Co-scheduling policy [6] generally results in good parallel job performance [5] on multiprocessor systems, and is a good policy for inter-processor coordination. However it is hard to be effectively implemented on NOWs where the communication overheads are high. In addition, no intra-processor coordination is provided.

We propose a scheduling scheme, called *self-coordinated local scheduling* based on the principle of co-scheduling. By preserving certain amount of computing power on each workstation, each local scheduler schedules independently the parallel task and local user jobs on its workstation. The coordination between parallel tasks running on different workstations is achieved, and the performance of both parallel and local jobs is guaranteed. To minimize the communication latency, the scheme uses a non-preemptive strategy to expedite the communication. Simulation results of four NAS benchmark programs indicate that it is an effective approach to scheduling parallel and local user jobs on NOWs.

2 Scheduling Scheme

We propose a scheduling scheme, called *self-coordinated local scheduling* for bulk synchronous applications. It is composed of two parts: computation scheduling which is scheduling parallel tasks when they are in computational phases, and the other, communication scheduling for scheduling tasks when they are in communication phases. A key issue in computation scheduling is to preserve a portion of power in each workstation for parallel tasks. We define the power weight of a workstation as its computing capability relative to the fastest workstation in a system. The value of the power weight is less than or equal to 1. Since the power weight is a relative ratio, it can also be represented by measured execution time. If $T(App, M_i)$ gives the execution time for executing program *App* on workstation M_i, the power weight can be calculated by the measured execution times as follows:

$$W_i(App) = \frac{\min_{i=1}^{m}\{T(App, M_i)\}}{T(App, M_i)}.$$ (1)

where m is the number of workstations in a NOW. (For detailed information about the power weight, the interested reader may refer to [8].) Each workstation first calculates its own capable power weight for parallel jobs:

$$\rho_i = W_i(1 - R_{kernel}(i) - R_{user}(i))$$ (2)

where W_i is the power weight of workstation i, $R_{kernel}(i)$ is the percentage of power for the kernel in workstation i, and $R_{user}(i)$ is the percentage of power for local user jobs in the workstation. We define the free power weight on workstation i, $(i = 1, ..., m)$, as:

$$F_i = 1 - R_{kernel}(i) - R_{user}(i).$$ (3)

Then the preserved power weight for parallel jobs in each workstation is determined by the minimum available power weight for parallel jobs among all the workstations:

$$\rho = min_{i=1}^{m}\rho_i.$$ (4)

The equivalent power preservation in each workstation for parallel tasks allows us to "simulate" co-scheduling in a virtual homogeneous system. The remaining percentage of the power in a workstation can be used for local user jobs.

Preserving the power in each workstation for parallel jobs does not guarantee coordination of parallel tasks because tasks need to be further locally scheduled in each workstation to ensure all the tasks will finish within a reasonable time period. This seems somewhat application dependent. Here we temporarily use an application program dependent parameter for the local scheduler, the size of parallel tasks, denoted by TS, which is measured by the number of floating point operations. For a given power of a workstation measured by the number of floating point operations per second, the time to finish a task on the relatively slowest workstation is

$$t_s = \frac{TS}{Pow(s) \times F_s},$$ (5)

where $Pow(s)$ is the power of the slowest workstation, and F_s is the free power weight in the slowest workstation; and the time to finish a task on workstation i is

$$t_i = \frac{TS}{Pow(i) \times F_i},$$ (6)

where $Pow(i)$ is the power of workstation i, and F_i is the free power weight of workstation i. If round-robin time-sharing fashion is used in the local scheduler of each workstation, the number of time slices used to finish a task on the slowest workstation is

$$N_{slice}(s) = \frac{t_s}{\delta_s},$$ (7)

where δ_s is the length of a time slice in the slowest workstation; and the number of time slices used to finish a task of the same size on workstation i is

$$N_{slice}(i) = \frac{t_i}{\delta_i},$$ (8)

where δ_i is the length of a time slice in workstation i, and $i = 1, ..., m$. Since the slowest workstation is the bottleneck of parallel jobs, if all the tasks finish within the time period of t_s in each loop, the performance would be optimal and equivalent to that in a dedicated NOW using co-scheduling. However, within t_s, there are $\frac{t_s}{\delta_i}$ time slices available in workstation i, which is larger than $N_{slice}(i)$. This means that all workstations except the slowest one have more time slices for additional processes. In order to make even distributions of time slices for parallel tasks and additional processes, we use (5), (6), (7) and (8) to quantify the time interval for a time slice assignment to a parallel task in workstation i, ($i = 1, ..., m.$). Therefore, if a time slice is given to the parallel task in workstation i within no less than a time quantum of $\frac{t_s}{\delta_i N_{slice}(i)} = \frac{t_s}{t_i}$, the local scheduler will ensure that within t_s, all the tasks in a local computation phase will finish. By (5) and (6), we further obtain

$$\frac{t_s}{t_i} = \frac{Pow(i) \times F_i}{Pow(s) \times F_s}$$ (9)

The time quantum in (9) is architecture dependent rather than application program dependent.

The scheduling scheme in the computational phase is as follows:

1. Determine ρ_i, the available power weight for parallel jobs on workstation i based on its local user's decision.
2. Broadcast $Pow(i)$ and F_i to other local schedulers.
3. Receive Pow and F from all other local schedulers.
4. Decide the minimum available power weight based on (4), and the relatively slowest workstation s.
5. The local scheduler in workstation i calculates its pace to assign a time slice to its parallel task by (9).

The communication phase is another essential section of parallel jobs. It has different requirements for schedulers. A message-passing can only be performed when it obtains both the CPU and the network. If a communicating task is given more chance to obtain the CPU, it would have a better chance to complete the communication phase as soon as possible. This would also increase the possibility of overlapping this communicating task with computational tasks on other workstations, and decrease the possibility of network contention. Motivated by this, we use such a scheduling scheme for communications:

1. The local scheduler is notified that the parallel task has been in communication phase (becomes a communicating task).
2. Set a larger time slice for this communicating task.
3. If the CPU is serving for a local user job, it does nothing until the local job finishes its time slice.
4. When the communicating task acquires the CPU, if the network is available, the above larger time slice is given to it, otherwise, the scheduler gives a time slice to a local job immediately.
5. The local scheduler is notified that the communicating task has finished its communication.
6. The local scheduler resumes its power preservation scheme.

Thus, the scheduler gives a larger time slice to a parallel task to complete the communication if the network is available, otherwise it is switched to process local user jobs. This would reduce unnecessary context switches and network contention. In the scheme, the CPU is switched to a local user job rather than to spin when the network is not available for the communication. This is because first, experiments reported in [4] indicate that an immediate switch strategy outperforms two-phase blocking strategy (spin and then switch) for coarse grained parallel jobs on relatively slow networks. Second, immediately yielding the CPU to other local sequential jobs increase the system utilization. Finally, if a spin is chosen, how long to spin is hard to determine because an optimal length is architecture- and application- dependent. The strategy of spin-until-it-gets seems to be an effective way, but it may lead to deadlocks.

3   Simulation Methodology

We designed and developed a simulator to perform event-driven simulation. The structure of the simulator is illustrated in Figure 1. The effect of computation scheduling was event-driven simulated. The communication effect was evaluated by a network simulator which simulated an Ethernet of 10Mbps bandwidth. We selected Ethernet because it is a popular network used to connect workstations.

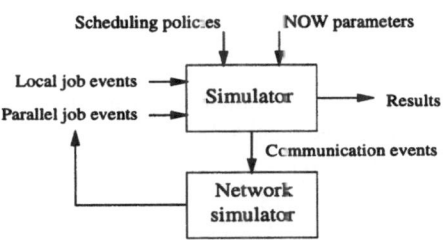

Fig. 1. The simulator structure.

The simulated heterogeneous NOW consisted of 7 types of Sun SPARCstations with different computing powers. We measured the power weight of each workstation (Table 1) by running 4 benchmark programs discussed later.

Table 1. The average power weight of 7 types Sun workstations.

S20-HS21	S20-HS11	S5-85	S20-50	S5-70	S10-30	Classics
A=1.0	B=0.790	C=0.562	D=0.461	E=0.436	F=0.374	G=0.239

We selected four programs from the NAS parallel benchmarks [2]: EP (Embarrassing Parallel), MG (Multigrid), IS (Integer Sort), and LU (LU decomposition). All of them follow the bulk synchronous model. The communication patterns of the 4 programs were classified into three types: *all-to-all* (IS, EP), *neighbors* (LU), and *transpose* (MG). The parallel job events were characterized by their computation sizes, communication patterns, the number of bytes to send in each communication, communication starting time, process arrival time, etc. The local workloads were only characterized by their starting time and computation sizes.

The scheduling policies in the simulator included a local scheduling scheme based on Unix (System V Release 4), co-scheduling using the matrix scheme in [6], and the self-coordinated local scheduling.

4 Performance Evaluation

The performance evaluation was done by simulating the execution of the 4 programs on NOWs with different workstation combinations. For space limitation, we only present results on 4 workstations of type B, C, F and G here. In the simulation, the following system parameters were used: the time slice was set 0.1 second; and a context switch took 200 μs. EP was iterated 10 times to increase its synchronization activities.

Table 2. The effect of varying the number of local jobs on the slowdowns of four programs using the self-coordinated local scheduling (*coordinated*) and local scheduling (*local*) in comparisons with co-scheduling.

# of local jobs	EP		IS		MG		LU	
	coordinated	*local*	*coordinated*	*local*	*coordinated*	*local*	*coordinated*	*local*
1	1.1	1.31	1.32	2.05	1.22	1.89	1.34	2.0
2	1.1	1.92	1.34	2.65	1.23	2.78	1.35	3.08
3	1.1	2.56	1.36	3.51	1.25	3.84	1.37	4.19
4	1.2	3.14	1.38	4.12	1.27	4.27	1.39	5.12
5	1.2	3.90	1.39	4.98	1.28	5.01	1.40	6.03

Table 2 presents the effect of varying the number of local user jobs on slow-down factors of the self-coordinated local scheduling and standard local scheduling. The slowdown is relative to the execution times of the four programs using co-scheduling on the corresponding dedicated NOW. The execution time of EP using self-coordinated local scheduling was very close to that using co-scheduling, differing by a factor of up to 1.2. The execution times of IS, MG and LU by self-coordinated local scheduling increased 22% to 34% in comparison with the times of co-scheduling when there was one local job. Besides scheduling skew and context switch overhead, there was another reason for the slowdown. The computation size of the three programs were dynamically changed as the programs proceeded. When the execution time of a computation size was close to a time slice, the execution pace was not well coordinated, because the time frame in the relatively slowest workstation for the task, t_s, in (5) became very small, and the scheduling became difficult in a tiny time space. However, with the increase of the number of local jobs, the slowdown of self-coordinated scheduling changed very slightly. In other words, the performance of parallel jobs was affected slightly by the change of local workload, and was guaranteed by the scheduling scheme. Meanwhile, the performance degradation of each local job was expected by the local user because he agreed to denote that amount of his workstation power to parallel jobs (see Step (1) in the scheme) and added the local workload by himself.

In contrast, when using local scheduling, the performance of parallel programs degraded significantly. The degradation increased almost proportionally with the increase of the number of local jobs. For example, the performance of IS decreased by a factor of 498% when 5 local jobs were executed with it.

Table 3. The breakdown of the execution times for EP and IS.

	EP				IS			
#	Comput.	Comm.	Sync.	Switch	Comput.	Comm.	Sync.	Switch
1	17.0	0.0007	2.88	0.06	28.3	6.42	4.29	0.09
2	18.6	0.0007	1.52	0.07	32.5	6.42	2.79	0.13
3	19.2	0.0007	0.89	0.08	33.8	6.42	1.37	0.15
4	19.9	0.0007	0.23	0.10	34.9	6.42	0.38	0.17
5	19.9	0.0007	0.23	0.10	34.9	6.42	0.37	0.18

Table 4. The breakdown of the execution times for MG and LU.

	MG				LU			
#	Comput.	Comm.	Sync.	Switch	Comput.	Comm.	Sync.	Switch
1	64.3	2.5	8.9	0.2	61.1	1.2	8.5	0.2
2	72.6	2.5	5.9	0.3	71.8	1.2	6.1	0.3
3	75.2	2.5	3.0	0.3	76.1	1.2	3.2	0.3
4	77.4	2.5	1.1	0.4	78.5	1.2	1.0	0.4
5	77.4	2.5	1.1	0.4	78.5	1.2	1.0	0.4

Table 3 and 4 list the breakdown of the execution times in simulated clock ticks for the four programs when the number of local jobs changes. The execution time includes computation time, communication time, synchronization time, and context switch time. The communication time includes the startup time (software overhead) and message transmission time (network latency). The synchronization time is that a workstation spends on waiting at a synchronization point in a program or in a synchronous send/receive protocol.

We further evaluate the efficiency of the communication scheduling, and compare it with two other policies. One is local scheduling, which means no special policy is used for scheduling communicating tasks. The other is either using the network if it is available or spinning for its turn. We call this policy, spinning scheduling. This policy may result in low system utilization, and deadlock if multiple parallel jobs are executed. Two communication patterns were studied, one is *all-to-all* and the other is *transpose* which is from IS and MG respectively. Table 5 presents the communication delays for two types of patterns changing with the problem sizes. For both patterns, our communication scheduling performed better than any of the two other schemes for different problem sizes.

Table 5. The communication delay of two types of communication patterns scheduled by the *Spinning* scheduling, the *Self-coordinated* scheduling and the *Local* scheduling.

Communication pattern	Problem size	Communication delay		
		Spinning	*Coordinated*	*Local*
all-to-all	2^4	6.5	5.2	14.9
	2^8	4.0	4.0	13.1
	2^{12}	6.5	4.5	11.6
transpose	32	8.0	6.3	10.4
	64	17.4	11.9	20.8
	128	35.0	12.6	18.1

5 Current Work

We propose a scheduling scheme for scheduling parallel tasks on non-dedicated NOWs. It guarantees the performance of both parallel and local jobs. The simulation results show its effectiveness. We are currently studying a way to preserve power in UNIX and how to effectively schedule a wider range of applications, and investigating adapting the communication scheduling in high speed networks.

References

1. R. H. Arpaci et al.: The interaction of parallel and sequential workloads on a network of workstations. Proceedings of ACM SIGMETRICS Conference, (1995)
2. D. Bailey et al.: The NAS parallel benchmarks. International Journal of Supercomputer Applications. 5(3) (1991) 63-73
3. M.-S. Chen and K. G. Shin: Subcube allocation and task migration in hypercube multiprocessor. IEEE Transactions on Computers. C-39(9) (1990) 1146-1153
4. A. C. Dusseau, R. H. Arpaci and D. E. Culler: Effective distributed scheduling of parallel workloads. Proceedings of ACM SIGMETRICS Conference, (1996)
5. A. Gupta, A. Tucker, and S. Urushibara: The impact of operating system scheduling policies and synchronization methods on the performance of parallel applications. Proceedings of ACM SIGMETRICS Conference. (1991) 120-132
6. J. Ousterhout: Scheduling techniques for concurrent systems. Proceedings of the 3rd International Conference on Distributed Computing Systems. October, (1982) 22-30
7. A. Tucker and A. Gupta: Process control and scheduling issues for multiprogrammed shared-memory multiprocessors. Proceedings of the 12th ACM Symposium on Operating Systems Principles. (1989) 159-166
8. X. Zhang and Y. Yan: Modeling and characterizing parallel computing performance on heterogeneous NOW. Proceedings of the Seventh IEEE Symposium on Parallel and Distributed Processing. IEEE Computer Society Press, October, (1995) 25-34

On Linear Schedules of Task Graphs for Generalized LogP-Machines

Welf Löwe   Wolf Zimmermann   Jörn Eisenbiegler

Institut für Programmstrukturen und Datenorganisation, Universität Karlsruhe, 76128 Karlsruhe, Germany, E-mail:{loewe|eisen|zimmer}@ipd.info.uni-karlsruhe.de

Abstract. We discuss linear schedules of task-graphs under the communication cost model of the LogP-machine. In addition to our previous work, we consider also non-constant parameters L, o and g, i.e. we introduce messages of different sizes into the LogP-model. The main results of this work are the following: (i) in the LogP-model, less communication in linear schedules does not necessarily imply a better performance. (ii) We give an upper time bound on the execution time of linear clusterings. (iii) We give an efficient algorithm which computes linear clusterings with a minimum number of clusters.

1   Introduction

The LogP-machine [CKP$^+$93] assumes a cost model reflecting latency of point-to-point-communication in the network, overhead of communication on processors themselves, and the network bandwidth. These communication costs are modeled with parameters $Latency$, $overhead$, and gap. The gap is the inverse bandwidth of the network per processor. In addition to L, o, and g, parameter P describes the number of processors. The parameters have been determined for several parallel computers [CKP$^+$93, CKP$^+$96, DI94, ELW97]. These works confirmed all LogP-based runtime predictions. Except the last, all works assume that only messages of the same size are sent. In contrast to our previous work on clustering and scheduling [ZL94, LZ95b, LEZ96] we assume that messages of different size can be sent within the same task graph, and sending one large message is cheaper than sending the same amount of data with several small messages. Like [ELW97], we assume that the parameters L, o, and g depend on the message size. We consider linear schedules of task graphs under the cost model of this generalized LogP-machine.

Other work on linear scheduling assume at least one of the following requirements.

- Overheads and gaps are ignored [PY90, GY93, YG94, LZ95a].
- Only special task graphs are considered [KSSS93, YG94, LZ95a].
- The task graphs must be coarse grained [LZ95b, ZL94].

In this paper we drop all of these requirements.

Papadimitriou and Yannakakis showed that finding an optimal clustering of a task graph G is NP-hard, even if $o = g = 0$ and $P = \infty$, [PY90]. They also showed that under the same assumptions approximations that guarantee a factor of two of the optimum $T_{opt}(G)$ cannot be found in polynomial time, unless P=NP. We can therefore not expect to find an efficient and optimal transformation. If $o = g = 0$ and $P = \infty$, Gerasoulis and Yang [GY93] find solutions guaranteeing $(1 + 1/\gamma(G)) \times T_{opt}(G)$ by linear clustering where $\gamma(G)$ is an adequate notion of granularity. In this paper, we generalize their definition.

For trees, they find the optimum if $\gamma(G) \geq 1$. For a subclass of trees, they find the optimum even for small computation times [YG94], both in polynomial time. We showed in [LZ95a] that an optimal solution by linear clustering with redundant computations can be found in polynomial time for tree-like communication structures, if $\gamma(G) \geq 1$, $o = g = 0$, and $P = \infty$. [LMZ97] generalizes these results for linear clusterings for $o, g > 0$.

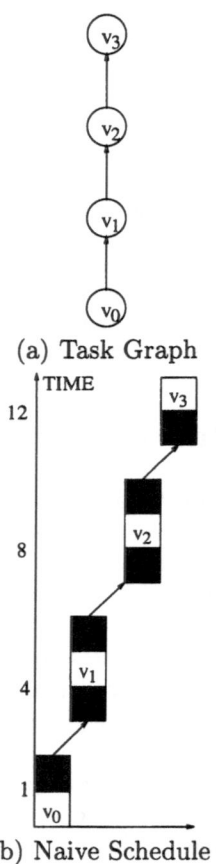

(a) Task Graph

(b) Naive Schedule

Fig. 1. A task graph where the bound in Lemma 1 is tight

Unfortunately, results on clustering and scheduling considering only latencies (i.e. $o = g = 0$) do not generalize for the LogP-machine in a straightforward way. Karp et al. [KSSS93] give optimal LogP-schedules for fork- and join-trees that significantly differ from the optimal schedules for the same structures proposed in [GY93, YG94, LZ95a]. We show in this paper that a performance of LogP-programs within a factor of $1 + 1/\gamma(G)$ of the optimal program can be guaranteed if the granularity $\gamma(G)$ is generalized to the LogP-machine according to [LZ95b].

Similar to [GY93] our approach runs in two phases: First, we cluster the task graph linearly, i.e. we determine the tasks which are computed on one processor. Second we introduce communication and determine the precise schedule. However, the proof of this result is different from [GY93]. Their proof is based on the following facts: (i) if each task is executed separately on one processor, then the execution time is at most $(1 + 1/\gamma(G)) \cdot T_{opt}(G)$. (ii) any linear clustering of G is no worse than (i). (iii) a schedule can be determined straightforwardly from a clustering.

At first glance, it seems that the same ideas might carry over to schedule for the LogP-machine. [ZL94] generalizes (i) for the cost model of the LogP-machine and shows that the approach can be chosen for coarse-grained task graphs. However, we show in this paper that (ii) need not be satisfied in the LogP-model, even if we assume constant message size. Instead, we show that the bound

$(1 + 1/\gamma(G)) \cdot T_{opt}(G)$ can be guaranteed. For proving this, the communication between two processors must be organized adequately. Therefore, (iii) is not as easy as in the case $o = g = 0$.

The paper is organized as follows: Section 2 introduces the basic definitions used throughout the paper. Section 3 proves (i) and shows that (ii) is violated. In Section 4 we prove the upper bound. This proof reorganizes communication and induces a schedule. In this paper we just consider the communication cost model of the generalized LogP-machine and assume that always enough processors are available. Section 5 introduces a processor-minimal linear clustering algorithm. Section 6 concludes our work.

2 Basics

A *task graph* is a directed acyclic graph $G = (V, E)$, where the vertices are sequential tasks and the edges are data dependencies between them. The set of *direct predecessors* $PRED_v$ of a vertex v in G is defined by $PRED_v = \{u \in V \mid (u, v) \in E\}$. The set of *direct successors* $SUCC_v$ is defined analogously. *Input vertices* are vertices without predecessors, *output vertices* are vertices without successors. $PATHS_v$ denotes the set of paths from any input vertex to v.

We assign the following costs to each vertex v: τ_v are the *computation costs* of vertex v, L_v is the *latency*, o_v is the *overhead*, and g_v is the *gap* of the messages sent by v. This is a generalization of the LogP-model to non-constant parameters. $G = (V, E, \tau, L, o, g)$ denotes the task graph (V, E) extended with these weight functions.

A *schedule* for G specifies for each processor P_i and each time step t the operation performed by processor P_i at time t, provided there starts one. The operations are the execution of a task $v \in V$ (denoted by the task), sending a message of a task v to another processor P_j (denoted by $send(v, j)$), and receiving a message of a task v from processor P_j (denoted by $recv(v, j)$). A schedule is therefore a partial mapping $s : \mathbb{N} \times \mathbb{N} \to OP$, where OP denotes the set of operations. s must satisfy some restrictions:

(i) Two operations on the same processor must not overlap in time. Computation of a task v requires time τ_v. Sending from v or receiving a message sent by v requires time o_v.

(ii) If a processor computes a task $v \in V$ then any task $w \in PRED_v$ must be computed or received before on the same processor.

(iii) If a processor sends a message of a task, it must be computed or received before on the same processor.

(iv) Between two consecutive send (receive) operations there must be at least time g_v, where v is the message sent earlier (received later).

(v) For any send operation there is a corresponding receive operation, and vice versa. Between the end of send operation and the beginning of the corresponding receive operation must be at least time L_v.

(vi) Any $v \in V$ is computed on some processor.

Condition (iv) is the same as described in [AISS95] for the LogGP-machine. We visualize schedules by Gantt-charts (cf. Figure 1). The x-axis corresponds to processors, the y-axis to time. We place a box at (i, t), if processor P_i starts at time t an operation. The vertical length of the box corresponds to the cost (i.e. τ_v if v is computed and o_v if it is send or receive operation). Send and receive operations are drawn as black boxes. White boxes denote the computation of a task. The *execution time* of a schedule is the time when the last operation is finished. $T_{\mathrm{opt}}(G)$ denotes the execution time of an optimal schedule, i.e. the minimal execution time of all schedules for G.

A *clustering* of G is a finite set $Cls(G) \subset 2^V$ of subsets of V such that for any $v \in V$ there is an $X \in Cls(G)$ with $v \in X$. A clustering is *linear* iff each $X \in Cls(G)$ is a path in G. The clustering is *naive* iff $|X| = 1$ for each $X \in Cls(G)$. A linear clustering $Cls(G)$ is *processor minimal* iff there is no other clustering $Cls'(G)$ of G with $|Cls'(G)| < |Cls(G)|$. Any schedule induces a clustering in the following way: for any processor P_i, let be $X \in Cls(G)$ where X is the set of tasks computed by P_i. The notion of *linear schedules* and *naive schedules* is then defined in the obvious way.

A sequential implementation of send and receive operations leads to the upper time bound $L_{\max}(u, v)$ for LogP-communication between adjacent tasks u and v of:

$$L_u + 2o_u + (|SUCC_u| - 1)\max(o_u, g_u) + \sum_{w \in PRED_v \setminus \{u\}} \max(o_w, g_w).$$

This result is a straightforward generalization of [ZL94].

The *granularity of a vertex* $v \in V$ relates computation costs to communication costs and is defined as

$$\gamma(v) = \frac{\min_{u \in PRED_v} \{\tau_u\}}{\max_{u \in PRED_v} \{L_{\max}(u, v)\}}$$

The granularity of G is defined as $\gamma(G) = \min_{v \in G} \{\gamma(v)\}$. G is *coarse grained* if $\gamma(G) \geq 1$, otherwise it is called *fine grained*.

3 Naive Clusterings

We first derive an upper time bound for executing naive clusterings (Lemma 1 and Corollary 2) and then show that the given bound is tight. Then we prove that there are task graphs with linear schedules that have a higher execution time than naive schedules.

Lemma 1 *Let* $G = (V, E, \tau, L, o, g)$ *be a weighted task graph,* $\gamma(G)$ *the granularity of* G, *and* $l_v = \max_{w \in PRED_v} \max_{\pi \in PATHS_w} \sum_{u \in \pi} \tau_u$ *be the weighted length of the longest path from an input vertex to a predecessor of* v *w.r.t.* τ. *Then there is a naive schedule* s *of* G, *such that any* v *starts at time* $t_v \leq (1 + 1/\gamma(G)) \cdot l_v$.

Proof: Let $v_1, \ldots, v_n$ be any topological sorting of G.

We prove the claim by induction on this topological sorting. Obviously, $t_{v_1} = 0$ can be chosen and the claim is satisfied.

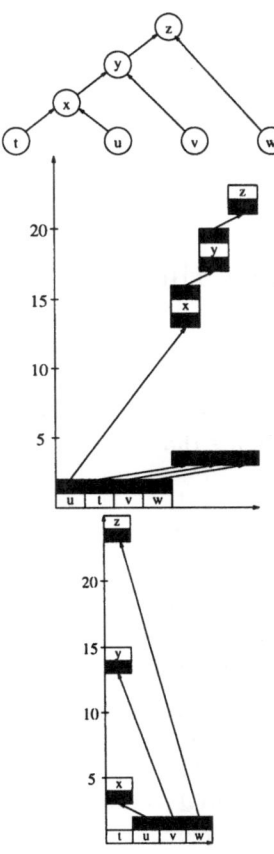

Consider v_i, $i > 1$. If v_i is an input vertex, the claim is proven as in the base case. Suppose $PRED_{v_i} \neq \emptyset$. It is easy to see that $l_{v_i} = \max\limits_{u \in PRED_{v_i}} l_u + \tau_u$. Let $w \in PRED_{v_i}$ be a vertex, s.t. $l_w + \tau_w + L_{\max(w,v_i)}$ becomes maximal. Since w is a predecessor of v_i, it occurs earlier in a topological sorting, and by induction hypothesis $t_w \leq (1 + 1/\gamma(G)) \cdot l_w$. Thus the computation of w is finished at time $t'_w \leq (1 + 1/\gamma(G)) \cdot l_w + \tau_w$. The worst scenario is that the last message sent by w is the message sent to v and this is the first message received by v. Observe that the other send and receive operations do not violate requirement (v) since they are sent earlier. We have already discussed in Section 2 that v can be computed at time $t_{v_i} \leq t'_w + L_{\max(w,v_i)}$. Thus

$$t_{v_i} \leq (1 + 1/\gamma(G)) \cdot l_w + \tau_w + L_{\max(w,v_i)}$$
$$\{\text{by definition of granularity}\}$$
$$\leq (1 + 1/\gamma(G)) \cdot l_w + \tau_w + \tau_w/\gamma(v_i)$$
$$\leq (1 + 1/\gamma(G)) \cdot (l_w + \tau_w)$$
$$\{\text{recursive characterization of } l_{v_i}\}$$
$$\leq (1 + 1/\gamma(G)) \cdot l_{v_i}$$

◇

Fig. 2. Example 2

Corollary 2 (Time Bound for Naive Schedules) *Let $G = (V, E, \tau, L, o, g)$ be a weighted task graph and $\gamma(G)$ the granularity of G. Then there is a naive schedule s with execution time $T(s) \leq (1 + 1/\gamma(G)) \cdot T_{\mathrm{opt}}(G)$.*

The following example shows that the bound given by Lemma 1 is tight.

Example 1 *Consider the task graph in Figure 1(a). Suppose for any vertex v, $\tau_v = 1$, $L_v = 1$, $o_v = 1$ and $g_v = 2$. Obviously with this definitions $\gamma(G) = 1/3$, and $l_{v_i} = i$, $i = 0, \ldots, 3$. The schedule in Figure 1(b) shows that vertex v_0 starts at time $0 = (1 + 1/\gamma(G))l_{v_0}$, vertex v_1 starts at time $4 = (1 + 1/\gamma(G))l_{v_1}$, vertex v_2 starts at time $8 = (1 + 1/\gamma(G))l_{v_2}$ and vertex v_3 starts at time $12 = (1 + 1/\gamma(G))l_{v_3}$.*

Gerasoulis and Yang [GY93] showed that linear clustering cannot increase the execution time of a task graph if $o = g = 0$. This result does not generalize to the case where $g \neq 0$. The following counter example shows that a linear clustering may delay the execution time w.r.t. the naive implementation.

Example 2 *Assume $L = 1$, $o = 1$, and $g = 10$, computation times of $\tau = 1$, and the task graph is given in Figure 2. The execution time of the naive schedule (23) is less than the execution time for a linear schedule (25) which sequentializes the tasks t, x, y, and z, see Figure 2. Observe that no linear schedule sequentializing t, x, y and z can have execution time less than 25.*

4   Execution Time Bounds for Linear Clusterings

Example 2 is the reason why the approach of [GY93] cannot be chosen to prove an upper time bound for linear clustering. In fact, due to the counter example it is not obvious from Lemma 1 or Corollary 2 that such a time bound can be established for linear schedules in general. This section shows that the bound in Lemma 1 is still valid for linear schedules.

Theorem 3 (Linear Clustering Time Bound) *Let $G = (V, E, \tau, L, o, g)$ be a weighted task graph, $\gamma(G)$ be the granularity of G, and l_v be the weighted length of the longest path from an input vertex to a predecessor of v w.r.t. τ. Then for any linear clustering $Cls(G)$ of G there is a linear schedule s of G inducing $Cls(G)$, such that any vertex v starts at time $t_v \leq (1 + 1/\gamma(G)) \cdot l_v$.*

Proof: Observe that a naive clustering is a special case of a linear clustering.

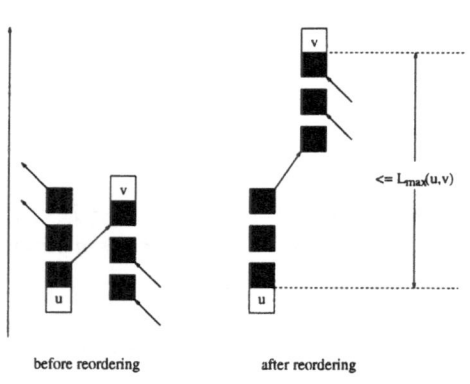

Consider two paths $\mathcal{U}$ and $\mathcal{V}$ of a linear clustering where the last vertex of $\mathcal{U}$ is connected to the first vertex of $\mathcal{V}$ by an edge. Then $\mathcal{U}$ and $\mathcal{V}$ can be concatenated to a path $\mathcal{W}$. It is easy to see that any linear clustering can be obtained by a sequence of concatenations of paths from the naive clustering.

We prove the theorem by induction on an arbitrary sequence of concatenations of two paths $\mathcal{U}$ and $\mathcal{V}$.

Fig. 3. Reordering of send and receive operations in $\mathcal{U}$ and $\mathcal{V}$.

INDUCTION BASIS: Because of Lemma 1, each task $v \in V$ can be computed at or before $t_v \leq (1 + 1/\gamma(G))l_v$ by a naive schedule.

INDUCTION HYPOTHESIS: Suppose for a linear clustering $Cls(G)$, there is a schedule s of G such that any $v \in V$ is computed at a time $t_v \leq (1 + 1/\gamma(G))l_v$.

We have to show that this also holds for any linear clustering $Cls'(G)$ constructed from $Cls(G)$ by concatenating two paths $\mathcal{U}$ and $\mathcal{V}$ of $Cls(G)$. We construct from s a linear schedule s' which induces clustering $Cls'(G)$. Let u be the last vertex of $\mathcal{U}$ and v be the first vertex of $\mathcal{V}$. Observe that $(u, v) \in E$ by definition of concatenation. The concatenation of $\mathcal{U}$ and $\mathcal{V}$ therefore saves communication time L_u and $2 \cdot o_u$. However, we have to show that the gap g_x (g_y) of the last send (receive) operation in $\mathcal{U}$ does not delay the first send (receive) operation in $\mathcal{V}$. A delay is only possible, if $\mathcal{U}$ contains at least two send ($\mathcal{V}$ contains at least two receive) operations. Otherwise obviously no delay by gaps occurs because we save the send (receive) operation by concatenation.

Claim: If $|SUCC_u| > 1$ or $|PRED_v| > 1$ the send (receive) operations of $\mathcal{U}$ ($\mathcal{V}$) can be reordered such that the send (receive) corresponding to edge (u, v) is the last in $\mathcal{U}$ (the first in $\mathcal{V}$) without violating any bound $t_x \leq (1 + 1/\gamma(G))l_x$.

Proof: (cf. Figure 3) Assume that send (receive) corresponding to edge (u, v) is not the last in $\mathcal{U}$ (the first in $\mathcal{V}$). By induction hypothesis and definition of $L_{\max}(u, v)$, $(1 + 1/\gamma(G))l_v \geq t_u + \tau_u + L_{\max}(u, v)$. The upper bound $L_{\max}(u, v)$ is only tight if all send operations of u are executed before the send corresponding to edge (u, v) and all receive operations of v are executed after the receive corresponding to edge (u, v). Hence, we may swap the last send (first receive) operations of $\mathcal{U}$ ($\mathcal{V}$) with the send (receive) operation corresponding to (u, v) without violating $t_v \leq (1 + 1/\gamma(G))l_v$. Since the last but one send (second receive) operation are executed sooner (later) after swapping, the corresponding receive (send) operation is not delayed. Hence, after reordering it still holds $t_x \leq (1 + 1/\gamma(G))$ for all $x \in V$.

Proof (Theorem 3 continued, cf. Figure 4):

The last send in $\mathcal{U}$ (first receive in $\mathcal{V}$) itself is saved by concatenation. Let t'_s be the starting time for the last send (t'_r the starting time for the first receive) operation and t''_s the starting time for the last but one send (t''_r the starting time for the second receive) operation[1] in $\mathcal{U}$ ($\mathcal{V}$) before path concatenation and after reordering. Observe that after reordering the last operation on $\mathcal{U}$ is the send operation to $\mathcal{V}$ and the first operation on $\mathcal{V}$ is the receive operation from $\mathcal{U}$. Let $'t_s$ be the starting time for the first send ($'t_r$ the starting time for the last receive) operation in $\mathcal{V}$ ($\mathcal{U}$) before path concatenation and after reordering. By concatenating $\mathcal{U}$ and $\mathcal{V}$ we save the last send (first receive) op-

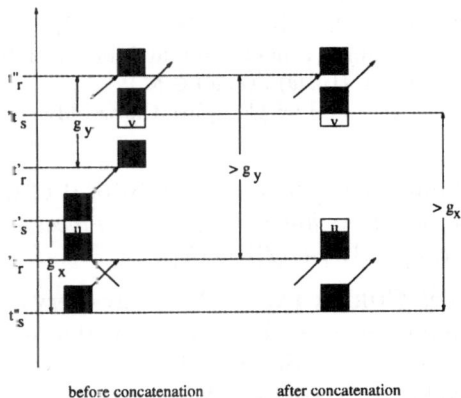

before concatenation   after concatenation

Fig. 4. Illustration of the proof of Theorem 3

---

[1] We identify paths with processors for simplicity.

eration in $\mathcal{U}$ ($\mathcal{V}$). Because of $t'_s \geq t''_s + g_x$ for a vertex x and $'t_s \geq t'_s$ ($t''_r \geq t'_r + g_y$ for a vertex y, $t'_r \geq' t_r$), the first send (second receive) operation in $\mathcal{V}$ is not delayed by the gap of the last but one send (last receive) operation in $\mathcal{U}$. Observe that the last but one send on $\mathcal{U}$ (second receive on $\mathcal{V}$) may send the message of a vertex different of u (receive an operand of a task later than v (cf. Figure 4). However, this is not important for the above reasoning. ◇

Remark: Observe that reordering the communication might increase the starting time of v in the new schedule. However, the claim in the proof of Theorem 3 guarantees that the amount of this increase does not become large. Without this reorganization of communication, the proof seems to be impossible. ◇

5 Processor Bounds for Linear Clusterings

Unfortunately, it seams to be impossible to compute a time minimal clustering of a task graph in polynomial time. Even worse, we cannot guarantee a lower upper time bound for any arbitrary linear clustered task graph than for the naive implementation. However, we may at least compute the processor minimal linear clustering of a task graph G. This can be done in polynomial time using an algorithm for computing a maximal matching of bipartite graphs.

Algorithm 1 Processor minimal clustering of a task graph G

INPUT: Task graph $G = (V, E)$
OUTPUT: processor-minimal clustering $Cls_{\min}(G)$ of G.

(1) $X := V$;
(2) $Y := \{\overline{v} : v \in V\}$;
(3) $B := (X, Y, \{(v, \overline{w}) : (v, w) \in E\})$;
(4) Compute a maximal matching M of B;
(5) $E' := \{(v, w) : (v, \overline{w}) \in M\}$;
(6) $Cls_{\min}(G) :=$ Compute the weakly connected components in $G' = (V, E')$;

Theorem 4 (Processor Minimal Clustering) *Let $G = (V, E)$ be a task graph. Algorithm 1 computes a processor minimal linear clustering $Cls_{\min}(G)$ in time $O(|V| + \sqrt{|V|} \cdot |E|)$.*

Proof: CORRECTNESS: Obviously every $v \in V$ is in exactly one weakly connected component of G'. First, we show that every weakly connected component of G' is a path in G. Because of $E' \subseteq E$, every weakly connected component of G' is acyclic. No vertex $v \in V$ has an in-degree larger than 1, otherwise M contains two edges $(u, \overline{v}) \in M$ and $(u', \overline{v}) \in M$. The latter is not possible because M is a matching. Analogously, there is no vertex $v \in V$ with an out-degree larger than 1. It follows that every weakly connected component in G' is a path in G. Hence, $Cls_{\min}(G)$ is a linear clustering.

It remains to show that $Cls_{\min}(G)$ is minimal. It is easy to see that $|V| = |Cls_{\min}(G)| + |M|$. Thus, $|Cls_{\min}(G)|$ is minimal iff $|M|$ is maximal. The latter is guaranteed after step (4).

COMPLEXITY: Steps (1)–(3) take time $O(|V| + |E|)$. Step (4) can be computed in time $O(\sqrt{|V|} \cdot |E|)$, see [HK73] (B contains $2 \cdot |V|$ vertices and $|E|$ edges). Steps (5)–(6) require time $O(|V|+|E|)$. Hence, Algorithm 1 is of time complexity $O(|V| + \sqrt{|V|} \cdot |E|)$. ◇

Therefore, Theorems 3 and 4 together imply that processor minimal linear schedule with execution time at most $(1 + 1/\gamma(G)) \cdot T_{opt}(G)$ can be obtained.

6 Conclusion

We have discussed linear schedules of task graphs G on the generalized LogP-machine. In contrast to previous work, we do not impose any restrictions on the task graphs considered. Our results differ from previous work in the following directions:

(i) We used a generalized LogP-cost model which takes into account messages of different sizes.
(ii) The upper bound of $(1 + 1/\gamma(G)) \cdot T_{opt}(G)$ can be established even for fine-grained task graphs.
(iii) We present an efficient algorithm that computes a processor-minimal linear clustering.

We showed that there is no way to prove (ii) using our previous approach for coarse-grained task graphs or the approach of [GY93], since it may happen that less communication leads to schedules with higher execution time (cf. Example 2).

Our approach allows for further optimization of task graphs without violating the results. E.g. we do not consider that two messages of two different tasks which are sent to the same processor may be merged. However, the performance guarantee in Theorem 3 cannot be improved. Example 1 shows that the bound given in Lemma 1 and Theorem 3 is also tight in this case.

Compilers for data-parallel languages or parallelizing compilers using results of dependency analysis and tiling usually compile source programs into target programs which correspond to linear schedules. Thus, Theorem 3 is applicable also to this case. This paper shows that the order of the communication is an important issue to improve performance. Therefore the question arises, how compilers can produce an adequate order of the communication.

References

[AISS95] A. Alexandrov, M. Ionescu, K. E. Schauser, and C. Scheimann. Loggp: Incorporating long messages into the logp-model – one step closer towards a realistic model for parallel computation. In *7th Annual Symposium on Parallel Algorithms and Architectures*, pages 95–105. ACM Press, 1995.

[CKP+93] D. Culler, R. Karp, D. Patterson, A. Sahay, K. E. Schauser, E. Santos, R. Subramonian, and T. von Eicken. LogP: Towards a realistic model of parallel computation. In *4th ACM SIGPLAN Symposium on Principles and*

Practice of Parallel Programming (PPOPP 93), pages 1–12, 1993. published in: SIGPLAN Notices (28) 7.

[CKP⁺96] D. Culler, R. Karp, D. Patterson, A. Sahay, K. E. Schauser, E. Santos, R. Subramonian, and T. von Eicken. LogP: A practical model of parallel computation. *Communications of the ACM*, 39(11):78–85, 1996.

[DI94] B. Di Martino and G. Ianello. Parallelization of non-simultaneous iterative methods for systems of linear equations. In *Parallel Processing: CONPAR 94 – VAPP VI*, volume 854 of *Lecture Notes in Computer Science*, pages 253–264. Springer, 1994.

[ELW97] Jörn Eisenbiegler, Welf Löwe, and Andreas Wehrenpfennig. On the optimization by redundancy using an extended LogP model. In *International Conference on Advances in Parallel and Distributed Computing (APDC'97)*, pages 149-155. IEEE Computer Society Press, 1997.

[GY93] A. Gerasoulis and T. Yang. On the granularity and clustering of directed acyclic task graphs. *IEEE Transactions on Parallel and Distributed Systems*, 4:686–701, June 1993.

[HK73] J. E. Hopcroft and R. M. Karp. An $n^{5/2}$ algorithm for maximum matchings in bipartite graphs. *SIAM Journal on Computing*, 2(4):225–231, 1973.

[KSSS93] R. M. Karp, A. Sahay, E. E. Santos, and K. E. Schauser. Optimal broadcast and summation in the LogP model. In *5th Annual ACM Symposium on Parallel Algorithms and Architectures*, pages 142–153. ACM, 1993.

[LEZ96] W. Löwe, J. Eisenbiegler, and W. Zimmermann. Optimizing parallel programs on machines with fast communication. In *9. International Conference on Parallel and Distributed Computing Systems*, pages 100–103, 1996.

[LMZ97] W. Löwe, M. Middendorf, and W. Zimmermann. Scheduling inverse trees under the communication model of the logp-machine. *Theoretical Computer Science*, 1997. Submitted, currently under revision.

[LZ95a] W. Löwe and W. Zimmermann. On finding optimal clusterings in task graphs. In N. Mirenkov, editor, *Parallel Algorithms/Architecture Synthesis pAs'95*, pages 241–247. IEEE, 1995.

[LZ95b] W. Löwe and W. Zimmermann. Programming data-parallel – executing process parallel. In P. Fritzson and L. Finmo, editors, *Parallel Programming and Applications*, pages 50–64. IOS Press, 1995.

[PY90] C.H. Papadimitriou and M. Yannakakis. Towards an architecture-independent analysis of parallel algorithms. *SIAM Journal on Computing*, 19(2):322 – 328, 1990.

[YG94] T. Yang and A. Gerasoulis. DSC: Scheduling parallel tasks on an unbounded number of processors. *IEEE Transactions on Parallel and Distributed Systems*, 5(9):951–967, 1994.

[ZL94] W. Zimmermann and W. Löwe. An approach to machine-independent parallel programming. In *Parallel Processing: CONPAR 94 – VAPP VI*, volume 854 of *Lecture Notes in Computer Science*, pages 277–288. Springer, 1994.

Rescheduling Support for Mapping Dynamic Scientific Computation onto Distributed Memory Multiprocessors [*]

Apostolos Gerasoulis[1] and Jia Jiao[2]

[1]Department of Computer Science, Rutgers University, New Brunswick, NJ 08903.
[2]Department of Computer Science, University of Arkansas, Little Rock, AR 72204.

Abstract. In this paper we discuss the approach of applying task graph rescheduling method to the efficient parallelization of irregular/dynamic problems. Two new rescheduling algorithms are proposed: One for task graph weight variation, the other for dynamic spawning of subgraphs. These algorithms are localized and incremental, with very low complexity. But for coarse grain task graphs they yield schedules that are competitive to global rescheduling from scratch. Experiments with the N-body problem and Vortex Sheet simulation show good speedups for these problems on the NCUBE-2s.

1 Introduction

Many real world scientific computing problems are irregular, and dynamic. Efficient parallelization of such problems are notoriously hard. As far as we know a complete integrated scheduling system that is sensitive to the application, where a user can select the appropriate scheduler for a given application and architecture, does not exist in the literature. In this paper, we report our results of applying compile-time scheduling/run-time rescheduling algorithms on applications that possess some of the properties such as slowly changing computation, iterative dynamic and local computation.

We consider two important classes of applications that have the property of *slowly changing computation* in time iterative steps: The N-body computation in Galaxy simulations and Vortex dynamics. One can take advantage of slowly changing computation to better guide run time scheduling algorithms. This property is commonly encountered in a wide range of numerical applications and is due to

[*] The work presented here was in part supported by ARPA contract DABT-63-93-C-0064 under "Hypercomputing and Design" project.

the fact that the stability of the computation requires sufficiently small time steps to compute accurately, [1, 6].

The Fast Multipole N-body Simulation: The N-body problem of simulating particle interactions is one of the most famous in scientific computing. The FMM method [7] is a low complexity approximation algorithm to solve the N-body problem. An example to illustrate the FMM irregular partitioning is shown in figure 1.

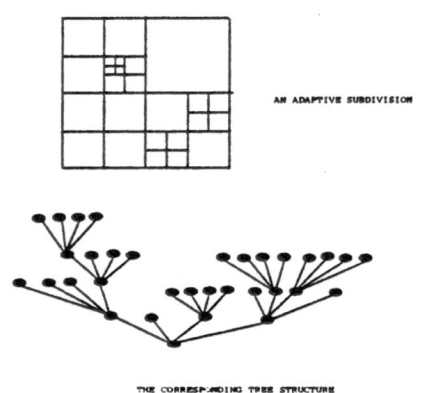

Fig. 1. A non-uniform adaptive subdivision

As we indicated, the task graph for the FMM algorithm is dynamic. Two kinds of changes are possible in the task graph:

During the N-body simulation, particles are moving across regions. If the initial space partitioning is kept unchanged the result will be a change in the task weights that correspond to each box.

After a significant number of iterations, the particle distribution may have shifted considerably. Although it's possible to retain the space partitioning and reschedule the task graph with weight changes only, it is usually more beneficial to conduct a repartitioning of space, by splitting the boxes with excessive concentration of particles. This will increase parallelism by new "spawning" of the task graph, i.e. new subtrees will be grown from the leaf nodes that are split or old subtrees will be deleted.

Vortex Sheet Roll-Up simulation: To gain deeper insight to the issue of the applicability of our scheduling algorithms we considered the vortex sheet roll-up calculation, [8]. We also use the FMM algorithm for this problem.

There are 2 types of configurations. For the first type, the tip of the sheet roll-up from the straight line. For the second type, there is global stretch of the sheet and vortex forming in the middle of the sheet. Dynamic point insertions are needed to interpolate the sheet, making the problem even more challenging. We show the 2nd type of configuration in figure 2

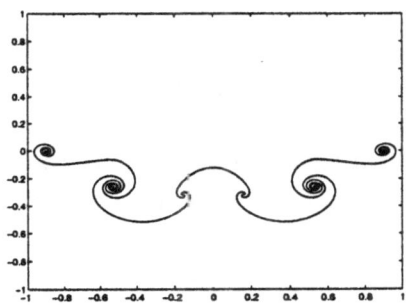

Fig. 2. A roll-up of type 2

2   Scheduling heuristics

Scheduling algorithms can improve the performance of many applications. Here we consider heuristics for dynamic and irregular applications. We use the macrodataflow model and task graphs to represent computation, [5]. We first describe the PYRROS system developed for static task graphs and then discuss a modification of PYRROS for dynamic task graphs.

The static PYRROS scheduling system: The PYRROS [5], scheduling and code generator tool takes as an input a weighted task graph, predicts its performance on a given parallel architecture and produces parallel code. The task and communication weights are estimated either at compile time using the number of operations and the volume of communication for task computation model. Our experimental evidence show that for coarse grain application with sufficient parallelism PYRROS predictions are very close to the actual execution time of the run time PYRROS system. Many real applications have been implemented using the PYRROS system on a variety of multiprocessors and reported excellent performance [3], [9].

The dynamic D-PYRROS scheduling system: We have modified the PYRROS system to be able to execute dynamic task graphs, [6]. The basic idea is to use an initial scheduling and then incrementally modify the schedule at run time to correct load imbalances. We have implemented two new scheduling heuristics that utilize work and communication counting at run time, which we briefly describe below.

A. Generalized Load Balancing for Weight Fluctuations:
We assume that the system is monitoring the parallel time, and as soon as the performance degrades to an unacceptable level, it counts the new task weights, and identifies a subgroup of tasks with large weight increases. Then it uses this information to re-map the task graph.

Load balancing is a topic that has been extensively studied in the literature. See for instance [10, 11, 12]. Traditional load balancing algorithms, e.g. self scheduling and task stealing algorithms [2], only consider work load among processors without regards to communication and data dependencies. This could result in an increase in the parallel time even though the load is balanced.

We have developed a "Generalized Load Balancing Heuristic", which detects potential parallel time increases before the load is transfered, and only carries out those load transfers that will not cause such problem. The algorithm uses the communication to computation ratio to decide if tasks should be moved or not in different processors, [6, 4]. We briefly describe our approach below:

1. Use a threshold to select N_c tasks with the largest increases in the weights.

2. For each task in N_c move tasks ordered after them in the same processor to another processor with light load provided certain conditions are satisfied.

If no conditions are imposed and only the load is used to move a task then this approach is similar to the traditional load balancing techniques. We have used two conditions to enhance the traditional approach. The first condition tries to avoid breaking dependence chains between two processors. The second conditions avoids the moving of tasks with high communication edges between them. The result of these two optimization is a global reduction of the parallel time rather than simply load balancing. More details are given in [6].

Since the algorithm only focus on the tasks with largest changes it is localized and has low complexity. For most cases the complexity is only a fraction of that of global scheduling. In particular, if the number of tasks with large weight changes is N_c and the average degree of the task graph is d, then the complexity of the algorithm is $O(N_c * d)$. This is because for each of the N_c "critical" tasks, only a constant number of "candidates", tasks considered to move, are examined, and the number of edges examined is about d times larger.

For most problems, including the N-body, $N_c << N$ and d is constant, N is the total number of tasks.

B. Incremental clustering for task graph spawning:
We use an incremental clustering method to handle cases when new subgraphs are grown from the original graph. We first cluster the new subgraphs locally using DSC, [3] and then merge the new cluster with the original schedule in an optimal way to reduce the parallel time. Although this does not ensure a global optimal, neither does the global PYRROS scheduler. However, if we are dealing with the specific case of "tree-spawning" subgraphs, meaning that the new subgraph is attached to one particular "spawning root" in the original graph, then we can prove that:

For coarse grain tree spawning task graphs the incremental clustering approach guarantees that the resulting clustering parallel time will not be worst than the global clustering parallel time.

The idea is roughly as follows: If the graph is coarse grain, then only linear clustering will be produced by the clustering algorithm [3]. So if we use local clustering on the subgraph, it will produce local linear clustering. On the other hand, if global clustering is used, because the graph is tree-spawning, there is only one connection point to the subgraph, and therefore the linear clustering on the subgraph will be localized. It can be shown that since the merge algorithm for the localized heuristic does the cluster merge with the global schedule by examining all the possible ways, it can produce a schedule no worse than the global one with linear clustering [6]. (Note that neither one can guarantee optimality, however, so the comparison is only between themselves).

Even if the graph is not strictly tree-spawning, our incremental algorithm has comparable to global clustering experimental performance, [6]. The complexity of this algorithm is proportional to the newly spawned subgraphs, not the whole graph, making it another localized heuristic. One central issue in the algorithm is the merging of new clusters into the original schedule. The algorithm needs to determine amongst 3 possible choices to find the best parallel time. However, this step has very little additional cost since parallel time information can be derived from the clustering step very quickly, [6].

3 Experimental performance results of the new heuristics

Application to the classic N-body problem: We consider the simulation of a non-uniform twin-galaxy test case with 10000 particles on 64 processors of an Ncube2s machine. In figure 3, we show the parallel time variation in time steps. Initially we use PYRROS to schedule the task graph and this schedule is good up to 100 time steps but start to degenerate afterward so rescheduling is called for. For this example the reduction of the parallel time of global rescheduling, shown in figure 3 at 145 step, was about the same as the two local heuristics described previously (reduced from 3.02 seconds to a range between 2.70-2.78 sec). However, the overhead for the global heuristic is much larger than the local heuristics, 8 sec. for global rescheduling, 0.7 sec. for graph spawning and 0.35 sec. for load balancing heuristics.

The Vortex Sheet Simulation: Next we consider the vortex sheet simulation and use both the load balancing and graph spawning local heuristics. If there is a global movement of the vortex sheet and no point concentration then the load balancing heuristic is used, but when the vortex sheet reaches a certain stability point and the vortices start forming the graph spawning heuristic is used.

We examine one example of vortex sheet roll-up of type 2 on 16 Nube2s processors. Initially, 800 blobs are lined up on a straight line.

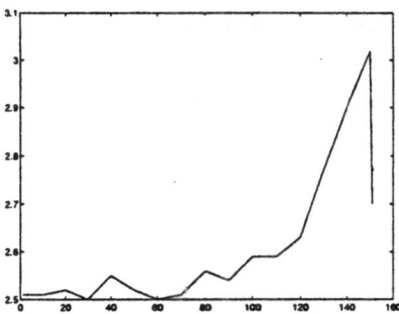

Fig. 3. Parallel time per iteration in the N-body test case

Later the middle section exhibits a large shifting pattern, which demand the use of generalized load balancing method. After about 180
iterations the pattern starts to stabilize but dense vortices are forming with concentration of new blobs, so we must use spawning and
incremental clustering to handle the performance loss. The evolution
of parallel time is shown in figure 4. If no rescheduling is attempted,
the parallel time will become 75% higher after 200 iterations. Otherwise rescheduling improves the performance as it can be seen at
steps 145, 180 and 190.

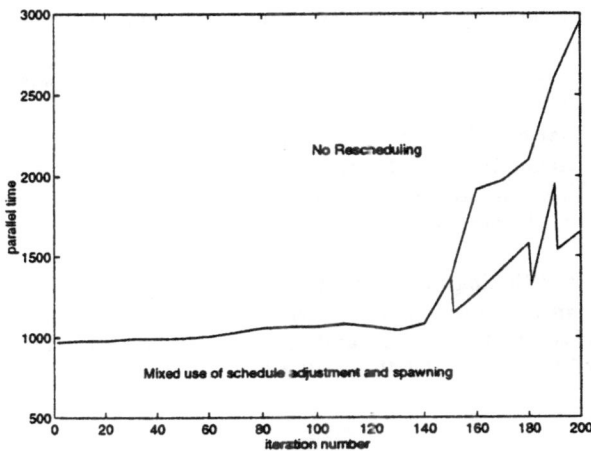

Fig. 4. Use of rescheduling to correct performance loss for the vortex sheet

We can also look at the speedup behavior and found that rescheduling improves the speedup performance. Details can be found in [6].

In conclusion, both examples show that rescheduling can be used to increase the speedup and reduce the parallel time. For such examples, with slowly changing and local behavior, local rescheduling heuristics are sufficient.

References

1. Jaswinder Pal Singh, Parallel Hierarchical N-Body Methods And their Implications for Multiprocessors, Ph.D. Thesis, Stanford U., Feb 1993.
2. Robert D. Blumofe, and Charles E. Leiserson. Scheduling Multithreaded Computations by Work Stealing. *In Proceedings of the 35th Annual Symposium on Foundations of Computer Science (FOCS)*, pages 356-368, Santa Fe, New Mexico, November 20-22, 1994. (90 KBytes, compressed postscript.) California, July 19-21, 1995. postscript.
3. T. Yang and A. Gerasoulis. DSC: Scheduling parallel tasks on an unbounded number of processors, *IEEE Trans. on Parallel and Distributed Systems.* Vol. 5, No. 9, 951-967, 1994.
4. A. Gerasoulis and T. Yang. On the granularity and clustering of directed acyclic task graphs. *IEEE Trans. on Parallel and Distributed Systems.*, Vol. 4, no. 6, June 1993, pp 686-701.
5. A. Gerasoulis, J. Jiao and T. Yang, Experience with Scheduling Irregular Scientific Computation, Proc. of 1st IPPS Workshop on Solving Irregular Problems on Distributed Memory Machines. 1-8, Santa Barbara, CA, April 1995.
6. Jia Jiao, Software support for parallel processing of irregular and dynamic computations, Ph.D. Thesis, Rutgers University, Oct. 1996.
7. Leslie Greengard, The rapid evaluation of potential fields in particle systems, Ph.D. Thesis, Yale University, 1987
8. R. Krasny, Computation of vortex sheet roll-up in the treffitz plane, *Journal of Fluid Mechanics*, Vol 184, pp. 123-155, 1987.
9. F. Chong, S. Sharma, E. Brewer, J. Saltz, Multiprocessor Runtime Support for Fine-Grained, Irregular DAGs. Parallel Processing Letters, December, 1995.
10. A. Y.Grama, V. Kumar and V. N. Rao, Experimental Evaluation of Load Balancing Techniques for the Hypercube. *Proceedings of Parallel Computing 1991, Spe. 1991, London*
11. H. D. Simon, Partitioning of Unstructured Problems for Parallel Processing, *Computing Systems in Engineering, Vol. 2, Number 2/3, pp. 135 - 148, 1991*
12. Bruce Hendrickson and Robert Leland, Multidimensional Spectral Load Balancing, *Proc. 6th SIAM Conf. Parallel Proc., 953-961, 1993*

Versatile Task Scheduling of Binary Trees for Realistic Machines

Cristina Boeres* and Vinod E. F. Rebello

Departamento de Computação, Universidade Federal Fluminense (UFF)
Niterói, RJ, Brazil
e-mail: boeres@dcc.uff.br

Abstract. In general, scheduling models only consider message latency as the sole dominant communication parameter. However, in many parallel systems, latency is negligible when compared to the CPU penalties associated with sending and receiving *communication events*. Our work considers a model in which this *CPU penalty* can also be a significant communication parameter. This paper focusses on analysing the effect of such a model on the scheduling of Full Binary Trees. We briefly describe a versatile, multi-stage scheduling approach that can be customised to classes of parallel systems according to their communication performance characteristics and which produces better makespans when compared with traditional techniques.

1 Introduction

It is well known that the scheduling problem is NP-complete in its general form and that a number of architectural and application-related characteristics have influenced the design of many scheduling heuristics [7, 9, 11]. With the development of distributed memory machines, more realistic models of parallel computation have been proposed where communication characteristics are represented. More recently however, the interface between a processor and the communication network has been identified as a potential *bottleneck* [5, 8]. Just as machines continue to evolve and their characteristics change, so must the models used to represent them. The scheduling model considered in this work defines not only the network delay or latency as a significant parameter, which has been given much attention previously, but also the overhead incurred when sending and receiving messages. This is not part of the delay but rather a cost in the form of a *CPU penalty* incurred on processors when communicating. Many scheduling models (e.g. [7]) fall into the class of the linear communication model, adding the overhead to the delay cost either for simplicity or when considering dedicated processors are used to perform communication. However, since the overhead is a blocking factor, the linear model may not reflect reality. The CLAUD model [4] is adopted as a more realistic model because it defines important characteristics

---

* This work was carried out while the author was with the Department of Computer Science, University of Edinburgh, Scotland, UK, and was being supported by a grant from CAPES, Ministry of Education, Brazil, under grant number 1553/91-8.

of applications and target architectures such as: the existence of *communication events* as tasks; the overheads associated with the *sending* and *receiving* of a message (denoted by λ_s and λ_r, respectively); and the delay τ which is incurred when a processor transmits a data item to an adjacent processor.

In this work, an input application is represented by a directed acyclic graph (DAG) $G = (V, E)$. The replication of tasks onto distinct processors is allowed, since task replication can be used to minimise communication costs in architectures with high communication costs [10]. The time that the processor is *blocked* characterises a *communication event* and can also be regarded as a "task". The existence of these special tasks depends on the scheduling of the computation tasks of the application. In parallel architectures with high overheads, the number of communication events can be reduced by bundling many small messages into fewer, larger ones through the clustering of tasks [1]. In this context, a cluster c of tasks is defined as a set of tasks of G, listed in accordance to their precedence relationship and with no communication occuring while the tasks in c are being executed. Messages to be sent by tasks in c to tasks in an immediate successor cluster c' are bundled and sent together. The dependencies between clusters represent both the dependency inherited from the original DAG G and the communication between the distinct processors to which the clusters have been allocated. This dependency relation and the set of all clusters defined are represented by a DAG called a *superdag*. The vertices in the superdag, the *supertasks*, correspond to clusters of tasks and the *superedges* are the communications between supertasks.

This paper gives a brief description of a *versatile* task scheduling approach followed by a discussion on the merits of two alternative partitioning strategies for Full Binary Trees together with experimental results.

2   A General Overview of the New Approach

The *multi-stage approach* (MSA) [1] performs a series of transformations on the input DAG until its schedule on the target machine is finalised. MSA aims to produce good schedules for general applications on a variety of target architectures, particularly those which have high communication overheads. The following sections briefly describe this new approach, but a more detailed description of each of the two stages of MSA can be found in [1].

I) Constructing a superdag – The first stage of MSA is comprised of two heuristics and produces a superdag with characteristics that are dependent on the target architecture and the input DAG. The first heuristic, the *replication* algorithm, adopts the *replication* principle (also used by Papadimitriou and Yannakakis [11]) to cluster tasks so that the number of communication events can be reduced. A cluster c is created by allocating v_i to a *virtual* processor p_i and replicating γ selected ancestors of v_i on to p_i, where γ is the *clustering factor*. The γ ancestors are chosen, to minimise the number of communication events, based on their value $e_l(v_i)$ which is the *latest* time at which task v_i can start without increasing the makespan of the resulting schedule. One advantage of using e_l-values is an increase in the number of candidate tasks that can be clustered with v_i (there tends to be larger groups of tasks with the same e_l-value than with the same earliest start time as defined in [11]). The result of the *replication* algorithm is a schedule containing sets of clearly defined parallel clusters, each set being executed during a *computational interval* or *band*. Between any two bands there exists a *communication interval* where only communication event tasks are scheduled.

Not all of the clusters produced by the *replication* algorithm may be needed in order to schedule the DAG G on the target machine. In order to select reasonably good candidates, a set of *rules* has been implemented in the second heuristic (the *superdag* algorithm [1]). This algorithm aims to construct a superdag with a granularity that matches the granularity of the target architecture. For each supertask, the immediate predecessors chosen are those which send the largest messages and have the smallest out degrees. These are necessary characteristics when the overhead is high. On the other hand, when the delay is the dominant communication parameter, the rules are parameterised in such a way that the immediate predecessors chosen are those which send the smallest messages.

II) Remapping onto physical processors – The superdag produced by the first stage will never contain more supertasks than the number of tasks in the original DAG. The number of supertasks may, of course, be greater than the number of available physical processors m, since the first stage of MSA does not take this parameter into account but rather allocates each supertask to a distinct virtual processor. Thus, the second stage of MSA transforms the given superdag so it can be executed efficiently on a fully connected network of m processors. This is achieved by exploiting the degree of parallelism within the superdag, i.e. the *parallel* supertasks scheduled in the same *computational* interval.

Given the schedule S and the superdag $G_S = (V_S, E_S)$, the supertasks are clustered by merging parallel supertasks and remapping them on to physical processors, if the number of supertasks in that band is greater than m. The merging is achieved in such a way that the number of communication events is minimised, even though an increase in the length of messages may be necessary.

3   Scheduling Full Binary Trees

The Full Binary Tree $B_n = (V_B, E_B)$ is a DAG with n tasks, in which the direction of the edges in E_B are towards the unique sink in V_B. In order to analyse the impact of the communication parameters on the scheduling of B_n, the *layer* and *subtree* partitions are each considered under the following models: the *delay* model (as described in [11]), where the communication cost is solely characterised by the communication delay τ; and the CLAUD model, where the overheads λ_s and λ_r are also defined as communication parameters together with τ (for brevity, whenever the overheads for sending and receiving are referred to as λ, it means that $\lambda_s = \lambda_r = \lambda$).

3.1 The *layer* partition

Based on the asymptotically optimal schedule of Papadimitriou and Yannaka-kis [11], Jung *et al.* [10] devised a scheduling strategy for B_n under the *delay* model by dividing it into layers. In order to incorporate the overhead λ, which is a blocking factor, the *layer* partition for B_n is formed by abstracting away knowledge of the communication parameters and making use of the *clustering factor* γ. Depending on the target model γ may assume different values, i.e. it may be assigned to λ, τ or even some other characteristic of the input *DAG*.

The tree B_n is partitioned into $h = \log_{\gamma+2}(n+1)$ layers, producing subtrees (supertasks) each comprised of $(\gamma + 1)$ tasks. The in-degree of each supertask (apart from the sources) is equal to $\gamma + 2$ and therefore, the resulting superdag is a $(\gamma+2)$-*ary tree*. Note also that each superedge is associated with unit weight since only the respective sink task in each subtree sends data to an immediate successor supertask.

The layer *partition and the* delay *model* – Since all communications between each layer take place in parallel, the cost (which is τ) is constant irrespective of the number of immediate predecessors allocated to distinct processors. Nevertheless, if more than one supertask from each layer is allocated to the same processor, then the computation time is increased. Therefore, the minimum makespan $h(\tau + 1) + (h - 1)\tau$ is achieved when there are at least $\frac{n+1}{\tau+2}$ processors.

The layer *partition and the* CLAUD *model* – The supertasks at each layer are allocated to m processors with each supertask being allocated to the same processor as one of its immediate predecessors. Note that the upper bound on the number of processors is the number of supertasks in layer $h - 1$, i.e. $(\gamma + 2)^{h-1}$. The makespan of this schedule, proved in [1], is expressed by

$$(\gamma+1)\cdot\sum_{i=k}^{h-1}\frac{(\gamma+2)^i}{m} + k(\gamma+1) + (k-1)[(\gamma+2)\lambda+\tau] + \frac{m}{(\gamma+2)^{k-1}}\lambda + \frac{(\gamma+2)^k}{m}\tau$$

where $k = \lceil\log_{\gamma+2} m\rceil$ is the layer in which the number of supertasks equals the number of processors.

3.2 The *subtree* partition

There exist scheduling algorithms [7, 9, 12] which tend to allocate the critical path of a *DAG* to a single processor. When such algorithms are applied to B_n, the allocation may be "inefficient" if the target machine has high overheads and particularly when they are much greater than the execution costs of the tasks. The *layer* partition provides subtrees as clusters of tasks, producing a coarser *DAG*. However, depending on the value of λ, the number of communication events which are incurred can be high since a $(\gamma + 2)$-ary tree is produced.

A partition for B_n is proposed in which the number of communication events is not dependent on the overhead parameter. The partition divides B_n *vertically* with each subtree being allocated to one of the m processors. For the sake of clarity, let $m = 2^k$, for some k. The $\frac{n+1}{2}$ sources of B_n are divided into m subsets so that $\frac{n+1}{2m}$ tasks are sources of m subtrees and with each subtree being allocated to a distinct processor. Each one of the m subtrees has its respective sink at level $k = \log m$. The expression for the makespan of a schedule produced by this partitioning strategy holds regardless of whether the communication cost is comprised of latency and/or overheads [1]. The makespan of the *subtree* partition, also proved in [1], is $\frac{n+1}{m} - 1 + (\log m - 1)(2\lambda + \tau + 1)$.

4 Experimental evaluations

MSA can produce a schedule for the Full Binary Tree which is equivalent to the *subtree* partition if the *clustering factor* γ has the value zero. For values other than zero, MSA produces the *layer* partition [1]. In this section, an analysis of the partitions discussed in this paper is carried out taking into account different target models.

Overheads are the only communication costs – When executing the *subtree* partition on m processors, the number of communication events that occur when two processors communicate is equal to two, regardless of the number of tasks, the value of λ and the number of processors m. On the other hand, in the *layer* partition with $\gamma = \lambda$, the binary tree is "transformed" into an $(\lambda + 2)$-ary tree corresponding to the fact that $(\lambda + 1)$ receiving events are carried out. Therefore, the higher the value of λ, the smaller the number of processors that can usefully be utilised to achieve the minimum makespan.

Figure 1 compares the makespans of the two partitioning strategies for B_{2047}. When $\lambda = 2$, the layer partition produces results very close to the *subtree* partition because of the low in-degree of each supertask. On the other hand, for larger overheads (e.g. when $\lambda = 30$), the schedules provided by the *layer* partition become worse when compared with those of the *subtree* partition.

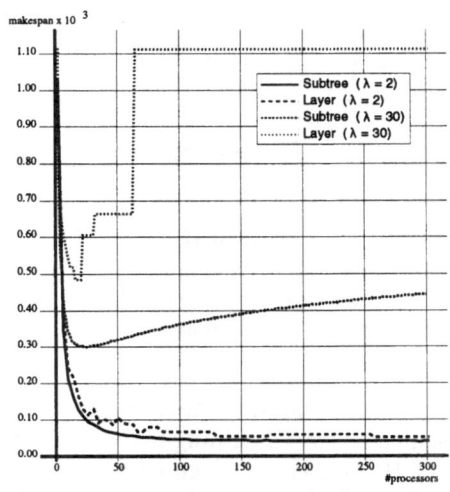

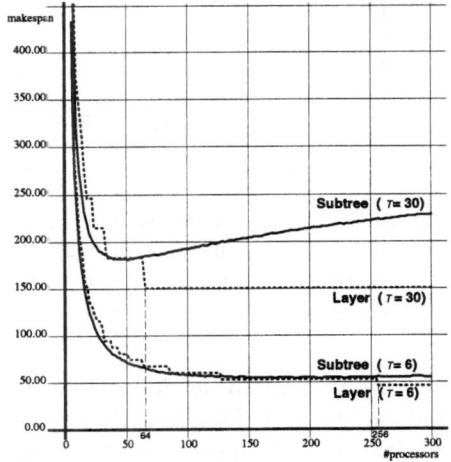

Fig. 1. The makespans of the *subtree* ($\gamma = 0$) and *layer* ($\gamma = \lambda$) partitions for B_{2047} with $\tau = 0$, when $\lambda = 2$ and $\lambda = 30$.

Fig. 2. The makespans of the *subtree* ($\gamma = 0$) and *layer* ($\gamma = \tau$) partitions for B_{2047} with $\lambda \simeq 0$, when $\tau = 6$ and $\tau = 30$.

The delay *model* – In order to compare the *subtree* and the *layer* partitions under the *delay* model, suppose that the number of available physical processors is at least $\frac{n+1}{\tau+2}$. Thus, the higher τ is, the smaller n (the number of tasks of B_n) must be so that the *subtree* partition produces a better schedule. The results confirm that the *layer* partition does provide schedules with smaller makespans than the *subtree* partition, particularly when $m \geq \frac{n+1}{\tau+2}$ (Figure 2 is an example for B_{2047}).

The CLAUD *model* – Both communication parameters are now considered significant and two cases are given attention in the experiments. Firstly, when $\lambda > \tau$ (the overhead is dominant), the clustering of tasks as in the *layer* partition is not wise since the in degrees can be large. Regardless of the number of processors, the best approach is the *subtree* partition, as seen in Figure 3. In the second case, when $\lambda \leq \tau/2$ (the factor of two allows a full subtree to be clustered into a supertask), the delay is dominant and although for $\gamma = \tau$ the respective superdag may have a high in degree, the values for λ are not high enough to hinder the benefits of partitioning B_n into layers. The best results were achieved when $m \geq \frac{n+1}{\tau+2}$ and $\gamma = \tau$ for the *layer* partition (Figure 4).

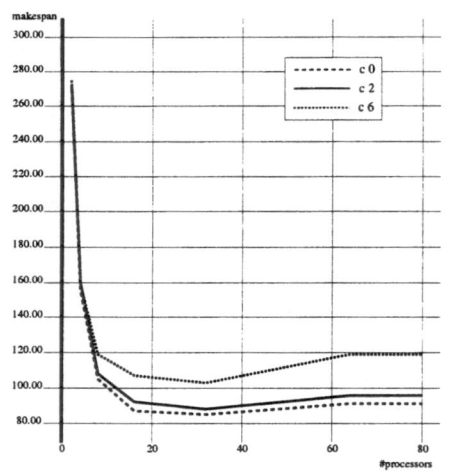
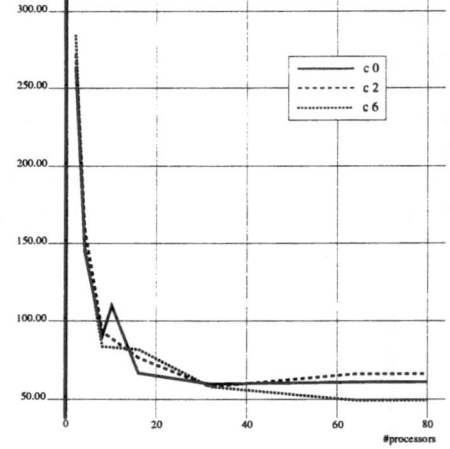

Fig. 3. Makespans of the *subtree* ($\gamma = 0$ corresponds to $c0$) and *layer* ($\gamma = 2, 6$ correspond to $c2$ and $c6$, respectively) partitions for B_{511}, $\tau = 1$ and $\lambda = 6$.

Fig. 4. Makespans of the *subtree* ($\gamma = 0$ corresponds to $c0$) and *layer* ($\gamma = 2, 6$ corresponds to $c2$ and $c6$, respectively) partitions for B_{511}, $\tau = 6$ and $\lambda = 1$.

Comparing MSA with other heuristics – In evaluating MSA and the *layer* and *subtree* partitions, two algorithms classified as list scheduling heuristics were implemented: Earliest Task First (ETF) of Hwang *et al.* [9] and the Mapping Heuristic (MH) of Rewini *et al.* [6]. The sending and receiving overheads were incorporated in the cost function that guides ETF. MH incorporates the cost of sending a message into its cost function, therefore representing the linear communication model. The schedules produced by MSA, ETF and MH were executed on a *simulated machine*, i.e. a tool capable of simulating the execution of given schedules on parallel machines with different characteristics.

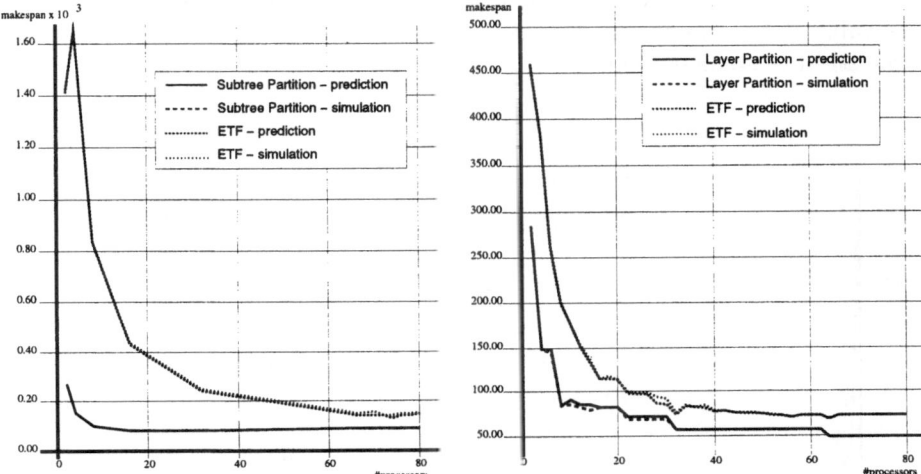

Fig. 5. Makespans of MSA (the *subtree* partition using $\gamma = 0$) and ETF for B_{511}, $\tau = 1$ and $\lambda = 6$.

Fig. 6. Makespans of MSA (the *layer* partition using $\gamma = \tau$) and ETF for B_{511}, $\tau = 6$ and $\lambda = 1$.

The best results produced by the *subtree* and *layer* partitions for B_{511}, are compared with ETF (see Figures 5 and 6). When $\lambda > \tau$, it is better to set the parameter γ to zero in the first stage of MSA (for the *subtree* partition). One of the problems with ETF is that due to local decisions, when the number of available processors is smaller than the number of source tasks, the source tasks which do not belong to the same subtree may be allocated to the same processor, i.e. ETF does not attempt to minimise the number of communication events. The second case examines the situation in which $\lambda < \tau$ and where MSA is executed with $\gamma = \tau$. In Figure 6, ETF's results are close to the *layer* partition of MSA, but only because the overheads are much smaller than the delay τ. The schedules generated by MSA and ETF for B_{511} were executed on the simulated machine considering a fully connected network. Both predictions are very accurate, but MSA gives better results than ETF for any number of processors.

To compare MSA with MH, experiments are carried out with the significant communication parameters being λ_s and τ. For B_n, the *sending event* tasks will be executed in parallel irrespective of the partitioning applied. Therefore, the best partitioning technique is the one that minimises the number of *sending event* tasks along the critical path of the corresponding superdag. In this case, the in degree of the $(\gamma+2)$-ary tree produced by the *layer* partition does not hinder the performance of the resulting schedule. First, experiments are performed with the delay τ as the only significant parameter (Figure 7). The best value of γ is that which produces supertasks containing Full Binary Trees (verified experiment-ally [1]). Figure 8 shows the results for the CLAUD model where λ_s and τ are significant. The times to completion of the simulation of MSA schedules are as predicted and therefore have been omitted from the respective figures. In all of the cases considering the *delay* model and CLAUD model, MSA produces better results then MH. Also, in the case of MH, Figures 7 and 8 together show that as the value of λ_s increases, so does the difference between the simulated execution time and the predicted one.

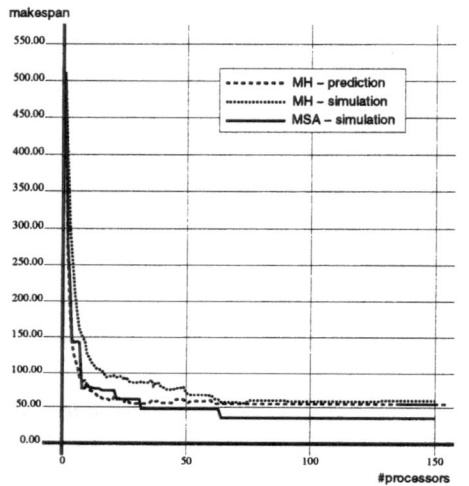

Fig. 7. Makespans of MSA with $\gamma = 6$ and MH, for B_{511} with $\tau = 6$ and $\lambda_s = \lambda_r = 0$.

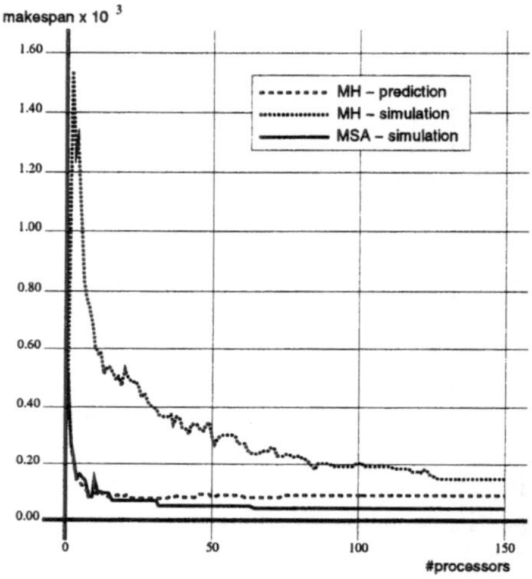

Fig. 8. Makespans of MSA with $\gamma = 6$ and MH, for B_{511} with $\tau = 1$, $\lambda_s = 10$ and $\lambda_r = 0$.

5 Concluding Remarks

Jung *et al.* [10] showed that the Full Binary Tree can be asymptotically optimally scheduled without replication, given that the target model has a single communication parameter – the delay τ. The *layer* partition provides a mechanism of transforming the input *DAG* onto a coarser one. However, when the *blocking* parameters λ_s and λ_r are considered, the partitioning of the Full Binary Tree into layers produces supertasks with in degree ≥ 2. In architectures with high overheads, the parallel execution of such superdags leads to a high communication cost due to the sending and receiving events. The *subtree* partitioning is proposed to overcome this problem and is provided by MSA when γ, the *clustering factor*, is zero.

Note that ETF and MH do not provide schedules that are completely unreasonable. In fact, for the *delay* model, both ETF and MH can provide better results than MSA when there is a small number of available physical processors and τ is very low. However, this is not true (under the *delay* model) for larger τ and, particularly, for the CLAUD model. In the latter case, by adding the startup overhead to the delay, ETF and MH specify that the startup cost can also be overlapped with task computation. This may be unrealistic and consequently, the execution times of the schedules may be much higher than their respective predictions.

This paper has evaluated the impact of the delay and the overhead communication parameters on the schedules of a certain class of regular *DAGs*, known as Full Binary Trees. Two types of partitions were analysed and the results showed that the partition which produced the best schedule depends on the characteristics of the target model. Our *multi-stage approach* displayed its versatility by tuning the input parameters that guide the clustering of tasks and bundling of messages so that a suitable schedule for the given underlying model is produced. An analysis of other regular structures, such as the diamond DAG and irregular graphs, and their partitioning strategies is described in [1].

References

1. C. Boeres. *Versatile Communication Cost Modelling for Multicomputer Task Scheduling Heuristics*. PhD thesis, Department of Computer Science, University of Edinburgh, Oct 1996.
2. C. Boeres, G. Chochia, and P. Thanisch. On the scope of applicability of the ETF algorithm. In *Proceedings of the 2nd International Workshop on Parallel Algorithms for Irregularly Structured Problems (IRREGULAR'95)*, Lecture Notes in Computer Science (LNCS 980), pages 159–164, Lyon, France, Sept 1995. Springer.
3. R.P. Brent. The parallel evaluation of general arithmetic expressions. *J. ACM*, 21:201–206, 1974.
4. G. Chochia, C. Boeres, M. Norman, and P. Thanisch. Analysis of multicomputer schedules in a cost and latency model of communication. In *Proceedings of the 3rd Workshop on Abstract Machine Models for Parallel and Distributed Computing*, Leeds, UK., April 1996. IOS press.
5. D. Culler *et al.* LogP: Towards a realistic model of parallel computation. In *Proceedings of the 4th ACM SIGPLAN Symposium on Principles and Practice of Parallel Programming*, San Diego, CA, May 1993.
6. H. El-Rewini and T.G. Lewis. Scheduling parallel program tasks onto arbitrary target machines. *J. Parallel Dist. Comput.*, 9:138–153, 1990.
7. A. Gerasoulis, S. Venugopol, and T. Yang. Clustering task graphs for message passing architectures. In *The International Conference on Supercomputing*, pages 447–456, Amsterdam, The Netherlands, June 1990.
8. F.W. Howell. Reverse profiling. In Innes Jelly, Ian Gorton, and Peter Croll, editors, *Software Engineering for Parallel and Distributed Systems*, pages 244–255. Chapman and Hall on behalf of IFIP, March 1996.
9. J-J. Hwang, Y-C. Chow, F.D. Anger, and C-Y. Lee. Scheduling precedence graphs in systems with interprocessor communication times. *SIAM J. Comput.*, 18(2):244–257, 1989.
10. H. Jung, L. Kirousis, and P. Spirakis. Lower bounds and efficient algorithms for multiprocessor scheduling of DAGs with communication delays. In *Proc. ACM Symposium on Parallel Algorithms and Architectures*, pages 254–264, 1989.
11. C.H. Papadimitriou and M. Yannakakis. Towards an architecture-independent analysis of parallel algorithms. *SIAM J. Comput.*, 19:322–328, 1990.
12. V. Sarkar. *Partitioning and Scheduling Parallel Programs for Multiprocessors*. Pitman, London, 1989.

Load Balancing Issues in the Prepartitioning Method

Jérôme Galtier

University of Versailles Saint-Quentin, Department of Computer Science, France
Electicité de France, Direction des Etudes et Recherches, France

Abstract. An original approach to the partitioning of 3D meshes (typically for the finite element method) is presented. Our technique applies on sub-domains defined by their polyhedrical boundary. It relies on the meshing of interfaces between sub-domains before meshing the domain itself.
Since this idea basically trades smoothness, small-size, and regularity of the interfaces for unbalance, we describe a fast, efficient, linear-time evaluation algorithm that correct this default. Its use is experienced with industrial benchmarks, and compared with other heuristic schemes.

Keywords. Mesh partitioning, load balancing, 3D finite element method, large irregular data structures

1 Introduction

Large 3D problems using the finite element method are well known to produce computations that require high CPU as well as large memory resources. In the recent years, this need has been addressed by the use of domain decomposition techniques that lead to parallel distributed computing [5,1,10,14,3]. As a result, a number of implementations have been done, and computations for large meshes that could be performed only on supercomputers in the past are now possible on a network of workstations.

Some industrials, however, have been deceived while applying these techniques to their problems. They realized that the preprocessing phase (including the mesh generation) was not parallel, and the data that really needed all these resources was hard to generate. *Indeed, the parallelization of a computation on a network of workstations with some data that can only be generated on a supercomputer is a kind of non-sense.* This fact probably explain that distributed supercomputers designers have met troubles selling their machines while the need for higher resources is still important.

In this paper, we address the problem of parallel mesh generation, and more specifically for meshes that are to be used in a domain decomposition scheme. Our method applies in the following framework:

- 3D finite element (i.e. tetrahedra) numerical method,

- Non-overlapping domain decomposition scheme,
- Polyedric domain Ω defined by its skin Γ, a list of points $\Gamma(P)$ and facets $\Gamma(F)$ that delimit the domain. We suppose all the facets to be conform (each of them shares exactly one side with another) and non-degenerated (having a non-zero area).

1.1 Main idea

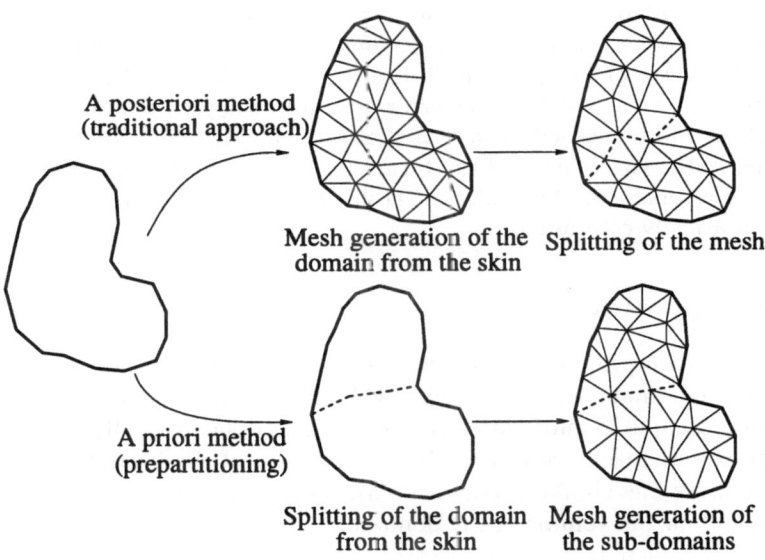

Fig. 1. Our splitting method

Our main idea is quite simple, and illustrated in 2D in Figure 1. Once we have the skin Γ of the domain, we generate directly a separator of it, namely, a list of facets that splits the domain in two. As a result, we obtain the skins Γ_a and Γ_b of the two sub-domains Ω_a and Ω_b. Then this procedure is applied recursively until the requested number of sub-domains are generated. This strategy can be identified as an *a priori* method, as shown in Figure 1. The classical approach that consists in splitting the domain after the mesh generation is designed as an *a priori* method. Naturally, on the one hand, our idea applies specifically to 3D domains for finite element discretizations, whereas classical mesh splitters can be applied to any kind of graph. On the other hand, as shown later, *a priori* methods can generate much larger data with smaller memory resources, and allow to improve drastically both the size and the smoothness of the separator.

1.2 Similar work

Other researchers have tried to implement *a priori* methods in the past. Two main approaches can be distinguished.

CAD/CAM systems. In many cases, the skin of a domain comes from a CAD/CAM system, in which an engineer has described the object. The user can also define a separating surface, so that the object is split in a number of sub-domains. The separator between the sub-domains is then created by the same surface mesh generator that has produced the external skin of the domain. Note that it is the user's responsibility to create well-balanced sub-domains. This approach is taken for instance in [9,2,4].

Underlying mesh. Löhner [13] has used a previously built mesh of the domain to build a separator. This technique is particularly relevant when, on an adaptive process, the mesh has to be rebuilt several times during the computation. In that case, the error estimator suggests a number of tetrahedra that will be needed in each area of the domain, which allows to estimate properly the size of each of the future meshed sub-domains. Then, the position of a separator that balance the load of each sub-domain can be easily automatically computed.

1.3 Our approach

The approach we propose in this paper requires no more information than the skin Γ of the domain, as defined previously. More precisely, the location of our separator is computed automatically (whereas it is done manually in CAD/CAM systems), and we take no advantage from any previously built mesh. As explained in section 2, we met similar problems to 3D mesh generation without having the same control on the volume of the domain.

Another problem that we have to face concerns load-balancing. An underlying mesh allows to evaluate the size (i.e. the number of tetrahedra) of each generated sub-domain. In a CAD/CAM system, it is the user's responsibility to balance properly the decomposition. In section 3 an algorithm is proposed that computes automatically the size of a domain only defined by its skin. The evaluation also allows to locate more precisely the place where the separator should be built. Comparisons between the prediction and the effective number of tetrahedra generated by some commercial mesh generators are given. Using this algorithm in the scheme presented in section 2, we show the improvement realized in terms of load balancing compared to other naïve heuristics.

2 Description of a prepartitioner

Part of the work presented here has been published in [6]. However, the presentation that follows is essential for a good understanding of the issues raised by our method. The basic steps of the algorithm are presented in Figure 2. It describes the split of one domain into two sub-domains. The process is then applied recursively until the requested number of sub-domains is obtained. In this section we consider the first step, the description of plane, as essentially heuristic. We discuss this point more precisely in Section 3.

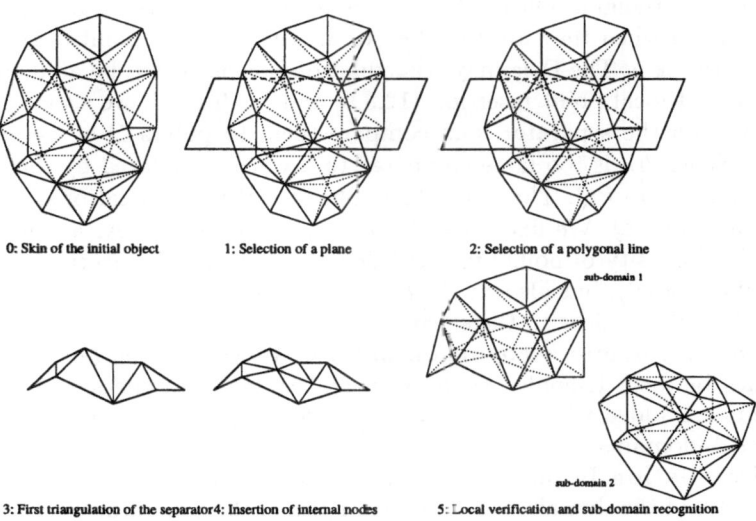

Fig. 2. Main steps of a prepartitioning process

Search for a connecting polygonal line This step consists in finding the intersection between the separator and the original skin. This polygonal line $\mathcal{L}$ is a set of edges $(e_i)_{1 \leq i \leq p}$ that may be disconnected. The two following properties have to be fulfilled:

- The intersection has to be conform to the skin, that is, every edge of the polygonal line has to belong to a triangle of the skin.
- The polygonal line has to be Delaunay-admissible with respect to the projective Delaunay criterium. More precisely:

$$(\forall e = [A, B] \in (\mathcal{L}))(\exists E \in (\Pi))(\forall f = [C, D] \in (\mathcal{L})) \begin{cases} d(E, A) = d(E, B) \\ d(E, C) > d(E, A) \end{cases}$$

We then form the projective Delaunay triangulation of the set of vertices $v_1 \ldots v_k$ included in $(\mathcal{L})$. ABC is a projective Delaunay triangle if the following is true:

$$(\exists P \in \Pi)(\forall f = [D, E] \in (\mathcal{L})) \begin{cases} d(P, A) = d(P, B) = d(P, C) \\ d(P, D) > d(P, A) \end{cases}$$

Triangulation of the separator - node insertion Since we only have the skin of the domain, the domain is "empty" and we have to add points to the set of points of $(\mathcal{L})$ to properly mesh the separator. In order to insert an adequate number of points, we associate an h_P parameter to each point P on the separator, the ideal size of a side of a triangle (or tetrahedron) connected to P. Initially, every point P of the separator also belongs to the skin. For any edge $[PQ]$ of the triangulation, new points are added between P and Q when the triangulation needs to be refined. Some of these new points are withdrawn when they are too close to another existing point of the separator or the skin. The Delaunay triangulation is constructed by an incremental process of local point insertion [12,8,6] and in the end, we scan all the h_P with P on the separator, and compare them to points Q of the domain's skin. When h_P is too large compared to h_Q, h_P is updated by the formula: $h_P = h_Q + \alpha \cdot PQ$. This formula emphasizes a geometrical progression between P and Q. We fixed $\alpha = 1.2$ for our experiments. As a consequence, the final density of points inserted depends both on the average side h_P of points P of $(\mathcal{L})$, but also on the h_P of points of the skin that do not belong to $(\mathcal{L})$, but are close to the separator.

Local validation and recognition process We control the value of dihedral angles between triangles of the skin and triangles of the separator. Some points are added when necessary. By a coloring process, we also identify the two sub-domains.

Global validation In this part, we check that no intersection occurs between triangles of the domain's skin and triangles of the separator.

3 Estimation of the size of a domain by its skin

We realized in the previous parts that the prepartitioning technique raise a difficult problem of load balancing. Related to this problem is the estimation of the number of tetrahedra that a mesh generator could generate on a domain defined by its skin.

In fact, the question is complex. Depending on the algorithm implemented, mesh generators can produce a varying number of tetrahedra. For instance, frontal-based algorithms [11] seem to generate less elements than Voronoï-based ones [7]. Also, the node insertion strategy appears to impact significantly the final amount of elements.

3.1 Idea of an evaluating algorithm and its requirements

In this section, we propose an algorithm that estimates the number of elements of a given domain defined by its skin, and locate the plane that half-split this domain. Not only this kind of evaluation could solve our load balancing problem, but it can help to normalize the amount of tetrahedra a mesh generator could produce. For this, the algorithm should satisfy the following requirements:

Simplicity. The algorithm should be simple and explicit enough so that the evaluation does not depend on the details of the implementation.

Theoretical justification. The evaluation has to rely on a fair approximation, that is theoretically justified.

Performance It should be performant enough to do some fast evaluations.

Realism. Compared to well known mesh generators, the result should be coherent.

Robustness. The algorithm is not allowed to fail, unless the skin is not consistent. In particular, any factor of numerical weakness should be clearly identified.

3.2 Instantaneous volume

As we saw in the prepartitioning process description, once a direction x_n, y_n, z_n (e.g. the principal inertia axis [1]) is given, the difficult point is to find out the constant c such that the plane

$$(\Pi_c) : x \cdot x_n + y \cdot y_n - z \cdot z_n = c$$

will split the domain Ω into balanced sub-domains Ω_+ and Ω_-. When the discretization parameter h is constant, balanced sub-domains correspond to equal-volume sub-domains. In that case, the function we compute is

$$vol(c) = \iiint_{\substack{(x,y,z)\in\Omega \\ x\cdot x_n + y\cdot y_n + z\cdot z_n \geq c}} dx \cdot dy \cdot dz = \sum_{M\in\Omega_+} dM.$$

The derivate $vol'(c)$ correspond to the area of (Π_c) that belongs to Ω.

---

[1] The principal inertia axis is the axis that contains the center of gravity of the object and along which the object has the smaller moment of inertia. The moment of inertia of a point P of weight w with respect to an axis D is $w \cdot d(D, P)^2$, and the moment of inertia of a discrete object is the sum of the moments of inertia of its points. Computing the principal inertia axis is equivalent to compute the eigenvector associated with the smaller eigenvalue of the matrix:

$$\begin{bmatrix} \sum_{s\in S}(y_s - y_g)^2 + (z_s - z_g)^2 & -\sum_{s\in S}(x_s - x_g)(y_s - y_g) & -\sum_{s\in S}(x_s - x_g)(z_s - z_g) \\ -\sum_{s\in S}(y_s - y_g)(x_s - x_g) & \sum_{s\in S}(z_s - z_g)^2 + (x_s - x_g)^2 & -\sum_{s\in S}(y_s - y_g)(z_s - z_g) \\ -\sum_{s\in S}(z_s - z_g)(x_s - x_g) & -\sum_{s\in S}(z_s - z_g)(y_s - y_g) & \sum_{s\in S}(x_s - x_g)^2 + (y_s - y_g)^2 \end{bmatrix}$$

where (x_s, y_s, z_s) are the coordinates of a point s in the set S of points of the object, and (x_g, y_g, z_g) are the coordinates of the center of gravity.

Theorem 1. *The function vol is a polynomial of degree 3 by segment.*

If c_1 and c_2 are such that no point of the skin belongs to (Π_c) for $c_1 \leq c \leq c_2$, the area given by vol'(c) is limited with segments that are defined by some affine functions of c. Therefore *vol'* is a polynomial of degree 2, and *vol* a polynomial of degree 3 between c_1 and c_2.

However, *vol'* results, on each segment $[c_1, c_2]$, from the combination of each face of Γ cut by (Π_c). Describing this list for each segment $[c_1, c_2]$ would be quite expensive: if n is the number of points of the skin, $n - 1$ such segment can appear. The algorithm we propose computes the instantaneous volume (at each c) in $O(n \cdot n_{cc})$ computational time, provided that the points $\Gamma(P)$ are already sorted along (x_n, y_n, z_n), and where n_{cc} is the number of connected components of Γ. We decompose the computation along the following steps, that are illustrated in 2D in Figure 3.

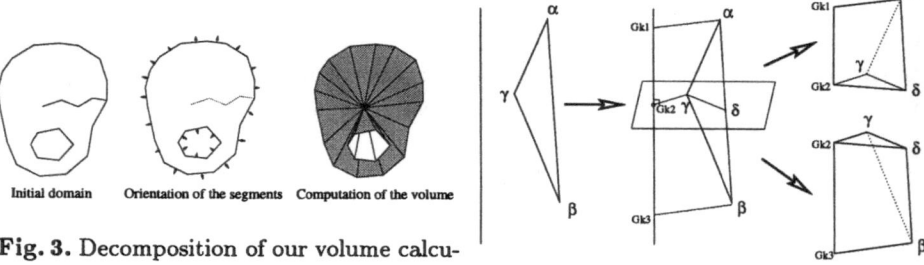

Fig. 3. Decomposition of our volume calculation.

Initial domain   Orientation of the segments   Computation of the volume

Fig. 4. Contribution of one face of the skin.

Orient the faces of the skin. The goal of this step is to reorder each face $[P_1, P_2, P_3]$ of $\Gamma(F)$ so that the normal $\frac{P_1 P_2 \wedge P_1 P_3}{\|P_1 P_2 \wedge P_1 P_3\|}$ point to the external part of Ω. By graph connection, the faces of the skin are classified into n_{cc} connected components in $O(n)$ time. Then, among each connected component, one face is oriented by a ray-launching algorithm (A ray that intersects a face of the component is defined. We compute all the faces of $\Gamma(F)$ that intersect the ray. Since we know that the faces always separate always separate the interior from the exterior part of the domain, a proper orientation can be computed for each intersected face.), which costs $O(n \cdot n_{cc})$. Then, the orientation of neighboring faces is derived, and step by step, all the faces of the skin are oriented.

Despite this case is not mentioned in our original framework, our algorithm can also handle a number of internal faces of Ω (faces that are given as a constraint for the mesh, but do not separate Ω from its external counterpart). Such a face is detected when one of its sides is shared with no other face - the reader will see that it is necessarily the case, otherwise the problem is not well defined from the beginning. Once it is removed, other internal faces can be detected, and removed. The process is applied until no other internal face is detected. For simplicity, we will now forget this particular case, and design by $\Gamma(F)$ the non-internal faces.

Compute the contribution of one face. Once all the faces $\Gamma(F)$ have been properly oriented, we can compute the contributions of them. Such contribution are stored in arrays ca, cb and cc, whose size is $n = |\Gamma(V)|$, and who are initialized to zero. First of all, the center of gravity $G(x_G, y_G, z_G)$ of the points $\Gamma(V)$ is computed. The idea is then to obtain, for each face $[P_1, P_2, P_3]$ of $\Gamma(F)$, and each Π_c, the area between:

- the orthogonal projection of G on (Π_c), and
- $\Pi_c \cap [P_1, P_2, P_3]$ if it is a segment. If this intersection is empty or contains the whole face $[P_1, P_2, P_3]$, the contribution of $[P_1, P_2, P_3]$ is null.

Let $\{\alpha, \beta, \gamma\}$ be a reordering of $\{P_1, P_2, P_3\}$ such that

$$x_\alpha x_n + y_\alpha y_n + z_\alpha z_n \geq x_\gamma x_n + y_\gamma y_n + z_\gamma z_n \geq x_\beta x_n + y_\beta y_n + z_\beta z_n,$$

and σ equal $+1$ if $[\alpha, \beta, \gamma]$ has the same orientation as $[P_1, P_2, P_3]$ and -1 otherwise. Let k_1, k_2 and k_3 be such that $\alpha \in \Pi_{k_1}$, $\gamma \in \Pi_{k_2}$ and $\beta \in \Pi_{k_3}$. Clearly, we have $k_1 \geq k_2 \geq k_3$. If $\Pi_{k_2} \cap [\alpha, \beta, \gamma]$ is a segment, γ is one of its endpoints (otherwise $[\alpha, \beta, \gamma]$ would be degenerated), and let δ be the the the other. Let finally G_k be the orthogonal projection of G on Π_k.

(k_1, k_2) **contribution.** If $k_1 = k_2$, we have $\delta = \alpha$, and the contribution of the segment is null. Otherwise, we compute the following oriented areas:

$$
\begin{aligned}
A &:= Area(G_{k_2}, \gamma, \delta) && in\ (\Pi_{k_2}) \\
A' &:= Area(mid(G_{k_1}, G_{k_2}), mid(\alpha, \gamma), mid(\alpha, \delta)) && in\ (\Pi_{\frac{k_1+k_2}{2}})
\end{aligned}
$$

For each k in $[k_1, k_2]$, the area of the triangle formed by G_k and $\Pi_k \cap [\alpha, \beta, \gamma]$ is given by $A(k) = ak^2 + bk + c$ with $(a, b, c) \in \mathbb{R}^3$ verifying:

$$
\begin{cases}
ak_1^2 + bk_1 + c = 0 & \text{since the area is zero when } k = k_1, \\
a\left(\frac{k_1+k_2}{2}\right)^2 + b\frac{k_1+k_2}{2} + c = A' & \text{since the area is A' when } k = \frac{k_1+k_2}{2}, \\
ak_2^2 + bk_2 + c = A & \text{since the area is A when } k = k_1,
\end{cases}
$$

which gives

$$
\begin{aligned}
a &:= \frac{2A-4A'}{(k_1-k_2)^2}, \\
b &:= \frac{-A(3k_1+k_2)+4A'(k_1+k_2)}{(k_1-k_2)^2}, \\
c &:= \frac{A(k_1^2+k_1k_2)-4A'k_1k_2}{(k_1-k_2)^2}.
\end{aligned}
$$

Then we modify the values of the array ca, cb and cc as follows:

$$
\begin{aligned}
ca[\alpha] &:= ca[\alpha] + \sigma a & ca[\gamma] &:= ca[\gamma] - \sigma a \\
cb[\alpha] &:= cb[\alpha] + \sigma b & cb[\gamma] &:= cb[\gamma] - \sigma b \\
cc[\alpha] &:= cc[\alpha] + \sigma c & cc[\gamma] &:= cc[\gamma] - \sigma c.
\end{aligned}
$$

(k_2, k_3) **contribution.** Similarly, when $k_2 > k_3$, we compute:

$$A'' := Area(mid(G_{k_2}, G_{k_3}), mid(\beta, \gamma), mid(\beta, \delta)),$$
$$a := \frac{2A - 4A''}{(k_3 - k_2)^2},$$
$$b := \frac{-A(3k_3 + k_2) + 4A''(k_3 + k_2)}{(k_3 - k_2)^2},$$
$$c := \frac{A(k_3^2 + k_3 k_2) - 4A'' k_3 k_2}{(k_3 - k_2)^2},$$

$$ca[\beta] := ca[\beta] - \sigma a \qquad\qquad ca[\gamma] := ca[\gamma] + \sigma a$$
$$cb[\beta] := cb[\beta] - \sigma b \qquad\qquad cb[\gamma] := cb[\gamma] + \sigma b$$
$$cc[\beta] := cc[\beta] - \sigma c \qquad\qquad cc[\gamma] := cc[\gamma] + \sigma c.$$

Total volume. Once this computation has been done for all the faces $\Gamma(F)$, and given the array k such that $P_i \in \Pi_{k[i]}$ and $k[i] \geq k[i+1]$ for $1 \leq i \leq n$ (i.e. all the points are sorted along x_n, y_n, z_n), the total volume of the object can be obtained by the following procedure (in Pascal):

```
volume := 0;
a := 0;   b := 0;   c := 0;
for i := 1 to n − 1 do
   begin
      a := a + ca[i];
      b := b + cb[i];
      c := c + cc[i];
      volume := volume + a (k[i]−k[i+1])³/3 + b (k[i]−k[i+1])²/2 + c(k[i] − k[i + 1]);
   end;
```

Remarks:

1. Since the number of faces of $\Gamma(F)$ is $O(n)$, the total construction of ca, cb and cc is in $O(n)$.

2. From a numerical point of view, the bottleneck of the algorithm is located in paragraph 3.2, where a division of type $\frac{1}{(k_1 - k_2)^2}$ occurs. This point can be easily detected and handled (for instance by removing the contribution of the faces when this factor is too big).

3. Thanks to geometrical properties, each part of the volume added in the above procedure is positive. As a result, we can detect instantaneous volumes, and more particularly, the plane Π_k that separates the domain into equal-volume sub-domains in $O(n)$, which answer our problem when the discretization of the domain is uniform.

The following paragraph will deal with non-uniform discretization.

3.3 Weighted volume

Not all the domains have a uniform discretization, especially in industrial problems. In that case, mesh generators estimate the ideal size $h[i]$ of one side of a triangle (or tetrahedron) connected to point P_i. Usually, $h[i]$ is the geometrical mean of all the sides of the faces of the skin connected to P_i. Then, the mesh is built so that it respects as well as possible $h[i]$ for each P_i in $\Gamma(P)$, and the discretization parameter $h[k]$ of internal points of the domain progress geometrically. It is clear that this requirement lets much home for various strategies, and as a result, the number of generated tetrahedron varies from one mesh generator to the other.

The first algorithm implementing a similar concept that we have experienced - called WVOL1 in the following - takes advantage from the fact that each addition of volume in the previous procedure is positive. The idea was then to weight the volume by local values of h, as done in the following:

$$wvolume := 0;$$
$$\dots$$
$$wvolume := wvolume + \frac{a\frac{(k[i]-k[i+1])^3}{3}+b\frac{(k[i]-k[i+1])^2}{2}+c(k[i]-k[i+1])}{(h[i]h[i+1])^{3/2}};$$
$$\dots$$

This idea tends to produce very unstable and unreliable weighted volumes. We identified two main reasons for this. We list them, along with the corresponding corrections to obtain our final estimator WVOL2.

Local weight $(h[i]h[i+1])^{3/2}$ is not reliable. Clearly, P_i can become to a very refined zone of the skin, and P_{i+1} to a very coarse one, so that the resulting weight is hazardous. A better solution consists in computing, for each segment $[k[i], k[i+1]]$, the geometrical mean of all the segments intersected by Π_k, with $k \in [k[i], k[i+1]]$. This can be done in linear time $O(n)$ by creating, along with ca, cb, and cc, two arrays clh and cnh updated during the step of paragraph 3.2 as follows:

$$cnh[\alpha] := cnh[alpha] + 2clh[\alpha] := clh[\alpha] + ln|\alpha\gamma| + ln|\alpha\beta|$$
$$clh[\gamma] := clh[\gamma] + ln|\beta\gamma| - ln|\alpha\gamma|$$
$$clh[\beta] := clh[\beta] - ln|\beta\gamma| - ln|\alpha\beta|$$
$$cnh[\beta] := cnh[\beta] - 2$$

The procedure becomes:

$$lh := 0; \quad nh := 0;$$
$$\dots$$
$$lh := lh + clh[i]; \quad nh := nh + cnh[i];$$
$$wvolume := wvolume + \frac{a\frac{(k[i]-k[i+1])^3}{3}+b\frac{(k[i]-k[i+1])^2}{2}+c(k[i]-k[i+1])}{exp(3lh/nh)};$$
$$\dots$$

$\Omega \cap \Pi_k$**may be disconnected for some k.** This is true especially when the geometry of Ω becomes complex. The result, in the previous procedure, is that h from disconnected regions are collapsed and should not be. If one of the region is much more refined than the others (usually the smallest in terms of area), the weighted volume is corrupted. The idea used to correct this default is to detect the zones of the skin where the domain splits in different regions along (x_n, y_n, z_n). By chance, the following theorem holds:

Theorem 2. *The zones where Ω splits along (x_n, y_n, z_n) are included are included in the set of 4-flip-flop points of $\Gamma(P)$ (defined in the following).*

We still suppose that all points P_i of $\Gamma(P)$ are sorted so that $k[i] \geq k[i+1]$. For one point P_i, a flip-flop is a face $[P_i, P_j, P_k]$ such that $(i > j$ and $i < k)$ or $(i < j$ and $i > k)$. One can see that the number of flip-flops of a point is even. A 4-flip-flop point is a point having more than four flip-flops (see Figure 5).

Usually, a simple domain has zero or one 4-flip-flop points along its principal inertia axis. Once those points have been located, a number of connected components of $\Gamma(P)$ are created using the following connecting rules:

(1) If i and j, $i > j$ belong to a common face, and no 4-flip-flop point l is such that $k[i] \geq k[l] \geq k[j]$, then i and j belong to the same connected component.

(2) Suppose all the points of one face $[P_\alpha, P_\beta, P_\gamma]$ belong to the same connected component, and all the points of one face $[P_a, P_b, P_c]$ belong to the same connected component. Suppose one oriented line $\mathcal{L}$ perpendicular to the axis x_n, y_n, z_n intersects $[P_\alpha, P_\beta, P_\gamma]$, and immediately after $[P_a, P_b, P_c]$, and intersects an even number of faces of the skin before or behind this event. Care is taken in the choice of $\mathcal{L}$ so that $\mathcal{L}$ intersects no side of a face of $\Gamma(F)$. Then the respective components of $[P_\alpha, P_\beta, P_\gamma]$ and $[P_a, P_b, P_c]$ are collapsed.

Since some quite complex geometries generated a large number of connected components, and given that this algorithm has only an estimation purpose, we restrict the number of final components to five by collapsing the smallest ones. The resulting algorithm, WVOL2, gives satisfying results, as we will see in the following paragraph.

3.4 Estimation results

To validate the algorithm, we used a set of benchmarks designed for the industrial CFD software N3S, provided by the *Laboratoire National d'Hydrolique* (LNH) of EDF. The geometry of the domains can be seen on the left part of Figures 8, 7, 6.

Case	I-DEAS Master Series^{TM}	GHS3D	WVOL2	Time (secs.)
TPN3S	8486	7295	7113.02	0.125
TPN3S*	215270	214301	204174	1.49
TRAVERS26	48144	34733	54734.8	0.8
COUV45	150109	147661	138689	2.72

Table 1. Comparisons between two mesh generators and our estimator.

TPN3S Turbulent flow through the abrupt widening of a pipe.

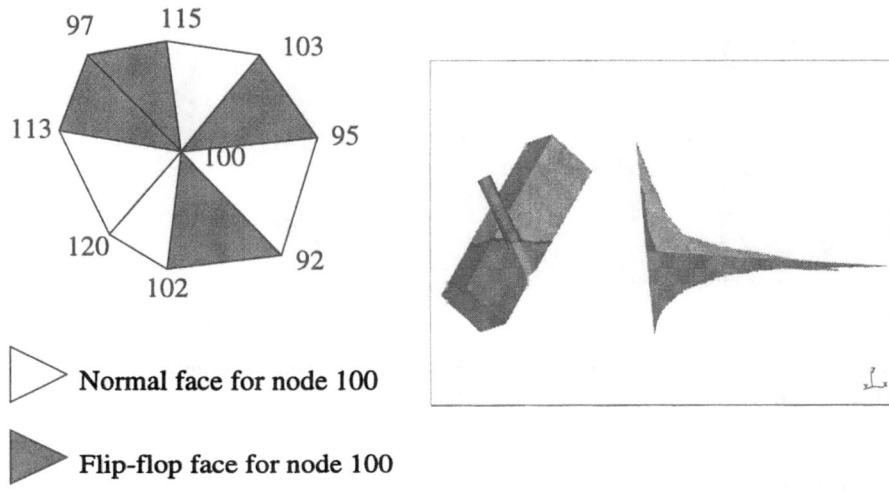

Fig. 5. A 4-flip-flop point.

Fig. 6. The TRAVERS26 domain along with its weighted area.

TPN3S* Same case, with a greater mesh refinement.

TRAVERS26 Flow disrupted by a shaft-like obstacle in the upper part of the tank of a nuclear power station.

COUV45 Flow through the tank of a 900 MegaWatts nuclear power station with simplified geometry, and two symmetry planes (1/8 of the tank is described).

In Table 1, we compare the generated number of tetrahedra on the previously mentioned domains with two mesh generators to the estimated number of tetrahedra given by WVOL2. Since one equilateral tetrahedron of side of length 1 has a volume of $\frac{\sqrt{2}}{12}$, the estimation is $\frac{12}{\sqrt{2}}wvolume$. We also give in Table 1 the time spent on a SUN ULTRA 1 in seconds to estimate the size of the domain. Even for large domains, this time remains small, and therefore acceptable. The difference observed between the estimation and the generated tetrahedra remains comparable to the difference between the two mesh generators, which is satisfying. However, in nearly all cases, the estimation is higher or smaller than the two obtained amounts of tetrahedra. Certainly, some progress has to be done to improve that point. We also note that in the case TRAVERS26, which presents a vicious domain splitting (and a 4-flip-flop point), the estimation is clearly too large.

In Figures 8, 7 and 6, we can see, on the left part, the geometries of our domains, and in the right part, having the principal inertia axis as a base, the curve of weighted areas. The weighted area is the derivate of the weighted volume, that give the estimated number of triangles that $\Pi_{k[i]} \cap \Omega$ should contain. This number is also a good approximation of the number of triangles that the prepartitioning algorithm will produce while splitting the domain along $(\Pi_{k[i]})$.

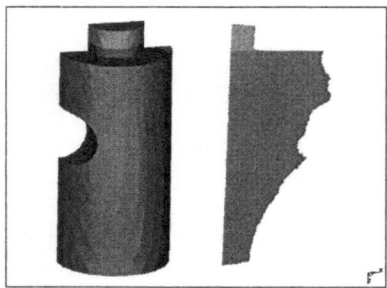 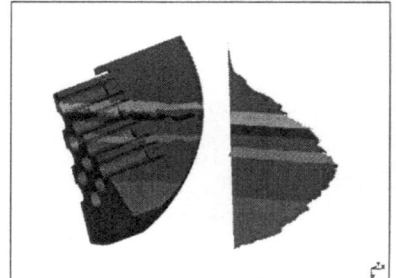

Fig. 7. The TPN3S domain along with its weighted area.

Fig. 8. The COUV45 domain along with its weighted area.

This number is:

$$w[i] := \frac{\frac{ak[i]^2}{4} + bk[i] + c}{exp(2lh/nh)}.$$

In the figures, the weighted area is cumulated separately for the different connected components; levels of gray allow to identify the connected components (in the left part) to their weighted area contribution (in the right part). Note that faces that belong to several connected components in the left part have colored in black. For instance, Figure 6 shows a domain with a very thin and refined arm, which appears as a darker zone. The contribution of this irregularity to the weighted volume remain reasonable. When this component is not distinguished from the other parts of the mesh, the estimator evaluates much more tetrahedra than this.

3.5 Efficiency as a splitting estimator

Are the algorithms WVOL1 and WVOL2 good in estimating recursively where a prepartitioner should split the domain? We present in this paragraph some experiments, based on the comparison with two heuristics.

MINMAX. Let k_1 (resp. k_2) be the smallest (resp. the largest) k such that $\Pi_k \cap \Omega \neq \emptyset$. The splitting plane is $(\Pi_{\frac{k_1+k_2}{2}})$.

MEDIUM. Let $k[i]$ be the previously defined array of sorted points along x_n, y_n, z_n. The splitting plane is $(\Pi_{k[\frac{n}{2}]})$.

In table 2 we present the final load balancing obtained for the splitting of our three largest cases into 16 sub-domains. The final mesh generator we have used in each sub-domain is GHS3D. We give the fraction of the size of the largest sub-domain over the total size of the domain, multiplied by 1600 so that the ideal sub-domain would obtain 100. We can see that heuristic approaches give good results for regular domains such as TPN3S*, but tend to be very

CASE	MINMAX	MEDIUM	WVOL1	WVOL2
COUV45	231	207	129	134
TPN3S*	194	113	110	131
TRAVERS26	519	341	322	131

Table 2. Load balancing for 16 sub-domains using various splitting estimators

hazardous while the geometry is becoming more complex. The TRAVERS26 case clearly exhibit foolish behaviors for all approaches excepted WVOL2. This last algorithm maintains a stable unbalance of 30%, whatever the domain is. Compared to the fluctuations observed on the two mesh generators in Table 1, this result is satisfying at first sight. Hence, we can expect that the unbalance remains stable as the number of sub-domains increases.

4 Conclusion

In this paper, efficient means to split a domain Ω defined by its only skin have been explored. Not only this technique can be expected to break down the maximal size of a mesh that can be generated by computers, but it also proves efficient as a splitting methods, in terms of

- size, smoothness of the interfaces,
- computational time and memory required for splitting.

Some other fields of investigation seem particularly promising for this technique, namely:

Adaptive methods. Two views of them can be discussed.

When the refinement is obtained by redivision of some elements, two main methods of re-balancing are applied. In one case, elements are moved one by one from a sub-domain to the other. Then, the original splitting has an unknown impact on the subdivisions after a few steps. In the other case, blocs of elements only are moved. If we suppose that these blocks are generated by a prepartitioner, certainly they will have smoother and smaller interfaces, and the original load unbalance will be masked by the unbalanced produced by adaption.

When the domain is re-meshed once an estimation of the error is done, the unbalance problem is even more easy, since, as we recalled in the beginning of the paper, the size of the sub-domains are easy to evaluate.

Physical shape of the interfaces adapted to the physical problem. We only mentioned prepartitioning along a plane during this article. Other kinds of shapes of the interface are possible theoretically (see [6]). Certainly the use of such adapted shapes could improve the performance of a parallel application. Naturally, in this case, the question of load-balancing is completely open.

Use of our volumic estimator to get even smaller interface. As we noticed, the weighted area notion that we have introduced in the previous lines is in fact an estimation of the size (in terms of triangles) that the prepartitioner will produce at a given position. Until now, no investigation has been done to take advantage from this in order to get even smaller interfaces.

Acknowledgments

I am particularly grateful to the departments MMN and LNH of EDF, who motivated these investigations, and provided industrial benchmarks to work with.

References

1. V. Aghoskov and V. Lebedev. The Poincaré-Steklov's operators and the domain decomposition in variational problems. *Computational Processes and Systems*, pages 173–227, 1985.
2. Geoffrey Butlin and Clive Stops. CAD Data Repair. In *Proceedings of the 5th International Meshing Roundtable*, pages 7–12, 1996.
3. T. Chan and T. Mathew. Domain decomposition algorithms. *Acta Numer.*, 8:, 1993.
4. H. S. de Cougny and M. S. Shephard. Surface meshing using vertex insertion. In *Proceedings of the 5th International Meshing Roundtable*, pages 243–256, 1996.
5. Q. Dinh, R. Glowinski, and J. Périaux. Solving elliptic problems by domain decomposition methods with applications. In G. Birkhoff and A. Schoenstadt, editors, *Elliptic Problem Solvers II*, pages 395–426. Academic Press, New York, 1984.
6. Jérôme Galtier and Paul Louis George. Prepartitioning as a way to mesh subdomains in parallel. In *Proceedings of the 5th International Meshing Roundtable*, pages 107–121, 1996.
7. P. L. George, F. Hecht, and E. Saltel. Automatic mesh generator with specified boundary. *Computer Methods in Applied Mechanics and Engineering*, 92:269–288, 1991.
8. Paul Louis George. Génération de maillages par une méthode de type Voronoï. Technical Report RR-1398, INRIA, 1991.
9. Gerhard Globisch. On an automatically parallel generation technique for tetrahedral meshes. *Parallel Computing*, 21:1979–1995, 1995.
10. D. Keyes and W. Groop. A comparison of domain decomposition techniques for elliptic partial differential equations and their parallel implementation. *SIAM J. Sci. Stat. Comput.*, 8:166–202, 1987.
11. R. Löhner and P. Parikh. Three-dimensional grid generation by the advancing-front method. *Int. J. Numer. Meth. Fluids*, 8:1135–1149, 1988.
12. F.P. Preparata and M.I. Shamos. *Computational geometry, an introduction*. Springer-Verlag, 1985.
13. Alexander Shostko and Rainald Löhner. Three-dimensional parallel unstructured grid generation. *Inter. J. Num. Meth. Engn*, 38:905–925, 1995.
14. P. Le Tallec. Domain decomposition method in computational mechanics. *Computational Mechanics Advances*, 1(2):121–220, 1993.

Design of Novel Load-Balancing Algorithms with Implementations on an IBM SP2

Sajal K. Das[1], Daniel J. Harvey[1], and Rupak Biswas[2]

[1] CS Dept., University of North Texas, Denton, TX 76203, USA
[2] NASA Ames Research Center, Moffett Field, CA 94035, USA

Abstract. In a distributed-computing environment, it is important to ensure that the processor workloads are adequately balanced. Among numerous load-balancing algorithms, a unique approach due to Das and Prasad defines a *symmetric broadcast network* (SBN) that provides a robust communication pattern among the processors in a topology-independent manner. In this paper, we propose and analyze three novel SBN-based load-balancing algorithms, and implement them on an SP2. A thorough experimental study with Poisson-distributed synthetic loads demonstrates that these algorithms are very effective in balancing system load while minimizing processor idle time. They also compare favorably with several other existing load-balancing techniques.

1 Introduction

To maximize the performance of a multicomputer system, it is essential to evenly distribute the load among the processors. In other words, it is desirable to prevent, if possible, the condition where one processor is overloaded with a backlog of jobs while another processor is lightly loaded or idle. The load-balancing problem is closely related to scheduling and resource allocation, and can be static or dynamic. A static allocation [8] relates to decisions made at compile time, and compile-time programming tools are necessary to adequately estimate the required resources. Dynamic algorithms [1, 5, 6, 7] allocate/reallocate resources at run time based on one or more system parameters. Determining which parameters to maintain and how to broadcast them are important design considerations.

In this paper, we consider general-purpose distributed-memory parallel computers in which processors (or nodes) are connected by a point-to-point network topology. These nodes communicate with one another using message passing. Responsibility for load balancing is decentralized, or spread among the nodes. Processor workload is determined by the length of the local job queue. The network is assumed to be homogeneous and that any job can be processed by any node. However, jobs cannot be rerouted once execution begins.

Recently, Das et. al. [3, 4] have suggested a different approach to dynamic load balancing, by introducing a logical topology-independent communication pattern called symmetric broadcast network (SBN). We refine this approach and propose three novel and efficient load-balancing algorithms, one of which is adapted for use on the hypercube architecture. Based on their operational characteristics, our SBN-based algorithms can be classified (e.g. [9]) as:

Adaptive: performance is adapted to the average number of queued jobs;
Symmetrically initiated: senders and receivers can initiate load balancing;
Stable: the network is not burdened with excessive load-balancing traffic;
Effective: system performance is not degraded when the algorithms operate.

The three proposed algorithms are implemented on an IBM SP2, using the
Message-Passing Interface (MPI). Performance is analyzed based on simulation
using Poisson-distributed synthetic loads under various metrics: total number
of jobs transferred, total completion time, message traffic per node, and maximum variance in node idle time. Empirical results of our extensive experiments
demonstrate that the load balancing achieved by the SBN approach is superior
to several other existing techniques such as Random [5], Gradient [7], Receiver
Initiated [6], Sender Initiated [6], and Adaptive Contracting [6].

2 General Characteristics of SBNs

A *symmetric broadcast network* (SBN) defines a communication pattern (logical
or physical) among the P processors in a multicomputer system. An SBN of
dimension $d \geq 0$, denoted as SBN(d), is a $d+1$-stage interconnection network
with $P=2^d$ processors in each stage. It is constructed recursively as follows:
- A single node forms the basis network SBN(0).
- For $d > 0$, SBN(d) is obtained from a pair of SBN($d-1$)s by adding a communication stage in the front and extra interprocessor links as follows: (a) node
  i in stage 0, is connected to node $j=(i+P/2)$ mod P in stage 1; and (b) node
  j in stage 1 is connected to the node in stage 2 that was the stage 0 successor
  of node i in SBN($d-1$).

An example of how an SBN(2) is formed from two SBN(1)s is shown in Fig. 1.

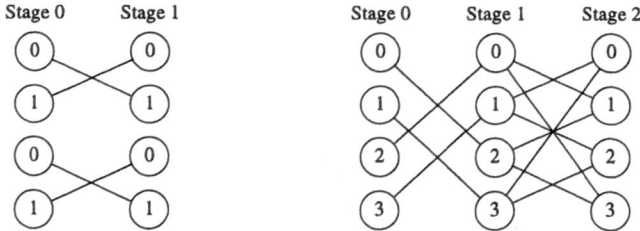

Fig. 1. Construction of an SBN(2) from a pair of SBN(1)s

The SBN approach defines unique communication patterns among the nodes
in the network. For any node at stage 0 as the source, there are $\log P$ stages
of communication with each node appearing exactly once. The successors and
predecessors for each node are uniquely defined by specifying the originating
node and the communication stage.

As an example, consider the communication pattern for SBN(3) shown in
Fig. 2 that is used for messages originating from node 0. In general, if n_x^s is the
corresponding node in the communication pattern for messages originating from
node x, then $n_x^s = n_0^s \oplus x$, where $\oplus$ is the exclusive-OR operator. Thus, all SBN
communication patterns can be derived from the template corresponding to the
one with node 0 as the root. The predecessor and successors to node n_0^s are:

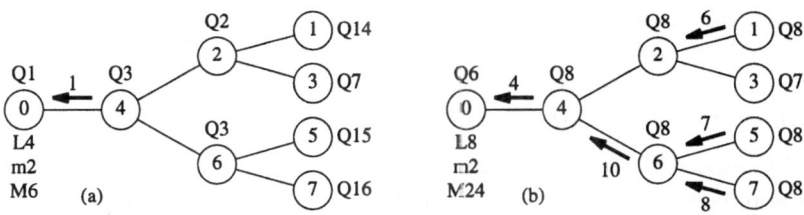

Fig. 2. An example of load balancing using the standard SBN algorithm

Predecessor: $\left(n_0^s - 2^{d-s}\right) \vee 2^{d-s+1}$, where $\vee$ is the inclusive-OR operator,
Successor 1: $n_0^s + 2^{d-s-1}$ for $0 \leq s < d$,
Successor 2: $n_0^s - 2^{d-s-1}$ for $1 \leq s < d$.

Figure 2 illustrates a possible SBN communication pattern, but many others can be easily derived based on network topology and application requirements. In [2], the SBN approach was adapted for use on the hypercube using a modified binomial spanning tree to ensure that all successor and predecessor nodes at any communication stage are adjacent nodes in the hypercube.

All SBN algorithms adapt their behavior to the system load. Under heavy (light) loads, the balancing activity is primarily initiated by processors that are lightly (heavily) loaded. This activity is controlled by two system thresholds: MinTh and MaxTh, the minimum and maximum system load levels. The system load level SysLL is the average number of jobs queued per processor. If a processor has a queue length QLen below MinTh, a message is initiated to begin load balancing. If QLen is larger than MaxTh, extra jobs are distributed through the network. If this distribution overloads other processors, load balancing is triggered. Algorithm behavior is affected by the values chosen for MinTh and MaxTh. For instance, MinTh must be large enough so that sufficient jobs can be received before a lightly-loaded processor becomes idle; however, it should not be too big so as to initiate unnecessary load balancing. If MaxTh is too small, it will cause an excessive number of job distributions. If it is too large, jobs will not be adequately distributed under light system loads. Moreover, once there is sufficient load in the network, very little load-balancing activity should be required.

Two types of messages are processed in the SBN approach. The first type are balancing messages that indicate unbalanced system load. These messages originate from an unbalanced node and are then routed through the SBN. As these balancing messages pass through the network, the cumulative total of queued jobs is computed to obtain SysLL. The second type of messages is for job distribution and used for two purposes. First, they are used to route the current SysLL through the network. Each node, upon receipt of such a message, updates its estimate of the average number of jobs queued per node in the network. The system thresholds, MinTh and MaxTh, are also updated. Second, job distribution messages are used to pass excess jobs from one node to another. This action can occur whenever a node has more jobs than its MaxTh. It can also be in response to a predecessor's need for jobs. This need is embedded in the load-balance messages as well as in the distribution messages that respond to these operations. To reduce message traffic, a node does not initiate additional load-balancing activity until all previous messages that have passed through the node have been completely processed.

3 SBN-Based Load-Balancing Algorithms

3.1 Standard SBN Algorithm

In the standard SBN algorithm, load-balancing messages are routed through the SBN from the source to the processors in the last stage. Messages are then routed back toward the original source with the total number of jobs in the system. The originating node thus has an accurate value of SysLL. Distribution messages are then sent to all nodes along with the SysLL. All nodes then update their local SysLL, MinTh, and MaxTh. Excess jobs are routed as part of this distribution to balance the system load. In addition, if a processor has QLen less than SysLL, the need for jobs is indicated during the distribution process. Successor nodes respond by routing back an appropriate number of excess jobs.

To illustrate the processing involved in a load-balancing operation, consider the SBN(3) in Fig. 2(a). The id and QLen for each node are shown. For example, node 6 has three jobs queued, indicated as Q3. The initial values of SysLL, MinTh, and MaxTh at node 0 are 4, 2, and 6, respectively (indicated as L4, m2, and M6). After a load-balancing request is sent through the SBN and then routed back to node 0, these values are updated to 8, 2, and 24, respectively, using:

$$\text{SysLL} = \lceil \text{TotalJobsQueued} / \text{P} \rceil,$$
$$\text{MinTh} = \min (\text{MinTh}, \text{SysLL}-1),$$
$$\text{MaxTh} = \text{SysLL} + 2^{\lfloor \text{SysLL} / \text{MinTh} \rfloor}.$$

Note that when load balancing is initiated, node 4 distributes half of its QLen, i.e., $\lfloor 3/2 \rfloor$ job, back to node 0. This is shown by the arrow in Fig. 2(a).

Distribution messages are then used to route excess jobs to the successor nodes or to indicate a need for jobs if the local QLen is less than SysLL. Jobs are routed back to the predecessor nodes when appropriate. Figure 2(b) shows the result of this distribution. The arrows indicate jobs routed between nodes. To load balance P processors, a total of $3P-3$ messages have to be processed.

3.2 Hypercube Variant

The SBN approach can be adapted for implementation on a hypercube topology, using the modified binomial spanning tree. A complete description of this hypercube variant is given in [2]. It operates in a manner similar to the standard SBN algorithm with the following differences:

- SysLL is computed when all balance messages arrive at the final node in the network. This is possible because there is a unique final node for every originating node. Distribution messages are then routed back to complete the load balancing. Since there are $P-1+\frac{P}{2}-1$ interconnections in the modified binomial spanning tree, a load-balancing operation requires $3P-4$ messages.
- Nodes in the SBN need to gather all balancing messages from their predecessors before routing the updated SysLL to the successors.
- The network topology is such that the number of predecessor and successor nodes vary at the different stages of communication.

3.3 Heuristic SBN Algorithm

Both previous algorithms process a large number of messages to accurately maintain SysLL. The heuristic version attempts to reduce the amount of processing by terminating load-balancing operations as soon as enough jobs are found that can be distributed. In general, this strategy reduces the number of messages; although, $O(P)$ messages are needed in the worst case.

In the heuristic SBN algorithm, a processor estimates SysLL by averaging QLen for the processors through which the balance message has passed. An appropriate number of jobs are then returned to the predecessor nodes as follows:

$$\text{ExJobs} = \begin{cases} 0 & \text{if QLen} < 3 \\ \lfloor \text{QLen} / 2 \rfloor & \text{otherwise.} \end{cases}$$

If ExJobs $= 0$ or if SysLL > 2 when ExJobs $= 1$, the balance message is forwarded to the next stage. Otherwise, the load balancing is terminated.

Job distribution is also performed differently in the heuristic SBN algorithm. For example, consider an SBN(3) that has a processor with MaxTh $= 15$ and QLen $= 24$. The number of jobs to be distributed is computed by dividing QLen by the total number of stages. Thus, six jobs are distributed in this case. SysLL is then set to 24−6=18. The processor that receives these jobs divides the number of jobs received by the remaining number of stages and adds the result to the SysLL stored at that node.

A significant advantage of the heuristic algorithm is that the balance messages do not have to be gathered until SysLL can be estimated. This reduces the interdependencies associated with the communication. If a particular processor fails, load balancing can still be accomplished for the remaining processors.

4 Experimental Results

The three SBN-based load-balancing algorithms have been implemented using MPI and tested with synthetically-generated workloads on the SP2 located at NASA Ames Research Center. The simulation program spawns the appropriate number of child processes and creates the desired network. The list of all process ids and an initial distribution of jobs is routed through the network.

In addition to the initial load, each node dynamically generates additional jobs during 10 job creation cycles. The number of jobs generated at each node during each cycle follows a Poisson distribution. By changing the parameter λ, both heavy and light system load conditions are dynamically simulated. Jobs are processed by "spinning" for the designated time period. The simulation terminates when all jobs have been processed. Three test runs are reported here:

Heavy System Load (cf. Fig. 3): Initially, 10 jobs per node are randomly distributed throughout the network. Jobs generated during execution are more than that the network can process. Job duration averages one second.

Transition from Heavy to Light System Load (cf. Fig. 4): Fifty jobs multiplied by the number of processors are distributed to a small subset of nodes as an initial load. A light load of jobs is generated as the load-balancing algorithms proceed. Job duration averages two seconds. Note that the initial load imbalance also needs to be corrected.

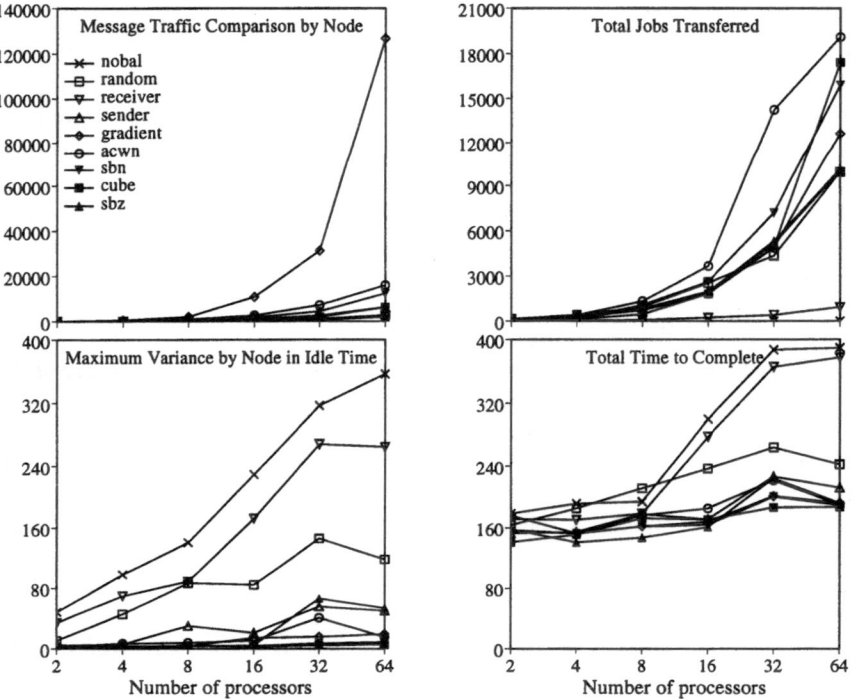

Fig. 3. Heavy system load

Light System Load (cf. Fig. 5): A small number of jobs are initially distributed to a small subset of nodes. A light load of jobs are created as the load-balancing algorithms operate.

The data and line charts in Figs. 3–5 measure the performance of the various load-balancing algorithms on an SP2, using the following variables:

Message Traffic Comparison by Node: Measures the maximum total number of load-balancing messages that were sent by any one of the nodes.

Total Jobs Transferred: Measures the total number of job transfers that occurred from one node to another.

Maximum Variance by Node in Idle Time: Measures the difference in processing time between the most busy node and the least busy node.

Total Time to Complete: Measures the total amount of elapsed time in seconds before all jobs are fully processed.

As expected, the program with no load balancing (*nobal*) performs by far the worst. The Random (*random*) algorithm, although significantly reducing the idle time, is less effective than the remaining algorithms. The Sender Initiated (*sender*) algorithm balances the load more evenly than *random*; however, the Receiver Initiated (*receiver*) algorithm does better only when the system load is light. For light to moderate loads, *receiver* generates more network traffic because all nodes poll neighbors to find jobs they can process. To overcome this deficiency, a time delay of one second has been introduced after a polling operation at

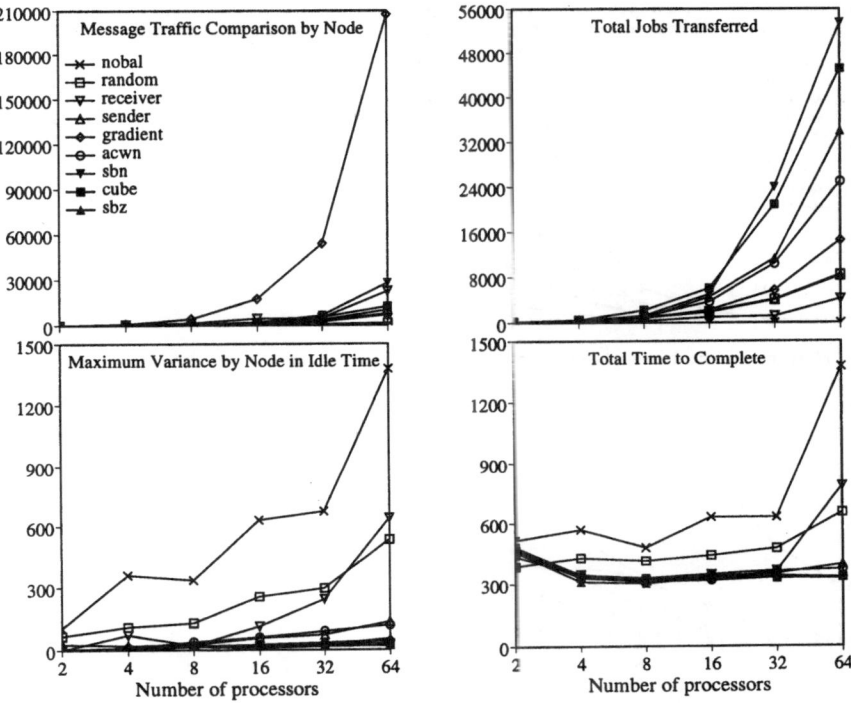

Fig. 4. Transition from heavy to light system load

the cost of increasing the idle time. At heavy system loads, *sender* can cause job thrashing. This has been overcome by reducing the number of job transfers that are done at high load levels; however, it can cause one or more nodes to remain lightly loaded. The Gradient (*gradient*) algorithm balances the load quite well without any of the above deficiencies. Unfortunately, lightly-loaded nodes can sometimes receive too many messages from overloaded nodes. Also significant communication is required to update neighbor node information, often resulting in excessive network traffic. The Adaptive Contracting (*acwn*) algorithm performs the best in periods of heavy system loads. However, as is true for *gradient*, the system traffic and the number of jobs migrated increase.

Both the standard SBN (*sbn*) algorithm and its hypercube variant (*cube*) are able to balance the system load more evenly than others. Their performance characteristics are very similar. Both require less message traffic than *gradient* but cause a higher number of job migrations, especially in light system loads. The heuristic SBN algorithm (*sbz*) performs well in minimizing idle time in light system loads. Although its performance during periods of heavy loads is relatively good, it does not balance the system load as well as *cube* or *sbn*. This is because its estimate of SysLL is not necessarily accurate. For light loads, *sbz* transfers more jobs than the other algorithms; however, it requires fewer messages than *gradient*, *sbn*, or *cube*. Overall, the empirical results demonstrate that the SBN-based approach to dynamic load balancing is an effective one.

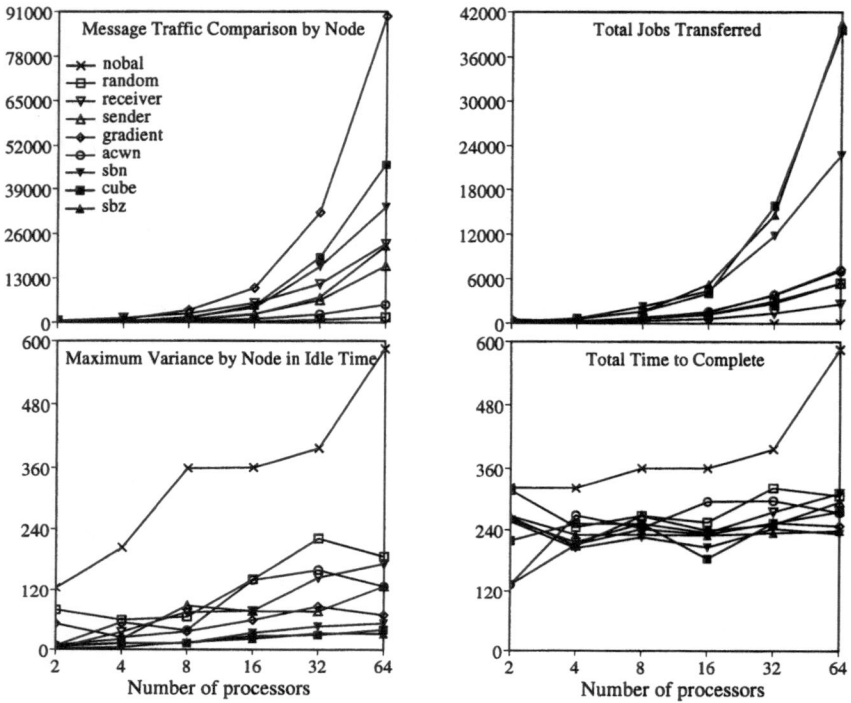

Fig. 5. Light system load

References

1. Cybenko, G.: Dynamic load balancing for distributed-memory multiprocessors. J. Parallel Distrib. Comput. **7** (1989) 279–301
2. Das, S., Harvey, D., Biswas, R.: Adaptive load-balancing algorithms using symmetric broadcast networks, NASA Ames Research Center Technical Report NAS-97-014 (1997)
3. Das, S., Prasad, S.: Implementing task ready queues in a multiprocessing environment. International Conference on Parallel Computing (1990) 132–140
4. Das, S., Yang, C., Leung, N.: Implementation of load balancing in multiprocessor systems using a symmetric broadcast network. International Conference of Parallel and Distributed Systems (1992) 589–596
5. Eager, D., Lazowska, E., Zahorjan, J.: Adaptive load sharing in homogeneous distributed systems. IEEE Trans. on Soft. Engrg. **12** (1986) 662–675
6. Eager, D., Lazowska, E., Zahorjan, J.: A comparison of receiver-initiated and sender-initiated adaptive load sharing. Perf. Eval. **6** (1986) 53–68
7. Lin, F., Keller, R.: The gradient model load balancing method. IEEE Trans. on Soft. Engrg. **13** (1987) 32–38
8. Sarkar, V., Hennessy, J.: Compile-time partitioning and scheduling of parallel programs. Scheduling and Load Balancing in Parallel and Distributed Systems (1995) 61–70
9. Shivaratri, N., Krueger, P., Singhal, M.: Load distributing for locally distributed systems. Computer **25** (1992) 33–44

Repartitioning of Adaptive Meshes: Experiments with Multilevel Diffusion *

Kirk Schloegel, George Karypis, Vipin Kumar
University of Minnesota, Department of Computer Science
(kirk, karypis, kumar) @ cs.umn.edu

Abstract. For a large class of irregular grid applications, the structure of the mesh changes from one phase of the computation to the next. Eventually, as the graph evolves, the adapted mesh has to be repartitioned to ensure good load balance. If this new graph is partitioned from scratch, it will lead to an excessive migration of data among processors. In this paper, we present a new scheme for computing repartitionings of adaptively refined meshes. This scheme performs diffusion of vertices in a multilevel framework and minimizes vertex movement without significantly compromising the edge-cut.

1 Introduction

For a large class of irregular grid applications, the computational structure of the problem changes in an incremental fashion from one phase of the computation to another. An example of such an effect results with the use of adaptive meshes. Eventually, as the graph evolves, it becomes necessary to correct the partition in accordance with the structural changes in the computation and to migrate a certain amount of computation between processors. Failure to do so will lead to load imbalance, which will bog down parallel run time.

Thus, for adaptive applications, we need a partitioning or repartitioning algorithm with the following constraints. It is fast. It is scalable. It balances the graph. It minimizes edge-cut. It minimizes vertex migration time. The use of parallel graph partitioners will result in a balanced graph with small edge-cuts, but will lead to excessive movement of data among processors. In this paper, we present a new scheme for computing repartitionings of adaptively refined meshes. This scheme performs diffusion of vertices in a multilevel framework and minimizes vertex movement without significantly compromising the edge-cut.

* This work was supported by NSF CCR-9423082, by Army Research Office contract DA/DAAH04-95-1-0538, by Army High Performance Computing Research Center cooperative agreement number DAAH04-95-2-0003/contract number DAAH04-95-C-0008, by the IBM Partnership Award, and by the IBM SUR equipment grant. Access to computing facilities was provided by AHPCRC, Minnesota Supercomputer Institute. Related papers are available via WWW at URL: http://www.cs.umn.edu/~karypis

In this paper, `TotalV` is defined as the number of vertices which change partitions as the result of partitioning or repartitioning. `MaxV` is defined as the maximum number of vertices which migrate into or out of any one partition as a result of partitioning or repartitioning.

2 Repartitioning Strategies: Review of Previous Work

In partitioning or repartitioning a dynamic graph, two strategies are available. The first is to simply partition the new graph from scratch. The advantage of this strategy is that edge-cut is minimized. The second strategy is to use the existing partition as input for a repartitioning algorithm and to attempt to minimize the difference between the original partition and the output partition. This strategy has the potential benefit of reducing `TotalV` by an order of magnitude or more over partitioning the modified graph from scratch.

The second strategy can be accomplished by the following *cut-and-paste repartioning* method. Excess vertices in an overbalanced partition are simply swapped into one or more underbalanced partitions in order to bring these partitions up to balance. However, while this method will optimize the vertex migration time, it will have an excessively negative effect on the edge-cut compared with more sophisticated approaches.

Another method which reduces edge-cut degradation over cut-and-paste repartitioning, while increasing `TotalV` only moderately, is analogous to diffusion from thermal dynamics. The concept is for vertices to move from overbalanced partitions to underbalanced partitions and to eventually reach balance, just as in the analogous case, uneven temperatures in a space cause the movement of heat towards equilibrium [1].

Directed diffusion is diffusion guided by a global view of the graph. One method of computing this view (hereafter referred to as the *diffusion solution*) involves minimization of the two-norm of this solution. Hu and Blake described a method which computes the diffusion solution while optimally minimizing its two-norm [3].

Walshaw, Cross, and Everett implemented JOSTLE, a combined partitioner and diffusion repartitioner. Their repartitioning scheme has two phases, a balancing phase based on the Hu and Blake method [3] and a refinement phase based on Kernighan-Lin refinement [5]. They obtained low `TotalV` results and low edge-cut degradation on mildly perturbed graphs [8]. Walshaw, Cross, and Everett later implemented JOSTLE-MD. This replaced the refinement phase of the original JOSTLE algorithm with multilevel refinement [2, 4]. The result was a decrease in edge-cut over the original algorithm at the cost of an increase in `TotalV` [7]. Unlike our schemes, JOSTLE-MD applies directed diffusion on the full graph. Thus, it misses the benefits of multilevel diffusion.

2.1 Multilevel Schemes for Graph Partitioning

The multilevel graph partitioning algorithm described in [4] finds high quality partitions. It has three phases, a coarsening phase, a partitioning phase, and a refinement, or uncoarsening, phase. During the coarsening phase, a sequence of smaller graphs are constructed from an input graph by collapsing vertices together. When enough vertices have been collapsed together so that the coarsest graph is sufficiently small, a partition is found using one among a variety of methods. Finally, the partition of the coarsest graph is projected back to the original graph by refining it at each uncoarsening level. Since now each uncoarsening level contains a finer graph, each subsequent graph has more degrees of freedom than the previous one had. These degrees of freedom can be used to decrease the edge-cut and increase the graph balance at each level.

Refinement is done in this scheme by a method based on the Kernighan-Lin partition algorithm [5]. Vertices are visited randomly. Each vertex visited is checked as to whether migrating partitions would

1. decrease the edge-cut while maintaining the graph balance, or
2. maintain the edge-cut and increase graph balance.

If so, the vertex is migrated. This process is repeated until it converges [4]. We define these two conditions as the *vertex migration criteria*.

3 Multilevel Diffusion

A repartitioning algorithm as a modification of the multilevel k-way partitioning algorithm implemented in MeIIS can be derived as follows. In the coarsening phase, only those pairs of nodes that belong to the same partition are considered for merging. Hence, the initial partitioning of the coarsest level graph is identical to the current partition of the graph that is being considered for repartioning.

We added a partition enforcement level to the refinement phase. The partition enforcement level is a constant which determines whether a partition is so overbalanced as to warrant forced vertex migration. During refinement, if the weight of the partition of the currently selected vertex is greater than the partition enforcement level and the vertex is a border vertex and at least one of the vertex's neighbor partitions is not overbalanced, then the vertex must migrate regardless of the consequences to edge-cut. If the vertex must migrate and more than one of its neighbor partitions are not overbalanced, then the vertex selects a partition to migrate to according to the vertex migration criteria described above. Thus, as the uncoarsening phase progresses, balance is automatically sought in conjunction with refinement. We use this repartitioning algorithm as a base algorithm and refer to it as R-MeIIS.

4 Multilevel Directed Diffusion

The R-MeIIS algorithm as described earlier can potentially balance any imbalanced graph. However, the diffusion of vertices from overbalanced partitions does not make use of the global information about the location of underbalanced partitions when determining vertex movement. It is possible to use the diffusion solution of the graph in order to direct the movement of vertices in R-METIS. That is, if the weight of a partition is greater than the enforcement level, then border vertices are migrated only in accordance with the diffusion solution. In essence, vertices flow directly from overbalanced partitions to underbalanced partitions with this global guidance. Thus, balancing is potentially quicker to converge than in R-METIS. This repartitioning algorithm will be referred to as Rf-METIS.

5 Results

We evaluated the performance of our repartitioning algorithms described above. These experiments were performed using 10 different graphs arising in finite element applications. All experiments were conducted on an SGI R10000 196MHz processor. The input graphs were partitioned by METIS and then the weights of some selected vertices were increased so as to overbalance and underbalance certain partitions.

Table 1 summarizes the results from these experiments. In particular, it shows the edge-cut, TotalV, MaxV, and run times for METIS, our multilevel repartitioner (R-METIS), our multilevel directed diffusion repartitioner (Rf-METIS), and a directed diffusion algorithm (DD-Repart). DD-Repart is essentially our implementation of the JOSTLE algorithm. The resulting value for each metric was normalized against the result from METIS. The means of 400 experiments are presented.

Table 1: Overview of Results

Algorithm	Edge Cut	TotalV	MaxV	Run Time
METIS	1.000	1.000	1.000	1.000
R-METIS	1.146	0.121	0.471	0.980
Rf-METIS	1.139	0.116	0.462	0.969
DD-Repart	1.455	0.173	0.504	2.880

Table 2: MDUAL, 4 Sources, 4 Sinks

Algorithm	Edge Cut	TotalV	MaxV	Run Time
METIS	34,317	252,747	5,499	15.0
Rf-METIS	35,262	36,237	2,912	16.1
DD-Repart	44,780	49,402	3,137	197.5

The R-METIS and Rf-METIS algorithms produced results within a few percent of each other for all four metrics. They both produced substantially lower results for TotalV and MaxV than the METIS partitioner while maintaining comparable edge-cuts. They also proved to be faster and more effective than the directed diffusion algorithm. In particular, the directed diffusion algorithm (DD-Repart) produced greater TotalV and MaxV results and larger edge cuts than either R-MℇℸS or Rf-MℇℸS.

Table 2 shows the results obtained on one particular experiment for MDUAL with a 128-way partition. The results from partitioning from scratch (MℇℸS), our multilevel directed diffusion repartitioning (Rf-MℇℸS), and the directed diffusion algorithm (DD-Repart) are compared on this table. Again edge-cut, TotalV, MaxV, and run time are compared. Here, however, only the unnormalized results of a single experiment are given.

6 Conclusions

Our multilevel directed diffusion repartioning algorithm is an excellent and robust repartitioning algorithm. It has been thoroughly tested on a variety of graphs for a wide variety of possible imbalance structures and has been found to be fast, scalable, and effective on them. Furthermore, it is highly parallel in nature. We have parallelized an optimized version of this algorithm described in [6]. In our early results we have obtained high-quality repartitionings of eight million vertex graphs in under three seconds using a 256-processor Cray T3D.

References

1. G. Cybenko. Dynamic load balancing for distributed memory multiprocessors. *Journal of Parallel and Distributed Computing*, 7(2):279–301, 1989.
2. Bruce Hendrickson and Robert Leland. An improved spectral graph partitioning algorithm for mapping parallel computations. Technical Report SAND92-1460, Sandia National Laboratories, 1992.
3. Y. F. Hu and R. J. Blake. An optimal dynamic load balancing algorithm. Technical Report DL-P-95-011, Daresbury Laboratory, Warrington, UK, 1995.
4. G. Karypis and V. Kumar. A fast and high quality multilevel scheme for partitioning irregular graphs. Technical Report TR 95-035, Department of Computer Science, University of Minnesota, 1995. Also available on WWW at URL http://www.cs.umn.edu/~karypis/papers/mlevel_serial.ps. A short version appears in Intl. Conf. on Parallel Processing 1995.
5. B. W. Kernighan and S. Lin. An efficient heuristic procedure for partitioning graphs. *The Bell System Technical Journal*, 1970.
6. Kirk Schloegel, George Karypis, and Vipin Kumar. Multilevel diffusion schemes for repartitioning of adaptive meshes. Technical Report TR 97-013, University of Minnesota, Department of Computer Science, 1997. http://www.cs.umn.edu/~karypis.
7. C. Walshaw, M. Cross, and M. G. Everett. Dynamic load-balancing for parallel adaptive unstructured meshes. *Parallel Processing for Scientific Computing*, 1997.
8. C. Walshaw, M. Cross, and M. G. Everett. Dynamic mesh partitioning: A unified optimisation and load-balancing algorithm. Technical Report 95/IM/06, Centre for Numerical Modelling and Process Analysis, University of Greenwich, London, UK, December 1995.

On the Embedding of Refinements of 2-dimensional Grids*

F. d'Amore[1], L. Becchetti[1], S.L. Bezrukov[2], A. Marchetti-Spaccamela[1],
M. Ottaviani[1], R. Preis[2], M. Röttger[2], and U.-P. Schroeder[2]

[1] Dipartimento di Informatica e Sistemistica, Università di Roma "La Sapienza",
Via Salaria 113, I-00198 Roma, Italy
[2] Fachbereich Mathematik/Informatik, Universität–GH Paderborn, Fürstenallee 11,
D-33102 Paderborn, Germany

Abstract. We consider the problem of constructing embeddings of 2-dimensional FEM graphs into grids. Our goal is to minimize the edge-congestion and dilation and optimize the load. We introduce some heuristics, analyze their performance, and present experimental results comparing the heuristics with the methods based on the usage of standard graph partitioning libraries.

1   Introduction

We consider the problem to embed large scale FEM graphs for the solution of partial differential equations into massively parallel computing systems. Roughly speaking, solving such equations with respect to a function F, say in two dimensions, requires to partition the domain of F into simple polygons (e.g. triangles or rectangles). Afterwards the value of the function F is computed in the nodes of the obtained partition. It turns out that accuracy requirements are not constant in the considered region but might vary considerably. This may lead to a partition of the area into polygons where the polygons sizes can be essentially different.

Such a partition can be viewed as a planar graph G whose nodes and edges correspond to the nodes and the sides of the polygons respectively. Each node represents a task for a processing element of the multiprocessor computing system. In order to minimize the running time the tasks have to be uniformly distributed among the processing elements. Furthermore, since the FEM requires to solve for each node x a difference equation involving x and its adjacent nodes, an information flow between the processing elements is caused which should also be minimized. These demands on the mapping can be expressed in the terms of graph embedding.

* This work was supported by the DFG-Sonderforschungsbereich 376 "Massive Parallelität: Algorithmen, Entwurfsmethoden, Anwendungen", the EC ESPRIT Long Term Research Project 20244 "ALCOM-IT" and the DFG Graduate Center "Parallele Rechnernetze in der Produktionstechnik", GRK 124/2-96.

Let $G = (V, E)$ and $H = (V', E')$ be finite graphs. An *embedding* of the *guest* graph G into the *host* graph H is a function $f : V \mapsto V'$ together with a routing scheme R_f which assigns to each edge $e = \{v_1, v_2\} \in E$ a path in H from $f(v_1)$ to $f(v_2)$. The *congestion of an edge* $e' \in E'$ is the number of paths in $\{R_f(e) \mid e \in E\}$ containing e'. The *edge-congestion of an embedding* is the maximum congestion of the edges of E'. The *dilation of an edge* $e \in E$ is the length of the path $R_f(e)$, the *dilation of an embedding* is the maximum length of the paths in $\{R_f(e) \mid e \in E\}$.

Many papers in the literature study the embedding of graphs with the goal to minimize both load and communication costs. However, most papers assume that either G is given or it belongs to a rather restricted class of graphs whose structural properties are exploited (see [MS] for an overview). A large amount of literature deals with the problem of finding a partitioning of a graph into k clusters of approximately the same size. Here the aim is to minimize the number of cut edges connecting nodes that belong to different clusters [DH,PSL]. In [DMT] the authors analyze the cost of implementing multigrid methods using parallel architectures. The multigrid methods define a hierarchy of graphs that need to be embedded which consist of the original fine grid and successively coarser grids. However, the authors did not consider the communication/load tradeoff of the embedding.

In this paper we examine the case where the host graph H representing the computing system is a grid (e.g. Intel Paragon and Parsytec GC are commercial grid-based systems), and study the embedding of quasi grids into H with the aim of minimizing the load, the dilation, and the edge-congestion. The *quasi grid* is defined as follows. Let R be a rectangular area on the 2-dimensional plane with sides parallel to the coordinate axes. By splitting this area with $x - 2$ horizontal and $y - 2$ vertical lines we get a $x \times y$ grid with $(x - 1) \cdot (y - 1)$ rectangular cells. Now for a cell C we define a *cell refinement* operation. The operation consists of splitting C into 4 subcells with one vertical and one horizontal line passing through the center of C. This results in a graph which has 5 new nodes and 4 new edges as shown with the thin lines in Fig. 1(a). Note that each edge of the original cell is now partitioned into 2 new edges. If two cells of the original graph which have to be refined have a common edge, we create the new node in their common edge just once (cf. Fig. 1(b)). The new subcells obtained after the cell refinement operation are allowed to be further refined. Applying the cell refinement operation a number of times we obtain a quasi grid S (cf. Fig. 2).

(a)   (b)

Fig. 1. (a) Refinement of a single cell and (b) of two neighboring cells.

Since the embedding problem is computationally hard [BU], we are interested in approximation algorithms. In Sect. 2, we introduce several heuristics and analyze them from the worst case point of view. We also present the results of preliminary experiments with the heuristics. In order to obtain fair results we compare the performance of the proposed heuristics with respect to the solution obtained by means of libraries for partitioning; these libraries exploit sophisticated algorithms that partition the nodes of a graph into clusters in order to minimize the load and the number of cut edges.

2 Algorithms

In this section we briefly present five heuristics for embedding a quasi grid S with m nodes into a $n_h \times n_v$ grid P and analyze their performance. Let $L^H(S, P)$ denote the load provided by heuristic H. A lower bound on the load is given by $L^{avg}(S, P) = \lceil m/(n_h \cdot n_v) \rceil$ and we define $R^H(S, P) := L^H(S, P)/L^{avg}(S, P)$.

Heuristic *Tile1*: This and the next two heuristics are based on partitioning the quasi grid S into boxes that correspond to the structure of the grid P. To describe *Tile1* we first introduce two orderings ϕ_h and ϕ_v of the nodes of S. We consider the nodes of S as points (x, y) on the plane, assuming that the origin of the coordinate system is the leftmost and bottommost node of S and its axes are parallel to the segments of S. For the nodes $(x_1, y_1), (x_2, y_2)$ of S we say that $(x_1, y_1) <_{\phi_h} (x_2, y_2)$ iff $y_1 < y_2$, or if $y_1 = y_2$ then $x_1 < x_2$. Similarly, we say that $(x_1, y_1) <_{\phi_v} (x_2, y_2)$ iff $x_1 < x_2$, or if $x_1 = x_2$ then $y_1 < y_2$. Now we partition the nodes into n_h sets $A_1, ..., A_{n_h}$. A_i consists of m_i' w.r.t ϕ_v consecutive nodes of S, with $\lfloor m/n_h \rfloor \le m_i' \le \lceil m/n_h \rceil$, $i = 1, ..., n_h$. Moreover, the nodes are partitioned into n_v sets $B_1, ..., B_{n_v}$ where B_j consists of m_j'' w.r.t. ϕ_h consecutive nodes of S, with $\lfloor m/n_v \rfloor \le m_j'' \le \lceil m/n_v \rceil$, $j = 1, ..., n_v$. The embedding defined by *Tile1* is the following: The nodes of $C_{ij} = A_i \cap B_j$ are mapped onto the node p_{ij} of P, with $i = 1, ..., n_h$ and $j = 1, ..., n_v$. (cf. Fig. 2(a)).
This heuristic is first of all designed to provide a small dilation and edge-congestion. It guarantees that the total load of each column (row) of the grid P is the same up to one. However, the loads of single processors can be essentially different.

Proposition 1. $R^{Tile1}(S, P) \le \min\{n_h, n_v\}$.

The first step of **Tile2** partitions the nodes in the same way as *Tile1*. Then each set A_i, $i = 1, ..., n_h$, is partitioned into sets C_{ij} of m_i^j consecutive nodes of A_i (w.r.t. ϕ_h), where $\lfloor m_i'/n_v \rfloor \le m_i^j \le \lceil m_i'/n_v \rceil$ and $j = 1, ..., n_v$ (cf. Fig. 2(b)).

Proposition 2. $R^{Tile2}(S, P) = 1$.

The heuristic **Tile3** involves an integer parameter d and uses the heuristic *Tile2* as a subroutine. First of all we partition the nodes of P into clusters C_{kl}:

$$C_{kl} = \{p_{ij} \mid (k-1)d + 1 \le i \le kd, (l-1)d + 1 \le j \le ld\},$$

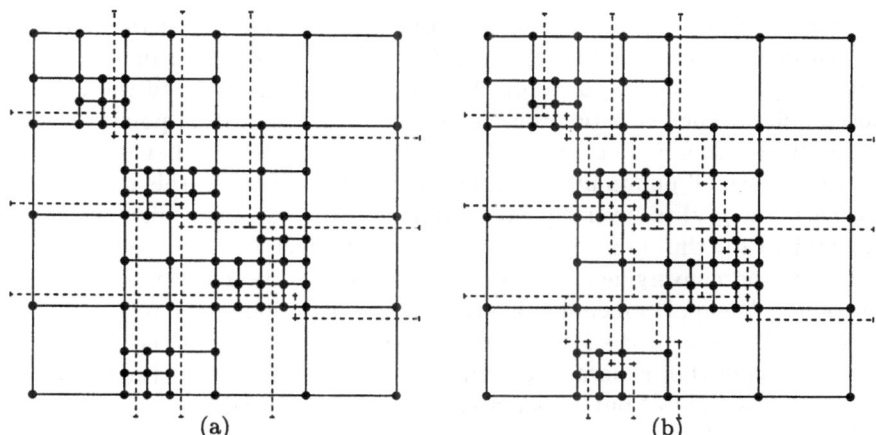

Fig. 2. The exemplified heuristics *Tile1* (a) and *Tile2* (b), where the guest graph is the 4×4 grid.

with $k = 1, ..., \lceil n_h/d \rceil$ and $l = 1, ..., \lceil n_v/d \rceil$. Furthermore, we partition the nodes of the quasi grid S into $\lceil n_h/d \rceil \cdot \lceil n_v/d \rceil$ blocks B_{kl} and map the nodes of B_{kl} onto the processors of the cluster C_{kl} using heuristic *Tile2*. The use of heuristic *Tile3* supposes that the unrefined quasi grid (that is, the grid obtained from S by considering its coarsest subgrid) coincides with, or is a subgraph of, P; if this is not the case a pre-embedding step is carried out. Details will be given in the full version of the paper.

The aim of the ***Pac-Man*** heuristic is to group nodes of S into $n_h \cdot n_v$ clusters in such a way that nodes of the same cluster lie as close to each other as possible. Under this approach we measure the nearness as the length of the shortest path between the corresponding nodes in S. We start by choosing randomly $n_h \cdot n_v$ seed-points of S. Then each seed-point tries in parallel to occupy nodes in its neighborhood; it stops when there is no node so that the already occupied area remains connected. Depending on the accrued clustering a new seed-point for the next iteration is computed. Namely, the center of each cluster is chosen as the seed-point for the next iteration. The algorithm terminates when all seed-points remain unchanged. Since there is no guarantee that this heuristic produces a balanced load, we have integrated a global load balancing step as post-processing. This step guarantees a totally balanced load without loss of the compactness of the previously computed clusters.

Heuristic ***Kohonen***: We adapt Kohonen's self-organizing maps [K] to compute an embedding that preserves topological relations between the nodes of the refinement. The general Kohonen process maps points of an euclidean space to adaptive elements called neurons in such a way that points of the space which are close to each other are mapped onto neurons which are close to each other. We represent the nodes of S as points of the 2-dimensional euclidean plane

E^2 and the grid P as the neural network. The Kohonen heuristic first assigns to each node v of P a point $p(v) \in E^2$ uniformly distributed in the rectangle $R = \{(x, y) \in E^2 \mid 1 \leq x \leq n_h, 1 \leq y \leq n_v\}$; let M denote the set of the chosen points. The heuristic proceeds by repeatedly relocating points as follows: randomly choose a node w of S, compute a point $n(w) \in M$ that is closest to w and then move all points $p(v) \in M$ in the direction of w. At-this the movement of each point $p(v)$ is inversely proportional to the euclidean distance between $p(v)$ and $n(w)$; the intensity of the movements decreases in time in order to guarantee the convergence. The process terminates after a certain number of iterations. The final embedding is given by the partition of the rectangle R into clusters $\{C(u) \mid u \in M\}$ induced by the Voronoi-diagram of the points of M. Since this algorithm generates an embedding which minimizes the communication costs (see [RMS]) but does not care about a balanced load, we additionally apply a partial load balancing procedure after an initial convergence phase every 300 iteration steps. In this procedure we compute the local load gradient of every node of the grid P by comparing the loads of all its neighbors. After that, for each node v of P, we move the point $p(v)$ in the direction of the local load gradient.

3 Partitioning Tools

Graph partitioning problems arise in many different applications, which leads to many heuristics based on different ideas. In general, graph partitioning can be viewed as an embedding of the guest graph into a complete graph, i.e., the load balance and the cut size are the major cost measures. Most applications require the load balance to be optimal, i.e., the number of nodes in each part differ at most by one. The goal is to minimize the cut size of the partition. The problem of constructing such a partition is known to be NP-complete even in the case of partitioning into two parts of the same size. Efficient heuristics have been designed in the last decades to construct partitions with very low cut sizes. Although they are based on some reasonable arguments for a low cut size, there is no guarantee that a method works well for all kinds of graphs.

Partitions with low cut sizes might be very useful for our embedding problem. In several previous studies (e.g. [BB,DMM]), graph partitioning was used in a first step to partition the graph in as many clusters as there are nodes in the host graph. A second step then performs the one-to-one embedding of the cluster-graph into the host graph. This two step strategy integrates the powerful partitioning methods into the embedding problem.

The efficiency of heuristics strongly depends on the implementation details. Several libraries like *Jostle* ([WCE]) by Walshaw, *Metis* ([KK]) by Karypis and Kumar, *Scotch* ([PR]) by Pellegrini or *Party* ([PD]) by Preis and Diekmann exist to solve the partitioning problem. A library like *Scotch* also encounters the embedding problem, but the cost function takes into account only the total sum of the dilation of all edges and not the maximum dilation or maximum edge-congestion. In a first approach, we use these libraries to compute good partitions of our graphs and compare the resulting loads and cut sizes to those heuristics described above.

4 Tests

In this section we will present some experimental results of the mentioned methods on two test graphs shown in Fig. 3. Both of them are subgraphs of the previously defined quasi grids.

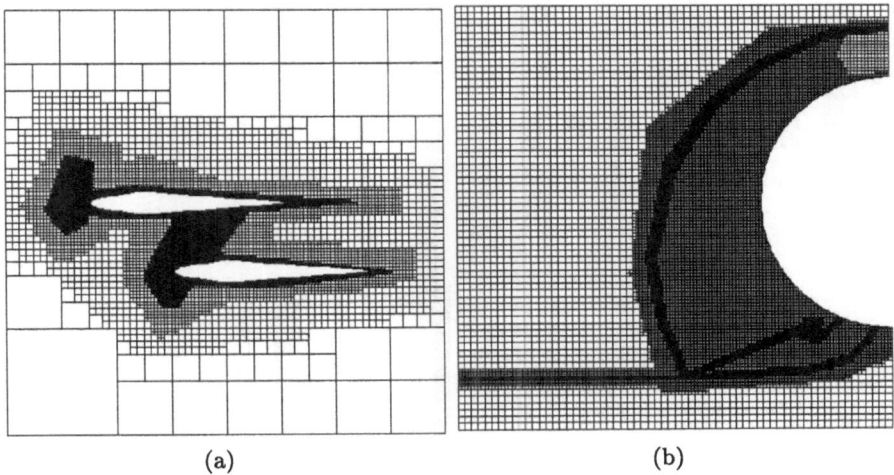

(a)   (b)

Fig. 3. (a) Graph *biplane* (21,701 nodes and 42,038 edges) and (b) graph *shock* (36,476 nodes and 71,290 edges).

The embedding is performed on an 8×8 grid as the host graph and the values for dilation, edge-congestion, load, and cut size are presented. A routing scheme for the edge-congestion is calculated in a sequential order for each path. To decide on a single path, the X/Y and the Y/X paths are considered and we choose the one with the lowest occuring edge-congestion along the path.

Our main cost criteria are the dilation and the edge-congestion which are shown in Fig. 4, (a) and (b). The results show that *Tile1* and *Tile3* have a very low dilation and that all *Tile*-heuristics have a low edge-congestion. Please note that the partitioning methods *Pac-Man*, *Party*, *Jostle*, and *Metis* are performing an embedding into a complete host graph. At this stage, we used the identical embedding on the grid to see how the dilation and congestion will be without optimization of the embedding. The maximum load, the cut edges and the CPU running times (sec) on a SUN Sparc20 workstation of the methods are shown in Fig. 4, (c), (d) and (e), respectively. A value of 100% refers to an optimal load balance. This or a just slightly unbalanced load is only guaranteed by the

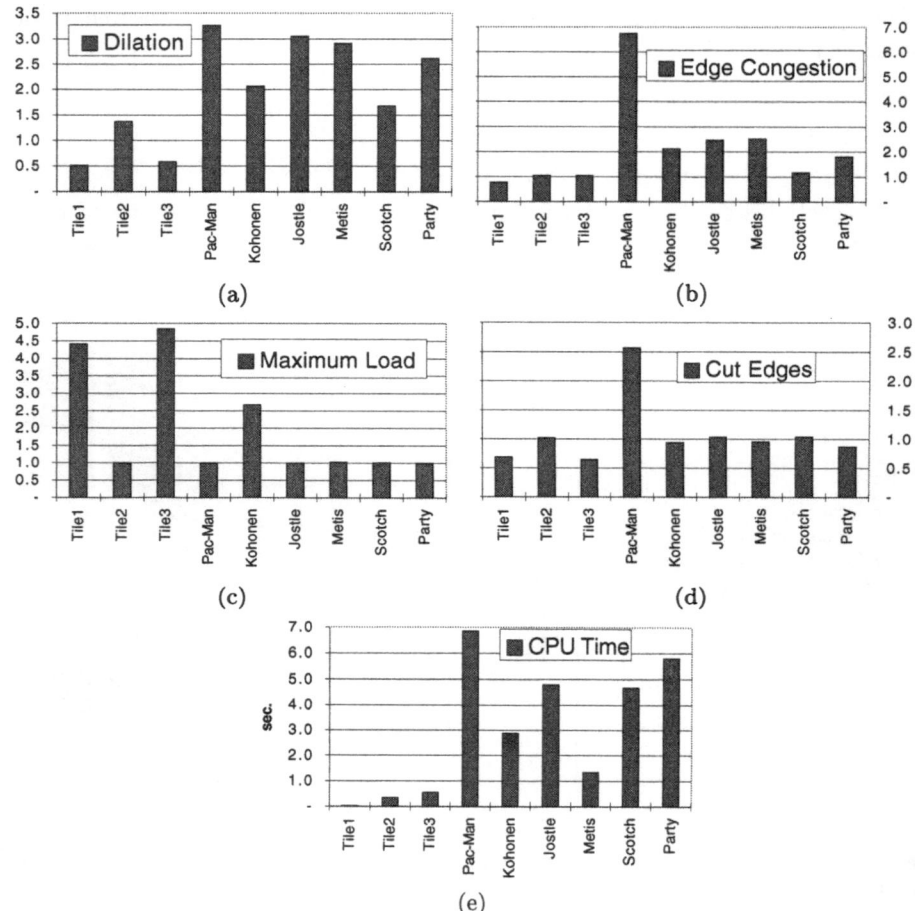

Fig. 4. Performances of tested algorithms: (a) Dilation, (b) edge-congestion, (c) maximum load, (d) cut size and (e) CPU time.

methods *Tile2*, *Jostle*, *Metis*, *Scotch*, *Party*, and *Pac-Man*. In case of *Tile1* and *Tile3*, the small values of the dilation imply high unbalance of the load.

The high values for the load for the heuristics *Tile1*, *Tile3*, and *Kohonen* are compensated by low cut sizes. In comparison with the other methods, *Party* seems to perform slightly better than the others. In general, the partitioning libraries produce a balanced load and very low cut sizes. This is a strong argument to integrate them in a two-step strategy for solving the embedding problem.

5   Conclusion

In this paper several heuristics for the off-line mapping of refinements of 2-dimensional grids are presented. They were all tested with two benchmark graphs considering a wide set of parameters. According to the results of the experiments, the *Tile*-heuristics seem to guarantee a good trade-off between load and dilation/edge-congestion and, in comparison with other partitioning tools, PARTY seems to provide good results with respect to cut-size and load.

Although the Kohonen method does not provide very promising results, being applied for the static embedding considered so far, it seems to be more suitable for dynamic embeddings than the other considered methods. This method can explore the already computed embedding to react on local changes of the guest graph without completely recomputing the whole embedding.

As to future work, we are going to modify and improve some of the presented heuristics and also to study the dynamic version of the problem. As to more theoretical aspects, we are completing the analytical study of our heuristics. One of the most interesting aspects is the trade-off existing between the different metrics considered, in particular between load, dilation, and edge-congestion, for which we already gained some preliminary results.

References

[BB]   M.J. Berger and S.H. Bokhari, *A Partitioning Strategy for Nonuniform Problems on Multiprocessors*, IEEE Trans. on Comp., C-36 (5), 570–580, 1987.

[BU]   S.L. Bezrukov and W. Unger, *On Refinement of 2-Dimensional Grids*, Preprint, 1995.

[DMM]   R. Diekmann, D. Meyer and B. Monien, *Parallel Decomposition of Unstructured FEM-Meshes*, Proc. of Irregular 95, LNCS 980, 199–215, 1995.

[DH]   W.E. Donath and A.J. Hoffman, *Lower bounds for the partitioning of graphs*, IBM J. Res. Develop. 17, 1973, 420–425.

[DMT]   S.E. Dorward, L.R. Matheson and R.E. Tarjan, *Toward efficient unstructured multigrid preprocessing*, Proc. of Irregular 96, LNCS 1117, 1996, 105–118.

[K]   T. Kohonen, *Self-Organization and Associative Memory*, 3rd edition, Springer Verlag, Berlin 1989.

[KK]   G. Karypis and V. Kumar, *A fast and High Quality Multilevel Scheme for Partitioning Irregular Graphs*, Tech. Rep. 95-035, Dept. of Computer Science, U. of Minnesota, 1995.

[MS]   B. Monien and I.H. Sudborough, *Embedding one Interconnection Network in Another*, Computing Suppl., 7, 1990, 257–282.

[PD]   R. Preis and R. Diekmann, *The PARTY Partitioning-Library User Guide - Version 1.1*, Tech. Rep. TR-RSFB-96-024, U. Paderborn, 1996.

[PSL]   A. Pothen, H.D. Simon and K.-P. Liou, *Partitioning sparse matrices with eigenvectors of graphs*, SIAM J. Matrix Anal. Appl.,11, 1990, 430–452.

[PR]   F. Pellegrini and J. Roman, *SCOTCH: A Software Package for Static Mapping by Dual Recursive Bipartitioning of Process and Architecture Graphs*, Proc. of HPCN, 1996, 493–498.

[RMS]   H. Ritter, T. Martinetz and K. Schulten, *Neural Computation and Self-Organizing Maps*, Addison Wesley, 1991.

[WCE]   C. Walshaw, M. Cross and M.G. Everett, *A Localized Algorithm for Optimizing Unstructured Mesh Partitions*, Int. J. Supercomputer Appl., 9, 1995, 280–295.

Dynamic Program Description as a Basis for Runtime Optimization *

Jörn Gehring

Paderborn Center for Parallel Computing (PC²), D-33095 Paderborn, Germany,
joern@uni-paderborn.de

Abstract. Dynamic load-balancing is an important topic for network-computing. In order to get the best performance out of a networked pool of dynamic resources, it is important to use as much information as possible for load-balancing decisions. Parallel programs are usually no "ex-and-hop" applications and a load-balancer that treats every execution as if it had never seen this application before, cannot achieve optimal results. We present an algorithm for generating formal descriptions of the dynamic execution behavior of distributed programs. These descriptions can then be used for a variety of tools, such as schedulers, load balancers, or routing systems.

1 Introduction

Currently there exists a variety of cluster management systems that support the distribution of sequential jobs on hundreds or thousands of networked computers. However, there is only rudimentary support for the distribution of parallel applications [6, 1]. Since communicating distributed programs need additional effort in scheduling, mapping, dynamic load-balancing, and message routing, much research has been done in these fields. However, most strategies either ignore the communication structure of the applications completely [3] or use a static representation [7]. The latter approach achieves better response times but it is still a simplification of the real problem. Fig. 1 depicts an example of three communicating processes. Looking at the dynamic program description, it is immediately clear that all three processes should be mapped onto one single machine. However, this cannot be derived from the static description. Therefore, newer policies

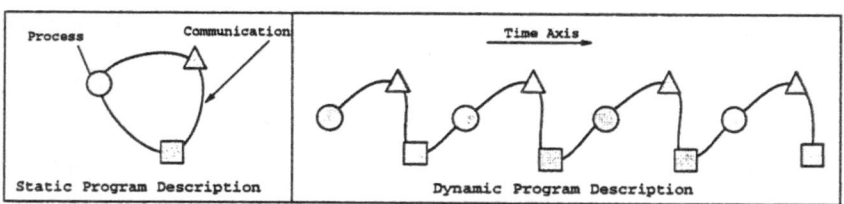

Fig. 1. Structure information of static vs. dynamic program description

---

* This work was supported by the Ministry for Science and Research of Northrhine-Westfalia, project *Metacomputing*

try to consider the dynamics of parallel applications as well. [9] and [2] are some examples of these. Both require dependency graphs which describe the dynamic behavior of the program during a certain time period. However, the user needs to provide these graphs for those sections of the algorithm that are worthwhile the effort of specific optimizations.

With the concept of heterogeneous metacomputing the impact of improper process assignments has increased essentially [5]. Consequently, the availability of formal descriptions of dynamic program characteristics is important for the performance of large WAN-distributed applications and thus for the acceptance of metacomputing at all.

In the following we present a concept for describing characteristics of the dynamic execution behavior of distributed communicating programs automatically. Since the general problem of extracting all features is known to be NP-complete, we reduce the complexity by identifying the characteristics usable for optimization purposes in Section 2. From this we derive an algorithm in Section 3 which provably finds all usable traits of a distributed message-passing program in polynomial time. This concept has been implemented in the context of the MARS [5] project which is part of the international *"Metacomputer Online"*[8] initiative.

2   Dependency Graphs and Phases

A trace of a parallel program can be represented by a directed graph with the nodes representing sequential computations and the edges standing for communications. In the following, we show how this representation can be used for deriving information about the internal structure of the parallel application.

2.1   Dependency Graphs

We consider a parallel program as a number of sequential computations interleaved by "critical" communications. These are blocking receive operations, blocking sends, waiting for termination of non-blocking sends/receives, or blocking collective communications (broadcast, gather, ...). For convenience, we also consider start and termination of a process as critical communications. The block of sequential statements between any sequence of two communications is called an *independent block*.

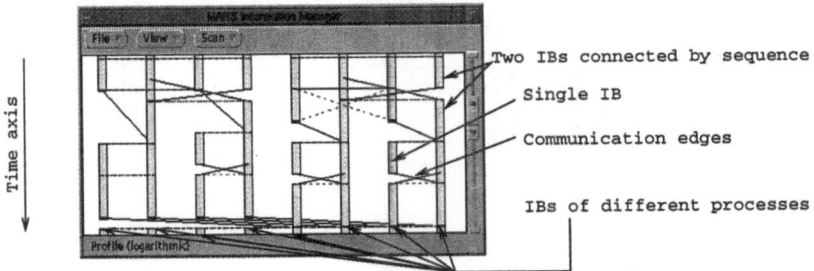

Fig. 2. Example for a dependency graph (sequence edges omitted for the sake of clarity)

Definition 1 *Let I be a sequence of instructions to be executed within a single process. I is called an "independent block" (or IB for short), iff the statements to be executed immediately before and after I are critical communications and there are no critical communications in I.*

Thus, all IBs are disjunct, no IB can start before all of its predecessors have finished, and once the execution of an IB has started, it can continue until completion. (See Fig. 2 for an example of a parallel sorting algorithm.)

Definition 2 *A directed graph is called a "dependency graph", if its nodes are independent blocks and for any two nodes α and β there is an edge from α to β iff at least one of the following conditions holds:*

1. *Both belong to the same process and β follows immediately after α (α "precedes" β). These edges are called "sequence edges". (Fig. 3.1)*
2. *The instruction after α is a blocking send which is received by the instruction directly before β. The latter is either a blocking receive or a wait for the termination of a non-blocking receive. (Fig. 3.2)*
3. *The instruction after α and that before β are both part of the same collective communication. (Fig. 3.3)*
4. *β waits for the termination of a blocking or non-blocking send initiated by a preceding IB and received immediately after α. (Fig. 3.4)*

In [5] we have shown, how dependency graphs can be collected from program runs without significant impact on the execution speed. Note that applications produce different dependency graphs in different execution runs. These have to be merged together for consolidating the statistical data of the application [4].

2.2 Phases

Our strategy is to perform offline analysis on dependency graphs in order to detect repeating patterns and store them into a knowledge base. In [5] we have shown, how this knowledge can be used for speeding up distributed programs by dynamic migration decisions. During the offline analysis we are looking for "phases" in the dependency graphs. A phase is a subgraph of a dependency graph that represents a set of instructions performing a closed subtask. It is considered to be of high quality, if it occurs frequently and covers a large amount of the program's overall resource usage. Concerning sequential or data parallel

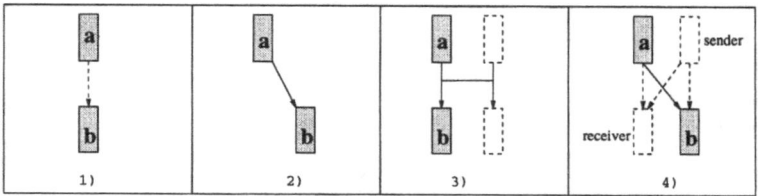

Fig. 3. Different edges of a dependency graph

programs, high quality phases are closely related to loop- and function-bodies and can therefore be detected during compile time. This is not possible for distributed MIMD programs, because the dynamic interaction between different modules can usually not be predicted by the compiler.

Since we are going to use phases for the prediction of runtime behavior, a phase which has started execution must not be interrupted by communications with IBs that are out of scope of this phase. I.e., the predecessors of an IB are either all inside the phase or all outside. This leads us to the following formal definition:

Definition 3 *Let $\mathcal{D} = (\mathcal{V}, \mathcal{E})$ be a dependency graph with $\mathcal{V}$ being the set of nodes and $\mathcal{E}$ being the set of edges. A connected subgraph $\mathcal{P} = (\mathcal{V}_\mathcal{P}, \mathcal{E}_\mathcal{P})$ of $\mathcal{D}$ is a "phase of $\mathcal{D}$", iff*

1. *For all $v, u \in \mathcal{V}_\mathcal{P}$: If $(v, u) \in \mathcal{E}$, then $(v, u) \in \mathcal{E}_\mathcal{P}$*
2. *For all $(u, v) \in \mathcal{E}_\mathcal{P}$, $u' \in \mathcal{V}$: If $(u', v) \in \mathcal{E}$, then $u' \in \mathcal{V}_\mathcal{P}$ and $(u', v) \in \mathcal{E}_\mathcal{P}$.*

Consequently, a single IB is also a phase, which we call an "atomic phase".

3 Detecting Phases

In the following we introduce an order on the quality of phases and describe a deterministic algorithm that extracts all valuable phases from a dependency graph. Although there may be a total of $2^{|\mathcal{V}|}$ phases in a dependency graph, we will prove that at most $O(|\mathcal{V}|^2)$ of them are usable for optimization purposes.

3.1 The Quality Function

As mentioned in Section 2.2, a phase is "good", if it describes a large part of the internal structure of the program. I.e., it occurs frequently and covers a high percentage of the overall resource usage. The algorithm presented in Section 3.3 constructs large phases out of smaller ones. Thus, the quality function has to make small frequently occurring phases attractive during the first iterations of the algorithm, while larger phases ought to become better to the end (see Sec. 3.4). This leads us to the following definition:

Definition 4 *Let $\mathcal{D} = (\mathcal{V}, \mathcal{E})$ be a dependency graph and $A \subseteq \mathcal{D}$ a phase. We define the "quality of A" $Q(A)$ to be $Q(A) := (\, Occ(A), |A| \,)$ with*

$$Occ(A) := \begin{cases} 0 & , \text{ if } A = \emptyset \\ |\{A' \subseteq \mathcal{D} \mid A, A' \text{ are isomorph}\}| & , \text{ else} \end{cases}$$

Two quality values are compared using lexicographical comparison.

Phase A is consequently considered to be of higher quality than B, if it appears more frequently or if A and B have the same number of instances but A is larger than B. In Section 3.4 we will show, how this definition directs the phase-detection algorithm on its search for valuable phases.

1) Let $i := 0$; $S_0 := \mathcal{V}$; $S_0^* := \mathcal{E}$
2) While $S_i^* \neq \emptyset$ Do
3)   Find $(A_i, B_i) \in S_i^*$ with $Q(A_i \oplus B_i) = \max_{(\alpha,\beta) \in S_i^*} \{Q(\alpha \oplus \beta)\}$
4)   Let $S_{i+1} := S_i \cup \{A_i \oplus B_i\}$
5)   Let $T_i := \{ \ (C, A_i \oplus B_i) \mid \ C \in S_i \backslash (A_i \oplus B_i) \ \wedge$
                       $[\exists \alpha \in C, \beta \in (A_i \oplus B_i) : (\alpha, \beta) \in \mathcal{E}] \ \} \quad \cup$
               $\{ \ (A_i \oplus B_i, C) \mid \ C \in S_i \backslash (A_i \oplus B_i) \ \wedge$
                       $[\exists \alpha \in (A_i \oplus B_i), \beta \in C : (\alpha, \beta) \in \mathcal{E}] \ \}$
6)   Let $U_i := \{(F, G) \in S_i^* \mid F \subseteq A_i \oplus B_i \wedge G \subseteq A_i \oplus B_i\}$
7)   Let $S_{i+1}^* := (S_i^* \cup T_i) \backslash U_i$
8)   Let $i := i + 1$
9) Let $I := i$

Fig. 4. The phase-detection algorithm

3.2 Creating Complex Phases

Complex phases are created by the concatenation of at least two simpler phases. Due to Def. 3 the union of two phases does not have to be a phase by itself. Other IBs that precede one phase but are not part the other one may have to be added. These additional IBs are covered by the following relation:

Definition 5 *Let A and B be phases of a dependency graph $\mathcal{D} = (\mathcal{V}, \mathcal{E})$ and let $\gamma \in \mathcal{V}$ be a node of $\mathcal{D}$. We say $\gamma \to (A \cup B)$ ("γ feeds $(A \cup B)$"), if γ has to be added to $A \cup B$ in order to make it a phase:*

$$\gamma \to (A \cup B) \iff \exists \, \alpha, \beta \in \mathcal{V} \text{ with } \alpha \in A \ \wedge \ \beta \in B \ \wedge \ (\alpha, \beta) \in \mathcal{E} \ \wedge \ (\gamma, \beta) \in \mathcal{E}$$

We can now define the concatenation of two phases A and B as:

$$A \oplus B := \begin{cases} \emptyset & , if \ \nexists \, \alpha, \beta \in \mathcal{V} \text{ with } \alpha \in A \ \wedge \ \beta \in B \ \wedge \ (\alpha, \beta) \in \mathcal{E} \\ A \cup B \cup \{\gamma \in \mathcal{V} \mid \gamma \to (A \cup B)\} & , otherwise \end{cases}$$

Thus, $A \oplus B$ is the smallest possible phase that includes $A \cup B$.

3.3 The Phase-Detection Algorithm

The phase-detection algorithm starts with a set S containing all atomic phases. It then successively merges the best pair out of all known phases and inserts it into S. Thus, after the algorithm has terminated, S will hold all extracted phases. In order not to do superfluous work, we exclude pairs lying entirely within a phase that has already been created. Fig. 4 gives a formal description of the algorithm which takes a dependency graph $\mathcal{D} = (\mathcal{V}, \mathcal{E})$ as input.

S_i contains all phases known after the i-th iteration and S_i^* holds all pairs of phases in S_i that may have to be considered during later iterations. Lines 5) and 6) may look somewhat complicated, but their purpose is simple. T_i includes all pairs to be added to S_i^* due to the creation of $(A_i \oplus B_i)$. U_i, on the other hand, incorporates all pairs having become superfluous.

3.4 A Closer Look at the Algorithm

Looking at Def. 4, it is obvious that $Occ(A)$ cannot increase, when phase A is extended. Furthermore, all pairs added to S^* in line 5) and 7) will be extensions of phases already contained in S_i^*. Thus, we can conclude that the greater the value of i is, the less often the newly constructed phases occur in $\mathcal{D}$. Since $Occ(A) \in \{0, \ldots, |\mathcal{V}|\}$ for any phase A, we have now shown that

$$\exists\, c_{|\mathcal{V}|}, \ldots, c_0 \in \{0, \ldots, I-1\} \text{ with } 0 = c_{|\mathcal{V}|} \leq c_{|\mathcal{V}|-1} \leq \ldots \leq c_0 = I \text{ and}$$

$$\forall_{j \in \{1, \ldots, |\mathcal{V}|\}} : \left[\forall_{k \in [c_j; c_{j-1}[} : Occ(A_k \oplus B_k) = j \right] \quad (1)$$

I.e., from iteration c_j inclusive to iteration c_{j-1} exclusive the algorithm creates only phases occurring exactly j times within the dependency graph.

The phases constructed in line 4) are connected subgraphs of $\mathcal{D}$. Since $\mathcal{D}$ is finite, there exist at most $n_j \leq |\mathcal{V}|$ disjunct phases $P_1, P_2, \ldots, P_{n_j}$ which occur precisely j times in $\mathcal{D}$ such that

$$Occ(P_1) = \ldots = Occ(P_{n_j}) = j \text{ and } \left[\forall_{k,l \in \{1, \ldots, n_j\}, k \neq j} : Occ(P_k \oplus P_l) < j \right] \quad (2)$$

During the c_j-th iteration, the algorithm constructs the so far largest phase appearing j times in $\mathcal{D}$. Because the quality function is defined to judge on the size of two phases, if their occurrence is identical, the algorithm will try to extend $A_{c_j} \oplus B_{c_j}$ during the next iteration. If this is not possible, because there is no phase C with $[(C, A_{c_j} \oplus B_{c_j}) \in S_{c_j+1}^* \lor (A_{c_j} \oplus B_{c_j}, C) \in S_{c_j+1}^*] \land Occ((A_{c_j} \oplus B_{c_j}) \oplus C) = j$, it tries extending another phase with the same number of instances that is not connected to $A_{c_j} \oplus B_{c_j}$.

Lines 6) and 7) remove all pairs of inner phases from S^* and therefore ensure that once a phase with occurrence j has been extended to its maximum, it will no longer be considered throughout iterations $[c_j; c_{j-1}[$. Thus, during the next iteration, the algorithm will continue extending the last phase or it starts inflating a completely new one. From Def. 4 we conclude that each IB $v \in \mathcal{V}$ belongs to at most one $P_i \in \{P_1, \ldots, P_{n_j}\}$ and during each iteration of $[c_j; c_{j-1}[$ at least one of these IBs will be assigned to its corresponding P_i. Lines 6) and 7) ensure that this IB will no longer be considered until at least iteration c_{j-1}. Since there are at most $|\mathcal{V}|$ IBs to assign, we have now shown that

$$\forall_{j \in \{1, \ldots, |\mathcal{V}|\}} : 0 \leq c_{j-1} - c_j \leq |\mathcal{V}| \quad (3)$$

Combining (3) and (1) we conclude that $I \leq |\mathcal{V}|^2$ and therefore the algorithm extracts $|S_I| \leq |\mathcal{V}| + I = O(|\mathcal{V}|^2)$ phases. Fig. 5 depicts an example run of the phase-detection algorithm. All IBs are the same and thus there is only one atomic phase with 12 instances. This phase is then iteratively extended until the algorithm has generated the last phase which is always the complete dependency graph.

Of course, the algorithm does not have to be implemented as it was described in Fig. 4. In [4] we show that it can be implemented with an overall time complexity of $O(|\mathcal{V}|^4)$. This sounds still very much, but experimental results have

964

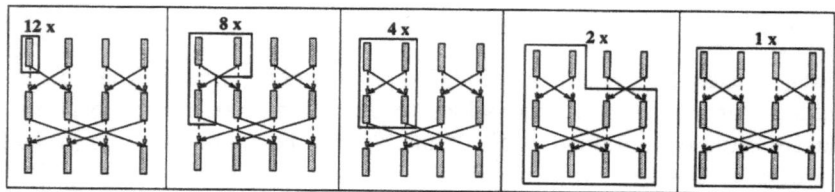

Fig. 5. Example run of the phase-detection algorithm

demonstrated that for two reasons the algorithm is usually much faster: First, the $\oplus$-operation most of the time produces larger phases than $A_i \cup B_i$ and second, not for every $j \in [1; |\mathcal{V}|]$ there exist A_i and B_i with $Occ(A_i \oplus B_i) = j$. Analyzing the dependency graph depicted in Fig. 5 for example, the algorithm generates only 26 occurrences of four valuable phases out of 12 IBs.

4 Experimental Results

Fig. 6 depicts the behavior of a distributed CG solver with different process mappings. We used two different clusters of four workstations each that were connected by a 32 Mbit/s wide area network. The bandwidth within the clusters was 155 Mbit/s. This configuration was used, because it demonstrates the absence of an optimal static mapping.

Both mappings shown in Fig. 6 are optimal for a particular range of problem sizes. Mapping A is optimal for problem sizes below 2^8 and mapping B is optimal for larger problems. Mapping A uses only one of the two clusters. Therefore, it is well suited for small problems, because the slow WAN connection cannot have any effect on the computation. For larger problems however, it is better to use all the computing power that is available. The optimal mapping for each problem size depends on the current network and computer configuration which changes dynamically in the shared environment. A process mapping tool that does not take advantage of any information about the program dynamics can only choose a mapping at random. Using dynamic program descriptions however, we first make a good guess according to the statistics stored in the knowledge base. I.e., if we have already seen a lot of runs with large problem sizes, we choose mapping B. Then, the application is monitored during the first few phases (iterations) and its behavior is matched against the knowledge base. From these

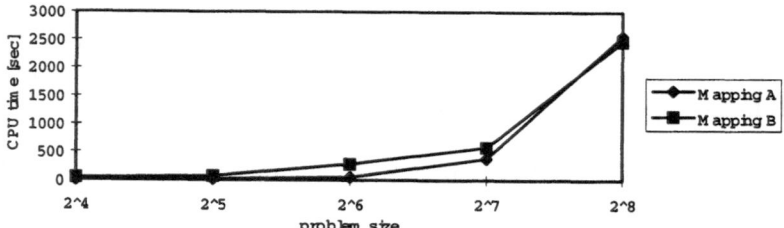

Fig. 6. Two different task mappings of a distributed CG solver

observations we decide, if B was the correct choice. Since the knowledge base provides estimations of the expected execution time, too, the load balancer can decide, if the application is likely to run long enough to be re-mapped. (More detailed experiments can be found in [5].)

5 Summary

We have demonstrated how characteristic features of a distributed program can be detected by analyzing dependency graphs. We presented a deterministic algorithm which provably detects all valuable characteristics within polynomial time. The algorithm considers results from previous runs of the same application and thereby produces better results each time the application is invoked. The extracted features are stored in a knowledge base which can be exploited by a variety of supporting tools like schedulers, load balancers, or routing systems for minimizing the response time. The consideration of characteristic phases enables these tools to adapt their optimization strategies to the dynamic communication behavior of the application. This is especially important for WAN-distributed applications arising in metacomputing environments.

References

1. M.A. Baker, C.G. Fox, and H.W. Yau. Cluster computing review. Technical report, Syracuse Univ., nov 1995.
2. W. Becker and G. Waldmann. Exploiting inter task dependencies for dynamic load balancing. In *Proc. HPCN-95*, pages 407–412. Springer LNCS 919, 1995.
3. T. Decker, R. Diekmann, R. Lüling, and B. Monien. Towards developing universal dynamic mapping algorithms. In *Proc. of the 7th IEEE Symposium on Parallel and Distributed Processing*, pages 456–459, 1995.
4. J. Gehring. Dynamic program description as a basis for runtime optimization. Technical Report PC2/TR-002-97, Paderborn Center for Parallel Computing (PC^2), 1997.
5. J. Gehring and A. Reinefeld. Mars - a framework for minimizing the job execution time in a metacomputing environment. *Future Generation Computer Systems*, 12(1996)(1):87–99, may 1996.
6. J.A. Kaplan and M.L. Nelson. A comparison of queueing, cluster and distributed computing systems. Technical memo, NASA, jun 1994.
7. Soo-Young Lee and J.K. Aggarwal. A mapping strategy for parallel processing. *IEEE Transactions on Computers*, C-36(4):433–442, apr 1987.
8. A. Reinefeld, R. Baraglia, T. Decker, J. Gehring, D. Laforenza, F. Ramme, T. Römke, and J. Simon. The MOL project: An Open, Extensible Metacomputer. In *Heterogenous computing workshop HCW'97 at IPPS'97*, 1997.
9. G.C. Sih and E.A. Lee. A compile-time scheduling heuristic for interconnection-constrained heterogeneous processor architectures. *IEEE Transactions on Parallel and Distributed Systems*, 4(2):175–187, feb 1993.

Workshop 16

Performance Evaluation and Prediction

Workshop 16:
Performance Evaluation and Prediction

Arndt Bode and Jack Dongarra

This workshop aims at bringing together people working in the different fields of parallel and distributed systems performance evaluation and prediction. It addresses methodologies, tools, modelling, and parallel and distributed applications. Especially welcome are contributions that bridge the gap between the fields of analytical modelling, performance measurement, parallel benchmarking and parallel program characterization. Topics of interest include analytical models, Petri net models, benchmarking, measurements, parallel simulation, queueing theory models, simulation models, tracing and analysis of traces, application optimization, application characterization, load modelling, monitoring, instrumentation, evaluation and tools and standardization.

The workshop attracted 15 submissions of which one was accepted as a distinguished, 5 as regular and 3 as short papers for the final workshop.

Indeed the work submitted covers a broad spectrum of techniques to measure the performance of distributed and parallel applications, software and hardware systems. Theoretical models of performance analysis are also covered.

The paper by Simon, Vieth and Weicker presents a case study on the SPEC CPU95 benchmarks applied to use several architectures.

Message-passing performance of parallel computers is investigated by Getov, Hernandez and Hey. The authors use the NAS Parallel Benchmarks and analyse the Cray-T3D and IBM-SP2. Also, the paper by Erich Strohmaier is based on Version 2.1 of the NAS Parallel Benchmarks. It introduces the technique of statistical performance modeling which it applies to several machines.

The paper by Lars Lundberg bridges the gap between theoretical work on the NP-hard finding of mappings with minimum completion time and the execution on a real multithreaded Solaris computer.

The simulation of the behavior of a rooting algorithms for an ATM network using distributed simulation techniques is proposed by Pham, Essmeyer and Fdida.

Another type of interconnection, namely multistage interconnection networks with crossbar switches for closely coupled multiprocessor systems, is evaluated using analytical models and queueing theory by Bouras, Garofalakis, Spirakis and Triantafillou.

The paper by J. Smith, "On Synchronisation in Fault-Tolerant Data and Compute Intensive Programs over a Network of Workstations", is partly analytical but also uses measurements obtained by a realistic fault tolerant computation.

Pahud, Guidec and Cornu have implemented a parallel program for wave propagation simulation on a Cray T3D computer. They report on an analytical model and measurements on the real machine.

Moreno, Cintra and Kofuji use a multilevel modelling methodology with Petri nets to predict the performance of prefetching and multithreading in bus-based multiprocessors.

The papers presented in this workshop show that the world of parallel and distributed systems is still heterogeneous regarding the hardware structure, the software structure and its applications. A general model or technique to be applicable must therefore contain flexibility and open endedness. Much work remains to be done, especially if one considers the performance of widely distributed and heterogenous networked systems.

Workload Analysis of Computation Intensive Tasks: Case Study on SPEC CPU95 Benchmarks

Jens Simon, Marco Vieth, and Reinhold Weicker[‡]

PC² – Paderborn Center for Parallel Computing, Germany
[‡] SNI – Siemens Nixdorf Informationssysteme AG, Germany
URL: www.uni-paderborn.de/pc2
{jens, ali}@uni-paderborn.de
weicker.pad@sni.de

Abstract. Several performance analysis tools have been developed with the drawback of dedicated hardware solutions or the compute intenseness of simulations. The modern microprocessors, with hardware support for counting of system hardware events, now make possible universal software tools for the performance analysis of complex application programs such as the SPEC benchmarks.

In this paper, we present a new method to determine system resource utilization (cache miss ratios, CPI values, branch miss predictions) of arbitrary programs, based on a sampling technique, combined with access to processor-internal event counter registers. We present the *sprof* tool set that is based on this method and enables also the detailed analysis of individual subroutines of a program, as they are executed over time. The high accuracy and the negligible overhead of the tool set is demonstrated. We used the SPEC95 benchmark suite, consisting of 8 integer and 10 floating-point intensive non-trivial programs that are commonly used to define the performance of workstations and servers. As an example, we present the analysis of a SPEC CPU95 benchmark program on different processor architectures.

1   Introduction

Several performance analysis tools have been developed. Some tools are based on hardware monitor devices which are very expensive, inflexible in handling and not usable in all system environments like parallel computers [SG94, GT95]. Other tools, which are implemented completely in software, are either very compute intensive or provide too few data for a detailed performance analysis. Thereby, simulation is the commonly used method to evaluate performance of different system components. With the increasing speed of CPUs, it appears that the memory-system performance is the crucial part of the computer architecture. Therefore several projects have developed cache and memory simulators (*CPROF* [LW94], *WARTS* [DF96]). However, today's simulation techniques are discouragingly slow. Simulation times can be two or three orders of magnitude slower than the runtime of the program on the original hardware. In a previous study, the simulation time of the SPEC benchmarks has been estimated as 17 months of processing time [GHPS93]. Therefore, new techniques have to be developed which overcome the disadvantages of these simulation methods and the inflexibility of extra hardware monitor devices.

The high-density integration technology of today's semiconductor industry enables new functionalities to be implemented directly on the processor chip. Almost all modern RISC microprocessors have hardware event counter registers which are accessible by system and user programs. The counter registers and new techniques, described later on, are now able to merge the benefit of hardware monitors and software profiling tools. Furthermore, standard profiling provides static performance statistics only. Usually, statistics are collected for the whole execution time of the programs. But in almost all cases a dynamic performance observation is required. Differences in the performance behavior caused by program phases can only be determined by a dynamic profiling tool. Also the assignment of performance statistics to certain parts of the program (subroutines, marked areas) is necessary for a detailed performance analysis.

The next section presents a short overview of the state-of-the-art profiling tools. Section 3 describes our new approach to a universal dynamic profiling tool *sprof*. Section 4 presents our analysis of SPEC benchmarks with the *sprof* tool set. The next section describes how the tool can be used to analyse workloads of tasks of parallel programs. The last section is a short conclusion.

2 Standard Profiling

The profiling tool, widely used for UNIX systems, is *prof* which produces an execution profile of a program [Pro]. The profile data is taken from a file which is created by programs compiled with a special compiler option. The option also links special libraries compiled for profiling. Moreover the profiled program is instrumented in such a way that for each external symbol (subroutine entry) the percentage of execution time spent between that symbol and the next is determined together with the number of times that routine was called and the number of milliseconds per call. An improvement of *prof* is *gprof* which propagates routine execution times along arcs of the call graph of the program [GKM83].

Generally, an execution profiler apportions statistic values, such as execution counts and execution times of a program, to its component parts. The grain of program decomposition of the profiling depends on the language in which the program is written. Information can be provided on statement level, routine level, or on individual objects. *prof* and *gprof* are designed for procedural languages with explicit control flows.

To avoid a runtime intensive code instrumentation a periodical execution inspection was developed where a logical partitioning of the program code in several consecutive blocks was introduced. A sampling technique is used to build a histogram of the performance data over these blocks. The program pointer is read periodically and the address of the program pointer is transformed to a block of the program code. In a post processing phase the entries of the blocks are assigned to appropriate subroutines of the inspected program. Subroutines can spawn one or more blocks, but a block does not necessarily have to be identified with a single subroutine. Two methods to determine the block sizes are applicable. The first method uses equal block sizes which allows a simple, constant time transformation between code address and block number. Due to the desired relation between code size of the functions and the block sizes, the number of blocks in the histogram table can get very large. The second method uses block sizes which are equal to the size of the code of the functions. These adapted block sizes lead to histogram tables with number of entries equal to the number of functions. The transformation between code address and block number is optimally done by binary search in logarithmic time.

A disadvantage of these tools is that only one statistic for the whole program is generated. A time-oriented presentation of execution time distributions is not possible. Also statistics of other events like cache misses or TLB misses is impossible. Superscalar microprocessors and several levels of memory are characteristics of modern computer systems with their sophisticated architectures [HP96]. But the standard tools do not consider resource utilization other than elapsed time.

3   Dynamic Profiling

Dynamic Profiling is a new approach to extend the standard profiling by dynamic analysis features which assign event counters to program sections during the runtime of the program. We have developed a set of highly portable and modular designed tools. Our tool-set can be adapted to other UNIX systems with minimal effort. It is flexible in the way that all possible events and a potential unlimited number of histograms can be handled. A user of the tool-set need not care about the actual number of physical event counters but he or she has only to select the necessary events out of the whole list of events. Thereby, sampling extents the physical counters to virtual event counters. With this method virtually all events can be counted simultaneously. In the following, this is called process event counter sampling. We introduce a profiling monitor similar to a debugging tool, which can profile nearly any program without the need of a program instrumentation. We show that the profiling tool set is very efficient with a small execution overhead of less than 2%.

3.1   Hardware Event Counter

Almost all modern microprocessors have on-chip integrated facilities for counting events [INTEL, MIPS, Hun95, WCNSH]. This facility avoids extra hardware equipment for event counting which is cost intensive and difficult to handle. The on-chip modules enable a very fast access to all necessary registers also on all processors of a parallel machine. Reading and setting the counter can be done without any additional overhead.

The Intel Pentium is able to count 41 events, the Mips R10000 has 31 events, and the PowerPC 604 handles 45 different events. Commonly used countable events are number of processor cycles, instructions completed, data and instruction cache misses, branch miss predictions, and so on. These events can be used to compute derived values like CPI (cycles per instruction), data cache misses per load/store instruction and further more.

Because of this broad functionality, the use of event counters will become a common usable tool for performance monitoring and optimization.

3.2 Event Counter Sampling

Often event counters can only be used in a restricted way. Typically modern microprocessors are only able to count two events concurrently. Furthermore, the counters are limited by their word size. For example, the event counters of the MIPS R10000 processor are 32 bit wide and will overflow after only 5.4 seconds assuming a 200 MHz clock and four events per cycle. A solution for both problems is to sample the events. After each sampling interval, the physical counters are read and reset for the next interval. Events not counted are assumed to be the same as in the interval before. We call the technique of the alternating counting and the extrapolation of event counter values *virtual event counters*. If short sampling intervals are used, the counters will never overflow.

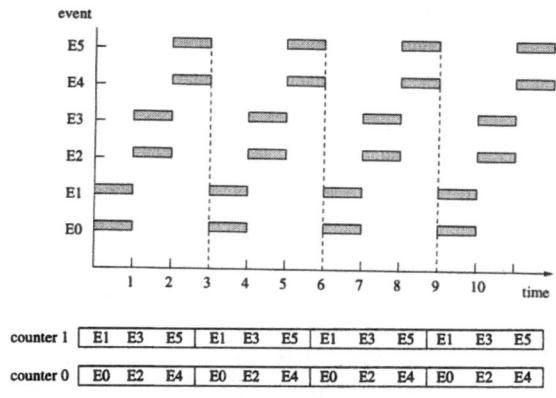

Fig. 1. Event counter sampling

Figure 1 depicts the process of virtual event counting. Six events are counted by virtually six event counters simulated by two physical counters.

3.3 Dynamic Symbol-Oriented Profiling

The standard profiling uses a single histogram with counters for equal-sized address blocks. The size of the histogram depends only on the code size and not on the execution time. But this solution is still not good enough for sequences of histograms, because the resulting data file would get much too large, especially for short histogram intervals. A solution to this problem is to have a single counter for each function. The drawback is to have a slightly longer time to find the counter for a given address. But even with a 1000 functions this can be done using binary search with only 10 comparisions in the worst case. It can be further limited to the functions of interest. Our study shows that only a fraction of the functions from the SPEC CPU95 benchmarks appears in the profiling statistic with a non-zero execution time.

3.4 The *sprof* Profiling Tool Set

The *sprof* tool set is implemented in two layers (Figure 2).

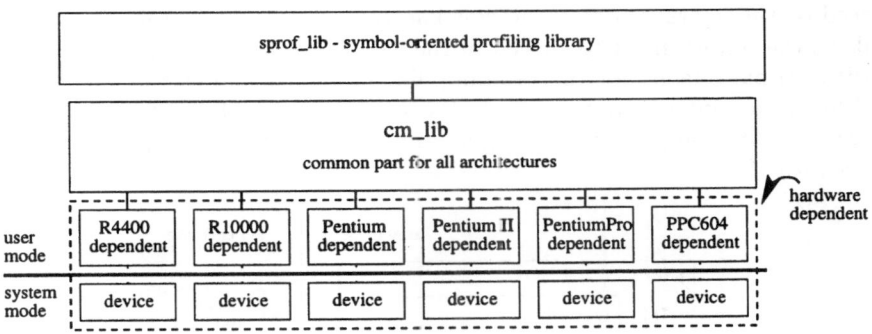

Fig. 2. The CPU monitor library cm_lib

- A hardware dependent layer:
  The layer is implemented as a UNIX device driver.
- A data processing and analysis layer:
  Consists of tools and libraries for processing phases.

Hardware Dependent Device Driver The hardware dependent layer is a device driver to access the event counters from user programs. Besides the handling of the system or kernel privileged accesses, the device driver also protects the counters from simultaneous accesses of different processes. While the general functionality is the same for all implementations, the details depend on the operating system and the processor. For example, the Linux kernel can be extended by modules during runtime, the SINIX kernel can be extended only at boot time. Also the parameters for the device driver slightly differ for the considered systems.

Data Processing and Analysis Data collection is done by user-level profiling with signals and interval timers. We tried several UNIX versions and found that today's signal mechanisms are reliable and efficient enough to be used for profiling. Under Linux, interval timers are the standard way to implement profiling.

sprof analyses files that have been generated during profiling and generates an output, either in extended *prof*-like textual form or graphically with the plotting tool *xmgr*. In both cases, measured counters can be combined together (with addition and division) to produce miss rates. Automatic endian conversation allows data post processing on any UNIX system.

Program Instrumentation Before a program can be profiled, it has to be instrumented first. This is done by linking a special library which supplies a "*main*" function to start and stop the profiling. The main-function of the analyzed program must be renamed. If it is not possible to make changes in the program, the startup function of the C-library must call the entry-function of the sprof-library instead of the main-function. Certainly, this process can be replaced by the instrumentation process known from standard profiling. Only two calls to the library must be placed in the startup code. The library needs some additional parameters for the profiling, like event list and size of the sampling and histogram interval and certainly the symbol table. This is done by the *preprof* tool, which prepares a profiling measurement by creating a profiling control file *a.out.prf*.

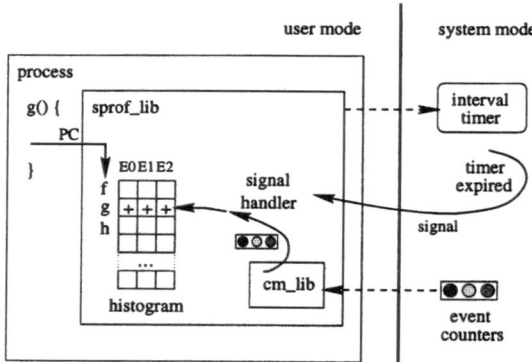

Fig. 3. Profiling with sprof_lib

sprof Library Figure 3 illustrates the information flow. First, an interval timer is programmed with the sampling interval. When it expires, the system invokes a signal handler. The handler reads the events and adds them to the histogram entry given by the interrupt position (PC, program counter). When the histogram interval expires, the histogram is written to a file (The histogram interval is measured in units of updates to the histogram).

sprof Monitor It is possible to profile a program without instrumentation. This is done with the UNIX *ptrace* system call, which allows debuggers to examine a child process. When the child process receives a signal, it is stopped and the parent process is asked for what to do. The parent can modify registers or memory areas and tell the child process to continue execution with any signal.

This functionality can be used to determine the program counter of the child process, as figure 4 shows.

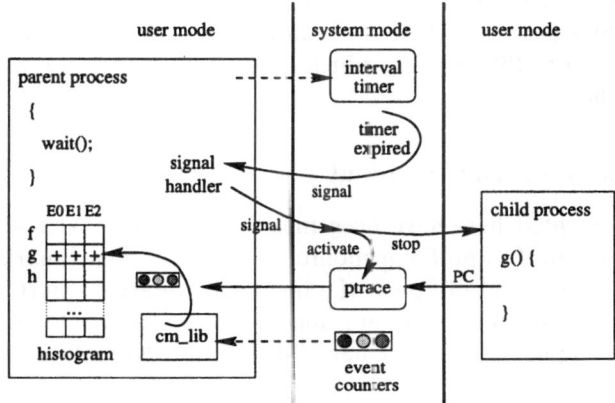

Fig. 4. Profiling with sprof_pt

The profiled program does not execute in the parent but in the child process. The parent sleeps most of the time until the interval timer expires. Then it sends a signal to the child process to activate the *ptrace* mechanism and to read the program counter.

Implementations The profiling tool is available for Mips R4400 (requires an extra hardware monitor developed by SNI) and Mips R10000, Intel Pentium and Pentium Pro, and for PowerPC 604. The hardware dependent part of the tool is implemented as a device driver integrated in the UNIX operating system. The event evaluation and post processing for data analysis and result presentation is equal for all hardware platforms.

4 Analysis of the SPEC CPU95-Benchmarks

In this section we analyse the SPEC benchmark suite with our tool-set and present a detailed performance study of one of the benchmark.

4.1 SPEC CPU95 Benchmark Suite

The Standard Performance Evaluation Corporation (SPEC) was formed to "establish, maintain and endorse a standardized set of relevant benchmarks that can be applied to the newest generation of high-performance computers" [Wei96]. The CPU benchmark set (SPEC CPU95) consists of non-trivial real-world application programs. It is broadly used to measure the processor and memory performance of UNIX systems [DR95]. Verification and reproducibility of the results are very important. This is achieved by strictly defined run rules and an exact hardware and software report of the system under test, also including all compiler flags, libraries and the operating system.

The SPEC CPU95 benchmarks consist of 18 integer and floating-point intensive programs (see Fig. 5). In the following we concentrate our study to the integer benchmark *129.compress* which consists of a sequence of compress and uncompress phases.

4.2 Measurement Environment

The results presented here have been measured on two UNIX servers (Tab. 1). We applied the *sprof_pt* profiling monitor tool to the benchmark *129.compress* with a sampling interval of $20ms$, the histogram interval length is 3000 or 500 times longer. All measurements are done with the same compilers CC4.0 and PyrC 5.0 and the same compiler flags "-O3 -Kold -Knoinline" and "-WM,-O3 -Kold -WM,-no_inlining".

Model Number	RM600-220/230/240	RM400-Exx
CPU	150 MHz R4400MC	200 MHz R10000
FPU	Integrated	Integrated
Number of CPUs	1	1
Primary Cache	16 kB I + 16 kB D	32 kB I + 32 kB D
Secondary Cache	4 MB (I+D) off chip	4 MB (I+D) off chip
Other Cache	None	None
Memory	128 MB	1024 MB
Disk Subsystem	2 GB, striped	1 x 4 GB
Other Hardware	SNI CPU monitor	None
OS & Version	SINIX-Y V5.43 B0043	SINIX-N V5.43 B0050
Compilers & Version	C-DS V1.1A, F77 1.4A	CC4.0 A 20, F77 1.4A and PyrC 5.0 (Nov. 96)
File System Type	UFS	UFS
SPECint95	2.4 (estimated on basis 92)	10.7

Table 1. R4400 and R10000 system description

4.3 Execution Time Overhead and Result Validation

We define overhead as the run time overhead of a benchmark run with profiling compared to a normal run. Figure 5 shows the profiling overhead for the benchmarks for two different histogram intervals on the R10000 system. The overhead is mostly under 2 percent, with only small differences between the two interval sizes.

One exception is gcc with an overhead of 6-7%. This is due to the high number of program starts of gcc in the benchmark. We compared the elapsed times reported by the SPEC control script, including 56 preparations for profiling and profiling setups for each gcc run. Without the preparations the gcc overhead would not be higher than for the other benchmarks.

We define correctness in terms of relative error of parameters from their ideal values. Since we do not know the exact values of miss rates for any of the

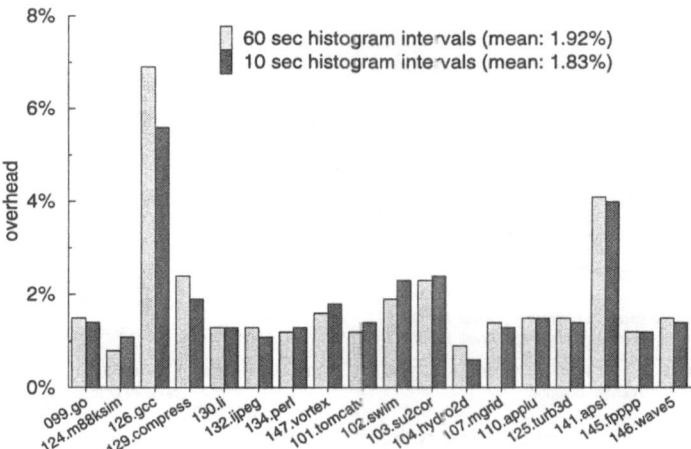

Fig. 5. Profiling overhead on R10000

SPEC benchmarks, we use a simple primary instruction cache miss generator to show the correctness. The program consists of two functions: a large one with a 32 kB code section and a small one with a 16 kB code section. Each section is surrounded by a loop. The big loop is executed N times while the small loop is executed $2 * N$ times to get the same amount of computation. Both functions are called 50 times alternately. On our R4400 system with 16 kB direct mapped cache, the 32 kB code causes a number of (32 kB/cache line size) cache misses for each run. On the other hand, the small code causes only a fraction of them, because of the loop code. We proved this theory by exact measurement of each function (also available in our profiling tool set) and got the following very reproduceable reference values for any number of loops (N) through the code sections:

	picm/inst	time
large_func	24.99%	75.03%
small_func	0.09%	24.97%

Then we measured the program with our statistical profiling tools *sprof_lib* and sprof_pt and analysed the accuracy for the large function.

Figure 6 shows on the left the relative error of picm/inst (primary instruction cache misses per instruction), compared to an exact value of 25% and the error of the time counter, compared to 75%. For a small number of loops, the picm error is high, because the runtime of the functions is very short and the profiling will collect many samples from the correct function, but without the exact cache miss value for it. The values are distributed to both functions. This is especially true for a large sampling interval of $80ms$. In all cases, the *sprof_pt* error is higher. For bigger N, both versions reach the exact value and the error gets smaller. The diagrams on the right show the same analysis for time. The error is much smaller, especially for *sprof_lib*.

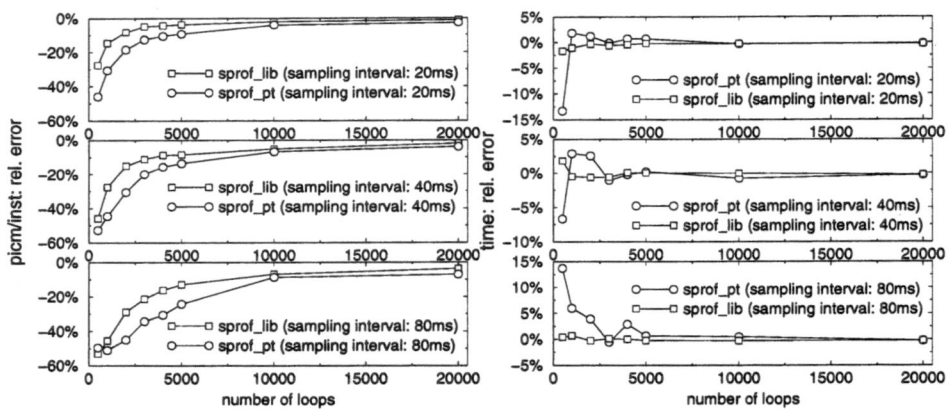

Fig. 6. Accuracy of primary instruction cache miss rates and time

4.4 Dynamic Analysis

In this section we analyse the dynamic behavior of the SPEC benchmark *129.compress* on both systems (R4400 and R10000). To get a more detailed analysis, we used a histogram interval length that is 50 times longer than the sampling interval. The setup phase of the benchmark fills a buffer with 14.5 million entries.

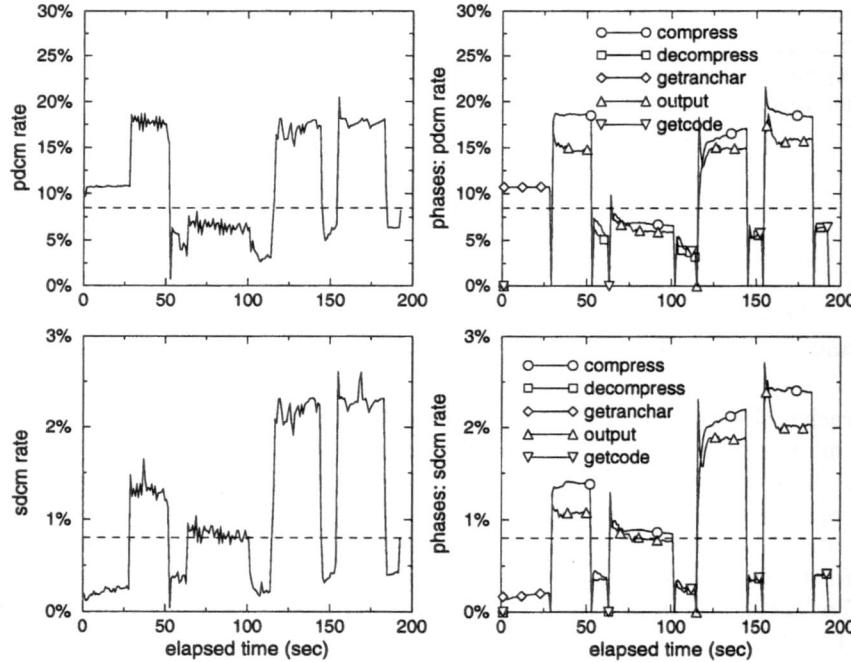

Fig. 7. 129.compress on R4400: Data Cache Miss Rates

Such a buffer is compressed and decompressed 25 times with the help of two additional buffers of the same size, using a modified Lempel-Zif compression method. The compress function uses a hash table with 69001 entries to find the next prefix and gets the code from a code table in case of success. Otherwise, another entry of the hash table will be checked.

Figure 7 shows the primary and secondary data cache miss rates on the R4400 system for the setup phase and for the first four compress and decompress phases. While the diagrams on the left show the rates for all functions, the right ones show the rates for the top 5 functions (sorted by execution time). To show things more clearly, we used the phase analysis function provided by our tool set. A phase is defined by a sequence of histograms, where a function does not change its running state (here we used the compress function). The diagrams show momentary values for phases, but not for single histogram intervals. This avoids too many "jump arounds" within phases. As expected, the rates of the compress function dominate together with output. The primary rate is 20%, the secondary nearly 2-2.5%. Decompress and getcode have much smaller rates. We have no explanation for the lower miss rates of the second compress phase.

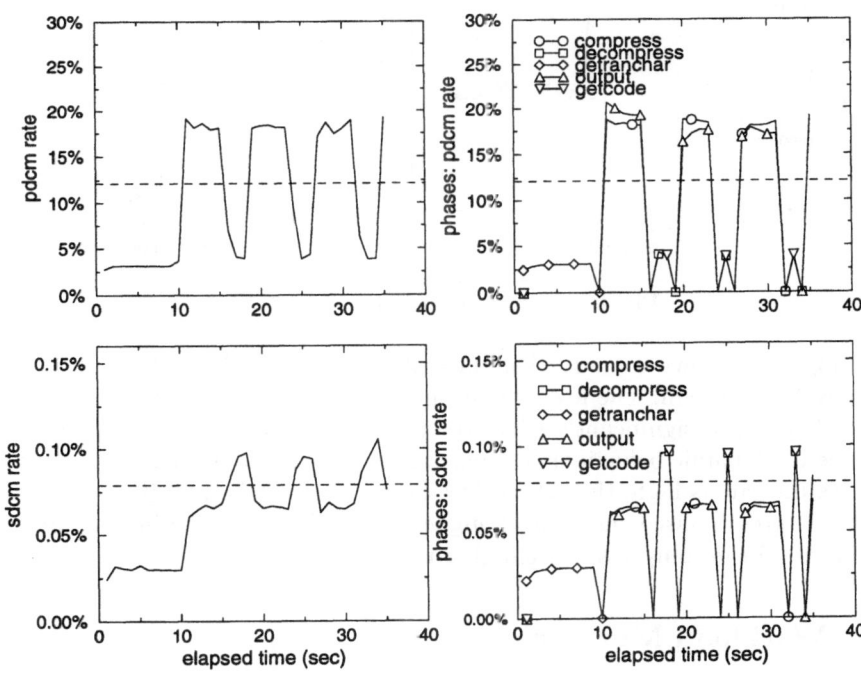

Fig. 8. 129.compress on R10000: Data Cache Miss Rates

Figure 8 shows the rates on the R10000 system. This system is much faster, not only due to the clock frequency but also because of 4-way speculative execution, branch prediction and non-blocking caches. So the same amount of work was completed in 35 instead of 198 seconds. The setup phase causes only 2.5%

picm/inst, compared to the R4400 system. While the primary rates of the other phases are nearly the same for both systems, this is not true for the secondary rates. decompress has a secondary rate of 0.15%, compress a smaller rate of only 0.05%.

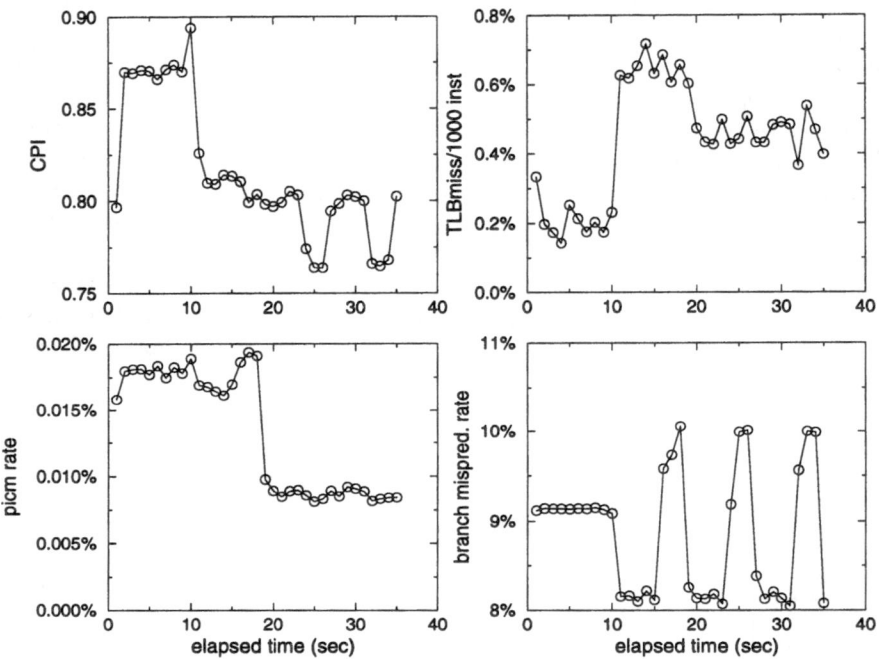

Fig. 9. 129.compress on R10000: Other rates

Figure 9 shows some further interesting rates simultaneously measured on the R10000 system. The low primary instruction cache miss rate (picm/inst) of 0.02% does not significantly effect the program execution time. It is interesting to see its dynamic behaviour. During the setup phase and the first (de)compress phase the rate is high, then the values drops down to the half (0.01%). The first (de)compress phase causes also a higher TLB miss rate. The branch misprediction rate for decompress is 2% higher than for compress.

5   Workload Monitoring

The presented method allows on-line as well as off-line analysis of programs. In the previous sections a number of off-line studies have been shown in detail. Beyond this, on-line profiling can be used for program monitoring. Our method is able to provide very detailed performance data which can be presented in a graphical user interface during program execution. Curve traces, bar diagrams as well as tachometer graphs are variants to present the measured performance data. But data evaluation during program execution has the potential to alter the measured parameters. The amount of occurring errors depends on the further overhead introduced by the data evaluation (e.g. replacement of blocks in the data and instruction caches).

A further application of our tool is the determination of complex load vectors of programs. Program loads are used to represent the workload of a piece of code in an abstract architecture model (e.g. execution time or resource demands). Generally, loads are determined by analyzing program code or by measuring the specific values. For a certain architecture the load can be measured by running or simulating the program. These load values can afterwards be used to predict the runtime of a program on an other architecture. Todays tools provide only a small set of values for representing program load and the major number of tools are using the runtime of the inspected program alone. With this simple load vectors an accurate performance prediction is not always possible, and more than ever for modern high complex system architectures.

The determination of workloads of program tasks is especially needed for load balancing activities in parallel systems. The optimal placement of tasks is driven by the load values of the tasks. In heterogeneous systems with different powerful computing nodes the accurate prediction of task runtimes is very important. We are developing a Workload Analyzer & Runtime Predictor (*WARP*) which will be integrated in a metacomputing environment [RBD+97]. A prototype of the workload analyzer is implemented for a PowerPC based parallel computer.

6 Conclusion

With the tools presented here, a dynamic event analysis is possible for all or a selectable number of subroutines of a program. While the tools do not allow a similarly exact association of events to individual statements of the source code as simulation-based tools [LW94], their overhead is much smaller: Only 2 percent execution time overhead is added, and with the exception of a small device driver, only normal services are used; no extra hardware is necessary. Therefore, even a modestly sophisticated programmer could use the tools to analyse his or her program's bottlenecks.

References

[DF96]   Joseph D. Darcy and Manuel Fähndrich.   Finding cache hotspots in SPEC95. *University of California at Berkeley, http://www.cs.berkeley.edu/~darcy/IRAM/spec95.hot/*, 1996.

[DR95]   Kaivalya Dixit and Jeff Reilly. SPEC95 questions and answers. *SPEC Newsletter*, 7:7 – 10, September 1995.

[GHPS93]   Jeffrey D. Gee, Mark D. Hill, Dinosios N. Pnevmatikatos, and Alan J. Smith. Cache performance of the SPEC92 benchmark suite. *IEEE Micro*, 13(4):17 – 27, August 1993.

[GKM83]   Susan L. Graham, Peter B. Kessler, and Marshall K. McKusick. An exe-
          cution profiler for modular programs. *Software Practice and Experience*,
          13:671 – 685, 1983.

[GL95]   Hui Gao and John L. Larson. Workload characterization using the cray
          hardware performance monitor. *Journal of Supercomputing*, 9:391 – 412,
          1995.

[GT95]   Aaron Goldberg and John Trotter. Interrupt-based hardware support for
          profiling memory system performance. *IEEE*, pages 518 – 523, August
          1995.

[HMMS96] M. Horowitz, M. Martonosi, T.C. Mowry, and M.D. Smith. Informing
          memory operations: Providing memory performance feedback in modern
          processors. In *Proceedings of the 23rd International Symposium on Com-
          puter Architecture*, pages 260 – 270, 1996.

[HP96]   John L. Hennessy and David A. Patterson. *Computer Architecture. A
          Quantitative Approach*. Morgan Kaufmann, 1996.

[Hun95]   Doug Hunt. Advanced performance features of the 64-bit PA8000. In
          COMPCON'95, 1995.

[INTEL]   Intel Corporation. *PentiumPro Processor User's Manual*, volume 1 – 3.
          1996.

[LW94]   Alvin R. Lebeck and David A. Wood. Cache profiling and the SPEC bench-
          marks: A case study. *IEEE Computer*, pages 15 – 26, October 1994.

[MIPS]   MIPS Corporation. *MIPS R10000 Microprocessor User's Manual*. 1995.

[Pro]   *PROF User's Manual*. UNIX Reference Manuals.

[RBD+97] A. Reinefeld, R. Baraglia, T. Decker, J. Gehring, D. Laforenza, F. Ramme,
          T. Römke, and J. Simon. The MOL project: An Open, Extensible Meta-
          computer. In *Proc. of Heterogeneous Computing Workshop HCW'97, IEEE
          Computer Science Press*, pages 17 – 31, 1997.

[SG94]   Ashok Singhal and Aaron J. Goldberg. Architectural support for perfor-
          mance tuning: A case study on the SPARCcenter 2000. *Proc. of 21th In-
          ternational Symposium on Computer Architecture*, pages 48 – 59, 1994.

[WCNSH]   E.H. Welbon, C.C. Chan-Nui, D.J. Shippy, and D.A. Hicks. Power2 per-
          formance monitor. PowerPC and POWER2. *Technical Aspects of the New
          IBM RISC System/6000*.

[Wei96]   Reinhold P. Weicker. A SPEC primer. *SPEC World Wide Web Site,
          http://www.specbench.org*, 1996.

[ZLTI96] Marco Zagha, Brond Larson, Steve Turner, and Marty Itzkowitz. Perfor-
          mance analysis using the MIPS R10000 performance counters. In *Preceed-
          ings Supercomputing'96*, 1996.

Statistical Performance Modeling: Case Study of the NPB 2.1 Results

Erich Strohmaier*

Computer Science Department, University of Tennessee, Knoxville, TN 37996

Abstract. With the results of version 2.1 a consistent set of performance measurements of the NAS Parallel Benchmarks (NPB) are available. Unchanged portable MPI code was used for this set of 269 single measurements. In this study we investigate how this amount of information can be condensed. We present a methodology for analyzing performance data not requiring detailed knowledge of the codes. For this we study several different generic timing models and fit the reported data. We show that with a joint timing model for all codes and all systems the data can be fitted reasonable well. The timing model also contains only a minimal set of free parameters. This method is usable in all cases where the analysis of results from complex application code benchmarks is necessary.

1 Introduction

The set of NAS Parallel Benchmarks (NPB) is one of the best accepted benchmarks for parallel processing [1]. End of 1995 a new version of this suite was released which asked for the first time for performance measurements of the unchanged MPI code which is available from the NASA Ames Research Center. In August 1996 a first set of such results was released [2]. It contained an almost complete set of measurements of 4 codes on 4 different systems for 3 different problem sizes. Such a homogeneous big set of performance data from application codes is the optimal starting point for any in depth analysis of the benchmark and systems in the set.

The number of 269 single measurements immediately brings up the question if and how this amount of information can be condensed. In previous studies we already showed that for the fully vendor optimized NPB 1.0 results a limited number of benchmarks would be sufficient and that Amdahl's Law describes these results very well [3].

In this paper we study to which extend this is true for the measurements of unchanged portable MPI code. We propose a generic timing model with a minimal number of free parameters and fit this model to the data using nonlinear regression. We evaluate the quality of our model by comparison with simpler models with more free parameters and careful examination of the statistical properties of the obtained fits[2]. A basic introduction to the statistics terminology used may be found in [4].

---

* e-mail: erich@cs.utk.edu

[2] All analysis discussed in this paper are done with the SAS statistical software package

2 Timing models for single systems

Starting point for our analysis are the measured execution times $T_{s,c}$ of a sample of n application codes c which are measured on m different systems s. These times are functions of the number of processors p and the problem size $\mathbf{n}$. For simplicity reasons we consider in the following only the dependency on p and not on $\mathbf{n}$.[3]

The measured times can be separated in times for different computational phases of the execution during which basic types of computational work j like parallel computation, serial computation or communication take place. We split each of these basic types of work j into a sum of products. One factor $u_i(p)$ of each term contains all the dependencies on p and is indexed independent of code or system. We normalize these functions such that $u_i(1) = 1$ or $u_i(1) = 0$. The other factors $t_{j,i}^{s,c}$ are parameters depending on code and system.

$$T_{s,c}(p) = \sum_{j=1}^{n} t_j^{s,c}(p) = \sum_{j=1}^{J} \sum_{i=1}^{I} t_{j,i}^{s,c} u_i(p) \tag{1}$$

We call u_i the "characteristic functions" as they contain all the dependencies of the performance on p. They therefor characterize the scaling behavior of the code with increasing processor number p. The set of the characteristic functions u_i reflects the typical algebraic form of timing relations for parallel computing.

We now continue by using the following general timing model for analyzing the measured data:

$$T_{s,c}(p) = \sum_{i=1}^{I} \delta_i^{s,c} u_i(p) \quad \text{with} \quad \delta_i^{s,c} = \sum_{j=1}^{J} t_{j,i}^{s,c} \tag{2}$$

If we analyze the performance data of a single code on a single system we will only be able to fit the sums $\delta_i^{s,c}$. With statistical methods we will not be able to gain information about the individual $t_{j,i}^{s,c}$ [5]. This can be achieved only by analyzing the results of a set of codes measured on a set of systems as done in section 5.

We now want to use equation 2 to analyze the performance results of an application without inspecting the code. For this it is critical to select a reasonable set of characteristic functions u_i such that the set can effectively characterize the execution times of parallel applications. This set of functions together with the number and quality of measurements will determine how many of the parameters $\delta_i^{s,c}$ can be fitted in a meaningful way. As characteristic functions for the further analysis we use the following set which is collected from different sources [6, 7, 5]

$$u_1 = \frac{1}{p^2}, \quad u_2 = \frac{1}{p}, \quad u_3 = \frac{log(p)}{p}, \quad u_4 = \frac{1}{\sqrt{p}}, \quad u_5 = 1, \quad u_6 = log(p), \quad u_7 = p \tag{3}$$

---

[3] In principal a similar approach as ours can be chosen to account for the dependency on the problem size $\mathbf{n}$.

3 Preparation of the data

Looking on the execution times of the NPB 2.1 we notice a big range in the measured times[4]. For class A problem size measured times range from 0.75 seconds up to 4873.7 seconds. This range of 4 magnitudes of orders poses a severe scale problem for any statistical method and a transformation of this scale is required for statistical reasons. We are using the following transformations.

One major source for the differences in measured times are the different number of processors used. As we are dealing with well parallelized codes multiplication of the measured times with the number of processor[5] has a smoothening effect on the data.

The variation in the floating-point operation count is one reason for the differences between the execution time of the different codes. Scaling measured times with the inverse of this operation count equalizes the scale for the different codes further.

Using processor with different computational power is another reason for differences in execution times. A multiplication of the measured times with peak performance r_{peak} provides additional correction of the scale.

Applying all three transformation in sequence the measured execution times $t(p)$ get transformed in a value $t'(p)$ which is given as

$$t'(p) = \frac{t(p) * p}{W_c} * r_{peak}. \tag{4}$$

t' is a dimensionless value which can be interpreted as inverse temporal efficiency. The values of t' now vary by a moderate factor of 4 to 7 for the three different problem size classes.

After removing two clear outliers from our sample of results we have for problem size class A and B on the average about 6 to 7 observations for each code on each system. This is sufficient for a statistical analysis. There are however almost no measurements for two of the systems in problem class C which is a clear limitation in the usability of this set of results.

A final inspection of the plotted performance data over the number of processors shows the following observations:

- Measurements for the Cray T3D often show unstable performance values over the number of processors. Performance values per processor can drop or rise by about 30% for measurements with similar processor numbers.
- The PowerChallenge Array shows clearly two regimes of operation associated with it's hierarchical architecture. Performance within a single SMP node tend to show super linear speedup while performance between SMP nodes can drop significantly.

These two facts are limits for our analysis as none of our characteristic functions from equations 3 can model such behavior effectively.

---

[4] see http://www.nas.nasa.gov/NAS/NPB/

[5] For our analysis we are using the number of active processors and not the number of allocated processors as done in the NPB report [2].

u_1	u_2	SSE	R^2	δ_1	δ_2
$log(p)/p$	$1/p^2$	26.13	0.99289	3.616	64.056
$1/p$	$1/\sqrt{p}$	26.40	0.99282	11.595	0.558
$1/p$	1	28.06	0.99237	14.066	0.027
$1/p$	$log(p)/p$	28.17	0.99234	6.443	2.422
$1/p$	$log(p)$	29.18	0.99206	14.516	0.005
$1/p$	p	35.41	0.99037	15.304	0.000

Table 1. All two parameter models with $R^2 \geq 0.99$ for the class A SP results on the Cray T3D. The total Sum of Squares (SST) is 3677.85. SSE: Sum of Squares of the remaining Error; R^2: coefficient of determination; δ_i are the fitted model parameter.

4 Results for individual fits

We start the analysis with two parameter models which are a compromise between the very limitations of one parameter models and the limited number of observations available for each analysis. We fit each possible timing model based on two characteristic functions separately for each code and each system to the data. As the number of observations for many cases is quite low (≤ 6) there are always several models which explain the same fraction R^2 of the total sum of squares in the model. As example we show in table 1 all meaningful combinations with an $R^2 \geq 0.99$ for the class A SP results on the Cray T3D. We have chosen this example as it contributes the most to the total Sum of Squares of this problem size class. For class A the Sum of Squares (SS) of SP on the Cray T3D is 3677.85 which is equivalent to 26.4% of the total Sum of Squares (SST) of this class.

Not only in this example but in most cases a precise selection of a single best model is not possible. There is however a clear trend to models containing $u_2 = \frac{1}{p}$ which is characteristic for parallel work. To find out which characteristic functions might be good candidates for a joint model for all codes and systems we fitted the same model to all combinations of code and systems and calculated the total SSE. In table 2 we show the SSE values for the 5 best models for each class together with the SSE value if taking the best individual model for each pair of code and system. Again most of the models contain $u_2 = \frac{1}{p}$. Models which contain a second function characteristic for limited parallelism $u_4 = \frac{1}{\sqrt{p}}$ or $u_3 = \frac{log(p)}{p}$ or for serial work $u_5 = 1$ tend to fit the data better then models including parallel overhead functions like $u_6 = log(p)$ or $u_7 = p$.

5 Joint timing model for all systems

We now proceed the analysis by making the additional assumption that the $t_{j,i}^{s,c}$ from equation 2 are the quotient of factors which only depend on the code $w_{j,i}^c$

u_1	u_2	SSE	R^2	u_1	u_2	SSE	R^2
Class A		SST = 13937.17		Class B		SST = 10769.02	
best	best	39.14	0.99719	best	best	85.17	0.99209
$1/p$	1	47.04	0.99663	$1/p$	$1/\sqrt{p}$	95.44	0.99114
$1/p$	$log(p)$	48.59	0.99651	$1/p$	1	96.95	0.99100
$1/p$	$1/\sqrt{p}$	52.39	0.99624	$1/p$	$log(p)$	99.15	0.99079
$1/p$	$log(p)/p$	80.87	0.99420	$1/p$	$log(p)/p$	101.51	0.99057
$log(p)/p$	$1/p^2$	103.91	0.99254	$log(p)/p$	$1/p^2$	124.63	0.98843

Table 2. The SSE values for the best 5 two parameter models used for all observations compared to the SSE value when using the best individual model for each pair of code and system. The best models for class C are the same as for class B.

(amount of work) or the system $r_{j,i}^s$ (power of the system).

$$t_{j,i}^{s,c} = \frac{w_{j,i}^c}{r_{j,i}^s} \tag{5}$$

The total execution time can now be written as

$$T_{s,c}(p) = \sum_{j=1}^{J}\sum_{i=1}^{I} \frac{w_{j,i}^c}{r_{j,i}^s} u_i(p) = \sum_{i=1}^{I} \delta_i^{s,c} u_i(p) \quad \text{with} \quad \delta_i^{s,c} = \sum_{j=1}^{J} \frac{w_{j,i}^c}{r_{j,i}^s} \tag{6}$$

By using this product representation we have introduced an additional degree of freedom for each characteristic function in equation 6. This follows as each $\delta_i^{s,c}$ is invariant if we multiply the values of $w_{j,i}^c$ and $r_{j,i}^s$ for all j by an arbitrary factor. This degree of freedom has to be fixed by an additional condition on the parameters $w_{j,i}^c$ and $r_{j,i}^s$. We choose for this study to fix one of the system parameters w^c equal to 1. This additional degree of freedom also implies that the absolute values of the parameters $w_{j,i}^c$ and $r_{j,i}^s$ by them self have no meaning as they can be manipulated by changing the normalization. Only the ratios of these parameters are invariant to such changes and can be interpreted in a safe way.

Analyzing the full set of results $T_{s,c}$ we can now fit values to the individual $w_{j,i}^c$ and $r_{j,i}^s$. The two sets of parameters work $w_{j,i}^c$ and speed $r_{j,i}^s$ together with the characteristic functions $u_i(p)$ fully describe the timing models for all codes on all systems included in the analysis.

Overall this product representation reduces the number of free parameters in the analysis effectively by a factor of $\frac{nm}{n+m-1}$ compared to fitting individual models for each pair of the m systems and the n codes. The number of free parameters is indeed quite small as we have for each "type of work" described by the characteristic functions only one parameter for each code and one for each system. This reduction in the free parameters represent the possible value of this model as it potentially can explain the same number of observations with less or even a minimal set of free parameters.

u_1	u_2	SSE	R^2	u_1	u_2	SSE	R^2
Class A		SST = 13937.17		Class B		SST=10769.02	
$1/p$	$1/\sqrt{p}$	130.28	0.99065	$1/p$	$log(p)/p$	134.73	0.98749
$1/p$	1	142.02	0.98981	$1/p$	$1/\sqrt{p}$	156.03	0.98551
$1/p$	$log(p)/p$	146.24	0.98951	$1/p$	1	179.27	0.98335
$1/p$	$log(p)$	148.84	0.98932	$1/p$	$1/p$	180.67	0.98322
$log(p)/p$	$1/p^2$	261.88	0.98121	$1/p$	$log(p)$	185.13	0.98281

Table 3. The 5 best two function models using the product representation from equation 5.

6   Results for the combined model

We now fit all possible timing models of the form of equation 6 based on two characteristic functions to all measurements. It turns out that for problem size class C because of the high number of missing measurements for two of the systems no analysis comparable to class A and B is possible. We compare the results show in table 3 to the results for fitting individual models in table 2.

The values of SSE are higher for the combined model. This was to expected as we now have only 14 free parameters instead of 32. The absolute increase compared to SST is quite small for each problem size class. This is a first strong confirmation that our factorization assumption from equation 5 works quite well.

We discuss now three of the overall best models in more detail. All three models contain $u_2 = \frac{1}{p}$ as first characteristic function. As second function they contain $u_5 = 1$ or $u_4 = \frac{1}{\sqrt{p}}$, $u_3 = \frac{log(p)}{p}$. This sequence of second functions is equivalent from going from serial work to better and better parallel execution.

In table 4 we show the actual fitted values for the 14 free parameters together with their asymptotic standard error for the class A and B. The parameters are fitted for the transformed t' from equation 4 and can be interpreted as the inverse of processor efficiencies and as code overhead factors. The absolute value of the parameters is however without any meaning. Only appropriate chosen ratios of them represent measurable values.

We notice that the standard errors for most parameters are in the range of 5% to 20% of the fitted value. For Class A only the second system parameter of the SGI PowerChallenge Array shows quite big error bars. This is certainly related to the previous mentioned special behavior of the measured data for this system. For class B the same is true for the second system parameter of the Intel Paragon. As an inspection of the measured data shows no special behavior of this system the most likely explanation of this large error is the small number of measurements for this system (16 out of 102).

For all systems the first system parameter varies only little between the three different models. If we interpret it as computational power then the IBM SP2 shows always performance efficiencies twice as large as the other systems. This

Parameter	Class A			Class B		
	$u_1 = 1/p$			$u_1 = 1/p$		
r_1^{IBMSP2}	1.	1.	1.	1.	1.	1.
$r_1^{Paragon}$	0.406 ± 0.026	0.370 ± 0.032	0.365 ± 0.041	0.401 ± 0.033	0.382 ± 0.038	0.381 ± 0.042
$r_1^{CrayT3D}$	0.425 ± 0.024	0.410 ± 0.031	0.466 ± 0.047	0.479 ± 0.035	0.541 ± 0.053	0.669 ± 0.102
$r_1^{PCArray}$	0.482 ± 0.031	0.408 ± 0.037	0.372 ± 0.041	0.457 ± 0.034	0.452 ± 0.041	0.445 ± 0.042
w_1^{BT}	4.572 ± 0.297	4.009 ± 0.365	3.714 ± 0.416	4.840 ± 0.390	4.563 ± 0.426	4.306 ± 0.413
w_1^{LU}	4.023 ± 0.249	3.506 ± 0.297	3.430 ± 0.367	4.227 ± 0.317	4.268 ± 0.386	4.309 ± 0.403
w_1^{SP}	5.355 ± 0.308	4.120 ± 0.337	3.521 ± 0.389	5.729 ± 0.402	5.166 ± 0.460	4.987 ± 0.475
w_1^{MG}	4.672 ± 0.293	3.970 ± 0.347	3.408 ± 0.373	4.049 ± 0.324	3.455 ± 0.375	3.161 ± 0.368
	$u_2 = 1$	$u_2 = 1/\sqrt{p}$	$u_2 = log(p)/p$	$u_2 = 1$	$u_2 = 1/\sqrt{p}$	$u_2 = log(p)/p$
r_2^{IBMSP2}	1.	1.	1.	1.	1.	1.
$r_2^{Paragon}$	1.538 ± 0.360	1.221 ± 0.233	1.012 ± 0.200	6.521 ± 18.08	2.902 ± 3.864	1.888 ± 1.829
$r_2^{CrayT3D}$	1.913 ± 0.266	1.106 ± 0.126	0.699 ± 0.079	1.053 ± 0.328	0.537 ± 0.143	0.335 ± 0.088
$r_2^{PCArray}$	2.351 ± 3.628	3.878 ± 4.918	4.037 ± 3.781	0.182 ± 0.086	0.343 ± 0.126	0.422 ± 0.132
w_2^{BT}	0.013 ± 0.006	0.172 ± 0.056	0.480 ± 0.114	0.004 ± 0.005	0.089 ± 0.044	0.271 ± 0.086
w_2^{LU}	0.033 ± 0.007	0.319 ± 0.064	0.577 ± 0.126	0.007 ± 0.006	0.059 ± 0.047	0.158 ± 0.087
w_2^{SP}	0.067 ± 0.007	0.753 ± 0.071	$1.535 = 0.154$	0.029 ± 0.009	0.306 ± 0.081	0.615 ± 0.155
w_2^{MG}	0.014 ± 0.007	0.235 ± 0.067	$0.750 = 0.132$	0.020 ± 0.007	0.260 ± 0.075	0.548 ± 0.143

Table 4. The fitted parameter with their asymptotic standard error for the best three combined models. The values shown are parameters for transformed t'. This means that system parameters are scaled by the peak performance and code parameters by the single processor floating point operation count.

is not always true in the second set of system parameters.

Looking on the first set of code parameters we see a larger influence of the chosen model on the parameter. This corresponds to the effect of the second functions which represent gradually different limited parallelism. The second code parameter increases as the second function changes to more parallel work. At the same time the first code parameter decreases. If the single processor floating point count which we used for scale transformation would accurately describe the amount of the total computational work then all code parameters should be equal. The values for BT, LU and MG seem indeed to be roughly equal. The values for SP are however consistently higher especially in the second parameter. This indicates that SP contains a substantial additional amount of computational work which is only partially parallelized.

A check of the statistical quality of the obtained fits shows that the correlation matrix of the parameters for the three models have typical values of about 0.5–0.7 for the first parameters and much smaller values for all other entries. The quantile-quantile plots for the errors are quite straight but show typically some outliers at the higher end of the curve. The maximum relative error of the predicted values is about 30% with only one value above 30% and the mean value of the relative error is only 7%.

7 Conclusions

In this paper we present a methodology for analyzing performance measurements without detailed knowledge of the used codes. It is based on the usage of generic timing models build with characteristic function which are typical for the algebraic form of timing equation in parallel computing. We use this methodology to analyze the NPB 2.1 results. Our results can be summarized as follows:

- Using a sequence of transformations solves the statistical scale problems.
- Analyzing each pair of system and code separately between 99.2% and 99.7% of the total Sum of Square (SST) can be explained with individual two parameter functions. This model has 32 free parameters for each class.
- Using a joint timing model with only 14 free parameter 99.1% of the SST of class A and 98.7% of the SST of class B can be explained by this model.
- Typical standard error for the fitted parameter are in the range of 10%. Only one parameter per class is not significantly different from 0.
- The maximum relative error of the predicted values is about 30% and the mean value of the relative error is 7%.
- The average efficiency of the SP2 processor is more than twice as high as for the other processors.
- The simulated CFD application SP contains a substantial amount of work which is not included in the single processor floating point counts.

This methodology for empiric modeling of performance measurements does not require detailed analysis of the implementations of the code. This makes this method to a good alternative in all cases where the analysis of results from complex application code benchmarks is necessary.

References

1. D. Bailey, J. Barton, T. Lasinski, and H. Simon (editors). The NAS parallel benchmarks. Technical Report RNR-91-02, NASA Ames Research Center, January 1991.
2. W. Saphir, A. Woo and M. Yarrow. The NAS parallel benchmarks 2.1 Results. Technical Report NAS-96-01, NASA Ames Research Center, August 1996.
3. Horst D. Simon and Erich Strohmaier. Statistical Analysis of NAS Parallel Benchmarks and LINPACK Results. In Bob Hertzberger and Guiseppe Serazzi, editors, *High-Performance Computing and Networking*, pages 626–633, May 1995.
4. Raj Jain. The Art of Computer Systems Performance Analysis. Wiley, 1991
5. Strohmaier, Erich. Extending the Concept of Computational Similarity for Analyzing Complex Benchmarks. Technical Report 43, Rechenzentrum der Universitaet Mannheim, April ,1995,
6. Vipin Kumar et al.. Introduction to Parallel Computing: Design and analysis of parallel algorithms. Benjamin/Cummings, 1994.
7. Jurgen Brehm and Patrick H. Worley and Manish Madhukar. Performance Modeling for SPMD Message-Passing Programs. Technical Report TM-13254, Oak Ridge National Laboratory, June 1996

A General Performance Model for Multistage Interconnection Networks *

C.J. Bouras, J.D. Garofalakis, P.G. Spirakis and V.D. Triantafillou

Department of Computer Engineering and Informatics,
and Computer Technology Institute,
P.O. Box 1122, 261 10 Patras, Greece
E-mail : $\{bouras, garofala, spirakis, triantaf\}$@cti.gr

Abstract. . In this paper we analyze the general case of Multistage Interconnection Networks (MINs), made of $k \times k$ switches with finite, infinite or zero length buffers (unbuffered). The exact solution of the steady state distribution of the first stage is derived for all cases. We use this to get an approximation for the steady state distributions in the second stage and beyond. In the case of unbuffered switches we reach the known exact solution for all the stages of the MIN. Our results are validated by extensive simulations.
Keywords: analytical models, queueing theory models, evaluation.

1   Introduction

Multistage Interconnection Networks (MINs) have attracted from the early '80s the attention of the designers of highly parallel multiprocessor systems with a large number of processors. MINs (which are packet-switched) have been adopted in the past in several machines ([2],[9]) and are expected also to play an important role in the development of high-speed networks based on Asynchronous Transfer Mode (ATM) . The performance of a MIN is of crucial importance, thus a lot of research has been dedicated to the study of how these networks perform under various conditions, through analytic techniques or simulation ([8], [10], [1], [6], [5], [7], [4]). Analytic results can be found for specific cases of MINs, which mainly rely on approximation methods.

The basic building block of the packet-switched MINs considered here, is a k-input, k-output ($k \times k$) switch grouped in stages. We examine MINs that provide a unique path from each source (processor) to each sink (memory module), which belong to the class of Banyan MINs [3]. Our work considers *general MINs*, that is, MINs made by switches with finite, infinite or zero length buffers (unbuffered), arbitrary switch size ($k \times k$) and variable injection rate p at the sources. Assuming that the traffic (requests for memory modules) is uniform,

* *This research was partially supported by the European Union ESPRIT Basic Research Projects ALCOM IT (contract no. 20244) and GEPPCOM (contract no. 9072) and the Greek Ministry of Education.*

that at each cycle a packet is generated with fixed probability p, and that packets are lost when they are attempted to be queued at a full buffer (relaxed blocking model), we derive for the general MIN, the exact steady-state distribution of queue lengths in the first stage, and of course exact formulas for the expected number of packets lost per cycle, and the mean queue length. We then use the results for the first stage and an operational approximation hypothesis to get the (approximate) distributions of the queue sizes of the second stage and beyond. Extensive simulations verify our results, as we discuss in Section 6. Our analysis, based on the theory of recurrence equations, explicitly provides the form of the queue length distribution, which is a linear mixture of geometrics.

2 Our Approach

2.1 The Model

MINs are packet switched and they are required to provide high bandwidth to support the communication between processors and memory modules. We consider that the network is built by switches connected by unidirectional lines. General MINs consist of a number of $k \times k$ switches (nodes) grouped into stages. A k-input, k-output switch, can receive packets at each of its k input ports and send them through each of its k output ports. In each output port there is a buffer. We assume that the buffers may be of infinite, finite or zero length (unbuffered switches).

If there is a unique path from each processor to each memory module then a MIN belongs to the class of Banyan Networks (BNs). We assume oblivious routing algorithms, i.e. algorithms in which the path of a packet through the network is fixed at the source node issuing it. The path can be encoded as a sequence of labels of the successive switch outputs of the path (path descriptor). Packets are generated at each processor by independent, identically distributed random processes. In our analysis we assume that each processor generates a packet with probability p at each cycle, and sends this with equal probability to any memory module (uniform access). The switches have a FIFO policy for their servers (outputs). Conflicts between packets simultaneously routed to the same output port are resolved by queueing the packet. Our analysis assumes that packets are lost when they are attempted to be queued at a full queue or in the case with unbuffered switches. In actual parallel machines, the sending processor is notified, in order to resubmit the packet later on. The service time of the output queues of each switch is assumed constant and equal to the network cycle time. The uniform access assumption allows us to represent any $k \times k$ switch as a system of k queues working in parallel, with a deterministic server each (of service time equal to 1). Any packet which enters any of the k inputs of the switch, goes with probability $1/k$ to any of the (output) queues of the switch. In our analysis we assume that the buffer length b includes the server (output). So, an unbuffered switch is referred with $b = 1$. We assume that arrivals happen at the end of each cycle (thus first the queue is served and then new packets arrive, if any). The routing logic at each switch is assumed to be fair, i.e. conflicts are randomly resolved.

2.2 The Equilibrium and Interstage Dependencies

Most authors that have used analytic approaches for the analysis of MIN's, have remarked the basic difficulty for any analytic approach. Except for the case of unbuffered switches ([8], [6]) in all other cases, the traffic flow between consecutive stages depends upon time, that is the distribution of packet arrivals at the second and the subsequent stages is not time independent, as is the case for the first stage which is fed by the independent "Bernoulli" processors. ([10], [6]). However, in [10] it is pointed out that the behaviour, say b_t, of a stage at time t depends mainly upon the present, a little bit $(b_{t-1}/4)$ upon the situation at time $t-1$, and is nearly independent $(b_{t-r}/4^r)$ from ancient events at time $t-r$. So, the dependency from history is exponentially desreasing. This last observation, together with the assumption that every stage of the MIN will reach an equilibrium (steady-state), leads to the marcovian approximation which we present in section 5: *The output queues of stage m that feed the stage m + 1, are assumed to operate like independent "Bernoulli" processors with a packet generation probability equal to their utilization.* Clearly, this hypothesis equates the dynamics of the output process of a stage with its "macroscopic" averages, ignoring any time dependency of its behaviour.

3   The General Recurrence Relation for the First Stage

Let C be the random variable denoting the number of packets arriving to an arbitrary output queue of an $k \times k$ switch of the first stage, at the end of a cycle and $x_{k,c} = \mathbf{Pr}(C = c)$. Some of these arriving packets may be lost due to a full queue.

Lemma 1. *The arrival process of packets at the output queues of the first stage of the network, is given by a Bernoulli distribution $B(p/k, k)$, where p is the fixed probability of a packet generated by a processor at each cycle. Therefore we have*

$$x_{k,c} = \begin{cases} \dbinom{k}{c} (\frac{p}{k})^c (1 - \frac{p}{k})^{k-c}, & for\ 0 \le c \le k \\ 0, & otherwise \end{cases} \tag{1}$$

Definition 2. Let $q^{(n)}$ be the number of packets in an arbitrary output queue at the end of the cycle n and let q be the steady state limit of $q^{(n)}$.

Definition 3. Let $v^{(n)}$ be the number of packets that are entering an arbitrary output queue at the end of cycle n and let v the steady state limit of $v^{(n)}$. It holds that $v^{(n)} \le C$ at each cycle n, when b is finite. If b is infinite, it is always true that $v^{(n)} = C$.

Definition 4. Let $p_j = \mathbf{Pr}(q = j)$, $j \geq 0$, be the distribution of q at the steady state. Also, $p_{0,1} = p_0 + p_1$

Lemma 5. *For $0 < m \leq min(b,k)$:*

$$\mathbf{Pr}(v^{(n)} = m) = \begin{cases} x_{k,m} & if\, q^{(n-1)} - \Delta(q^{(n-1)}) < b - m, \\ (x_{k,m} + x_{k,m+1} + \cdots + x_{k,k}) & if\, q^{(n-1)} - \Delta(q^{(n-1)}) = b - m, \\ 0 & otherwise \end{cases}$$

(2)

where $\Delta(q^{(n)})$ is the departure of a packet from an arbitrary output queue at the end of cycle n, if any.

Also for $m = 0$, $\mathbf{Pr}(v^{(n)} = 0) = x_{k,0}$ for any $q^{(n-1)}$.

Theorem 6. *The steady state flow balance equations are : $(p_{0,1} = p_0 + p_1)$*

$$\begin{aligned} p_0 &= p_{0,1} x_{k,0} \\ p_1 &= p_{0,1} x_{k,1} + p_2 x_{k,0} \\ p_2 &= p_{0,1} x_{k,2} + p_2 x_{k,1} + p_3 x_{k,0} \\ &\vdots \\ p_k &= p_{0,1} x_{k,k} + p_2 x_{k,k-1} + \cdots + p_{k+1} x_{k,0} \end{aligned}$$

(3)

while for $k \leq j < b$, the general recurence holds :

$$\begin{aligned} p_j x_{k,0} &= p_{j-k+1}(x_{k,k}) + p_{j-k+2}(x_{k,k-1} + x_{k,k}) + \cdots + \\ & p_{j-2}(x_{k,3} + x_{k,4} + \cdots + x_{k,k}) + \\ & p_{j-1}(x_{k,2} + x_{k,3} + \cdots + x_{k,k}), \quad k \leq j < b \end{aligned}$$

(4)

The same equation (4) holds for $j = b$, in the case of finite buffers, or unbuffered switches $(b = 1)$. (Proof in full paper).

4 Solution of the First Stage

The characteristic equation for the above recurence relation (4) is, for $b \geq j \geq k$ ($b < \infty$ or $b = \infty$) : $F(y) = 0$, where

$$F(y) = x_{k,0} y^{k-1} - (x_{k,2} + x_{k,3} + \ldots + x_{k,k}) y^{k-2} - \ldots - (x_{k,k-1} + x_{k,k}) y - x_{k,k}$$

CASE 1: $F(y)$ has distinct roots $R_1, \ldots, R_{k-1}$. Then the steady-state probabilities are

$$p_j = A_1 R_1^{j-1} + A_2 R_2^{j-1} + \cdots + A_{k-1} R_{k-1}^{j-1} \qquad (5)$$

where $A_1, A_2, \ldots, A_{k-1}$ are constants that can be derived from the initial conditions

$$\begin{aligned} p_{0,1} &= A_1 + A_2 + \cdots + A_{k-1} \\ p_2 &= A_1 R_1 + A_2 R_2 + \cdots + A_{k-1} R_{k-1} \\ &\vdots \\ p_{k-1} &= A_1 R_1^{k-2} + A_2 R_2^{k-2} + \cdots + A_{k-1} R_{k-1}^{k-2} \end{aligned}$$

(6)

together with $p_{0,1} = p_0 + p_1$, $\sum_{n=0}^{b} p_n = 1$, ($b < \infty$ or $b = \infty$) and the equations (3) for $p_0, p_1, \ldots, p_{k-2}$.

CASE 2: $F(y)$ has at least one multiple nonzero root. Then the system is unstable, that is $\lim_{n \to \infty} q^{(n)} = \infty$

Theorem 7 (stability criterion). *A steady state queue size distribution exists if and only if $F(y)$ has distinct roots.*

The cases of instability should occur only when $b = \infty$ (infinite buffers) and $p = 1$. However, applying our method for networks with switches 2×2, 3×3 and 4×4, we never faced the above CASE 2.

4.1 Switches with finite buffers

By applying (5) for $k = 2$, we get one root R_1, which is, given that $x_{2,0} = (1 - p/2)^2$ and $x_{2,2} = p^2/4$: $R_1 = \frac{x_{2,2}}{x_{2,0}} = (\frac{p}{2-p})^2$ The constant A_1 is given by: $A_1 = \frac{1-R_1}{1-R_1^b}$ The steady state probabilities are :

$$p_0 = A_1 x_{2,0}, p_1 = A_1(1 - x_{2,0}), p_j = A_1 R_1^{j-1}, 2 \le j \le b \quad \text{(for } p < 1) \quad (7)$$

or

$$p_0 = 1/4b, p_1 = 3/4b, p_j = 1/b, 2 \le j \le b \quad \text{(for } p = 1) \quad (8)$$

By an easy calculation, the *mean number of packets in an output queue of the first stage is*

$$E(q) = \sum_{j=0}^{b} j p_j = p + \frac{p^2[1 - p_b(1 - p + b)]}{4(1 - p)}, \quad \text{for } p < 1 \text{ and } b > 1 \quad (9)$$

and $E(q) = \sum_{j=0}^{b} j p_j = \frac{b+1}{2} - \frac{1}{4b}$, for $p = 1$ and $b > 1$

It is worth pointing out that for $b \to \infty$, we get $p_b \to 0$ much faster, thus equation (9) agrees with the known formula of [6] for the infinite buffer case (equation [19]).

For the *mean number of packets lost in a cycle at an output queue of the first stage* we have:

$$\underline{\text{for } p < 1} : E(\text{packets lost in one cycle}) = \begin{cases} (p^2/4)p_b, & b > 1 \\ p^2/4, & b = 1 \end{cases} \quad (10)$$

$$\underline{\text{for } p = 1} : E(\text{packets lost in one cycle}) = \begin{cases} 1/4b, & b > 1 \\ 1/4, & b = 1 \end{cases} \quad (11)$$

4.2 Switches with infinite buffers

In this case we have $b = \infty$, thus for $k = 2$, we get $x_{2,0} = (1 - p/2)^2$, $x_{2,2} = p^2/4$ and the root $R_1 = (\frac{p}{2-p})^2$. The difference is in the constant A_1 which is now : $A_1 = 1 - R_1$

The steady-state probabilities are :

$$p_0 = 1 - p, p_1 = A_1(1 - x_{2,0}), p_j = A_1 R_1^{j-1}, j \ge 2 \quad \text{for } p < 1 \quad (12)$$

For $p = 1$ we don't have steady-state probabilities, since this is an instability case. Equations (12) are in agreement with [6], since they provide the known result: $E(q) = p + p^2/4(1 - p)$

4.3 Unbuffered switches

For the general case ($k \times k$ switches), we have two balance equations :

$$p_0 = p_0 x_{k,0} + p_1 x_{k,0} = x_{k,0}$$
$$p_1 = p_0(x_{k,1} + x_{k,2} + \cdots x_{k,k}) + p_1(x_{k,1} + x_{k,2} + \cdots x_{k,k}) \qquad (13)$$
$$= 1 - x_{k,0}$$

Since $x_{k,0} = (1 - p/k)^k$, we have

$$p_1 = 1 - (1 - p/k)^k \qquad (14)$$

Equation (14) is exactly the equation $P_{m+1} = 1 - (1 - P_m/k)^k$ of [8] and [6], when $m = 0$. We may remark here, that the above authors, derive this equation for all the stages of the network. This is an evidence that our approximation for the stages beyond the first stage (section 5) is valid even for the cases when $b > 1$. Easily, we get

$$E(q) = 1 - x_{k,0} = 1 - (1 - p/k)^k$$
$$E(lost) = p - 1 + x_{k,0} = p - 1 + (1 - p/k)^k \qquad (15)$$

The last equation is the same with (10) for $b = 1$, when $k = 2$.

5 Subsequent Stages and Network Performance

In accordance to the remarks stated in Section 2.2, we assume now the following approximation hypothesis :

Hypothesis : The output queues of stage m that feed stage $m + 1$, are assumed to operate like processors with a packet generation probability $p_{(m)}$ such that

$$p_{(m)} = \text{utilization of an output queue of stage } m \text{ (and } p_{(0)} = p)$$

This hypothesis equates the dynamics of the output process of a stage with its "macroscopic" averages.

Definition 8. Let $p_{j,i}$ = the steady state probability of finding j packets in an output queue of stage i of the network.

Suppose that we have a network with L stages. Our approximation scheme is iterative and is described in following Algorithm I.

```
p(0) := p
FOR i = 1 TO L DO
BEGIN
    Set p := p(i-1)
    Calculate x_{k,0}, x_{k,1}, ..., x_{k,k},
    Evaluate p_{0,i}, p_{1,i}, ..., p_{b,i}, from equations (5),(6)
```

Evaluate $E(q)$, $E(lost)$ for stage i

$p_{(i)} := 1 - p_{0,i}$

END

CALCULATE NETWORK PERFORMANCE MEASURES
(BANDWIDTH, AVERAGE TRANSIT TIME etc.)

This approximation scheme has the following nice properties :

- It provides an *exact* solution for all stages for unbuffered networks as we commented in Section 4.3
- It approximates not only the average measures such as $E(q)$ and $E(lost)$, but also the distribution itself of the queue sizes, with a maximum relative error in all cases, less than 5%. Higher errors are observed only in cases where the absolute values are very small and the simulation experiments count only a few respective events (e.g. lost packets when p is small).

6 Comparison with Simulation Results and Discussion

We performed extensive simulations to validate our results. The simulations verify our analysis for the first stage and the subsequent ones for all different cases (unbuffered, infinite and finite buffers). Moreover, they prove that the hypothesis introduced has a strong physical sence.

The comparison of the analytic results with the simulation experiments, confirms the exact solution of the first stage for all classes of MINs studied and the fact that the algorithm for the next stages presents an exact solution for all stages in the case of unbuffered network ($b = 1$). Our approximation predicts cumulative performance measures (such as mean queue length) with very small relative error. As far as the steady state distribution of queue sizes is concerned, we approximate the largest steady state probabilities with a very good accuracy, in all stages. For the low-valued probabilities (p_b, p_{b-1}) we odserve a small absolute error and a greater relative one. This error is caused probably due to the fact that the blocking phenomena that relate to these probabilities happen rarely, thus they are encountered a few times by the simulation of the network. The relative maximum relative error of 5% observed for the above probabilities, could cause a relative error of about 10% for the mean number of lost packets per queue, for the stages beyond the first since it depends mainly on those small probabilities. It is interesting to note that under our analysis networks with 2×2 switches seem to perform better than the 3×3 switches, with respect to the mean number of packets lost per queue.

- For networks with 2×2 switches :
  For low traffic ($p \leq 0.4$) buffers of size 3 are sufficient to allow only a small fraction (of about 0.0001) of the packets to be lost per queue. The buffer size becomes $b = 8$ for moderate to heave traffic ($0.4 < p \leq 0.8$) and $b = 15$ for very heavy traffic ($0.8 < p \leq 0.9$), respectively in order to keep the losses at the same low level.

- For networks with 3×3 switches :

  The buffers should be respectively of length $b = 4, b = 10, b = 18$ in order to get the same proportion of lost packets per cycle.

We expect that this tendancy - as k increases the mean number of packets lost increases - also holds for networks with greater k. The small fraction of lost packets implies that resubmission of those packets from the processors will not increase the input traffic noticaably. Thus, one can use our analysis to predict the performance of actual networks where lost packets are resubmitted later.

References

1. C. Bouras, J. Garofalakis, P. Spirakis and V. Triantafillou, *Queueing Delays in Buffered Multistage Interconnection Networks*, Proc. of the ACM Sigmetrics Conference 1987, pp. 111-121
2. A. Gottlieb, R. Grishman, C. P. Kruscal, K. P. McAuliffe, L. Rudolph, M. Snir, *The NYU Ultracomputer-Designing an MIMD Shared Memory Parallel Computer*, IEEE Trans. Computers, Vol. C-32, No. 2, Febr. 1983, pp. 175-189
3. G.F. Goke, G.J.Lipovski *Banyan Networks for Partitioning Multiprocessor Systems*, Proc. 1st Ann. Symp. on Computer Architecture, 1973, pp. 21-28
4. J. Garofalakis, P. Spirakis *The performance of Mutlistage Interconnection Networks with Finite Buffers*, Proc. of the ACM SIGMETRICS Conference, 1990, short paper.
5. R.R. Koch *Increasing the size of a Network by a constant factor Can Increase Performance by More Than a Constant Factor*, IEEE Symp. on Found. of Comp. Sc. (FOCS 88), pp. 221-231
6. C.P. Kruskal, M. Snir *The performance of multistage interconnection networks for multiprocessors*, IEEE Trans. Comp., vol. C-32, Dec 1983, pp. 1091-1098
7. C.P. Kruskal, M. Snir, A. Weiss *The Distribution of Waiting Times in Closed Multistage Interconnection Networks*, IEEE Trans. on Computers, vol. 32, 1988, p. 1337-1352
8. J.H.Patel, *Performance of processor-memory interconnection for multiprocessors* IEEE Trans. on Computing, vol. C-30, 1981, pp. 771-780
9. G. Pfister, M. C. Brantley, D. A. George, S. L. Harvey, W. J. Kleinfelder, K. P. McAuliffe, E. A. Melton, V. A. Norton, J. Weiss, *The IBM Research Parallel Processor Prototype (RP3): Introduction and Architecture*, Proc. 1985 Int. Conf. Parallel Processing, pp. 764-771
10. R. Rehrmann, B. Monien, R. Luling, R. Diemann, *On the Communication Throughput of Buffered Multistage Interconnection Networks*, ACM SPAA'96, pp. 152-161

Simulation of a Routing Algorithm
Using Distributed Simulation Techniques

C.D. Pham   J. Essmeyer*   S. Fdida

Laboratoire LIP6
Université Pierre et Marie Curie
4 place Jussieu 75252 Paris Cedex 05
e-mail : {pham,essmeyer,fdida}@masi.ibp.fr

Abstract. This paper describes our experiment in simulation of a routing algorithm in ATM networks using distributed simulation techniques. These techniques are a promising tool for performance evaluation of large and complex systems that can not be handled sequentially. The simulations are performed on a CM-5 and the results show that interesting speedups can be achieved when compared to a sequential execution. However, they also raise the problem of optimal partitioning and load balancing in communication network models where communication costs represent the main overhead.

1 Introduction

Discrete event simulation often represents the only way to study the behavior of a system. This is because analytical methods may be unsuitable when the size and the complexity of the system require simplifications that reduce the usefulness of the study. Simulation is more likely to allow a complete study since the level of detail is not theoretically limited. However, the flexibility of simulation is paid by a large amount of computation time and a model can turn out very quickly to be very complicated as soon as it is a little realistic. Therefore, there has been a growing interest over the last decade in bringing the available power of parallel machines to the simulation task.

In this work, we report the speedup obtained by distributing the simulation of a complex ATM network model on a Connection Machine (CM-5) of the *Centre National de Calcul Parallèle en Science de la Terre* (CNCPST). Performance evaluation of large communication networks is a challenge because such evaluation requires the execution of millions of events. Distributed simulation of these models is also difficult because the overhead per event must be kept small enough in order to obtain significant acceleration over a sequential simulation. Previous works by Earnshaw and Hind [2], Mouftah and Sturgeon [8] have also investigated the communication network simulation problem but they are limited to the simulation of simple transfer of packet/cell between nodes with uniform

---

* This work was conducted while the author visited the LIP6 (previously MASI) laboratory for a 6 months period as part of his Master degree.

routing. The speedup reported in [2] is almost linear for the most favorable case when compared to an optimized version of the parallel simulator. Our work includes the routing algorithm described in [6] and connection management. The simulation of a routing algorithm implies to simulate (i) the mechanism that consists in constructing and updating (on cost change such as a link failure) the routing table and (ii) the flow of cells transported by the network.

This paper is organized as follows: the next section presents some backgrounds in distributed simulation. Section 3 presents the ATM network model and the routing algorithm. Section 4 details our conservative simulation kernel. The preliminary speedup measures are presented in Sect. 5. Conclusion and future works are then given in Sect. 6.

2 Distributed simulation background

The distributed approach usually assumes that the system to be simulated can be spatially partitioned into disjoint sub-systems. Each sub-system is simulated by an associated *Logical Process* (LP) on a dedicated processor. To model the interactions that may exist in the real system messages are exchanged and times-tamped by the sending LP according to its *Local Virtual Time* (LVT). At the receiving side, they are put in a local *Future Event List* (FEL) to be processed. Other methods to parallelize a simulation exist (see [3, 4]) but their applications are limited.

The distribution of virtual clocks and event lists requires special care to ensure that the distributed simulation is *correct* and produce the same output when compared to a sequential one. In fact, correctness is achieved if *causality constraints* are maintained. Now, since the different LPs may advance at different rates synchronization problems are likely to occur. For instance, we are confronted to a *causality error* when a message arrives at a receiving LP and is outdated, or old, according to the LVT. These problems arise the need of an explicit synchronization mechanism that are traditionally classified in two categories: *conservative* and *optimistic*.

The conservative approach [1] only processes events that can definitely be considered safe, i.e. that processing an event would not result in further causality violations. Safety is guaranteed by forcing the LP to block when bad decisions can be taken. The advantage is to obtain a correct simulation at any time but this scheme introduces deadlocks that must be avoided by sending *null-messages* relying on *lookahead* ability in the purpose of artificially propagating and advancing the simulation time. An alternative that consists of detecting the deadlock and breaking it also exists. On the other side, optimistic approaches [7] do not search for safety but provisions are made to *roll back* to an earlier coherent state when they occur. Periodic check-pointing and *anti-messages* to cancel bad computations are then needed as a counterpart of more freedom. In addition, a *Global Virtual Time* (GVT) is required to monitor the simulation progress and to reclaim memory used by obsolete information.

Experimental studies have shown that neither conservative nor optimistic approaches can be used for a large variety of applications. Instead, a lot of variations on the original scheme are defined to take into account specific problems because the performances of a distributed simulator are tightly linked to the application. In some cases, a distributed simulation may perform worse than a sequential one. In this work, we choose a conservative approach as the underlying synchronization method for our kernel. This choice is dictated by two main reasons. First, some thoughts suggest that communication network models are not well-suited for optimistic scheduling since they have a very low computation granularity that would propagate bad computations very quickly. And second, a conservative method is much more simple than an optimistic one and therefore is more likely to be efficiently implemented by a small team.

3 The ATM network model

We propose to study the topology illustrated by Fig. 1. Cells are generated by 7 traffic sources and are sent to the corresponding sinks by intermediate ATM switches. Connection mode is also modeled so traffic sources send a setup message to their connected switch indicating the address of the destination before sending any data cell.

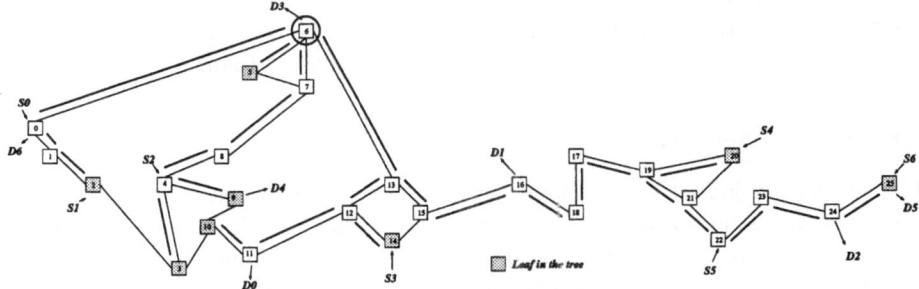

Fig. 1. Routing network topology.

All along the path, a new entry is added in the connection table of each intermediate ATM switches for this new connection. In particular, this table stores the VCI/VPI (Virtual Channel Identifer/Virtual Path Identifer) translation for the connection. Upon reception of the setup message, the destination can either send back a connect message or a reject message. For sake of simplicity, we will assume that the destination never refuses a call demand and that the source can always send cells to its associated sink.

The routing function consists in calculating the best path for the packets/cells from a source to a sink [10]. This calculation is done by considering either implicit criterion such as the number of hops or explicit one given by the user (Quality Of Service). The routing function works in two phases: (i) constructs new routing tables (done at initialization) and (*ii*) updates these tables to reflect the network changes (link failures or dynamic routing).

In this paper, we are interested in the routing algorithm proposed in [6]. This algorithm belongs to the *vector-distance* category and is said by the authors to have a quick recovery from failure property. In Fig. 1, the additional bold links represent the tree constructed for a given destination $D3$ (the tree is noted $G(D_3)$). The distributed routing algorithm works by exchanging cost information between switches to construct local routing tables. At the end of the first phase, each switch is supposed to have a complete routing table that indicates for each possible destination the best output link. Finally, for each possible destination D_i, a different tree $G(D_i)$ rooted by D_i can be constructed with the edges representing the best link an intermediate node should take to reach D_i. When the first phase is completed, the switches can begin to send cells from sources to sinks according to their routing table. The second phase consists in experimenting link failures to study the recovery behavior of the routing algorithm.

4 The conservative simulator

The conservative simulator is written in C++ using an object oriented design. The PVM (*Parallel Virtual Machine*) package [9], and especially the CM-5 implementation that uses the underlying low-level CMAML message passing library instead of the high-level CMMD [5], provides the communication facilities between processors. All simulations are performed on a 32 nodes partition.

4.1 The kernel structure

The simulation kernel we developed is a unified object-oriented framework for both sequential and parallel conservative simulation. Therefore, the same source code can either be run sequentially or in a distributed manner. This method has the advantage to allow fair speedup measures, easy debugging and automatic modification of the two versions. An abstract base class CSimulatedObject is defined for the purpose of deriving child classes for each simulated object such as switch, source and link. In particular, the routing functionality is implemented in the CRoutingSwitch class derived from the CSwitchATM class that only provides basic switching functionalities. The main part of the simulator is defined in the CSimulation class that contains all the simulated objects, the Future Event List (FEL) and the Local Virtual Time (LVT) for an LP. At this point, there is no difference yet between the sequential and the distributed simulation. For conservative distributed simulation, Input Channels (IC) are introduced in order to handle messages from the outside. For this purpose the CConservativeLP class is derived from CSimulation, and therefore inherits the simulated objects, the FEL and the LVT, and implements the management of the ICs. This means that a sequential simulation only needs the CSimulation class while the distributed one uses CConservativeLP instead. In both cases, the simulated objects remain the same and they do not know whether they are working sequentially or in a distributed environment.

The management of the ICs and the FEL are the core of the simulator. Since messages arrive on several ICs, one has to scan all the ICs to determine the messages that are safe for processing. Instead of scanning on a per event basis, all messages that are considered safe are transfered to the FEL and transformed into events in one step. Each IC i handles a virtual clock h_i that ticks according to the timestamp of the last received message. In this way, it is possible to determine the minimum of the h_i, called the Sure Future Timestamp (SFT), in order to decide whether a message is safe, i.e. its timestamp is less than the SFT, or not safe, i.e. its timestamp is greater than the SFT. Events produced by an LP are stored inside the FEL that determines whether the destination resides in a different LP on a different processor or in the same LP. This last case can happen when the mapping is not one switch to one processor as explained in the next section.

Null-messages for deadlock avoidance are sent when all safe events have been processed, i.e. the simulation virtual time has advanced to the value of SFT. One null-message is sent on each output link relying on the lookahead provided by the link propagation delay. This delay represents the minimum timestamp increment between 2 successive messages.

4.2 Initialization of the simulation

The programming model for the CM-5 is the master-slave model. At initialization, a master program spawns (with the pvm_spawn function) all the LPs on the nodes of the partition and checks for the end of the simulation. The call to pvm_spawn must be unique so that all the slaves reside in one CM-5 process in order to achieve fast communication.

A mandatory input file must be provided to the simulator which describe the topology of the network to be simulated. Each slave reads this input file to build the simulated objects. It is possible to specify an optional mapping file that describes the mapping of the network components (switches) into LP. The default mapping is to assign one LP per switch that is simulated on one physical processor. With a mapping file, several network components can be grouped in one LP. In this case, all objects in the LP are simulated sequentially and conservative synchronization is only needed for external messages.

5 Speedup measures

To measure the efficiency of our distributed simulator we compare the overall execution time of a sequential version running on one node of the parallel machine to the execution time of a parallel version. In a first attempt, we assign one physical processor to one switch of the network model. In a second step, several switches are simulated on the same physical processor with a simple aggregation method that groups several adjacent switches together. Finally a sequential-like simulation is realized for all the switches in the same physical processor whereas the conservative synchronization mechanism is used between switches on different processor. In this way, we want to reduce the communication cost between adjacent LPs since switches in the same LP do not need to exchange remote message nor null-messages. Figure 2 illustrates the 10 processor mapping. In all cases, traffic sources and sinks are encapsulated in the same LP of the switch they are connected into.

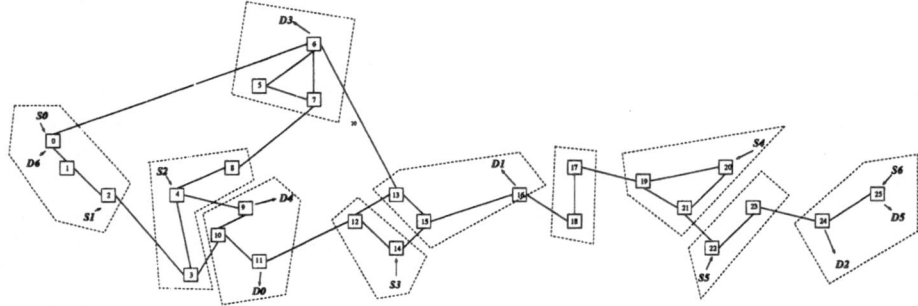

Fig. 2. 10 processor mapping.

Figure 3 shows these execution times for the simulation of 100000 time slots. This represents more than 6 millions events simulated (approximately 2 millions cells exchanged between the switches). For the one to one mapping, the resulting speedup is 4.52. This speedup results in only 0.17 for the efficiency (defined as the ratio between the speedup and the number of processors used). This modest speedup, when compared to those obtained by previous works, is linked to the complexity of the model. In particular, the routing algorithm, instead of being uniform, creates load imbalances that result in a high null-message ratio between LPs that do not exchange real messages. Also, the general network topology of our tests contributes to lower the degree of parallelism. Figure 3 also shows that unfortunately no better speedup can be obtained with less processors. Therefore it appears that the aggregation method fails to reduce the communication cost between LPs but the reason for these poor performances now becomes clear in the light of the results. Let us take the LP that aggregates switches 0, 1 and 2 to illustrate our explanation.

When aggregating adjacent switches on the same physical processor, the number of external links in an LP often remains the same for the case where an LP simulates one switch or for the case where the aggregated LP simulates several switches (2 in the example). This has two consequences. First, the number of null-messages sent by the LP does not change significantly so the overhead for these extra messages is not reduced. Second, the number of external remote messages remains the same since the number of cells sent on a given link is not affected by the logical mapping. The aggregation of switches has the consequence of increasing the load of the processor, because it simulates several switches in-

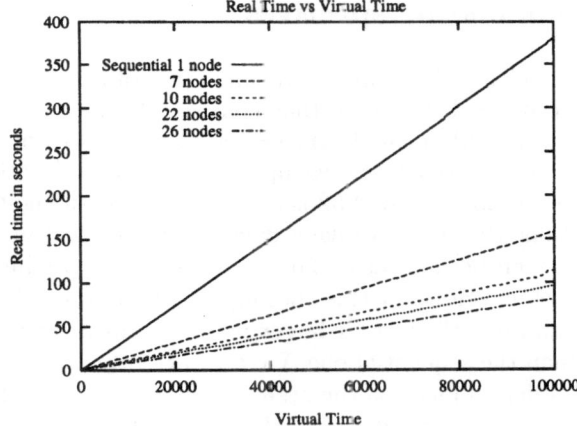

Fig. 3. Overall execution time of parallel version when *nproc* is varied.

stead of one, while keeping the same overhead for sending external messages. These results point out the difficult problem of optimal partitioning and load

nproc	real time	speedup	efficiency
1	380	1.0	1.0
7	158	2.41	0.34
10	115	3.30	0.33
22	97	3.92	0.18
26	84	4.52	0.17

Table 1. Execution time, speedup and efficiency as *nproc* is varied

balancing in distributed simulation of communication networks. In this particular application where the ratio between real computation and communication overhead (packing and unpacking messages) is very low, putting several switches that are not adjacent is useless because the level of external synchronization will remain the same. Little gain can only be achieved for small cycles such as the one introduced by switches 5, 6 and 7 in Fig. 2 where we can see that the number of external links is reduced from 4 (case when switch 6 is simulated alone) to 3 (case when switches 5, 6 and 7 are simulated together). In most cases, aggregation can improve the efficiency but not the overall speedup. Clearly, a difficult trade-off should be done when aggregating switches between the communication cost reduction and the load increase of the processor. If the computational power of a single processor increases, it should be possible to find a partitioning that maximizes the efficiency while keeping the speedup very close to the one to one mapping speedup. On the CM-5, this phenomenon can not be observed because the processing power of a single processor is quite low (SPARC at 32Mhz).

6 Conclusion and perspectives

In this paper we presented a distributed simulation of a real routing algorithm. Given a realistic network topology that consists of 26 switches, a speedup of 4.52 can be achieved with a simple one switch to one processor mapping. Using less processors provides no better speedup since the communication cost remains roughly the same. A speedup of 4.52 is a lot and ... not enough! It is very good because simulations that should take 4 days can be realized in less than a day. Now it is not enough because using 26 processors for a speedup of 4.52 is only 0.17 for the efficiency. However this speedup can be improved: we believe that better speedup can be achieved for larger networks since the parallel version is more scalable than the sequential one. We are currently porting the kernel on a CRAY T3E in order to increase the system size. Nevertheless this work shows that the use of a parallel computer can help to push the design of communication networks because interesting problems can be simulated in a tolerable amount of time.

Acknowledgment

Part of this work is supported by the French defense research institute Direction de la Recherche Et de la Technologie (DRET).

References

1. Chandy, K. M., Misra, J. : Distribution Simulation: A Case Study in Design and Verification of Distributed Programs. Trans. on Soft. Eng., 5(5) (May 1979) 440–452.
2. Earnshaw, R. W., Hind, A. : A Parallel Simulator for Performance Modeling of Broadband Telecommunication Networks. Proc. of the WCS'92 1992, 1365–1373.
3. Ferscha, A., Tripathi, S. K. : Parallel and Distributed Simulation of Discrete Event Systems. CR-TR-3366, Dept. of Comp. Science, Univ. of Maryland.
4. Fujimoto, R. M. : Parallel Discrete Event Simulation. Comm. of the ACM, 33(10) (October 1990) 31–53.
5. Hoppe, H. C., Ossadnik, P., Stüttgen, W. : CM-PVM: An Efficient Implementation of PVM 3.3 for the CM-5. Proc. of The EuroPVM'95, 52–58.
6. Jaffe, J. M., Moss, F. H. : A Responsive Distributed Routing Algorithm for Computer Networks. IEEE Trans. on Comm., 30(7) (July 1982) 1758–1762.
7. Jefferson, D. R. : Virtual Time. ACM Trans. on Prog. Lang. and Sys., 7(3) (July 1985) 405–425.
8. Mouftah, T., Sturgeon, R. T. : Distributed Discrete Event Simulation for Communications Networks. IEEE JSAC, 8(9) (December 1990) 1723–1734.
9. Geist, A. and al. : PVM 3 User's Guide and Reference Manual. (May 1993).
10. Schwartz, M., Stern, T. : Routing Techniques used in computer communication networks. IEEE Trans. on Comm., 28(4) (April 1980) 539–552.

Message-Passing Performance of Parallel Computers

Vladimir Getov[1,2], Emilio Hernández[3], Tony Hey[2]

[1] School of Computer Science
University of Westminster, Harrow Campus, London, UK
[2] Department of Electronics and Computer Science,
University of Southampton, Southampton, UK
[3] Departamento de Computacion,
Universidad Simon Bolivar, Apartado 89000, Caracas, Venezuela

Abstract. In this paper we investigate some of the important factors which affect the message-passing performance on parallel computers. Variations of low-level communication benchmarks that approximate realistic cases are tested and compared with the available Parkbench codes and benchmarking techniques. The results demonstrate a substantial divergence between message-passing performance measured by low-level benchmarks and message-passing performance achieved by real programs. In particular, the influence of different data types, the memory access patterns, and the communication structures of the NAS Parallel Benchmarks are analysed in detail, using performance measurements on the Cray-T3D in Edinburgh and the IBM-SP2 in Southampton.

1 Introduction

Generally speaking, the time taken for communications represents a substantial part of the performance penalty that one has to pay when using message-passing parallel computers. In most cases, the communication subsystems on such machines work in a *pipeline* fashion, and therefore, the communication time can be modelled as a simple linear function of the message length [9]:

$$t_{com} = t_0 + t_{trans} n \qquad (1)$$

where t_0 is the message startup time, t_{trans} is the transmission time per byte, and n is the message length in bytes.

Perhaps the simplest measure of message-passing performance is the *ping-pong* or *echo* benchmark between a pair of nodes, which has been described by several authors [9,5,2]. In this test one node sends a message to the other, which in turn, receives the message and sends it back immediately to the first node. Half the time for this test is recorded as the time to send a message of the given length. The message-passing performance can be characterised by extending Hockney's performance description for vector pipelined processors over the above model using the asymptotic performance, r_∞, and the half-performance length, $n_{\frac{1}{2}}$, parameters [5].

$$t_{com} = \frac{n_{\frac{1}{2}}^c + n}{r_\infty^c} \tag{2}$$

Although the communication subsystems of tightly coupled parallel computers have been improving very quickly over the last few years, this same simple model is also valid for multistage interconnection networks based on high-speed communication switches [10].

Different sets of low-level benchmarks may be defined, depending on the aims of the particular evaluation exercise. If the benchmarking goal is *to compare* the performance of different parallel computers, the low-level benchmarks are necessary to mark off the bounds of calculation and message-passing performance. The communication patterns which outline the message-passing performance bounds may be grouped into four categories [4]: single point-to-point messages; single broadcasts; point-to-point transmissions from all nodes; broadcasts from all nodes (multibroadcasting). If, however, the benchmarking goal is *to estimate* the application performance, the low-level benchmarks must provide much more accurate and sometimes application-specific time measurements. There have been attempts to relate message-passing performance of real applications to the results of the existing low-level communication Parkbenchmarks [8]. However, it is difficult to predict the performance of a particular communication pattern, for instance, a circular shift, by the currently available low-level Parkbenchmarks, as they have been designed for performance comparisons and not for performance predictions. One of the main reasons for this is the fact that the communication time depends not only on the message length, but also on the current workload of the interconnection network [2], because the network itself cannot be considered as a single resource. It has input and output ports for each node in the distributed memory system, all of which may operate simultaneously, and it is therefore fallacious to draw conclusions about the aggregate message-passing performance from simple ping-pong measurements. Such tests highlight only the case of a single message in isolation and could not be useful as a basis for performance predictions of higher level benchmarks and application codes.

In order to tackle this problem, a list of performance models for frequently used communication patterns can be employed to estimate the message-passing performance of an application, but such a list may be very large. A more pragmatic alternative is to extract the communication structure from the application kernel and to evaluate the asymptotic communication parameters by varying the message length for different numbers of processors. In this article we follow the latter approach in order to demonstrate the substantial divergence between message-passing performance measured by the existing low-level benchmarks and message-passing performance achieved by real programs. The next section shows results related to transmitting data elements other than bytes, which is a common requirement in scientific computing. Section 3 focuses on the performance effects of data movements associated with message-passing, while section 4 presents results related to real communication structures, extracted from the NAS parallel benchmarks [1]. Finally, section 5 discusses the conclusions to the study.

2 Alternative data types

Modern processors can load/store a 64-bit word with a single instruction. The memory access speed depends on the data type and is usually optimised for double precision (64-bit) words. In order to investigate how the choice of data type affects the message-passing performance, the COMMS1 low-level benchmark was modified to send contiguous and non-contiguous double precision elements, rather than bytes. Such performance figures are closer to those a real scientific application would yield. Table 1 shows measurements made with the original COMMS1 and a version of COMMS1 modified to send double precision data type elements. The experiments were conducted on a Cray T3D and an IBM SP2. The T3D used for the experiments has 150MHz 21064 Alpha processors, each with 64MB local memory, 8KB instruction cache and 8KB data cache, while the operating system was UNICOS 8.0.4 with a cf77 6.2.1 Fortran compiler. The compiler directives used were "-dp -Oscalar3". Meanwhile, the IBM SP2 nodes used were "thin 1" type nodes, each with a 66MHz Power2 processor, 128MB RAM (64bit memory bus), 32KB instruction cache and 64KB data cache, with an AIX 4.1.4 operating system and a xlf 3.2.5 Fortran compiler. The codes were compiled with "-O3 -qstrict" compiler directives.

	r_∞^c	r_∞^c (MB/s)	$n_{1/2}^c$	$n_{1/2}^c$ (Bytes)	Startup Time
Orig. (SP2)	28.592 MB/s	28.592	2844.935 B	2844.935	99.501 μsec
Modif. (SP2)	4.188 Mdp/s	33.504	897.039 dp	7176.312	214.174 μsec
Orig. (T3D)	28.466 MB/s	28.466	1966.929 B	1966.929	69.098 μsec
Modif. (T3D)	3.709 Mdp/s	29.672	387.048 dp	3096.384	104.359 μsec

Table 1. Bandwidth, $n_{1/2}^c$, and startup time reported by COMMS1, using bytes (B) and double precision elements (dp). The modified version transmits double precision elements.

Even though the bandwidth measured when transmitting bytes is similar on the T3D and the SP2, there is a key difference in message-passing performance between these two machines when the ping-pong benchmark uses double precision elements. Normalised measures of both r_∞^c and $n_{1/2}^c$, in Mbyte/s and bytes respectively, change when 64-bit words are transmitted.

3 Memory Access Patterns

Data movements related to message-passing are very common in distributed-memory parallel applications. For instance, data are frequently moved from non-contiguous locations to the communication buffer when a row of a matrix stored by columns is transmitted, and vice versa. In order to test the communication bandwidth when the data are non-contiguous, several variants of the COMMS1 benchmark were created. These variants use either DO loops to explicitly transfer data from the original non-contiguous location to the message buffer, or the proper message-passing library support in order to make such data transfers. In this way, we are not only testing the impact of the data movements themselves, but also the ability of different message-passing library implementations to perform such operations efficiently. The stride used in all examples was 16 double precision elements – a stride big enough to load a different cache line into cache for every element copy, in the architectures under consideration.

The Message-Passing Interface (MPI) provides a mechanism to specify non-contiguous data on the basis of *derived data types* [7] and the user can declare the shape and size of the object to be transferred using *data type constructors*. A *data type descriptor* of the object must have been previously created, so that when an MPI_SEND or an MPI_RECV is executed, the data is taken from the specified location according to the data type descriptor information. The software/hardware implementation may have to copy the original object into a buffer before sending it out, relying on the information provided in the data type descriptor. The variants of the MPI version of COMMS1 used to conduct the experiments are sketched in the following boxes:

(1)
```
Non-contiguous data in master and slave (MPI_Datatype solution)
/* Master */                        /* Slave */
MPI_TYPE_VECTOR(stride=16)          MPI_TYPE_VECTOR(stride=16)
...                                 ...
MPI_SEND(BUFFER)                    MPI_RECV(BUFFER)
MPI_RECV(BUFFER)                    MPI_SEND(BUFFER)
```

(2)
```
Non-contiguous data in master and slave (DO-loop solution)
/* Master */                        /* Slave */
DO I = 1,N                          MPI_RECV(BUFFER)
  BUFFER(I) = MATRIX(16,I)          DO I = 1,N
ENDDO                                 MATRIX(16,I) = BUFFER(I)
MPI_SEND(BUFFER)                    ENDDO
MPI_RECV(BUFFER)                    DO I = 1,N
DO I = 1,N                            BUFFER(I) = MATRIX(16,I)
  MATRIX(16,I) = BUFFER(I)         ENDDO
ENDDO                               MPI_SEND(BUFFER)
```

(3)
```
Non-contiguous data in master only (MPI_Datatype solution)
/* Master */                        /* Slave */
MPI_TYPE_VECTOR(stride=16)
...
MPI_SEND(BUFFER)                    MPI_RECV(BUFFER)
MPI_RECV(BUFFER)                    MPI_SEND(BUFFER)
```

(4)
```
Non-contiguous data in master only (DO-loop solution)
/* Master */                        /* Slave */
DO I = 1,N
  BUFFER(I) = MATRIX(16,I)
ENDDO
MPI_SEND(BUFFER)                    MPI_RECV(BUFFER)
MPI_RECV(BUFFER)                    MPI_SEND(BUFFER)
DO I = 1,N
  MATRIX(16,I) = BUFFER(I)
ENDDO
```

(5)
```
Contiguous data in master and slave
/* Master */                        /* Slave */
MPI_SEND(BUFFER)                    MPI_RECV(BUFFER)
MPI_RECV(BUFFER)                    MPI_SEND(BUFFER)
```

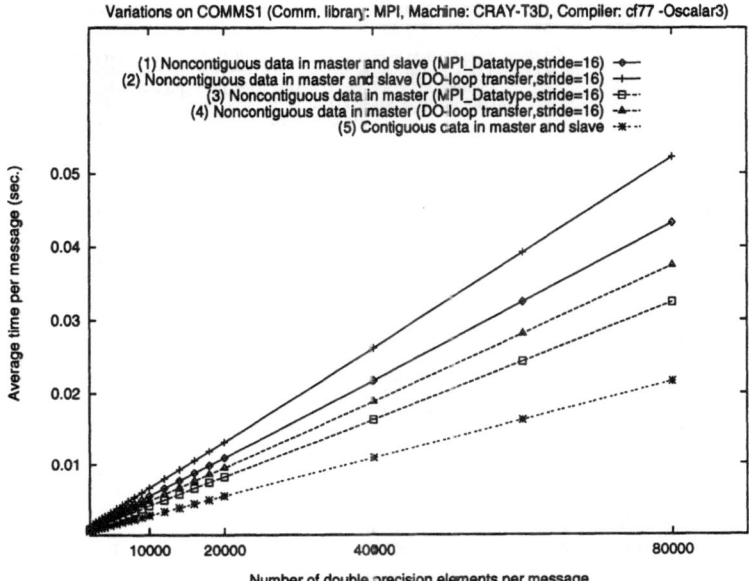

Fig. 1. COMMS1 on CRAY-T3D

The standard COMMS1 benchmark with MPI calls as provided in the Parkbench suite follows the implementation scheme of variant 5. The results obtained from all of the MPI variants on the Cray T3D and the IBM-SP2 are shown in Figures 1 and 2, respectively. The general conclusion that can be drawn from these results is that there is a very significant difference between the raw communication bandwidth and the communication bandwidth with associated data movements. On the SP2, the ratio between the raw communication bandwidth and the slowest version which performs data movements is 5.43, while on the T3D, the same ratio is 2.42. Meanwhile, the ratio between the best implementation of the non-contiguous case and the best implementation of the contiguous case is 4.50 on the SP2 and 2.00 on the T3D. It is interesting that on the SP2, the best option for transmitting non-contiguous data is by means of a DO loop (a performance ratio equal to 1.20), while on the T3D the fastest COMMS1 implementation was the one which uses library calls (here the same performance ratio was 0.83). This demonstrates the importance of the quality of the communication library in terms of performance.

4   Communication Structures

We have used the NAS parallel benchmarks (NPB), version 2.0 [1], in order to verify the validity of the ping-pong benchmark for predicting which communication pattern will execute faster on different machines. The NPB suite consists of two kernels (FT and MG) and three compact applications (LU, BT ans SP).

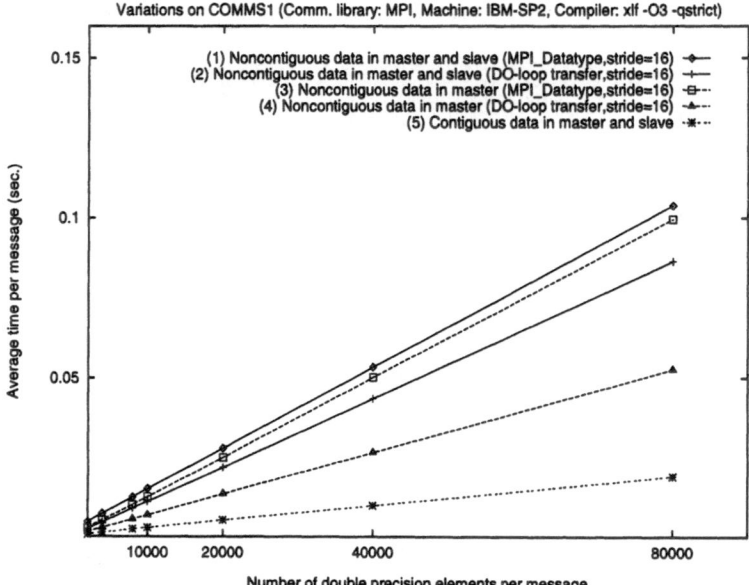

Fig. 2. COMMS1 on IBM-SP2

Our experiments were based on the compact applications, in which the communication skeletons were extracted by commenting out the computation codes of these benchmarks, leaving only the communication functions and the data movements associated with message-passing.

The LU benchmark is a compact application which finds a finite difference solution of the 3-D compressible Navier-Stokes equations. The implementation utilises a block-lower-triangular block-upper-triangular approximate factorisation of the original scheme. A particular characteristic of the algorithm is that it sends a relatively large number of small messages (40 bytes each). Consequently, this benchmark would penalise machines whose interconnect subsystem has a high startup time. The BT benchmark solves three sets of uncoupled systems of equations which are block tridiagonal with 5x5 blocks, while the SP benchmark is similar to BT, but solves a scalar pentadiagonal system. The code structures of the BT and the SP benchmarks are so similar because the systems of equations are solved using Gaussian elimination, without pivoting, in both cases. In contrast to LU, these programs have a coarse-grained communication scheme. In other words, the LU communication kernel should benefit from low latency interconnect subsystems, while the BT and SP communication kernels would execute faster on networks with a greater bandwidth.

Several experiments concerning the communication structure of the NPB kernels were performed on the Cray T3D and the IBM SP2. Figure 3 compares the communication overhead on the T3D and the SP2 for LU, BT and SP by dividing the communication times shown into the "pure" communication time and time spent by explicit data movements related to message-passing. The

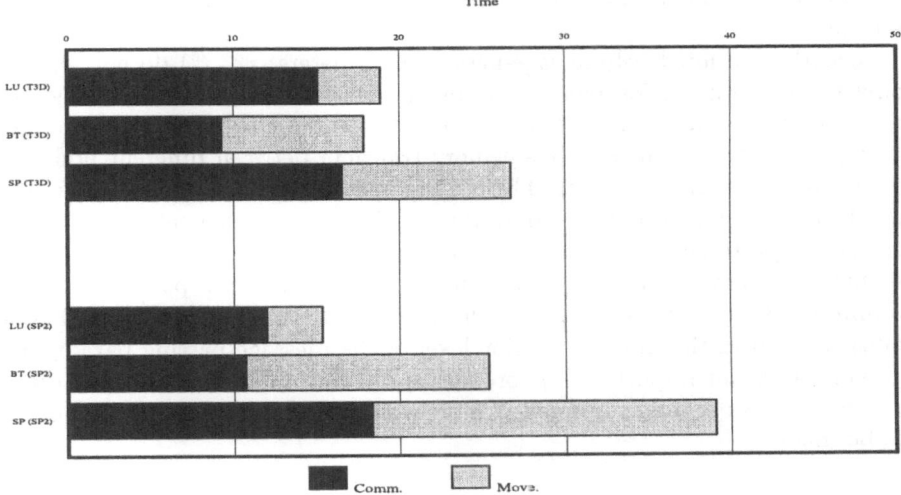

Fig. 3. Comparison of communications in T3D and SP2 (16 processors).

communication skeleton of LU runs marginally faster on the SP2 than on the T3D, while the SP and BT communication skeletons execute faster on the T3D.

The measurements shown in Table 1 were obtained on the same platforms and these COMMS1 results seem to contradict the observed behaviour of the LU, BT and SP communication skeletons on the T3D and the SP2. Apart from the bandwidth and startup time, many other factors not measured by COMMS1, play an important role in message-passing performance. These include the network contention, the presence of collective communication functions, the fact that messages are sent from different memory locations, etc. The execution of the communication skeletons indicates that there is not a substantial difference in message-passing performance between the SP2 and the T3D. The main feature in favour of the T3D, however, is the shorter time spent in data movements associated with message-passing, rather than the communication time itself.

5 Conclusions

The performance analysis of communication subsystems, based on the performance reported by ping-pong benchmarks, has been of great value for comparisons of the message-passing performance across different machines. However, predicting communication performance in terms of performance parameters provided by existing Parkbench codes is not just a function of low-level parameters, measured by simple communication benchmarks. Previous studies have shown that the analytic expression (2) fits well into parallel performance data [6,3]. Unfortunately, the fitted values of parameters turn out to be very different from those determined by the low-level benchmarks. Therefore, in most cases the existing low-level communication benchmarks are not suitable for performance estimation.

Clearly, the low-level message-passing parameters (r_∞^c, t_0^c) do not represent only the communication bandwidth and latency of the message-passing channels, but as our analysis and measurement results show, they are also aggregate parameters for memory-to-memory transfers between different nodes of a distributed memory computer. The message-passing performance, therefore, depends upon such factors as the data type of the message elements and the memory access patterns, as well as the storage location of the data to be transferred within the memory hierarchy, in addition to the hardware parameters of the communication subsystem. Hence, the trade-off between efficiency and portability, as well as the implementation level of the message-passing paradigm are also of significant importance. A broader spectrum of low-level benchmarks and parameters has to be considered if more accurate performance predictions are to be made.

6 Acknowledgments

Special thanks go to the staff of the Edinburgh Parallel Computer Centre and Computing Services at the University of Southampton, for their assistance in operating the Cray-T3D and the IBM-SP2 installations.

References

1. D. Bailey, T. Harris, W. Saphir, R. van der Wijngaart, A. Woo, and M. Yarrow. The NAS Parallel Benchmarks 2.0. TR-NAS-95-020, NASA Ames RC, Dec. 1995.
2. T. Dunigan. Performance of the Intel iPSC/860 and Ncube 6400 Hypercubes. *Parallel Computing*, 17:1285–1302, 1991.
3. V. Getov. 1-Dimensional Parallel FFT Benchmark on SUPRENUM. *Lecture Notes in Computer Science*, 605:163–174, 1992.
4. V. Getov and C. Jesshope. Simulation Facility of Distributed Memory System with 'Mad Postman' Communication Network. *Lecture Notes in Computer Science*, 487:224–233, 1991.
5. R. Hockney. Performance Parameters and Benchmarking of Supercomputers. *Parallel Computing*, 17(10/11):1111–1130, 1991.
6. R. Hockney. Computational Similarity. *Concurrency: Practice and Experience*, 7(2):147–166, 1995.
7. Message Passing Interface Forum. MPI: A Message-Passing Interface Standard. *International J. Supercomputer Applications*, 8(3/4):169–414, 1994.
8. Parkbench Committee. Report - 1: Public International Benchmarks for Parallel Computers. *Scientific Programming*, 3(2):101–146, 1994.
9. D. Reed and D. Grunwald. The Performance of Multicomputer Interconnection Networks. *IEEE Computer*, 20:63–73, June 1987.
10. Z. Xu and K. Hwang. Modeling Communication Overhead: MPI and MPL Performance on the IBM SP2. *IEEE Parallel and Distributed Technology*, 4(1):9–23, 1996.

Prefetching and Multithreading Performance in Bus-Based Multiprocessors with Petri Nets

[1]Edward D. Moreno, [1]Sergio T. Kofuji, [1,2]Marcelo H. Cintra
(edmoreno, kofuji)@lsi.usp.br and cintra@csrd.uiuc.edu

[1]*LSI-EPUSP, University of Sao Paulo, Brazil*
[2]*CSRI-UIUC, University of Illinois, Urbana-Champaign, USA*

Abstract. The large latency of memory accesses is a major obstacle in obtaining high processor utilization in large scale shared-memory multiprocessors. Access to remote memory is likely to be slow, compared to the ever-increasing speeds of processors. Thus, any scalable architecture must rely on techniques that can cope with the large latency of memory accesses to reduce/hide/tolerate remote-memory-access latencies.

In this paper, we shall consider two architectural techniques, that address the latency problem: Prefetching and Multithreading. We also intend to develop and analyse such techniques, using simple but useful analytical models that predict the performance benefits achievable on bus-based multiprocessors.

First, we study the effects of various parameters such as latency, bandwidth, degree of prefetching on speed-up and network utilization of the system. Then, using a multilevel modeling methodology for Petri Nets, we show that multithreaded architectures have higher processor utilization, but offer a higher load to the interconnection network and the memory system.

1. DSPN MODEL TO HARDWARE-BASED PREFETCHING

We have constructed one simple DSPN (Deterministic and Stochastic Petri Net) model to Fixed Sequential Prefetching for a bus-based shared memory multiprocessor. In figure 1, a rectangular box indicates a transition with fixed delay time, a single line represents a transition with zero delay and the fill-black rectangular box corresponds to an exponentially distributed stochastic transition.

Figure 1.a corresponds a processing element. Initially, place P_1 contains N_p=4 processors, which are executing instructions. The transition T1 defines the mean rate of pcycles *(processor cycle)* by access. Using a 80 MHz, 65 MIPS processor and supposing that each instruction needs two accesses to data, each processor requires data with a mean rate of 1.62 access/cycle. The data can be founded in its local cache (T_4) or upon its prefetching buffers (T_5). Transitions T_4 and T_5 are a random switch and its probabilities are associated to cache hit rate and buffers hit rate. These values are strongly based on the sharing pattern and locality property of a given application [1, 4, 7]. The values used for these probabilities are 0.95 and P_b=0.5-0.9. We assume that mean access time to caches (T_2) and buffers (T_3) are exponentially distributed with a mean value of 1 and 2 *pcycle* respectively to cache and buffers.

If a cache miss happens, then the request will be transmitted to interconnection network, represented by figure 1.b. Here, are showed two branches that correspond to external access replied by the other cache on the system (T_7) or by the main memory (T_{10}). Assuming a good degree of sharing, the probability associated to T_7 is defined as 0.75. Thus, the probability of T_{10} is 0.25.

To analyze the network contention, each part consists of two places: P_6/P_8 and P_7/P_9 associated with waiting time to access the bus and transmitting time, respectively. The access time upon caches (T_8) or main memory (T_{11}) are exponentially distributed with mean time of 1 pcycle. Once the request has been serviced by the cache/memory, it must be transmitted by the interconnection network. This process is represented by transitions T_9 and T_{12}. which have a deterministics distribution whose delay depends on the characteristics of the network and on the characteristics of prefetching (degree of prefetching and probabilities associated to prefetched data[1]).

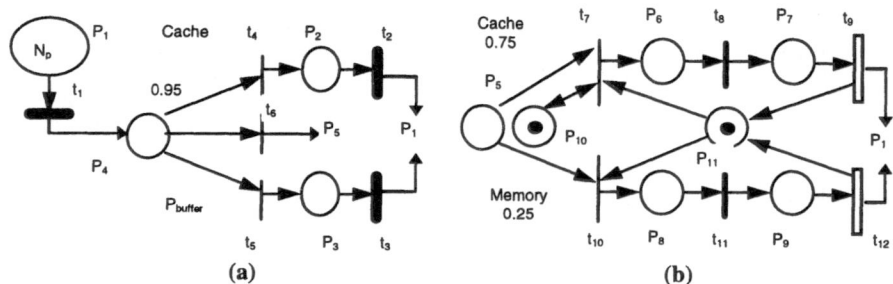

Figure 1. DSPN Model to Hardware-Based Prefetching

The main key of this model appears on the values associated to T_9 and T_{12} transitions, which take into account the possibility of utilization of the network. The effects of the interconnection network depend on its latency and bandwidth. In this work, such delays are: $T_9 = L_c + FB*(B_{size}/T_{net})*(1.0+D_g*P_f)$ and $T_{12} = L_m + FB*(B_{size}/T_{net})$.
In the previous relation: (i) $L_c = 4$-64 pcycles, if the transfer is from cache-to-cache and $L_m = 8$-128 pcycles, if the transfer if memory-to-cache, since the memory access are more slower than cache access. (ii) the FB factor is the relation between the processor's frequency and the network's bandwidth (In this case, we have chosen the characteristics of the SBus - 50 MBytes/s). (iii) D_g is the degree of prefetching. It is varied between 0 and 10. (iv) P_f is the probability associated to prefetched data, (v) B_{size} is the transfer unit size (cache block size). We have chosen 32 bytes. And, (vi) T_{net} is the bus width (SBus = 32 bits).

2. DSPN MODEL TO MULTITHREADING

2.1 A Bus-Based Multiprocessor Model

To evaluate the performance of a multithreaded architecture we must accurately evaluate the long-latencies observed in the system. In this work we consider a popular

---

1. (P_f, P_b) are probabilities associated to prefetching. P_f: determines the number of consecutive blocks to current block, that will be transmitted by the interconnection network when a cache miss occurs. The state of these blocks should be valid and shared. P_b: is associated to the use of prefetched data store in the prefetches buffers (e.g. the hit rate on prefetching buffers).

configuration of a bus-based multiprocessor with N processors and M memory modules, connected by a common bus. In this system, the contexts requesting a memory access are put in a queue to access the bus, which can only service one request at a time. Once the bus is granted to a context, a data request packet is sent to one of the memory modules and the requested for data is enqueued at this module. All memory modules service only one request at a time and to send the responses back to the processors they must compete to access the bus. Figure 2.a shows a Petri net model for such a system.

In this figure, place P_3 represents the shared bus and place P_5 represents the memory pool with M modules. For simplicity, in this model we consider that the memory requests are uniformly distributed across the memory modules. Place P_1 represents the contexts requesting memory access and the firing of transition t_6 removes a context from the memory system after its request is completed. Transitions t_1 and t_5 represent the contention for using the bus for the data request and reply, respectively. Transitions t_2, t_4 and t_6 have fixed firing delays and represent the packet transmission time (t_2 and t_6) and the actual memory access time (t_4). The values used for these times (in pcycles) are typical of a modern multiprocessor system: $t_2=t_6=5$ and $t_4=20$.

With the purpose to simulate the impact of bus contention on multithreading, we have used a multilevel modeling methodology. In this case, is necessary to transfer the behaviour of this memory system to the higher level system, we must compute the average overall access times for different number of contexts. We then closed the model presented in figure 2.a, by connecting transition t_6 to place P_1, and used the simulator RP_SIM [2] to compute the average access time for 1 to 32 tokens in place P_1, representing 1 to 32 contexts requesting memory access. This data is then used in the multithreaded processor model (see fig. 2.b).

We observed that modeling in detail the bus and memory system is highly desirable because the effective memory access time is very sensitive to the number of contexts contending for the bus. In our simulations, we found that the effective access time varies from 30 cycles, with only one context, to about 320 cycles with 32 contexts. Thus, a simple model of the multithreaded multiprocessor that approximates the memory access time by a single value for all loads will lead to misleading results.

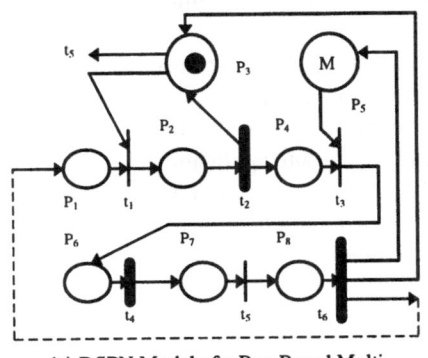

(a) DSPN Model of a Bus-Based Multi processor with M memory modules.

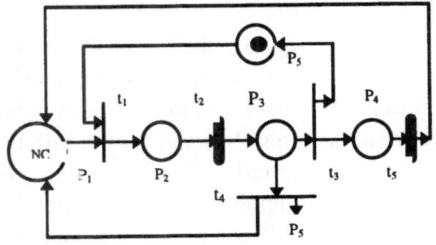

(b) DSPN Model of a Multithreaded Scalar Processor with NC resident contexts and implementing a interleaving multithreading model. A Multilevel Model.

Figure 2. DSPN Model to Multithreading in a Bus-Based Multiprocessor

2.2 A Multilevel Model of a Multithreaded Architecture

In this work we developed a Petri net model of an interleaved multithreaded architecture [3], which we believe to be most flexible and better performing multithreading model for traditional scalar processors. In this model, an instruction from a different context is issued and executed at each cycle and if the instruction is a memory request the context is marked blocked, otherwise the context remains ready in the task pool. The Petri net model for this processor is shown in figure 2.b.

In figure 2.b, place P_1 represents the task pool with NC ready contexts, P_5 represents the single issue slot available each cycle and place P_4 represents contexts requesting memory access. The system considered has an off-chip cache and the firing of transition t_4 indicates that either the instruction was not a memory operation or the instruction was a memory operation that could be satisfied from the cache. Transition t_3, on the other hand, represents a memory operation that could not be satisfied from the cache. Transitions t_3 and t_4 are a random switch and the probability associated with the firing of transition t_3 is the cache-miss rate observed for a given application. To model a multiprocessor system with P processors, we need only replicate P times the Petri net model in figure 2.b and use the sum of tokens in all places P_4 as the total number of contexts contending for the memory. In this case, the union of places P_4 corresponds to the load place.

The times involved in the model of figure 2.b are 1 cycle, fixed, for t_2 and the memory access times for the bus-memory system, for t_5. To model the timing behaviour of the memory system, the firing delay of transition t_5 is adjusted according to the number of tokens in all places P_4 at each simulation cycle to reflect the memory access time obtained from the bus and memory system model.

Using the Petri net model of figure 2.b, along with the values obtained of multilevel model (see fig 2.a), we evaluated the processor utilization for multiprocessor systems of various sizes and different degrees of multithreading (the number of resident contexts allowed by the hardware). The configurations studied are representative of current shared-memory multiprocessors, with up to 32 processors. For all configurations studied, the number of memory modules used is the same as the number of processors in the system. The processors used have 1 (no multithreading), 2, 4 or 8 resident contexts.

To assess the gains obtained with multithreading, we then performed a series of simulations to compare multithreaded architectures with different degrees of multithreading but the same overall number of contexts. The cache miss rates used are similar to the rates observed for the MP3D and LocusRoute applications [4,8] from the SPLASH benchmarks: 7% and 0.6% respectively. We have chosen these applications because they represent common parallel scientific applications with bad cache behaviour (Mp3d) and good behaviour (LocusRoute).

3. SIMULATION RESULTS

We now present results from our simulation studies. The detailed simulation model is built on a RP_SIM simulator [2, 5]. To evaluate the performance of the system, we must consider the system when it is in the steady state. All the experiments have been performed to 100000 time units and a confidence interval of 95%.

3.1 Performance of Prefetching

In our analysis, the key parameters are: Latency, Bandwidth, Degree of Prefetching and the probabilities (P_f, P_b) associated to prefetching [5]. We have used the speed-up (Sp) and the bus contention percentage (Bc%) as metrics of evaluation. The speed-up at a particular degree prefetching is taken as execution time without prefetching divided by the execution time with prefetching. The Bc (%) is taken as the waiting time to access the network divided by the total time.

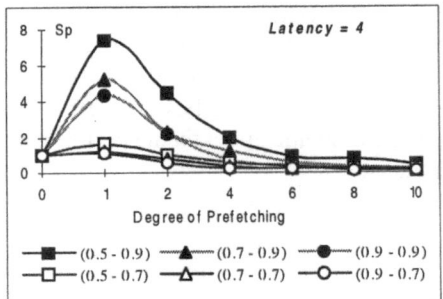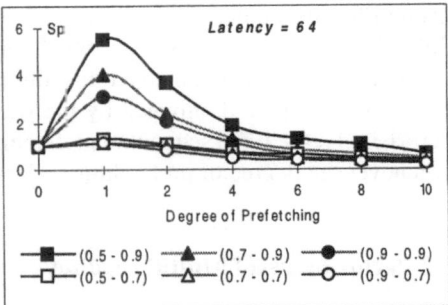

Figure 3. Speed-up to Hardware-Based Prefetching in a Bus-Based Multiprocessor

The behaviour of a system with these metrics is shown in figures 3 and 4 respectively. In figure 3, the improvements on speed-up only are seen in values where the curve is above 1. Thus, can be seen that for degree of prefetching varying among 1 to 4, there is a reasonable reduction of time. Hereafter, the speed-up diminishes considerably for the case in which there is a stronger contention on the network, since the processors are transmitting a larger quantity of data at each access (see figure 4).

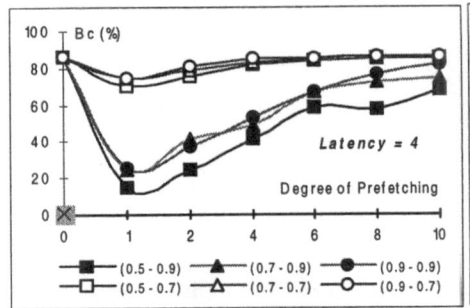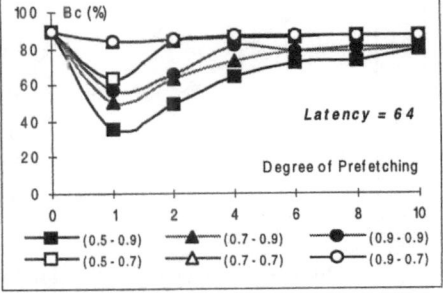

Figure 4.Bus Contention in a Bus-Based Multiprocessor with FSP

According to figure 3, prefetching to a great number of data, (i.e. larger than 4 consecutive blocks) does not improve the system's performance, regardless of the application program used, or of its sharing pattern (P_f, P_b).

When P_f=0.5 and the maintenance of data prefetched and stored into the buffers is relatively high (P_b = 0.9), this means that the data prefetches are still valid or updated in the moment which are requested, i.e. does not execute any invalidated actions initiated by cache coherence protocols. In this case, Pf=0.5 & Pb=0.9, a great speed-up is obtained: 7.3 and 5.5 when the latency is 4 and 64 pcycles, respectively with degree of prefetching equal to one. Thus, figure 3 shows that, fixed sequential prefetching, would ameliorate the total execution time even on systems with medium latencies = 64 pcycles.

As it is seen in figure 4, the bus contention with non-prefetch is about to 90%. This value means that the contention in this system is high and it become the key aspect of bottleneck in bus-based multiprocessors.

Using FSP, the contention bus is diminishes up to 18% (latency =4 pcycles) and up to 35% (latency =64 pcycles) for degree equal to one. For degree between 1 and 4, the contention is dominated by the transmitting time, and when degree is superior to 4, the contention is again dominated by the waiting time to access the bus. Therefore, not surprisingly, the fixed sequential prefetching, has a dramatic effect on the performance whenever the degree of prefetching, is larger than 4.

3.2 Performance of Multithreading

The processor utilization (P_u) versus number of contexts in the system for different degrees of multithreading is shown in figure 5. From this figure we verify that the increase in the level of multithreading improves the overall processor utilization. Also, for the same configuration, we observe that the processor utilization decreases as the size of the system is increased. This latter dependence reflects the bus and memory contention accurately modeled in the Petri net model of figure 2 (particularly in fig. 2.a), and which is not accurately modeled in [6].

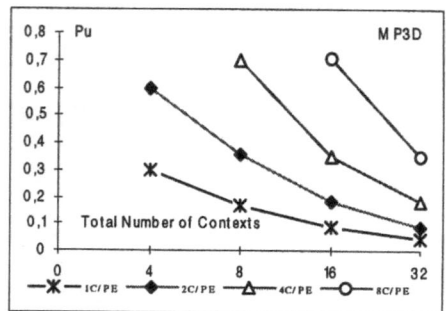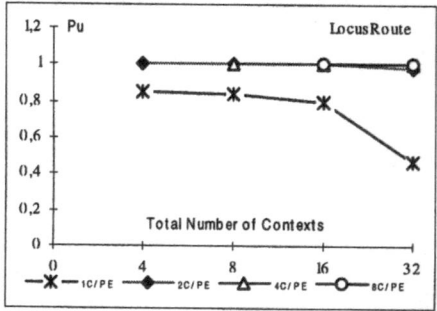

Figure 5. Processor Utilization for different degrees of Multithreading.
The cache miss rates are similar to the rates observed in the (a) MP3D program and (b) LocusRoute application.

Comparing the curves from figure 5.a with the those in figure 5.b, we notice that the performance improvements from multithreading are highly dependent on the cache behaviour of the applications. For programs with good cache behaviours (figure 5.b), the utilization of a single-context processor is already high, the utilization of a two-context processor is almost 1 and the incremental gains from multithreading degrees of more than 4 become marginal. Applications with bad cache behaviours (figure 5a), on the other hand, seem to require higher degrees of multithreading to achieve a high processor utilization.

A side-effect of using multithreading in multiprocessor systems is the increase in the load offered to the network. This happens because processors with C resident contexts can have up to C outstanding memory requests, requiring larger buffers to hold the requests and a higher network and memory bandwidth to serve ore requests in the same time. To verify this effect on the bus-memory system, we observed the effective memory access times for the configurations studied. The observed memory access times are shown in figures 6 for the Mp3d and LocusRoute applications.

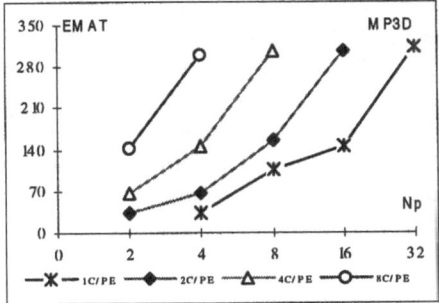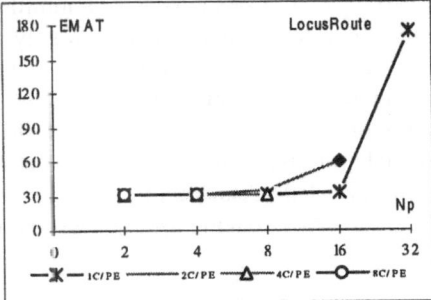

Figure 6.Effective Memory Access Time in pcycles on Multithreading

Analyzing figure 6, we observe that indeed systems with higher degrees of multithreading experience higher delays for the memory accesses. Again, we observe that applications with good cache behaviours tend to be less sensitive to effects of higher degrees of multithreading. For a program with bad cache behaviour (figure 6.a) we notice that the effective memory access time is very close to the maximum observed time for the bus and memory subsystem with 32 contexts (see section 2). This result shows that for these applications multithreading, and thus multiple outstanding memory requests, together with a high cache miss, can lead the memory system to its maximum load.

4. CONCLUDING REMARKS

Prefetching: We have described and analyzed one simple prefetch strategy that reduces the influence of medium latencies on the system. Thus, sequential-prefetch caches could reduce the influence of medium latencies and it can be have better system performance than a no-prefetch caches The traffic on the bus is the key factor of bottleneck in bus-based multiprocessors and limit the benefits of prefetching when the probabilities P_f and P_b are inferior to (0.5, 0.5).

Multithreading: We have investigated two important aspects of multithreading in a bus-based multiprocessor: processor utilization and contention in the shared bus. We have shown that multithreaded architectures with 2 to 8 contexts have better processor utilization than single-thread with the same overall number of contexts. This increase in the processor utilization can be exploited to speed up parallel applications and to increase the throughput of multiprogrammed multiuser systems.

Differently from previous work with analytical models, we have modeled in detail the behaviour of the bus and the memory system, and the contention for their access. We have verified that using multithreaded processors increases the contention for the bus and the overall memory access time. Thus, when designing multithreaded architectures with high degrees of multithreading, we must account for the increase in the network traffic and design the memory and interconnection system with sufficient bandwidth.

In this work, we have used a multilevel modeling methodology for Petri nets that allowed us to efficiently and accurately solve the multithreaded multiprocessor model. The flexible approach for handling multiple-level models allows us to easily modify both models and even replace the subsystems and their models without changing other parts of the model. Some extensions to the work presented in this paper are then to model different interconnection networks and memory systems and different multithreaded architectures, such as Simultaneous Multithreading, which is more likely to be used in the next generation of superscalar processors.

5. REFERENCES

1. Bianchini, R., Beng-Hom, L. Evaluating the performance of multithreading and prefetching in multiprocessors. In *Journal of IEEE Parallel & Distributed Technology.* April (1997)
2. Cintra, M., Ruggiero, W. A tool for modeling and simulation of computer architecture using Petri nets. In *Proceedings of the VII Brazilian Symp. on Computer Architecture (SBAC-PAD)*, Brazil, Aug. (1995) 581- 594
3. Laudon, M., Gupta, A., Horowitz, M. Interleaving: A multithreading technique targeting multiprocessors and workstations. In *Proceedings of the 6th International Conference on Architectural Support for Programming Languages and Operating Systems.* Oct., (1994) 308-318
4. Lenosky, D., Weber, W. Scalable shared-memory multiprocessing. *Edit. Morgan Kaufmann Publishers*, San Francisco, CA, (1995)
5. Moreno, E., Kofuji, S. Performance evaluation of the fixed sequential prefetching on a bus-based multiprocessor: preliminary results. *Proceedings of ISPAN'96 Intl. Symp. Parallel Architectures, Networks and Algorithms.* June (1996) 487-493
6. Nemawarkar, S., Govindarajan, R., Gao, G., Agarwal, V. Performance evaluation of latency tolerant architectures. *In Proceedings of the 4th Intl. Conf. on Computing and Information.* May (1992) 183-186
7. Tullsen, D., Eggers, S., Levi, H. Simultaneous multithreading: maximizing on-chip parallelism. *In proceedings of the 22nd Ann. Intl. Symp. on Computer Architecture*, June, (1995) 392-403
8. Weber, W., Gupta, A. Exploring the benefits of multiple hardware contexts in a multiprocessor architecture: preliminary results. *In proceedings of the 16th Ann. Intl. Symp. on Computer Architecture*, June (1989) 273-280

On Synchronisation in Fault-Tolerant Data and Compute Intensive Programs over a Network of Workstations

J.Smith

Department of Computing Science, The University of Newcastle upon Tyne Newcastle upon Tyne, NE1 7RU UK
jim.smith@newcastle.ac.uk

Abstract. An application structured as a fault-tolerant bag of tasks adapts easily to changing resources. To be represented by a single bag of tasks, a computation must decompose into purely independent tasks. The work summarised here investigates performance of structuring approaches applicable where this ideal is not possible, partly through analysis and partly through measurements of a realistic fault-tolerant computation.

1 Introduction

Where applicable, the well known "bag of tasks" organisation for parallel computations has proved popular particularly in networked environments due to its transparent load balancing property. Examples include seismic computations [1] and materials science [13]. Typically the computation is controlled by a single master process and the data manipulated by the computation is located on a single disk with all I/O being performed by the master. A fault-tolerant version of this structure [6, 2, 10] allows cheap recovery since only the particular task affected by a machine failure needs to be recovered.

It is possible to increase capacity and bandwidth of storage at a single machine using RAID techniques [5], but in some computations the data manipulated outstrips the capacity of a single machine either in terms of volume or bandwidth requirements. It is then necessary to distribute data over multiple machines and then valuable gains may be made by taking advantage of computation structure to optimise I/O [7]. To ensure application fault-tolerance however a further mechanism is required, such as checkpointing.

It is possible however to avoid the single node bottleneck in a bag of tasks computation but it is necessary to take measures to ensure consistency of access to the distributed disk based state. An approach deriving from queued transaction processing [9] is demonstrated in [12] where computation state is persistent and located in a shared distributed object store and a recoverable queue [4] serves as a fault-tolerant bag of tasks. Writes to the shared state are enclosed in an atomic action (transaction), and abort of that action leads to rollback of writes enclosed in the failed action.

One of the computations implemented in the earlier work is dense Cholesky factorisation which does not decompose into purely independent tasks. In the earlier work an algorithm is employed directly from [8, §6.3.8.]. The matrix operands are partitioned into blocks and each task entails computation of a single block of the output. The order of computation starting at the top left and proceeding down block columns and from left to right is represented in the queue, but it is necessary to delay an accessing slave until a block of the matrix becomes available. This is achieved through synchronisation flags, employed within the operations of the distributed matrix object itself.

Employing an additional synchronisation mechanism together with a single queue is one approach. An alternative requires no such synchronisation mechanism and instead relies on global barrier type synchronisation points to ensure that data is ready when required. A computation consists of a sequence of queues of which one is current at any time. All slaves wait till the current queue is actually empty before attempting to dequeue from the next in the sequence. The *dequeue* operation returns a status which allows the caller to distinguish between the situation where the queue is empty and that where entries remain but are all locked by other users. The resulting structure which supports a parallel loop programming style is similar to that of CALYPSO [3], but operating on disk based state.

One issue in the choice between such possible structures is the ease of programming at the application level, but important also is the achievable performance. This is clearly specific to each application. The work here investigates the performance of alternative realisations of Cholesky factorisation. The algorithm referred to above is used as the basis of two alternative realisations.

Single queue Where all tasks are stored in a single queue, they are all enabled at the start of the computation so it is necessary to employ a synchronisation mechanism separate to the queue so that dependencies may be satisfied. Use of an array of atomic flags is described in [12].

Homogeneous It is possible to avoid the need for synchronisation flags by loading tasks into a number of separate queues. In the structure considered here blocks on the diagonal are computed serially and all blocks in the same block column below the diagonal are computed in parallel.

Because of the dynamic nature of the computation structure it is not simple to predict parallel performance precisely. However where I/O is invariant with the number of slaves it is possible to characterise a computation by its single slave time and bounds on the minimum parallel time. Queue and synchronisation flag access cost is assumed negligible. If tasks are independent, a lower bound is the sum of all communications with the shared store and an upper bound the same plus the longest of all the task computation components. An alternate lower bound is reached when each task is computed by a separate slave. If there are inter-task dependencies then the computation may be modelled as if there are barriers to allow an upper bound to be obtained. Algebraic expressions for these bounds are derived in [12] for the single queue based realisation of Cholesky factorisation to allow extrapolation of the predicted performance to allow for upgraded hardware. Similar analysis for the alternate structure described here is detailed in [11]. Space restrictions here accommodate only the results.

The implementation of the single queue structure on a network of HP9000 based machines was described in [12]. For this work the application was ported to a network of 133 MHz Pentium machines, and the alternative application structure implemented in the same environment.

The Pentium machines have 32 Mbytes main memory and 256 Kbytes secondary cache. Hard disks are connected via fast SCSI 2 controllers. Most of the machines have 1 Gbyte IBM Pegasus disks, though the available scratch space on these disks is limited. Two of the machines have also a pair each of MAXTOR 540SL disks, providing an aggregate 2 Gbytes of storage. All the machines are running Linux version 2.0.23. The machines are connected by both a LinkBuilder FMS 100 Stackable Fast Ethernet Hub from 3Com and a ForeRunner ASX-200WG ATM switch with 155Mbit/s links. The following configurations are used.

Fast The object store is located on the IBM disk connected to a single machine and communications between slaves and store is via fast ethernet.

ATM The object store is distributed between the four MAXTOR disks. On each of the two machines hosting these disks, the two local MAXTOR disks are managed as a RAID-0 pair through the *md* [14] software. Communications between slaves and store is via ATM.

In each case the slaves are located on separate machines from those hosting the shared object store.

For block sizes above 250, the low level transfer rates for local memory to remote memory and to and from disk are found to be roughly constant. While the read and write cost for the IBM disk have nearly equal cost, the cost of writes to the RAID configuration is rather higher than the cost of reads. Clearly in the ATM configuration the limiting I/O rate should be double that offered by a single node, but a single slave can only use the bandwidth of a single node.

The computation rates vary for the different matrix primitives. In the overall computations however, matrix multiplication dominates and as tuned this primitive runs at about 27 Mflop/s for a block size above about 30. The experimental configuration is summarised below.

Configuration			Fast	ATM
Computation	(Mflop/s)		27	
Network	(Mbyte/s)		4.6	9.0
Disk	(Mbyte/s)	*read*	2.8	5.5
		write	2.8	2.4

2 Comparison

Figure 1 compares the maximum performance for the two computation structures for a matrix size of 4800^2 in the **fast** configuration. Upper and lower bounds on maximum performance calculated from the data in the table above are shown and so are a number of experimental measurements. The computation rate assumes an overall operation count of $\frac{n^3}{3}$. The performance is greatest for the single queue configuration and it can be shown analytically that this is generally true [11].

Figure. 2 shows the performance for the larger example of a 15000^2 matrix in the ATM configuration. Again both derived and measured performance are plotted. Here

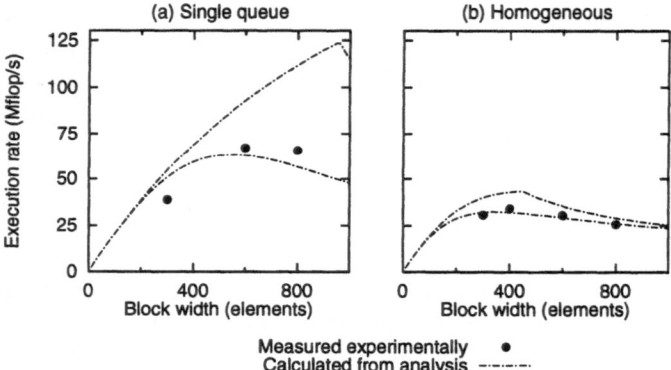

Fig. 1. Maximum performance of parallel Parallel Cholesky factorisation of a 4800^2 matrix for different synchronisation structures in the **fast** configuration.

though, five slaves do not exhaust the available bandwidth, so rather than the bounds on limiting performance, a lower bound on expected performance using just five slaves is plotted in each case. For the single queue this is obtained by dividing the single slave time by five. The approach is similar where multiple queues are used, but each phase of the computation is treated separately. It is seen that even the fault-tolerant

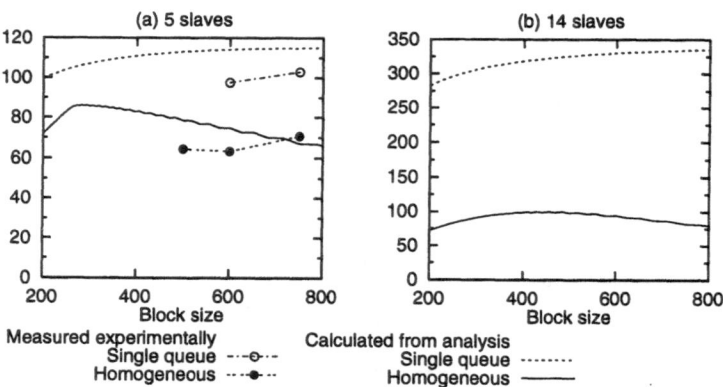

Fig. 2. Performance of parallel Cholesky factorisation of a 15000^2 matrix for different synchronisation structures in the **ATM** configuration.

implementations achieve performance which is close to that predicted, so that even with the assumptions made, the modelling process is not unrealistic.

Significantly better performance can be obtained using the single queue configuration particularly when the number of slaves is high. A simple modification to the homogeneous structure is to include computation of a block on the diagonal in the same task with the block immediately to the left. The new structure is found to perform better than the homogeneous structure, but only approaches the performance of the single queue structure for small block size, of about 250, and for a small number of slaves [11]. Intuitively the single queue structure offers least restriction to parallelism so long as the cost of synchronisation is small.

Overall the evidence suggests that the single queue approach cannot be abandoned for performance reasons.

Acknowledgements

The support of all the Arjuna team is gratefully acknowledged, and in particular collaboration with S. Shrivastava in earlier experiments which this work builds upon and the assistance of M. Little, G. Parrington and S. Wheater with implementation issues relevant to this work.

References

1. G. S. Almasi and A. Gottlieb. *Highly Parallel Computing*. Benjamin/Cummings, 2nd edition, 1994. ISBN 0-8053-0443-6.
2. D. E. Bakken. *Supporting Fault-Tolerant Parallel Programming in Linda*. PhD thesis, The University of Arizona, Aug. 1994.
3. A. Baratloo, P. Dasgupta, and Z. M. Kedem. CALYPSO: A novel software system for fault-tolerant parallel processing on distributed platforms. In *4th International Symposium on High Performance Distributed Computing*. IEEE, Aug. 1995.
4. P. A. Bernstein, M. Hsu, and B. Mann. Implementing recoverable requests using queues. *ACM SIGMOD*, pages 112–122, 1990.
5. P. M. Chen, E. K. Lee, G. A. Gibson, R. H. Katz, and D. A. Patterson. RAID: high-performance, reliable secondary storage. *ACM Computing Surveys*, 26(2):145–185, June 1994.
6. T. Clark and K. P. Birman. Using the ISIS resource manager for distributed, fault-tolerant computing. Technical Report 92-1289, Cornell University Computer Science Department, June 1992.
7. J. M. del Rosario and A. Choudhary. High performance I/O for parallel computers: Problems and prospects. *IEEE Computer*, pages 59–68, Mar. 1994.
8. G. H. Golub and C. F. V. Loan. *Matrix Computations*. John Hopkins University Press, second edition, 1989. ISBN 0-8018-3772-3.
9. J. Gray and A. Reuter. *Transaction Processing: Concepts and Techniques*. Morgan Kauffman, 1993.
10. K. Jeong. *Fault-Tolerant Parallel Processing Combining Linda, Checkpointing, a nd Transactions*. PhD thesis, New York University, Jan. 1996.
11. J. Smith. *Fault Tolerant Parallel Applications Using a Network Of Workstations*. PhD thesis, University of Newcastle upon Tyne, 1996. Forthcoming.
12. J. A. Smith and S. Shrivastava. Performance of data and compute intensive programs over a network of workstations. *Theoretical Computer Science*, 1997. To appear in special issue for Euro-Par'96 papers.
13. V. S. Sunderam, G. A. Geist, J. J. Dongarra, and R. J. Manchek. The PVM concurrent computing system: Evolution, experiences, and t rends. *Parallel Computing Vol. 20(4)*, pages 531–546, 1993.
14. M. Zyngier. *md*. ftp://sweet-smoke.ufr-info-p7.ibp.fr/pub/Linux/, Apr. 1996. version 0.35.

Performance Analysis of a Parallel Program for Wave Propagation Simulation

Michel Pahud, Frédéric Guidec, Thierry Cornu

Swiss Federal Institute of Technology Lausanne
Parallel Computing Research Group
EPFL-DI-LITH
CH-1015 Lausanne
E-mail: [pahud,guidec,cornu]@di.epfl.ch

1 Introduction

During the last decade, performance prediction has been repeatedly quoted as a key factor to developing parallel systems [1, 2, 3]. Predicting the performance of a parallel program as a function of the number of processors and of the problem size is crucial for choosing the best hardware configuration and for tuning various parameters.

This paper presents a method for achieving performance analysis for parallel irregular applications. The model is closely related to the *Bulk Synchronous Programming* (BSP) model [4]. It is based on the measurement of basic communication and computation routines. The computational workload of each processor and the load imbalance are modeled analytically.

The method is used for predicting the performances of ParFlow++, an irregular, parallel radio-wave propagation algorithm.

2 Wave propagation with ParFlow++

The ParFlow simulation method is based on the so-called Lattice Boltzman Model. It describes a physical system in terms of motion of fictitious, microscopic particles over a lattice. In the current ParFlow model, it is assumed that waves do not penetrate buildings: wall points are perfectly reflecting points that return any incident wave with opposite sign. Indoor points may therefore be ignored in the computation.

ParFlow++[i] is a data-parallel C++ implementation of the ParFlow method that implements these optimizations. It maintains a list of references to *active points*, that is, outdoor grid points that have already been reached by the wave. At each step of the simulation, values need only be calculated for active points, and the propagation of outgoing flows is only required from active points to their neighbors.

---

[i] For a full description of ParFlow++ see the paper *'Object-Oriented Parallel Software for Radio Wave Propagation Simulation in Urban Environment'*, also published in these proceedings.

The simulation zone is split into horizontal strips called partitions, which are then allocated to processors based on a round-robin policy. Communications take place at the borders between adjacent partitions, where wave flows and activation status are propagated. Balancing of the workload is achieved by increasing the number of partitions allocated per processor. Yet, increasing the number of partitions increases the communication costs. Therefore, a trade-off must be found in the number of partitions per processor. Predicting the optimal number of partitions is one of the goals of the performance model developed in the next section.

3 Performance model

Our performance model focuses on the behavior of the algorithm while performing the actual wave propagation simulation. The computation time of one iteration is assumed to be that needed by the most heavily loaded processor, and the computation workload of a processor is assumed to be proportional to the number of active points handled by this processor. This number is not necessarily identical for each processor and increases at each iteration step. The cost of the communication itself is assumed to be proportional to the number of partition borders handled by one processor. This cost is constant over the iterations since the volume of data transferred only depends on the number of points per partition border.

To model the behavior of ParFlow++ on a new parallel architecture, only two measurements are required: the computation time for each active point, and the communication time per partition border. The computation time per active point is obtained by timing one iteration of a sequential ParFlow++ execution, when all grid-points are active. The communication time is obtained by timing the point-to-point communication time of a message of the same size as a partition border. Contention is neglected in a first approximation. The measured communication time includes the cumulated time taken by the PVM communication routine, by the procedures responsible for packing and unpacking the data contained in a border, and finally by the call to the encapsulating C++ methods.

To model local computation, we evaluate the number of active points handled by a processor at a given time as a function of the number of processors and of the number of partitions per processor. The number of active points is first predicted for a void area (i.e., without buildings). When buildings are present the actual number of active points is obtained by subtracting all indoor points, since they are not processed in the ParFlow++ implementation. To avoid considering the actual location of buildings on the grid, we assume that they are uniformly distributed.

4 Results

The ParFlow++ performance model was validated on the Cray T3D multi-processor architecture. Figure 1(a) compares the measured and the theoretical computation times for a 16-processors system working on a 1000 × 1000 simulation zone (namely, a district of the city of Geneva), with one partition per processor. Communication times are not considered here. Each curve shows the computation time spent by one of the processors as a function of the iteration step i.

Figure 1(a) confirms that the model fits quite well to the actual results. Discrepancies between them show the impact of the non uniform distribution of buildings.

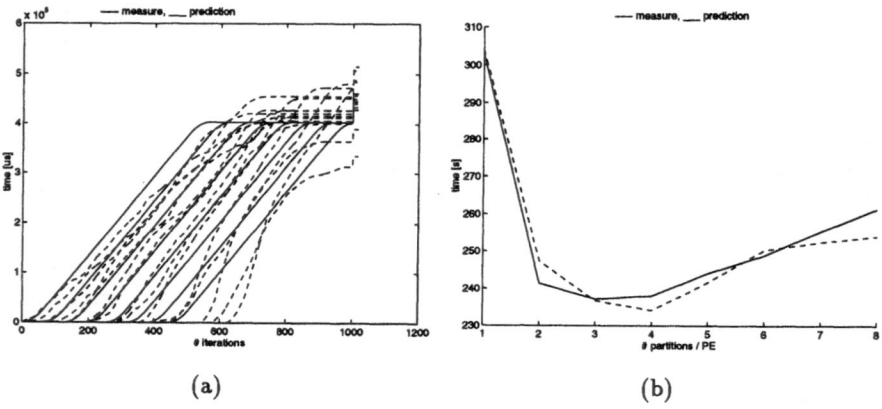

Fig. 1. Workload distribution (a) and execution times (b) of ParFlow++ running on 16 processors. Dashed curves show actual measurements and plain curves show predicted values.

Although the load of individual processors is predicted only imperfectly, this prediction can still be used very efficiently in the estimation of the total execution time. Figure 1(b) shows the predicted and measured execution times for the district of the Geneva city, as a function of the number of partitions per processor. The communication time is taken into account. Both the predictions and the measurements clearly show the trade-off between a good load-balancing and reduced communication costs. The optimal partitioning for this problem size is actually four partitions per processor, while the prediction indicates three partitions. The performance difference between the predicted and measured option is quite small. The discrepancies between the two curves are probably due to the irregularity of the distribution of buildings. Without *a priori* information on building locations, three partitions per processors would actually be the best bet.

The evolution of the measured and predicted speedups for the same area was also compared, as a function of the number of processors used. In this case, only one partition per processor was used. The quality of the prediction was very satisfactory: prediction error varied between 0 and 6%.

5 Conclusion

The performance model for a parallel, irregular application presented is based on the BSP model [4], known for its ability to represent regular algorithms. It was tested on an irregular simulation problem and was able to provide a valid prediction for executions on the Cray T3D. It can be used for scalability analyses, such as speedup prediction, or to seek optimal trade-offs in issues such as data partitioning. Prediction relies on only a few basic measurements: the timing of point-to-point communications and of elementary sequential computations.

Future work will include the modeling of other parallel irregular algorithm with a similar methodology. Parallel optimization algorithms like branch & bound, tabou search and evolutionary algorithms are good candidates for such investigations. On other multi-processor architectures (distributed memory machines such as the Intel Paragon, shared memory architectures like the SGI Origin 2000, and networks of workstations) performance prediction will also be investigated to check the generality of the modeling method. In the long term, our objective is the development of a generic prediction tool, that should process annotated parallel irregular programs automatically and predict their performances on any target parallel platform.

References

1. T. Fahringer. *Automatic Performance Prediction for Parallel Programs on Massively Parallel Computers*. PhD thesis, Technical University Vienna, Austria, 1993.
2. A. Gupta and V. Kumar. Analyzing scalability of parallel algorithms and architectures. *Journal of Parallel and Distributed Computing*, 22(3):379–391, Sept. 1994.
3. W. Kuhn. Performance prediction and benchmarking results from the ALPSTONE project. In *Proceedings of the International Conference and Exhibition on High-Performance Computing and Networking (HPCN Europe'96)*, number 1067 in Lecture Notes in Computer Science, pages 763–769, Brussels, Belgium, Apr. 1996. Springer Verlag.
4. L. G. Valiant. Bulk-synchronous parallel computers. In M. Reeve and S. E. Zenith, editors, *Parallel Processing and Artificial Intelligence*, Chichester, UK, 1989. Wiley.

Bounding the Minimal Completion Time of Static Mappings of Multithreaded Solaris Programs

Lars Lundberg

Department of Computer Science, University of Karlskrona/Ronneby,
Soft Center, S-372 25 Ronneby, Sweden, email: Lars.Lundberg@ide.hk-r.se

1 Introduction

A multithreaded Solaris program can be executed in parallel on a multiprocessor. Previous experience show that, using the default scheduling algorithm, threads are frequently relocated from one processor to another [5]. After each such relocation, the code and data associated with the relocated thread have to be moved to the cache of the new processor. In order to avoid this problem, one can map threads to processors using the *processor_bind* directive [2]. The major problem with such static mappings is that one can easily end up with an unbalanced load. The problem of finding a mapping of threads to processors which results in minimum completion time is NP-hard [1]. It is even difficult to determine if a certain mapping is close the optimal case or if it is worth-while to look for other mappings.

Previous results [4] show that, based on certain information about the program, one can obtain a tight bound on the minimal completion time using static mappings, i.e. it is always possible to find a mapping with a completion time less than or equal to the bound. This makes it possible to determine if a certain mapping is close to the optimal case or if it is worth-while to look for other mappings. In this paper, we present a set of tools which make it possible to obtain such a bound, using an ordinary uni-processor workstation.

2 Method Overview

Figure 1 shows the steps used for bounding the minimum completion time of a static mapping of a multithreaded Solaris program P using a multiprocessor with k processors. Bold boxes indicate the parts developed by us. The routines in the Solaris thread library are overloaded with an instrumented thread library. For each call to a thread routine, a number of values are recorded, e.g. the identity of the thread making the call, the identity of the routine (e.g. *thr_create, thr_exit* and *sema_wait*) and the local process time when the call is made.

In order to bound the minimum completion time for a multiprocessor with k processors, we use an ordinary uni-processor workstation. The uni-processor execution of the multithreaded program, using the instrumented thread library, results in a list of recorded values. This list is then restructured into the dynamic program behaviour format, which is the information needed in order to calculate the bound. The bound calculation tool takes the dynamic program behaviour of the monitored program and the number of processors (k) in the multiprocessor system for which we are going to calculate the performance bound.

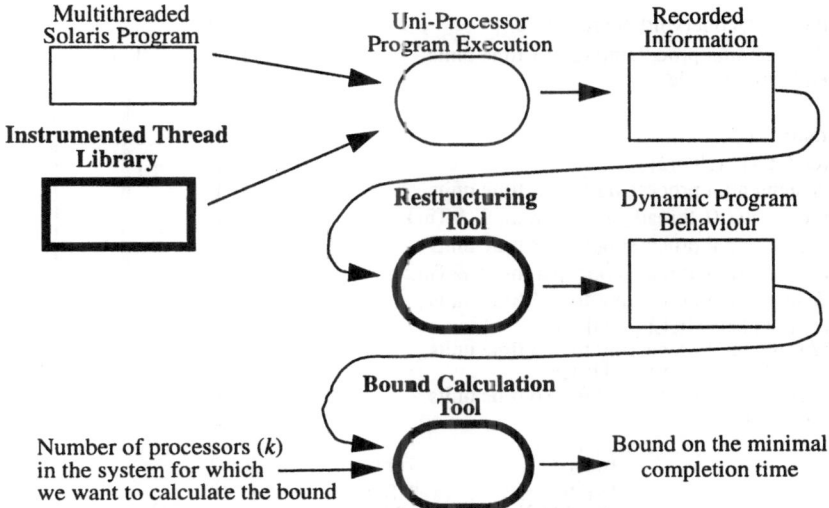

Figure 1: The tools which enable us to calculate a bound on the minimal completion time of any static mapping of a multithreaded Solaris program.

3 Calculating the Bound

The upper left part of figure 2 shows a multithreaded Solaris program, containing three threads: main, Thr1 and Thr2. The upper right part of the figure shows a graphical representation of the execution of this program using one processor for each thread. The lower part of figure 2 shows a textual representation corresponding to the graphical representation; *work(t)* denotes sequential processing for *t* time units. Each thread is represented as a list of synchronization events separated by periods of sequential processing. These lists represent the dynamic program behaviour (see figure 1). The lists correspond to the sequence of events during the monitored execution. Therefore, the lists are completely deterministic. Indeterministic events, e.g. *mutex_trylock* are modelled by the corresponding deterministic events. If a *mutex_trylock* succeeds in the monitored execution, this is modelled as a *mutex_lock*, otherwise the event is not modelled at all.

The table below shows the recorded information generated during the monitored uni-processor execution of the multithreaded program in figure 2.

Process local time	Thread Id	Thread routine	Parameters
T2	main	thr_create	Thr1
T2+T3	main	thr_create	Thr2
T2+T3+T4	main	thr_join	Thr1
T2+T3+T4+T1	Thr1	thr_exit	--
T2+T3+T4+T1+T1	Thr2	thr_exit	--
T2+T3+T4+T1+T1+T5	main	thr_join	Thr2
T2+T3+T4+T1+T1+T5+T6	main	thr_exit	--

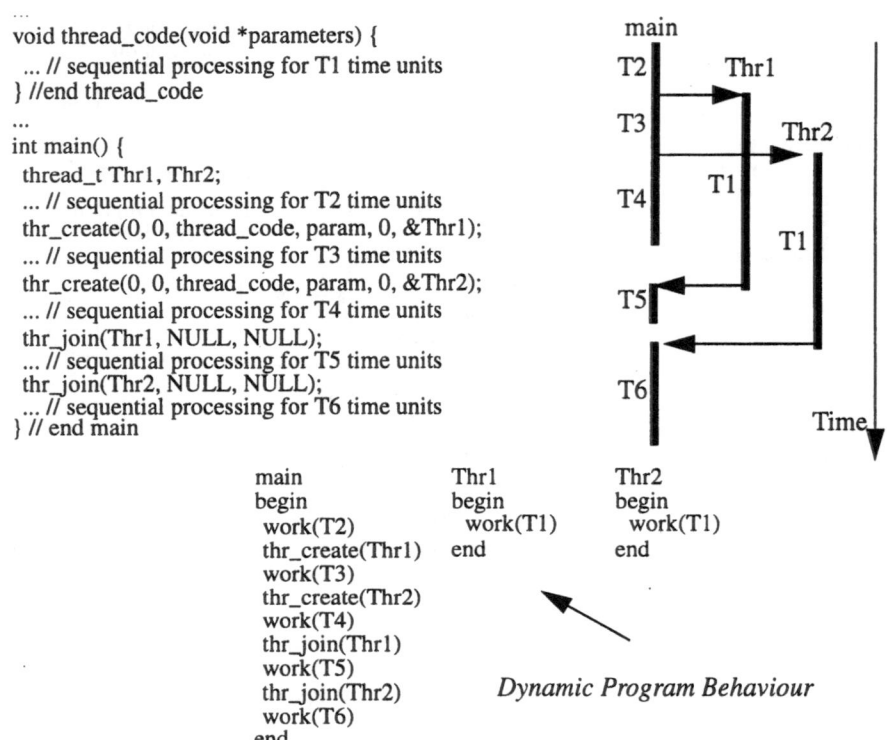

```
void thread_code(void *parameters) {
    ... // sequential processing for T1 time units
} //end thread_code
...
int main() {
    thread_t Thr1, Thr2;
    ... // sequential processing for T2 time units
    thr_create(0, 0, thread_code, param, 0, &Thr1);
    ... // sequential processing for T3 time units
    thr_create(0, 0, thread_code, param, 0, &Thr2);
    ... // sequential processing for T4 time units
    thr_join(Thr1, NULL, NULL);
    ... // sequential processing for T5 time units
    thr_join(Thr2, NULL, NULL);
    ... // sequential processing for T6 time units
} // end main
```

```
main                    Thr1          Thr2
begin                   begin         begin
  work(T2)                work(T1)      work(T1)
  thr_create(Thr1)      end           end
  work(T3)
  thr_create(Thr2)
  work(T4)
  thr_join(Thr1)
  work(T5)
  thr_join(Thr2)              Dynamic Program Behaviour
  work(T6)
end
```

Figure 2: Three representations of a multithreaded Solaris program. Upper part: the source code and a graphical representation of the execution using one processor for each thread. Lower part: a set of lists corresponding to the behaviour of the program.

The monitored execution of the multithreaded program is done on a uni-processor system using one LWP (Light Weight Process) for the entire program. The scheduling of threads to LWPs is non-preemptive, i.e. if there is only one LWP, a running thread cannot be preempted by another thread in the same program. This means that thread switching only occurs at a call to a routine in the thread library, e.g. *thr_exit*, *sema_wait* and *mutex_lock*. We monitor all these calls and record the time when they occur. Therefore, it is possible to restructure the recorded information into the lists (one list for each thread) representing the dynamic program behaviour.

4 Method Example

We are going to bound the completion time of a parallel implementation of an algorithm for generating prime numbers. A number of filters form a line with a number generator feeding numbers into the line. There is a prime number associated with each filter. Each filter filters out numbers which are divisible by its prime number, e.g. the first one filters out all even numbers. If a filter cannot divide a number, that number is forwarded to the next filter. Therefore, prime numbers reach the end of the line. When a prime number reaches the end of the line a new filter is created. The number generator stops by generating the number 3,000,000. The filter chain is cut up into contiguous subchains containing 500 filters each. Each subchain is executed by a Solaris thread. The maximum number of such subchains, which is reached at the end of the execution, is 44, i.e. $n = 44$.

The program was executed on a uni-processor version of a Sun Sparc Center 1000, using the instrumented thread library. Based on recorded values we calculate the bound. Figure 3a shows the bound in the interval ($1 \leq k \leq 8$).

Using the *processor_bind* routine, a simple mapping was implemented. The first $\lfloor n/k \rfloor$ threads are executed by processor zero (thread zero is the first subchain, thread one is the second, and so on). Similarly the next $\lfloor n/k \rfloor$ threads are executed by processor one. The last $n - (k - 1) \times \lfloor n/k \rfloor$ threads are executed by processor k-1. We refer to this as contiguous mapping, since the chain of threads is cut up into k contiguous subchains, and each subchain is mapped to a processor.

Figure 3b shows the completion time using contiguous mapping and the upper bound. The figure shows that the completion time using contiguous mapping is above the upper bound when the number of processors (k) is less than 6. Consequently, in this interval we know that contiguous mapping is not optimal, and it is thus worth-while to look for other mappings.

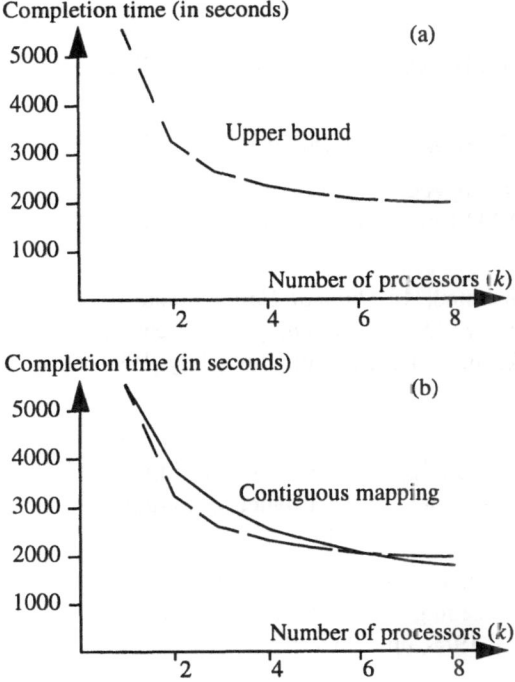

Figure 3: Comparing the completion using contiguous and round-robin mapping with the upper bound.

5 Conclusions

The technique is based on the assumption that the behaviour of the multi-threaded program is (almost) deterministic. There are programs for which this assumption is reasonable, e.g. programs which multiply matrices of a certain size and programs which perform parallel image processing by splitting the image into fixed sized pieces.

The idea of using information from monitored uni-processor executions in order to estimate the multiprocessor performance of a parallel program is not new. The same idea has been used for parallel Ada and Fortran programs [3][6]. However, in these cases the goal was to predict the speedup of a the parallel program, i.e. no performance bounds were calculated in these projects.

The technique and tools described in this paper show that it is possible to integrate a theoretical performance bound in a real parallel programming environment, making the bound accessible to practitioners. The applicability of the result has been demonstrated by comparing the completion time of a real multithreaded Solaris program with the completion time bound. The bound itself, which was obtained in a previous study [4], is optimal in the sense that it cannot be improved within the definition of the problem. The tool used for obtaining the necessary information require no modification of the source code. Except for a minimal recording overhead, the behaviour of the multithreaded program is not affected by the recordings.

References

[1]   M. Garey and D. Johnson, *Computers and Intractability*, W.H. Freeman and Company, 1979.

[2]   B. Lewis and D. J. Berg, *Threads Primer*, Prentice Hall, 1996.

[3]   L. Lundberg, *Predicting the Speedup of Parallel Ada Programs*, in Proceedings of the Ada-Europe 1992 Conference, Amsterdam, June 1992, Springer-Verlag, pp. 257-274.

[4]   L. Lundberg and H. Lennerstad, *An Optimal Upper Bound on the Minimal Completion Time in Distributed Supercomputing*, in Proceedings of the 1994 ACM International Conference on Supercomputing, July, 1994, Manchester, England, pp. 196-203.

[5]   L. Lundberg, *Multiprocessor Performance Evaluation of Billing Gateway Systems for Telecommunication Applications*, in Proceedings of the ISCA International Conference on Parallel and Distributed Computing Systems, Dijon, France, September, 1996, pp. 225-231.

[6]   K. So, A. S. Bolmarich, F. Darema and V. A. Norton, *A Speedup Analyser for Parallel Programs*, in Proceedings of the 1988 International Conference on Parallel Processing, August 1988, pp. 126-129.

Workshop 17

Instruction-Level
Parallelism

Workshop 17: Instruction-Level Parallelism

D. K. Arvind

Research in Instruction-level Parallelism, or ILP for short, is concerned with architectural innovations to expose concurrency between the elemental operations in a processor, and compilation methods which exploit this concurrency in an efficient manner. Modern commercial processors, especially the ones at the high-performance end, have been influenced by this research in a manner which has been largely transparent to the users. And ILP research will continue to play a seminal rôle in defining the computation structures for the billion-transistor processors of the future.

The essence of an ILP architecture is its ability to sustain multiple instructions at any one time. This is achieved by identifying independent instructions and either issuing them simultaneously or in rapid succession. This process can be achieved statically as in the case of Very Large Instruction Word (VLIW) architectures, where the onus is on the compiler to recognise and issue in lockstep unrelated instructions whose types match the concurrent architectural resources. Superscalar architectures achieve this dynamically at run-time through the use of specialised hardware.

A limiting factor for ILP is the serialisation imposed by the control flow in programs, and the data dependences between pairs of instructions. Branch predictors mitigate the effect of control dependences by speculating on the path to be taken at a branch, and provide additional mechanisms for undoing the damage and returning to a consistent state should this choice be later shown to be mistaken.

Collapsing true data dependences is a far more onerous task. Lipasti and Shen propose two variations on their *value prediction* method called *source operand value prediction* and *dependence prediction* which allow instructions to execute before their data dependences are resolved. Their method hinges on the observation of "value locality" in load instruction addresses, and in dependence relationships between dynamic instructions.

Lee et. al. suggest a method for improving the accuracy of multiple branch prediction for superscalar architectures. In this case, the accuracy of prediction of successive branches depends on the unresolved previous branches. Any error in prediction can lead to a cascade of mispredictions. Therefore, multiple branch predictions suffer from interferences by employing a global history scheme. The approach in this paper is to devote separate Branch History Registers and Pattern History Tables to each predicted branch which reduces interferences due to mispredictions.

More accurate multiple branch speculation leads to greater ILP in the programs but at a cost of increased pressure on the hardware resources. Filho and Fernandes have studied the impact of speculation depth on the performance of superscalar processors and identify potential architectural bottlenecks.

The emphasis in the papers so far has been for the architecture to dynamically choose the appropriate instructions to execute in a given cycle. In contrast the following set of three papers rely on the static analysis of the programs to extract a schedule of operations to utilise the concurrent architectural resources efficiently.

Software pipelining is an effective technique for scheduling loops but at a price of increased pressure on the registers. Fernandes et. al. consider a method for alleviating this by replacing part of the register file in a VLIW architecture with queues. In the case of loops, the scheduler must ensure that computations which produce and consume data values do so in the right order through a named queue.

Software pipelining assumes a timing model of the VLIW target when producing the schedules. The quality of these schedules degrades if the timing information is uncertain, as in the case of memory instructions using caches. Ding et. al. employ cache reuse data to inform the scheduler to assign longer latencies to loads which habitually miss, and a default hit latency to the rest of the load instructions.

For schedulers across basic blocks, the degree of ILP in a program is sensitive to the nature and size of groupings of basic blocks. Banergia et. al. suggest partitioning the program into a number of *Treegions*, which are single-entry acyclic regions with multiple paths of control and free of merge points.

The final set of papers deal with the problem of memory latency which will become increasingly significant with higher ILP and relatively slower memory speeds. González and González rely on the observation that the source operands of load/store instructions are highly predictable. This is used to compute the effective address of memory references so that the pipeline can speculatively execute past predicted loads and stores. Chou et. al. present a detailed analysis of simultaneous multithreading in superscalar architectures for realistic workloads. Another approach to attenuate the effects of memory latency is to overlap computation and memory access operations. Jones and Topham present a study into the theoretical limits of *access decoupling* for a model with maximum ILP and perfect dependency analysis.

The Performance Potential of Value and Dependence Prediction

Mikko H. Lipasti and John P. Shen

Department of Electrical and Computer Engineering
Carnegie Mellon University, Pittsburgh PA, 15213

Abstract. The serialization constraints induced by the detection and enforcement of true data dependences have always been regarded as requirements for correct execution. We propose two data-speculative techniques–source operand value prediction and dependence prediction–that can be used to relax these constraints to allow instructions to execute before their data dependences are resolved or even detected. We find that inter-instruction dependences and source operand values are easily predictable. These discoveries minimize the per-cycle instruction throughput (or IPC) penalty of deeper pipelining of instruction dispatch and result in average integer program speedups ranging from 22% to 106%, depending on machine issue width and pipeline depth.

1 Introduction

There are two restrictions that limit the degree of IPC that can be achieved with sequential programs: control flow and data flow. Control flow limits IPC by imposing serialization constraints at forks and joins in a program's control flow graph. Data flow limits IPC by forcing data-dependent pairs of instructions to serialize. Examining the extent of these limits has been a popular and important area of research (e.g. [1][2][3]). However, in light of the energies focused on eliminating control-flow restrictions on parallel instruction issue, surprisingly little attention has been paid to eliminating data-flow restrictions on parallel issue.

In [5], Lipasti et al. introduce the notion of value locality and demonstrate Load Value Prediction, or LVP, for predicting the results of load instructions at dispatch. In [5], they generalize the LVP approach to all instructions to allow one to exceed the classical dataflow limit.

Detecting data dependences between multiple instructions in flight is an inherently sequential task that becomes very expensive combinatorially as the number of concurrent in-flight instructions increases. Olukotun et al. argue convincingly against wide-dispatch superscalars because of this very fact [7]. Wide dispatch is difficult to implement and has adverse impact on cycle time because all instructions in a dispatch group must be simultaneously cross-checked.

One obvious solution to the problem of the complexity of dependence detection is to pipeline it to minimize impact on cycle time. In Section 3 we propose a pipelined approach to dependence detection that facilitates implementation of wide instruction dispatch. However, pipelined dependence checking aggravates the cost of branch mispredictions by delaying resolution of mispredicted

Table 1. Benchmark set

Benchmark	Run Length	BHT Mispred	BTB Mispred	RAS Mispred
go	79.6M	12.0 % (1.7%)	8.5% (0.8%)	0.0% (0.0%)
m88ksim	107.0M	2.7 % (0.4%)	4.3% (0.5%)	0.0% (0.0%)
gcc	181.8M	5.1 % (0.6%)	8.7% (1.1%)	3.9% (0.0%)
compress	39.7M	5.9 % (0.7%)	0.0% (0.0%)	0.0% (0.0%)
li	56.8M	3.0 % (0.4%)	5.3% (0.5%)	12.1% (0.3%)
ijpeg	92.1M	2.7 % (0.4%)	0.6% (0.1%)	18.7% (0.1%)
perl	50.1M	2.4 % (0.3%)	11.1% (1.2%)	4.1% (0.1%)
vortex	153.1M	0.6 % (0.1%)	1.6% (0.2%)	11.4% (0.1%)

branches. In Figure 1, we see the IPC impact of pipelined dependence checking on a 16-dispatch machine with an advanced branch predictor and no other structural resource limitations (refer to Section 2 for further details). Lengthening dispatch to two or three pipeline stages severely increases the number of cycles during which no useful instructions are dispatched and increases CPI (decreases IPC) dramatically, to the point where sustaining even 2-3 IPC becomes very difficult.

We propose to alleviate these problems in two ways: by introducing a scalable and speculative approach to dependence detection called dependence prediction and also by exploiting a modified approach to value prediction called source operand value prediction [6]. Fundamental to these is the notion that maintaining semantic correctness does not require rigorous enforcement of source-to-sink data-flow relationships or even exact detection of these relationships before we start execution. Rather, dynamically adaptive techniques for predicting values as well as dependences allow early issue of instructions, before their dependences are resolved or even known. We find that the dependence relationships between instructions are quite predictable, and propose dependence prediction for capturing and exploiting this value locality to allow early issue of instructions in wide-dispatch machines. Furthermore, we find that combining value and dependence prediction leads to significant performance increases.

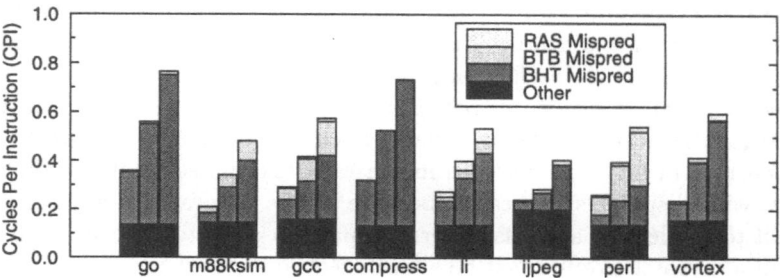

Fig. 1. Branch Misprediction Penalty. The approximate contribution of RAS, BTB, and BHT mispredictions to overall CPI is shown for single-cycle dispatch (left bar), 2-cycle (middle bar) and 3-cycle (right bar) pipelined dispatch.

2 Experimental Framework

To evaluate the performance potential of the dependence prediction and source operand value prediction, we implement a flexible emulation-based simulation framework for the PowerPC instruction set. The simulation framework is built around the PSIM PowerPC functional emulator that is distributed as part of the GDB debugger, and accurately models branch and fetch prediction, dispatch width constraints, and all branch misprediction and data dependence delays for realistic instruction latencies.

We selected the SPEC95 integer benchmark suite for our study, since it is readily available, widely used, and well-understood. Table 1 shows the run length and BHT (branch history table), BTB (branch target buffer), and RAS (return address stack) branch misprediction rates. The BHT, BTB, and RAS misprediction rates are shown with respect to total number of predictions and total number of completed instructions. The benchmarks are compiled for PowerPC GCC version 2.7.2 at full optimization. PSIM emulates user-state and NetBSD library code, but does not account for supervisor-state execution. All of the benchmarks are run to completion with reduced input sets and/or fewer iterations than in the SPEC95 reference runs.

The machine model used in our simulations has a canonical four-stage pipeline: fetch, dispatch, execute, and complete. The width of the fetch and dispatch stages can be varied arbitrarily, while the execute stage has unlimited width. The latency of the dispatch stage can also be varied from one to three cycles, while the latency of the execute stage is instruction-dependent and is summarized in Table 2. All functional units are pipelined, all architected registers are dynamically renamed, and instructions are allowed to execute out-of-order subject to data dependences. However, branches are executed in program order, and loads must wait until the addresses of all preceding stores are known. If an alias to a store exists, the load is delayed until the store's data becomes available and is forwarded directly to the load (in effect, the processor renames memory).

Our model uses a modern *gshare* branch predictor [4] with a 256K entry branch history table (BHT) with 2 bits per entry that is indexed by the XOR of the 18-bit branch history register and the branch instruction address. The RAS and the untagged, direct-mapped BTB both have 1024 entries. Fetch and branch prediction both occur during the fetch stage of the pipeline, while instructions are fetched from a dual-banked instruction cache with line size equal to the

Table 2. Instruction Latencies

Instruction Class	Issue Latency	Result Latency
Integer Arithmetic and Logical	1	1
Integer Multiply	1	3
Integer Divide	1	10
Load/Store	1	2
Branch(pred/mispred)	1	0/1

Table 3. Machine Model Parameters

Parameter	Value
Branch Predictor	3-wide gshare(16)
Fetch and Dispatch Width	4,8,16
Completion Width and Instruction Window	Unrestricted
Instruction and Data Cache	Perfect

specified fetch width (this configuration is described as interleaved sequential in [8]). Up to three conditional branches can be predicted per cycle with the gshare predictor. The key parameters for the machine model used in our simulations are summarized in Table 3.

Additional simulation results for a variety of machine models as well as floating point benchmarks were omitted from this paper, but are available in [9].

3   Pipelined Dispatch Structure

In this section, we describe a pipelined dispatch structure that facilitates wide dispatch by reducing the circuit complexity and cycle-time demands imposed by simultaneous cross-checking of data dependences within a large dispatch group. In this scheme, dependence checking is divided into two pipeline stages. During the first stage, all destination registers within a dispatch group are identified and renamed, and the rename mappings are written into the dependence-resolution buffer (DRB) and the mapping file (MF). During the second stage, all source registers are identified and their rename mappings are looked up in the DRB and the shadow mapping file (MF'). In our microarchitecture, all register writes are renamed to slots in a value silo. The value silo is used to scoreboard, hold, and forward the results of instructions until they are ready to complete and write back into the architected register file.

As shown in Figure 2, during the first pipeline stage P1 of pipelined dispatch, all instructions in a fetch group allocate value silo slots for their destination operands, and then write the (register number, value silo slot) mapping tuples into their dependence-resolution buffer (DRB) entries. At the same time, the value silo slot numbers are written into the mapping file (MF), a table indexed

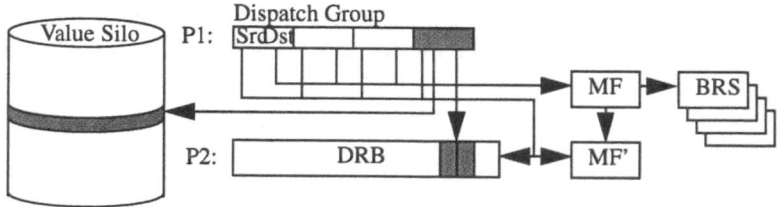

Fig. 2. Pipelined Dispatch Structure. During stage P1, all instructions in a fetch group write destination register rename mappings into the DRB and the MF. During stage P2, the instructions search the DRB and MF' for source register rename mappings.

by the register number. If a dispatch group contains more than one write to the same architected register, arbitration logic selects the last write before a taken or predicted-taken branch. During the second pipeline stage P2, all instructions in a fetch group search ahead in the DRB for a register number matching each of their input registers (the DRB is multi-ported and content-addressable). If multiple matching entries are found, the closest one (i.e. the most recent definition) is selected. If no matching entry is found, the shadow mapping file (MF') entry for the register is used instead. MF' summarizes the register-to-value silo mappings for all previous fetch groups, and is a one-cycle-delayed copy of the mapping file MF. If no register-to-value-silo mapping exists, the appropriate MF' entry will instead point to the architected register file. At the end of P2, all the instructions in the fetch group know where in the value silo they can find their input operands, and can check the scoreboarded valid bits to see if they are available.

Whenever a predicted branch occurs within a dispatch group, a snapshot of the mapping file MF that includes all register writes through the branch is pushed onto a branch recovery stack (BRS). Any instruction following a taken or predicted-taken branch within a fetch group is discarded and prevented from writing into either the DRB or the MF. When a branch misprediction is resolved, any instructions that are newer than the branch are discarded along with their value silo slots, and fetching starts over from the actual destination of the mispredicted branch, while the MF snapshot corresponding to that branch is retrieved from the branch recovery stack.

As described here, instruction dispatch is pipelined into two stages. However, it is easy to envision even deeper pipelining of this process. Hence, we simulate the performance effects of and present results for one-, two-, and three-stage dispatch pipelines.

4 Dependence Prediction and Recovery

Figure 1 illustrates the detrimental performance effects of a pipelined dispatch structure, which increases the number of cycles between a branch misprediction and the detection of that misprediction, hence aggravating the misprediction penalty and severely limiting performance. To alleviate these effects, we propose

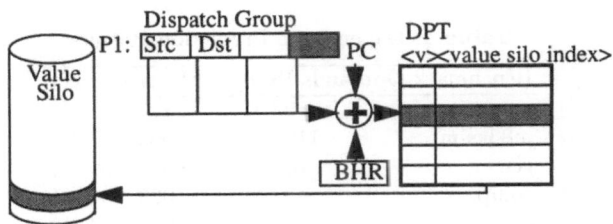

Fig. 3. Dependence Prediction Mechanism. During stage P1, the source operand position, PC, and branch history register (BHR) are hashed together to index the dependence prediction table (DPT), which predicts the value silo entry that contains the source operand. During P2 the prediction is verified.

dependence prediction that can frequently short-circuit multi-cycle dispatch by predicting the dependences between instructions in flight and speculatively allowing instructions that are predicted to be data ready to execute in parallel with exact dependence checking.

As shown in Figure 3, dependence prediction is implemented with a direct-mapped dependence prediction table (DPT) with 8K entries indexed by hashing together the instruction address bits, the gshare branch predictor's branch history register (BHR), and the relative position of the operand (i.e. first, second, or third) being looked up. Each DPT entry contains a numeric value which reflects the relative index of that input operand's source in the value silo. This relative index is used to check the value silo to see if the operand is already available. If all of the instruction's predicted input operands are available, the instruction is permitted to issue early, after the first dispatch cycle. In the second (or third, in the three-cycle dispatch pipeline) dispatch cycle, exact dependence information becomes available, and the earlier prediction is verified against the actual information. In case of a mismatch, the DPT entry is replaced with the correct relative position, and the early issue is cancelled.

The total number of operands predicted, average number of predictions per instruction, and percentage of correct predictions are shown in Table 4. We find that for most benchmarks, the DPT achieves a respectable hit rate. For two benchmarks–go and perl–the dependence prediction hit rates were rather low. This behavior can be attributed to the unpredictable branch behavior of these three benchmarks, since unpredictable branches can lead to unpredictable dependence distances when there are multiple definitions reaching a use. As seen in Figure 1 and Table 1, both go and perl have high BTB misprediction rates, while go has a high BHT misprediction rates.

In Figure 5 we show the effect of dependence prediction on IPC for dispatch widths of four, eight, and sixteen, and dispatch latencies of one, two and three cycles. Without dependence prediction, the best performance is obtained with single-cycle dispatch, which sustains about 2.8 IPC in the worst case (go), and 4.8 IPC in the best case (m88ksim) with 16-wide dispatch. Lengthening dispatch to two and three cycles degrades go to asymptotic IPC of 1.7 and 1.3, respectively, while reducing ijpeg (which is now the best performer) to 3.7 and 2.5

Table 4. Dependence Prediction Results

Benchmark	Operands Pred	Per Instr	Correct
go	89.6M	1.126	38.0%
m88ksim	113.4M	1.060	77.4%
gcc	74.2M	0.958	59.7%
compress	40.6M	1.023	87.3%
li	49.8M	0.878	72.4%
ijpeg	92.3M	1.001	71.9%
perl	47.3M	0.944	48.2%
vortex	120.7M	0.788	71.3%

IPC. Furthermore, wider dispatch provides rapidly diminishing returns, hence eroding incentive for building processors with dispatch widths exceeding four. Fortunately, dependence prediction is able to alleviate these trends by reducing the average dispatch latency. For both two- and three-cycle dispatch, dependence prediction significantly elevates the sustainable IPC and brings it much closer to the single-cycle case. Furthermore, wider dispatch again harvests greater IPC, restoring incentive for building wider superscalar processors. Three benchmarks–compress, li, and ijpeg–behave particularly well, eliminating nearly all of the performance penalty induced by two- and three-cycle dispatch.

5 Source Operand Value Prediction and Recovery

A complementary approach for reducing the adverse performance impact of pipelined dispatch involves a variation on previous work on value prediction [6]. In earlier work, the destination operands (i.e. results) of instructions were predicted via table-lookup at fetch/dispatch, and then speculatively forwarded directly to dependent instructions. The shortcoming of this approach is that dependence relationships must be detected before values can be forwarded to dependent instructions. To overcome this problem, we propose predicting the values of source operands, rather than destination operands, hence decoupling value-speculative instruction dispatch entirely from dependence detection. As in the earlier work, we predict only floating-point and general-purpose register operands, and not condition registers or special-purpose registers.

Source operand value prediction is illustrated in Figure 4. As in [6], we use a value prediction table (VPT) to keep track of past operand values, and exploit the value locality [5] of operands to predict future values. In our experiments, the VPT is direct-mapped, 32KB in size, and is indexed by hashing together the instruction address bits and the relative position of the operand (i.e. first, second, or third) being looked up. Source operand value prediction also uses a direct-mapped classification table (CT) similar to the one proposed in [6] for classifying the predictability of source operands and deciding whether or not the operands should be predicted. In our experiments, the CT is direct-mapped, has 8K entries with a 2-bit saturating counter at each entry, and is indexed by hashing together

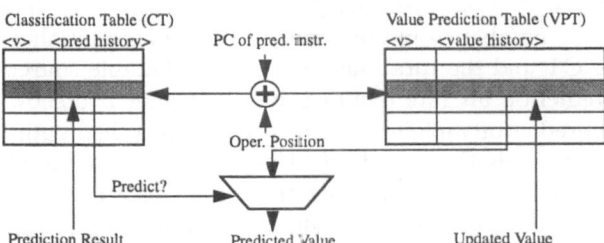

Fig. 4. Source Operand Value Prediction Mechanism. The source operand position and PC are hashed together to index the VPT and CT. The prediction and value histories are updated at completion.

Table 5. Source Operand Value Prediction Results

Benchmark	Value Locality	CT Pred Hit Rate	CT Unpred Hit Rate	Dep Pred Hit Rate
go	45.3%	77.0%	83.7%	42.1%
m88ksim	56.1%	92.8%	89.6%	89.6%
gcc	40.9%	78.0%	89.6%	63.0%
compress	42.4%	97.5%	98.8%	94.6%
li	33.7%	76.9%	92.9%	75.4%
ijpeg	35.2%	91.6%	95.9%	81.2%
perl	44.5%	76.4%	84.3%	54.4%
vortex	32.9%	83.3%	93.9%	82.5%

the instruction address bits and the relative position of the operand being looked up.

When all of the input operands of an instruction are classified as predictable, the instruction is permitted to issue early, after the first dispatch cycle (instructions with unpredictable source operands may still end up executing sooner than without value prediction, in cases where an operand that is predicted is on a critical path). Once dispatch finishes and exact dependence information becomes available, the instruction waits for its verified operands to become available in the value silo (operands in the value silo become verified when the instructions that generate them have validated all of their input operands) and then compares them against its predicted operands. If they match, the result operands of the instruction are marked verified, and the instruction is allowed to complete in program order. If they don't match, the instruction re-executes with the correct operands. Just as in [6], this results in a one-cycle misprediction penalty, since the instruction in question as well as all of its dependents do not execute with their correct inputs until one cycle later than if there had been no prediction.

Table 5 summarizes the value locality, classification hit rates, and dependence prediction hit rates for each of our benchmarks. Value locality (column two), as defined in [5], is the ratio of the dynamic count of source operands that are predictable with the VPT mechanism and the dynamic count of all source operands. The predictable hit rate (column three) is the ratio of the number of predictable source operands that were identified as such by the CT and the total number of predictable source operands. Similarly, the unpredictable hit rate (column four) is the ratio of the number of unpredictable source operands that were identified as such by the CT and the total number of unpredictable source operands. The dependence prediction hit rate (column five) is included to show the interaction between value prediction and dependence prediction. When both types of prediction are used, operands that are deemed unpredictable by the CT are relegated to dependence prediction. We see that the dependence prediction hit rates are better across the board than the ones shown in Table 4, indicating that the techniques are mutually synergistic. We also note that the value locality numbers are similar to those reported earlier [6], while the CT hit rates are somewhat better.

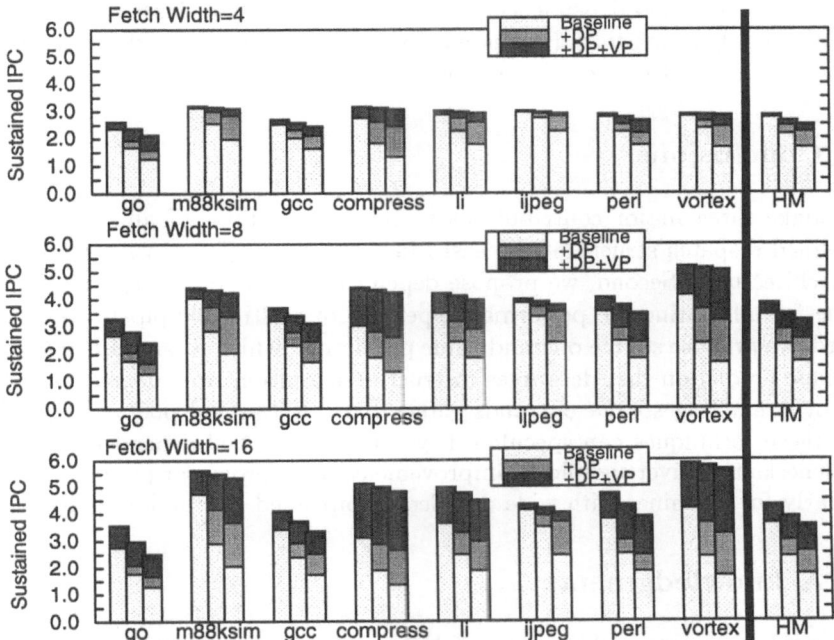

Fig. 5. Effect of Dependence and Value Prediction. The sustained IPC for dispatch widths of 4, 8 and 16 is shown for single-cycle dispatch (left bar), two-cycle dispatch (middle bar), and three-cycle dispatch (right bar). Each stacked bar shows cumulative IPC attainable with dependence prediction (+DP) and value prediction (+DP+VP).

The former is not surprising, since source operands should be no more or less predictable than destination operands, while we attribute the latter improvement to the larger CT size used in these experiments.

In Figure 5 we show the effect of dependence prediction and value prediction on IPC for various dispatch widths of four, eight, and sixteen, and dispatch latencies of one, two and three cycles. The best performance, obviously, is obtained with value prediction and single-cycle dispatch, which sustains 3.6 IPC in the worst case (go), and 5.9 IPC in the best case (vortex) with 16-wide dispatch. Lengthening dispatch latency to two and three cycles degrades go to 3.0 and 2.8 IPC, respectively, while reducing vortex to 5.8 and 5.6 IPC. Value and dependence prediction improve performance significantly over the baseline in all cases, and wider dispatch harvests even greater additional IPC, restoring incentive for building wide-dispatch processors.

We see that with dependence and value prediction, virtually all of the performance penalty associated with pipelined dispatch has been eliminated, allowing even three-cycle dispatch to nearly match the performance of single-cycle dispatch. Even the worst case benchmark (go) only degrades by 17% from single-cycle to two-cycle dispatch, while the best case (vortex) degrades by only 2% for two-cycle and 4% for three-cycle dispatch. Furthermore, three-cycle dispatch with value and dependence prediction can usually at least match, and frequently clearly outperform (compress, vortex, m88ksim, li), single-cycle dispatch without value or dependence prediction.

6 Conclusions

We make three major contributions in this paper. First of all, we propose a pipelined dispatch structure that eases the implementation of wide-dispatch microarchitectures. Second, we propose dependence prediction, a speculative technique for alleviating the performance penalty of multi-cycle pipelined dispatch. Third, we propose source operand value prediction, which is a modified approach to value prediction that decouples instruction execution from dependence checking by predicting source operands rather than destination operands. We show that these techniques can speculate beyond data-flow and dependence detection bottlenecks to deliver significant improvements in uniprocessor performance, particularly for machines with wide and deeply pipelined instruction dispatch.

7 Acknowledgments

This work was supported in part by ONR grant N00014-96-1-0928. We gratefully acknowledge the generosity of the Intel Corporation for donating numerous fast Pentium Pro-based workstations for our use. We also wish to thank the authors of the PSIM functional emulator for their generosity in making this tool publicly available.

References

1. Riseman, E., Foster, C.: The inhibition of potential parallelism by conditional jumps. IEEE Transactions on Computers, pages 1405-1411, December 1972.
2. Lam, M., Wilson, R.: Limits of control flow on parallelism. In Proceedings of ISCA-19, pages 46-57, June 1992.
3. Sazeides, Y., Vassiliadis, S., Smith, J.E.: The performance potential of data dependence speculation & collapsing. In Proceedings of MICRO-29, December 1996.
4. McFarling, S.: Combining branch predictors. Technical Report TN-36, Digital Equipment Corp, June 1993.
5. Lipasti, M.H., Wilkerson, C., Shen, J.P.: Value locality and load value prediction. In Proceedings of ASPLOS-VII, October 1996.
6. Lipasti., M.H., Shen, J.P.: Exceeding the dataflow limit via value prediction. In Proceedings of MICRO-29, December 1996.
7. Olukotun, K., Nayfeh, B.A., Hammond, L., Wilson, K., Chang,K.: The case for a single-chip multiprocessor. In Proceedings of ASPLOS-VII, October 1996.
8. Conte, T., Menezes, K., Mills, M., Patel, B.: Optimization of instruction fetch mechanisms for high issue rates. In Proceedings of ISCA-22, pages 333-344, June 1995.
9. Lipasti, M.H., Shen, J.P.: Approaching 10 IPC via superspeculation. Technical Report CMU-MIG-1, Carnegie Mellon University, 1997.

An Enhanced Two-Level Adaptive Multiple Branch Prediction for Superscalar Processors

Jong-bok Lee, Wonyong Sung and Soo-Mook Moon

School of Electrical Engineering
Seoul National University
Seoul 151-742, Korea

Abstract. We propose an enhanced multiple branch predictor using per-primary address branch histories. Using this scheme, the interferences among different branches are reduced, enhancing the average prediction accuracy. Also, since the branches predicted each cycle do not suffer from successive dependencies, predictions are generated in parallel. This scheme results in higher average branch prediction accuracy than the previous global history scheme under the same implementation cost. For two branch predictions, the prediction accuracy of integer benchmarks varies between 92.0 and 96.9 percent. For floating point benchmarks including *nasa*7, the accuracy is between 94.8 and 95.8 percent.

1 Introduction

In order to increase the fetch bandwidth of a superscalar processor, we need to predict more than one branch and to fetch multiple non-consecutive basic blocks in a single cycle. Yeh and Patt developed *Two-Level Adaptive Branch Prediction* and extended it to predicting multiple branches per cycle [1]. Several variations of the scheme have been introduced, yet all of them use the global history register in common. The idea of the global history register makes each branch to share the same space for storing its prediction history, thus causing interferences. In addition, branches which need to be simultaneously predicted each cycle have successive dependencies in the mechanism of prediction, respectively. Hence, the prediction accuracy is gradually degraded as the number of simultaneously predicted branches is increased.

In order to overcome these shortcomings, this paper proposes an enhanced *Two-Level Adaptive Multiple Branch Prediction using a Per-primary Address Branch History Table*. Only those branches that are predicted simultaneously share the same space, thus reducing interferences. Unlike the previous scheme, the predicted values for multiple branches can be generated in parallel at each cycle. Several hardware configurations of the per-primary address history scheme are introduced, simulated, and compared with the previous scheme. The performance is evaluated by conducting an empirical study on a subset of SPEC benchmark suite using the trace-driven simulation. The proposed scheme further enhances the branch prediction accuracy of integer and floating point programs.

The rest of this paper is organized as follows. Section 2 briefly describes *Two-Level Adaptive Branch Prediction* and the previous multiple branch prediction schemes. Section 3 proposes the per-primary address history scheme. Section 4 presents the simulation environments and results. Finally, a conclusion follows in Section 5.

2 Previous Work

2.1 Two-Level Adaptive Branch Prediction

Dynamic branch predictions use run-time execution history. Among them, *Two-Level Adaptive Branch Prediction* is known to present the highest prediction accuracy [2] [3] [4]. *Two-Level Adaptive Branch Prediction* uses two major data structures, the Branch History Register (BHR) and the Pattern History Table (PHT), as shown in Figure 1(a). The BHR is used to record the history of taken

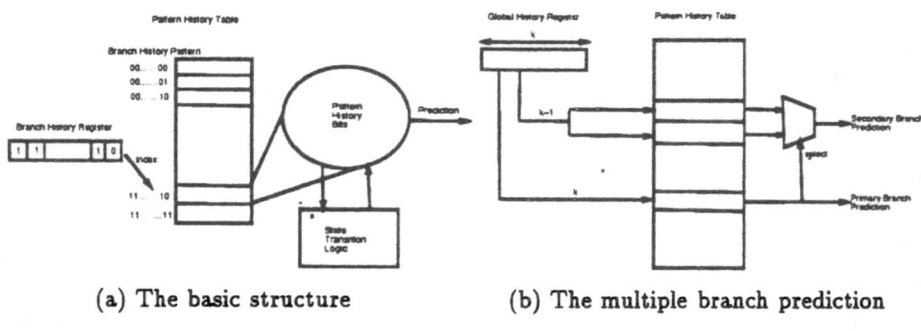

(a) The basic structure   (b) The multiple branch prediction

Fig. 1. Two Level Adaptive Multiple Branch Prediction.

and not-taken branches. For each possible pattern in the BHR, a pattern history is recorded in the PHT. Prediction decision logic interprets the two pattern history bits to make a branch prediction.

2.2 The Previous Multiple Branch Predictions

The multiple branch prediction proposed by Yeh and Patt extends the global history scheme of *Two-Level Adaptive Branch Prediction*. The extended global history scheme makes the prediction of an immediately following branch and also extrapolates the predictions of subsequent branches as shown in Figure 1(b) [1]. Since a global BHR and a global PHT are employed in this scheme, it is called

Two-Level Adaptive Multiple Branch Prediction Using a Global History Register and a Global Pattern History Table (MGAg).

The global history scheme used in Yeh's multiple branch prediction has several disadvantages. The prediction of a branch instruction is interfered by the history of other branches due to the use of the single global history register. Another problem is the mechanism of successive prediction. When two branches are predicted, the prediction of the secondary branch is based on the yet unresolved prediction value of its primary branch. Similarly, for three branch predictions, the prediction of a tertiary branch is based on unresolved predictions of both the primary and the secondary branches. Therefore, a misprediction at the prior stage may produce successive mispredictions at that cycle. Moreover, predicted values of multiple branches only can be generated sequentially by this mechanism. This might affect the cycle time of the processor, since table lookup for the prediction already requires considerable amount of time.

Another method of predicting multiple branches per cycle was proposed, where a tree-like subgraph of the control flow graph was employed [5]. In this scheme, multiple branches are predicted indirectly through the path prediction of a subgraph. Instead of not storing the condensed history of all branches, the subgraph history pattern is stored. However, the reported branch prediction accuracy is not higher than the study of Yeh or proposed multiple branch predictor, which employs 2-bit saturating updown counter for each PHT entry.

3 Multiple Branch Prediction Using a Per-primary Address Branch History Table

In order to reduce interferences in the first level of the branch history information, one history register is associated with each distinct conditional primary branch, which changes the structure of the Branch History Table (BHT). For two branch predictions where the branch history length is k bits, each entry of the new BHT is composed of a BHR with the length of $2k$ bits. In addition, separate PHTs are employed for each primary branch address. Each entry of PHT contains a single 2-bit saturating up-down counter like the original design.

Primary branches are predicted by accessing the BHT and the PHT using the branch address. The first half k bits in the history register are used to index into the PHT. In order to predict the secondary branch, the BHT and the PHT are also accessed by the primary branch address. This time, the second half k bits in the history registers are used to index into the PHT. When three branches are predicted at each cycle, each entry of the BHT is composed of $3k$ bits and last k bits are used for the prediction of the tertiary branch. In summary, since the branch addresses of the secondary and the tertiary branch are not known at the time of prediction, the primary branch address is used for the access of the BHT and the PHT. This multiple predictor is named as *Two-Level Adaptive Multiple Branch Prediction using a Per-primary Address Branch History Table and Per-primary Address Pattern History Tables* (MPAp). Figure 2 depicts the prediction mechanism of the MPAp scheme, in case of two branch predictions.

In the global history scheme, the prediction of a branch is influenced by the

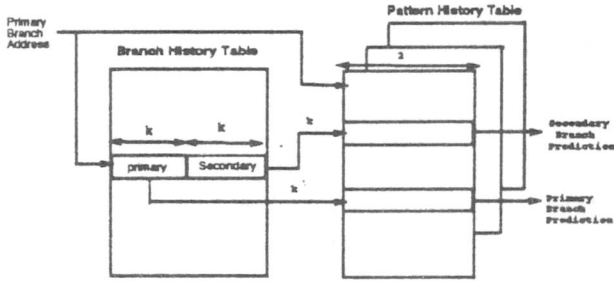

Fig. 2. The MPAp scheme with two branch predictions.

history of other branches since all branch predictions are based on the single global history register. However, only two secondary branches share a single BHR associated with each primary branch address for two branch predictions in our scheme. For three branch predictions, the two secondary branches and the four tertiary branches are shared, respectively. Although small interferences may exist among the branches which share a BHR at the same cycle, this substantially reduces branch interferences than the global history scheme.

Another advantage of this scheme is related to the hardware mechanism of prediction. Unlike the global history scheme, this method independently performs the prediction of multiple branches once the primary branch address is known. Therefore, the predictions of multiple branches can be generated in parallel.

4 Experimental Results

We compare the prediction accuracy of the per-primary address scheme with the previous global history scheme considering various hardware configurations and implementation costs.

4.1 Experimental Environments

We use the trace-driven simulation using ten programs in SPEC benchmarks. Four integer programs are *eqntott, espresso, xlisp*, and *gcc*. Six floating point programs are *nasa7, doduc, spice2g6, tomcatv, matrix300*, and *fpppp*. These programs are compiled by C and Fortran 77, with compiler optimizations turned on. The tracing system is based on SPARCstation 2 [6]. To produce the instruction traces of SPEC benchmark programs, Shadow is used [7]. Each benchmark is traced for 10 million instructions and fed into the multiple branch predictor. In order to obtain the instruction traces uniformly from the wide range of each benchmark, 2 million instructions are sampled for five times to trace up to the first 50 million instructions.

The Branch Address Cache (BAC) has 1024 entries with the set associativity of 4. The configuration of the Branch History Table (BHT) is also 1024-entry and 4-way set associative, utilizing the LRU (Least-Recently-Used) algorithm for the replacement policy. The selected BHR length ranges from 4 to 14 bits. Finite number of tables selected from 1 to 256 are employed for the PHT. We allowed a maximum number of 16 instructions to be fetched from the instruction cache per cycle. When a basic block is large as in most floating point benchmarks, the percentage of performing multiple predictions is low.

The previous multiple branch predictor to compare is *Two-Level Adaptive Multiple Branch Prediction using a Global Branch History Register and Per-primary Address Pattern History Tables* (MGAp) [1]. For the clarification, the MPAp and the MGAp will be also called the per-primary address history scheme and the global history scheme, respectively.

4.2 Results

Prediction Impact of BHR Lengths and Number of PHTs Figure 3(a) shows the average prediction accuracy of integer programs obtained by using the previous global history and the proposed per-primary address history scheme. Two branches are predicted, and the branch history lengths range from 4 to 14 bits. Each curve shows the prediction accuracy for three different numbers of PHTs, which are 1, 16, and 256.

The performance of the global history scheme is sensitive to the BHR length and the number of history tables. When the BHR length is only 4-bit and a single PHT is employed, the average prediction accuracy is below 78 percent. This scheme requires branch history length over 14 bits or more than 256 PHTs to reduce the interference for effective overall performance. The average prediction accuracy for integer programs ranges from 77.7 to 96.5 percent. On the contrary, the prediction accuracy of the per-primary address history scheme is not as sensitive to the BHR length and the number of PHTs as the global history scheme. Using the 4-bit BHR and a single PHT, the prediction accuracy of the per-primary address history scheme outperforms that of the global history scheme by more than 14 percent. The average prediction accuracy for integer programs ranges from 92.0 to 96.9 percent.

Figure 3(b) shows the average prediction accuracy of floating point programs with the same hardware configuration as above. By the nature of floating point benchmarks, the curves of the global history scheme are less sensitive to the branch history lengths and the number of PHTs than in case of integer programs. The average prediction accuracy of the global history scheme ranges from 87.4 to 95.5 percent. Using a 4-bit BHR and a PHT, the prediction accuracy of the per-primary address history scheme outperforms that of the global history scheme by more than 7 percent. The prediction accuracy ranges from 94.8 to 95.8 percent, which is lower than the results of Yeh and Patt. This is because we

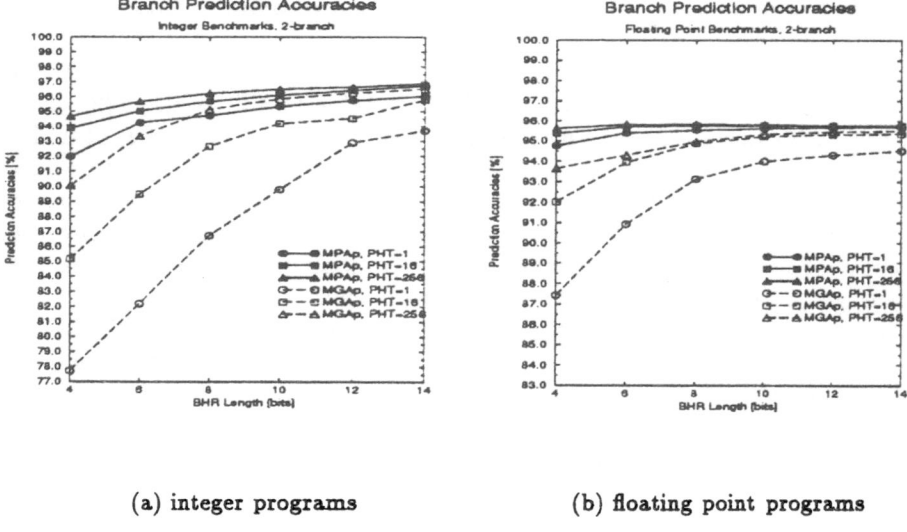

(a) integer programs   (b) floating point programs

Fig. 3. The prediction accuracies of MPAp and MGAP scheme.

have included *nasa7*, which degrades the average performance of floating point benchmarks from around 98 percent down to 95.

The simulation was repeated for three branch predictions per cycle. The prediction accuracy of the global history scheme ranges between 76.2 to 95.6 percent for integer benchmarks. The per-primary address history scheme ranges from 91.1 to 95.6 percent. The prediction accuracy of three branch predictions is less than that of two about one to three percent. Except for the degradation of prediction accuracy, the curves show similar characteristics to the results of two branch predictions.

For floating point benchmarks, the prediction accuracies of the global history scheme ranges from 82.0 to 95.0 percent. The per-primary address history scheme ranges between 94.3 to 95.0 percent. In case of the global history scheme, the maximum degradation in the prediction accuracy between two and three branch predictions is as much as 5.4 percent. For the per-primary address history scheme, the difference is within only 1.3 percent.

Cost Effectiveness The simplified hardware cost estimate functions for the global history (MGAp) and the per-primary address history scheme (MPAp) are described in Table 1. Figure 4(a) shows the comparison of the most effective configuration of the global and the per-primary address history scheme when two branches are predicted with an implementation cost of 128 K bits. For the global

Table 1. Branch predictor configurations and their estimated costs.

b is the number of entries in the BHT.
m is the number of branches (2 or 3) per cycle.

scheme same	BHR length	number of PHTs	hardware cost
MGAp(h,p)	h	p	$h+p\times 2^h\times 2$
MPAp(h,p)	h	p	$b\times h\times m+p\times 2^h\times 2$

history scheme, the most cost-effective configuration is MGAp(8,256). For the per-primary address history scheme, the most cost-effective one is MPAp(10,16). Most of per-primary address history scheme outperforms the global history

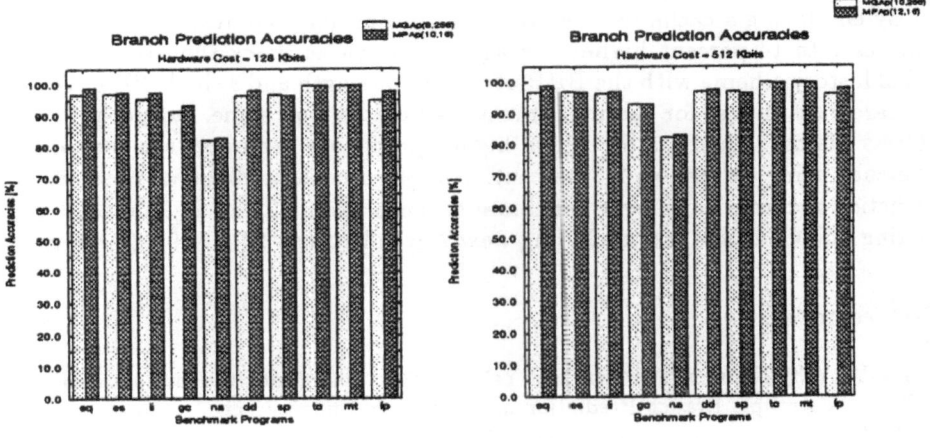

(a) The implementation of 128K bits   (b) The implementation of 512K bits

Fig. 4. Comparison of MGAp and MPAp with the same implementation cost.

scheme. Especially, *eqntott*, *xlisp*, *gcc*, and *fpppp* brings prediction accuracies higher than 2.0 percent. However, the results of *spice2g6* are lower, and *tomcatv* and *matrix300* are comparable. For these floating point benchmarks, the increased number of pattern history tables in the global history scheme substantially reduces the possibility of pattern history interference. The average prediction accuracies of the global history scheme are 95.1 and 94.9 percent for integer and floating point programs, respectively. For the per-primary address history scheme, they are enhanced to 96.1 and 95.7 percent.

Figure 4(b) compares the cost effectiveness of the global and the per-primary address history scheme given a higher hardware budget of 512 K bits for the costs of both BHRs and PHTs. For the global history scheme, the most cost-effective configuration is MGAp(10,256). For the per-primary address history scheme, the most cost-effective one is MPAp(12,16). For the global history scheme, the average prediction accuracies for integer and floating point benchmarks are 95.9 and 95.3 percent, respectively. For the per-primary address history scheme, they are enhanced to 96.4 and 95.7 percent.

5 Conclusions

We have proposed a variation of the multiple branch prediction named as the per-primary address history scheme where the interferences among branches are reduced and the predictions of subsequent branches do not depend on previous unresolved predictions, thus enhancing the average prediction accuracy.

By employing a cache for the BHT, the per-primary address history scheme achieves 7 to 14 percent higher average prediction accuracy than the previous global history scheme with the BHR of only 4-bit length and a single PHT. When the hardware budget for the implementation cost is kept same, our per-primary address history scheme still brings higher prediction accuracy, the maximum difference being as high as 2.7 percent. For two branch predictions, the average prediction accuracy of integer benchmarks ranges from 92.0 to 96.9 percent. For floating point benchmarks including *nasa7*, it is between 94.8 to 95.8 percent.

References

1. Tse-Yu Yeh, Deborah T. Marr, and Yale N. Patt, "Increasing the Instruction Fetch Rate via Multiple Branch Prediction and a Branch Address Cache," in *International Conference on Supercomputing '93*, 1993, pp. 67–76.
2. T-Y Yeh and Y.N. Patt, "Two-Level Adaptive Branch Prediction," in *The 24th ACM/IEEE Intl Sym and Workshop on Microarchitecture*, Nov. 1991, pp. 51–61.
3. T-Y Yeh and Y.N. Patt, "Alternative Implementations of Two-Level Adaptive Branch Prediction," in *Proceedings of the 19th International Symposium on Computer Architecture*, May. 1992, pp. 124–134.
4. K. So S-T Pan and J.T. Rameh, "Improving the Accuracy of Dynamic Branch Prediction Using Branch Correlation," in *Proceedings of the 5th International Conference on Architectural Support for Programming Languages and Operating Systems*, Oct. 1992, pp. 76–84.
5. Simonjit Dutta and Manoj Franklin, "Control Flow Prediction with Tree-Like Subgraphs for Superscalar Processors," in *"Proc. 28th Annual International Symposium on Microarchitecture "*, 1995, pp. 258–263.
6. *The SPARC Architecture Manual*, Prentice-Hall, Inc., 1992.
7. *Introduction to SHADOW*, Sun Microsystems. Inc., Jul. 1989.

The Effect of the Speculation Depth on the Performance of Superscalar Architectures

Eliseu M. C. Filho
Edil S. T. Fernandes
Department of Systems and Computer Engineering
COPPE/Federal University of Rio de Janeiro
P.O. Box 68511 21945-970 Rio de Janeiro,RJ Brazil
e-mail: {eliseu,edil}@cos.ufrj.br

Abstract

Speculative execution is a key concept to increase the performance of superscalar processors. Given accurate branch prediction mechanisms, the efficiency of speculative execution is mainly determined by the speculation depth. In this work, we evaluate the pressure of speculative execution on the resource requirements of a typical superscalar architecture.

1   Introduction

Speculative execution improves processor performance by anticipating the execution of instructions beyond the boundaries of basic blocks. On the other hand, there is a cost imposed by the cancellation of instructions along mispredicted paths. Current dynamic branch prediction mechanisms [1] yield an average accuracy as high as 85%, and there are new predictors that can achieve a near 100% accuracy [2]. Given these accuracy ratios, the cost of mispredictions are significantly reduced, and the performance gain becomes mainly determined by the ability to anticipate the execution of instructions across unresolved branches. This ability is characterized by the *speculation depth* — the number of unresolved branches beyond which the instruction dispatch mechanism stalls. Current superscalar machines exhibit different values for the speculation depth: it is two branches in the PowerPC 604 [3], four branches in the PowerPC 620 [4] and in the R10000 [5] and 16 branches in the SPARC64 [6].

The speculation depth has implications on the architecture's configuration. The resources in the execution unit should be dimensioned to accommodate the flow of dispatched instructions resulting from a certain value of the speculation depth. This is an important factor that should be taken into account to obtain a well-balanced architecture. This work quantifies the demands on the machine's resources for different speculations depths. Section 2 of this paper describes the superscalar model considered in the study. Section 3 shows the experiments to assess the resource requirements. The paper concludes in Section 4 with some remarks.

2   Superscalar Model and Experimental Framework

The superscalar model here adopted is representative of existing superscalar microarchitectures. The model is organized as a pipeline with seven stages: *fetch, predict, decode, dispatch, issue, execute* and *writeback.* The fetch stage transfers instructions from the cache memory into a fetch buffer. The predict stage transfers instructions from this buffer into an instruction queue (i-queue), and predicts the result of conditional branches by using a branch target buffer [1]. The dispatch stage sends decoded instructions, in order, from the i-queue to *reservation stations.* The maximum number of instructions that can be dispatched on a cycle will be referred to as the *dispatch width.* The issue stage schedules instructions dynamically using a scheme based on the Tomasulo algorithm [7]. Finally, the writeback stage sends the results of completed instructions to the register file and to the reservation stations. A reorder buffer and a future register file [8] are employed to support speculative execution. Branch instructions are executed, in order, by a specific branch unit. In the case of a misprediction, the branch unit stalls the dispatch stage, recovers the correct architectural state and then resumes instruction dispatch.

In this work we have used a trace-driven simulator to reproduce the operation of our superscalar model. This simulator accepts programs compiled for the SPARC Version 7 architecture [9]. The traces were generated by a simulator of a four-stage pipelined scalar implementation of the SPARC architecture. The instruction latencies adopted are similar to those in the PowerPC 604. We have used eight integer and six floating-point applications from the SPEC92 and SPEC95 benchmarks as the test programs. The traces used in the experiments cover the execution of 20 million of each test program.

3   Effect on the Resource Requirements

In architectures like the one considered here, the number of reservation stations and the size of the reorder buffer determine the maximum dynamic instruction window size. In order to avoid excessive dispatch stalls, these two architectural parameters should be properly adjusted to the flow of instructions resulting from a certain combination of the speculation depth and dispatch width. In another work [10], we present this balance for various machine configurations. Due to the lack of space, here we present only the results concerning the number of reservation stations, for a machine configuration which dispatches four instructions per cycle, with two integer units and one memory unit.

Instruction dispatch can be stalled due to: 1) all reservation stations are occupied, 2) the reorder buffer is full, 3) the speculation depth is insufficient, 4) a mispredicted branch was found 5) the instruction queue is empty. Dispatch stalls due to an insufficient speculation depth occur whenever the number of pending branches is equal to the speculation depth and a new branch instruction arrives for dispatch. We have measured the individual contribution of each one of these stall components.

In Figure 1 we show the occurrence of dispatch stalls when the number of reservation stations per integer unit (*integer reservation stations*) is changed and the number of reservation stations attached to the memory unit (*memory reservation stations*) is fixed at a large value. Dispatch stalls are expressed as the percentage of *zero-dispatch cycles* within the total cycle count. Each stacked bar gives the percentage of zero-dispatch cycles for a certain number of integer reservation stations. Each group of stacked bars corresponds to a different value of the speculation depth. The percentages are the geometric mean of the individual values presented by the integer test programs.

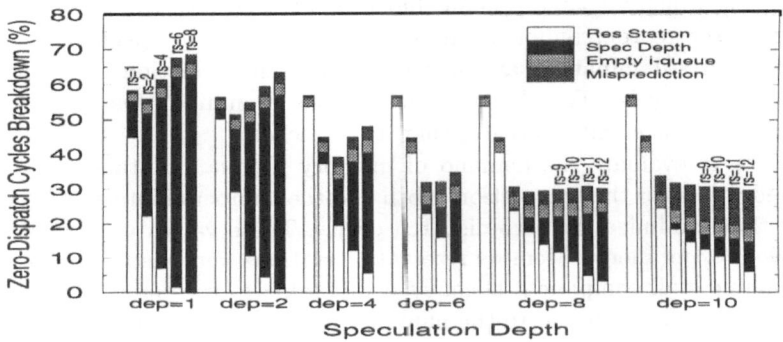

Fig. 1. Varying the number of integer reservation stations

For speculation depths less than or equal to four branches, we observe a high number of dispatch stalls, ranging from 40% to 70% of the total cycle count. The contribution from branch mispredictions does not exceed 4% of the total cycle count, while dispatch stalls caused by an empty i-queue accounts for at most 4%, even for deep speculations.

There is an inverse relationship between the stall components related to the speculation depth and to the number of integer reservation stations. For example, with a four-branch speculation depth, dispatch stalls caused by the lack of integer reservation stations drops from 54% to only 6% when the number of reservation stations is increased. On the other hand, stalls due to an insufficient speculation depth increases from 0.1% to 30%, keeping the total number of stalls still high. This happens because the inclusion of more integer reservation stations increases the number of instructions being dispatched per cycle, including branch instructions. As more branches are dispatched within a time interval, the speculation depth is more frequently reached.

To compensate this effect, one would simply increase the speculation depth. However, with a deeper speculation, more instructions would be dispatched within the same time interval, and the number of dispatch stalls caused by the lack of integer stations would increase again. The balance between the speculation depth and the number of integer stations can not be obtained by simply fixing a value for one of these parameters and then choosing the appropriate value for the other one. Both parameters should be dimensioned together, until the number of dispatch stalls is minimized. Figure 1 indicates that the number of stalls is minimized for a speculation depth of 8 branches and 6 integer reservation stations. Another minimum was obtained with a speculation depth of 10 branches, but in this case 12 integer reservation stations were required.

Observe that a deep speculation does not always contribute to the reduction of stalls. This comes from the fact that the average dynamic instruction window size increases with the speculation depth, allowing more instructions to be ahead of the mispredicted branch within the window. This increases the number of cycles during which dispatch remains blocked, waiting until the branch reaches the head of the reorder buffer. From a certain value of the speculation depth, this effect starts dominating, avoiding any significant reduction in the number of zero-dispatch cycles. For all the configurations examined, this happens with speculation depths equal or greater than 10 branches.

In order to evaluate the demand of memory reservation stations, we have attributed values for the speculation depth and the number of integer reservation stations that minimize the zero-dispatch cycles. These values were taken from the previous experiments. Memory reservation stations were then added until the corresponding stall component becomes comparable to the components related to the speculation depth and to the number of integer reservation stations. The results are shown in Figure 2. Stalls caused by an insufficient number of memory reservation stations are above 10% of the total cycles when there are less than 14 of these stations. In [10] we have also evaluated machines that can perform two memory accesses simultaneously. In this case, six memory reservation stations are sufficient to keep dispatch stalls to a ratio below 10%.

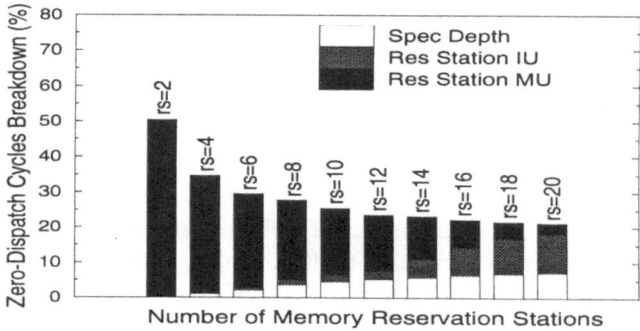

Fig. 2. Varying the number of memory reservation stations.

4 Conclusions

The cost of speculating across a limited number of branches can be much higher than the cost related to speculations along wrong paths. Dispatch stalls involved with state recovery does not exceed 12% of the total cycles, while in a configuration with a two-branch depth (like the PowerPC 604), the percentage of dispatch stalls caused by an insufficient speculation depth is as high as 30%. Nevertheless, there is a specific value of the speculation depth from which the number of stalls related to state recovery starts dominating, limiting the performance. In the majority of the configurations investigated, this threshold occurred for a 10-branch speculation depth.

The results presented here demonstrate that the potential benefit of increasing the speculation depth can be lost if the resources are not properly balanced. In [10], we have also assessed the role of the speculation depth on the cycle count of the speculative model, relative to a similar but non-speculative model. For integer programs, we have observed an average reduction of 18% in the cycle count for a two-branch speculation depth (as in the PowerPC 604), and of 40% for a speculation depth of four branches (as in the PowerPC 620 and R10000). An average reduction of 48% (and of almost 60% for some integer benchmarks) was obtained with a speculation depth of eight branches.

5   References

[1] Lee, J. K. F., A. J. Smith, *Branch Prediction Strategies and the Branch Target Buffer Design*, IEEE Computer Vol. 17 N. 1, September 1980, pp. 261-294.
[2] Yeh, T.-Y., Y. Patt, *Two-Level Adaptive Training Branch Prediction*, Proc. of the 24th Annual International Symposium on Microarchitecture, 1991, pp. 51-61.
[3] Song, S. P., M. Denman, J. Chang, *The PowerPC 604 RISC Microprocessor*, IEEE Micro Vol. 14 N. 5, October 1994, pp. 8-17.
[4] Diep, T. A., C. Nelson, J. P. Shen, *Performance Evaluation of the PowerPC 620 Microarchitecture*, Proc. of the 22th International Symposium on Computer Architecture, 1995, pp. 163-175.
[5] MIPS Inc., *The R10000 Microprocessor User's Manual*, 1995.
[6] Williams, T., N. Patkar, G. Shen, *SPARC64: A 64-b 64-Active-Instruction Out-of-Order-Execution MCM Processor*, IEEE Journal of Solid State Circuits Vol. 30 N. 11, November 1995, pp. 1215-1226.
[7] Tomasulo, R. M., *An Efficient Algorithm for Exploiting Multiple Arithmetic Units*, IBM Journal of Research and Development Vol. 11 N. 1, January 1967, pp. 25-33.
[8] Smith, J. E., A. R. Pleszkun, *Implementing Precise Interrupts in Pipelined Processors*, IEEE Transactions on Computers Vol. 37 N. 55, May 1988, pp. 562-573.
[9] Sun Microsystems, *The SPARC Architecture Manual, Version 7*, Mountain View, CA, 1987.
[10] *The Effect of the Speculation Depth on the Performance of Superscalar Architectures*, Tech Rep ES415/96, Department of Systems and Computer Engineering, COPPE/UFRJ.

Allocating Lifetimes to Queues in Software Pipelined Architectures

Marcio M. Fernandes[1], Josep Llosa[2], Nigel Topham[1]

[1] Edinburgh University, UK
[2] Universitat Politècnica de Catalunya, Spain

Abstract. Software pipelining is an effective technique for increasing the throughput of loops in superscalar or VLIW machines, however it generates high register pressure, which in some cases requires the introduction of spill code into the schedule. Large multi-ported register files present significant problems in the construction of scalable VLIW systems, which has lead us to investigate architectures in which part of the register file is replaced by queues. We believe that this organization has distinct advantages in terms of hardware complexity, silicon area, instruction name space, and scalability. Queues also represent a natural mechanism for communication between clusters of functional units in a partitioned VLIW system. In this paper we present an overview of this approach, along with some experimental results suggesting it as being a feasible organization.

1 Introduction

Instruction-level parallelism (ILP) is a family of processor and compiler design techniques that speed up program execution by causing individual machine operations to execute in parallel. Decisions about which operations should be executed in a given cycle and a given functional unit can be taken either at compile time or at run time, depending on the architecture model in use. In Very Long Instruction Word (VLIW) machines the compiler provides information as to which operations are independent of one another, so the hardware knows without further checking which operations can execute concurrently.

The scheduling of operations plays a major role in achieving near optimal performance from an ILP machine. One of the scheduling schemes that can be employed is software pipelining, with the objective of initiating successive loop iterations before prior iterations had completed [2]. Modulo scheduling is a class of software pipelining algorithms in which all loop iterations have the same schedule of operations [9]. Most software pipelining schemes assume an architectural model in which arithmetic operations are all register-register operations

---

[0] This work has been supported by research grants from Capes (Brazil), British Council and Ministry of Education of Spain under Acciones Integradas grant no. 1016, and also UK EPSRC under grant no. K19723
[1] Department of Computer Science, mmf,npt@dcs.ed.ac.uk
[2] Department d'Arquitectura de Computadors, josepll@ac.upc.es

and data is transferred between registers and memory using load and store instructions. The lifetime of a value is the time span from the reservation of a register to hold the value up to the last moment when the value is used. Lifetimes often exceed the initiation interval, which means multiple live values from a single instruction must coexist. Early designs proposed alternative register file organizations to deal with the problem. The Polycyclic architecture ([9]) uses a delay element, implemented as a queue with shift capabilities between every pair of communicating functional units, often resulting in a full cross-bar. This queue organization facilitates write operations by means of a write pointer and compacting non-empty locations, however it requires a book-keeping function to determine the exact address of a value being read. The Cydra 5 architecture ([12]) relies on a large number of registers and provides a mechanism to perform a sort of register renaming, which also helps to avoid code size explosion, a scheme called *rotating register file*. It requires a bank of registers between every pair of communicating functional units, which also leads to a full cross-bar. In addition to the problem of overlapped lifetimes, advances in technology have increased the parallelism available in a microprocessor through a larger number of functional units, which in turn increases register pressure dramatically [7], requiring once again new register file organizations. Assuming that a single register file is not able to support the high register pressure generated by modulo scheduled loops for large numbers of functional units, we believe that some sort of register file partitioning might be a reasonable alternative. Thus, a processor composed of clusters of functional units and private register files could be used as a starting point for a new hardware scheme. However, simply reorganizing the processor in this way can not guarantee a solution for the whole problem as inter-cluster communication delays can impose a severe performance penalty. To effectively take advantage of this concept a more elaborated register file organization and scheduling mechanism should be employed.

In a modulo scheduled loop the register values used to hold data referring to the same operation in different loop iterations have the same lifetime, but with the start times offset by the initiation interval. Therefore, if two computations produce values with lifetimes of equal length, and their start times are different, then the production order of their respective values will exactly match the consumption order of the values. Under this condition the computations can name a shared queue as the common destination for their result values. Thus, sets of lifetimes of the same length could be stored in the same queue, simplifying register access and reducing register name pressure. Further investigations have shown that this constraint can be relaxed under certain strictly defined conditions to permit lifetimes of *different* lengths to share the same output queue.

We are currently investigating the possibility of designing a scalable VLIW architecture comprising clusters of functional units and private register files implemented as queue structures, which in turn may also be used for inter-cluster communication. As the number of queues will be finite the code partitioning and scheduling process will involve an element of queue allocation similar in some ways to conventional register allocation. Overall, we believe that the use of queues has distinct advantages in terms of hardware complexity, silicon area, name space, and scalability. This paper presents the current status of our research, together with some of our initial experimental results and conclusions.

2   Using Queues to Organize Register Files

We show in [4] that the register file area needed to store enough registers and to provide sufficient access to those registers in a software pipelined loop is proportional to the cube of the number of functional units. This result clearly shows that is impractical to rely on a large multi-port register file to hold live values in a VLIW machine using modulo scheduling techniques if scalability of parallelism is the goal. It may even be the case that a shared multi-ported register file is not the most area-efficient storage scheme for the moderate degrees of ILP found in superscalar microprocessors.

This paper proposes a partitioned register file in which individually address-able registers are replaced by queues. In terms of similarities with other systems we understand that it resembles the Polycyclic machine only in which concerns writing values to a queue. The rotating register file employed by the Cydra 5 architecture could be viewed as a queue organization in which every distinct lifetime is allocated to a distinct queue, however that would require an unaccept-able number of machine resources. The remainder of this paper is devoted to demonstrating that queues can reduce the register pressure generated by mod-ulo scheduled loops in a VLIW machine, incorporating the following advantages over conventional organizations:

- *Hardware complexity and silicon area:* The access to a queue of registers is simpler than the access to a conventional register file as there is no need to select the register to be read or written to. Instead a value is always written on the last position in the queue and read from the first position, which can be controlled by means of two pointers. We expect that this organization might reduce the hardware complexity, and consequently the silicon area required.
- *Name Space:* We show in [4] that the number of registers required by a modulo scheduled loop is proportional to the number of functional units and to the pipelining degree, which increases the pressure on the name space as the machine scales up. In our queue register file model a data value is not allocated to a specific register location but instead to a specific queue, which implies that the name space problem is shifted from distinct register locations to distinct queues. We have found through experimental analysis that using a queue register file may reduce dramatically the pressure on the name space, as shown in Sect. 4.
- *Register Allocation:* The problem of register allocation, either considering a conventional register file [10] or a partitioned one [6] has been pointed out by several authors as being a non trivial task. We have developed a simple and efficient strategy to allocate data values to queues that we understand as being simpler than most of the techniques described in the literature.

- *Code Generation:* Kernel-Only ccde is a scheme that avoids code size explosion [11], which may be implemented if a queue register file is used along with support for predicate execution.
- *Inter-Cluster Communication:* It is well known that the efficiency of the inter-cluster communication system is a major issue to be addressed when designing clustered architectures. We believe that register queues may be used for this purpose, implementing a sort of asynchronous communication between clusters, with no need of extra instructions to move data values.

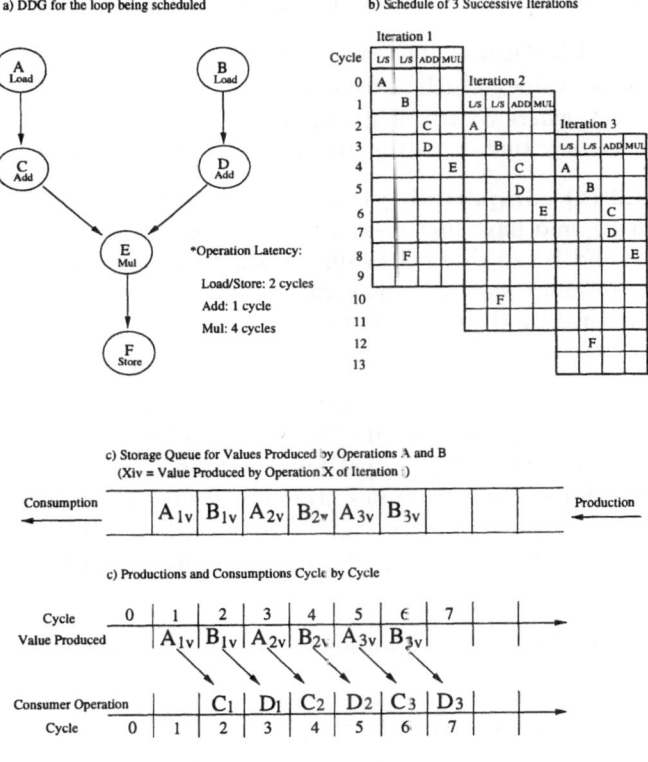

Figure 1. Allocating Registers to a Queue

To illustrate some of the ideas presented in this section we take the *data dependence graph (ddg)* of a given innermost loop (Fig. 1a) and the corresponding modulo schedule for 3 successive loop iterations (Fig. 1b). Assuming that a queue register file is being used, Fig. 1c shows the data flow in one of the storage queues, which contain values produced by successive executions of operations A and B. It can be seen that the production order of such values matches the consumption order required by operations C and D, i.e., the first element in the queue is always the value required by the next operation to be executed.

3 Queue Compatibility Condition

The ability to minimize the number of queues required by a modulo scheduled loop is critical to the use of a queue register file. We have developed a condition to check if two lifetime values can share the same storage queue. We also show how this condition can be evaluated through a simple and practical compile-time test. Due to space limitations we have ommited the theorem proof, which can be found in a technical report ([4]).

In a modulo-scheduled loop each computation generates a new value every Initiation Interval (II) cycles. Each value has a fixed lifetime which begins at some start-point and terminates at some end-point within the schedule.

Definition 3.1 (Lifetimes). On each iteration of a loop every computation **a** produces a new value which exists over a period defined by the pair $\langle S_a, S_a + L_a - 1 \rangle$, where S_a is the start-point and $S_a + L_a - 1$ is the end-point of that value. We say that L_a is the lifetime of computation **a**.

Definition 3.2 (Q-compatibility). Let two computations **a** and **b** have start-points S_a and S_b, and have lifetimes L_a and L_b such that $L_a \geq L_b$. The values produced by **a** and **b** can share the same destination queue if the relative order in which they produce values is identical to the relative order in which those values are consumed by their successor computations, and their start-points are different.

It is now necessary to formulate a simple way of determining the compatibility of any pair of computations. We do this by formulating a proposition which encapsulates our definition of Q-compatibility and then we prove that there exists a simple relationship between lifetimes, start-points and Initiation Interval which can be used in a scheduler to determine Q-compatibility. We now formulate a proposition based on Definition 3.2 which provides us with a formal criteria for queue compatibility.

Proposition 3.3. *The two computations* **a** *and* **b** *are Q-compatible if, and only if:*

$$\forall_{i,j \geq 0} : a_i > b_j \Rightarrow a_i + L_a > b_j + L_b \tag{1}$$

$$\wedge \ a_i < b_j \Rightarrow a_i + L_a < b_j + L_b \tag{2}$$

$$\wedge \ a_i \neq b_j \tag{3}$$

This proposition, although an accurate formulation of Definition 3.2, cannot be used directly when scheduling a loop as it contains universal quantifiers. These imply a large, possibly unbounded, search space for i and j. The following theorem defines an alternative, and computationally efficient, test for Q-compatibility.

Theorem 3.4 (Exact Compatibility Test:). *Two computations* **a** *and* **b**, *with start-times* S_a *and* S_b, *and lifetimes* L_a *and* L_b *such that* $L_a \geq L_b$, *are Q-compatible if and only if* $L_a - L_b < (S_b - S_a) \mod II$.

4 Experimental Evaluation

In order to obtain quantitative data regarding modulo scheduled loops for a hypothetical VLIW machine, an experimental scheduling framework has been built. The basic algorithm used in this framework is *Iterative Modulo Scheduling (IMS)* [8]. The scheduler assumes the existence of a simple VLIW machine, comprising of some fully pipelined functional units connected to either a *multi-ported register file (RF)* or a *register file organized by means of queues (QRF)*. Three machines configurations have been considered, as shown in Table 1.

To evaluate the effectiveness of queues as an alternative to conventional registers all eligible innermost loops from the Perfect Club Benchmark were scheduled, totalling 1258 loops. The optimizations and data dependence analysis were performed by the ICTINEO compiler [1], which supplied the input data set used by our framework. Due to space limitations we only briefly present some of the experimental results obtained, which can be found in [4].

Functional unit type	Operation latency	Issue rate	Number of functional units		
			machine A	machine B	machine C
load/store	2	1/~	2	2	4
add/subtract	1	1/~	1	2	4
multiply	4	1/~	1	2	4

Table 1. Functional units for three target machine configurations

Number of Queues Required The graphics presented in Fig. 2 shows the fraction of loops, from the set of 1258 loops considered, that can be scheduled employing only a given number of queues. The results show that with a fixed number of 32 queues it is possible to schedule most of the loops regardless the number of functional units, suggesting that number as being the size of the name space required, which is considerably smaller than that required by conventional register file organizations. It also shows a tolerable increase in the required number of queues as more functional units are used, suggesting that the scalability of the model is not constrained by this resource.

Number of Storage Positions Required In Fig. 3 it is shown the total number of queue positions required to schedule a given fraction of the loops. It can be seen that it is possible to schedule over 90% of all the loops using no more than 64 queue positions. It may be worth at this point to make a rough comparison between this figures and the register requirements when using a conventional register file organization. Similar analyses performed by other groups [7, 3, 5] found that it is possible to schedule around 90% of all the loops with 32 registers, which may suggest that their schemes are more efficient regarding this aspect. In spite of our beliefs that the possibly lower complexity of a queue register file may compensate this difference, we are currently working in a number of alternatives to improve this figure.

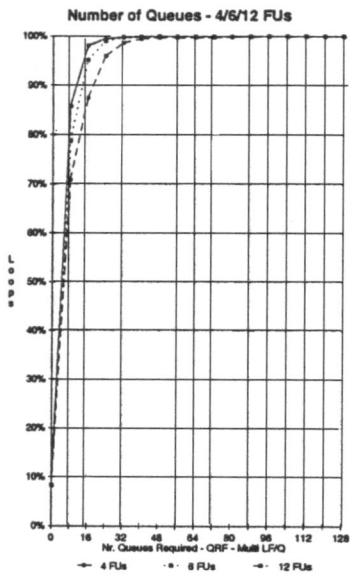
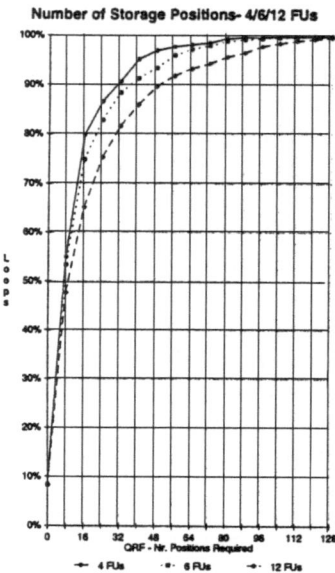

Figure 2. Number of Queues Required   **Figure 3.** Queue Capacity Required

Loops that Benefit from Greater Parallelism We found that significant speedups can be attained for around 70% of the loops when more functional units are employed, which justify the use of aggressive hardware configurations. In most of the cases the number of extra queues required for that falls between 0 and 15, which we understand as being a good prospect in terms of scalability.

5 Conclusions

We have investigated alternative register file organizations to address the high register pressure generated by a modulo scheduled loop. A register file organized by means of queues has been considered, and a number of quantitative data regarding machine resources was obtained from a preliminary evaluation. We have observed that the number of distinct queues required to schedule the benchmark loops is around 32 for configurations up to 12 functional units, which is less than other schemes reported in the literature. We have found that the total number of bits of queue storage is larger than that required by a conventional register file but we believe that the silicon area requirements will remain significantly lower. The small differences found between machine resources required by distinct number of functional units suggests that there is an advantage in terms of scalability, which is not the case of systems that relies on conventional register files or cross-bar organizations.

1073

We are currently working in a number of improvements on the proposed model, including loop unrolling to maximize functional units utilization, introduction of copy operations to deal with the problem of simultaneous writes of the same value to distinct queues, allocation of loop invariant and a hardware complexity model for the queue register file. We are also working on a new machine model organized by means of clusters composed of functional units and a private register file, which in turn communicate among each other through a bidirectional ring of queues. Finally, an enhanced machine model should be employed in the near future, increasing the level of details and assuming a finite number of machine resources, which may lead to the use of other techniques like graph coloring and the introduction of spill code.

References

1. E. Ayguadé, C. Barrado, J. Labarta, J. Llosa, D. Lopez, S. Moreno, D. Padua, E. Riera, and M. Valero. Ictineo: Una herramienta para la investigacion en paralelismo a nivel de instrucciones. In *VI Jornadas de Paralelismo*, July 1995.
2. A. Charlesworth. An approach to scientific array processing: The architectural design of the AP120B/FPS-164 family. *Computer*, 14(9), 1981.
3. A. Eichenberger, E. Davidson, and S. Abraham. Minimum register requirements for a modulo schedule. In *Proceedings of the MICRO-27 - The 27th Annual International Symposium on Microarchitecture*, November 1994.
4. Marcio M. Fernandes, Josep Llosa, and Nigel Topham. Using queues for register file organization in VLIW architectures. Technical Report ECS-CSG-29-97, Edinburgh University, February 1997.
5. R. Huff. Lifetime-sensitive modulo scheduling. In *Proceedings of the SIGPLAN'93 - Conference on Programming Language Design and Implementation*, 1993.
6. J. Janssen and H. Corporaal. Partitioned register file for TTAs. In *Proceedings of the MICRO-28 - The 28th Annual International Symposium on Microarchitecture*, November 1995.
7. J. Llosa, M. Valero, E. Ayguadé, and J. Labarta. Register requirements of pipelined loops and their effect on performance. In *2nd International Workshop on Massive Parallelism: Hardware, Software and Applications*, October 1994.
8. B. Rau. Iterative modulo scheduling. *The International Journal of Parallel Processing*, February 1996.
9. B. Rau and C. Glaeser. Some scheduling techniques and an easily schedulable horizontal architecture for high performance scientific computing. In *14th Annual Workshop on Microprogramming*, October 1981.
10. B. Rau, M. Lee, P. Tirumalai, and M. Schlansker. Register allocation for software pipelined loops. In *Proceedings of the ACM SIGPLAN'92 - Conference on Programming Language Design and Implementation*, June 1992.
11. B. Rau and P. Tirumalai M. Schlansker. Code generation schema for modulo scheduled loops. In *Proceedings of the MICRO-25 - The 25th Annual International Symposium on Microarchitecture*, December 1992.
12. B. Rau, D. Yen, W. Yen, and R. Towle. The Cydra 5 departmental supercomputer. *Computer*, January 1989.

Treegion Scheduling for Highly Parallel Processors

Sanjeev Banerjia and William A. Havanki and Thomas M. Conte

Department of Electrical and Computer Engineering
North Carolina State University
Raleigh, North Carolina 27695-7911
(919)-515-7983
{sbanerj,wahavank,conte}@eos.ncsu.edu

Abstract. Instruction scheduling is a compile-time technique for extracting parallelism from programs for statically scheduled instruction-level parallel processors. Typically, an instruction scheduler partitions a program into regions and then schedules each region. One style of region represents a program as a set of decision trees or *treegions*. The non-linear nature of the treegion allows scheduling across multiple paths. This paper presents such a technique, termed *treegion scheduling*. The results of experiments comparing treegion scheduling to scheduling for basic blocks and across "simple linear regions" show that treegion scheduling outperforms the other techniques.

1 Introduction

The performance of statically-scheduled, instruction-level parallel (ILP) processors depends on compiler techniques that extract parallelism from programs. In order to extract large amounts of ILP from non-scientific, integer programs, instruction scheduling must be performed across basic blocks [1], [2]. Schedulers typically group together basic blocks which may execute together into *regions* and then schedule each region. Regions are either *linear* (containing a single path of control) or *non-linear* (containing multiple paths of control).

The grouping process (*region formation*) is often done using profile information [2], [3]; if program behavior differs from this information, performance can suffer [4]. Other problems may arise due to *merge points*, instructions to which control can flow from multiple instructions. If an instruction is speculated above a merge point, it must be duplicated along all paths that join at the merge point. Merge points also add complexity to dynamic recompilation techniques [5].

One region that is resistant to unpredictable execution and that does not include merge points is a *treegion*, a tree-shaped subgraph of a program's control flow graph (CFG). This paper describes treegions and how they can be scheduled and is organized as follows. Section 2 defines treegions and introduces treegion scheduling via an example. Section 3 presents experimental results for treegion scheduling and compares the results with scheduling for basic blocks and "simple linear regions". Section 4 describes related work in non-linear regions, and Section 5 concludes with comments on future work and a summary.

2 Treegions

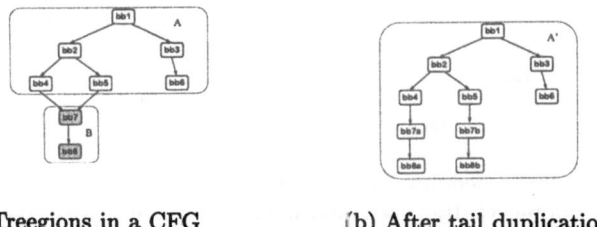

(a) Treegions in a CFG   (b) After tail duplication

Fig. 1. Figure (a) shows the CFG broken into two treegions A and B. Figure (b) shows how the two treegions can be combined into one treegion A' with tail duplication.

A treegion is a rooted tree subgraph of a CFG. An example of a CFG partitioned into treegions is shown in Figure 1(a). The size and number of treegions in a CFG are determined by the CFG topology, not profile information. However, heuristics using profile information can guide methods to expand treegions; tail duplication on basic blocks 7 and 8 results in the CFG shown in Figure 1(b). Many of the procedures used with superblocks [3] may be applied to treegions.

Treegion formation begins at each entry node of a CFG. Nodes encountered while traversing from each entry node are absorbed into a treegion until merge points are encountered, each of which becomes the root of a new treegion. This process continues until every node is in some treegion.

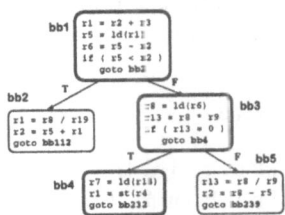

Fig. 2. A sample CFG. The emphasized basic blocks are a possible preferred path.

Figure 2 shows a sample CFG. Figure 3(a) shows a schedule formed from the CFG using the successive retirement scheduling algorithm [6] (the example machine is a two-issue processor with universal functional units and unit latency). This schedule retires the exits from the *preferred path*[1] in sequential order and performs speculation only along that path. Program execution along the preferred path { bb1, bb3, bb4 } takes seven cycles (cycles 0–6), assuming there

[1] The preferred path is the most frequently executed path within a region as indicated by profile information or static heuristics.

1076

Cycle #	ALU-1	ALU-2
0	r1 = r2 + r3	
1	r5 = ld(r1)	*r1 = st(r4)*
2	r6 = r5 - r2	blt bb2,r5,r2
3	r8 = ld(r6)	
4	r13 = r8 * r9	
5	bne bb5,r13,0	r7 = ld(r13)
6	goto bb232	
7	bb2: r1 = r8 / r19	
8	goto bb112	r2 = r5 + r1
9	bb5: r13 = r8 / r9	r2 = r8 - r5
10	goto bb239	

(a) Successive retirement

Cycle #	ALU-1	ALU-2
0	r1 = r2 + r3	*r1a = r8 / r19*
1	r5 = ld(r1)	r1 = st(r4)
2	r6 = r5 - r2	*r2 = r5 + r1a*
3	blt bb2,r5,r2	r8 = ld(r6)
4	r13 = r8 * r9	*r13 = r8 / r9*
5	bne bb5,r13,0	r7 = ld(r13)
6	goto bb232	
7	bb2: goto bb112	*r1 = r1a*
8	bb5: goto bb239	r2 = r8 - r5

(b) Treegion scheduling

Fig. 3. Sample CFG schedules. Underlined instructions are speculated above their control-dependent branches. Italicized instructions have had register renaming performed.

are no cache misses and perfect branch prediction. Program execution along the path { bb1, bb3, bb5 } takes eight cycles (cycles 0–5,9,10).

Figure 3(b) is a schedule formed from the CFG using *treegion scheduling*. The priority function used is the number of treegion execution paths through the operation [4]. Unlike successive retirement, operations from other paths ("off-paths") become intermingled into the schedule, so that operations from multiple paths are scheduled to execute together. Compile-time register renaming is used to allow speculation of operations above their control-dependent branches, preserving live-out register values. If the preferred path is executed at run-time, this schedule again takes seven cycles to execute. However, the execution time of the path { bb1, bb3, bb5 } has been reduced from eight to seven cycles.

One strength of treegion scheduling is that by scheduling multiple paths in parallel, a high-performance schedule for a preferred path can be generated without unduly penalizing off-paths. This characteristic hedges against poor performance when the executed path differs from the compile-time preferred path. In this respect, treegion scheduling is similar in spirit to the speculative hedge heuristic [4] of superblock scheduling.

3 Experimental results

Experiments were conducted to gauge the effectiveness of treegion scheduling using the SPECint95 benchmark suite. Classic optimizations and a profiling run using training inputs were applied to the benchmarks before scheduling for treegions, "simple linear regions"[2] (*SLRs*), and basic blocks using the LEGO compiler, a research ILP compiler developed at N.C. State University. Scheduling was performed for two statically-scheduled machine models: an eight-issue processor with universal functional units, EIGHT-AGGR, and one with a mix of four integer/branch, two memory, and two floating-point units, EIGHT-CONS. Instructions are unit latency except loads (2 cycles), floating-point multiply (3 cycles), and floating-point divide (9 cycles). Program performance was measured by using the profile count and schedule height of each region to estimate

---

[2] Simple linear regions are built like superblocks, but without tail duplication.

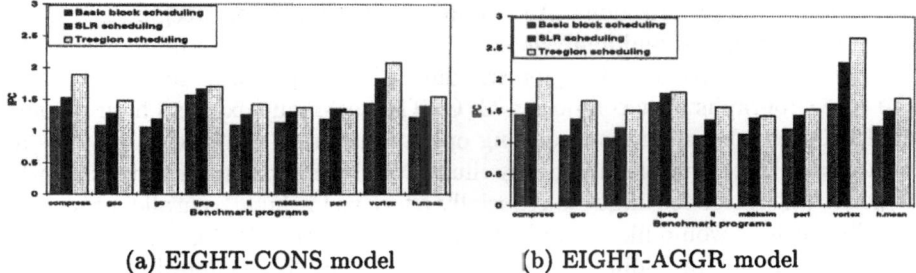

(a) EIGHT-CONS model   (b) EIGHT-AGGR model

Fig. 4. Performance of basic block scheduling, SLR scheduling and treegion scheduling for the two machine models. h.mean denotes harmonic mean.

execution time. The effects of instruction and data caches were ignored. Useful instructions completed per cycle (IPC) was the performance metric used. Instructions added due to renaming were not used in computing IPC.

Figure 4 presents the results. In every case, treegion scheduling yielded higher performance than basic block scheduling, and about the same as or better than SLR scheduling. The treegion schedule performed worse than the SLR schedule for **perl** under EIGHT-CONS because of aggressive speculation, which extends the preferred path schedule by speculating more off-path operations. The IPC improvements are larger with EIGHT-AGGR because the flexibility of the model permitted the treegion scheduler to fill more empty slots in the schedule with off-path operations. This illustrates that treegion scheduling yields the most benefit on highly parallel processors.

4 Related work

Hsu and Davidson's decision tree scheduling (DTS) [7] is the predecessor of the work presented here. DTS schedules along multiple paths within a decision tree, inserting instructions into branch delay slots and using guards to control write-back of speculated instructions. The VLIW project at IBM Research embellished Nicolau's percolation scheduling [8], using them to implement a VLIW compiler [9]. The heart of the IBM VLIW machine is a *tree instruction*, which has the ability to evaluate multiple branches in one clock cycle. The initial work in VLIW architectures was based on a single-path scheduling algorithm called trace scheduling [2]. The Trace Scheduling-2 algorithm is an extension of the original trace scheduling algorithm that schedules along multiple paths simultaneously [10]. Hyperblock scheduling also schedules multiple paths in parallel [3] by removing branches from the instruction stream entirely through if-conversion.

5 Concluding remarks and acknowledgements

There are issues related to treegions that merit further research. The use of if-conversion and tail duplication could eliminate merge points and allow for the formation of larger treegions. Also, different heuristics for treegion scheduling need to be identified and analyzed.

This paper introduced treegion scheduling, which performs scheduling across the tree subgraphs that compose a CFG. The technique extracts high amounts of ILP by scheduling and speculating operations along multiple paths. The advantages of treegion scheduling were illustrated by comparing treegions to other regions. The latter technique is especially effective for highly parallel processors.

The authors would like to thank Scott Mahlke of the CAR Group at Hewlett-Packard Labs for providing the optimized SPECint95 benchmarks used in this paper, and Kishore Menezes and Sumedh Sathaye for discussions that greatly improved the quality of this paper. The comments from the anonymous referees are also appreciated. This work was supported by IBM, Hewlett-Packard, and the National Science Foundation under grants MIP-9696010, MIP-9625007, and GER-9454175.

References

1. G. S. Tjaden and M. J. Flynn, "Detection and parallel execution of independent instructions," *IEEE Trans. Comput.*, vol. C-19, pp. 889–895, Oct. 1970.
2. J. A. Fisher, "Trace scheduling: A technique for global microcode compaction," *IEEE Trans. Comput.*, vol. C-30, no. 7, pp. 478–490, July 1981.
3. S. A. Mahlke, *Exploiting instruction level parallelism in the presence of branches.* PhD thesis, Department of Electrical and Computer Engineering, University of Illinois at Urbana-Champaign, Urbana, IL, 1996.
4. B. L. Deitrich and W. W. Hwu, "Speculative hedge: regulating compile-time speculation against profile variations," in *Proc. 29th Ann. Int'l Symp. on Microarchitecture* [11].
5. T. M. Conte and S. W. Sathaye, "Dynamic rescheduling: A technique for object code compatibility in VLIW architectures," in *Proc. 28th Ann. Int'l Symp. on Microarchitecture*, (Ann Arbor, MI), Nov. 1995.
6. C. Chekuri, R. Johnson, R. Motwani, B. Natarajan, B. Rau, and M. Schlansker, "Profile-driven instruction level parallel scheduling with application to superblocks," in *Proc. 29th Ann. Int'l Symp. on Microarchitecture* [11], pp. 58—67.
7. P. Y. T. Hsu and E. S. Davidson, "Highly concurrent scalar processing," in *Proc. 13th Ann. Int'l Symp. Computer Architecture*, (Tokyo, Japan), June 1986.
8. A. Nicolau, "Percolation scheduling: a parallel compilation technique," Technical report TR-678, Department of Computer Science, Cornell University, Ithaca, NY, May 1985.
9. K. Ebcioğlu, "Some design ideas for a VLIW architecture for sequential-natured software," in *Proceedings of the IFIP Working Group 10.3 Working Conference on Parallel Processing*, (Pisa, Italy), pp. 3–21, North Holland, 1988. (published as *Parallel Processing*, M. Cosnard, et al., (eds.).).
10. J. A. Fisher, "Global code generation for instruction-level parallelism: Trace Scheduling-2," Tech. Rep. HPL-93-43, Hewlett-Packard Laboratories, June 1993.
11. *Proc. 29th Ann. Int'l Symp. on Microarchitecture*, (Paris, France), Dec. 1996.

Modulo Scheduling with Cache Reuse Information*

Chen Ding[1] and Steve Carr[2] and Phil Sweany[2]

[1] Dept. of Computer Science, Rice University, Houston TX 77005-1892
[2] Dept. of Computer Science, Michigan Technological University, Houghton MI 49931

Abstract. Software pipelining for instruction-level parallel computers with non-blocking caches usually assigns memory access latency by assuming either all accesses are cache hits or all are cache misses. We contend setting memory latencies by cache reuse analysis leads to better software pipelining than either an all-hit or all-miss assumption. Using a simple cache-reuse model, our software pipelining optimization achieved 10% improved execution performance over assuming all-cache-hits and used 18% fewer registers than required by an all-cache-miss assumption. We conclude that software pipelining for architectures with non-blocking cache should incorprate a memory-reuse model.

1 Introduction

In modern processors, main-memory access time is at least an order of magnitude slower than processor speed. To tolerate main memory latency, hardware designers use *non-blocking* caches which allow cache accesses to continue when misses occur [2]. This, in turn, allows an instruction scheduler to overlap more operations with memory accesses, possibly hiding main-memory latency. Thus, a significant increase in instruction-level parallelism (ILP) can be achieved.

Even though memory latency is variable with non-blocking caches, instruction schedulers typically either assume that all memory accesses are cache hits or that all are cache misses. Assuming all hits reduces register lifetimes keeping register pressure to a minimum. However, significant penalties are incurred when a cache miss occurs. Assuming all cache misses tolerates the latency of a cache miss better, but may increase register pressure significantly. Knowing the latency of a memory operation allows the instruction scheduler to get the best of both schemes. Previous work has used profiling to determine memory latencies for architectures with non-blocking caches [1] or has used software prefetching instructions to hide main-memory latency [6]. We report here on an experimental evaluation of modulo scheduling using cache-reuse analysis [5] to choose the proper latency for loads and stores on machines with non-blocking caches and no prefetching instructions.

---

* This work was partially supported by the National Science Foundation under grants CCR-9409341 and CCR-9308348, as well as a grant from Digital Equipment.

2 Experiment

To evaluate our contention that taking advantage of memory reuse information can improve software pipelining's efficiency, we compiled and simulated 120 Fortran loops in which our software pipelining implementation [7] used one of three different memory latency policies, namely 1) all loads are cache hits, 2) all loads are cache misses, 3) each load is either always a cache hit or always a cache miss, as determined by memory reuse analysis [5]. When reuse anlaysis determines that an array reference has any reuse, we assume the reference is always a cache hit; otherwise, we assume it is always a cache miss. Our hypothesis is that software pipelining in which the load latency is determined using this model yields better execution performance than pipelining with an all-hit latency policy, and though it should lead to slightly poorer execution performance than pipelining with an all-miss latency policy (assuming an infinite number of registers) the reuse policy leads to significantly fewer registers required for the loop.

2.1 Machine Model

The hypothetical superscalar architecture we chose has two integer and two floating point functional units; each may issue an instruction in each cycle should data dependences allow. Latency for integer instructions is two cycles; latency for floating point instructions is four cycles. Only one load or store can be issued per cycle. All loads and stores use an integer unit and the cache hit latency is two cycles.

Because one parameter we wished to investigate was register usage, we could have chosen a fixed, relatively small number of registers similar to current ILP machines and "measured" register pressure as part of execution time, since spilling would necessarily degrade loop performance. However, to separate register concerns and loop performance concerns, we assume 256 integer and 256 floating point registers, thus ensuring no spill code and allowing evaluation of register pressure effects by a direct measurement of registers required to generate software pipelined code for the loop.

Our cache model is an 8K, direct-mapped cache with 32-byte lines. The cache is non-blocking and allows up to 6 outstanding misses to occur in parallel. The penalty for a cache miss is an additional 25 cycles. On each miss, two consecutive 32-byte lines are brought into the cache.

We simulated loop behavior by resetting the simulator for each outermost loop construct. Thus, although we only pipelined innermost loops we counted all nested loops in our simulation results. However, we did not simulate non-loop code.

2.2 Test Programs

We software pipelined 107 Fortran innermost loops from three SPEC programs (hydro2d, su2cor, swm256) and an additional 13 loops from Fortran kernels, yielding a total of 120 innermost loops. For 45 loops there was no difference in any of the pipelined schedules. We used iterative modulo scheduling [7] to pipeline the loops, and restricted our attention to loops with no control flow or function calls. Thus, we pipelined only single-block loops in this study.

2.3 Results

Table 1 summarizes results for the performance, in terms of execution cycles, of the 75 loops tested. Column one shows "normalized" execution time for code compiled with an all-hit latency policy. Column two shows the same computation for the all-miss latency policy and column three lists results of code compiled with reuse information. We normalized execution cycles so that whichever of the three compiled codes (hit, miss, reuse) required the fewest cycles was set to 100, and the other two were normalized with that value. Table 1's values represent the unweighted average of these normalized execution values for all 75 loops. As expected, loops pipelined with latencies set by reuse required fewer cycles, on average, than those compiled with latencies set by a cache-hit assumption (roughly 10% less). Based only on execution cycles, we also expected reuse to perform slightly worse than cache-miss, due in part to our over-simplified model of cache behavior that assumes each static load is either always a hit or always a miss. We anticipated that this would lead to a small performance penalty, but, in fact, virtually all of the roughly 8% degradation in performance between cache miss and reuse policies can be attributed to our over-simplified reuse model, as discussed in Section 2.4. Finally, the summary data shows that miss was not always the best performance policy. Several loops showed better performance with reuse than cache miss. We discuss this unexpected result as well in Section 2.4.

Cache Hit	Cache Miss	Reuse
123	104	112

Table 1. Summary Performance Numbers — Normalized

In our experiment, the reuse version of a loop improved on the hit version for 17 of the 75 loops tested, while hit never did better than reuse. Thus, all of the roughly 10% average performance improvement was found in less than one fourth of the loops. In fact, for 13 loops, the reuse-compiled code was more than 20% faster than the code compiled with hit latencies. The largest difference was a factor of 2.61. The other 58 loops all produced the same results when compiled with hit latencies or reuse latencies. In contrast, while miss resulted in better schedules 34 times out of 75, reuse outperformed miss 19 times, by as much as 41% in one instance. To obtain the roughly 8% average improvement of cache miss to reuse then, cache miss had to be significantly better than reuse for some loops and, in fact, this is what we found. Although reuse outperformed miss by at least 20% only twice, miss was more than 20% faster than reuse for 19 of the 75 loops. The maximum penalty of reuse for any loop was 69%.

Table 2 shows register requirements of the pipelined loops. While compiling with reuse required about one register more on average than compiling with hit latencies, it took 6 fewer registers than were required by assuming miss latency.

This represents a 17.9% register savings over schedules that assume miss latency. For architectures with moderate numbers of registers this can be a considerable factor in deciding between using miss latencies and reuse information. When we restrict ourselves to those loops in which miss provided at least 20% better execution performance the difference is even greater. For those loops, reuse required an average of 31.6 registers while miss required 40.9, a savings of 22.8%.

Cache Hit	Cache Miss	Reuse
33.7	40.8	34.6

Table 2. Summary of Registers Required

2.4 Discussion

Our basic premise was that compiling with reuse information would allow both more efficient pipelined loops than would compiling with hit latency and fewer registers required than would compiling with miss latency. Our experimental evidence certainly suggests that this is true.

To understand the reason for the degradation of reuse compared to miss, consider the definition of self-spatial reuse. Self-spatial reuse occurs because an entire cache line is brought into the cache on a single miss. If we assume stride-1 access of data (accessing adjacent data items on successive loop iterations) then self-spatial reuse leads to one miss every N loop iterations, where N is the number of adjacent data elements brought into the cache at once. This is quite different from our compiler's assumption that *every* access is a hit when we have self-spatial reuse.

Investigation of the 34 loops for which miss led to more effective pipelined schedules showed that each exhibited self-spatial reuse. Many included several self-spatial reuse loads. This means that, in our machine model, each self-spatial reuse load incurs a 25-cycle penalty every 8 loop iterations (since we bring 8 data items into the cache for each miss.) resulting in the observed performance degradation of reuse.

Perhaps more puzzling is the fact that for 19 loops the schedule generated with reuse information required fewer cycles than were needed using miss latency. Our intuition suggested that miss should never yield a worse schedule, but it did. Closer investigation of the loops in question showed that, for each, software pipelining "overhead" of prelude and postlude as well as preconditioning (used when a loop is unrolled for modulo variable expansion [4]) was significantly greater for the miss schedule than the reuse schedule. Thus, performance improvement is not an advantage of the reuse model but, rather, a side effect of code generation strategy.

2.5 Refinements

The above experimental data indicates that scheduling with reuse information can achieve performance equivalent to all-cache-miss with lower register pressure if we properly handle references exhibiting self-spatial reuse. To overcome this problem, we suggest two possible hardware modifications. First, hardware could be modified to prefetch the next cache line on a hit or miss if that line were not already in the cache. Alternatively, load instructions could be modified to stream a specified number of consecutive cache lines into the cache to reduce the negative effects of self-spatial accesses.

3 Conclusion

In this paper, we have demonstrated experimentally that using reuse information while software pipelining is effective. On our benchmark suite we produce on average 10% better schedules than an all-cache-hit assumption (a factor of 2.61 better on one loop) and on average we use 18% fewer registers than an all-cache-miss assumption. Even though all-cache-miss sometimes out performs reuse, it does so at the cost of 23% more registers. We refer the reader to the extended version of this paper for more details on this experiment [3].

Given that the cycle time of cache-miss latencies is increasing, software pipelining methods must eliminate performance degradation caused by these latencies. The methods presented here are an important step in eliminating the latency problem.

References

1. ABRAHAM, S., SUGUMAR, R., WINDHEISER, D., RAU, B., AND GUPTA, R. Predictability of load/store instruction latencies. In *Proceedings of the 26th International Symposium on Microarchitecture (MICRO-26)* (Austin, TX, December 1993), pp. 139–152.
2. CHEN, T.-F., AND BAER, J.-L. Reducing memory latency via non-blocking and prefetching caches. In *Proceedings of the Fifth International Conference on Architectural Support for Programming Languages and Operating Systems* (Boston, Massachusetts, 1992), pp. 51–61.
3. DING, C., CARR, S., AND SWEANY, P. Software pipelining with cache-reuse information. Tech. Rep. 96-07, Michigan Technological University, Sept. 1996. *ftp://cs.mtu.edu/pub/carr/modulo.ps.gz.*
4. LAM, M. Software pipelining: An effective scheduling technique for VLIW machines. *SIGPLAN Notices 23*, 7 (July 1988), 318–328. *Proceedings of the ACM SIGPLAN '88 Conference on Programming Language Design and Implementation.*
5. MCKINLEY, K. S., CARR, S., AND TSENG, C.-W. Improving data locality with loop transformations. *ACM Transactions on Programming Languages and Systems 18*, 4 (1996), 424–453.
6. MOWRY, T. C., LAM, M. S., AND GUPTA, A. Design and evaluation of a compiler algorithm for prefetching. In *Proceedings of the Fifth International Conference on Architectural Support for Programming Languages and Operating Systems* (Boston, Massachusetts, 1992), pp. 62–75.
7. RAU, B. Iterative modulo scheduling. In *Proceedings of the 27th International Symposium on Microarchitecture (MICRO-27)* (San Jose, CA, December 1994), pp. 63–74.

Memory Address Prediction for Data Speculation *

José González and Antonio González

Departament d'Arquitectura de Computadors
Universistat Politècnica de Catalunya
C. Jordi Girona 1-3, 08034 Barcelona (Spain)
Email: {joseg,antonio}@ac.upc,es
Tel:+ 34 4 4016988
Fax: + 34 4 4017055

Abstract. Data speculation refers to the execution of an instruction before some logically preceding instructions on which it is data dependent. Data speculation implies some form of prediction of the data required by the speculative executed instruction and a recovering mechanism in case of misspeculation. This paper shows that load/store instructions are very good candidates for speculative execution since their effective address is highly predictable. We propose a novel technique called Memory Address Prediction (MAP) that implements speculative execution of load/store instructions in an out-of-order processor. The cost of this mechanism is mainly the addition of an address prediction table since the misprediction recovery hardware is already present in many current microprocessors for other purposes. The mechanism is evaluated for the SPEC95 benchmark suite showing significant performance gains.

1 Introduction

The performance of current processors heavily rely on the exploitation of Instruction Level Parallelism (ILP). The effectiveness of this technique is limited by the necessity to obey the data dependences existing among instructions.

There have been very few proposals trying to overcome the limitations imposed by having to obey data dependences. In the same way as control dependent instructions can be speculative executed before the branches on which they depend, data dependent instructions can do so. What is needed for the latter is the ability to predict the value required by the data dependent instruction and a recovery mechanism for misspeculated instructions.

In this paper, we propose a data speculation mechanism that is based on the observation that the source operands of load/store instructions are highly predictable.

* This work has been supported by the Spanish Ministry of Education under grant CYCIT TIC 429/95 and the Direcció General de Recerca of the Generalitat de Catalunya under grant 1996FI-03039-APDT.

The proposed Memory Address Prediction mechanism (MAP) identifies which load/store instructions are highly predictable and issues them speculatively, as well as those instructions that depend on them. In case of misprediction, the misspeculated instructions are re-executed. The mechanism is evaluated for an out-of-order processor with two different memory ordering techniques: a conventional total disambiguation scheme and a more aggressive partial disambiguation mechanism.

2 Data speculation and related work

Data prefetching schemes [2] have certain similarities with the MAP in the sense that they predict the effective address of future load/store operations and bring the data into cache if not yet present. However, data prefetching does not execute any instruction speculatively.

Data speculation is a family of techniques that try to avoid the ordering imposed by data dependences. Data speculation allows the speculative execution of some instructions before some other instructions on which they are data dependent. Data speculation could be applied to values that flow either through memory or registers.

The most remarkable proposals on data speculation are: [1][3][5][9] [10][11]. The main differences between the mechanism proposed in this paper and previous work are:

- The mechanism proposed is this work predicts the address of memory instructions as [1] [3][5]. However in our proposal the instructions that depend on the predicted load are issued speculatively meanwhile such previous proposals do not perform speculative execution of those instructions.
- In [11] the effective address of load instructions is predicted and the load and subsequent dependent instructions are speculatively issued as is done in our work. But in our proposal the effective address of store instructions is also predicted. Besides, in [11] a perfect memory disambiguation scheme is considered, whereas in this paper we study the performance of address prediction for two realistic memory disambiguation schemes, where the prediction of the effective address of stores plays an important role in order to achieve an significant performance improvement.
- Regarding [9][10] the difference with our method is that their predict the result of an operation whereas we speculate with load/store instructions predicting their effective address. In addition, we will show in this paper that memory addresses are more predictable than memory values.

3 Motivation

In [6] it is shown that even with unlimited resources and perfect control speculation, the performance of current architectures would not be much higher than a hundred IPC (instructions per cycle) for many programs, and in some cases it

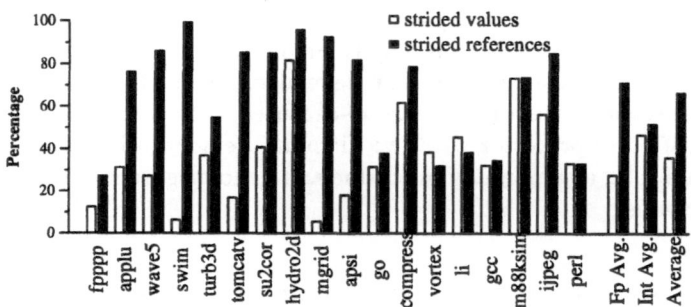

Fig. 1. Percentage of strided references compared with percentage of strided values.

would be just a few tens of IPC. Data speculation allows to go beyond this barrier. Data speculation is a powerful technique to increase the ILP of a program.

The mechanism proposed in this paper is based on the observation that the effective address of many load/store instructions are highly predictable. In particular, it tries to identify those load/store instructions whose effective address in successive executions is either the same or equal to the previous one plus a constant value. We will refer to both as strided references.

To validate our hypothesis on the predictability of load/store effective address we have run all the SPEC95 benchmarks and measured the percentage of strided references. The experimental framework is described in section 5.1. For the results in this section, each benchmark was run until the first billion load instructions or until completion if it happened earlier. Strided references are captured by means of a history table with 1K entries. This table is direct-mapped, non-tagged and it is indexed with the ten least-significant bits of the load/store instruction address. For each entry, the table stores the last effective address and the last observed stride.

Figure 1 shows the percentage of dynamically executed load/store instructions that exhibit strided references. It can be seen that strided references are very common in the SPEC95 suite: In average, they represent about 70% of all memory references. In both integer and floating point applications strided references are very frequent although the percentage is higher in floating-point applications. For comparison, Figure 1 also shows the performance of the load value prediction scheme proposed in [9], extended to account for strided values for both load and store instructions. Although strided values are quite frequent they are much less common than strided references. We conclude that memory addresses are more predictable than memory values.

4 Memory Address Prediction

The fact that the effective address of memory references is highly predictable can be used to speedup processor performance in different ways (e.g. prefetching

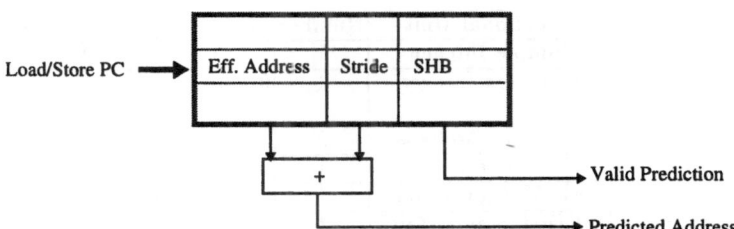

Fig. 2. Memory History Table (MHT)

mechanisms). We propose to use memory address prediction to speculate on the effective address of unresolved load/store instructions and execute them speculatively as well as the instructions that depend on them.

We have investigated the use of memory address prediction in the context of an out-of-order execution processor with in-order retirement to support precise exceptions [6]. Two different memory ordering mechanisms have been considered:

- Total disambiguation: In this scheme, a load can be issued as soon as there are not preceding stores with unknown effective address and the load operands are available. A store is issued as soon as its effective address can be computed and all previous loads and stores have already been issued.
- Partial disambiguation: This memory ordering scheme is based on the mechanism implemented in the HP PA8000 [4] and the ARB proposed in [7]. In this scheme a load or store can be issued as soon as its operands are ready, without being fully disambiguated with previous references. When a store founds that a load has already been performed to the same address from a succeeding instruction with no intervening stores in between, a recovery action is initiated.

In both schemes data can be forwarded from a previous resolved store to a subsequent dependent load.

The MAP mechanism is implemented by means of a table called Memory History Table (MHT), as it is shown in Figure 2. It is a 1024-entry, direct-mapped table that is indexed with the least-significant bits of the instruction address and does not contain tags. Each entry stores the following information:

- Effective Address: This is the last effective address seen by that load/store instruction.
- Stride: This field contains the last stride observed for that load/store instruction. The length of this field is 4 bytes.
- Stride History Bits (SHB): This field is used to assign confidence to the prediction. It is implemented by means of a two-bit up-down saturated counter. The prediction is determined by the most-significant bit of this field.

The MAP works as follows: At the decode stage, the corresponding MHT entry is read and the predicted effective address is computed. If the SHB most

Functional Unit	Number	Latency	Repeat Rate
Simple Integer	1	1	1
Complex Integer	1	9 multiply	9
		67 Divide	64
Effective Address	2	1	1
Simple FP	1	2	1
FP Multiplication	1	2	1
FP Divide and SQRT	1	21 divide	21
		35 SQRT	35

Table 1. Functional Units and instruction latency.

significant bit is set, then a correct prediction is assumed and the instruction is considered to be ready-to-issue to the load/store buffer.

Speculatively executed load/stores must be verified. This is done by issuing them to the address computation unit when their operands are ready. The computed effective address is compared against the predicted effective address and in case of a mismatch a recovering action is initiated. At this time the SHB field is updated and in case of a misprediction, the stride field is set to the new value.

The recovery for mispredicted loads and stores is implemented in a different way. For loads, we assume the recovering mechanism proposed in [9]. When a load is mispredicted, all subsequent instructions that depend on such load are re-executed. This mechanism cannot be used for stores because the destination of a store is not a register but a memory location, and in consequence, dependent instructions are not known until the correct address is calculated. For mispredicted stores we assume the same recovery mechanism used to recover from mispredicted branches (a pipeline flush).

Since the store misprediction penalty is higher than that of loads, a store instruction is executed speculatively only when its operands are unknown when it is decoded.

5 Performance evaluation

This section studies the effectiveness of the MAP in the context of a superscalar out-of-order execution processor for the two memory ordering schemes discussed in the previous section.

5.1 Experimental Framework

We have developed a simulator of a superscalar processor with out-of-order execution that resembles some of the latest microprocessors. The execution of an instruction consists of the typical stages: fetch, decode, issue, execute, write-back and graduate (or commit). Branch Prediction is performed by means of a

	No MAP		MAP	
	Total disambiguation	Partial disambiguation	Total disambiguation	Partial disambiguation
104.hydro2d	1.14	1.21 (6.1%)	1.21 (6.1%)	1.26 (10.5%)
107.mgrid	1.68	1.68 (0%)	1.82 (8.3%)	1.82 (8.3%)
110.applu	1.20	1.20 (0%)	1.22 (1.7%)	1.23 (2.5%)
146.wave5	0.95	0.99 (4.2%)	1.02 (7.4%)	1.02 (7.4%)
124.m88ksim	0.91	0.91 (0%)	1.14 (25.3%)	1.18 (29.7%)
129.compress	1.15	1.28 (11.3%)	1.31 (13.9%)	1.31 (13.9%)
130.li	0.92	1.02 (10.9%)	1.03 (12.0%)	1.06 (15.2%)
134.perl	0.91	1.03 (13.2%)	0.98 (7.7%)	1.06 (16.5 %)
Avg. improvement		5.7%	10.3%	13.0%

Table 2. Instruction completion rates. In brackets it is shown the percentage improvement over the total disambiguation scheme with no memory address prediction.

2048 entry Branch History Table with a 2 bit up-down saturated counter per entry.

The results presented in this section assume the following configuration. The size of the reorder buffer is 32 entries. There are two separate physical register files for integer and FP data. The size of both files is 64 registers. The processor has a lookup-free data cache [8] that allows up to 16 pending misses to different cache lines. The cache size is 8Kb, and it is direct-mapped with 32-byte line size. Cache hit latency is 2 cycles and the penalty for a cache miss is 20 cycles. An infinite L2 cache is assumed. Table 1 shows the number of functional unit and their latency.

Our experimentation methodology is trace-driven simulation. The object code, previously compiled with full optimization for a DEC AlphaStation 600 5/266 with and Alpha AXP 21164 processor, is instrumented using the Atom tool [12]. Because of the detail at which simulation is carried out the simulator is slow, so we have simulated 50 million of instructions for each benchmark after skipping the first 100 million of instruction. Eight SPEC'95 benchmarks (4 FP and 4 Integer) has been selected for this study.

5.2 Results

Table 2 shows the IPC (instructions committed per cycle) achieved by the two memory disambiguation schemes with and without the MAP mechanism proposed in this paper. It is shown in brackets the percentage improvement over the total disambiguation scheme without memory address prediction. In average, the improvement achieved by MAP is around 10

The improvement due to MAP depends on a number of issues in addition to the percentage of strided accesses. It depends on the accuracy of the memory address predictor. The data in Figure 3a represents the percentage of memory ref-

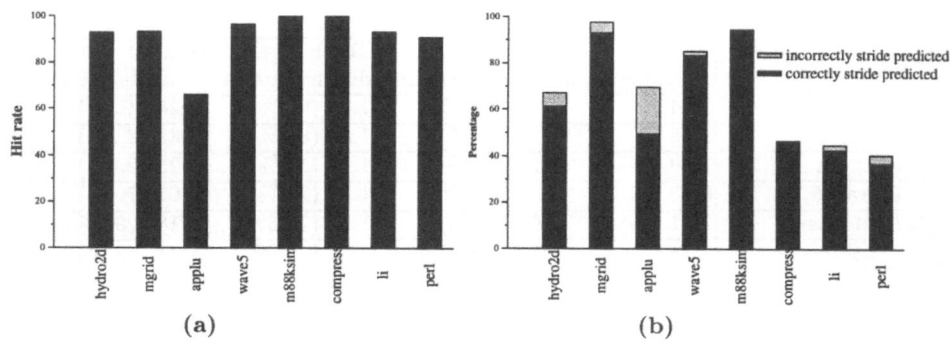

Fig. 3. Percentage of: a) correctly predicted loads and stores. b) correctly predicted loads

erences that are correctly predicted either as strided references or as not strided references. Obviously, a better prediction will imply more accurate speculation and less number of mispredictions. The performance also depends highly on the percentage of memory references that are predicted to be strided references and the number of these predictions that are correct. This is shown in Figure 3b for loads. A higher percentage of strided references means more speculation. On the other hand, a higher number of misprediction implies a higher penalization due to recovering.

For instance, 110.applu is the program that experiences the lowest improvement with the MAP. In Figure 1, it can be observed that it has a quite high number of strided accesses but the predictor exhibits a rather low hit ratio (Figure 3a). In particular, this translates in a high percentage of loads that are predicted to be strided but they are not actually, as it can be seen in Figure 3b.

On the other hand, 124.m88ksim shows the highest improvement. This benchmark has a high percentage of strided accesses, most of them predicted correctly as it is shown in Figure 3b.

We can conclude that memory address prediction for data speculation can be an interesting mechanism to be included in future microprocessors as a way to overcome the ordering imposed by true dependences. Its hardware cost is not negligible mainly because of the large memory history table that it requires but it may be affordable for next generation machines.

The objective of this paper is to demonstrate the potential benefits of a new technique for data speculation. Detailed evaluation of different prediction schemes and their associated hardware cost is not the aim of this paper.

6 Conclusions

We have presented a novel technique for data speculation based on the observation that the effective address of many load/store instructions is highly predictable. By predicting their address, these instructions and the instructions dependent on them can be executed speculatively before their source operands are known, allowing the processor to go beyond the limits imposed by having to obey the true dependences.

We have observed an average performance gain of about 10%, and in some cases it was much higher (up to 30%). We have also shown that memory addresses are more predictable than memory values. Thus speculation based on memory addresses may be more effective than speculating on memory values. However, both techniques could be combined to performed a more aggressive speculation. We plan to investigate this issue as an extension to this work. We have also shown that data speculation based on memory address prediction is more effective than speculation based on partial memory disambiguation and both types of speculation can be combined to obtain an average gain of about 13%.

References

1. T.M. Austin, G.S. Sohi Zero-Cycle Loads: Microarchitecture Support for Reducing Load Latency. Proc. of Int. Symp. on Microarchitecture, pp 82-92, 1995.
2. T-F. Chen and J-L. Baer: A Performance Study of Software and Hardware Data Prefetching Schemes Proc of the Int. Symp. Computer Architecture, pp. 223-232, 1994.
3. R.J. Eickemeyer and S. Vassiliadis: A Load Instruction Unit for Pipelined Processors. IBM Journal of Research and Development, 37(4), pp. 547-564, July 1993
4. M. Franklin and G.S. Sohi: ARB: A Hardware Mechanims for Dynamic Reordering of Memory References IEEE Transactions on Computers, 45(6), pp. 552-571, May 1996.
5. M.Golden and T.N. Mudge:Hardware Support for Hiding Cache Latency. Technical Report #CSE-TR-152.93. University of Michigan, 1993.
6. J.L Hennessy and D.A. Patterson: Computer Architecture. A Quantitative Approach. Second Edition. Morgan Kaufmann Publishers, San Francisco 1996.
7. D. Hunt: Advanced Performance Features of the 64-bit PA-8000 Proc. of the CompCon'95, pp. 123-128, 1995.
8. D. Kroft: Lockup-free Instruction Fetch/Prefetch Cache Organization Proc. of the Int. Symp. on Computer Architecture, pp. 81-87, May 1981.
9. M.H. Lipasti, C.B. Wilkerson and J.P. Shen: Value Locality and Load Value Prediction Proc. of the 7th. ACM Conf. on Architectural Support for Programming Languages and Operating Systems, Oct. 1996.
10. M.H. Lipasti and J.P. Shen: Exceeding the Dataflow Limit via Value Prediction. Proc. of Int. Symp. on Microarchitecture, 1996
11. Y. Sazeides, S. Vassiliadis and J.E. Smith: The Performance Potential of Data Dependence Speculation & Collapsing. Proc. of Int. Symp. on Microarchitecture, December 1996.
12. A. Srivastava and A. Eustace: ATOM: A system for building customized program analysis tools Proc of the 1994 Conf. on Programming Languages Design and Implementation, 1994.

A Realistic Study on Multithreaded Superscalar Processor Design

Yuan C. Chou, Daniel P. Siewiorek, and John Paul Shen

Department of Electrical and Computer Engineering
Carnegie Mellon University
Pittsburgh, PA 15213, U.S.A.
{yuanchou,siewiorek,shen}@ece.cmu.edu

Abstract. Simultaneous multithreading is a recently proposed technique in which instructions from multiple threads are dispatched and/or issued concurrently in every clock cycle. This technique has been claimed to improve the latency of multithreaded programs and the throughput of multiprogrammed workloads with a minimal increase in hardware complexity. This paper presents a realistic study on the case for simultaneous multithreading by using extensive simulations to determine balanced configurations of a multithreaded version of the PowerPC 620, measuring their performance on multithreaded benchmarks written using the commercial P Threads API, and estimating their hardware complexity in terms of increases in die area. Our results show that a balanced 2-threaded 620 achieves a 41.6% to 71.3% speedup over the original 620 on five multithreaded benchmarks with an estimated 36.4% increase in die area and no impact on single thread performance. The balanced 4-threaded 620 achieves a 46.9% to 111.6% speedup over the original 620 with an estimated 70.4% increase in die area and a detrimental impact on single thread performance.

1 Introduction

Simultaneous multithreading is a recently proposed technique in which instructions from multiple threads are dispatched and/or issued concurrently in every clock cycle. To aggressively exploit instruction-level parallelism on a wide-issue superscalar, a large number of functional units are necessary to achieve good performance even though the average utilization of most of the functional units is low. This is due to the unsmooth parallelism profile of most programs. By executing instructions from multiple threads concurrently, simultaneous multithreading has the potential effect of smoothing the aggregate parallelism profile and thereby making better use of the functional units to improve the latency of a multithreaded program or the throughput of a multiprogrammed workload, making it an attractive microarchitectural technique.

Recent studies in simultaneous multithreading [9, 10] have demonstrated an average throughput of 5.4 instructions per cycle (IPC) on a suite of multiprogrammed workloads each comprising of 8 SPEC benchmarks. While these results are very promising, the chosen workloads may not be very representative of realistic environments. Other studies [6, 7, 8, 14] have demonstrated the potential of simultaneous multithreading on multithreaded applications. However, these research have assumed the use of a custom threading paradigm. We believe that for simultaneous multithreading to be widely adopted in industry, it must be able to take advantage of the multithreaded programs that have been written for symmetric multiprocessors (SMPs). These programs are usually written in a threads package supported by the operating system. Current operating systems such as OSF/1, Mach, Solaris, AIX, Windows NT and

Windows 95 all support multithreading and provide programmers with threads application programming interfaces (APIs).

This paper attempts to present a more realistic examination of the case for simultaneous multithreading by evaluating the performance of realistic simultaneous multithreaded superscalar processor models on multithreaded applications written using a commercial threads API and estimating the complexity required to implement these processors in terms of die area. Towards this end, we choose a real superscalar processor, the PowerPC 620 [1], as the baseline microarchitecture, and the IBM AIX 4.1 P Threads API, which is based on the emerging POSIX 1003.1c industry standard for a portable user threads API, as the threads API for creating our multithreaded applications. In the course of our study, we also attempt to understand the bottlenecks of a realistic simultaneous multithreaded processor and answer questions such as the number of threads the processor should support, the fetch and dispatch strategy the processor should employ and the functional unit mix the processor should adopt in order to achieve an efficient and balanced design. This is in contrast to previous studies in which execution resources are predetermined (usually arbitrarily expanded from a non-multithreaded design) before support for simultaneous multithreading is added to this baseline processor.

The rest of this paper is organized as follows. Section 2 gives an overview of the methodology employed in this research, as well as a description of the simulation environment and the benchmarks used. Section 3 describes the PowerPC 620 and the parameters studied in designing a balanced multithreaded 620. Section 4 presents the effects of varying these parameters and the resulting configurations of balanced multi-threaded 620s while Section 5 presents an estimation of the extra complexity required to implement these processors. Section 6 briefly mentions related work. Finally, Section 7 concludes this paper.

2 Methodology

To evaluate the performance of simultaneous multithreaded superscalar processor models, we first identify the resources of the baseline PowerPC 620 that have to be replicated or modified in order to support simultaneous multithreading, and we also identify the major parameters that affect the performance of the multithreaded processor. The PowerPC 620 is chosen as the baseline microarchitecture because it represents the current state-of-the-art in superscalar processor design and because we already have a cycle-accurate trace-driven simulator for this microarchitecture [2]. Simulations are then performed using five multithreaded benchmarks written using the IBM AIX 4.1 P Threads API and based on the simulation results, we identify balanced configurations of the multithreaded processors.

2.1 Simulation Environment

We modified the original PowerPC 620 simulator to incorporate the additional features required to support simultaneous multithreading. In addition, we also developed a tool that automatically generates PowerPC traces from multithreaded C/C++ programs written using the IBM AIX 4.1 P Threads API. This trace generation tool works by instrumenting the assembly files of the original C/C++ multithreaded program. In addition to generating the instruction and data traces, it also generates additional thread synchronization information that is used by the simulator to ensure that the synchronization structure of the original multithreaded program is preserved during simulation.

2.2 Benchmarks

Five multithreaded benchmarks representing CAD, multimedia and scientific applications are used in the simulations. We believe that these applications are realistic workloads for current and future high performance computers and therefore are good target applications for simultaneous multithreaded processors. The first three are integer benchmarks while the last two are floating-point benchmarks. All benchmarks are simulated till completion. These benchmarks are briefly described in Table 1.

TABLE 1. Description of benchmarks.

Benchmark	Description	Dynamic Inst Count
Router	Multithreaded version of ANAGRAM II [3].	14.1M
MPEG	Parallel decoding of four MPEG-1 video streams.	26.4 M
Matrix Multiply	Parallel integer matrix multiply [4].	5.1 M
FFT	P Threads implementation of SPLASH-2 [5] FFT.	22.9 M
LU	P Threads implementation of SPLASH-2 [5] LU.	16.6 M

3 The Multithreaded PowerPC 620

The PowerPC 620 [1,2] is used as the baseline processor in developing our simultaneously multithreaded processor models.

3.1 Replicated Resources

In the P Threads API, a thread is defined as a basic unit of CPU utilization that consists of a program counter, a register set, and a stack space, that shares its address space with other peer threads. To allow instructions from multiple threads to exist in the processor at the same time, the program counter, the GPR and FPR files, the Condition Register (CR), the Link Register as well as its Link Register Stack and Shadow Link Registers, the Count Register as well as its Shadow Count Register are replicated. To simplify branch misprediction and exception recovery, the reorder buffer is also replicated.

Although the GPR, FPR and CR rename buffers can be shared among threads, this will result in a multiplicative increase in the number of read and write ports to these buffers which can seriously impact machine cycle time. Therefore, we choose to replicate these rename buffers rather than to share them among threads.

Depending on the instruction fetch strategy employed, the number of instruction cache ports may or may not need to be increased. The instruction queue can be replicated or shared among threads. We choose to replicate it because it reduces the complexity of the dispatch mechanism.

3.2 Shared Resources

The functional units and their reservation stations are shared by all threads, as are the instruction and data caches and their memory management units. The bus interface unit and the L2 cache interface are also shared by all threads. Since all threads share the same code section, the branch target address cache (BTAC) is shared. We also choose to share the branch history table (BHT). In Section 4.5, we show that simultaneous multithreading has little impact on the hit rates of the shared caches and the prediction accuracy of the BHT.

3.3   Variable Parameters

Having identified the replicated and shared resources, we identify four major parameters that can be varied to achieve a balanced multithreaded 620. They are: 1) the number of threads supported, 2) the instruction fetch strategy and bandwidth, 3) the instruction dispatch strategy and bandwidth, and 4) the functional unit mix. The effects of varying these parameters are presented in the next section.

4   Experimental Results

In order to prune the design space to a manageable size (which is necessary since each simulation run takes about 2 days to complete on our fastest CPUs), we adopted the following design space exploration methodology. We first examine the effects of different fetch and dispatch strategies while assuming an expanded functional unit mix, and we select combinations that provide good cost-performance trade-offs. Next, assuming these selected instruction fetch and dispatch strategies, we gradually scale back the functional unit mix, eliminating those units that do not contribute to better performance. As the number of instructions executed by each benchmark remains constant for all processor configurations, the results are presented in terms of the average number of instructions completed per cycle (IPC). We do not attempt to quantify the impact of our microarchitectural changes and additions on machine cycle time. In all the tables presented in this section, the numbers shown are IPC numbers. "1T" means that configuration supports one thread (i.e. it is non-multithreaded) while "2T" and "4T" represent configurations that support two and four threads respectively.

4.1   Number of Threads

In Section 3.1, the resources that have to be replicated for each thread supported by the processor are described. In addition, the number of threads supported also has a direct impact on the number of buses connecting the functional units to the rename buffers and reorder buffers. Moreover, since the main advantage of a multithreaded processor over a multiprocessor is its ability to share functional units to reduce the amount of resources required to achieve the same performance level, increasing the number of threads supported increases the number of replicated resources relative to the number of shared resources, thus diminishing this advantage. For this reason, in our study, we only consider supporting two threads and four threads. The one threaded case is included for comparison purposes.

4.2   Fetch and Dispatch

The number of threads from which instructions are fetched in each cycle (F) can be varied, from fetching from one thread to fetching from all threads. Although the number of instructions fetched from a thread can also be varied, we choose to retain the 620's design of fetching at most four instructions per thread. To fetch from multiple threads in each cycle, we model the instruction cache as being multi-ported. In terms of dispatch strategy, we assume that instructions are dispatched from all threads in every cycle and we vary the number of instructions dispatched from each thread per cycle (**D**).

TABLE 2. Functional unit configuration of the baseline 620.

Functional Unit Type	#Units	#Reservation Station Entries	Issue Latency
Single-Cycle Integer Unit (SCIU)	2	2	1
Multi-Cycle Integer Unit (MCIU)	1	2	3-8 (multiply) 37 (divide)
Floating Point Unit (FPU)	1	2	1 (multiply-add) 18 (divide) 22 (square root)
Load/Store Unit (LSU)	1	3	1
Branch Unit (BRU)	1	4	1

The functional unit mix of the original 620 is shown in Table 2. In studying the effects of our fetch and dispatch combinations, we double the number of SCIUs, MCIUs and FPUs, and we add a second execution pipeline to both the LSU and the BRU. Since simultaneous multithreading increases the dispatch rate, we double the number of reservation station entries of each functional unit. The number of SCIUs, MCIUs, and FPUs can easily be increased but not the LSU. This is because of the cost and complexity of having multiple data cache ports and because of the complexity of the 620's alias detection mechanism (the 620 allows loads and stores to execute out-of-order). The second execution pipeline allows up to two loads or two stores to be issued and executed in every cycle. In the enhanced BRU, the two execution pipelines share a common reservation station. Within a thread, branch instructions are issued in-order.

TABLE 3. Effects of varying instruction fetch and dispatch strategy (IPC).

<F, D>	Router			MPEG			Matrix Multiply			FFT			LU		
	1T	2T	4T	1T	2T	4T	1T	2T	4T	1T	2T	4T	1T	2T	4T
<1, 2>		1.80	1.95		2.54	2.99		1.63	1.81		1.73	2.04		1.75	1.80
<1, 4>	1.29	1.92	1.98	1.89	2.67	3.00	1.19	1.61	1.74	1.18	1.75	2.08	1.11	1.75	1.79
<1, 8>	1.30	1.93		1.89	2.66		1.20	1.61		1.18	1.76		1.11	1.76	
<2, 2>		1.83	**2.19**		2.63	**3.60**		1.62	**1.80**		1.76	**2.25**		1.76	**1.96**
<2, 4>		**1.97**	2.26		**2.91**	3.69		**1.61**	1.77		**1.78**	2.27		**1.77**	1.97
<2, 8>		1.97			2.93			1.61			1.80			1.78	
<4, 2>			2.24			3.62			1.82			2.26			1.97
<4, 4>			2.32			3.73			1.75			2.28			1.97

F = number of threads instructions are fetched from in every cycle
D = number of instructions dispatched per thread in every cycle

Table 3 shows the effects of the various fetch and dispatch combinations on the performance of the multithreaded 620. For the 2T processor, fetching from two threads in every cycle and dispatching 4 instructions/thread (<F, D> = <2, 4>) is a good trade-off. For the 4T processor, if the total dispatch bandwidth is limited to 8 instructions/cycle (i.e. dispatch 2 instructions/thread/cycle), fetching from 2 threads/cycle almost matches the performance of fetching from 4 threads/cycle except on the *Router* benchmark. Increasing the total dispatch bandwidth to 16 instructions/cycle results in a moderate performance gain on the *Router* and *MPEG* benchmarks. However, dispatching 16 instructions/cycle may be unrealistic given current technology. Therefore, we conclude that for the 4T processor, fetching from two threads every cycle and dispatching 2 instructions/thread (<F, D> = <2, 2>) is a good trade-off. Finally, we note that dispatching 8 instructions/thread/cycle for both the 1T and 2T processors results in little performance gain because we limit the fetch bandwidth to 4 instructions/thread. Fetching more than 4 instructions/thread may require fetching from multiple cache lines (since the typical basic block length is 4-5 instructions for integer programs), thus demanding more sophisticated fetching mechanisms [15].

4.3   Functional unit Mix

In our study of instruction fetch and dispatch strategies, we have deliberately assumed a rich functional unit mix. However, not every one of the functional units may be well utilized and the number of functional units may possibly be scaled back with little or no sacrifice in performance.

Table 4 shows the effects of scaling back the number of functional units. Case A is the expanded functional unit mix assumed earlier. In Cases B and C, we scale back the number of SCIUs by one and two units respectively. In Cases D and E, we remove the second MCIU and the second FPU respectively. In Case F, we remove the second LSU execution pipeline, while in Case G, we remove the second BRU execution pipeline. Examining this table, we observe that for both the 2T and the 4T processors, removing the second LSU execution pipeline and removing two SCIUs each results in significant performance degradation, while eliminating the second BRU execution pipeline has little impact on performance. Eliminating the second MCIU only degrades the performance of the 4T processor on the *Matrix Multiply* benchmark. Removing the second FPU degrades the performance of both the 2T and the 4T processors on the *FFT* benchmark by less than 4%. Since a FPU is expensive to implement (see Table 6), the second FPU is difficult to justify.

Based on these observations, in Case H (<S, M, F, L, B> = <3, 1, 1, 2, 1>), we eliminated the second MCIU, the second BRU execution pipeline, the second FPU as well as the fourth SCIU, resulting in a functional unit mix that is a good trade-off for both the 2T and the 4T processors. For comparison purposes, in Case I, we set the functional unit mix and number of reservation entries to that of the original 620. The performance of the 2T and 4T processors are much lower than in Case H, indicating that the functional unit mix and reservation station entries of the original 620 must be suitably enriched for simultaneous multithreading to be effective. At this point, we make two observations. First, among functional units, the LSU is a bottleneck in the multithreaded processors. In fact, when the rich functional unit mix of Case A is assumed, our simulation statistics show that the dual execution pipeline LSU is the most saturated functional unit of both the 2T and 4T processors. Second, in addition to the second LSU execution pipeline, only an additional SCIU need to be added to the functional unit mix of the 620 for effective multithreading. This indicates that simultaneous multithreading is making more efficient use of the execution resources due to the smoothing of the aggregate parallelism profile.

TABLE 4. Effects of varying functional unit mix (IPC).

Case <S, M, F, L, B>	Router			MPEG			Matrix Multiply			FFT			LU		
	IT	2T	4T	IT	2T	4T	IT	2T	4T	IT	2T	4T	IT	2T	4T
A <4, 2, 2, 2, 2>	1.29	1.97	2.19	1.89	2.91	3.60	1.19	1.60	1.80	1.18	1.78	2.25	1.11	1.77	1.96
B <3, 2, 2, 2, 2>	1.29	1.96	2.19	1.88	2.86	3.58	1.19	1.61	1.82	1.18	1.78	2.24	1.11	1.78	1.96
C <2, 2, 2, 2, 2>	1.29	1.92	2.16	1.84	2.63	3.09	1.19	1.61	1.80	1.17	1.75	2.21	1.11	1.78	1.94
D <4, 1, 2, 2, 2>	1.29	1.96	2.19	1.87	2.89	3.58	1.19	1.60	1.65	1.18	1.78	2.25	1.10	1.76	1.95
E <4, 2, 1, 2, 2>	1.29	1.96	2.19	1.89	2.91	3.60	1.19	1.61	1.80	1.14	1.72	2.17	1.11	1.78	1.92
F <4, 2, 2, 1, 2>	1.24	1.86	2.04	1.77	2.69	3.24	1.13	1.59	1.66	1.15	1.72	2.12	1.07	1.57	1.62
G <4, 2, 2, 2, 1>	1.28	1.96	2.19	1.87	2.86	3.56	1.19	1.61	1.82	1.18	1.78	2.25	1.11	1.76	1.96
H <3, 1, 1, 2, 1>	1.28	1.95	2.19	1.85	2.81	3.47	1.19	1.60	1.66	1.14	1.72	2.18	1.10	1.75	1.92
I Original 620	1.20	1.77	1.82	1.64	2.24	2.67	1.16	1.57	1.64	1.08	1.51	2.00	1.06	1.55	1.52

<S, M, F, L, B> = <#SCIUs, #MCIUs, #FPUs, #LSU pipelines, #BRU pipelines>

4.4   Comparison of Multithreaded 620 vs. Original 620

To evaluate the effectiveness of simultaneous multithreading, the performance of the balanced multithreaded 620s is compared to that of the original 620. The balanced 2-threaded (2T) 620 fetches instructions from both threads in every cycle, dispatches and completes up to four instructions per thread per cycle (<F, D> = <2, 4>). In addition to the baseline 620's functional unit mix, it has a dual-execution pipeline LSU and an additional SCIU (<S, M, F, L, B> = <3, 1, 1, 2, 1>). It also has twice as many reservation station entries in each functional unit. The balanced 4-threaded (4T) 620 fetches instructions from two threads in every cycle, dispatches and completes up to two instructions per thread per cycle (<F, D> = <2, 2>), and has the same functional unit mix and number of reservation station entries as the balanced 2T 620. Also included in the comparisons is an expanded but still single-threaded 620. The expanded 620 dispatches and completes up to four instructions per cycle (<F, D> = <1, 4>), and has the same functional unit mix and number of reservation station entries as the balanced 2T and 4T 620s. Table 5 presents the results of the comparison. The expanded 620 improves upon the performance of the original 620 by 3.8% to 12.8%. The balanced 2T 620 shows a 41.6% to 71.3% improvement, while the balanced 4T 620 shows a 46.9% to 111.6% improvement.

Although simultaneous multithreading improves the performance of multithreaded programs, it can also degrade the performance of single-threaded programs since the dispatch bandwidth is now partitioned among multiple threads. To highlight the impact of simultaneous multithreading on single-thread performance, the performance of the balanced 2T and 4T processors are measured on two SPEC92 single-threaded benchmarks, *eqntott* and *tomcatv*. As shown on the right half of Table 5, the 2T processor has the same performance as the expanded 620 since it dispatches 4 instructions/thread every cycle (like the original and expanded 620s) and has an identical functional unit mix. In contrast, because the 4T processor dispatches only 2 instructions/thread in every cycle, its performance on these single-threaded programs suffers. On the *eqntott* benchmark, its performance is 21.2% worse than the original 620. This leads to the observation that if separate instruction queues are assumed for each thread and the total dispatch bandwidth is limited to 8 instructions/cycle, the 2T processor is a better trade-off than the 4T processor when the target applications of the processor comprise both multithreaded and single-threaded applications.

TABLE 5. Balanced multithreaded 620s vs. original 620 on multithreaded and non-multithreaded benchmarks (IPC).

| | Multithreaded Benchmarks | | | | | Single-threaded | |
	Router	MPEG	Matrix Multiply	FFT	LU	eqntott	tomcatv
Original 620	1.20	1.64	1.13	1.08	1.06	1.32	1.00
Expanded 620	1.28	1.85	1.19	1.14	1.10	1.35	1.01
	+6.7	+12.8	+5.0	+5.6	+3.8	+2.3	+1.0
Balanced 2-threaded 620	1.95	2.81	1.60	1.72	1.75	1.35	1.01
	+62.5	+71.3	+41.6	+59.3	+65.1	+2.3	+1.0
Balanced 4-threaded 620	2.19	3.47	1.66	2.18	1.92	1.04	0.99
	+82.5	+111.6	+46.9	+101.9	+81.1	- 21.2	- 1.0

4.5   Effects on Caches and Branch Prediction

We also studied the impact of simultaneous multithreading on the hit rates of the shared instruction and data caches and the prediction accuracy of the shared branch history table (BHT) when the multithreaded benchmarks are run. Due to limited space, we briefly state that simultaneous multithreading has negligible impact on the hit rate of the shared instruction cache. It degrades the hit rate of the data cache by less than 2% on all benchmarks for both the 2T as well as the 4T processor. It also has little impact on the shared BHT. The prediction accuracy for both the 2T and 4T processors are within 3% of that of the original 620.

5   Complexity Estimation

The commonly cited advantage of a multithreaded processor over a multiprocessor is that the former requires less resources to achieve a given level of performance because of the sharing of execution units. In this section, a first-order estimation of the extra complexity required to implement the balanced multithreaded 620s is made.

The estimation is made by determining from the die photo (not shown due to space limitations) of the original 620 the die areas of the various shared and replicated resources. We make the simplistic assumption that the area of the replicated resources scale linearly with the number of replications unless otherwise stated. Since the number of reservation station entries of the functional units are doubled in the 2T, 4T and Expanded 620s, we estimate the die areas of these functional units to be 25% larger. Although the instruction cache of the 2T and 4T processors are assumed to be dual-ported in our simulations, here we assume that they occupy the same die area as the data cache which is two-way interleaved. We assume that the die area of the enhanced LSU is twice that of the original LSU. We also assume that the die area occupied by the dispatch and completion unit of the 2T 620 is twice that of the original 620 while that of the 4T 620 is four times larger.

The results of the estimation are shown in Table 6. The units of area are relative square units. The balanced 2-threaded 620's die area is estimated to be 36.4% larger than that of the original 620, while that of the balanced 4-threaded 620 is estimated to be 70.4% larger. For comparison purposes, the expanded 620 is estimated to be 12.6% larger than the original 620. Overall, the die area increases appear to be more significant than implied in previous studies but is still reasonable.

TABLE 6. Comparison of die areas (in relative square units).

	Original 620	Expanded 620	Balanced 2T 620	Balanced 4T 620
Instruction Cache + MMU	26	26	31	31
Data Cache + MMU	31	31	31	31
Bus + L2 Interface + Perf. Monitor + PLL + COP JTAG	19	19	19	19
SCIU	8	10	15	15
MCIU	6	7.5	7.5	7.5
FPU	18	22.5	22.5	22.5
LSU	9	18	18	18
BRU	5	6.5	6.5	6.5
GPR and Rename Buffers	3	3	6	12
FPR and Rename Buffers	6	6	12	24
Dispatch + Completion Unit	16	16	32	64
Total	147	165.5 +12.6	200.5 +36.4	250.5 +70.4

6   Related Work

Other than the studies mentioned in Section 1, [11] studied the effects of limited fetch bandwidth and instruction queue size on a 2-threaded superscalar processor with unlimited issue width and unlimited functional units while [12, 13] studied simultaneous multithreaded processors that dynamically interleave VLIW instructions.

7   Conclusion

The cost/performance trade-offs of both the 2-threaded and 4-threaded processors when executing multithreaded benchmarks are reasonable although much less impressive than reported in previous studies using multiprogrammed workloads. The dispatch bandwidth and load/store unit limitations result in the diminishing returns of the 4T processor. Because of the partitioning of the dispatch bandwidth among multiple threads, the 4-threaded processor actually degrades the performance of non-multi-threaded programs.

In conclusion, we believe that the future of simultaneous multithreading depends on the widespread adoption of the multithreaded programming paradigm. If this programming paradigm is widely adopted, resulting in a proliferation of multi-threaded applications, simultaneous multithreaded processors are an efficient means of exploiting the thread and instruction level parallelism of these applications. On the other hand, if single-threaded applications continue to dominate or if future single-threaded processors are able to effectively utilize the available dispatch bandwidth, the future of simultaneous multithreading may be less promising.

References

[1] D. Levitan. T. Thomas, and P. Tu, "The PowerPC 620 Microprocessor: A High Performance Superscalar RISC Microprocessor", in *Spring CompCon 95 Proceedings*, pages 285-291, 1995.

[2] T. A. Diep, C. Nelson and J. P. Shen, "Performance Evaluation of the PowerPC 620 Microarchitecture", in *Proceedings of the 22nd Annual International Symposium on Computer Architecture*, pages 163-175, 1995.

[3] J. M. Cohn, D. J. Garrod, R. A. Rutenbar, and L. R. Carley, "KOAN/ANAGRAM II: New Tools for Device-Level Analog Placement and Routing", in *IEEE Journal of Solid-State Circuits*, Vol. 26, No. 3, March 1991.

[4] J. Boykin, D. Kirschen, A. Langerman, and S. LoVerso, "Programming Under Mach", Addison-Wesley, 1993.

[5] S. C. Woo, M. Ohara, E. Torrie, J. P. Singh, and A. Gupta, "The SPLASH-2 Programs: Characterization and Methodological Considerations", in *Proceedings of the 22nd Annual International Symposium on Computer Architecture*, pages 24-36, 1995.

[6] H. Hirata, K. Kimura, S. Nagamine, Y. Mochizuki, A. Nishimura, Y. Nakase, and T. Nishizawa, "An Elementary Processor Architecture with Simultaneous Instruction Issuing from Multiple Threads", in *Proceedings of the 19th Annual International Symposium on Computer Architecture*, pages 136-145, 1992.

[7] M. Gulati and N. Bagherzadeh, "Performance Study of a Multithreaded Superscalar Microprocessor", in *Second International Symposium on High-Performance Computer Architecture*, pages 291-301, 1996.

[8] M. Loikkanen and N. Bagherzadeh, "A Fine-Grain Multithreading Superscalar Architecture", in *Proceedings of PACT '96*, pages 163-168, 1996.

[9] D. M. Tullsen, S. J. Eggers, and H. M. Levy, "Simultaneous Multithreading: Maximizing On-Chip Parallelism", in *Proceedings of the 22nd Annual International Symposium on Computer Architecture*, pages 392-403, 1995

[10] D. M. Tullsen, S. J. Eggers, J. S. Emer, H. M. Levy, J. L. Lo, and R. L. Stamm, "Exploiting Choice: Instruction Fetch and Issue on an Implementable Simultaneous Multithreading Processor", in *Proceedings of the 23rd Annual International Symposium on Computer Architecture*, pages 191-202, 1996.

[11] G. E. Daddis and H. C. Torng, "The Concurrent Execution of Multiple Instruction Streams on Superscalar Processors", in *International Conference on Parallel Processing*, pages I 76-83, 1991.

[12] R. G. Prasadh and C. Wu, "A Benchmark Evaluation of a Multi-Threaded RISC Processor Architecture", in *International Conference on Parallel Processing*, pages I 84-91, 1991.

[13] S. W. Keckler and W. J. Dally, "Processor Coupling: Integrating Compile Time and Runtime Scheduling for Parallelism", in *Proceedings of the 19th Annual International Symposium on Computer Architecture*, pages 202-213, 1992

[14] M. Bekerman, A. Mendelson, and G. Sheaffer, "Performance and Hardware Complexity Trade-offs in Designing Multithreaded Architectures", in *Proceedings of PACT '96*, pages 24-34, 1996.

[15] T. M. Conte, K. N. Menezes, P. M. Mills, and B. Patel, "Optimization of Instruction Fetch Mechanisms for High Issue Rates", in *Proceedings of the 22nd Annual International Symposium on Computer Architecture*, pages 333-344, 1995.

Acknowledgments

This research is supported by the NSF under grant MIP-9403473 and by the ONR under Contract N00014-96-1-0347.

A Limitation Study into Access Decoupling

G.P. Jones and N.P.Topham*

Department of Computer Science,Edinburgh University,Edinburgh,U.K.

Abstract. This paper presents a study into the theoretical limits of a latency hiding technique called *access decoupling*. Access decoupling is effective at hiding memory latency for low ILP and conservative dependency analysis [9,12,13]. We assess if this result still applies for maximum ILP and perfect dependency analysis.
We find that access decoupling with a basic decoupled memory model fails to hide latency. However, when the memory system is optimised to capture temporal locality exposed by decoupling, sensitivity to memory latency is almost removed. The optimised memory is also found to be a powerful bandwidth filter.

1   Introduction

Access decoupling is a latency hiding technique which tries to hide memory latency by overlapping computation and memory access operations [1,5,10,16]. Studies into access decoupling have shown that it can successfully hide memory latency [9,12,13] when ILP is low and data dependency analysis is conservative.

However, one can predict that future latency hiding technology will be stressed by relatively slower memory speeds and increased ILP. Increased ILP will place greater pressure on memory systems and require higher sustained bandwidth. One can also predict that improvements in compiler technology and data dependency analysis may offer greater opportunities for latency hiding.

In this paper we describe our study into the theoretical limits of latency hiding through decoupled execution with maximum ILP and perfect dependency analysis. We have developed a model of decoupled execution that allows us to simulate the execution of scientific Fortran applications on an abstract decoupled machine. To focus on the limits of decoupling, our model had unlimited address computation resources and ideal out-of-order (o-o-o) execution.

We have found that a basic memory model for access decoupling fails to hide large memory latency. If however, we *optimise* our decoupled model to include a mechanism that utilises the temporal locality exposed by decoupling, sensitivity to memory latency is almost completely removed. This mechanism is also found to act as a powerful bandwidth filter. Our results provide useful information about the necessary requirements of future access decoupled machines.

This paper is organised in the following way. In section 2 we discuss some of the previous work in decoupling. In section 3 we describe the latency hiding

* This work was spupported by U.K. EPSRC under grant K19723

model and introduce β, an architecture independent measure of the efficiency of a latency hiding technique. In section 4 we describe our decoupled model and in section 5 discuss our simulation techniques. In section 6 we present our results. Finally in section 7 we conclude with a summary of our findings and a discussion of future work.

2   Background

Access decoupling is an asynchronous data prefetching technique which tries to hide memory latency by overlapping computation and memory access operations. Central to all decoupled machines [1,5,10,16] is an architectural visible address unit (AU) and data unit (DU); these units are responsible for performing, respectively, the memory accesses and data computations in a program. They each have their own program counter allowing the AU to run ahead of the DU. The degree to which the AU is ahead of the DU is called the *slippage*. The units communicate with each other and with memory via queues.

The early decoupled machines like the ZS-1 [5] and PIPE [10] differed in how they split the instruction stream. The ZS-1 had a single instruction stream with a splitter whilst PIPE had separate instruction caches for the access and execute unit. The ZS-1, unlike PIPE, also included a data cache. More recently decoupled machines like the DAE [1], MISC [14] and ACRI [2] have appeared. To increase slippage DAE includes specialised hardware for efficient address generation. This hardware is effective at reducing DU stall time and increasing cache utilisation. The MISC [14] architecture, derived from PIPE, has four asynchronous units each with their own instruction cache but common data cache. The ACRI machine included an additional control unit responsible for computing conditional branches and dispatching instructions to the AU and DU.

Decoupling has gained currency in superscalar architectures like the MIPS R10000 [17]. The R10000 is able to support a decoupled mode of operation through o-o-o execution and a separate access instruction queue. The R10000 can decouple address and execute operations even though there is no architecturally visible AU and DU.

3   A Model for Latency Hiding

The use of memory hierarchies in high performance architectures is a consequence of the need to balance the cost, capacity and performance benefits of different memory technologies. Latency hiding techniques try to hide the latency of the slowest level of the memory hierarchy so that all accesses are perceived by the CPU to occur at the speed of the fastest element. The difference between the speeds of the fastest and slowest memory we refer to as the *memory differential*. In assessing the efficiency of any latency hiding technique we focus on the extent to which the memory differential can be hidden. In this section we introduce a new term called β which is an architecture independent measure of the efficiency of a latency hiding technique. We define β as the *fraction of the memory differential which is hidden from a memory access* and use it to quantify the *latency hiding efficiency*.

A generic model of a memory hierarchy is shown in figure 1. The element χ is used to denote any number of levels and types of memory (e.g. queues, buffers, caches) in the memory hierarchy. Since χ characterises the type of memory system we will refer to it as an χ memory system. First we define the following terms for an χ memory system.

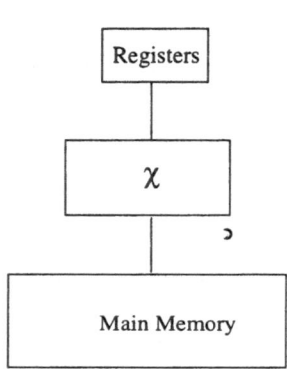

Fig. 1. χ Memory System   **Fig. 2.** Access Decoupled model

- c_M, c_m are, respectively, the execution cost of a main memory and register access.
- δc is the memory differential. It is simply given by $c_M - c_m$.
- H_r, H_w are, respectively, the cost of a read and write hit in χ.
- M_r, M_w are, respectively, the cost of a read and write miss in χ.
- $\overline{p}$ is the average perceived access time for the χ memory system.
- δl is the average hidden memory latency. It is given by

$$\delta l = \delta c - (\overline{p} - c_m) = c_M - \overline{p}$$

- α_r and α_w are respectively, the read and write hit ratio for χ.

The efficiency of a latency hiding technique is the fraction of the memory differential hidden by χ. This is given by :

$$\beta = \frac{\delta l}{\delta c} \tag{1}$$

The average perceived access time is given by

$$\overline{p} = (1 - w)[\alpha_r H_r + (1 - \alpha_r)M_r] \\ + w[\alpha_w H_w + (1 - \alpha_w)M_w] \tag{2}$$

where w is the proportion of write accesses. We know that for any memory system $H_r, H_w \geq c_r$ and $M_r, M_w \geq c_M$. Substituting into equations 1 and 2 we find

$$\beta \leq (1 - w)\alpha_r + w\alpha_w \qquad (3)$$

When $\alpha_r = \alpha_w$ equation 3 reduces to $\beta \leq \alpha$. This shows that β for any memory hierarchy is bounded by the hit ratio for any memory hierarchy.

As we can compare different latency hiding techniques using β, it is possible to predict the miss ratio required by a cache in order to attain the same level of latency hiding through decoupling (see section 6).

An alternative and equivalent way of expressing β is $(1 - w)\beta^{read} + w\beta^{write}$ where $\beta^{read} \leq \alpha_r$ and $\beta^{write} \leq \alpha_w$. β^{read} and β^{write} are the latency hiding efficiency for, respectively, a read and write access.

4   A model of Access Decoupling

In our *decoupled model*, based on the ACRI machine [2] and illustrated in figure 2, there are three asynchronous superscalar units, a control unit (CU), an address unit (AU) and a data unit (DU). Each of the units can schedule dynamically and perform o-o-o issue and completion. The bulk of the computational work is performed on the AU/DU pair. The CU is responsible for dispatching instructions and computing conditional branches ahead of the AU/DU pair. This is a technique called *control decoupling*. A discussion of this technique is beyond the scope of this paper.

The *decoupled memory* is capable of sending and receiving data from the AU/DU pair and main memory. The AU and DU can both fetch and write data into the decoupled memory but only the AU can access main memory. We avoid an implementation description of the decoupled memory instead concentrating on the semantics necessary to support decoupling.

A *decoupled load* to the DU is performed by way of a split instruction executed on the AU and DU. The AU initiates the load by computing the load address and sending it, via the decoupled memory, to main memory. Once the requested data is returned it waits in the decoupled memory until it is *fetched* by the DU. The AU is also capable of issuing decoupled self loads in a similar manner. A *decoupled store* is performed by the AU sending an address to the decoupled memory where it waits until the DU computes the data. The decoupled store *completes* when the address and data are paired and sent to main memory.

The decoupled memory guarantees that RAW [1] hazards will be resolved, provided the AU sends load and store addresses to main memory in program order. Load addresses are, however, allowed to overtake each other. The decoupled memory detects RAW hazards by matching load and store addresses. When a hazard is detected the load is *suspended* until the store completes. In this paper we investigate two different decoupled memory models for handling RAW conflicts:

---

[1] Our simulation technique allows us to remove false dependencies so we do not need to consider WAW and WAR data hazards (see section 5).

1. In our first memory model each load waits for completion of any outstanding writes to the same address before going to main memory. We refer to this as the *basic decoupled memory* since it relies only on the slippage between the AU and DU to hide the main memory latency.
2. In our second memory model, information about *future* access patterns is used to *cache* data close to the DU. When a store datum is received from the DU it is written to main memory and a copy is kept in the decoupled memory. Note the data is only cached when there are suspended loads. We refer to this as the *optimised decoupled memory*.

An advantage of the optimised over the basic decoupled memory model is that it acts as a *bandwidth filter* reducing required memory bandwidth. In section 6 we investigate the required bandwidth for the optimised memory model.

5   Simulations

The purpose of our experiments was to find the limitations of access decoupling when ILP is maximised. We used a technique [7], which allowed us to :

- simulate the execution of the program at the source level. This meant we could concentrate on the high level semantics of access decoupling without bringing in issues of code generation.
- simulate an idealised decoupled machine with unlimited resources, idealised o-o-o execution and perfect data dependency analysis. This meant our machine had *maximum ILP* and *optimal slippage*. The latter is a consequence of the unlimited addressing resources and the perfect data dependency analysis removing false synchronisation points.

There was no speculative execution but operations from different loop bodies could execute in parallel. In our experiments we simulated four programs from the PERFECT club suite [6]. We recognise that this represents a small subset of programs but we justify our approach in the following way. Researchers typically only simulate *part* of a program's total execution time because *full* simulations can be prohibitive. We have found that some programs do need to be simulated in full because there are identifiable phases in the program which exhibit differing behaviour [7]. We therefore chose to perform full simulation of a small number of programs to completion.

We selected benchmarks from the PERFECT club to represent varying degrees of vectorisation and also to span known degrees of decoupling. Table 1 shows the four selected benchmarks. This table gives the reported proportion of vectorised operations (VO) [15] and the decoupling efficiency (DE) [13]. The other columns give the measured total number of operations, the total number of loads, a breakdown by unit of the ILP [2], the fraction of loads (FOL) and the fraction of operations (FOP). The measurements were made with $\delta c = 0$ and a floating point cost of 5 cycles.

---

[2] These ILP measurements are high but are within the bounds of previous studies [8]

Program	Reported		No. of ops (*10⁹)	No. of loads (*10⁸)	Measured								
	VO (%)	DE (%)			ILP			FOL(%)			FOP(%)		
					cu	au	du	cu	au	du	cu	au	du
ARC2D	91.1	99	6.797	13.139	302	1255	2155	8	0.1	91.9	8	34	58
TRFD	69.8	98	1.857	4.419	3	1067	1676	0	0.1	99.9	0.1	38.9	61
FLO52Q	91.5	84	2.244	4.626	30	156	224	7	0	93	7	38	55
QCD2	4.2	19	1.072	1.126	5	7	10	8	2	90	22	33	45

Table 1. Characteristics of benchmark programs

6 Experimental Results

The three major findings of our experiments into the effectiveness of access decoupling as a latency hiding technique under the pressure of maximum ILP are :

1. access decoupling with an optimised decoupled memory is potentially a powerful latency hiding and bandwidth filtering mechanism.
2. β_0 for access decoupling is comparable to that of a copy-back cache with hit ratios between 88% and 99%.
3. access decoupling with a basic decoupled memory can not hide large memory latency, even with optimal slippage.

To provide the most favourable conditions for latency hiding the execution cost of AU loads and address computations were set to 1 and 0 cycles respectively. The floating point and CU load latency were 1 cycle. Communication between units carried no cost.

Figure 3 shows the variation of *relative increase in execution time* [3] as a function of the memory differential. It can be seen that, even with optimal slippage, access decoupling using the basic decoupled memory model is sensitive to increases in memory differential. Only TRFD has sufficient parallelism to hide the latency. For FLO52Q, a memory differential of 80 cycles increases the program execution time by a factor of 12.

Figure 4 shows the effect introducing the optimised decoupled memory. It can be seen that all four programs show little variation with increases in the memory differential. For FLO52Q, a memory differential of 80 cycles only results in a relative increase of 0.6% in the execution time. Figure 5 shows that if the floating point latency increases to 5 cycles, effectively slowing down the DU, the sensitivity to memory latency decreases still further.

Figure 6 shows the simulation results for a more realistic case in which AU loads are decoupled and AU address computations take a single cycle. The results show that all the programs still remain insensitive to increases in the memory differential.

---

[3] The relative increase is given by

$$\frac{T(l) - T(0)}{T(0)}$$

where $T(l)$ and $T(0)$ are the times to execute the program when the memory differential is l and 0 respectively.

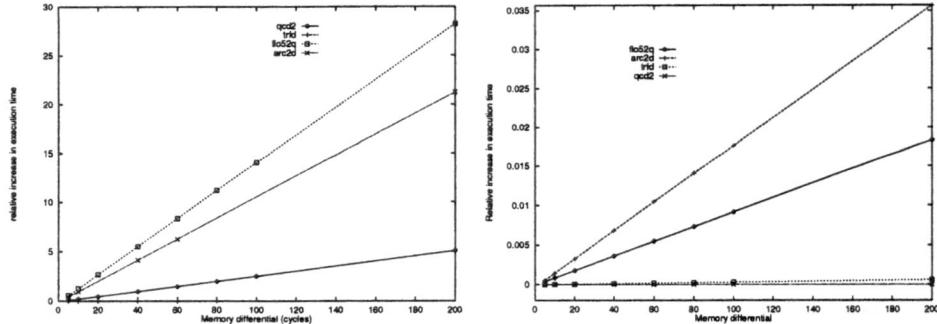

Fig. 3. Basic decoupled memory model; FP latency = 1

Fig. 4. Optimised decoupled memory; FP latency = 1 cycle

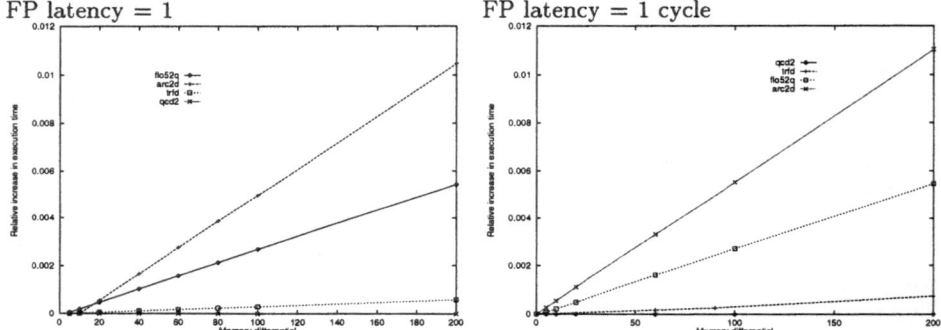

Fig. 5. Optimised decoupled memory; FP latency=5 cycles

Fig. 6. AU and DU with decoupled loads; FP latency=5 cycles

One possible criticism of our findings is that the partitioning of the code has fortuitously biased our experiments. If most of the loads on a program's critical path were executed on the CU then our experiments would show positive results for access decoupling (recall that CU loads perceive a fixed cost of 1 cycle). To answer this criticism we moved all loads previously executed on the CU to the DU. Figure 7 and 8 show that whereas three of the programs are still insensitive to the memory differential the increase in QCD2's execution time is comparable to results for a basic decoupled memory model (see figure 3).

The explanation for QCD2's poor results can be found in the routines **LADD**, **LMULT** and **PRANF** where scalar loads previously executed on the CU dominate the critical path. These loads, executed on entry to the routines, can not decouple and perceive all of the memory differential. The three routines are most frequently called from inside a while loop in the **CHOOS** routine. By inlining the three routines it is possible to hoist the loads out of the loop. The effect of this simple optimisation can be seen in figure 8. The results show clearly that QCD2 is once again insensitive to increases in the memory differential.

We can conclude that under conditions of maximum ILP, access decoupling with the optimised decoupled memory is still effective at hiding memory latency. We have shown that higher floating point latencies improve latency hiding and that our results are independent of the partitioning of loads between CU and DU.

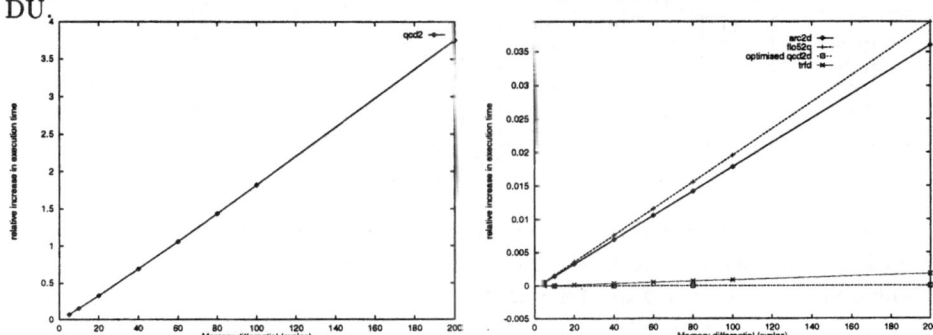

Fig. 7. Relative increase in execution time with no CU for QCD2

Fig. 8. Relative increase in execution time with no CU

The optimised decoupled memory also acts as a bandwidth filter by reducing the number of accesses that need to go to main memory. This finding is reinforced by the results in table 2 which show the distribution of the decoupled loads in the programs. The table shows the percentage of loads which are *coupled*(loads perceive the full memory differential), *fully decoupled*(loads perceive a single cycle latency) and *partially decoupled* (loads perceive part of the memory differential). The column labelled 'cached' is the percentage of decoupled loads that are cached in the optimised decoupled memory (see section 4). Since cached decoupled loads will not access main memory this column also shows the bandwidth filtering. It can be seen that, with the exception of TRFD, between 77% and 98% of operations are satisfied in the decoupled memory.

In TRFD almost 50% of loads are either partially decoupled or coupled. This phenomenon is due to induction variables serialising frequently executed loops. The induction variables can be removed using generalised induction variable elimination [4]. The effect of this optimisation can be seen in the row labelled 'optimised TRFD'. The increase in parallelism allows the AU to slip further ahead of the DU, increasing the number of fully decoupled loads from 14.3% to 49.8%.

Clearly the unlimited resources in our decoupled model makes the levels of bandwidth filtering unrealisable. For the four programs we analysed our experiments show that the maximum capacity of the decoupled memory is between 65% and 75% of the DU working set size. Although this size of capacity could not reasonably be implemented, we still expect high filtering for more realistic degrees of ILP.

Our motivation for deriving β was to provide a way of comparing the latency hiding efficiency of different techniques. Since caches are the most widely used technique for hiding latency we have compared access decoupling against a copy back (CB) cache. A full derivation of β for a copy back cache is discussed in [7].

In our experiments we computed β for our decoupled machine by measuring the average perceived load latency. Table 3 shows a comparison of β against the

Program	Decoupled load			
	cached	Full	Partial	Coupled
	(%)	(%)	(%)	(%)
ARC2D	97.5	1.9	0.0	0.6
QCD2	77.3	1.8	0.4	20.5
FLO52Q	89.6	3.5	0.4	6.4
TRFD	35.8	14.3	25.3	24.5
optimised TRFD	49.4	49.8	0.0	0.8

Table 2. Decoupled loads for benchmark programs

program	$\delta c = 5$			$\delta c = 60$	
	β	hit rate		β	hit rate
		$b = 8$	$b = 2$		$b = 8$
ARC2D	99.3	99.8	99.7	99.5	99.8
FLO52Q	96.6	99.1	98.2	94.3	96.6
QCD2	81.8	95.1	90.4	81.8	89.2
TRFD	70.7	92.1	84.6	80.1	88.2
optimised TRFD	99	99.7	99.5	99.3	99.6

Table 3. β versus CB cache hit rate for different block size

CB cache hit rate required to achieve an equivalent degree of latency hiding. The term b in columns 3,4 and 6 is used to denote the cache block size. The values in the columns labelled β are for our decoupled machine.

For example in ARC2D when the memory differential is 60 cycles β is 99.5%. For comparable latency hiding efficiency, a CB cache with block size 8 words would require a hit rate of 99.8%. It is clear that the comparable hit rates are high, in the range 88% to 99%. The table also shows that when the memory differential is reduced to 5 cycles the comparative CB hit rates increase. When the block size is reduced to 2 words the comparative hit rates are similar to those when $\delta c = 60$ and $b = 8$.

7 Conclusion

This paper has described a study into the theoretical limits of latency hiding through decoupled execution when ILP is pushed to a maximum. To focus on the limits of decoupling, our machine model had unlimited address computation resources, ideal o-o-o execution and perfect data dependency analysis.

We have shown that when ILP is a maximum and dependency analysis is perfect, access decoupling with an optimised decoupled memory is a powerful latency hiding and bandwidth filtering technique. The optimised decoupled memory uses information about future access patterns to cache data close to the AU and DU. However, unlike a typical cache where resident data may be accessed, the cached data in the decoupled memory will definitely be used. In Smith's paper [11] a similar technique was discussed but no empirical evidence was given to support its use. The SUNDER architecture employed a similar technique through the use of pending store and load queues in the prefetch engine [3].

Our study has also shown that access decoupling with the basic decoupled memory model is not capable of hiding memory latency.

We can conclude that if future designers are able to obtain the levels of ILP and dependency analysis outlined in this paper, access decoupling could still be an effective latency hiding technique.

We are currently investigating whether these results apply to ILP levels projected for the next decade. We are also investigating an implementation of the optimised decoupled memory with a data cache.

References

1. A. Berrached, L.D. Coraor, and P.T. Hulina. A Decoupled Access/Execute Architecture for Efficient Access of Structured Data. In *Proc. of the 26th Hawai Int. Conf. on System Sciences*, volume 1, pages 438–47, Los Alamitos, CA, USA, Jan 1993. IEEE.
2. P. Bird, A. Rawsthorne, and N.P. Topham. The Effectiveness of Decoupling. In *Proc. Int. Conf. on Supercomputing*, Tokyo, Japan, May 1993.
3. Tzi cker Chiueh. Sunder : A Programmable Hardware Prefetch Architecture for Numerical Loops. In *Proc. Supercomputing '94*, pages 488–497, Los Alamitos, CA, USA, Nov 1994. IEEE Comput. Soc., ACM , SIAM, IEEE Comput. Soc. Press.
4. R. Eigenmann, J. Hoeflinger, L. Zhiyuan, and D. Padua. Experience in the Automatic Parallelisation of Four Perfect-Benchmark Programs. CSRD 1193, Center for Computing Research and Development., University of Illinois at Urbana-Champaign, Urbana Illinois., 1992.
5. J.E. Smith et al. The ZS-1 Central Processor. In *Proc. of the 2nd Int. Conf. on Architectural Support for Programming Languages and Operating Systems*, October 1987.
6. M. Berry et al. The Perfect Club Benchmarks, Effective Performance Evaluation of Supercomputers. Techreport 827, CSRD, University of Illinois, Urbana-Chmpaign, Urbana, Illinois., May 1989.
7. G. Jones. Evaluating the Limits of Access Decoupling using the Latency Hiding Model. Technical report, Edinburgh University, 1997.
8. M. Kumar. Measuring Parallelism in Computation Intensive Scientific/Engineering Applications. *IEEE Transactions on Computers*, 37(9):1088–1098, Sept. 1988.
9. L. Kurian, P.T. Hulina, and L.D. Coraor. Memory Latency Effects in Decoupled Architectures. *IEEE Trans. on Computers*, 43(10):1129–39, Oct. 1994.
10. M.K.Farrens and A.R.Pleszkun. Implementation of the PIPE Processor. *IEEE Computer*, pages 65–70, Jan 1991.
11. J.E. Smith. Decoupled Access/Execute Computer Architectures. *ACM Trans. on Computer Systems*, 2(4):289–308, Nov. 1984.
12. J.E. Smith. A Simulation Study of Decoupled Architecture Computers. *IEEE Trans. on Computers*, C-35(8):692–701, Aug. 1986.
13. N.P. Topham, A. Rawsthorne, C.E. McLean, M.J.R.G. Mewissen, and P.Bird. Compiling and Optimising for Decoupled Architectures. In *Proc. of Supercomputing '95*, San Diego, Dec. 1995. ACM press.
14. G. Tyson, M. Farrens, and A.R. Pleszkun. MISC : A Mutiple Instruction Stream Computer. In *Proc. of the 25th Ann. Sym. on Microarchitecture*, Portland, Oregon, Dec 1-4 1992.
15. S. Vajapeyam, G.S. Sohi, and W-C Hsu. An Empirical Study of the CRAY Y-MP Processor using the PERFECT Club Benchmarks. In *Proceedings of the 1991 ACM Int. Conf. on Supercomputing*, pages 170–179, New York, 1991. ACM, ACM press.
16. Wm A. Wulf. An Evaluation of the WM Architecture. In *Proc. Int. Symp. on Computer Architecture*, Gold Coast, Australia, May 1992.
17. K.C. Yeager. The Mips R10000 Superscalar Microprocessor. *IEEE micro*, 16(2):28–41, April 1996.

Workshop 18

Parallel and Distributed Database Systems

Workshop 18:
Parallel and Distributed Database Systems

Andreas Reuter

I will start with some observations on the state of both distributed and parallel database technology and the relations between the two, and then try to put the papers in this year's Euro-Par workshop into the context of the evolution of those technologies.

The technology of distributed database systems has been around for more than ten years now, which makes it an "old" technology by the standards of our discipline. For quite some time, its application was confined to a small number of system classes, where scalability, reliability and very high- performance OLTP are the key issues. Now that distributed databases are being used in environments that are (substantially) different from what the technology was originally designed for, new problems need to be solved – as well as some old problems that were put aside as being "less important".

Similar observations can be made for parallel databases, although this field is a bit younger and more active judged by the ongoing development efforts. Parallel database technology is an offspring of distributed databases, focussing on performance and treating distribution as a technical rather than an application issue. Consequently, these two lines of research and development have been pursued independently, at least for a while. Meanwhile, it has become obvious that distributed databases on the one hand and parallel databases on the other have a number of structurally similar problems, although the instances of those problems may be characterized by different parameters (bandwidth, latency, dynamic vs. static mapping, etc.). Originally, the differences were emphasized, resulting in research into the same topic, here under the heading of "distributed databases", and under "parallel databases" there. A deeper understanding of the underlying structural issues has fostered a different attitude: increasingly, researchers try to understand the abstract problem, for example the principles of data partitioning, first, and then explore systematic mappings onto a range of parallel and distributed architecture, run-time systems, and so on. In the same vein, it has been discovered that techniques which originally were thought to be "clever tricks", optimizing system performance under special constraints such as data placement, process and thread control etc., are in fact facets of the much larger (and more complicated) problem of static and dynamic load balancing.

And finally, new application areas such a multi-media databases require solutions which consist of a mix of components, some of which could be characterized as distributed technology, while others are decidedly parallel – all instantiated from the same set of primitives.

This year's workshop on parallel and distributed databases reflects this trend of convergence with respect to the two lines of research. The two sessions (the first consisting of the first four, and the second of the last three papers) are *not* organized into one covering the distributed systems, and the other one parallel systems – as one might have expected. Rather, each session focuses on problem classes which are equally important in a parallel and in a distributed execution environment.

Session 1 covers the increasingly important subject of load balancing. The paper by Manegold et al. makes the point that query evaluation and optimization is closely intertwined with load balancing – an observation which will have important consequences for the design of future optimizers. The second paper focuses on a special type of queries, recursive queries, which need very efficient dynamic load balancing in order to realize the performance potential of parallel architectures.

The other two papers discuss the problem of data partitioning in databases with structured schema objects, where the structure can be exploited to anticipate certain access patterns. It turns out that data partitioning, put into the proper perspective, is but one facet of static load balancing.

The second session is a little less coherent with respect to the subject areas, but all papers address the issure of performance measurement and optimization. About the first paper by Christodoulakis et al. the reader may wonder why it is presented in this workshop at all. On the surface, it says little about either distribution or parallelism. On second thought, however, it fits quite nicely, because high-performance multi-media systems will be both distributed and parallel, one way or the other. And of course, we have learned how deeply application-specific benchmarks can influence the architecture of database systems.

The second paper by Hameurlain and Bonneau deals with load balancing, but at a different level than those in the first session. It adopts a fairly formal view in order to model the behaviour of a run-time system of a distributed and/or parallel database system. Based on a number of detailed assumptions, the model allows to exploit locality dynamically – which is the key issure for optimization in all kinds of data-parallel computations.

The last paper by Nafjan und Kerridge is a little outside the scope of this workshop – strictly speaking. Most of the paper deals with genetic algorithms and ways to improve their breeding success; joint processing seems to be but a non-trivial example. Considering, however, the general state of the art in database query optimization, it is perfectly justified to explore all kinds of new (and hopefully better) ideas.

All in all, the workshop presents parallel and distributed databases as a rapidly maturing technology with virtually unlimited applicability – in fact, as *the* success story of large scale parallel and distributed processing in the commercial domain.

Load Balanced Query Evaluation in Shared-Everything Environments* **

Stefan Manegold[1]   Johann K. Obermaier[1]   Florian Waas[2]

[1]Humboldt-Universität zu Berlin
Institut für Informatik
10099 Berlin, Germany
⟨lastname⟩@dbis.informatik.hu-berlin.de

[2]CWI, P.O.Box 94079
1090 GB Amsterdam
The Netherlands
flw@cwi.nl

Abstract. In this paper, we present *data threaded execution*, a new strategy to exploit both, *pipelining* and *intra-operator parallelism* in shared-everything environments. Data threaded execution is intuitive, straightforward to implement, but resistant against workload estimation errors and resistant against the *discretization error* of processor scheduling, that conventional strategies suffer from. Furthermore, data threaded execution minimizes *startup* and *shutdown execution delays*. Simulation results show that data threaded execution outperforms conventional strategies significantly due to the better utilization of parallel processing resources.

1   Introduction

Parallel processing in database systems is one of the keys to the required performance improvements of modern database applications [DG92]. In general, parallelism for the evaluation of database queries is classified into three main categories: *inter-query*, *inter-operator*, and *intra-operator parallelism*. Inter-operator parallelism with no execution dependencies between operators is called *bushy parallelism*. With a producer/consumer relationship between operators, we speak of *pipelining parallelism*. Recently, the use of inter-operator parallelism has been investigated [CLYY92, SD90, SYT93, SE93, WA91]. Pipelining parallelism is of particular interest.

The major problem with the usage of pipelining parallelism are the dependencies between operators, i.e. the performance of the pipelining execution is dominated by the slowest operator. Hence, it is important to predict the workload of the operators as precisely as possible to determine the optimal degree of parallelism for each operator. There are two main sources of errors: failures in the prediction of the operators' work (*execution skew*) and the *discretization error* i.e. there is no discrete processors-to-operators assignment such that every operator reaches its optimal degree of parallelism [SE93, WFA95]. Minimizing the discretization error by using more processes than processors, as a straightforward solution, adds the unacceptable overhead of process context switching.

---

* supported by the German Research Council under contract DFG Fr 1142/1-1
** full version of this paper is available as [MOW96]

An additional problem with the dependencies between operators are *startup* and *shutdown execution delays* [GHK92, WFA95]. Processors assigned to operators at the end of a pipeline are idle at the beginning of the computation, whereas processors assigned to operators at the begin of the pipeline are idle towards the end of the execution.

In this paper, we focus on the issue of load balanced execution of pipelining segments (*PS*) [CLYY92, SD90, SYT93]. We assume that an optimizer has already generated a tree-shaped query plan and partitioned the plan in PSs with the following characteristics: (1) Only the last operator of each segment may be a blocking operator, all other operators are non-blocking operators. The optimizer tuned the size of each segment so that (2) all necessary tables can be loaded in main memory and (3) all processing then can be done in main memory. To achieve this, the optimizer splits a sequence of non-blocking operators into multiple segments if necessary.

All segments are evaluated one after another according to the producer/consumer data dependencies between them. We do not consider parallel evaluation of data independent PSs, as this obtains no performance improvements [SYT93]. Evaluation of a segment proceeds in two phases: In the first phase all inner relations of joins in the segment are loaded by parallel I/O and the (hash) indices are built in parallel. In the second phase all tuples of the outer relation are piped through selections, projections, or probe phases of joins.

The contribution of this paper is *data threaded execution (DTE)*, a new parallelization strategy for efficient evaluation of the second phase of PSs on a shared-everything system. DTE allocates processing threads not to operators, but to data streams. Thus, DTE subsumes intra-operator parallelism and conventional pipelining parallelism. As additional advantages it includes load balancing, is resistant against various kinds of skew and discretization error, and avoids startup and shutdown execution delays.

The remainder of the paper is organized as follows. In Section 2, we present the pipelining query execution. Data threaded query evaluation is described in Section 3. A simulation model and a comparative performance evaluation is given in Section 4. Section 5 concludes the paper.

2 Evaluation of Pipelining Segments

To show how the different execution strategies work, we chose a rather simple example here. Of course, all strategies presented are also applicable to much more complex queries, consisting of arbitrary non-blocking operators.

We model a flight-information-system. The relation Connections consists of the attributes from and to that represent airports. Each tuple (A,B) denotes non-stop flights from A to B. Table 1 shows a sample instance of Connections. We ask for connections from JFK to SBA with two stop-overs, i.e. with three single non-stop flights. We call this query JFK2SBA-query. A possible query evaluation plan for this query is depicted in Figure 1. R_i are instances of Connections, I_i are intermediate results, θ_i are the selection and join predicates, respectively.

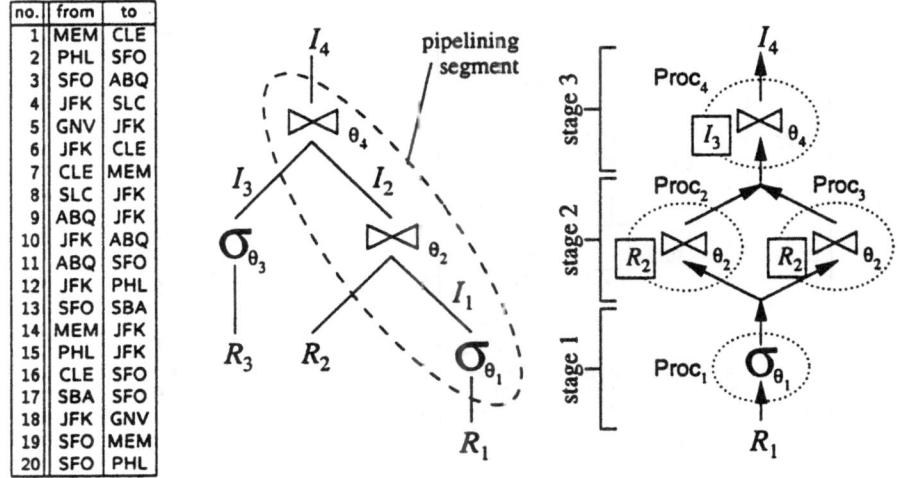

no.	from	to
1	MEM	CLE
2	PHL	SFO
3	SFO	ABQ
4	JFK	SLC
5	GNV	JFK
6	JFK	CLE
7	CLE	MEM
8	SLC	JFK
9	ABQ	JFK
10	JFK	ABQ
11	ABQ	SFO
12	JFK	PHL
13	SFO	SBA
14	MEM	JFK
15	PHL	JFK
16	CLE	SFO
17	SBA	SFO
18	JFK	GNV
19	SFO	MEM
20	SFO	PHL

Table 1. Connections   **Fig. 1.** QEP   **Fig. 2.** PE

In the *pipelining execution model (PE)* each operator forms its own processing *stage*. Inter-operator parallelism is achieved by executing each stage on its own distinct set of processors. When using more processors than stages to evaluate a PS, intra-operator parallelism within stages becomes possible. The first problem to solve, is to determine the optimal degree of parallelism within each stage. Let p be the number of processors and w the work, i.e. the total sequential processing time, of the whole segment, then the optimal parallel execution time is w/p. Let n be the number of stages and w_i the work of stage i ($w = \sum_{i=1}^{n} w_i$). To achieve the optimal parallel execution time, we have to assign $\overline{p_i}$ processors to each stage i such that (1) $\sum_{i=1}^{n} \overline{p_i} = p$, (2) the parallel execution time of the stage is not longer than that of the whole segment (i.e. $w_i/\overline{p_i} \leq w/p$), and (3) the processors in the stage are never idle (i.e. $w_i/\overline{p_i} \geq w/p$). We call this *processor allocation problem (PAP)*. The resulting equation system

$$w_i/\overline{p_i} = w/p \quad \Leftrightarrow \quad \overline{p_i} = (w_i/w) \cdot p, \qquad i \in \{1, \dots, n\}$$

does not have integer solutions for $\overline{p_i}$, in general. An algorithm to find integer approximations $p_i \geq 1$, which provide minimal response time and which fulfill $\sum_{i=1}^{n} p_i = p$, is described in the full paper [MOW96].

A first-come-first-served policy is used to distribute the tuples, that are to be processed by one stage, to the processors that participate in the execution of the particular stage. Data partitioning is not necessary. Each join needs only one shared hash table, as all accesses during the probe phase are read-only, i.e. there are no conflicts. Figure 2 depicts the pipelining execution of the JFK2SBA-query on 4 processors. The shared hash tables are represented by boxes attached to the joins.

Table 2 shows a sample schedule for the JFK2SBA-query executed with PE on 4 processors. Each row represents one unit of time, called *tick*. We assume that performing a selection (S) or the probe phase of a join (P) takes one tick, whereas performing the build phase of a join (B) takes 3 ticks. Tuples of R_1

tick	stage 1 (σ_{θ_1}) processor 1	stage 2 ($\bowtie_{\theta_2}$) processor 2	processor 3	stage 3 ($\bowtie_{\theta_4}$) processor 4
1	$S_{\theta_1}(1)$ o			
2	$S_{\theta_1}(2)$ o	ⓐ		
3	$S_{\theta_1}(3)$ o			
4	$S_{\theta_1}(4) \to 4.1$ •		ⓐ	
5	$S_{\theta_1}(5)$ o	$P_{\theta_2}(4.1,R_2)$		ⓐ
6	$S_{\theta_1}(6) \to 6.1$ •	$B(4.1 \bowtie R_2.8)$:		
7	$S_{\theta_1}(7)$ o	(JFK,SLC,JFK)	$P_{\theta_2}(6.1,R_2)$	
8	$S_{\theta_1}(8)$ o	$\to 4.1.1$ •	$B(6.1 \bowtie R_2.7)$:	
9	$S_{\theta_1}(9)$ o	ⓑ	(JFK,CLE,MEM)	$P_{\theta_4}(4.1.1,I_3)$ o
10	$S_{\theta_1}(10) \to 10.1$ •		$\to 6.1.1$ •	ⓑ
11	$S_{\theta_1}(11)$ o	$P_{\theta_2}(10.1,R_2)$	$B(6.1 \bowtie R_2.16)$:	$P_{\theta_4}(6.1.1,I_3)$ o
12	$S_{\theta_1}(12) \to 12.1$ •	$B(10.1 \bowtie R_2.9)$:	(JFK,CLE,SFO)	
13	$S_{\theta_1}(13)$ o	(JFK,ABQ,JFK)	$\to 6.1.2$ •	ⓑ
14	$S_{\theta_1}(14)$ o	$\to 10.1.1$ •	$P_{\theta_2}(12.1,R_2)$	$P_{\theta_4}(6.1.2,I_3)$
15	$S_{\theta_1}(15)$ o	$B(10.1 \bowtie R_2.11)$:	$B(12.1 \bowtie R_2.2)$:	$B(6.1.2 \bowtie I_3.1)$:
16	$S_{\theta_1}(16)$ o	(JFK,ABQ,SFO)	(JFK,PHL,SFO)	(JFK,CLE,SFO,SBA)
17	$S_{\theta_1}(17)$ o	$\to 10.1.2$ •	$\to 12.1.1$ •	$\to 6.1.2.1$ ⊙
18	$S_{\theta_1}(18) \to 18.1$ •	ⓑ	$B(12.1 \bowtie R_2.15)$:	$P_{\theta_4}(10.1.1,I_3)$ o
19	$S_{\theta_1}(19)$ o	$P_{\theta_2}(18.1,R_2)$	(JFK,PHL,JFK)	$P_{\theta_4}(10.1.2,I_3)$
20	$S_{\theta_1}(20)$ o	$B(18.1 \bowtie R_2.5)$:	$\to 12.1.2$ •	$B(10.1.2 \bowtie I_3.1)$:
21		(JFK,GNV,JFK)		(JFK,ABQ,SFO,SBA)
22		$\to 18.1.1$ •		$\to 10.1.2.1$ ⊙
23				$P_{\theta_4}(12.1.1,I_3)$
24	ⓒ			$B(12.1.1 \bowtie I_3.1)$:
25		ⓒ	ⓒ	(JFK,PHL,SFO,SBA)
26				$\to 12.1.1.1$ ⊙
27				$P_{\theta_4}(12.1.2,I_3)$ o
28				$P_{\theta_4}(18.1.1,I_3)$ o

Table 2. Sample schedule (PE)

are identified by their number (cf. Tab. 1). To demonstrate how PE works, we describe the processing of one tuple throughout the PS: In stage 1, tuple 10 (JFK,ABQ) satisfies θ_1, and entails ($\to$) tuple 10.1 (JFK,ABQ). This is forwarded (•) to stage 2, where it finds two join partners from R_2 ((ABQ,JFK), (ABQ,SFO)). Hence, tuples 10.1.1 (JFK,ABQ,JFK) and 10.1.2 (JFK,ABQ,SFO) are built and forwarded to stage 3. Tuple 10.1.1 has no join partner in I_3, thus its processing is cancelled (o). Tuple 10.1.2 finds (SFO,SBA) as join partner in I_3, so that the final output tuple 10.1.2.1 (JFK,ABQ,SFO,SBA) is built (⊙).

In our example, we have $(w_1,w_2,w_3) = (20,29,17)$ and $w = 66$. The exact solution of the PAP is $(\overline{p_1},\overline{p_2},\overline{p_3}) = (1.21,1.76,1.03)$. The best approximated processor assignment is $(p_1,p_2,p_3) = (1,2,1)$. This results in minimal execution times of 20, 14.5, and 17 ticks for σ_{θ_1}, $\bowtie_{\theta_2}$, and $\bowtie_{\theta_4}$, respectively. Thus, the total execution time of PE using 4 processors cannot be less than 20 ticks for the whole segment. This shows, that PE cannot reach the ideal execution time of $66/4 = 16.5$ ticks due to the discretization error. But, as Table 2 shows, the actual execution time is even worse (28 ticks). This results from two other shortcomings of PE: At the beginning, processors 2, 3, and 4 are idle as they have to wait for the tuples being produced by the previous stages (*startup execution delay*, ⓐ). For the same reason, processors 2 and 4 are idle before they finish their work (ⓑ). At the end, processors 1, 2, and 3 are idle until processor 4 has finally finished (*shutdown execution delay*, ⓒ).

3   Data Threaded Execution (DTE)

The performance of PE suffers mainly from idle time. This problem is a consequence of load balancing by static assignment of processors to stages. The key
idea of our approach is to assign the available processors dynamically to the data.
This leads to a much more efficient resource utilization without any additional
overhead. In contrast to PE, we gather all operators of a PS into one stage and
assign all processors to this stage. Obviously, this *avoids* the PAP completely.

As it is not possible to perform two operators on the same tuple in parallel,
we switch from operator parallelism to data parallelism. Data parallelism covers
both, intra-operator and inter-operator parallelism. We create only one thread
per processor to avoid context switching and scheduling overhead. Each thread
is able to perform all operations within the active PS.

Evaluation of a PS proceeds as follows:
The input tuples for the PS are provided in
a single queue that all threads can access.
Each thread takes one tuple at a time from
this queue and guides it the way through all
the operators of the PS by subsequently calling the procedures that implement the operators. A tuple does not leave the thread (and
thus the processor) during its way through
the PS, until it has been processed by the last
operator or it failed to satisfy a selection or
join predicate. As soon as one tuple has left
a thread, this thread is able to process the
next input tuple from the queue. In the case
that one tuple finds more than one partner in
a join (i.e. the operator produces more than

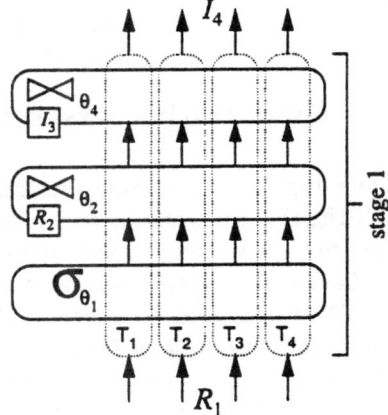

Fig. 3. DTE

one output tuple from one input tuple), the thread has to process all these tuples
first, before it can proceed with the next input tuple from the queue. Figure 3
depicts the data threaded execution of the JFK2SBA-query on 4 processors.

Table 3 shows one possible sample schedule for the JFK2SBA-query executed
with DTE on 4 processors. We use the same notation as in Table 2. There are
no data dependencies between the threads, as no tuple is forwarded (•) from
one thread to another. Thus, all threads start their processing simultaneously
without any idle time, and none of them is idle until it finishes its work, i.e. there
is no startup execution delay. A minimal shutdown execution delay (1 tick, ©)
cannot be avoided. This (nearly) optimal resource utilization reduces the total
execution time from 28 ticks (PE, cf. Tab. 2) to 17 ticks. Thus, in contrast to
PE, DTE (nearly) reaches the minimal execution time of 16.5 ticks.

In DTE, load balancing between the processors is automatic and dynamic,
as each thread can process the next input tuple as soon as it has finished the
processing of the former tuple. Thus, all processors are working as long as there
are input tuples in the queue. DTE optimizes resource utilization, and as no
overhead is needed to achieve this, DTE minimizes the execution time.

tick	thread 1		thread 2		thread 3		thread 4	
1	$S_{\theta_1}(1)$	o	$S_{\theta_1}(2)$	o	$S_{\theta_1}(3)$	o	$S_{\theta_1}(4) \rightarrow 4.1$	
2	$S_{\theta_1}(5)$	o	$S_{\theta_1}(6) \rightarrow 6.1$		$S_{\theta_1}(7)$		$P_{\theta_2}(4.1,R_2)$	
3	$S_{\theta_1}(8)$	o	$P_{\theta_2}(6.1,R_2)$		$S_{\theta_1}(9)$		$B(4.1 \bowtie R_2.8)$:	
4	$S_{\theta_1}(10) \rightarrow 10.1$		$B(6.1 \bowtie R_2.7)$:		$S_{\theta_1}(11)$	o	$\mid$(JFK,SLC,JFK)	
5	$P_{\theta_2}(10.1,R_2)$		$\mid$(JFK,CLE,MEM)		$S_{\theta_1}(12) \rightarrow 12.1$		$\rightarrow 4.1.1$	
6	$B(10.1 \bowtie R_2.9)$:		$\rightarrow 6.1.1$		$P_{\theta_2}(12.1,R_2)$		$P_{\theta_4}(4.1.1,I_3)$	o
7	$\mid$(JFK,ABQ,JFK)		$P_{\theta_4}(6.1.1,I_3)$	o	$B(12.1 \bowtie R_2.2)$:		$S_{\theta_1}(13)$	o
8	$\mid \quad \rightarrow 10.1.1$		$B(6.1 \bowtie R_2.16)$:		$\mid$(JFK,PHL,SFO)		$S_{\theta_1}(14)$	o
9	$P_{\theta_4}(10.1.1,I_3)$	o	$\mid$(JFK,CLE,SFO)		$\mid \quad \rightarrow 12.1.1$		$S_{\theta_1}(15)$	o
10	$B(10.1 \bowtie R_2.11)$:		$\rightarrow 6.1.2$		$P_{\theta_4}(12.1.1,I_3)$		$S_{\theta_1}(16)$	o
11	$\mid$(JFK,ABQ,SFO)		$P_{\theta_4}(6.1.2,I_3)$		$B(12.1.1 \bowtie I_3.1)$:		$S_{\theta_1}(17)$	o
12	$\mid \quad \rightarrow 10.1.2$		$B(6.1.2 \bowtie I_3.1)$:		$\mid$(JFK,PHL,SFO,SBA)		$S_{\theta_1}(18) \rightarrow 18.1$	
13	$P_{\theta_4}(10.1.2,I_3)$		$\mid$(JFK,CLE,SFO,SBA)		$\mid \quad \rightarrow 12.1.1.1$	o	$P_{\theta_2}(18.1,R_2)$	
14	$B(10.1.2 \bowtie I_3.1)$:		$\mid \quad \rightarrow 6.1.2.1$	o	$B(12.1 \bowtie R_2.15)$:		$B(18.1 \bowtie R_2.5)$:	
15	$\mid$(JFK,ABQ,SFO,SBA)		$S_{\theta_1}(19)$	o	$\mid$(JFK,PHL,JFK)		$\mid$(JFK,GNV,JFK)	
16	$\mid \quad \rightarrow 10.1.2.1$	o	$S_{\theta_1}(20)$	o	$\mid \quad \rightarrow 12.1.2$		$\rightarrow 18.1.1$	
17	⊚		⊚		$P_{\theta_4}(12.1.1.2,I_3)$	o	$P_{\theta_4}(18.1.1,I_3)$	o

Table 3. Sample schedule (DTE)

4 Quantitative Assessment

The implemented simulation framework models the structure of operators, the CPUs, the bus system, and even synchronization effects of the queuing mechanisms. As various experiments showed, our framework achieves characteristic behavior even in speed-up and scale-up.

We investigate right-deep PSs consisting of joins, only. Each join consumes materialized relations (either base relations or intermediate results) as its left input, and the results of the preceding join as its right input. Hence, queries are determined by a few parameters: The number of joins altogether, the number of tuples of the right-most input relation and the selectivities of each single join.

The *augmentation factor (AF)* denotes the ratio of the number of input tuples an operator consumes of the outer (piped) relation to the number of produced output tuples. In case of selections the augmentation factor equals the conventional selectivity. In case of joins the augmentation factor equals to $|R_I \bowtie R_O|/|R_I|$, where R_I denotes the inner relation and R_O the outer one.

Within the simulation, the respective number of output tuples produced for one single input tuple is implemented as a normal distributed number with the given AF as mean. As a consequence, the sizes of all inner relations are given implicitly and thus we do not need to model attribute values. To obtain stable results we took the arithmetic middle of at least 25 runs. The size of the right-most input relation was chosen between 10^3 and 10^5 tuples.

To examine the impact of discretization error and various kinds of data skew, separately, the respective *critical* parameter is variable in each experiment, while all other parameters provide *optimal adjustments for PE*. In the final experiment, all parameters are chosen randomly to give an estimation of the average case.

The first experiment examines the impact of discretization errors. Consider a query consisting of 4 joins with an AF of 1.0 each. Whenever the number of CPUs is a multiple of 4, PE is optimal and DTE yields only poor savings ($\leq 6\%$)

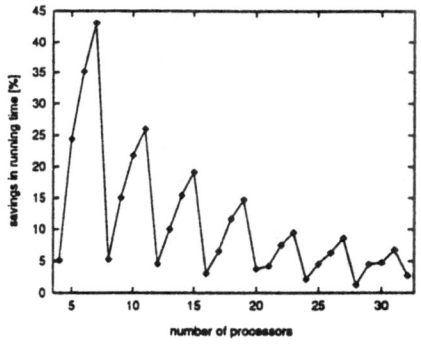

Fig. 4. Discretization errors

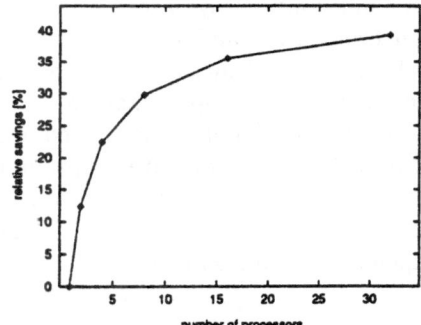

Fig. 5. Relative performance (query mix)

compared to PE that result from startup and shutdown execution delays. Contrary, in presence of discretization error DTE yields savings up to 43% (Fig. 4). Obviously, the impact of the discretization error decreases when the number of processors increases, because of the decreasing ratio of work one single processor performs to the complete work, e.g. moving from 4 to 5 processors can save 20% running time while moving from 31 to 32 can save at most 3.6%, irrespective of the query.

The following experiments examine the case where the actual execution diverges from the assumed one, the static scheduling was based on. Again, consider queries involving 4 joins with an AF of a_i for the i-th stage, where a_i is a normal distributed random number with mean 1.0. As a measure of deviation we introduce the *augmentation skew* $q = \sum_{i=1}^{n} (1 - a_i)^2$ where n is the number of stages. Our experiments show, at a skew of more than 0.35, DTE on 8 CPUs is faster than PE on 12 CPUs. For 12 and 16 CPUs, this effect already occurs at an augmentation skew of 0.225.

For a given number p of processors a query with at most p joins is generated randomly, i.e. the largest queries involve 33 base relations. The AFs are chosen randomly, too, and vary between 0.25 and 1.75. We ran 5600 different queries where each was evaluated at least 5 times with both strategies. The important observation with this experiment is that discretization error and data skew intensify each other. DTE provides savings up to nearly 40% and more than 25% at an average (Fig. 5). Note, that these results do not contradict to the previous experiments, where the amount of total work was constant and the number of processors was variable.

A detailed description of the simulation model, query configuration and further experiments can be found in [MOW96].

5 Conclusion

This paper addresses the topic of load balanced query execution in parallel database systems. We presented *data threaded execution*, a new technique for parallel query execution in shared-everything environments. Compared to conventional execution methods, DTE provides substantial advantages: (1) Startup and shutdown delays are minimized, (2) no discretization error arises, (3) less synchronization and inter-process communication is needed, (4) implicit load balancing establishes almost linear speedup, and (5) DTE is resistant to estimation errors during optimization.

In various simulations, we compared DTE with conventional pipelining execution (PE). In opposite to previous approaches we did not limit our considerations to idealized query parameters, but also considered configurations that cause execution and data skew. In each case, DTE outperforms the conventional pipelining execution strategies.

References

[CLYY92] M.-S. Chen, M. Lo, P. S. Yu, and H. C. Young. Using Segmented Right-Deep Trees for the Execution of Pipelined Hash Joins. In *Proc. Int'l. Conf. on Very Large Data Bases*, Vancouver, BC, Canada, August 1992.

[DG92] D. J. DeWitt and J. Gray. Parallel Database Systems: The Future of High Performance Database Systems. *Communications of the ACM*, 35(6), June 1992.

[GHK92] S. Ganguly, W. Hasan, and R. Krishnamurthy. Query Optimization for Parallel Execution. In *Proc. ACM SIGMOD Int'l. Conf.*, San Diego, CA, USA, June 1992.

[MOW96] S. Manegold, J. K. Obermaier, and F. Waas. Load Balanced Query Evaluation in Shared-Everything Environments (Extended Version). Technical Report HUB-IB-70, Humboldt-Universität zu Berlin, Institut für Informatik, September 1996. ftp://ftp.dbis.informatik.hu-berlin.de/pub/papers/techreports/HUB-IB-70.ps.gz.

[SD90] D. A. Schneider and D. J. DeWitt. Tradeoffs in Processing Complex Join Queries via Hashing in Multiprocessor Database Machines. In *Proc. Int'l. Conf. on Very Large Data Bases*, Brisbane, Australia, August 1990.

[SE93] J. Srivastava and G. Elsesser. Optimizing Multi-Join Queries in Parallel Relational Databases. In *Proc. Int'l. Conf. on Parallel and Distr. Inf. Sys.*, page 84, San Diego, CA, USA, January 1993.

[SYT93] E. J. Shekita, H. C. Young, and K. Tan. Multi-Join Optimization for Symmetric Multiprocessors. In *Proc. Int'l. Conf. on Very Large Data Bases*, Dublin, Ireland, August 1993.

[WA91] A. N. Wilschut and P. M. G. Apers. Dataflow Query Execution in a Parallel Main-Memory Environment. In *Proc. Int'l. Conf. on Parallel and Distr. Inf. Sys.*, page 68, Miami Beach, FL, USA, December 1991.

[WFA95] A. N. Wilschut, J. Flokstra, and P. M. G. Apers. Parallel Evaluation of Multi-Join Queries. In *Proc. ACM SIGMOD Int'l. Conf.*, San Jose, CA, USA, May 1995.

Exploring Load Balancing in Parallel Processing of Recursive Queries *

Sérgio Lifschitz[1][**], Alexandre Plastino[2][***] and Celso C. Ribeiro[1][†]

[1] Pontifícia Universidade Católica do Rio de Janeiro, Departamento de Informática
Rua Marquês de São Vicente 225, Rio de Janeiro, RJ 22453-900, Brasil
[2] Universidade Federal Fluminense, Departamento de Ciência da Computação
Praça do Valonguinho s/n, Niterói RJ 24210-130, Brasil

Abstract. We study a *dose-driven* dynamic load balancing strategy to evaluate database recursive queries. This proposal aims at obtaining a good workload balance with full use of the available resources, with different kinds of skew considered. The strategy is intended for general recursive queries and preliminary computational results illustrate its efficiency when applied to the particular case of linear transitive closure.

1 Introduction

Recent work on load balancing, mostly on join processing, has confirmed its importance when one wants to achieve good performances during the evaluation of parallel database queries. We are interested here in more complex queries, such as the recursive ones. In this case, the work due to a task cannot be previously determined and, consequently, no method can define at the outset the tasks to be executed in parallel in order to balance the workload at each processor.

We claim that the set of tasks to be executed should be assigned to each site dynamically during the query evaluation process, with any workload imbalance controlled, avoided if possible. This way, we may take into account different kinds of skew: intrinsic (related to the input data distribution), partition (due to the algorithm to be performed or to the parallel strategy considered) [WDJ91] and also, what we call here a concurrency skew, due to either a multi-user (or multi-transaction) or an heterogeneous processing environment.

The main contributions of these paper are: (1) the introduction of a dynamic load balancing strategy for processing recursive queries, that deals well with different skew reasons and keep all processing sites active most of the time and (2) an initial implementation of the strategy in a parallel environment, here applied to the simple, yet important, case of linear transitive closure queries.

The paper is organized as follows. In Section 2, the general strategy is explained and some related work is presented. Next, in Section 3, its specialization

---

* Work partially done in LNCC, Laboratório Nacional de Computação Científica.
** E-mail: lifschit@inf.puc-rio.br. Supported in part by CNPq under grant 300048/94-7.
*** E-mail: plastino@dcc.uff.br.
† E-mail: celso@inf.puc-rio.br. Supported in part by CNPq under grant 302281/85-1.

when applied to the transitive closure query is discussed, with preliminary experimental results also given. Conclusions and future work are found in Section 4.

2 Dose Driven Strategy

We present a dynamic task-oriented demand-driven parallel processing strategy for dealing primarily with recursive queries. It is dynamic, in the sense that the workload is assigned to each site during the evaluation process, and demand-driven because new tasks are allocated to a site when it becomes idle and asks for new tasks to process. Our focus is on determining and controlling the execution when tasks are being distributed, in order to avoid load imbalance at most, rather than correcting it later whenever it occurs. Tasks may have variable sizes so to better tune which and how many tasks are to be run at each site. Thus, our strategy is called *dose-driven*, which stands for a method that aims at obtaining an even workload distribution with variable-sized tasks corresponding to doses that are dinamically assigned to the parallel sites.

A task-oriented demand-driven strategy has been suggested previously in [LT92] for balancing the load during the processing of join queries. These are clearly less complex than recursive ones and the strategy proposed is guided by a specific algorithm (hash join), which limits the way tasks are defined. Also, we consider load balancing as part of the whole strategy and not only for redistribution purposes when an uneven workload assigned to parallel sites is detected.

In the case of recursive queries, existing works appear in the context of (datalog) rule programs, where the general framework for processing recursive queries is known as *data reduction* (or *rule instantiations* paradigm [WO93]. The main idea is to parallelize the query evaluation by assigning subsets of the rule instantiations (which are obtained by appending arithmetic predicates like hash functions to some of the rules) among the sites, such that each site evaluates the same program but with less data.

A load balancing method is proposed in [WO93] where a list of alternative parallelization strategies (a set of different restricting predicates) could be used to change the strategy dinamically whenever a workload imbalance is detected. However, a better performance cannot be guaranteed with the new strategy chosen and there is no simple way to implement this mechanism. In [DSH+94], a more sophisticated parallel strategy is proposed, which includes a *predictive* protocol for detecting potential uneven processing at each site and a *correction* algorithm that balances the load. The problem here is that some considerations about load imbalancing that may not be true in practice are made. Indeed, it is considered that a larger local database in a given site implies more work in the future and this is not always the case, as for the join product skew. Moreover, the concurrency skew is not considered.

In our dose driven strategy, we keep the rule instantiations partitioning method but each site is able to evaluate any of the restricted versions and not only those assigned initially. In our case, each subset of instantiations is considered a task. There are many ways to determine these tasks and the strategy will define which site will execute a given task. Basically, these could be subsets of constants in the active domain of the database, or generated by horizontal and/or vertical partitioning of database relations. As so, we have all the flexibility needed to guide the evaluation strategy and control uneven workload in different skew conditions.

In order to illustrate the applicability of our proposed strategy and its tasks generation method, we have adapted it to the case of the transitive closure query. We compare the behavior of our approach to that proposed in [AJ88] - called here *AJ* - which is a static distribution strategy that does not include load balancing techniques and corresponds to the data reduction paradigm applied to the linear transitive closure query. Further details can be found in the full paper [LPR96].

3   Experimental Results

In this section, we give the specialization of our strategy for the evaluation of the transitive closure of a binary relation, say R, usually defined as follows:

$$r_1 : Tc(x, y) : - R(x, z), Tc(z, y).$$
$$r_2 : Tc(x, y) : - R(x, y).$$

The evaluation of the Tc relation may be understood as the computation of all successors of all nodes in the relation's corresponding direct graph. Thus, we may define the tasks to be executed as the computation of all successors of a subset of constants in R. In its linear definition, as above, the transitive closure evaluation can be executed in parallel with no communication during the evaluation process (known as *pure parallelization*), as long as R is replicated through all sites. This property of recursive programs is called *decomposability* [WO93].

The number of tasks t that will be generated must be bigger than p, the number of processing sites available. We could determine each of the tasks as follows: considering an order in the constants set taken into account by the query, a task will be the pair (i, j), where i is the i-th constant in the domain and j represents the number of constants belonging to the task itself.

One of the sites, the coordinator, controls the distribution of tasks to all p sites. There are two phases: in the first one, the coordinator distributes p tasks, exactly one allocated to each site. In the second and last phase, it is time for the dynamic and adaptive distribution of the remaining $t - p$ tasks. As long as there are still tasks to be distributed, the coordinator site sends a new task to a site upon request as soon as it becomes idle. When all tasks have been assigned to the processing sites, the evaluation has reached its end.

We have made preliminary implementations for both methods *DD* and *AJ*, so to investigate their behavior in different skew situations. We have done all implementations on the IBM 9076 SP/2 machine. All programs were coded in C/SQL (with a complete Open Ingres DBMS available at each node). To make full use of the parallel environment, MPI (Message Passing Interface) was chosen.

In order to better illustrate our experiments, we have chosen two input binary relations among all those randomly generated. Relation R1 has 5,000 tuples

(150,000 in the closure) and R2 10,000 tuples (1,000,000 in the closure). Both R1 and R2 correspond to 1,000 nodes graphs, R1 being acyclic, R2 cyclic. In particular, we see that the closure of R2 is the 1,000 nodes complete graph.

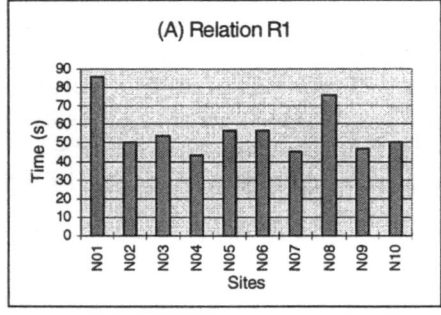
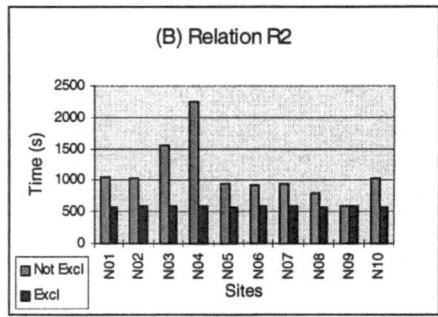
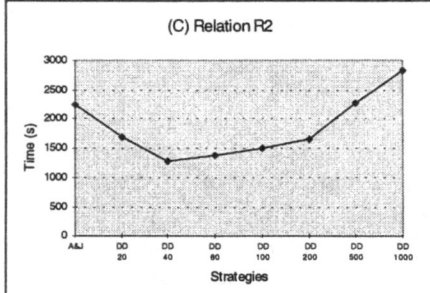
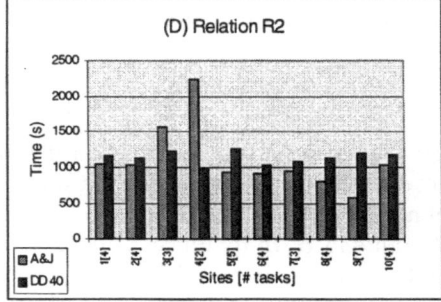

Fig. 1. (A) *AJ* for R1: exclusive environment (B) *AJ* for R2: exclusive and non-exclusive environments (C) *DD* parallel time for R2 and (D) *DD* distribution of 40 tasks

To illustrate the effects of partition skew, here related to the unknown workload associated to each task assigned to a processing node, we show in Figure 1(a) bar coded graphics representing the evaluation of the transitive closure query by the *AJ* strategy in a single-user (exclusive) parallel environment. Times are given in seconds and N01, N02, ... N10 are the 10 sites used. As we can see, there is a strong workload imbalance. For example, site N01 has taken twice as much the time N04 has processed its job.

We consider now a non-exclusive environment where there are other processes - accessing the database or not - running concurrently on the same sites. Figure 1(b) shows *AJ* algorithm on relation R2 in this situation. An uneven processing time has occurred, although the work at each site was equivalent.

We have also tried out the *DD* strategy with distinct total number of tasks, ranging from 20 up to 1000 (one task corresponding to exactly one attribute constant) tasks. In Figure 1(b), it can be seen that the parallel time of the *AJ* strategy for relation R2 is 2240 seconds and in Figure 1(c), with 20 and 40 tasks to be dynamically allocated to the sites, *DD* has obtained parallel processing times of 1676 and 1263 seconds respectively. One could think that a continuous increase on the number of tasks would imply even better results. However, as shown in Figure 1(c), as the number of tasks increase, the parallel time keep its ascending curve, where it gets even worse than the AJ algorithm. The problem is that the query processing work, when partitioned in a set of tasks, has a fixed cost per task that is intrinsically sequential.

In Figure 1(d), we observe the actual distribution of tasks that occurred for algorithm DD with 40 tasks, which has obtained the best parallel time before. In the horizontal axis, $I[J]$ indicates that site NI has performed J tasks. So, we see that N04 has executed only 2 tasks, as its external load was high, while site N09 was responsible for 7 tasks, almost 20% of the total number of tasks. There is not only a gain in the efficiency of the parallel query processing but also it is shown that a good workload balancing was obtained.

4 Final Comments and Future Work

There are many interesting points to discuss and further explore. First, we believe that an increasing number of tasks is valid while the sum of fixed costs related to every task does not offset the gain in performance obtained by the DD strategy. It is still an open question if there is any fixed cost variation with respect to the size of the tasks and this must be better investigated.

Some other results that we have obtained [LPR96] point out that, if the minimization of the total parallel time is the goal to be achieved, not always the best workload balance corresponds to the best parallel strategy. It deserves further investigation but so far we believe that these results are due to the fixed cost mentioned above. We will also implement variable-sized tasks during the evaluation process, probably with a non-blind strategy in mind, that enable a fine tuning of tasks sizes.

References

[AJ88] R. Agrawal and H.V. Jagadish "Multiprocessor Transitive Closure Algorithms" *Procs. Intl. Symp. Database in Parallel and Distributed Systems*, 1988, pp 56–66.

[WDJ91] C.B. Walton, A.G. Dale and R.M. Jenevein "A Taxonomy and Performance Model of Data Skew Effects in Parallel Joins", *Procs VLDB*, 1991, pp 537-548.

[DSH+94] H.M. Dewan, S.J. Stolfo, M.A. Hernandez and J-J. Hwang, "Predictive Dynamic Load Balancing of Parallel and Distributed Rule and Query Processing", *Procs. ACM-SIGMOD Intl. Conf. on Management of Data*, 1994, pp 277-288.

[LT92] H. Lu and K-L. Tan, "Dynamic and Load-Balanced Task-Oriented Database Query Processing in Parallel Systems", *Procs. Intl. Conf. on Extending Data Base Technology*, 1992, pp 357–372.

[LPR96] S. Lifschitz, A. Plastino and C.C. Ribeiro, "Exploring Load Balancing in Parallel Processing of Recursive Queries" *Technical Report MCC37, PUC-Rio, Departamento de Informática*, November 1996.

[WO93] O. Wolfson and A. Ozeri, "Parallel and Distributed Processing of Rules by Data-reduction", *IEEE Trans. Knowledge and Data Eng.* 5(3), 1993, pp 523–530.

Use of a Semantically Grained Database System for Distribution and Control Within Design Environments

Caetano Traina Junior[1,4], João Eduardo Ferreira[2,4], Mauro Biajiz[3,4]

[1] University of São Paulo at São Carlos-Brazil - caetano@icmsc.sc.usp.br
[2] São Paulo State University at Rio Claro-Brazil - jeferreira@ifqsc.sc.usp.br
[3] Federal University at São Carlos-Brazil - mauro@dc.ufscar.br
[4] Development and Research Institute on Database and Information Management - São Carlos

Abstract This paper presents a technique to share the data stored in an object-oriented database aimed at designing environments. This technique shares data between two related databases, called the Original and Product databases, and is composed of three processes: data separation, evolution and integration. Whenever a block of data needs to be shared, it is spread into both databases, resulting in a block on the original database, and another into the Product database, with special links between them controlled by the Object Manager. These blocks do not need to be maintained identical during the evolution phase of the sharing process. Six types of links were defined, and by choosing one, the designer control the evolution and reintegration of the block in both databases. This process uses the composite object concept as the unit of control. The presented concepts can be applied to any data model with support to composite objects.

Key Words: Distributed Databases, Information Sharing, Data Integration, Composite Objects, Integration Semantics, Design Environments.

1. Origin and Needs of the Problem.

The main characteristic of the distribution of Conventional Databases [1], [2] is the simultaneous availability of data to all users. In such environments, the conflicts caused by concurrency to access data are significantly increased. To prevent data inconsistency, algorithms are specified which try to schedule conflicting transactions [7]. There is also an Integrity Control, in order to reduce and/or eliminate the effects of system failures [5]. Although these controls reduce the throughput of the system, they offer the user simultaneous access to shared data in a consistent manner.

In Non-conventional Databases [3], [6], distribution characteristics are different from those of a Conventional Database. Although there may be situations where the system must deal with data concurrency, this feature is another one among other specific distribution requirements for these environments, such as: capacity to perform lengthy transactions (which may take months); restrict a portion of the work exclusively to a given designer while sharing other portions to several designers; and the capacity of the database of to bear conflicts, since these would be solved externally. In these environments, data distribution also involves dealing with concurrency, version and partition [9], insofar as the occurrence of most varied forms of data copying is concerned.

The distribution needs in such environments are greater than those of the traditional relational one, so the term **"Data Sharing"** will be used in this paper. Distribution of data is one of the possibilities of Data Sharing. The proposed **Data Sharing Model** deals with the partitioning of the database, supplying different kinds of links between distributed/replicated partitions, controlling the evolution of each partition; and accomplishing an eventual reintegration of the partitions.

2. Composite Objects in the Data Sharing Model .

To treat a database as a huge collection of data, we need a new paradigm to define how to group data, partitioning it into referential units which, herein, we shall call **Sharing Divisions**. A Sharing Division will be the element used by the system to establish the links between divisions in the database itself, or with other databases in the shared environment.

When a portion of data is devised under the paradigm of object orientation, the data will consist of a collection of objects gathered into one unit, which can be called a **Composite Object**. This term is used here whit a someway different meaning of how it is used in object models.

Within the scope of object-oriented Data Models, the term Composite Object has usually been employed to indicate that two or more objects are associated to each other, and that this association is represented by a reference to the other object by at least one of the objects involved. This association does not always carry the same meaning as that of the real composition. For example, if there is an object *person who lives* in an object of type *residence*, this normally turns *people* to be considered a composite object because it has a reference to another object. *Persons*, however, are not in fact composed of *residences*, so this situation is called herein as "Relationships". The term **"Composite Object"** is used to characterize situations where objects are really *"composed of"* others, as when one states that a *building* is *"composed of"* rooms and *corridors*. Composition includes the idea of grouping, which is fundamental to the Model proposed herein.

Figure 1a) shows the relationship *"is part of"* that exists between a **composite object** and its **parts**. Usually the set of parts forms an object with distinct properties, so to treat the set of parts as another object eases the management of the database. The term **Object Colony** is defined as being "The Set of Objects which Composes a Composite Object according to a given aspect". For instance, a printed circuit board design can be composed of a collection of electronic components (a colony of electronic components), or by their wiring draft (a colony of wires).

Figure 1b) shows the implicit relationships between the objects involved in an occurrence of the Composition Abstraction: a **composite object** *constricts* a **colony**, where the **part objects** which are *part of* that composite object *inhabit*. Colonies have a type, as any other modelling element. Objects always inhabit exactly one colony, the biggest of all being the database itself.

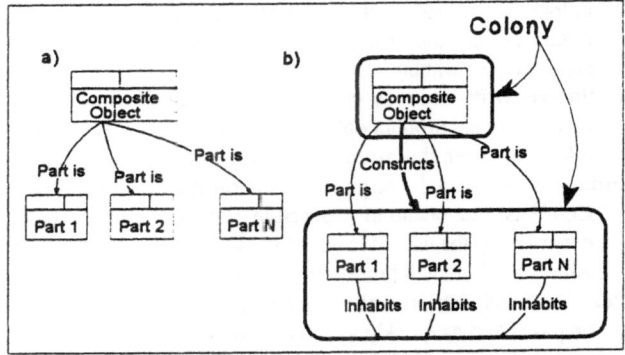

Figure 1: Composite Objects and Colonies that constrict objects.

Colonies establish a relationship of existential dependence between itself and each object inhabits it. One Composite Object can constrict more than a Colony, each one according to one aspect. A colony is the modelling element that embodies the concept of a Sharing Divisions.

The database itself is a colony of the system controlled type "Global", of which there must exist exactly one instance: the database itself, called the "Global colony". The composition abstraction defines a hierarchy, establishing the "context" in which each object is created, composed, how each one can be reached, and its access right parameters.

According to the data evolution and to the application semantics, a database could sometimes be divided. Each colony becomes a unit of data to the Sharing Operation, able to be duplicated and possibly placed into different databases. Every time a separation occurs, types of links are established between corresponding colonies. Such separation may duplicate, partition, or duplicate/partition the data of the original database between the two resulting databases.

3. Types of Links Between Databases in a Sharing Operation.

The execution of a Sharing Operation over one set of colonies is called a "Data Sharing Process". It is composed of three phases: **separation, evolution** and **reintegration**. For each Data Sharing Process the *data separation phase* creates the partitions or duplications of colonies of the database, where a *link* is defined between each corresponding colony in the "original" and in the "product" databases. This link continues to exist while the original and product databases exist, during the *evolution phase*. Such a link will define and circumscribe the set of read and write operations allowed over the objects that inhabit each colony of each database. When (or if) the two databases need to be merged, these links are used again to drive the *reintegration phase*.

We consider the following six types of links as the most important:

read only: only read operations are allowed both in the original and the product databases;

snapshot: that colony is frozen in the original database, and it can be updated in the product database;

isolated: the original database can continue to be updated, and the product database has a copy of the colony as it was in the original database at the time of the separation operation;

Type of Link	Original Database	Type of link
read only	read	read
snapshot	read	read/write
isolated	read/write	read
exclusive	read/write	read/write
independent	read/write	read/write

Figure 2: Types of links between the colonies of the original and product databases

exclusive: write and read operations are allowed in both databases, however, only the original database or the product database will be taken as for the integration process, getting rid of the other. The user chooses one at the time of integration;

independent: both the original and the product databases can evolve in an independent manner. Conflicts that occur in a possible reintegration process should be avoided and/or solved externally by the user. It is possible to select a type of integration among the available choices, where conflicts will be resolved or avoided by an externally supplied integration algorithm;

"on-line": both databases allow read and write operations, and any modification in a database will be passed on as quickly as possible to the other database. This is the "traditional" situation.

4. Sharing Operation.

The user specifies the set of colonies to be separated, and the type of link between each original and product colony. The access to each colony and its inhabitant objects is controlled by an Access Rights Register (ARR), whether or not a sharing process had occurred. The ARR stores the allowable operation to create, delete, modify or lock the several elements of the model, such as objects, attributes, colonies, etc. All the attributes of each ARR need not have the same access rights.

The separation phase of a Sharing Operation modifies the ARR of each shared colony. Also, for each shared colony the separation process must treat all the composite objects under the database's composition hierarchy. The sharing operation also maintains an additional Sharing Register for each participating colony.

4.1. Separation Phase.

The separation phase reproduces the objects. attributes, interrelationships, and composite objects subordinated to the composition hierarchy. It also generates the appropriate ARR of each separated colony in the original and product databases, adding the appropriate links to each access operation allowed. It is also necessary to evaluate the interrelationships, i.e. the relationships between an object that inhabits a colony involved in the Sharing Context, and objects that does not inhabit these colonies. These relationships are called Inter-relationships, while those that occur between objects that inhabit colonies of the same Sharing Context are called Intra-relationships.

Each ARR attribute must be independently analysed. For example, if a colony is marked read-only before separation, this colony will have read-only rights in both databases, regardless of the chosen link. If the field in question has access rights to read and write operations, there will be six possible types of links between the original and product databases.

4.2. Evolution Phase.

During further development of the database, operations in each colony of the original and product databases will be performed restricted by the access rights defined by the ARR, as if no Sharing Operation were taking place. This phase will go on until the user decides either to reintegrate the databases or to cancel the sharing link, thus making each database autonomous.

In the evolution phase, an additional operation should be performed over the usual operations of the database: the recording of the objects deleted in any shared colony in both databases. This recording will be needed in the reintegration phase, because the original or the product database could be updated, in line with the list of objects eliminated in the Evolution phase.

4.3. Integration Phase.

The integration phase depends on data generated during the separation phase, kept in both the original and the product databases. To reintegrate each colony, it uses the previous ARR which existed at the time immediately prior to the separation phase of the database, the current ARRs, and the type of link established. A special procedure is executed, which is based on each type of link, and on the updates made over each pair of colonies.

It must be noted that the user is responsible for solving any possible conflicts due to conflicting updates in the evolution phase. With the independent link it is impossible to assure atomicity and, therefore, the serialization in the integration process. This is due to the non-maintenance of the temporal order of the operations, so various sharing operations may generate different results. However, this kind of evolution in a distributed database system aimed at project support is frequently needed, and the possible conflicts almost always have a semantic origin, understandable and solvable by the design team.

5. Modelling of a Sharing Operation.

The concepts presented herein are developed in a system developed according to the SIRIUS [4] data model, and implemented by means of emulators for this model, using both an Oracle and an Access RDBMS as the object repository [8]. SIRIUS is a formal object oriented data model based on the parametrization of data abstractions. SIRIUS supports four Data Abstractions: Classification, Generalization, Aggregation, and Composition. The composition abstraction is used together with the classification abstraction to complete the operation of object creation, so objects are created inside specific "places" of the database: the colonies. The type of an object dictates the unique type of the colony it can inhabit. This gives a semantic and physical significance to colonies, because it simplifies the process of managing, sharing and insulating information in the database.

A colony of a system-controlled type is created in each database to store data of the pending sharing operations involving this database. Each colony of this type stores data from one sharing operation, so one database can participate in many independent sharing operations. Each such colony stores objects that mirror the composite objects that constrict the shared colonies, with their respective ARR, redefined as a combination of the initial ARR (initial_reg) and the type of link of the databases, generating two new ARRs for each participating colony: one register for the colony in the original database (reg_share_original), and one register for the colony in the product database(reg_share_product).

In addition, another register is created to hold the type of link between the original and product databases (reg_link), in which each attribute can freely specify one of the possible links. Thus, three registers are needed in one Sharing Operation for each shared colony in each database: reg_initial, reg_link and reg_share_original in the original database; and reg_initial, reg_link and reg_share_product in the product database.

6. Conclusion.

The concept of data sharing presented in this paper has a different connotation from that of a conventional distribution data model. The types of links between the original and product databases establish the data sharing semantics aimed at supporting a computer-aided design environment. This semantic also supports the coexistence of data in an "on-line" environment, which is today the predominant characteristic of distributed databases.

In this paper we have sought to transform a complex problem, that of data sharing in object-oriented databases, into a system-treatable sharing of data between several professionals, mainly considering the case of engineering project development environments. The result has proved to meet a fairly wide scope of needs within the bounds of the target application. The partitioning and modularity sought for the proposed solution enable us to state that even the needs which are not met by the solution adopted herein can be much more easily analysed and resolved, since it is possible to more accurately define the location where any new operation or alteration of operations already concluded must be performed.

Acknowledgment

The authors want to thank the USP Physics Institute at São Carlos, were the work was developed.

7. Bibliography.

1. Bernstein, P.A.; Shipman D.W.; Rothnie Jr.,J.B. - "Concurrency Control in a System for Distributed Database (SDD-1)", ACM Transactions on Database System 5:1, 18-25, 1980.

2. Bernstein, P.A.; Goodman, V. - "Concurrency Control in Distributed Database Systems", Computing Surveys 13:2, 186-221, 1981.

3. Bertino, E.; Lorenzo, M. - "Object Oriented Database Systems" International Computer Science Series, Addison - Wesley - 1993.

4. Biajiz, M. - "Development of a Meta-Model of Object Guided Data Models" (in portuguese) , doctorate thesis presented to the Institute of Physics of the University of São Paulo at São Carlos.

5. Ceri, S.; Pelagatti,G. - "Distributed Database: Principles and Systems". MacGraw-Hill, New York, 1984.

6. Chorafas-93 Dimitris N Chorafas-"Manufacturing Databases and Computer Integrated Systems" 1993,320p. Crc Press.Inc./Lewis Publishers.

7. Eswaram, K.P., et al. - "The Notions of Consistency and Predicate Locks in a Relational Database System", Communications of the ACM 19:11, 624-634, 1976.

8. Ferreira, J.E. - "Composite Object Sharing between Object Oriented Databases" (in portuguese) , doctoral thesis presented to the Institute of Physics of the University of São Paulo at São Carlos.

9. Kim, Won, et al. - "A Distributed Object Oriented Database System Supporting Shared and Private Databases", ACM Transactions in Information Systems, Vol. 9, no. 1, 1991.

Method Transformations for Vertical Partitioning in Parallel and Distributed Object Databases

Gajanan S. Chinchwadkar and Angela Goh

School of Applied Science,
Nanyang Technological University, Singapore 639798
gajanan@booch.sas.ntu.ac.sg, asesgoh@ntu.edu.sg

Abstract. Vertical partitioning is a useful technique for performance improvement in parallel and distributed object oriented database systems (**POBSs and DOBSs**). Arising from partitioning, there is a need to modify methods in a user defined schema for correct execution in the partitioned domain. A partitioning scheme should also support application transparency in order to make schema definitions and vertical partitioning independent. We have followed an approach of method transformations for dealing with these problems. A complete taxonomy for the method types arising from vertical partitioning of a class is presented with transformations for each of these types. The role of the transformations in determining the cost of methods is also briefly presented.

Keywords Vertical partitioning, Parallel object database, Distributed object database, Object oriented database

1 Introduction

The popularity of object oriented database systems (**ODBSs**) has inspired researchers to find techniques for improving their performance. The growing performance demands from modern applications such as CAD, CAM and multimedia will make the use of POBSs and DOBSs inevitable in the near future. Support for parallelism and distribution has been the focus of many ODBSs projects. Class partitioning is often used as a means to achieve better performance in such systems. Various reported ways of class partitioning in ODBSs include vertical, horizontal, mixed and path partitioning [4, 5].

In this paper we consider the problem of vertical partitioning. In particular, we discuss a way to handle methods in a vertically partitioned ODBS which is useful in both POBSs and DOBSs. Most of the reported vertical partitioning work has been carried out in the context of DOBSs. A form of vertical partitioning has also been used in a POBS [8]. In relational databases vertical partitions are defined as subsets of attributes in a relation [7]. The task of partitioning algorithms is to find the best subsets with respect to certain performance criteria. This definition of vertical partitions does not hold good in ODBSs since a class consists not only of attributes but also of methods. Hereafter, we refer to the definition of vertical partitions as **Partitioning Scheme** [5] and the procedure for partitioning as **Partitioning Algorithm**.

One way to vertically partition a database class is to partition the methods in the class into different sets and insert attributes in these sets afterwards [1, 2]. The other way is to partition the attributes first and insert the methods afterwards [4]. In both cases, there exist some methods which do not find all the required attributes in a single partition. Further, we assume that the object oriented (**OO**) data model allows complex methods [2]. After partitioning the class, a complex method should invoke the other method from the appropriate partition. Thus, the execution paths of methods change depending upon the partitioning scheme. An idea of providing transparency to the applications using method transformations has been introduced in [6]. However, they considered only a few (and simple) types of methods. In this paper we present method transformations for several (and general) types of methods. The main contributions of the paper include presentation of a unified view of the vertical partitioning problem, a taxonomy of the possible method types, transformations for all these types and a brief discussion on the transformation based determination of method costs.

We assume an OO data model which supports features such as **objects**, **unique object identifiers (OIDs)**, **encapsulation**, **classes**, **simple** and **complex methods** and **ISA** and **class-composition** hierarchies [2]. A **containing class** accommodates complex attributes which can be instantiated to point to the objects of **contained classes**. The parallel model is based upon the shared nothing architecture. User queries arrive at a node designated as the **root** which distributes them among different methods. These methods are executed on the nodes where the classes containing them reside. Each method returns results, if any, to the root.

In the next section, a review of related work is presented. In Sect. 3, a unified view of the vertical partitioning problem is covered. Method transformations are described in Sect. 4. An example and applications of method transformations are discussed in Sect. 5. The paper is concluded in Sect. 6.

2 Related Work

Several fundamental issues related to partitioning an ODBS are described in [4]. It defines various method types, transparency, types of method accesses and existence/access constraints. Vertical, horizontal and path partitioning schemes are defined for ODBSs in [5]. For vertical partitioning, the approach suggested is to partition the attributes in the first place and associating methods with each partition afterwards. Method-induced partitioning schemes and a discussion about application transparency using method transformation has been provided in [6]. It presents transformations for four methods: to create/destroy objects and read/write attributes. Algorithms for generating method based vertical partitions are proposed in [1, 2]. The use of vertical partitioning in POBS is reported in [8] wherein each class is vertically partitioned into OID-attribute pairs and all partitions corresponding to a class are mapped onto a single processor.

3 Unified View of Vertical Partitioning

The schema in Fig.1 contains classes DEPT, EMPLOYEE and DATE. Class DEPT contains two simple attributes DName and DNo and a complex attribute DHead. Class EMPLOYEE contains two simple attributes EName and Salary and one complex attribute DateBirth. Class DATE contains three simple attributes. First three methods in class DEPT retrieve attributes. Method m4 invokes a simple method m7 in class EMPLOYEE. Method m5 is a complex method which invokes a complex method m9 in class EMPLOYEE (which in turns invokes a simple method m13 in class DATE). It also accesses the attribute DName for comparing it with the department name passed as an argument. Suppose attributes in DEPT are to be partitioned in two sets DEPT1{DName, DNo} and DEPT2{DHead}. Methods m1 and m3 can be associated with the first partition. Methods m2 and m4 can be associated with the second partition. The partitions become new classes DEPT1 and DEPT2 and contain **local methods**[5]. Method m5 accesses attributes from different partitions so it is associated with class DEPT' and is referred to as a **global method**[5]. The complex attributes D1 and D2 of class DEPT' point to the partitions DEPT1 and DEPT2 respectively. The class-composition hierarchy rooted at DEPT' represents the logical equivalent of the class DEPT.

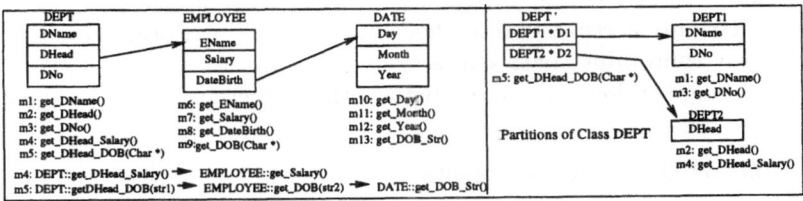

Fig. 1. Department Schema and Partitions of Class DEPT

The main components of the query processing architecture of Fig. 2 are the query processor and the partition catalogue. A query targeted at the user defined schema is submitted to the query processor which identifies and invokes the methods needed to complete the query. It locates relevant global and/or local methods by referring to the partition catalogue. These methods may access attributes or invoke methods from other partitions. Hence they need transformations for correct execution in the partitioned domain.

A unified view of the vertical partitioning problem is shown in Fig. 3. The user defined class acts as an input to the partitioning processor and the class in the partitioned domain is its output. The partitioning processor consists of a partitioning algorithm and a transformation processor. The partitioning algorithm forms subsets of attributes and methods. It also generates the partition

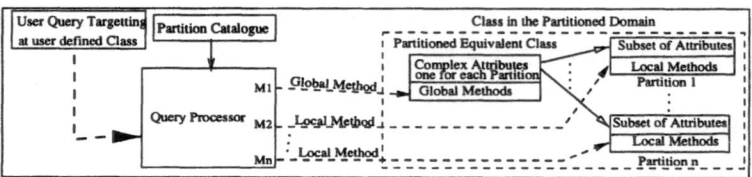

Fig. 2. Query Processing Architecture

catalogue. The transformation processor transforms the methods so that they

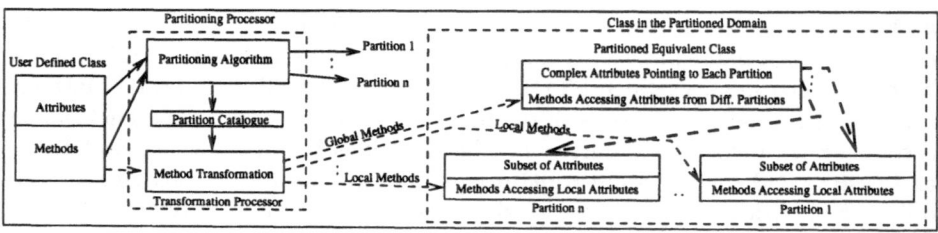

Fig. 3. Unified View Of Vertical Partitioning Problem

can execute correctly in the partitioned domain. This transformation involves replacing attribute accesses and method invocations in the original method definitions with appropriate path expressions. The transformation processor uses the partitioning catalogue to determine whether a method is local or global.

4 Method Transformations

First, we present some assumptions and notations. Without loss of generality it can be assumed that a method performs some computation, accesses one or more attributes of the same class and invokes other methods. For brevity, we do not show the arguments and return values of methods in the following discussion. For any class $C(I, M)$, $I = \{a, b, \ldots, n\}$ is the set of attributes and $M = \{m_1, m_2, \ldots, m_k\}$ is the set of methods. In more general terms, the partitioning scheme described above can be expressed as, $V = \{C_1(I^1, M^1), C_2(I^2, M^2), \ldots, C_p(I^p, M^p)\}$, where p is the number of partitions. Class $C'(I', M')$ (similar to DEPT' in Fig.1) represents the root of class-composition hierarchy equivalent to $C(I, M)$ such that,

1. $I' = \{A_1, \ldots, A_p\}$ and $M' \subseteq M$. Each A_j is a complex attribute which is bound to class $C_j(I^j, M^j)$, $1 \leq j \leq p$. Each method from M' is a global method.
2. $I^j \subseteq I$ and $M^j \subseteq M$. Each class $C_j(I^j, M^j)$ is a vertical partition, where $1 \leq j \leq p$. Each method from M^j is a local method of the partition.

An object O of class $C(I, M)$ is denoted by O_C and retrieval of attribute
"a" from this object is denoted by $O_C \cdot a$. Paths consisting of multiple dots have
been interspersed by "|". The subscripts of "|" indicate the class of OIDs which
are retrieved by the path preceding the "|" symbol. For example the expression,
$c' = O_{C'} \cdot A_r|_{C_r} \cdot r|_{D'} \cdot A_c|_{D_c} \cdot c$, indicates that, complex attribute A_r of object
$O_{C'}$ is retrieved, resulting in an object of class C_r. From the object of class C_r,
attribute r which gives object of class D' is retrieved and so on. Finally, attribute
c of the object of class D_c is retrieved. $X_C(a, b, c)$ denotes a simple method X
belonging to class C that accesses attributes a,b and c. $X_C(a, b, c, Y_*)$ denotes a
complex method X of class C that accesses attributes a,b and c of the same class
and invokes method Y_*. Y_* may be a method of class C or other class. Hence, '*'
should be substituted by name of the class to which method 'Y' is bound. When
there are two methods belonging to unspecified classes in the same expression,
"**" is used to indicate the placeholder of the second class. 'X' and 'Y' are
replaced by 'L' when referring to local methods and by 'G' when referring to
global methods. $O_C \rightarrow X_C(a, b, c)$ indicates invocation of a method X on object
O of class C. In some expressions, method execution paths are denoted by Z^1
and Z^2. The symbol "=" (to be read as "expands to"), is used to denote the full
paths corresponding to Z^1 and Z^2. The left and right hand sides of the symbol
"$\rightsquigarrow$" (to be read as "transformed to") indicate the method execution paths in
the user defined and the partitioned schema, respectively.

For simplicity, a complex method is shown to be invoking only one method.
But the transformations can easily be extended to cover methods that invoke
multiple methods. The OO model allows simple and complex methods. Vertical partitioning causes local and global methods. As a result, several types of
methods exist in a vertically partitioned ODBS. We present these types and
corresponding method transformations below. We assume that all methods are
safe [6]. Our work differs from [6] in that, it assumes a general method structure
(simple and complex) and it covers several types of methods.

4.1 Transformation of Local Methods

Local Simple Methods: Let C_r be a partition of $C(I, M)$ and contains attributes a,b,c and method L of $C(I, M)$. Then,

$$O_C \rightarrow L_C(a, b, c) \rightsquigarrow O_{C'} \cdot A_r|_{C_r} \rightarrow L_{C_r}(a, b, c) \qquad (1)$$

This is the simplest case in which only the method invocation path changes.

Local Complex Methods Invoking Local Simple Methods: Let attributes
a, b and method L^1 belong to class $C(I, M)$. After partitioning, say attributes a,
b, and method L^1 belong to class C_a. Method L^1 invokes another simple method
L^2 which accesses attributes c and d. Then,

$$O_C \rightarrow L^1_C(a, b, L^2_*(c, d)) \rightsquigarrow O_{C'} \cdot A_a|_{C_a} \rightarrow L^1_{C_a}(a, b, Z^1) \qquad (2)$$

There are two ways in which Z^1 can be expanded depending upon whether L^2
belongs to class C or another class.

Case1 (L^1 and L^2 belong to the same class): Before partitioning, attributes c, d and method L^2 belong to $C(I, M)$. After partitioning, they belong to partition C_r. In (2) above, '*' should be replaced by C. Therefore,
$Z^1 = O_{C'} \cdot A_r|_{C_r} \to L^2_{C_r}(c, d)$.

Case2 (L^1 and L^2 belong to different classes): Before partitioning, attributes c, d and method L^2 belong to $D(I, M)$. After partitioning, attributes c, d and method L^2 belong to class D_d. In (2) above, '*' should be replaced by D. L^1 invokes other method L^2 which accesses attributes c and d. Assume that r is a complex attribute of C(I,M) with D(I,M) as its domain and it belongs to class C_r after partitioning. Therefore, $Z^1 = O_{C'} \cdot A_r|_{C_r} \cdot r|_{D'} \cdot A_d|_{D_d} \to L^2_{D_d}(c, d)$.

If C(I,M) is not a containing class of D(I,M) then D_d should be reachable through $O_{D'}$. Therefore, $Z^1 = O_{D'} \cdot A_d|_{D_d} \to L^2_{D_d}(c, d)$.

Local Complex Methods Invoking Local Complex Methods: Let attributes a, b and method L^1 belong to $C(I, M)$. After partitioning, they belong to class C_a. Then,

$$O_C \to L^1_C(a, b, L^2_*(c, d, L^3_{**}(\cdots))) \rightsquigarrow O_{C'} \cdot A_a|_{C_a} \to L^1_{C_a}(a, b, Z^1) \qquad (3)$$

Similar to (2), there are two ways in which Z^1 can be expanded.

Case1 (L^1 and L^2 belong to the same class): Before partitioning, attributes c, d and methods L^2, L^3 belong to $C(I, M)$. After partitioning, attributes c, d and method L^2 belong to class C_c and L^3 belongs to C_j. In (3) above, both '*' and '**' should be replaced by C.
$Z^1 = O_{C'} \cdot A_c|_{C_c} \to L^2_{C_c}(c, d, Z^2)$ and $Z^2 = O_{C'} \cdot A_j|_{C_j} \to L^3_{C_j}(\cdots)$.

Case2 (L^1 and L^2 belong to different classes): Before partitioning, attributes c, d and method L^2 belong to $D(I, M)$ and L^3 belongs to $E(I, M)$. After partitioning, attributes c, d and method L^2 belong to class D_d and method L^3 belongs to E_e. Hence in (3) above, '*' should be replaced by D and '**' should be replaced by E. Assume that r is a complex attribute of C(I,M) with D(I,M) as its domain and it belongs to class C_r after partitioning. Similarly, q is a complex attribute of D(I,M) with E(I,M) as its domain and it belongs to class D_q after partitioning. Therefore, $Z^1 = O_{C'} \cdot A_r|_{C_r} \cdot r|_{D'} \cdot A_d|_{D_d} \to L^2_{D_d}(c, d, Z^2)$
and $Z^2 = O_{C'} \cdot A_r|_{C_r} \cdot r|_{D'} \cdot A_q|_{D_q} \cdot q|_{E'} \cdot A_e|_{E_e} \to L^3_{E_e}(\cdots)$.

If D(I,M) is not a containing class of E(I,M), L^3 will be reachable through $O_{E'}$. Therefore, $Z^2 = O_{E'} \cdot A_e|_{E_e} \to L^3_{E_e}(\cdots)$.

Similarly, if C(I,M) is not a containing class of D(I,M) Z^1 should be reachable from $O_{D'}$. In that case again Z^2 will take two forms depending upon whether D(I,M) is a containing class of E(I,M) or not.

4.2 Transformation of Global Methods

Global Simple Methods: Let attributes a, b and c and method G belong to class $C(I, M)$. After partitioning, say attribute a belongs to class C_a, attribute b belongs to C_b and attribute c belongs to C_c. Then,

$$O_C \to G_C(a,b,c) \rightsquigarrow O_{C'} \to G_{C'}(a',b',c') \tag{4}$$

where, $\quad a' = O_{C'} \cdot A_a|_{C_a} \cdot a$, $b' = O_{C'} \cdot A_b|_{C_b} \cdot b$ and $c' = O_{C'} \cdot A_c|_{C_c} \cdot c$.
Since attributes a,b and c do not belong to the class C' their retrieval is equivalent to invocation of methods to retrieve them in their respective classes.

Global Complex Methods Invoking Global Simple Methods: Let attributes a, b and method G^1 belong to class $C(I,M)$. After partitioning, attribute a belongs to class C_a and attribute b belongs to class C_b.

$$O_C \to G_C^1(a,b,G_*^2(c,d)) \rightsquigarrow O_{C'} \to G_{C'}^1(a',b',Z^1) \tag{5}$$

where, $\quad a' = O_{C'} \cdot A_a|_{C_a} \cdot a$ and $b' = O_{C'} \cdot A_b|_{C_b} \cdot b$.
There are two ways in which Z^1 can be expanded depending upon whether G^2 belongs to class C or another class.

Case1 (G^1 and G^2 belong to the same class): Before partitioning, attributes c, d and method G^2 belong to C(I,M). After partitioning, attribute c belongs to class C_c and attribute d belongs to C_d. In (5) above, '*' should be replaced by C. Therefore,
$Z^1 = O_{C'} \to G_{C'}^2(c',d')$, where, $c' = O_{C'} \cdot A_c|_{C_c} \cdot c$ and $d' = O_{C'} \cdot A_d|_{C_d} \cdot d$.
Case2 (G^1 and G^2 belong to different classes): Attributes c, d and method G^2 belong to $D(I,M)$. After partitioning attribute c belongs to class D_c and attribute d belongs to class D_d. In (5) above, '*' should be replaced by D. Assume that r is a complex attribute of C(I,M) with D(I,M) as its domain and it belongs to class C_r after partitioning. Therefore, $Z^1 = O_{C'} \cdot A_r|_{C_r} \cdot r|_{D'} \to G_{D'}^2(c',d')$,
where, $c' = O_{C'} \cdot A_r|_{C_r} \cdot r|_{D'} \cdot A_c|_{D_c} \cdot c$ and $d' = O_{C'} \cdot A_r|_{C_r} \cdot r|_{D'} \cdot A_d|_{D_d} \cdot d$.
If C(I,M) is not a containing class of D(I,M) then D_d should be reachable through $O_{D'}$. Therefore, $\quad Z^1 = O_{D'} \cdot A_d|_{D_d} \to G_{D_d}^2(c',d')$, where, $c' = O_D' \cdot A_c|_{D_c} \cdot c$ and $d' = O_D' \cdot A_d|_{D_d} \cdot d$.

Global Complex Methods Invoking Global Complex Methods: Let attributes a, b and method G^1 belong to class $C(I,M)$. After partitioning, attribute a belongs to class C_a and attribute b belongs to class C_b. Then,

$$O_C \to G_C^1(a,b,G_*^2(c,d,G_{**}^3(\cdots))) \rightsquigarrow O_{C'} \to G_{C'}^1(a',b',Z^1) \tag{6}$$

where, $a' = O_{C'} \cdot A_a|_{C_a} \cdot a$ and $b' = O_{C'} \cdot A_b|_{C_b} \cdot b$.
Similar to (5), there are two ways in which Z^1 can be expanded.

Case1 (G^1 and G^2 belong to the same class): Before partitioning, attributes c, d and methods G^2, G^3 belong to $C(I,M)$. After partitioning attribute c and d belong to classes C_c and C_d, respectively. In (6) above, both '*' and '**' should be replaced by C. Therefore, $Z^1 = O_{C'} \to G_{C'}^2(c',d',Z^2)$ and
$Z^2 = O_{C'} \to G_{C'}^3(\cdots)$, where, $c' = O_{C'} \cdot A_c|_{C_c} \cdot c$ and $d' = O_{C'} \cdot A_d|_{D_d} \cdot d$.

Case2 (G^1 and G^2 belong to different classes): Before partitioning, attributes c, d and method G^2 belong to $D(I, M)$ and method G^3 belong to $E(I, M)$. Hence in (6) above, '*' should be replaced by D and '**' should be replaced by E. After partitioning, attribute 'c' belongs to class D_c and attribute 'd' belongs to class D_d. Assume that 'r' is a complex attribute of C(I,M) with D(I,M) as its domain and it belongs to class C_r after partitioning. Similarly, 'q' is a complex attribute of D(I,M) with E(I,M) as its domain and it belongs to class D_q after partitioning. Therefore, $Z^1 = O_{C'} \cdot A_r|_{C_r} \cdot r|_{D'} \rightarrow G^2_{D'}(c', d', Z^2)$ and $Z^2 = O_{C'} \cdot A_r|_{C_r} \cdot r|_{D'} \cdot A_q|_{D_q} \cdot q|_{E'} \rightarrow G^3_{E'}(\cdots)$, where,

$c' = O_{C'} \cdot A_r|_{C_r} \cdot r|_{D'} \cdot A_c |_{D_c} \cdot c$ and $\quad d' = O_{C'} \cdot A_r|_{C_r} \cdot r|_{D'} \cdot A_d |_{D_d} \cdot d$.

If D(I,M) is not a containing class of E(I,M), G^3 will be reachable through $O_{E'}$. Therefore, $Z^2 = O_{E'} \rightarrow G^3_{E_e}(\cdots)$. Similarly, if C(I,M) is not a containing class of D(I,M) Z^1 should be reachable from $O_{D'}$. In that case again Z^2 will take two forms depending upon whether D(I,M) is a containing class of E(I,M) or not.

There is a possibility of hybrid types such as global(local) complex methods invoking local(global) complex methods. We do not elaborate upon these types due to space constraints.

5 Example and Applications

In Fig. 1, methods of class DEPT can be classified into three groups: local simple methods such as m1, m2 and m3; a local complex method invoking local simple method such as m4; and a global complex method such as m5 which invokes another complex method m9. We explain the transformations for method m5 in the following discussion. Suppose that class EMPLOYEE is partitioned as $EMPLOYEE_1$(EName) and $EMPLOYEE_2$(Salary, DateBirth). Method m9 accesses attribute EName to compare this name with the argument which is passed to it. Class DATE is not partitioned. In the non partitioned class definition the execution path of method m5 is as follows. It accesses attribute DName and invokes method m9 in class EMPLOYEE. Method m9 in turn accesses attribute EName of class EMPLOYEE and invokes method m13 of class DATE. In the partitioned domain, method m5 becomes a global method of DEPT'. It retrieves attribute DName using a path DName' and invokes the global complex method m9 of class EMPLOYEE' and so on. The transformation of m5 can be expressed as,

DEPT → **get_DHead_DOB(DName, EMPLOYEE** → **get_DOB(DATE** → **get_DOB_Str())))**

$\rightsquigarrow$ **DEPT'** →**get_DHead_DOB(DName', Z^1)**, where,

DName' = **DEPT'** $\cdot$ |$_{DEPT_1}$ $\cdot$ **get_DName()** and

Z^1 = **DEPT'** $\cdot$ **D2**|$_{DEPT_2}$ $\cdot$ **DHead**|$_{EMPLOYEE'}$ $\cdot$ **E2**|$_{EMPLOYEE_2}$ → **get_DOB(EName', Z^2)** where,

EName' = **DEPT'** $\cdot$ **D2**|$_{DEPT_2}$ $\cdot$ **DHead**|$_{EMPLOYEE'}$ $\cdot$ **E1**|$_{EMPLOYEE_1}$ $\cdot$ **get_EName()** and

Z^2 = **DEPT'** $\cdot$ **D2**|$_{DEPT_2}$ $\cdot$ **DHead**|$_{EMPLOYEE'}$ $\cdot$ **E2**|$_{EMPLOYEE_2}$ $\cdot$ **DateBirth**|$_{DATE}$

$\rightarrow$ **get_DOB_Str()**.

In a vertically partitioned ODBS several paths need to be traversed for a single method invocation. In addition to the IO cost of explicit retrievals each method invocation can also add to IO cost if the method accesses attributes. Each "·"(dot) in the transformation rule can result into communication depending upon the placement of partitions in a multiple processor environment. In DOBSs, vertical partitioning aims at minimizing irrelevant attribute accesses, communication cost and providing more concurrency [3]. In POBSs, it aims at minimizing irrelevant attribute accesses and achieving high retrieval parallelism [8]. The IO and communication costs reflected through the transformations are useful for further study related to partitioning and placement strategies. This is important in the case of global methods when the partitions are placed in different nodes and considerable data shipping are needed. The transformations allow existing applications to run on a partitioned schema without rewriting them [6] and enable independence of schema design and class partitioning.

6 Conclusions

Vertical partitioning in ODBSs is difficult due to the presence of methods [2]. We have presented a unified view of the vertical partitioning problem and a set of transformation rules for various method types. This work is useful in developing vertically partitioned POBSs, DOBSs and centralized ODBSs. More detailed analysis of vertical partitioning requires a cost model incorporating both IO and communication costs. Our present research includes the development of such a cost model for POBSs. Other related problems include designing of partition catalogues and query processors and finding various partition placement strategies.

References

1. L. Bellatreche, et al. Vertical Fragmentation in Distributed Object Database Systems with Complex Attributes and Methods, Proc. DEXA'96, Sept(1996).
2. C. I. Ezeife and K. Barker. Vertical Class Fragmentation in a Distributed Object Based System, Tech. Report 94-03, Comp. Sci. Dept. Uni. of Manitoba, (1994).
3. C. I. Ezeife. Class Fragmentation in a Distributed Object Based System, Ph.D. Thesis, Comp. Sci. Dept., Uni. of Manitoba, (1995).
4. K. Karlapalem, et al. Issues In Distribution Design of Object Oriented Databases, In Distributed Object Management by M. T. Ozsu et. al. Morgan Kaufman Publications, (1994), 148-164.
5. K. Karlapalem and Q. Li. Partitioning Schemes for Object Oriented Databases, Proc. of RIDE-DOM'95, (1995).
6. K. Karlapalem, et al. Method Induced Partitioning Schemes in Object Oriented Databases, Proc. of 16th Intl. Conf. on Distributed Computing Systems, Hongkong, May(1996).
7. M. T. Ozsu and P. Valduriez. Principles of Distributed Database Systems, Prentice Hall, (1991).
8. A. K. Thakore, S. Y. W. Su and H. Lam. Algorithms for Asynchronous Parallel Processing of Object-Oriented Databases, IEEE TKDE, 7(3), June(1995), 487-504.

Benchmarking and Performance Tuning of Multimedia Servers *

Peter Triantafillou, Stavros Christodoulakis and Theodora Magoulioti

MUltimedia Systems Institute of Crete, MU.S.I.C.
Department of Electronics and Computer Engineering
Technical University of Crete, T.U.C.
Chania, Crete, Greece **

Abstract. In this paper we study the issues related to the development of a multimedia server benchmark. Multimedia benchmarks can be used to compare the performance of existing multimedia servers and to fine-tune the performance of servers under development. The modeling of multimedia servers includes the modeling of databases, multi-user workloads, user behavior, and the definition of the relevant performance metrics. The proposed benchmark has been implemented and has been used to fine-tune the performance of a multimedia server, called *KYDONIA*. We present the results of testing *KYDONIA* and representative instances of fine-tuning *KYDONIA*'s performance.

1 Introduction

Benchmark techniques have gained significant popularity and global acceptance which is widely reflected in the daily market life. As computer technology and computer products evolve nowadays, more and more the consuming society looks for realistic performance benchmarks, so as to have the correct attitude about the computer market. However, benchmarking is a difficult process because the goal of evaluation is typically ill-defined: End-users, sometimes even designers, either don't know, or can't specify exactly what result they expect. Often, they don't specify the architectural variants to consider, and often the metrics and workload they expect users to use are ill-defined. Moreover, they rarely clarify which kind of model and evaluation method best suit the evaluation problem.

In this paper, we study issues of systematic performance evaluation of large scale multimedia information systems, including DAVIC-compliant systems ([9], [20]). Such systems are typically geographically distributed and are centered on powerful multimedia servers whose high performance depends on exploiting parallel I/O mechanisms. We study the typical functionalities and requirements of the computer applications demanding multimedia access, the impact of user behavior on system performance, as well as the impact of data properties. The component performance metrics that most completely express the user's attitude about the overall system's efficiency were then determined. The objective is to

* This work was done in the context of and supported by the ESPRIT Long Term Research Project HERMES (project number 9141).
** Authors address: P.O. 133 Chania, 731 10 Crete, Greece. Phone: +30-821/64803, 69737, 69738. FAX: +30-821/64846. E-mail address: {peter,stavros,dora}@ced.tuc.gr

obtain benchmark results in order to evaluate the performance of the multimedia servers. To this end, it is important to mention that performance results depend significantly on the system architecture, which always must be revealed in full detail before characterizing the system performance ([18]).

The remainder of the paper is organized as follows. Section 2 describes the benchmark model. In Section 3 we show how we used the model to fine-tune the performance of the *KYDONIA* multimedia server ([5]). The corresponding results are described and analyzed in the same section. Section 4 concludes the paper and gives a short description of future work.

2   The Benchmark Model

We have developed and implemented a benchmark model in order to evaluate the performance of large scale multimedia information systems. The performance modeling of databases, multi-user workloads, and user behavior, as well as the definition of the performance metrics are the topics that must be addressed. The benchmark consists of three main parts, namely the *Database Characterization*, the *Workload Characterization*, and the *Performance Metrics Definition*. The parts are analyzed in the following sections.

2.1   Database Characterization

In general, database characterization involves the specification of the number of objects residing in the database, the characterization of database objects, and finally, the definition of the parameters of the scheme organization. In this paper we do not refer to scheme organization issues since it has been long studied by others ([3], [13]). However, we insist on object characterization analysis, which consists of two parts, namely *object type information*, and *object replica information*. Fig. 1 gives an overview of the database model.

Object Type Information We classify the objects into two types, namely *atomic* and *composite*.

An object of atomic type may be a *delay-sensitive (ds)*, or a *non-delay-sensitive (nds)* object. The *ds* object type is the generic data type which can be used to describe time-dependent objects, such as video and audio sequences. Objects of this category consist of a sequence of media quanta (such as video frames and audio samples), which have to be presented in specified time intervals. A fundamental characteristic of the *ds* objects is their large storage demands. Compression schemes are employed in order to limit them ([12], [15]). Different compression schemes correspond to different consumption data rate requirements. High data rates are typical of the *ds* objects. Video and audio objects are currently the *ds* instances studied here.

As *nds* we call the general data type which can be used to describe traditional textual and numeric data objects. Images are objects which also belong to this category. In general, they are characterized by simple attributes. For example, the attributes maintained within a text object is -likely- the text size and information concerning its structure and content. In addition, the information that image objects may hold includes size, resolution and compression schemes ([1], [17]). Images and text objects are representative classes of *nds* objects.

Atomic objects can be combined in a fashion which usually requires synchronization. For example, a slide presentation must synchronize audio (music and commentary) with images. Thus, we define composite objects to consist of two or more atomic objects. Their basic feature is the number and type of atomic objects, as well as their interrelationships in time and space (e.g., atomic objects can start together, finish together, or start based on a lapse ([14], [19])).

Object Replica Information Replication may be desirable for performance and availability reasons, as well as for supporting different object qualities (e.g., multiple video resolutions). Consequently, we must first specify the number of copies for each database object, and second, we must characterize each copy based on its *compression, size,* and *popularity* information.

The *ds* objects can be compressed by scalable compression algorithms, which create versions with different sizes ([16]). On the other hand, uncompressed objects (such as text) may exist in the database. We distinguish, therefore, two object categories, namely: *uncompressed* and *compressed.* Compressed objects may be characterized as having *single,* or *multiple* compressions. The JPEG and the MPEG standards are the compression schemes that typically are used for images and audio/video objects, respectively. We have also adopted multiple compression schemes which can also be used to implement the interactive VCR operations easily, because of the fact the multiple compression schemes can offer *low, medium, high,* and *full* data rate demands, simultaneously ([16]).

An object's type and compression information determine its size in bytes. A real database has objects of variable sizes. If, for instance, we consider the movie on demand service, then the subtitles, along with the compressed audio and video objects are representative database objects which are characterized by variable sizes. In general, we can consider that subtitles are small objects when compared with an audio object, which is a medium size object when compared to a video object. We define, therefore, three categories of objects based on their sizes: *small* objects, *medium* objects, and *large* objects. The lower and upper bounds of the ranges that contain small, medium and large objects, respectively, are given as parameters.

The frequency of access of real database objects varies significantly. Objects then are classified into two categories, namely *hot* and *cold.* The users access the objects of each class either uniformly, or according to an access pattern distribution. Also, the object's popularity varies with time.

2.2 Workload Characterization

In order to test the performance of multimedia information systems, we have to model actual multi-users workloads. The workloads denote representative tasks or service requests, which characterize the system's applications. To this end, we propose a workload model which consists of three tasks, namely *functionality characterization, workload types characterization,* and *user population characterization.* Fig. 2 overviews the workload model.

Functionality Characterization The design of a multimedia benchmark should allow the testing of all multimedia services that the system under test supports, such as video on demand ([9]). The services provide typical functionalities to

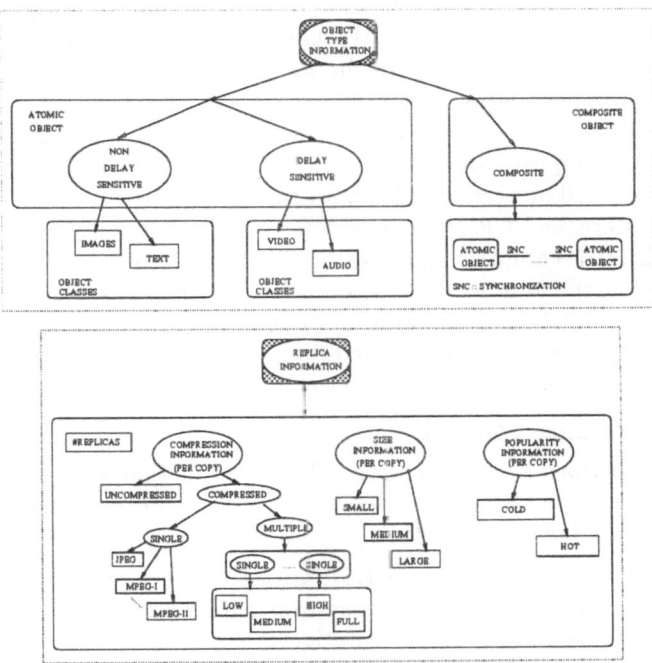

Fig. 1. An overview of the database model.

users. A benchmark needs to check whether typical service functionalities are provided. In our model we include the recommendation of $DSM\text{-}CC$ ([10]), which has also been adopted by $DAVIC$ ([9]), for defining the system's functionality. In this paper we consider the following functionalities:

1. Storing and retrieving various media types (e.g., text, audio, video, etc.).
2. Performing typical VCR operations (i.e., playback, FF, FB, stop, etc.).

Depending on whether user requests access atomic or composite objects, they are classified into two types, namely *atomic* and *composite* request types, respectively. The atomic requests are further classified into *nds* and *ds* requests, depending on whether they access *nds*, or *ds* database objects, respectively.

Workload Types Modeled workloads consist of user sessions, each of which contains a number of requests. We distinguish two main types of multi-user workloads, namely *simple* and *complex*. Simple workloads are characterized by sessions that consist of only a single type of atomic requests (i.e., either *nds*, or *ds* requests). Complex workloads are classified into *atomic* and *composite* workloads. An atomic complex workload contains atomic requests of any type (i.e., both *nds* and *ds* request type), while a composite and complex workload contains a set of composite requests.

User Characterization Users enter the system either with an exponential distribution arrival rate (in an open system), or according to the rules of a closed system. They create sessions which contain requests which typically follow a non-uniform access distribution. We have chosen to model the "*a/b* rule", in which *a* percent of accesses go to the most popular *b* percent of the data, and the *Zipf's Law* distribution ([4], [24]).

To model the user's behavior, we must first decide which atomic/composite object (and if composite the constituent atomic objects and their synchronization requirements) will be accessed. Then for each request we must decide (i) whether it is a read/store, or VCR type request, (ii) on the object's type, (iii) whether the accessed object is compressed, or not, and (iv) on the object's size. Furthermore, we must define the number of sessions and the number of requests in each session which the workload contains. Finally, in addition to the arrival rate of users, in each session we define an arrival rate for its requests and (perhaps) a user's think time, in between the requests of a given session.

2.3 Performance Metrics

The system performance evaluation is based on measurements of numerous aspects of efficiency, which mainly depend on service specific features. The particular metrics are now presented.

Response Time and Startup Latency An aspect of response time can be defined as the cumulative time elapsed from submission of a request until reception of the whole answer. Representative requests involve the retrieval and/or the processing of traditional data. The usefulness of this definition depends on data length, since it is more suitable for the relatively smaller and *nds* objects. Indeed, considering requests that involve relatively larger objects, or *ds* requests, we define response time as the time elapsed from submission of a request until the reception of a partial reply that can temporarily satisfy the user. For instance, a user who submits a playback video request realizes the server's response upon reception of enough video frames to begin its playback; not upon completion of the whole video playback. This is the *start up latency* metric defined, as the delay a user experiences from submission of a request until reception of the first units of data. Finally, it is important to measure the delay a user experiences from submission of a VCR request until system reaction to the user demands. Thus we define one more significant metric, called *reaction time*.

Throughput The throughput of a system is defined as the number of requests that are submitted and completed during a time unit. For example, concerning the movie on demand service, we can monitor the number of movies submitted and completed per hour (*mph*). The throughput typically expresses the maximum number of concurrent users that can be supported by the system. Various measurements of throughput can be reported depending on the response time restrictions enforced, the intra-/inter-object synchronization requirements enforced, the resolution requirements enforced, and at last, the system's price.

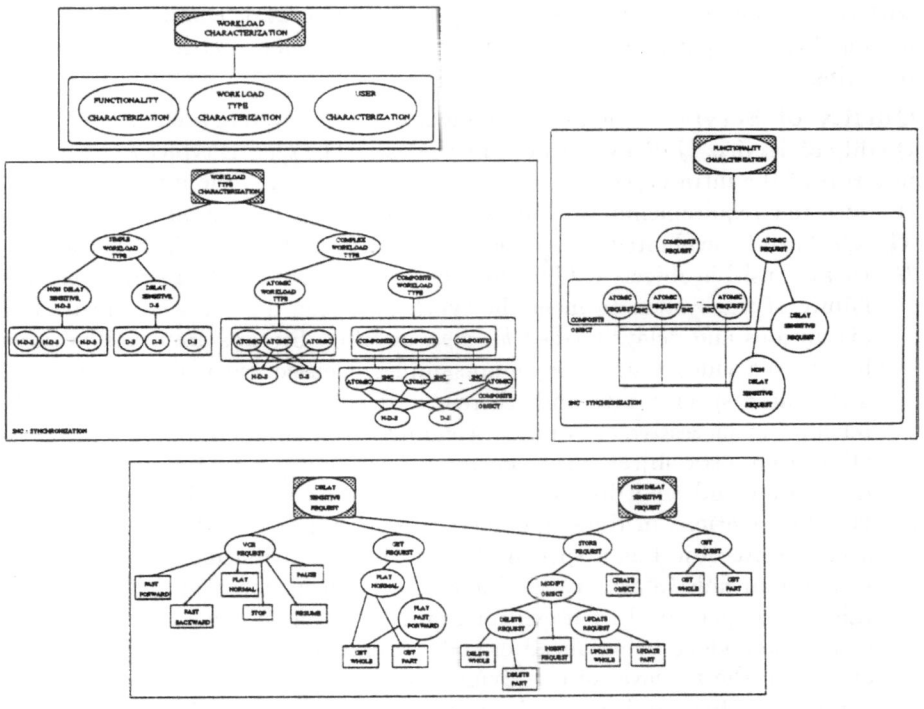

Fig. 2. An overview of the workload model. Functionality Characterization: Modeling of a single request. Workload Type Characterization: Modeling the resulting workload.

Throughput under Restrictions Apart from pure throughput reporting, several hybrid metrics can be defined. If, for example, we consider the response time metric, we can compute the percentage of requests that are actually serviced under user response time or startup latency constraints. A two-second delay was generally considered to be appropriate for "simple" commands, which are defined as those the user believes they can be done without any substantial processing and for which the user is not willing to wait. An example for a "simple" command is deleting an object. VCR requests are characterized by different user response time constraints, etc. User synchronization and presentation resolution constraints can also be considered.

Price/Throughput The cost of constructing and maintaining the system which can be either the hardware, software, or communication cost can be combined with the measured throughput that the priced configuration enables. The combination of throughput and associated price produces a new metric which is worth reporting.

Quality of Service A significant indicator of the system's efficiency is the Quality of Service, QoS, experienced by the user. The QoS is typically expressed in terms of the intra-object synchronization requirements, the inter-object synchronization requirements, and at last, the object presentation requirements. The QoS can be evaluated according to the function $QoS = \mathcal{F}(\mathcal{H}, \mathcal{S}, \mathcal{R})$ where,

$\mathcal{H}$: Delay-sensitive objects exhibit intra-object synchronization requirements. A failure of the system to meet the synchronization demands is perceived as disruptions and delays termed *hiccups*, or *jitter*. Hiccups are usually caused by delayed video frames, or audio samples. Consequently, data units must arrive at a specific rate and, therefore, deadlines are associated with the retrieval of successive data units. A cumulative characterization, $\mathcal{H}$, of deviation from intra-object synchronization requirements can be established via monitoring and reporting the total number of missed deadlines.

$\mathcal{S}$: The interrelations in time and space among atomic objects that constitute a composite object is a special characteristic contributing in QoS metric. If a system fails to achieve the inter-object synchronization demands, then this can be perceived by users, for example, as audio being ahead of video, or inversely video being ahead of audio, Consequently, deadlines are associated with the retrieval of qualifying data units, as previously identified. A cumulative characterization, $\mathcal{S}$, of deviation from inter-object synchronization requirements can be established via monitoring and reporting the total number of missed deadlines.

$\mathcal{R}$: This factor refers to video and image presentations on user display devices. Users request screen dimensions $SD_1 \times SD_2$ that can be evaluated as the product $SD_1 * SD_2$. However, the system supports the resolutions $R_1 \times R_2$ that can be evaluated as the product $R_1 * R_2$. The greater the product $R_1 * R_2$, the better the resolution that the users experience. The bigger the difference $SD_1 * SD_2 - R_1 * R_2$, the worse the resolution that the users experience. The mean (averaged over distinct requests) percentage of difference experienced is the quantity $\mathcal{R}$.

3 Performance Tuning of the KYDONIA Server

We currently use the benchmark presented above to observe and fine-tune *KYDONIA* ([5]). *KYDONIA* is a multimedia information server designed and implemented in MUSIC/TUC with the aim to investigate design issues and develop technology for the multimedia information systems of the future. *KYDONIA*'s main features are: storage management for multimedia objects, real-time pumps towards multiple clients over ATM, multimedia DBMS functionality, multimedia object modelling, text and video access methods and browsing techniques. *KYDONIA* serves as a testbed for the research and development of multimedia information systems that support distributed multimedia applications, and with respect to this, a series of benchmarking tests takes place in order to improve overall server performance. In the following we present some of the benchmarks carried out along with the results.

Benchmarks For our performance tests, we assume
- an MPEG-2 compression scheme with a data rate of 1.5 megabits per second per video stream. Based on this data rate, object sizes range from 576 kilobytes for a 3-second video to 11.25 megabytes for a 60-second video. For our experiments, we assume all objects are 150-second videos.
- that the server stores enough one-copy videos to fill available space.
- *ds* workloads. We also assume that all objects are requested in their entirety and played sequentially. In reality, some percentages of the time users will deviate from sequential viewing to perform VCR operations, such as pause, fast backward and fast forward. Due on space constraints, we ignore them in this paper.
- that accesses to videos in the server follow a uniform distribution.

All measurements have been performed on the *KYDONIA* server (avoiding issues such as network bandwidth and client caching). Typical video on demand servers maintain strict performance constraints to guarantee adequate presentation quality; the data rate demands delivery of a video block as often as every ($block\_size\_in\_bytes \div data\_rate\_demands\_in\_bytes$) secs. The peak number of streams that can be supported without hiccups in presentation are the results of our experiments.

Results We show the performance of *KYDONIA* when qualifying videos have been stored in $m(=1,2,3)$ *Seagate ST51080N* of size 1GB and nominal bandwidth TR_{disk} equal to 5MB/s each disks and the video data block size varies in values of 1, 2, 3, 4, 5, and 8MB. The disks connect to individual SCSI controllers of nominal bus bandwidth 10MB/s each. The data has been placed on the disks according to the *fine-grained striping* technique. In fine-grained striping, the striping unit is typically small. As a consequence, if there are m disks, then every retrieval involves all the m disk heads, and the m disks behave like a single disk with bandwidth $m * TR_{disk}$. The following table shows the results on the throughput measurements. We obviously expect that as the number of disks increases and the data block size increases, throughput also increases.

Data Block Size (in MB)	Throughput		
	$m=1$	$m=2$	$m=3$
1	14	17	15
2	15	26	28
3	17	28	34
4	17	31	36
5	18	31	40
8	19	35	46

Discussion KYDONIA supports a maximum of 19 *smooth* streams (i.e., without hiccups) when using a single disk, and a maximum of 35 and 46 smooth streams when using 2 and 3 disks, respectively. Since multiplying by two and three the number of the involved disks definitely increases by a factor of two and three the disk bandwidth, the throughput doesn't follow a related increase! This is because in fine-grained striping the size of the striping unit decreases as the number m of the involved disks increases, and this finally results in effective disk bandwidth decrease per disk!

Furthermore, by examining the design of the server closely and paying attention in *KYDONIA*'s scheduling mechanism, we found that the current scheduler presents some important limitations. In particular, *KYDONIA* has incorporated a scheduling algorithm which is based on deadlines in order to meet the associated deadlines of the delay-sensitive objects. The deadlines are computed by taking into accounts *only* the following two factors; the consumption rate and the video data block size. Factors such as the load of the system are not taken into account. In addition, although the basic idea of the algorithm is to order all requests for disk access according to their associated deadlines (in increasing order), the server also performs SCAN scheduling in order to reduce the cost of positioning seeks (it takes place at *KYDONIA*'s disk access module). This is obviously against philosophy of a deadline-based scheduling algorithm. The above result in missed deadlines[1]. In additional, we have seen that if for example all requests have the same deadline, some of them miss their deadline because of the fact that service rounds have interleaved (i.e., the block $j + 1$ of a stream is retrieved before the block j of another stream). Efforts are currently underway to re-examine very closely the scheduling of delay-sensitive requests in the KYDONIA server. Particularly we will implement the concept of *round* scheduling using an efficient algorithm like SCAN to process requests within a round. Round-based scheduling is known for solving such problems.

4 Conclusions and Future Work

We have developed a multimedia benchmark which consists of three parts, namely: *Database Characterization*, *Workload Characterization* and *Performance Metrics Definition*. Synthetic databases and multi-user workloads are realistically approximated by the first two parts, while the third part defines the performance metrics.

The benchmark can be used not only to evaluate the performance of a system in comparison to other similar systems, but also to fine-tune the performance of a given system. With respect to this, the proposed benchmark was used to fine-tune the *KYDONIA* server.

With respect to future work, our objective is to use the benchmark model to further fine-tune *KYDONIA*. In particular, we plan to incorporate in *KYDONIA* our work on striping techniques ([6], [23]), which will greatly improve its performance, and new scheduling algorithms, which will avoid the inefficiences of EDF-like algorithms. We will also incorporate our results on optimal data placement on multi-zones disks, disk arrays, and tertiary storage libraries to fine-tune those parts of *KYDONIA* managing these devices ([8], [21], [22], [23]). Finally, we will incorporate batching techniques to improve the performance.

---

[1] This can be avoided if the display of each stream is delayed until the end of the current round instead of starting the display as soon as the first block arrives in memory (as in currently done in KYDONIA).

References

1. Anastasiadis. M., et al.: State of the Art: Image Compression Schemes. MU.S.I.C. Tech Report (1995) (in greek)
2. Böckle, G., et al.: Structured Evaluation of Computer Systems. IEEE Computer (June 1996) 45-50
3. Cattel. R., Skeen J.: Object Operations Benchmark. ACM Transactions on Database Systems (March 1992)
4. Chervenak. A.L.: Tertiary Storage: An Evaluation of New Applications. Doctor of Philosophy, Dept of Computer Science, Univ. of California, Berkeley (1994)
5. Christodoulakis. S., et al.: The KYDONIA Multimedia Information Server. To appear in ECMAST '97.
6. Christodoulakis. S., Zioga. F.: Fundamentals of Striping and Placement of Delay-Sensitive Data on Disks. (submitted for publication)
7. Christodoulakis. S., Triantafillou. P.: Research and Development Issues for Large-Scale Multimedia Information Systems. ACM Computing Surveys (December 1995)
8. Christodoulakis. S., Triantafillou. P., Zioga F.: Principles of Optimally Placing Data in Tertiary Storage Libraries. VLDB '97 (accepted to appear)
9. DAVIC 1.0 Specification (1995-1996)
10. ISO/IEC JTC1/SC29/WG11: Digital Storage Media Command & Control. (1996)
11. Quinn, S., R., Sitaram, M.: State of the Art: Testing, Testing. Byte (1996) 97-101
12. Gall, D.L.: MPEG: A Video Compressed Standard for Multimedia Applications. Communications of the ACM (vol. 34) (no. 4) (April 1991)
13. Gray, J.: The Benchmark Handbook for Database and Transaction Processing Systems. 2nd edition, edited by Jim Gray Digital Equipment Corporation, Morgan Kaufmann Publishers, Inc., California
14. ISO/IEC CD 13522-5: Information Technology-Coding of Multimedia and Hypermedia Information. Part 5: MHEG Subset for Base Level Implementation (1995)
15. ISO/IEC 13818-3: Information technology - Generic coding of moving pictures and associated audio information -. Part 3: Audio (1995)
16. Keeton. K., Katz, R.: Evaluating Video Layout Strategies for a High Performance Storage Server. ACM/Verlag Multimedia Systems(3) (1995) 43-52
17. Kontogiannis. P., et al.: State of the Art: Image Formats. MU.S.I.C. Tech. Report (1996) (in greek)
18. Magoulioti. T., et al.: Benchmarking and Performance Aspects in Multimedia Information Systems. MU.S.I.C./HERMES, Tech. Report (1996)
19. Steinmetz, R., Nahrstedt, K.: Multimedia: Computing, Communications and Applications. Innovative Technology Series (1995)
20. Triantafillou, P., Christodoulakis, S.: Digital Libraries: A Survey of Developments in Required Technology. MU.S.I.C./HERMES Tech. Report, no. 5 (October 1996)
21. Triantafillou, P., Christodoulakis, S., Georgiadis, C.: Optimal Data Placement on Disks: A Comprehensive Solution for Different Technologies. IEEE Transactions on Knowledge and Data Engineering (conditionally accepted)
22. Triantafillou, P., Papadakis, T.: On-Demand Data Elevation in a Hierarchical Multimedia Storage Server. VLDB '97 (accepted to appear)
23. Triantafillou, P., Faloutsos, C.: Optimal Striping for Parallel I/O in Modern Applications. Parallel Computing journal, Special issue on Parallel Data Servers and Applications (accepted to appear)
24. Zipf. G.: The psycho-biology of language: An introduction to dynamic philology. Boston: Houghton Mifflin, 1935, Cambridge, MA:MIT Press (1965)

Database Program Mapping
onto a Shared-Nothing Multiprocessor Architecture:
Minimizing Communication Costs

Sophie Bonneau, Abdelkader Hameurlain

Institut de Recherche en Informatique de Toulouse, Université Paul Sabatier
118, route de Narbonne, 31062 Toulouse cedex, France
E-mail: {bonneau, hameur}@irit.fr

Abstract. This paper deals with the minimization of inter-task and inter-processor communication costs in the context of one SQL query mapping onto a shared-nothing architecture, as far as the parallel decisional query processing is concerned. After setting both the models of the application handled and the target multiprocessor architecture, a study aimed at minimizing both inter-processor transfer times depending on the interconnection network topology, and the impact of pipeline starting/closing (start-up) times, is mainly presented.

1 Introduction

In the context of the compilation of a database program on a parallel architecture, the present paper focusses on the particular problem of mapping tasks making up a SQL query onto shared-nothing parallel architecture processors.

The mapping problem, generally as well as for many special cases known to be NP-complete, has been intensively studied within the database community, further to parallel DBMS development, for these last few years [7, 8, 10]: typically, the double optimization criterion of the proposed resolution methods (such is also our case) is the response time (or makespan as well) minimization and the system throughput maximization, and the allocation constraint, processor load partitioning or balancing.

Nevertheless, these heuristics are all distinguished by not taking into account communication costs, indeed estimated on the basis of inter-processor and inter-task communication times, inherent in the target parallel architecture specificity and the application type features, while these contribute to the increase in the response time [2]: first, a shared-nothing parallel system may be characterized by a sizeable inter-processor communication cost overhead, and second, the considerable communication (data and control) volume exchanges of the decisional SQL query bound tasks generate significant inter-task communication cost. On the one hand, inter-task communication times can be improved by reducing communication volume: at the mapping level, it has been achieved by introducing allocation constraints which facilitate propagating the number of processors and partitioning attributes [3] to avoid data redistributing whose cost may significantly exceed a task execution cost, and force mapping of the tasks in charge of reading data from disk, at the very place where data they use are stored. However, on the other hand, inter-processor communication times depend on both the target multiprocessor architecture interconnection network topology, and the available different parallelism (partitioned and pipeline) types taken into account in the application mapping process.

This is the reason why the problem of inter-processor communication cost minimization is here particularly focussed on, within the context of one SQL query static mapping onto a shared-nothing parallel architecture [5]. In Section 2, a model for the application to be mapped and the target parallel system, are described. In Section 3, our study of inter-processor communication cost optimization is summed up, as well as its consequences on the mapping process to be designed. Finally, Section 4 concludes with a synthesis of the work achieved and an overview of future related prospects.

2 Problem Modelization

The application to be mapped is a *decisional SQL query dependence graph* [9]: this is an oriented graph whose nodes represent tasks (i.e. Scani, Buildi and Probei[1]), and edges, information (data and control messages) communication and/or time-related dependence links. The communication link between two tasks I and J is either of the "precedence" or of the "pipeline" type: in the former case, J will be executed and will receive information from I only when I has completed its execution, whereas, in the latter case, I and J exchange data flow in a producer-consumer manner. Only tasks Buildi and Probei are constrained by a precedence type communication. In addition, this graph is valuated by each task local response time, each task couple communication cost (see Section 3), and the number of processors required to execute each task, which is estimated by the cost evaluator [4]. The communication mode of each bound task couple is also known (i.e. broadcast, distribution or propagation).

The target *shared-nothing parallel architecture* is characterized by the number of its processors, and its interconnection network topology (i.e. point-to-point, multi-stage or fully connected). It is to be noted that communications, depending on the network topology and the required data partitioning on each memory space, may generate a sizeable parallelism overhead.

3 Inter-processor Communication Cost Minimization

3.1 Inter-processor Communication Time Definition and Minimization Approach

Nowadays, trying to minimize inter-processor communication times on shared-nothing parallel architecture, still seems to be relevant. Multi-stage or fully connected type networks have, indeed, uniform communication times unlike point-to-point networks, and new technologies (i.e. optic fibre, co-axial cable support) cannot account for a minimal Communication Time/CPU Time ratio yet, for the number of MIPS has not improved in the same proportion as commmunication latency and rate have decreased.

Generally, the inter-processor communication time between two processors refers to

---

1. i.e. tasks respectively reading tuples from disk, building and probing the hash table of a simple-hash join node.

$T_{comm} = T_{transfer} + T_{start\text{-}up}$, where $T_{transfer}$ is the information (data and control) transfer time between two processors, and $T_{start\text{-}up}$, the information receiving/sending (start-up) CPU time overhead. On the one hand, in order to minimize the former component while mapping, when a point-to-point topology is at stake, "bringing nearer" emitter and receiver processors involved in communication can be attempted by using a distance matrix. On the other hand, as the latter component can prevail in target parallel system enabling pipeline parallelism to be exploited (such is our case), and because our application data are transferred in flow, minimizing of pipeline start-up time impact is essential, and based on the following observation: as a rule, in order to maximize pipeline potential chains, it is attempted to allocate two different processor sets to each pipeline communicating task couple. Well, the communication cost (estimated from data communication volume) between two of these tasks can sometimes exceed their execution cost (based on local response time) because some tasks such as Scan, Join, Grouping, don't always handle all the data received [2]. In this case, if only one processor is supposed to be allocated to each task, it is to be noted that the pipelined task couple response time mapped onto two different processors, is less than the pipelined task couple response time mapped onto the same processor, only if the communication cost is relatively less than the task local response times. As a consequence, taking into account partitioned parallelism (i.e. assigning several processors to each task) sets the question of whether it is always relevant to systematically allocate two different processor sets to pipelined tasks. Our resolution method consists, for each pipelined task couple, in analyzing associated costs to determine whether it should be better, in terms of mapped task couple response time, to allocate onto the same processor set or onto two different processor sets.

3.2 Resolution Method Mainspring

Let two tasks I and J, with respective local response times $t_i>0$ and $t_j>0$, and pipeline communication cost $C_{i,j}>0$, which respectively require $n_i>1$ and $n_j>1$ processors. Depending on the inter-task communication mode and the number of processors assigned to each of them, the following cases can be distinguished:
1st case (see Figure 1):
* $n_i = n_j$ (for instance, I and J are both assigned 2 processors.)
* communication mode: propagation
2nd case (see Figure 2):
* $n_i \neq n_j$ (for instance, I and J are respectively assigned 2 and 3 processors.)
* communication mode: distribution or broadcast
Our method consists in attempting to *formally evaluate* this *task couple response time* for each of the mapping configurations [1], under the following assumptions: i/ The couple tasks to be mapped will be scheduled in a balanced and symmetric way [6], ii/ all processors are identical.

3.3 Formal Study Consequences and Interpretation

The following conclusions can be drawn from the response time study of a task couple

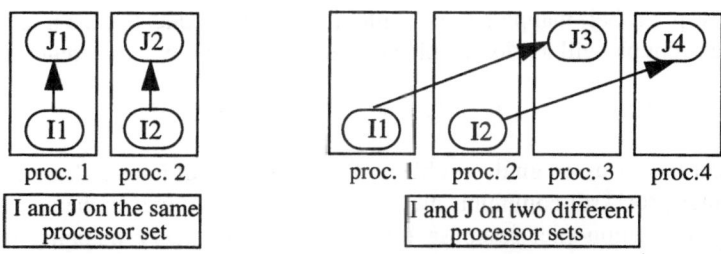

Fig. 1. Possible mapping configurations of pipelined tasks propagating

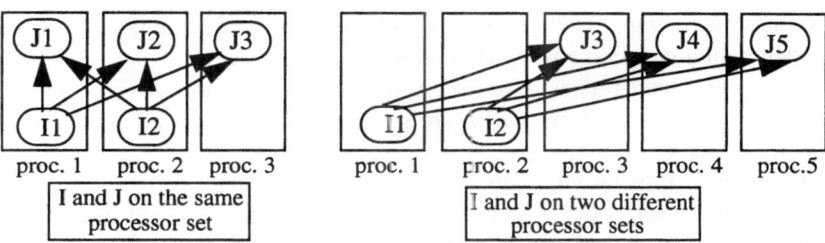

Fig. 2. Possible mapping configurations of pipelined tasks distributing or broadcasting

in each mapping configuration, depending on communication mode, towards the application dependence graph:

i/ tasks distributing or broadcasting (e.g. Scani→Buildj, Scani→Probej and Probei→Buildj if the number of processors and partitioning attributes were not propagated), should always be mapped onto two different processor sets, to exploit pipeline parallelism. In this case, it can, indeed, be proved that the response time of the task couple mapped onto two different processor sets will always be less than the response time of the task couple mapped onto the same processor set, whatever the communication cost (start-up) significance with regard to task execution costs [1].

ii/ tasks propagating (e.g. Probei→Buildj if the number of processors and partitioning attributes were propagated), should be mapped onto two different processor sets, to favour pipeline parallelism, when the communication cost is relatively less than the task execution cost. On the contrary, multi-programming execution would better be favoured by mapping tasks onto the same processor set. Decision relativity can be settled by the following fast calculation:

$$\text{let } R = max(t_i, t_j) - (t_i + t_j) \text{ (negative by definition for } t_i, t_j > 0)$$

1st case. $C_{i,j} > |R|$: the communication cost will compensate saving due to pipeline parallelism, i.e. the communication cost will inhibit saving due to pipeline execution. Therefore, task multi-programming execution would better be favoured by mapping onto the same processor set.

2nd case. $C_{i,j} \leq |R|$: saving due to multi-programming execution is compensated by the communication cost. Pipeline execution would better be favoured by mapping tasks onto two different processor sets.

It is to be noted that the pipeline communication format Probei→Probej is not deliberately considered in this work: the data localization allocation constraint [5] binds

Probei (resp. Probej) mapping to be Buildi (resp. Buildj) one. So, the mapping clue for such a task couple will a priori be physical allocation onto two different processor sets.

4 Conclusion

This paper has focussed on the problem of minimization of inter-task and, more particularly, inter-processor communication costs, which contributes to the optimization criterion minimization (i.e. response time) of one SQL query mapping onto shared-nothing parallel architecture. After describing a problem modelization, this study and an adequate resolution method have been presented.

So it has been emphazised that the mapping process could mainly act on the inter-processor communication cost by attempting to:

i/ minimize inter-processor transfer time for point-to-point networks, by choosing to allocate sets of processors which are nearer each other to communicating tasks, thanks to a distance matrix,

ii/ minimize pipeline communication start-up time impact, by working particularly on dependence graph pipeline communications. For each pipelined task couple, this handling determines whether it would be better, in terms of mapped task couple response time and depending on the communication mode, to allocate onto the same processor set or onto two different processor sets.

Finally, the next goal is to integrate these results into the PSA (Parallel Scheduling Algorithm) scheduling and simultaneous mapping algorithm [5].

References

1. BONNEAU, S. et al., "Placement d'un programme bases de données sur une architecture multiprocesseur à mémoire distribuée: minimisation des coûts de communication", Rapport de Recherche, No. IRIT/97-08-R, Lab. IRIT, Janvier 1997, 23 pages.
2. ENGLERT, S., et al., "Parallelism and its price: a case study of NonStop SQL/MP", Sigmod Record, vol. 24, n°4, Dec. 1995, pp. 61-71.
3. HAMEURLAIN, A., et al., " An Optimization Method of Data Communication and Control for Parallel Execution of SQL Queries ", Intl. Conf. on Database and Expert Systems Applications, LNCS, n°720, Prague, Sept. 6-9, 1993, pp. 301-312.
4. HAMEURLAIN, A., et al.," A Cost Evaluator for Parallel Database Systems ", 6th Intl. Conf. on Database and Expert Systems Applications, DEXA'95, London, 4-8 Sept. 1995, LNCS, n°978, pp. 146-156.
5. HAMEURLAIN, A., et al., "Scheduling and Mapping for Parallel Execution of Extended SQL Queries", 4th Intl. Conf. on Information and Knowledge Management, ACM Press, Baltimore, Maryland, 28 Nov. - 2 Dec. 1995, pp. 197-204.
6. HASAN, W., "Optimization of SQL queries for parallel machines", dissertation fot the degree of Doctor of Philosophy, Stanford University, Dec. 1995.
7. LO, Y.L., et al., "Scheduling queries for parallel execution on multicomputer DataBase Management System", 7th Intl. Conf. on Database and Expert Systems Applications, DEXA'96, Zurich, Sept. 1996, LNCS, n°1134, pp. 698-707.
8. MAHIOUT, A., et al., "Modéliser les dépendances entre les tâches data-parallèles pour le placement et l'ordonnancement automatique", 6ièmes Rencontres Francophones du parallélisme, Lyon, Juin 1994, pp. 37-40.
9. SCHNEIDER, D., et al.,"Tradeoffs in Processing Complex Join Queries via Hashing in Multiprocessor Database Machines", Proc. of the 16th VLDB Conf., Brisbane, Australia 1990, pp. 469-480.
10. WOLF, J. L., et al., "A Hierarchical Approach to Parallel Multiquery Scheduling", IEEE Transactions on Parallel and Distributed Systems, 1995, vol. 6, n°6, pp. 578-589.

Large Join Order Optimization
on Parallel Shared-Nothing Database Machines
Using Genetic Algorithms

Khalid A Nafjan
Department of Computer Science,
University of Sheffield, Sheffield UK
K.nafjan@dcs.shef.ac.uk

Jon M Kerridge
Department of Computer Studies
Napier University, Edinburgh, UK
J.kerridge@dcs.napier.ac.uk

Abstract This paper proposes the use of genetic algorithms (GAs) for optimizing the sequence of large joins execution on parallel shared-nothing database architectures. In order to measure the suitability of this method we compare the GA that we have specifically developed for this problem with previously proposed GAs. Experimental results show that our GA was able to outperform its counterparts. We also compare the performance of our GA with some known heuristics that were employed for optimizing joins in parallel queries. It turned out that for smaller number of relations, heuristics were able to produce query execution plans as good as those of GAs. However when the number of relations increases, GAs outperform heuristics.

1. Introduction

Although Shared-nothing(SN) parallel database architecture has its own drawbacks, it has emerged as the best choice for managing large data-intensive applications. This is due to that architecture's high level of scalability and throughput. In such data-intensive applications, queries that involve very large joins are not uncommon. For example, we have designed an Object-Oriented interface to a SN database machine which requires joining all the relational tables to materialize objects from relations [Nafj97]. To provide real-time response to these kinds of queries, it is crucial to investigate ways to improve performance of query optimizers.

Since join operators are the most expensive, they became the primary target for any query optimizer. Early work in query optimization for centralised database systems has used simple enumerative strategies that search all or most of the solution space [Seli79]. Typically, this approach is exhaustive and might be acceptable for small number of joins only (i.e. < 7 tables). With the required number of joins becoming larger, the solution space increases rapidly. To tackle this problem, several researchers have been studying the use of other techniques such as heuristics and randomized algorithms [Swam88,Ioa90,Shim94]. On the other hand, optimization of large joins in parallel queries is rather more complex than that of centralised databases due to the influence of other factors such as data distribution and processor allocation, which increase the solution space much more. In this paper we shall examine the use of GAs for optimizing parallel queries with large number of joins on shared-nothing database systems. GAs have the ability to find good quality solutions in problems with large solution spaces without the need to evaluate every single solution. Further, unlike GA methods used with centralised databases which was originally developed for the traveling sales person problem, our method is specifically designed for query optimization. We compared the performance of our GA (We called it PREFIX) with these methods [Benn91,Stei93]. experimental results show that our GA was able to outperform its counterparts. We have also compared its performance with some famous heuristics that were employed for optimizing parallel queries. It turned out that for smaller number of tables, heuristics were able to produce query execution plans as good as those of GAs. However when the number of relations increased, GAs were able to outperform heuristics. We consider the use of left deep join trees. We believe that any drawn conclusion for left deep trees can be generalized to other solution spaces.

2. GAs- what are they

GAs are search algorithms which are usually used for problems with very large search spaces. They differ from other standard search algorithms (e.g. Hill-climbing) in that the search is conducted using a population of individuals. Each individual (also called chromosome) represents a possible solution to the problem. Each possible solution is allocated a fitness value which determine its quality. Before a GA can be applied, a fixed-length representation to the chromosome should be devised. The GA attempts to find the optimal (or near-optimal) solution to the problem by genetically breeding the population of solutions over a series of generations using mainly two mating operators called crossover and mutation. Individuals are selected to participate in those two operators based on their fitness values. Basically a crossover operator works by selecting two chromosomes (parents) and swaps parts of them to produce new offspring which share some features taken from each parent chromosome. The mutation operator alters one or more bit values at randomly selected locations in chromosomes that are also selected randomly. At the end of the mating process, all or most of existing chromosomes are replaced by new offspring which in turn constitute a new population. The newly created population becomes the next generation and the optimization process is repeated until the best chromosome in the population reaches the anticipated solution or a predefined number of generations has been achieved. For a detailed discussion of GAs, the interested reader is referred to many text sources, e.g., [Gold89,Mich94].

3. Encoding Join Processing Trees using GAs

Join processing trees are the traditional abstraction of sequence using which joins are executed. Basically, each join processing tree is encoded as a single chromosome. The chromosome representation adopted here is a permutation of an ordered-based encoding which is a vector of bits containing a number of distinct integers. Each integer value corresponds to a relation in the join processing tree. In an example of a query that consist of 4 relations to be joined, the chromosome might by represented by the integer list [1,0,3,2]. This is interpreted as a join processing tree as shown in Figure 1. The value in the first location of the chromosome corresponds always to the base building relation and the rest of values represent the probing relations. From the sequence of values in the chromosomes, joins are performed in the following order: Relations 1 and 0 are joined first and resulting intermediate relation of the join (I_1) is joined with relation 3. Again, the resulting intermediate relation of this stage (I_2) is further joined with relation 2 to produce the final answer to the query (I_3). Obviously, for a chromosome to be valid join processing tree encoding, it must not contain any duplicated or absent relations.

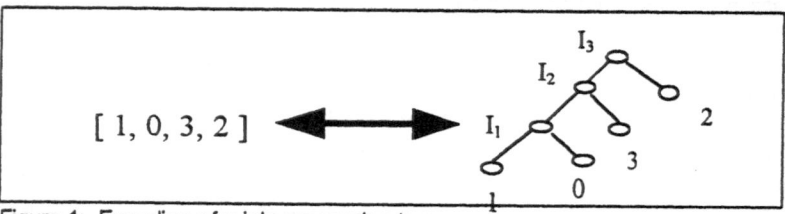

Figure 1: Encoding of a join processing tree

4. PREFIX Crossover Operator

In this section we describe the PREFIX crossover operator that have been specifically designed for join ordering problem. The basic idea is based on the observation of Grefenstette in [Gref87]. They have found that to successfully design a crossover for a certain problem, heuristic information about the nature of the problem should be encompassed in the crossover operator. Basically, the PREFIX crossover builds offspring by choosing a sub-sequence from a starting cut point of each of the two chromosomes and prefixing it to the other one. For example, the two parent chromosomes C_1 and C_2 consisting of 9 relations would produce offspring in the following way.

$$C_1 = [1\,2\,3\,9\,8\,6\,5\,7\,4] \ , \ C_2 = [8\,6\,5\,4\,3\,2\,1\,7\,9]$$

We firstly choose randomly the cut point at each chromosome. For instance, cut points at location 4 and 5 for C_1 and C_2 respectively with sub-sequence of length 3 will produce the sub-sequences S_1 and S_2:

$$S_1 = [9\,8\,6] \ , \ S_2 = [3\,2\,1]$$

Each sub-sequence is then prefixed to the front of the other chromosome, that is S_1 will be prefixed to C_2 and S_2 will be prefixed to C_1 as follows

$$C_1' = [\underline{3\,2\,1}\ 1\,2\,3\,9\,8\,6\,5\,7\,4] \ , \ C_2' = [\underline{9\,8\,6}\,8\,6\,5\,4\,3\,2\,1\,7\,9]$$

The first value in each chromosome C_1' and C_2' is copied into the front of the offspring O_1 and O_2.

$$O_1 = [3,.....,...] \ , \ O_2 = [9,.....,...]$$

Successive values from C_1' and C_2' are then copied one by one to offspring O_1 and O_2 respectively, provided that any value that is duplicated or produces Cartesian product will not be copied. The final results of offspring O_1 and O_2 will be:

$$O_1 = [3\,2\,1\,9\,8\,6\,5\,7\,4] \ , \ O_2 = [9\,8\,6\,5\,4\,3\,2\,1\,7]$$

The crossover is further enhanced by self-learning capabilities. The general principle is that the crossover learns which location should be favored for cut points and at what location the sub-sequence should be prefixed rather than doing that randomly. In order to accomplish this, an additional punctuation value is added to each chromosome. The present cut point location is saved in this punctuation . If the choice of this cut point leads to an improved offspring then the cut point information in the parent chromosome will be propagated to the offspring to be used in further generation. Otherwise a null value is copied and a random value will be used. Choosing the point where the sub-sequence is prefixed depends on how good the new population is when compared to the old one. If the improvement satisfies a predefined target (e.g. 20% of chromosomes have been improved) then the same prefix location is used for next generation, otherwise it will be incremented by one.

5 Experimental Results

Performance experiments presented here are divided into two parts. In the first one, the performance of the PREFIX crossover is compared with former crossovers that were used with centralised query optimizers, namely, PMX, ORDER, and CYCLE crossovers. In the second part,. the performance of two heuristics called minimum intermediate and minimum cost described in [Swam88] are compared against the performance of our GA.

5.1 Comparison between crossovers.

Figure 2 illustrates the relative average results of 20 queries runs for 400 generations. In the initial generation, the CYCLE and ORDER operators start with similar cost but much higher than the other two crossovers. However at the 400th generation the three crossovers other than the

PREFIX, end with similar costs. Table 2 indicates the cost of the best plan found at the 1st ,200th and last generation for each crossover operator.

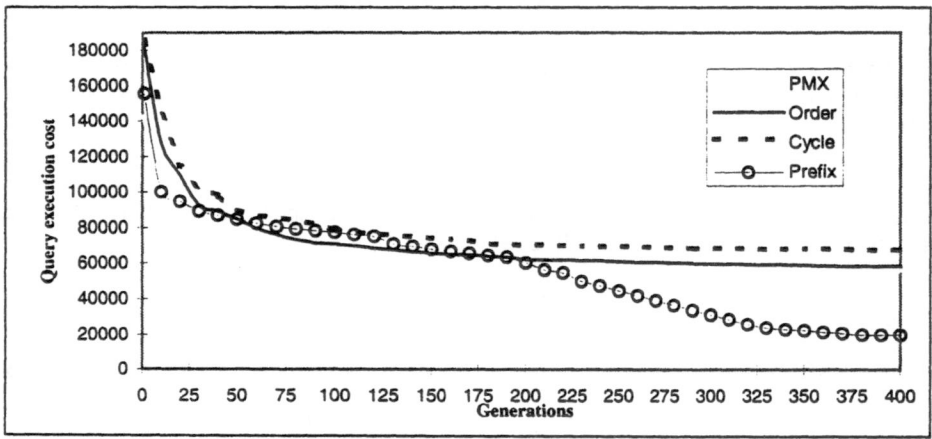

Figure 2: Comparing crossover operators

From the table we can see that apart from the PREFIX operator, other operators show very little improvement between generation 200 and 400. This is due to the fact that these operators concentrate their efforts in choosing valid plans that do not contain cartesian products rather than improving best plans from former generations.

Method	1st Generation	200th Generation	Last Generation	% Improvement
Prefix	155600	58962	19900	780%
Cycle	185700	70399	67400	275%
Order	181300	62259	58200	311%
PMX	141800	60187	57100	248%

Table 1: Maximum and minimum costs of crossovers

5.2 Heuristics

Figure 3 shows how the performance of our GA and the two heuristic change as the number of tables in the query is increased. The two heuristics perform well and similar to the GA as long as the number of tables is not very large. However there was a sharp deterioration in heuristics performance when the number of tables has exceeded 19. This is due to the fact that the solution space and the percentage of bad plans increase dramatically at this point. It is also clear that the GA is the only algorithm that exhibits stable performance with larger number of tables.

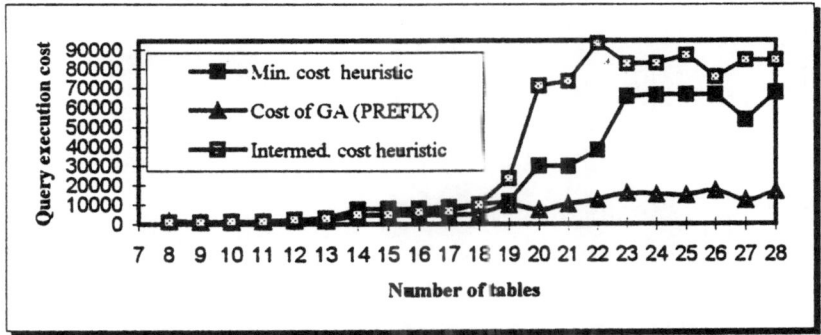

Figure 3: Sensitivity to number of tables between GAs and Heuristics

6. Conclusions and Future Work

We have studied the performance of a GA that we have developed for optimizing join orders in parallel shared-nothing database systems. We have compared our PREFIX crossover with other GAs that were proposed for query optimization in centralised database systems. Our empirical results have demonstrated much better query execution plans when this algorithm is used. This is mainly because our algorithm has the ability to employ data knowledge from previous runs and that it does not allow invalid plans to appear in the genetic population. We have also compared this algorithm with two heuristics. The results show that for small number of relations, the two heuristics can obtain similar query execution plans to our GA. However when the number of relations increases, GA outperforms these heuristics considerably.

Future work involve expanding the crossover operator reported here to other solution spaces, namely right deep, segmented right deep and bushy trees. For right deep trees, the same crossover operator can be used without the need to make any changes, except for the fitness function to reflect the cost model of such solution space. Segmented right deep and bushy trees are rather different. Some modifications to the present crossover would be inevitable due to the way joins sequence are processed in these solution spaces.

References

[Benn91] K.Bennett, M.Ferris, and Y.Ioannidis, *A Genetic Algorithm for Database Query Optimization*, Computer Science Technical Report No.1004,Wisconsin university, 1991.

[Gold89] D. E. Goldberg, *Genetic Algorithms in Search, Optimization and Machine Learning*, Addison Wesley, MA, 1989.

[Gref87] J. Grefenstette. *Incorporating Problem Specific Knowledge into Genetic Algorithms.* IN L Davis, editor, GAs and Simulated Annealing, Pitman Publisher, London 1987.

[Ioan90] Y. E. Ioannidis, and Y. Kang, *Randomized Algorithms for Optimizing Large Join Queries*, Proceedings SIGMOD 90, May 1990.

[Mich94]Z. Michalewicz, *GAs + Data Structures=Evolution Programs*, Springer-Verlag, 1994.

[Nafj97] K. Nafjan and J. M. Kerridge, *Toward an Object-oriented Interface to a Parallel Shared-nothing Database Machine*, Submitted for publication.

[Seli79] P. G. Selinger et al, *Access Path Selection in a Relational DBMS*, Proceedings of ACM SIGMOD- International Conference on Management of Data, May,1979.

[Stei93] M. Steinbrunn, G. Morekotte, A. Kemper, *Optimizing Join Orders*, Technical Report MIP-9307, University of Passau,1993.

[Swam88] A. Swami and A. Gupta, *Optimization of Large Join Queries*, SIGMOD Record, vol. 17. 1988.

Workshop 19

Symbolic Computation

Workshop 19: Symbolic Computation

Manuel Hermenegildo

The Call for Papers

The call for papers for this workshop invited submissions on all topics related to parallel symbolic computation. Symbolic computation was defined in its broadest sense, comprising all programming and computation paradigms that deal with symbolic data, or that deal with exact numeric data rather than floating-point values. Functional programming, logic programming, constraint programming, computer algebra, theorem proving, parallel search, and AI-related paradigms were explicitly included. Specific topics of interest mentioned included parallel execution models, parallel algorithms, parallel compilation and run-time techniques *as applied to the specific domain of symbolic computation*, including parallelizing recursion, speculative parallelism, manipulation of complex parallel data structures, parallel search, parallel language constructs, and static, dynamic or hybrid resource control (heap management, load management, data placement, granularity, etc.). Papers reporting on significant applications were also particularly welcome.

The Reviewing Process

A total of 15 papers were submitted, representing 13 countries (Austria, France, Germany, Hungary, Israel, Italy, Japan, the Netherlands, Spain, Switzerland, Ucraine, United Kingdom, and the United States). 5 submissions were accepted as regular papers, 3 as short papers, and 7 were rejected. No papers were moved to other workshops.

The Program

The final selection of papers covers a reasonable subset of the topic areas, with special emphasis on parallel implementation of logic and functional programming languages. A recurrent topic is implementation on distributed systems. In particular, within logic programming, two papers deal with the correct implementation in distributed systems of built-in predicates for pruning the search and collecting solutions, and with improving the efficiency of distributed unification by means of automatically inferred type information. The other paper presents a technique for optimizing task granularity for concurrent, committed-choice dialects. In the functional programming area, attention has focussed on the overly neglected issue of good data distribution.

In the context of constraint satisfaction problems, the accepted paper presents heuristics for the implementation of parallel tree search in distributed systems.

The paper on term rewriting describes a parallel implementation technique for order-sorted conditional term rewriting systems with promising initial speedup results. Within computer algebra, the accepted paper deals with the parallelization of arbitrarily long integer multiplication and division, which are fundamental in all other algebraic computations. The paper parallelizes the Karatsuba method for multiplication and division. The remaining paper deals with theoretical foundations, presenting several models of fine-grain parallelism.

An overview of the topic of parallel logic program implementation can be found in [1]. See also [3] (in this volume) for a tutorial introduction to the issues faced by parallelizing compilers in these paradigms. An introduction to parallel functional programming can be found in [2]. A comprehensive overview of parallel computer algebra can be found in several papers in [4].

References

1. J. Chassin and P. Codognet. Parallel Logic Programming Systems. *Computing Surveys*, 26(3):295–336, September 1994.
2. K. Hammond. Parallel functional programming: An introduction. In *Proc. First Int. Symposium on Parallel Symbolic Computation (PASCO'94)*, volume 5 of *Lecture Notes Series in Computing*, pages 181–193. World Scientific, September 1994.
3. M. Hermenegildo. Automatic parallelization of irregular and pointer-based computations: Perspectives from logic and constraint programming. In M. Griebl C. Lengauer and S. Gorlatch, editors, *Euro-Par'97*, Lecture Notes in Computer Science. Springer-Verlag, 1997.
4. H. Hong, editor. *Proc. First Int. Symposium on Parallel Symbolic Computation (PASCO'94)*, volume 5 of *Lecture Notes Series in Computing*. World Scientific, September 1994.

Using the Parallel Karatsuba Algorithm for Long Integer Multiplication and Division

Tudor Jebelean

RISC-Linz, A-4040 Linz, Austria
Jebelean@RISC.Uni-Linz.ac.at

Abstract. We experiment with the sequential and parallel Karatsuba algorithm under the paclib system on a Sequent Symmetry shared-memory architecture, obtaining efficiency 88% on 9 processors for the controlled-depth version, and 71% on 18 processors for the scalable version. Moreover, we use the Karatsuba algorithm within *long integer division*, by a recent divide-and-conquer technique. The parallel version exhibits modest speed-ups (2.3 on 3 processors and 3.4 on 9 processors), however the combined speed-up over the classical sequential method is more than 10 at 500 words.

Introduction

Fast methods for operations over long integers are very important for applications like cryptography and arbitrary precision arithmetic. Since FFT is only useful for very long operands [12, 1], the practical method of choice is Karatsuba multiplication. Moreover, this algorithm is easy to parallelize: experiments have been reported in [12] on shared memory, [3] on workstation network and [1, 2] on distributed memory.

Under the paclib [4] computer algebra system on a Sequent-Symmetry architecture with 20 processors (Intel 386, 16 MHz), we experiment with the "controlled depth" parallel Karatsuba algorithm. The speed-up ranges from 2 (at 30 words) to almost 3 (at 100 words) on 3 processors, while on 9 processors it ranges from 2 (20 words) to 7.4 (250 words) to 8 (500 words). This is better than the speed-up on 9 processors of [2] (6.5 at 250 words and 7.5 at 500 words), and almost 2 times better than the results reported in [3].

We also experiment with a scalable algorithm using up to 18 processors. For 500 words the efficiency ranges between 93% (3 processors) and 71% (18 processors). The speed-up is significantly higher than in [12] on 12 processors (100 words: 6.8 vs. 3.0, 200 words: 8.8 vs. 3.9, 300 words: 9.5 vs. 4.7 on 300).

Long integer division has the same theoretical complexity as multiplication ([10], p. 275), however using Karatsuba method involves a loss of 15 to 30 times in speed, which leads to a break-even point of several thousand words (see [11, 8]). Also, parallelization of integer division is difficult: theoretical parallel algorithms have been designed (see [13] for a survey), but practical implementations are realized mostly for VLSI design [14] and on systolic architectures [6]. *Exact* division has been parallelized on 2 processors [11], and the method presented below can be used for further increasing the speed [5].

Here we use a novel technique from [8], which allows to use Karatsuba multiplication for division with a slow-down of only a factor of two. Embedding the parallel "fork-join" Karatsuba method into division does not yield good speed-ups, because most of the calls to the multiplication are for small-length operands. For 500 words, the speed-up on 3 processors is 2.3 (efficiency 77%) and on 9 processors is 3.44 (efficiency 38%). However, the combined speed-up over the classical sequential algorithm is quite significant.

1 Multiplication

The Karatsuba multiplication algorithm [9], [10], p. 258, is a recursive divide-and-conquer technique which replaces a multiplication by three multiplications of half-length operands. In practice the threshold for stopping the recursion is determined experimentally (6 words of 29 bits in our implementation). Table 1 shows the timings (averaged over 100 runs) of the classical algorithm (column 2), the Karatsuba algorithm (column 3) and the speed-up (column 4).

As the previous authors, we use the straight-forward "fork-join" parallelization of the Karatsuba scheme: the three multiplications involved are performed in parallel. For the "controlled-depth" approach we obtained the timings shown in Table 1. Columns 5, 8 show the absolute time, columns 7, 10 show the speed-up over the sequential Karatsuba algorithm, and columns 6, 9 show the combined speed-up over the sequential classical algorithm. The efficiency surpasses 75% on 3 processors already at 40 words, and on 9 processors at 150 words. At 500 words the efficiency is about 90%.

In order to obtain a **scalable algorithm**, we give up the control over the recursion depth and we use instead a parallelization threshold of 24 words (experimentally determined). The algorithm scales well (efficiency over 72%) until 18 processors only for the length of 500 words - see Fig. 1.

Table 1. Multiplication (up) and division (bottom) in milliseconds.

length (words of 29 bits)	classic ($\mathcal{C}$)	seq. Karatsuba ($\mathcal{K}$) abs.	vs.$\mathcal{C}$	parallel Karatsuba					
				on 3 processors			on 9 processors		
				abs.	vs.$\mathcal{C}$	vs.$\mathcal{K}$	abs.	vs.$\mathcal{C}$	vs.$\mathcal{K}$
40	65	37	1.76	15	4.33	2.47	8	8.13	4.63
100	398	158	2.52	57	6.98	2.77	25	15.92	6.32
200	1,585	480	3.30	168	9.43	2.86	66	24.02	7.27
500	9,862	2,034	4.85	704	14.01	2.89	255	38.67	7.98
76/40	70	55	1.27	45	1.56	1.22	47	1.49	1.17
196/100	447	278	1.61	175	2.55	1.59	161	2.78	1.73
396/200	1,790	881	2.03	454	3.94	1.94	365	4.90	2.41
996/500	11,255	3,825	2.94	1,659	6.78	2.31	1,111	10.13	3.44

Fig. 1. Speed-up of the scalable multiplication algorithm.

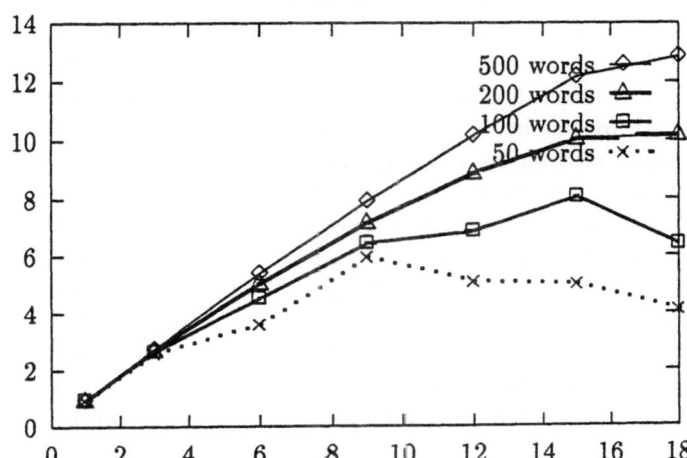

2   Division

The classical division algorithm (see [10] p. 237) consists of a series of successive updates of the dividend A by subtracting the divisor B multiplied by the current digit of the quotient Q, and right-shifted. This process is represented pictorially by the parallelogram in Fig. 2. The final value of A will be the remainder R. The scheme introduced in [8] starts by splitting the quotient Q into its high-order part Q_H of length q_H and its low-order part Q_L of length q_L such that $q_L \leq q_H \leq q_L + 1$, and also the divisor B into its high order part B_H of length b_H and its low-order part B_L such that $q_H + 3 \leq b_H$. Computing the high-order part Q_H of the quotient would normally require to perform the updates in the upper half of the parallelogram of Fig. 2 (areas 1 and 2). However, Krandick [11] proves that in most cases it is enough to update only 3 words below the lowest digit of A needed for the lowest quotient digit to be computed (i.e. down to the vertical line crossing area 1). This allows to split the updates of the dividend into 4 parts as shown by 1, 2, 3, 4 in Fig.2, and to perform them in the order of

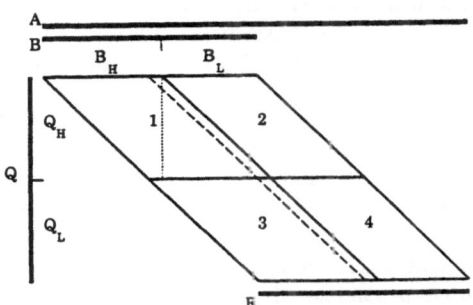

Fig. 2. Organization of the dividend updates $(Q, R \longleftarrow A : B)$.

the numbering: parts 2 and 4 by Karatsuba multiplication, and parts 1 and 3 by the same recursive technique. Under a certain threshold (15 words in our implementation), the recursion is replaced by classical division. This method has Karatsuba-like complexity, and the number of digit-multiplications is only 2 times higher than in Karatsuba multiplication (for details see [8]).

The speed-up can be increased by using the parallel "fork-join" version of the Karatsuba multiplication algorithm presented in the previous section. Table 1 shows the results of the experiments using the "controlled depth" algorithms on 3 and 9 processors. As expected, the efficiency of the parallel division is much lower than the one of multiplication, because the Karatsuba multiplication is called many times, and mostly with short-length operands. However, the combined speed-up over the classical sequential algorithm is quite significant: on 9 processors it ranges from almost 3 at 100 words, to more than 10 at 500 words.

Further details on the algorithms, timings, and the behavior of the parallel implementation. are given [7], which is available on the internet.

References

1. G. Cesari and R. Maeder. Parallel 3-primes FFT algorithm. In J. Calmet and C. Limongelli, editors, *DISCO'96*, pages 174–182. Springer LNCS 1128, 1996.
2. G. Cesari and R. Maeder. Performance analysis of the parallel Karatsuba multiplication algorithm for distributed memory architectures. *J. of Symbolic Computation*, 21:467–473, 1996.
3. B. Char, J. Johnson, D. Saunders, and A. P. Wack. Some experiments with bignum arithmetic. In H. Hong, editor, *PASCO'94*, pages 94–103. World Scientific, Singapore, 1994.
4. H. Hong, W. Schreiner, and A. Neubacher. The design of the SACLIB/PACLIB kernel. *J. of Symbolic Computation*, 19(1–3):111–132, 1995.
5. T. Jebelean. Exact division with Karatsuba complexity. Technical Report 96-31, RISC-Linz, December 1996.
6. T. Jebelean. Integer and rational arithmetic on MasPar. In J. Calmet and C. Limongelli, editors, *DISCO'96*, pages 162–173. Springer LNCS 1128, 1996.
7. T. Jebelean. Applications of the parallel Karatsuba algorithm to long integer multiplication and division. Technical Report 97-08, RISC-Linz, http://www.risc.uni-linz.ac.at/library, February 1997.
8. T. Jebelean. Practical integer division with Karatsuba complexity. In W. Kuechlin, editor, *ISSAC'97*. ACM Press, 1997.
9. A. Karatsuba and Yu Ofman. Multiplication of multidigit numbers on automata. *Sov. Phys. Dokl.*, 7:595–596, 1962.
10. D. E. Knuth. *The art of computer programming*, volume 2. Addison-Wesley, 1981.
11. W. Krandick and T. Jebelean. Bidirectional exact integer division. *Journal of Symbolic Computation*, 21:441–455, 1996.
12. W. Kuechlin, D. Lutz, and N. Nevin. Integer multiplication on PARSAC-2 on stock microprocessors. In *AAECC-9*, pages 216–217. Springer LNCS 539, 1991.
13. S. Lakshmivarahan and S. K. Dhall. *Analysis and design of parallel algorithms: Arithmetic and matrix problems*. McGraw-Hill, 1990.
14. E. E. Swartzlander, editor. *Computer Arithmetic*, volume 2. IEEE Computer Society Press, 1990.

Towards Full Prolog on a Distributed Architecture [1]

Lourdes Araujo
Fac. Matemáticas (Dpto. Sistemas Informáticos y Prog.)
Universidad Complutense de Madrid
Madrid 28040, Spain
lurdes@dia.ucm.es

Abstract. This paper presents an implementation of some essential side-effects of Prolog: *cut* and *findall*, on a distributed memory system. Although the techniques proposed herein are valid for any distributed memory implementation, they are advantageous in those based on re-computation, such as PDP (Prolog Distributed Processor), a model for Independent_AND/OR parallel execution of Prolog. The key idea to implement the *cut* predicate is to exploit as much parallelism as possible, but in such a way that the computation of a branch of the search tree which cannot be pruned by a *cut* is never delayed to control computations depending on a *cut*, (i.e. to analyze the pruning of the branch of these computations and to kill them). The model proposed for the *findall* predicate reduces the communication as much as possible.

1 Introduction

PDP (Prolog Distributed Processor) [4, 5] is a recomputation-based model for the exploitation of independent_AND/OR parallelism. The development of parallel systems to implement Prolog usually begins with the design of the part corresponding to the pure language (horn clauses) and then it is extended with side-effect predicates. PDP is not an exception. So, the purpose of this paper is to introduce some essential side-effect predicates, namely *cut* and *findall*. Although the techniques proposed herein are valid for any distributed memory implementation, they turn out to be advantageous in those based on recomputation, such as PDP. Recomputation allows PDP to exploit combined parallelism as a set of independent computations (see [4, 5] for details). Thanks to this, the problem of gathering solutions inherent to AND_parallelism is overcome. Another advantage of recomputation, already discussed in the literature [9], is that it allows to introduce side-effect predicates without sacrificing parallelism exploitation. What this paper will show is that side-effects can be incorporated to PDP without introducing modifications to the model.

The *cut* predicate prunes all the branches on the right of the one corresponding to the clause in which the *cut* occurs. If OR_parallelism is exploited, branches of the same predicate containing a *cut* may be executed by different processes. Accordingly, a process working on a branch which is not the leftmost in the search tree can find its corresponding solution before the process working on the leftmost branch reaches the *cut*. Therefore, maintaining the exploitation of OR_parallelism in the presence of a *cut* requires the control of the processes

---

[1] Supported by the projects TIC95-0433.

which execute each non-leftmost branch in order to be able to stop them if necessary. This mechanism may be too expensive on a distributed memory system, for which stopping a process requires an exchange of messages. A number of solutions have been proposed for parallel systems [7, 10, 9]. However, they require a large amount of communication on a distributed memory system. Another model by Ali [2] consists in constraining the OR_parallelism exploitation to the cases outside the scope of a *cut*. This "constrained" approach seems to be suitable for PDP because it avoids communications overhead, but, at the same time, many opportunities of parallelism may be lost. Accordingly, a more ambitious approach has been devised. The key idea of this approach is to exploit as much parallelism as possible, but in such a way that a *safe* computation (a computation which can not be pruned by a *cut*) is never delayed.

This paper also presents an implementation of the *findall* predicate. As it is well-known, *findall(X, G, L)* constructs a list L consisting of all of the bindings of X for which the goal G holds. Some solutions have been proposed for this predicate [6, 1]. These approaches assume multiple processes sharing access to some special structure (a *set* [6] or a *findall tree* [1]). Hence, they are not appropriate for a distributed implementation. The model proposed herein reduces the communication as much as possible by distributing the control of the execution of the *findall* predicate.

The rest of the paper proceeds as follows: Section 2 presents an execution model overview of PDP; section 3 describes the implementation of *cut*; section 4 discusses the implementation of the *findall* predicate; and section 5 draws the main conclusions of this work.

2   Execution Model Overview

The execution model —described in full detail in Ref. [5]— has been implemented as an extension of the Warren Abstract Machine (WAM) [13]. PDP has been devised to run on a pool of processors organized within a hierarchy of clusters, each consisting of a *scheduler* and a set of *workers*. Schedulers are responsible for the distribution of pending work among idle workers.

The goals and clauses to be executed in parallel (*parallel goals* and *parallel clauses*) are annotated in the program. Independent AND_parallelism is exploited by following a *fork-join* approach, which is an extension for distributed memory systems of the one followed in the RAP model [11]. OR_parallelism is exploited by following a recomputation approach [3]; a processor environment is reconstructed by recomputing the query without backtracking, following the socalled *success_path*, i.e. the sequence of clauses which have succeeded until the last choicepoint with pending parallel alternatives, obtained from the parent processor (the one finding the parallel clause). Recomputation allows the exploitation of OR_under_AND parallelism in a very natural way. The PDP approach to exploit OR_under_AND parallelism [4] is designed to create, in an automatic and decentralized way, an independent computation for each solution.

The computation of a goal in PDP is called a *task*. Two types of tasks are distinguished: *OR_tasks* and *AND_tasks*. AND_parallelism and OR_parallelism are exploited by AND_tasks and OR_tasks respectively. They are created by the task finding the annotation of parallelism, which is their *parent task*. An OR_task computes solutions to the query by exploring a portion of the search tree. An AND_task computes a solution to a goal which belongs to a parallel call and gives the result to its parent task. In this way, the parallel execution of a program defines a *task tree*. The model supports combined parallelism in a very natural way. As a result, the execution of the search tree is automatically distributed among tasks, so that no specific task is in charge of the distribution. The model is outlined as follows:

- The program execution begins as an OR_task (the root of the task tree), which performs a sequential computation until a parallel call or a parallel clause is reached.
- The execution of a *parallel call* is carried out by the creation of an *AND_task* for each independent goal. These AND_tasks receive from its parent task a goal and the computed answer substitution restricted to the variables of the goal. Each AND_task computes its goal, returns the *local solution* to the parent task and finishes. The parent task waits for the answer to each independent goal and it is in charge of synchronizing the reception of those answers.
- The execution of a *parallel procedure*, i.e. a procedure with clauses annotated with OR_parallelism, is carried out by the creation of a new *OR_task*. This receives the *success_path* of the predecessor OR_task and recomputes the query following this path. After the recomputation, the execution continues as usual.
- If OR_parallelism appears under AND_parallelism, the OR_tasks arising from an AND_task have to re-execute the parallel call in order to find new solutions to it. If this were done blindly, the result would be the simple *repetition of solutions*. To avoid this, it has been introduced a rule, called *combination rule* which decides which branch is explored to solve each independent goal. The *ancestor goal* of an OR_task arising from an AND_task is defined as the goal executed by this AND_task. The combination rule fixes the solution to the goals on the left of the ancestor goal and combines them with *every* solution of the remaining goals.

The results of the implementation of this model [5] have proved that OR_parallelism exploitation provides a linear speedup for high granularity programs. For some programs presenting both kinds of parallelism PDP achieves a greater speedup than the product of the speedups achieved by exploiting each kind of parallelism separately. The reason is that the exchange of messages required in the exploitation of AND_parallelism is avoided in PDP when OR_under_AND parallelism is exploited.

3 Cut Implementation

The model relies upon the distinction between *safe* tasks, i.e. computations of branches which can not be pruned by *cut*, and *speculative* tasks, i.e. computations that can be pruned by *cuts*. The principle under which this approach has been designed is never to delay a safe task because of controlling speculative tasks, i.e. because of analyzing if these tasks are computing a branch which has been pruned by a cut and killing them. The model can be outlined with the following points:

- OR-parallelism is exploited and all the branches of a predicate are executed simultaneously. Branches in the scope of a *cut*, i.e. which can be pruned by a cut, are executed as speculative tasks.
- If a safe task fails, the task corresponding to the execution of the next branch, in a depth-first, left-to-right order, becomes safe – by receiving a message from the previous safe task.
- If a safe task succeeds, the speculative tasks executing branches arising from the same node are killed.
- A solution reached by a speculative task is not given as an output but it is stored until the task becomes safe or it is killed.

Figure 1 shows the scheme of a possible case in the execution of the goal p in the following program:

```
p :- s, t, !, u.     s :- s1.        t :- t1.
                     s :- s2.        t :- t2.
                     s :- s3.        u. s1. s2. s3. t1. t2.
```

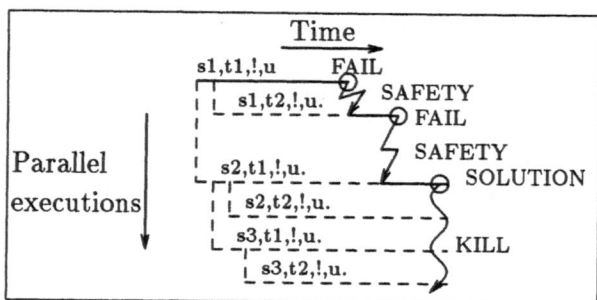

Fig. 1. Parallel implementation of *cut*.

If the computation corresponding to $s1$ and $t1$ does not fail, the remaining computations (speculative) would be useless because of the *cut* and they would be killed. Figure 1 shows the case in which the first task fails when executing

$t1$, and then informs the following speculative task $(s1, t2, !, u)$ about its new safe status. This task also fails and transforms the next task $(s2, t1, !, u)$ into *safe*. This task does not fail and when the solution is reached the remaining speculative computation are killed.

The implementation of *cut* in PDP leads to distinguish between safe and speculative tasks as follows:

- **speculative AND_task:**
  An AND_task is *speculative* in these two cases: (a) if the task executes a goal on the left of a *cut* in a parallel call, for instance b in a, $(b\&!\&c)$, or (b) if the task arises from another speculative task. Since the solution to the goals on the left of a *cut* in a parallel call are fixed to the first solution encountered, in case (a), any the speculative OR_task that the AND_task might create is killed as soon as the solution to the goal is reached, or it is made safe as soon as the AND_task fails.
  Notice that, in spite of being speculative, an AND_task always computes the first solution to its goal and gives it to its parent task. What the speculative character determines is that every task it creates is also speculative.
- **speculative OR_task:**
  Any OR_task arising from a speculative AND_task or from another speculative OR_task is also *speculative*, and exploits parallelism by means of speculative tasks. The solution obtained by a speculative OR_task is stored until the task becomes safe or it is killed.

PDP, which was initially implemented on a transputer network, has now been ported to a workstation network using PVM (*Parallel Virtual Machine*) [8], a software package that allows a heterogeneous network of parallel and serial computers to appear as a single concurrent computational resource. Nevertheless, the execution times obtained are not optimal since the resources have been shared with other processes. The current results have been obtained with 12 processes running on three SUN SPARC 1 workstations (20 MHz).

The above presented scheme has been compared with the other one in which the exploitation of parallelism is constrained to the part of the program outside the scope of a *cut* (the method used in ref. [2]). The benchmarks are a synthetic program, which presents coarse grain parallelism; the *mastermind* program, with eight digits to choose and different secret codes – mm1([1,5,0,4,3]), mm2([2,2,2,2,2]), mm3([3,3,4,5,1]), mm4([5,4,3,2,0]) –; the *number* program, previously used by Hausman [10], with five digits to choose and different queries – number1(3,62), number2(3,65), number3(3,56), number4(3,88) –; and the Chat-80 natural language query system, with the query *db5* used by Hausman (db5 *20 means that the query was run twenty times). Table 1 shows the results for these programs. The speedup obtained for these programs with the unconstrained approach is only slightly lower then the one obtained by Hausman [10] with 4 workers in the Aurora system, which presents some shared memory. According to the results the unconstrained approach seems to be better. For some programs, such as number, the constrained approach impedes the parallelism exploitation.

1178

program	Sequential	Constrained	PDP
synthetic	7.10	7.10(1)	2.81(2.53)
mm1	63.80	48.72(1.31)	35.64(1.79)
mm2	25.16	21.50(1.17)	16.55(1.52)
mm3	151.45	101.64(1.49)	55.68(2.72)
mm4	165.94	109.89(1.51)	57.79(2.87)
number1	24.07	24.05(1)	20.23(1.19)
number2	49.32	49.41(1)	28.51(1.73)
number3	38.11	37.72(1.01)	34.50(1.10)
number4	21.54	21.60(1)	16.20(1.33)
db5*20	3.7	3.7(1)	2.74(1.35)

Table 1. Comparison of approaches for implementing *cut* (time in seconds and speedups in brackets.)

4 Findall Implementation

The PDP implementation of *findall* introduces some differences in the exploitation of OR_parallelism. One of them is the fact that a solution corresponding to a *findall* predicate does not have to be sent to the output process, but to the one executing the *findall*. A second key point is to detect when all solutions to the *findall* predicate have been collected.

Let us call *F_task* the task executing the *findall* predicate. This task may be either an OR_task or an AND_task. Thus, the extension of the model to implement *findall* introduces two new types of tasks: *findall_OR_tasks* and *findall_AND_tasks*. A findall_OR_task computes a solution to the goal of the *findall* predicate (*findall goal*) instead to the query. It receives a success path which starts at the findall goal and leads to a new solution to it. The obtained solution is sent to the F_task instead of the input_output process, as it would be the case of a normal OR_task. If this task finds parallelism it exploits it by further findall_tasks. A findall_AND_task is similar to an AND_task, except for the fact that it exploits parallelism by further findall_tasks. As in the case of the implementation of *cut*, AND_tasks are merely transmitters of the character (speculative or findall) they have. It is in the OR_tasks that this character is translated into an actually different behavior.

The *findall* execution approach is outlined in the following points:

- A task executing a *findall* predicate enters a new execution mode, *findall_mode*, in which its behavior is adapted to the execution of *findall*.
- The OR_parallelism and the AND_parallelism of the findall goal is exploited by means of findall_OR_tasks and findall_AND_tasks respectively.
- In findall_mode a task receives solutions until the completion of the *findall* predicate. Two strategies have been investigated for this detection, which will be discussed later.

– Each solution is received along with its computed success path, which allows sorting the solutions. Finally the computation mode is changed to normal mode.

The parent task needs to know when all the solutions to *findall* have been found. This happens when every task taking part in the computation has failed. Two strategies have been checked to detect this event. In *Strategy 1* each task taking part in the execution of *findall* collects the failures of its offspring tasks. If the task fails before all its offspring tasks, it informs the remaining of them that their new parent task is its own parent task. In *Strategy 2* failure is directly reported to the F_task. In this strategy the F_task needs to knows how many tasks participate in the execution of *findall*. Therefore, each task reporting a solution or a failure must also inform about how many tasks it created. The implementation of the first strategy has the advantage of decentralizing the process, but the disadvantage of introducing more exchange of messages whenever a task fails before all its offspring tasks. For the second strategy it is the other way around. Table 2 shows the results of implementing each of these strategies, as well as

program	sequential (prog, fail)	parallel (prog, fail)	findall strategy a	findall strategy b
queen(6)	1.7	1.37(1.24)	1.5(1.13)	1.5(1.13)
queen(8)	39.0	14.68(2.65)	17.0(2.29)	15.40(2.53)
mm5	379.89	102.90(3.69)	118.13(3.21)	116.92(3.25)
mm6	356.99	99.65(3.58)	110.14(3.24)	109.85(3.25)
mm7	388.31	112.78(3.44)	121.10(3.20)	119.48(3.25)
number5	58.97	17.33(3.40)	23.57(2.50)	23.02(2.56)
number6	82.17	25.97(3.16)	32.28(2.54)	31.90(2.57)
number7	138.60	43.60(3.18)	54.84(2.53)	54.09 2.56)

Table 2. Comparison of the strategies for the detection of the *findall* completion (time in seconds and speedups in brackets.)

the time to collect all solutions in a 'program, fail' manner. The latter allows to estimate the overhead introduced by the findall mechanism (around 15% in both approaches). The programs used as benchmarks are the *queen* program; the mastermind program, with four digits to choose and different secret codes – mm5([1,1,2]), mm6([0,3,1]), mm7([3,3,2]) –; and the *number* program, with five digits to choose and different queries – number5(3,15), number6(3,5), number7(3,3). Results show slightly shorter execution times for strategy b. They seem to indicate that the difference increases with the program size and the number of solutions to collect. The reason is the smaller number of messages exchanged with strategy b. The number of messages exchanged by both strategies is the same unless a parent task fails before their offspring tasks.

5 Conclusions

This paper has presented an implementation of some important side-effects in PDP. The implementation reveals that the parallelism exploitation model of PDP turns out to be advantageous to include the side-effects to the extent that the model need not be altered in any way.

An "unconstrained" *cut* implementation model, which exploits all parallelism appearing in the program, has been compared with an approach, already proposed in the literature, which constrains OR-parallelism exploitation to those parts of the program outside the scope of any *cut*. The model is based upon the principle of never to delay safe computations because of controlling speculative ones. The results obtained implementing both models on a network of workstations reveal that the unconstrained approach is clearly advantageous.

The *findall* predicate has also been implemented in such a way that the control of the execution is decentralized. Two strategies have been tried for detecting when all solutions have been collected. It turned out that the best strategy is that in which failures are reported directly to the task executing the *findall* predicate instead of the parent tasks.

References

1. Ali, K. A. M., Karlsson, R. *A Novel Method for Parallel Implementation of findall.* Proc. Int. Conf. on Logic Programming (1989), pp. 235-245.
2. Ali, K.A.M. *A Method for Implementing Cut in Parallel Execution of Prolog.* Research Report SICS R87001 (1987).
3. Araujo, L., Ruz, J.J. *OR-Parallel Execution of Prolog on a Transputer-based System.* Transputers and Occam Research: New Directions. IOS Press (1993), pp. 167-181.
4. Araujo, L., Ruz, J.J. *PDP: Prolog Distributed Processor for Independent_AND\OR Parallel Execution of Prolog.* Proc. Int. Conf. on Logic Programming (1994), pp. 142-156.
5. Araujo, L., Ruz, J.J. *A Parallel Prolog System for Distributed memory.* The Journal of Logic Programming, 33(1), (1997), pp. 49-79.
6. Carlsson, M., Danhof, K., Overbeek, R. *A Simplified Approach to the Implementation of AND-parallelism in an OR-parallel Environment.* Proc. Int. Conf. on Logic Programming (1988), pp. 1565-1577.
7. Calderwood A., Szeredi, P. *Scheduling Or-parallelism in Aurora - the Manchester scheduler* Proc. Int. Conf. on Logic Programming (1989), pp. 419-435.
8. Geist, A., Beguelin, A., Dongarra, J., Jiang, W., Manchek, R., Sunderam, V. *PVM: Parallel Virtual Machine* MIT Press (1994).
9. Gupta, G., Santos Costa, V., *Cuts and Side-effects in AND-OR Parallel Prolog.* The Journal of Logic Programming 27(1), (1996), pp. 45-71.
10. Hausman, B. *Pruning and Speculative Work in OR-Parallel PROLOG.* PhD thesis, SICS (1990).
11. Hermenegildo, M., *An abstract Machine Based Execution Model for Computer Architecture Design and Efficient Implementation of Logic Program in Parallel.* PhD thesis, U. of Texas at Austin (1986).
12. Muthukumar, K., Hermenegildo, M., *Complete and Efficient Methods for Supporting Side-effects in Independentbackslash Restricted And-parallelism* Proc. Int. Conf. on Logic Programming (1989), pp. 80-97.
13. Warren, D.H.D., *An Abstract Prolog Instruction Set.* Tech. Note 309, SRI International, (1983).

Improving Distributed Unification through Type Analysis

Evelina Lamma[1], Paola Mello[2], Cesare Stefanelli[1], Pascal Van Hentenryck[3]

[1] DEIS, Università di Bologna, Viale Risorgimento 2, 40136 Bologna, Italy
{elamma,cstefanelli}@deis.unibo.it
[2] Istituto di Ingegneria, Università di Ferrara, Via Saragat, 44100 Ferrara, Italy
pmello@ing.unife.it
[3] Dept. of Computer Science, Brown University, Box 1910, Providence RI 02912, USA
pvh@cs.brown.edu

Abstract. In distributed implementations of logic programming, data structures are spread among different nodes and unification involves sending and receiving messages to access them. Traditional implementations make remote data structures accessible to other processes by sending messages which carry either the overall data structure (*infinite-level copying*) or only remote references to these data structures (*zero-level copying*). These fixed policies can be far from optimal on various classes of programs and may induce substantial overhead. The purpose of this paper is to present an implementation scheme for distributed logic programming which consists of tailoring the copying level to each procedure. The scheme is based on a consumption specification which describes the way the procedure "consumes" its arguments locally. The consumption specification (or an approximation of it) can be automatically obtained through a static analysis inspired by traditional type analyses. The paper also describes a high-level distributed implementation that uses the consumption specification to avoid unnecessary copying and to request data structures globally. Experimental results for a network of workstations show the potential of the approach.

1 Introduction

In distributed implementations of logic programming, data structures are spread among different nodes and unification involves sending and receiving messages to access them. Traditional implementations make remote data structures accessible to other processes by sending messages which carry either the overall data structure (*infinite-level copying*) or only remote references to these data structures (*zero-level copying*). An intermediate approach is possible in some systems, where data structures are copied up to a certain level. For instance, an approach based on replication of data is used in [4], where each message carries the infinite-level copying of each argument of the goal to be solved. Ichiyoshi et al. [12] introduced the copying level concept in the implementation of the KL1 language on the MultiPSI machine where a fixed level of copying can be "a-priori" used for data structures.

It should be clear that any fixed policy can be far from optimal and may induce substantial overhead on various class of programs. A high copying level may cause overhead due to the transmission of unnecessary information; a low copying level might cause, instead, further messages to be exchanged for accessing a remote structure in order to perform unification (remote dereferencing).

The purpose of this paper is to present a novel implementation scheme that, informally speaking, tailors the copying level to each argument of each procedure in the distributed program. The scheme is based on a consumption specification which describes how a procedure "consumes" its arguments locally. The implementation uses this specification to improve both the sending of arguments to goal processes and the remote dereferencing phase possibly occurring during clause head unification. The copying phase is improved by avoiding the copy of subterms that are not used by the called procedure. The remote dereferencing phase is improved by requesting, from a remote node, entire subterms instead of accessing them piece by piece, thus reducing the amount of communication. Consumption specifications are expressed in terms of a simple generalization of tree-grammars [5] (or type graphs [8, 17]). Consumption specifications (or approximations of them) can be obtained automatically through a static analysis inspired by traditional type analyses, making the approach fully transparent for programmers.

The paper also describes a high-level implementation of the scheme based on attributed variables following the lines suggested by [7]. The implementation uses a blackboard for communication. Experimental results show the potential of this approach.

2 Consumption-based Distributed Unification

In our distributed model, a number of concurrent processes cooperate in solving the query. Processes communicate via message passing, since data structures can be spread over different nodes of the underlying architecture. In the following we focus our discussion on the implementation of distributed unification. The main difference between distributed and sequential implementations of unification is that the first demands message management for data exchange.

The put phase in the standard Warren Abstract Machine (WAM) [1, 18] loads the arguments in the registers but, in a distributed implementation, the put phase consists of preparing a message to be sent to the process responsible for executing the goal. For a given argument, this *message preparation* phase works as follows. If the argument is an atomic value, the value is included in the message. If the argument is an unbound variable, a remote reference to this variable is created and inserted in the message. If the argument is a structure, different policies can be found in the literature [9, 15, 16, 12]:

- the argument is copied in its entirety in the message; this is called *infinite-level copying*;
- a remote reference to the structure is included in the message; this is called *zero-level copying*;

— the structure is copied in the message up to a certain level with remote references to the rest of the structure; this is called *k-level copying*.

Following the adopted policy of copying, the arguments are packed in the message that is sent to the remote node responsible for the procedure execution.

The standard WAM **get** phase is also modified in the distributed case. In particular, before performing the head unification, the remote process executes a *message reception* phase to extract the arguments from the message. In addition, for efficiency reasons discussed later, the message may not contain all the necessary information needed for unification and therefore some remote references may be encountered when extracting the arguments from the message. In this case, it might be necessary to request to the appropriate remote processes the needed data structures. This task, i.e., requesting the value of a remote reference to a process, is known as *remote dereferencing* and is performed by using message-passing. Also, when performing remote dereferencing, it is possible to follow several policies of copying. In particular, it is possible to request either all the referenced data structure or only a part of it. The two solutions (and any intermediate case) have the same advantages and inconvenients of the policies described for the message preparation phase.

None of the previously described strategies (infinite-level copying, zero-level copying, or k-level copying) is the most appropriate for all programs. Infinite-level copying has the problem of sending too much information, i.e., copying terms that are not used. Zero-level copying has the problem of sending many small messages, increasing the communication between processes. K-level copying combines the advantages and inconvenients of zero- and infinite-level copying. In addition, the choice of the copying policy depends not only on the structure of the application, but also on the architecture type (in particular, the communication cost), since it affects the cost of the various operations. An analysis of the architectural factor in the choice of the copying policies may be found in [2].

The main contribution of this paper is to present a novel implementation scheme that, informally speaking, tailors the copying and dereferencing level to each argument of each procedure in the distributed program. Consider the traditional Prolog program for list concatenation [13]:

```
append([],L,L).
append([F|T],S,[F|R]) :- append(T,S,R).
```

and the top-level goal **top :- append(LongList, List, Res)** where **LongList** is a long list of complex data structures. **append/3** needs the spine of the list for the first argument but does not need to inspect the elements of the list. It also may not need to inspect its second and third arguments. As a consequence, the top-level goal should be compiled to prepare a message containing a copy of the spine of the list with a remote reference to its various elements for the first argument. Remote references to the two other arguments must be included in the message as well. This makes sure that the message preparation phase copies only what is really needed by **append/3** and that **append/3** is executed without requiring any remote dereferencing. An infinite-level policy would copy the whole list and thus potentially many irrelevant structures, while zero-level copying would require to dereference the remote list element by element, introducing notable communication cost. We are interested in driving unification by the minimal level needed for term inspection, which we call consumption.

2.1 The Consumption Specification

In order to specify how a procedure minimally "consumes" its arguments, our implementation makes use of a consumption specification. Consumption specifications are simple enough to be provided by programmers but we also show in Section 4 how to automatically obtain them (or an approximation of them) through a static analysis inspired by traditional type analyses, making the approach transparent to programmers. In practice, a consumption specification describes a superset of the type of the procedure, as it will be clear in Section 4.

A consumption specification is expressed as a tree grammar [5] extended with an additional terminal Remote. This additional terminal simply specifies zero-level copying. The rest of the specification identifies what part of the term is consumed locally by the predicate. For instance, the consumption specification for append/3 is append(T_1,T_2,T_3) *where*

```
T₁ ::= [] | cons(Remote,T₁).
T₂ ::= Remote.
T₃ ::= Remote.
```

It specifies that append/3 consumes the spine of the first argument and that the two other arguments are not consumed. Consumption specifications can be rather complex. The consumption specification for the program

```
process([]).
process([s(D)|R]) :-
     process(R).
process([c(D)|R]) :-
     process(R).
```

is given by process(T) *where*

```
T  ::= [] | cons(T₁,T).
T₁ ::= s(Remote) | c(Remote).
```

It specifies that process/1 consumes all the elements of the list but limited to the main functor of each element.

Consumption specifications are expressive enough to associate different level of copying not only with each procedure argument but also with distinct subtrees of a given argument.

2.2 Exploiting the Consumption Specification

The consumption specification is exploited in the distributed implementation during message preparation, message reception, and remote dereferencing.

During message preparation, consumption specifications are used to include in the message only those parts of the data structures consumed by the called procedure. For instance, a call to append/3 requires for each argument a copy instruction which inserts in the message the appropriate part of the argument. In case of a term t with consumption specification T ::= [] | cons(Remote,T), the copy instruction inserts inside the message only the spine of the list, i.e., only the reference to each element of the list.

When receiving a message the consumption specification is used to request the remote data structures that are necessary for the procedure to execute locally. The main interest of consumption specifications in this context is the ability of requesting the needed data structures globally instead of element by element. Note also that these data structures are requested at the procedure level before executing any clause.

3 A High-level Implementation

To experimentally verify our approach, we implemented a distributed logic language on top of SICStus Prolog [14] using the Linda library and attributed variables.

Attributed variables. Our high-level implementation was inspired by [7] and uses attributed variables to implement the communication variables, i.e., the variables occurring in a remote goal. Attributed variables introduced by Le Houitouze [11] are variables associated with an attribute (i.e., a term) and unification of these variables can be specified by programmers. Our high-level implementation attaches an attribute rem(Process,Id,Bound,Type) where the pair (Process,Id) uniquely identifies a communication variable, Bound specifies if the variable is bound or unbound and Type is the type associated with the variable.

The blackboard structure. The message sending is achieved by writing the message onto a blackboard implemented via the Linda library. There is a server process which handles the blackboard. Prolog client processes can write (using out/1), read (using rd/1), and remove (using in/1) data (i.e., Prolog terms) to and from the blackboard. Partial bindings of the communication variables (determined by the consumption specification) are inserted in a message and posted on the Linda blackboard.

The consumption specification. The consumption specification is represented by Prolog facts of the kind type(t,term).. The copy of terms in their blackboard representation is performed until a terminal remote is reached, thus avoiding unnecessary communication overhead. Notice that the consumption specification is also used for run-time type checking when constructing the message.

Message preparation. Consider the preparation of a message containing the variable X, associated with the consumption specification T described in the previous paragraph. Assume that X is bound to the list `[c([1,2,...,200])]`. In this case, the element of the list is copied but copying is limited to the main functor. A new communication variable X1 is created and locally bound to `[1,2,...,200]`. Only the structure `[c(X1)]` is then posted into the blackboard. In fact, the message inserted in the blackboard contains the binding for the variable X:

$$\texttt{msg\_binding(rem(1,2,bound,t),[c(rem(1,3,bound,list))])}$$

where the pair `(1,2)` identifies X and the pair `(1,3)` identifies X1.

Message reception. After receiving a message, the argument values are extracted in order to perform head unification and goal evaluation. This implies the building of a local structure starting from the blackboard representation of the arguments contained in the message. A new local structure `[c(X2)]` is created, where X2 is a new attribute variable with the same identifier of X1. Notice that, during the head unification, some parts of the data structure may not be locally present. In this case, *remote dereferencing* is automatically raised by the unification of attribute variables. For instance, assume that `g([c([1|_])])` is the head of the clause. This implies the unification of the attribute variable X2 with `[1|_]`. Therefore the unification handler (i.e.,`verify_attr/3`) is called and a request for a remote dereferencing of X2 is sent to the appropriate process.

4 Static Analysis

In this section, we sketch how the consumption specification can be obtained by a static analysis enhancing traditional type analyses. We convey the main ideas behind the approach. Recall that the consumption specification describes a superset of the type of the procedure. Consider the list concatenation code. A goal-independent type analysis produces the result:

$$\texttt{append(T,Any,Any)} \ \textit{where} \ \texttt{T} \ ::= \ \texttt{[]} \ | \ \texttt{cons(Any,T)}.$$

which is essentially the consumption specification we showed previously (replace Any by Remote). These type analyses have been investigated extensively in the literature (e.g., [6, 8, 17]). Type analysis of logic languages, and Prolog in particular, is of primary importance for high performance compilers. In the sequential case, type analysis has been applied in order to improve indexing, to specialize unification and to produce more efficient code for built-in predicates.

We exploit type analysis to automatically obtain the consumption specification. To this purpose, we can profitably use a system like GAIA [17], where type analysis is based on abstract interpretation [3] and type graphs [8]. The appealing feature of GAIA is its good trade-off between accuracy and efficiency and the ease with which type analysis can be combined with other analyses that can be useful in our case, such as, for instance, mode analysis.

To obtain effective consumption specifications, it is necessary to refine the result of the type analysis. Consider the program

```
p([]).                              q([]).
p([F|Ta]) :- q(Ta).                 q([F|Ta]) :- q(Ta).
```

The standard analysis, and GAIA in particular, would produce the results p(T) and q(T) where T ::= [] | cons(Any,T)., which are superset of the types of the procedures.

Notice that an efficient implementation of distributed unification should not send the whole list to p/1, but only its first cons cell with arguments which are remote references to the head and tail of the list. In this respect, the consumption specification for p/1 is $p(T_1)$ where T_1 ::= [] | cons(Remote,Remote).. This consumption specification is a larger superset of the type (i.e., what the analysis would return if the goal q(Ta) is omitted in the clause for p/1). To determine the consumption specification from the type, traditional type analyses should be enhanced by annotating each functor with the predicate in which it occurs. Then, the widening operator for type graphs [17] is applied only for types related to the same predicate.

When enhanced in this way, the type analysis for the above program now determines the results $p(T_1)$ and $q(T_2)$ where

$$T_1 ::= []_p | cons_p(Any, T_2).$$
$$T_2 ::= []_q | cons_q(Any, T_2).$$

From these results, it is not difficult to obtain the consumption specification for p/1, since the type T_1 now indicates that p only accesses the first cons cell locally. To this purpose, it suffices to substitute each occurrence of Any with Remote in type T_1 and any reference to other types (e.g., T_2) with Remote as well, thus obtaining the consumption specification for p and q:

$$T_1 ::= [] | cons(Remote, Remote).$$
$$T_2 ::= [] | cons(Remote, T_2).$$

5 Experimental Results

This section presents the experimental results obtained by running some example programs on the system described in Section 3. These results give some indication of the practical usefulness of the consumption specification approach. Obviously, the high-level implementation cannot achieve the performance of a specific abstract machine. For this reason, in [10] we sketched a low-level implementation based on the WAM, suitably extended in order to perform the distributed unification driven by the consumption specification.

To compare the consumption specification approach with traditional copying policies, we executed each program with different policies of copying. In particular, we compared both zero- and infinite-level copying policies with the "optimal" policy driven by the consumption specification determined by the static analysis sketched in Section 4. All programs have been executed with the same granularity of execution, i.e., we allocated each procedure onto a different node, although in some cases this is not the best possible allocation strategy (we are not interested in obtaining the maximal performance, but only in the comparison of the different solutions for copying the arguments). Tables 6.1-6.4 show the results obtained by running the examples on a network of SUN workstations (Sparcstation2) connected over an Ethernet network. The results are presented in terms of complexity of the data structures involved. All the presented programs work on lists and are executed with lists of different length.

As a first benchmark, consider the **append/3** program. We have considered for this program both zero- and infinite-level copying and the consumption specification requiring to send the spine of the list for the first argument and a remote reference for both the second and third arguments of the goal. Table 6.1 shows the results obtained with the first two arguments being lists of 10, 30, and 50 integer elements. Times are in milliseconds. As shown in Table 6.1, the gain achieved with the consumption specification approach increases with the length of the lists.

length	zero-level copy (0)	∞-level copy (∞)	consumption copy (*c*)	0/*c*	∞/*c*
10	122	56	24	5.08	2.33
30	365	80	32	11.40	2.50
50	585	112	40	14.62	2.80

Table 6.1: **append/3** with lists of integers

Table 6.2 shows the timings of the usual **reverse/2** [13] program with zero-, infinite-level copying, and the consumption specification described in Section 2. Benchmarks have been executed with a list of 10 elements each one being, on its turn, a list of 10, 30, 50 integer values. Zero-level copying is very inefficient on this benchmark, because it requires a remote dereference for each element of the list, i.e., 10 remote dereference requests for all cases. ∞-level induces to copy the list twice from the top-level goal to **reverse/2** and from **reverse/2** to **reverseAcc/3**.

length	zero-level copy (0)	∞-level copy (∞)	consumption copy (*c*)	0/*c*	∞/*c*
10	1020	240	208	4.90	1.15
30	1123	544	496	2.27	1.09
50	1210	800	568	2.13	1.40

Table 6.2: **reverse/2** with one list of lists of increasing size

Tables 6.3 and 6.4 show the results of the **keysort/2** and **quicksort/2** programs [13]. **keysort/2** is a variant of **quicksort/2**, and sorts a list of strings on the basis of the first character of each string. The results reported in Table 6.3 are obtained by sorting a list of 10 strings, each one of increasing length (i.e., 10, 30, 50 characters), and by adopting zero-, infinite-level copying and the copying policy driven by the consumption specification, that for **keysort/2** is **keysort(T$_1$,T$_1$)** *where*

$T_1 ::= [] \mid cons(T_2, T_1).$
$T_2 ::= cons(char, Remote).$

It specifies that `keysort/2` consumes the first character of each string. The `split/4` procedure in `keysort/2` splits the list of strings on the basis of the first character of first string and its consumption specification is `split(T_2,T_1,T_1,T_1)` which requires to send to `split/4` only the first character of the first argument and the first character of each (nested) string for the other arguments.

length	zero-level copy (0)	∞-level copy (∞)	consumption copy (c)	0/c	∞/c
10	4710	1640	816	5.77	2.01
30	4713	4184	832	5.66	5.02
50	4816	7360	848	5.67	8.68

Table 6.3: `keysort/2` with one list of strings of increasing size

Finally, Table 6.4 shows the results obtained by executing the `quicksort/2` program on a list of 10 characters. Differently from `keysort/2`, for this program the consumption specification requires to send to `split/4` the whole string.

length	zero-level copy (0)	∞-level copy (∞)	consumption copy (c)	0/c	∞/c
10	620	480	380	1.63	1.26

Table 6.4: `quicksort/2` with one list of 10 characters

6 Conclusions

Traditional distributed implementations of logic programming make remote data structures accessible to other processes by sending messages which carry the overall data structures or only remote references to these data structures. These fixed policies can be far from optimal on various classes of programs and may induce substantial overhead. This paper has presented an implementation scheme for distributed logic programming which consists of tailoring the copying level for each argument of procedures. The scheme is based on a consumption specification which describes the way each procedure "consumes" its arguments locally. The implementation scheme uses the consumption specification to avoid unnecessary copying and to request data structures globally. Moreover, the consumption specification (or an approximation of it) can be automatically obtained through a static analysis inspired by type analysis, as shown in Section 4.

The paper has also described a high-level implementation of the scheme, built on top of SICStus Prolog by using both the attributed variable and the Linda libraries. The results obtained with some example programs running on a network of workstations show the viability of the approach. The system is flexible with respect to the copying specification, and has been used to test different copying policies for the goal arguments. In particular, it showed an effective performance improvement when adopting the consumption specification obtained by the static analysis of the program. This is a high-level implementation suitable for giving the idea of the usefulness of the technique. A significant gain in term of performances could be achieved by a low-level implementation, by pushing the implementation of critical operations down to C, at WAM level.

References

1. H. Aït-Kaci: Warren's Abstract Machine. The MIT Press (1991).
2. A. Ciampolini, E. Lamma, P. Mello and C. Stefanelli: Multilevel Copying for Unification in Parallel Architectures. Proc. Second Euromicro Workshop on Parallel and Distributed Processing, IEEE Computer Society Press (1994) 518-525.
3. P. Cousot and R. Cousot: Abstract Interpretation and Application to Logic Programs. Journal of Logic Programming, **13** (1992) 103-180.
4. G. Frosini, P. Corsini and L. Rizzo: Implementing a parallel Prolog interpreter by using Occam and Transputers. Microprocessors and Microsystems, **13** (1989) 271-279.
5. F. Gecseg and M. Steinby: Tree Automata. Akademiai Kiado, Budapest (1984).
6. N. Heintze and J. Jaffar: A Finite Presentation Theorem for Approximating Logic Programs. Proc. 17th ACM Symp. on Principles of Programming Languages (1990) 197-209.
7. M. Hermenegildo, D. Cabeza and M. Carro: Using Attributed Variables in the Implementation of Concurrent and Parallel Logic Programming Systems. Proc. Int'l Conf. on Logic Programming ICLP95, The MIT Press (1995).
8. G. Janssens and M. Bruynooghe: Deriving Descriptions of Possible Values of Program Variables by Means of Abstract Interpretation. Journal of Logic Programming, **13** (2-3)(1992) 205-258.
9. P. Kacsuk and M. Wise (eds.): Implementations of Distributed Prolog. J. Wiley and Sons (1992).
10. E. Lamma, P. Mello, C. Stefanelli and P. Van Hentenryck: Consumption-based Distributed Unification in parallel architectures. Proc. APPIA-GULP-PRODE96, M. Martelli and M. Navarro eds., San Sebastian (S) (1996).
11. S. Le Huitouze: A New Data Structure for Implementing Extensions to Prolog. P. Deransart and J. Maluszunski eds, Proc. Programming Language Implementation and Logic Programming, Springer-Verlag (1990) 136-150.
12. K. Nakajima, N. Ichiyoshi, K. Rokusawa and Y. Inamura: A new external reference management and distributed unification for KL1. ICOT editor, Proc. Int'l Conf. FGCS-88 (1988) 904-913.
13. L. Sterling, E. Shapiro: The Art of Prolog, MIT Press (1986).
14. Swedish Institute of Computer Science, SICStus Prolog User's Guide, S. Kista (1990).
15. S. Taylor: Parallel Logic Programming Techniques. Prentice-Hall International Editions (1989).
16. E. Tick: Parallel Logic Programming. MIT Press (1991).
17. P. Van Hentenryck, A. Cortesi, and B. Le Charlier: Type Analysis of Prolog using Type Graphs. Journal of Logic Programming, 22 (3) (1995).
18. D.H.D. Warren: An abstract Prolog instruction set. Technical Report TR 309, SRI International (1983).

Static Granularity Optimization of a Committed-Choice Language Fleng

Takuya Araki and Hidehiko Tanaka

School of Engineering, the University of Tokyo,
7-3-1 Hongo, Bunkyo-ku, Tokyo 113, Japan.
{araki,tanaka}@mtl.t.u-tokyo.ac.jp

Abstract. The committed-choice language Fleng can extract much parallelism easily even from irregular programs using dataflow synchronization. However, there is a large overhead because the granularity of execution is very fine. If granularity of a program is coarsened, such an overhead can be reduced. This can be attained by fusing several goals into one goal, but this may cause deadlock. In this paper, we propose a safe goal fusion algorithm that statically optimizes granularity of a Fleng program. We implemented the algorithm and evaluated it on a parallel computer PIE64. The evaluation shows that enough speedup can be attained by this method.

1 Introduction

Committed-choice languages can extract much parallelism easily even from programs with irregular parallelism such as symbolic computation. This is achieved by single assignment variable and dataflow synchronization.

However, the granularity of execution is so fine that overheads of parallel execution, such as context switching, synchronization, and communication, is large. Therefore, to reduce these overheads is the key of efficient implementation of these languages.

Because the main source of the overhead is fine grain execution, the overhead can be reduced by making granularity coarse. However, it is not easy to make granularity coarse, because that may cause deadlock.

In this paper, we present a safe goal fusion algorithm. The basic idea of the algorithm is derived from separation constraint partitioning [4]. We implemented the algorithm for a committed-choice language Fleng [3] and evaluated its effectiveness on a parallel computer PIE64.

2 Committed-choice language Fleng

Fleng is a kind of parallel symbol processing language. Its ancestor is Prolog and the syntax is similar to it. However, the semantics is very different from Prolog, because a Fleng program doesn't backtrack. GHC and KL1 [8] are other examples of committed-choice languages.

The characteristics of Fleng are:

- execute all goals in parallel
- synchronize using single assignment variable

These make it possible to extract much parallelism easily.

Take a short program for example:

$$\underbrace{\text{foo}(A, R)}_{head} :- \underbrace{\text{add}(A, 1, B), \text{mul}(B, 2, R)}_{body}$$

This is a definition of a *predicate* foo that executes R = (A + 1) * 2. A definition of a predicate consists of *clauses*. In this case, it consists of only one clause. A Clause is separated by ":-"; the left side of a clause is called a "*head*", and right side is called a "*body*".

The unit of computation is called a *goal*. If an initial goal foo(A,R) is given, this goal is rewritten into add(A,1,B) and mul(B,2,R). These two goals are executed in parallel. But in this case, mul cannot be executed before the value of B is calculated. Thus, the execution of mul is *suspended* if B is not bound. When B is bound by add, mul is *activated*. To realize this mechanism, variables are *single assignment variables* and the value of the variable cannot be changed.

Branch can be expressed as follows:

```
foo(true,R):- R = 1.
foo(false,R):- R = 0.
```

This program consists of two clauses. This program means, if the first argument is true then executes R = 1, if it is false then executes R = 0. Like the previous example, if the first argument is not bound, the goal is suspended. When it is bound, the goal is activated and branches according to the value.

Arithmetic operations are defined such as:

```
add(#A,#B,R):- compute(+,A,B,R).
```

Here, "#" means that the predicate waits for the variable with "#" to be bound. compute is executed without suspension and it can be compiled into a few assembly codes. The input arguments of compute should be bound, which is guaranteed by "#"s of the head in this case. If there is a sequence of computes, they are executed sequentially.

3 Goalfusion

3.1 Problem of goalfusion

We propose *goalfusion* as a coarsening method of Fleng programs. Goalfusion is a method that fuses goals into one goal, but the method cannot always be applied.

For example:

```
foo(U,V,R,S):- add(U,U,R), mul(V,V,S).
add(#A,#B,R):- compute(+,A,B,R).
mul(#A,#B,R):- compute(*,A,B,R).
```

This program executes R = U + U and S = V * V. add and mul are fused into one goal:

```
foo(U,V,R,S):- add_mul(U,V,R,S).
add_mul(#U,#V,R,S):- compute(+,U,U,R), compute(*,V,V,S).
```

In this case, add and mul are fused into add_mul, and the overhead of goal invocation and synchronization is reduced. However, this transformation is *wrong*.

In the original program, if foo(1,Tmp,Tmp,S) is called, it is rewritten into add(1,1,Tmp) and mul(Tmp,Tmp,S). Then Tmp is bound to 2 by add, S is bound to 4 by mul.

But in the transformed program, foo(1,Tmp,Tmp,S) is rewritten into add_mul (1,Tmp,Tmp,S); add_mul(1,Tmp,Tmp,S) waits for Tmp to be bound forever. Thus, transformed program causes deadlock. That means this transformation changed the semantics of the program, so such a transformation should be avoided.

Why the fusion of add and mul causes deadlock? Figure 1 shows the dataflow graph of the original program and the transformed program.

In the dataflow graph of the original program, mul depends on add through Tmp, which is an argument of foo. This dependency does not always exist. This kind of dependency is called a *potential dependency* [4]. This dependency makes a *cyclic dependency* after goalfusion. A Cyclic dependency means that the goal depends on its own output, which causes deadlock. The original program does not cause deadlock because it does not have any cyclic dependencies.

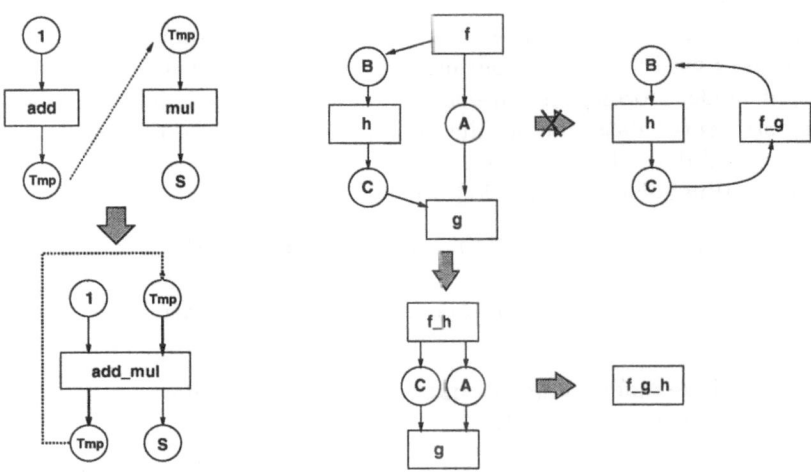

Fig. 1. Dataflow graph   Fig. 2. Algorithm of goalfusion

3.2 Algorithm of goalfusion

To avoid deadlock, cyclic dependencies should be avoided. So the basic idea of the algorithm is: *two goals can be fused only when there is no indirect dependency between these goals.* This is because indirect dependencies always make cyclic dependencies when these two goals are fused.

Consider figure 2 for example; it is a dataflow graph of a program. A, B and C are variable and f, g and h are goals. In this program, f and g cannot be fused because g indirectly depends on f through h. If f and g are fused, the indirect dependency makes a cyclic dependency.

On the other hand, f and h are safely fused into f_h. Though h depends on f, the dependency is direct. The direct dependency disappears after fusion because the dependency exists inside of the fused goal.

Then f_h and g can be fused into one goal likewise.

Brief sketch of the algorithm is as follows:

1. Select one clause.
2. Analyze dependencies in the clause (described in section 3.3).
3. Fuse two goals if there are no indirect dependencies between them.
4. Repeat 2-3 until any two goals cannot be fused.

Here, potential dependencies and dependencies which go through one or more other goals are treated as indirect dependencies.

3.3 Dependency analysis

In order to fuse two goals, it is necessary to guarantee that there is no indirect dependency between these goals. In this section, we will describe how to analyze dependencies in a program.

Mode inference In a Fleng program, it is not known from syntax whether an argument is used as input or output. To analyze dependencies precisely, input/output mode information is needed. [9] describes a mode inference method including structured data, but we did simple mode inference that does not include structured data. The algorithm is borrowed from [9].

The algorithm is based on the following ideas:

1. If a variable is output of a goal, the variable is input of other goals that share the variable.
2. If a variable is input of all goals except one goal, the variable is output of the goal.

Variables whose mode cannot be inferred are treated as both input and output to analyze dependencies safely.

Detection of dependency In this section, we will describe how to detect dependencies from each program element.

Body goals: If a body goal cannot be reduced until an argument variable is bound, the goal directly depends on the variable (and the variable is input of the goal); and if an argument variable is guaranteed to be bound after a goal is reduced (by `compute` or `=`), the variable directly depends on the goal (and the variable is output of the goal).

In addition, a body goal may make dependencies from any input variables to any output variables, because sub goals of the goal may make dependencies.

Of course, sub goals may not make dependencies; but in order to detect potential dependencies safely, dependencies from all input variables to all output variables should be assumed.

If the definition of a body goal is known, the information is propagated using inter-clause analysis described below.

Unification: Unifying variable with atomic value does not make a dependency. However, unifying variable with variable should be treated to make dependencies of both directions. For example, `A = B` is treated to make dependencies from `A` to `B` and that from `B` to `A`. Actually `A` does not depend on `B` and vice versa; but `A = B` may be used as a path of dependencies.

Unifying variable with structured data is more complex. For example, `A = {B,C}` (`{B,C}` means a vector which consists of two elements, `B` and `C`) should be treated to make dependencies from `A` to `B` and `C`, and those from `B` and `C` to `A`, because this unification may be also used as paths of dependencies.

For example:

```
foo(...):- A = {B,C}, bar(A,A1), A1 = {X,Y}, ...
bar({B,C},A1):- add(B,B,X), mul(C,C,Y), A1 = {X,Y}.
```

In the definition of `foo`, `X` depends on `B` and `Y` depends on `C`. These dependencies cannot be detected unless dependencies from `B` and `C` to `A` and those from `A1` to `X` and `Y` are assumed.

Head: The caller site (i.e. head) may make dependencies between arguments. The first example whose body goals cannot be fused is of this type.

While analyzing dependencies, a head and body goals are alike in the sense that both may make dependencies between arguments. So a head can be treated in the same way as a body goal during mode inference and dependency analysis. However, the inferred mode of a head is reversed to what is seen from the outside.

For example:

```
foo(U,V,R,S):- add(U,U,R), mul(V,V,S).
```

This program is same as the first example whose body goals cannot be fused.

Figure 3 shows the dataflow graph of this program including the head, which is represented in italic. Inferred mode of the head `foo` is: `R` and `S` are input and `U` and `V` are output (The mode of `foo` seen from outside is: `U` and `V` are input and `R` and `S` are output).

This dataflow graph shows that there are indirect dependencies from `R` to `V` and from `S` to `U`, which indicates that `add` and `mul` cannot be fused.

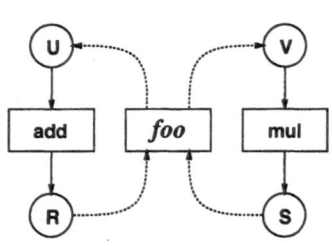

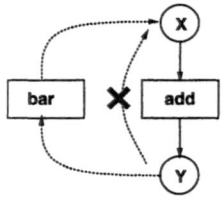

Fig. 4. Removal of a potential dependency that is contradicted by a certain dependency

Fig. 3. Dependencies through a head

Removal of contradicted dependencies Analyzed dependency may contain dependencies that cannot exist.

For example:

```
foo(...):- bar(Y,X), add(X,X,Y),...
```

In the dependency analysis, a dependency from Y to X is detected from a body goal **bar**; but there cannot be such a dependency. Because **add** makes a dependency from X to Y certainly, if Y depends on X, the program is in deadlock. If the given program is right and runs without deadlock, there cannot be such a dependency (Fig. 4).

In this study, we used the method described in [4]: *remove potential dependencies that are contradicted by certain dependencies.* Here, dependencies that consist of only direct dependencies (and dependencies that are propagated as certain dependencies by inter-clause analysis) are treated as certain dependencies. Other dependencies are treated as potential dependencies.

3.4 Inter-clause analysis

We described analysis within a clause so far. We will describe inter-clause analysis in this section.

There are two directions of passing information in inter-clause analysis: from a callee to a caller (bottom-up) and from a caller to a callee (top-down). But in the current implementation, top-down information propagation is not implemented to avoid complexity, though it can be implemented almost in the same way as bottom-up method.

And there are two kinds of information to pass: input/output modes and dependencies between variables. Dependencies are classified into certain dependencies and potential dependencies. This classification is used to remove potential dependencies that is contradicted by certain dependencies.

If the information varies according to the conditional branch, the information should be merged to be safe.

Dependencies are analyzed from the top goal and the call tree is traversed in the depth-first order. The goal already analyzed is memorized and analysis is stopped when the goal was already analyzed; it avoids infinite loop while analyzing (mutually) recursive goals.

For example:

```
foo(U,V,R,S):- add(U,U,R), mul(V,V,S).
bar(...):- foo(U,V,R,S),...
```

The information analyzed in foo is propagated to bar as follows.

The input/output mode of foo can be inferred using the method described above. However, because the analyzed mode is what is seen from body goals, the mode needs to be reversed. In this case, U and V are input, and R and S are output.

The dependencies to pass is what is analyzed without head. This is the dependencies made inside of foo. In this case, certain dependencies from U to R and V to S are analyzed. There are no other potential dependencies.

It is notable that the propagated dependency information indicates that dependencies from U to S and from V to R do not exist. It cannot be known only from the mode information.

4 Implementation and evaluation

4.1 Implementation

We implemented this method as a preprocessor of a compiler. It inputs a Fleng program and outputs a Fleng program whose granularity is coarsened. The preprocessor is about 6,000 lines of a Fleng program.

And the Fleng compiler is extended to manage conditional branch within a goal in order to fuse goals that branches at heads. This branch is described as (Cond --> Then; Else). This branch assumes that all variables needed to evaluate Cond are bound, and it is compiled into a simple branch operation like procedural languages.

In addition, if a goal can be executed without suspension, it is inlined.

An example of transformation is as follows:

```
abs(A,R):-
  greater(A,0,IsGt), abs1(IsGt,A,R).
abs1(true,A,R):- R = A.
abs1(false,A,R):- sub(0,A,R).
```

This program is transformed into:

```
abs(A,B):- C = 0, greater_abs1(A,C,B).
greater_abs1(#A,#B,C) :-
  compute(>,A,B,D),
  ((D == true)--> C = A; E = 0, compute(-,E,A,C)).
```

Table 1. Benchmark programs

Program	Code size	Input size	Binary size (byte)		Compilation time (sec)	
			Original	Optimized	Original	Optimizing
queen	55	9-Queens	8925	7918	3.58	7.04
primes	39	2000	7076	7367	2.87	5.00
qsort	31	10000	4365	5373	2.55	14.1
fme	1175	queen	194110	170054	135	539

4.2 Evaluation

We evaluated this method on the Parallel Inference Engine PIE64 [1]. PIE64 is a parallel computer that is designed to execute Fleng programs efficiently. PIE64 has 64 processor elements and is a distributed shared memory machine, that is, the address space is global. The processor of PIE64 supports multi-context processing with which latency of communication can be hidden. The interconnection network supports automatic load balancing, and the speed of the network is relatively high.

Execution time used to calculate relative speed is an average of 3 executions excluding garbage collection time.

The granularity of the transformed program is made to be as coarse as possible though the implementation allows to specify the granularity of the program. This is because even the coarsest granularity was finer than the optimal granularity.

Compilation time was evaluated on a workstation (Sun Ultra 1).

We selected "queen", "primes", "qsort" and "fme" as benchmark programs. Queen is a program that solves N-Queens problem. Primes is a prime number generator that uses Eratosthenes' sieve. Qsort is a quick sort program. Fme is a program that expands macros of Fleng program; it is a practical Fleng application. The input of fme is queen whose macros are not expanded. These programs except queen have relatively small parallelism.

Table 1 shows binary size and compilation time of benchmark programs. Binary size was expected to increase because new predicate definition was added. However, it did not increase very much and even decreased in two programs. This is because the optimizer removes predicates that became unused after goalfusion. Compilation time increased 2 to 6 times. We think this increase is reasonable considering the complex dataflow analysis.

Figure 5–8 show the speedup by granularity optimization. X-axis of the graph shows the number of processors, and Y-axis of the graph shows the relative speed normalized by the speed of the original program with one processor. The optimized programs except fme are 2 to 9 times faster than the original programs irrespective of number of processors. Optimized fme ran about 1.2 times faster than the original program. We think the limited speedup is due to the difficulty of global analysis of a large program.

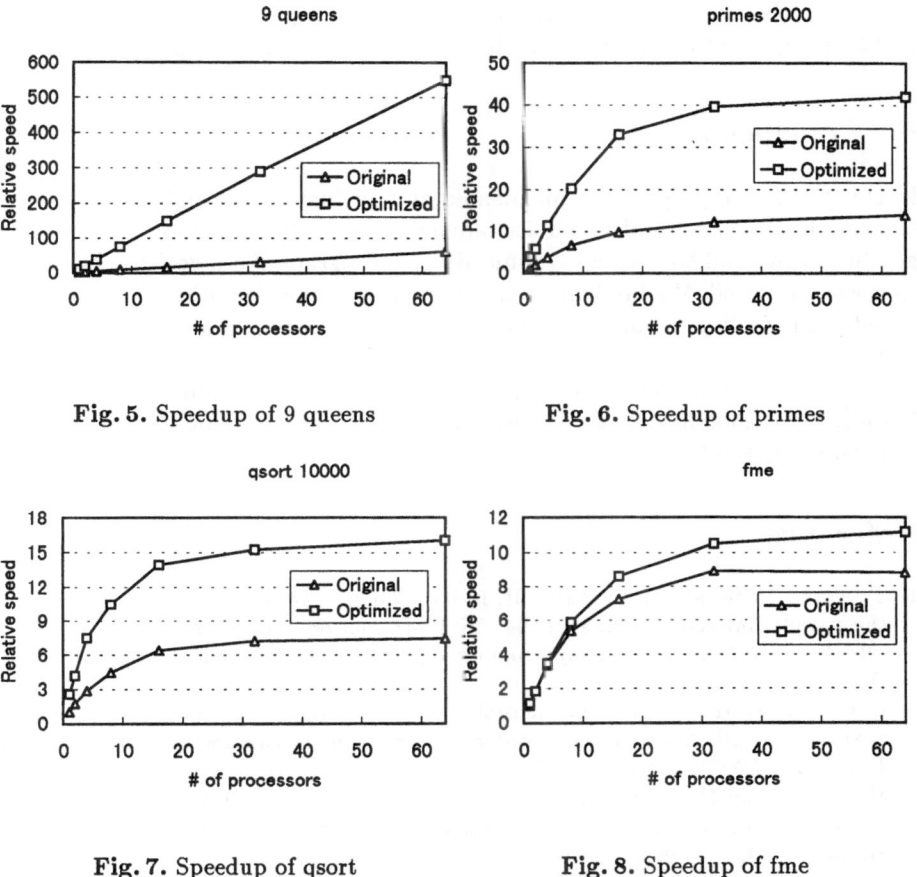

Fig. 5. Speedup of 9 queens

Fig. 6. Speedup of primes

Fig. 7. Speedup of qsort

Fig. 8. Speedup of fme

5 Related work

Execution model of lenient functional languages such as Id is similar to that of committed-choice languages. Same kind of optimization as our work is also studied in the field of functional languages [5, 6, 7, 4]. Separation constraint partitioning (SCP) [4] is the latest algorithm of Id to make large threads from dataflow graphs; it strongly influenced our study.

The basic algorithm of our study is substantially same as SCP. SCP uses dataflow graphs as input, but our algorithm runs on the Fleng source code level. That makes it possible to process other source code level optimizations simultaneously. For example, if goalfusion enables any goals to be inlined, they are inlined. In addition, a transformed program is correct after every step of our algorithm, though it is not true in SCP.

On committed-choice languages, there is a study of sequentialization of programs [2]. In this study, they limit the programs to a class called "fully moded, feedback free". Feedback free means that there is no dependency through a head. It seems that their limitation is very strict; especially to limit programs to "feedback free" makes it impossible to write many kinds of programs.

6 Conclusion

In this paper, we proposed, implemented and evaluated a granularity optimization method of a committed-choice language Fleng. In this study, we used the goalfusion method to coarsen granularity of a program. An algorithm of goalfusion was proposed. We implemented and evaluated the algorithm. Several times speedup in small programs and about 1.2 times speedup in a realistic program were attained.

The reason of relatively small speedup in a realistic program seems to be due to difficulty of global analysis. Future work will focus on developing a granularity coarsening method for larger programs.

References

1. T. Araki, Y. Hidaka, H. Nakada, H. Koike, and H. Tanaka. System integration of the parallel inference engine PIE64. In *Workshop on Parallel Logic Programming attached to International Symposium on Fifth Generation Computer Systems 1994*, Dec. 1994.
2. B. C. Massey and E. Tick. Sequentialization of parallel logic programs with mode analysis. In *4th International Conference on Logic Programming and Automated Reasoning. LNCS 698*, pages 205–216. Springer-Verlag, 1993.
3. M. Nilsson and H. Tanaka. Fleng Prolog - the language which turns supercomputers into Prolog machines. In *Logic Programming '86, LNCS 264*, pages 170–179. Springer-Verlag, 1989.
4. K. E. Schauser, D. E. Culler, and S. C. Goldstein. Separation constraint partitioning – a new algorithm for partitioning non-strict programs into sequential threads. In *POPL '95*, pages 259–271. ACM, 1995.
5. K. E. Schauser, D. E. Culler, and T. von Eiken. Compiler-controlled multithreading for lenient parallel languages. In *FPCA '91, LNCS 523*, pages 50–72. Springer-Verlag, 1991.
6. K. R. Traub. Multi-thread code generation for dataflow architectures from non-strict programs. In *FPCA '91, LNCS 523*, pages 73–101. Springer-Verlag, 1991.
7. K. R. Traub, D. E. Culler, and K. E. Schauser. Global analysis for partitioning non-strict programs into sequential threads. In *LFP '92*, pages 324–334. ACM, 1992.
8. K. Ueda and T. Chikayama. Design of the kernel language for the parallel inference machine. *The Computer Journal*, 33(6):494–500, December 1990.
9. K. Ueda and M. Morita. A new implmentation technique for flat GHC. Technical report, ICOT, 1990. TR-560.

Distributed Arrays in the Functional Language Concurrent Clean

Pascal R. Serrarens

Computer Science Institute
University of Nijmegen, The Netherlands
email: pascalrs@cs.kun.nl

Abstract. In this paper, we show how distributed arrays can be implemented in Concurrent Clean without extensions to the language. We introduce the notion of remote values, associating a value at a remote processor with its location. By combining remote values with arrays we can build distributed arrays, having the same flexibility as standard arrays. With the example of matrix-vector multiplication, we show that our distributed arrays are well suited to implement parallel algorithms with little overhead.

1 Introduction

At this moment, one can say that efficient sequential implementations of functional languages exist. Still, parallel functional programming is not very popular. Perhaps some features for effective parallel programming are still missing. In this paper we present an important construction for parallel programming in the functional language Concurrent Clean [PvE93], namely distributed arrays: arrays of which the elements are distributed over a number of processors. With Concurrent Clean, these arrays can be built efficiently without new language constructions. We continue with the work initiated by Kesseler [Kes95] and present a toolkit for working with distributed arrays.

In Section 2 we introduce remote values: values located on other processors. Section 3 then describes how distributed arrays can be built with remote values and gives some functions for them. In Section 4 we show an example of a matrix-vector multiplication on a mesh-shaped distributed memory computer model. Section 5 shows some results obtained with distributed arrays. Finally, Section 6 describes related work and Section 7 concludes the paper.

2 Remote Values

In Concurrent Clean it is possible to introduce parallelism by annotating expressions with the P-annotation [PvE93] [Kes94]. In the expression {| P at p |} f x, the expression f x is evaluated on the processor identified with p, where p has type ProcId. If the expression is not completely evaluated, the result is only partially returned to the root processor. The rest of the result is send when it is needed, which is not very efficient. In the rest of the paper we assume that all results are completely evaluated, so they are always send in one piece.

The P-annotation seems to be less suited for data-parallel programming because of two problems: first, it is not possible to determine the location of the result of the remote expression, and second, the result of the expression is communicated to the root processor very easily, as soon the remote expression is used, it gets evaluated and the result is send back to the root processor.

In his paper Kesseler [Kes95] used a new primitive function arg_id to solve the first problem: it returns the ProcId of its argument. For the second problem he used an ordinary datatype for representing remote values. We will also use such a datatype, but we add a new argument of type ProcId representing the location of the remote value:

```
:: R x = Remote x ProcId
```

With this datatype we have solved both problems: the location is stored with the value and can be determined easily. The Remote constructor protects the value x for being used directly.

We only have to make sure that the physical location of the remote value corresponds with the location stored in the datatype. This can be accomplished by making Remote an abstract datatype and providing a set of functions on it, each fulfilling the property above.

2.1 Moving Data and Information

The first three functions we introduce respectively copy data to, from and between processors:

```
putRemote :: ProcId a → R a
putRemote p x = Remote y p
where
    y = {| P at p |} x

getRemote :: (R a) → a
getRemote (Remote x p) = x

copyTo :: ProcId (R a) → R a
copyTo p2 (Remote x p1) = Remote y p2
where
    y = {| P at p2 |} x
```

The last function is important. Concurrent Clean provides implicit communication: when a value is needed at a different processor, the runtime system copies the value to the new location. But many parallel algorithms describe in detail how and when data is moved between processors, making sure that the communication overhead is as small as possible. Therefore, we provide a function for explicitly moving data between processors.

The last function is this section returns the location of a remote value. Together with copyTo, it can be used to align data on the same processor.

```
locationOf :: (R a) → ProcId
locationOf (Remote x p) = p
```

2.2 Sending Work

The key functions on remote values are the apply-like functions: they will send a function to the remote value and apply it there:

```
rap :: (a → b) (R a) → R b
rap f (Remote x p) = Remote y p
where
    y = {| P at p |} f x


rap2fst :: (a b → c) (R a) (R b) → R c
rap2fst f (Remote x p1) (Remote y p2) = Remote z p1
where
    z = {| P at p1 |} f x y
```

When applying a binary function, we have a choice where to leave the result: it may be the location of either argument, depending on the situation. Therefore we provide two functions: rap2fst and rap2snd which leave the result on the location of their first and second argument respectively.

3 Distributed Arrays

Using the remote values from the previous section we can build an array with elements located on remote processors. A distributed array is simply a type synonym for an array of remote values:

```
:: DArray a :== { R a }
```

Arrays in Clean are quite similar to lists, apart from some array-specific constructions. The type of arrays is denoted as { x }[1]. Clean provides ZF-expressions for manipulating arrays and list. The syntax of these expressions is adapted from mathematics for describing sets. Some examples of ZF-expressions are:

```
{ f e \\ e ← l }
{ g e1 e2 \\ e1 ←: a1 & e2 ←: a2 }
{ h e1 e2 \\ e1 ←: a1, e2 ←: a2 }
```

In the first example, the function f is applied to all elements of list l. In the second example the elements are drawn form arrays a1 and a2 simultaneously and g is applied to them. In third example is similar, but now the generators are nested. Each element of a1 is then combined with all elements of a2. Selecting an element from position i of array a is denoted as a.[i].

---

[1] In fact, Clean provides three kinds of arrays: lazy, strict and unboxed, with types { x }, { !x } and { #x } respectively [vG96]

3.1 Basic Operations on Distributed Arrays

Distributed arrays are created by combining two arrays: an array with the elements and an array for the locations of those elements. These two arrays are combined with putRemote:

```
createDArray :: { Procld } { a } → DArray a
createDArray ps a = { putRemote p e \\ e ←: a & p ←: ps }
```

Many functions on distributed arrays are straightforward since they are basically array manipulations. We give the implementation of mapping a function over a distributed array and zipping two distributed arrays with a function.

```
map_Dist :: (a → b) (DArray a) → DArray b
map_Dist f dx = { rap f rx \\ rx ←: dx }
```

```
zipwithFst_Dist :: (a b → c) (DArray a) (DArray b) → DArray c
zipwithFst_Dist f dx dy = { rap2Fst f rx ry \\ rx ←: dx & ry ←: dy }
```

Similar to the functions rap2fst and rap2snd in Section 2, we have two zipwith versions: zipwithFst_Dist and zipwithSnd_Dist.

The current implementation of distributed arrays is not the most efficient. In case of the map_Dist function, the function f is applied to every element using rap, resulting in a message. When we have n elements in our array, a map will need n messages. In the future we plan a broadcast system to reduce this to a single message.

3.2 Rearranging Distributed Arrays and Communication

Kesseler [Kes95] did not use an explicit copy function to move data between processors. This was achieved by rearrangements of the array:

```
implRotateL_Dist :: Int (DArray a) → DArray a
implRotateL_Dist s dx = { dx.[(i + s) mod n] \\ i ← [0 .. n - 1] }
where
    n = size dx
```

The advantage of this is that it saves communication: when two rotations are performed after each other, the elements are copied only once. However, it may also lead to too little communication. Take for example the following expression:

```
map_Dist ((+) 1) (implRotateL_Dist 1 dx)
```

The rotation suggests that communication takes place, but the map function uses the locations stored in the distributed array, which did not change by the rotation. Therefore, the rotation will not be performed. This problem may be very hard to detect. When the replacement of subarrays is explicit, we do not have this problem. The subarrays are copied explicitly with the copyTo function:

```
rotateL_Dist :: Int (DArray a) → DArray a
rotateL_Dist s dx =
    { copyTo (locationOf dx.[i]) dx.[(i + s) mod n] \\ i ← [0 .. n - 1] }
where
    n = size dx
```

3.3 Folding Distributed Arrays

Fold-like functions combine the elements of compound datastructures using a binary function. The best choice for a parallel fold is a tree shaped fold, which make an effective use of parallelism. However, the location of the result is not determined and the communication distances will not be minimal. An option is are left-to-right-fold, which behaves like a left-fold and leaves the result at the same location of the rightmost element. Its complement is right-to-left-fold, similar to a right-fold and leaving the result at the leftmost location.

```
foldlr_Dist :: (a a → a) (DArray a) → R a
foldlr_Dist f dx = foldlr 1 dx.[0]
where
    foldlr i r
        | i >= size dx = r
        | otherwise    = foldlr (i + 1) (rap2snd f dx.[i] r)
```

4 An Example: Matrix-Vector Multiplication

In this section we will implement the matrix-vector multiplication algorithm by Kumar et al. [KGGK94]. The algorithm assumes a mesh-shaped processor network. The matrix is distributed blockwise over the network, while the vector is distributed blockwise over the rightmost column of the network. In the algorithm, the vector is first distributed over the processors, then all processors perform a local matrix-vector multiplication and at the end, the local results are summed, leaving the result vector in the rightmost column of the processor network.

We need two kinds of distributed structures: a distributed matrix and a distributed vector, with types:

```
:: DMatrix :== {{ R Matrix }}
:: DVector :== { R Vector }
```

For the broadcast of the vector, the algorithm prescribes a two-step mechanism, to get a minimum communication overhead. In the the first step the vector is copied to the diagonal of the network:

```
copyToDiagonal :: DVector DMatrix → DVector
copyToDiagonal dv dm =
    { copyTo (location of dm.[i,i]) dv.[i] \\ i ← [0 .. size dv - 1] }
```

Table 1. Performance measurements for the matrix-vector multiplication. $t =$ running time in seconds, $s =$ speedup compared to sequential code

matrix size	sequential	4 processors		16 processors		64 processors	
	t	t	s	t	s	t	s
512^2	6.2	1.7	3.7	0.7	8.8	1.2	5.2
2048^2	97.6	24.7	3.9	6.7	14.6	3.0	33.0
8192^2	1555.5	390.1	4.0	98.9	15.7	27.1	57.5

In the second step the vector is broadcasted over the columns. The result of this step is a distributed vector for each row of processors:

```
distributeVector :: DVector DMatrix → { DVector }
distributeVector dv dm =
    {{ copyTo (locationOf dm.[r,c]) dv.[c] \\ c ← [0..nc - 1] } \\ r ← [0..nr - 1] }
where
    nr = size dm
    nc = size dm.[0]
```

Next, every processor evaluates its own local matrix-vector multiplication. For that we use zipWithFst_2Dist, a 2-dimensional variant of zipWithFst_Dist:

```
localMatMulVec :: DMatrix { DVector } → { DVector }
localMatMulVec dm dv = zipWithFst_2Dist matVecMult dm dv
```

Every row of the distributed vector is summed from left to right, with the result ending up in the rightmost column. We do this for every row in the array. We assume that the addition operator (+) is defined for vectors:

```
accumSums :: { DVector } → DVector
accumSums dv = { foldr1_Dist (+) row \\ row ←: dv }
```

5 Performance Measurements

Our implementation of the matrix-vector multiplication was tested on a 64-node transputer network. We compared the code against sequential code with no overheads for parallelism. Table 1 shows good speedups, because the local matrix-vector multiplications take most of the time.

Another test case the conjugate gradient algorithm. The sequential version compared well to C and Haskell [Ser96]. The parallel implementation ran on a network of 4 Macintosh II computers connected by Localtalk and Ethernet. Using many map-like operations, we had lots of communication, so worse speedups could be expected, but table 2 shows that the results were encouraging. We

Table 2. Performance measurements for the conjugate gradient algorithm.

matrix size	sequential	4 Localtalk		4 Ethernet	
	t	t	s	t	s
400^2	22.1	67.9	0.3	25.5	0.9
1600^2	175.9	152.7	1.2	83.1	2.1
3600^2	590.9	306.9	1.9	200.7	2.9
6400^2	1386.9	552.7	2.5	416.1	3.3

expect that these times will improve when broadcasting as described in section 3.1 is implemented.

The performance for matrix multiplication is close to the results obtained by Kesseler. For multiplication of a matrix of size 256×256, the running times in seconds on 4 and 16 processors were 4.0 and 13.3 for our implementation, compared to 4.0 and 15.5 for Kesseler. Our times are slightly better, because the explicit movement of data gives less parallel overhead.

6 Related Work

The remote values in this paper are similar to network objects [BNOW94]. A network object consists of a local and a remote object. Invocation of a method of the local object will send a message to the remote object which then invokes its corresponding method. Network objects do not differ from normal objects for the programmer, while we need the special apply function rap for remote values.

Various other functional languages provide some kind of distributed array, but they all provide it as a primitive language construction. The primitive parallel datatype of NESL [BCH+93] are sequences. It uses a construction similar to ZF-expressions for the application of a function over a sequence.

Skil [BK96] tries to have maximal flexibility with its pardata construct. It allows any distributed data structure, as long as it is composed of identical data structures place on each processor. Nested parallelism is not possible.

SCL [DGTY95] give the ParArray for data-parallelism. It also provides configuration skeletons [Col89] to determine the distribution and alignment of the data. No direct control over locations of values is possible in NESL and Skil.

Kuchen and Geiler [KG91] suggest distributed arrays in favor of list- and tree-like structures. Every processor has an array with links to non-local distributed arrays elements. In this way processors have easy access to array elements at other processors, but is more complex to manage than our solution.

7 Conclusions and Future Work

In this paper we showed how an efficient implementation of remote values can be implemented in Concurrent Clean. Apart from the standard P-annotation, we did not need any other extension of the language. Remote values are used to build and efficient implementation distributed arrays, an important construction for parallel computing. We showed that good speedups are possible with our distributed arrays.

We will continue improving the implementation of distributed arrays, with emphasis on minimal communication costs. This will include broadcasting functions and values over the network. The possibility of using remote values in other datastructures will also be a topic of further research.

References

[BCH⁺93] G.E. Blelloch, S. Chatterjee, J.C. Hardiwck, J. Sipelstein, and M. Zagha. Implementation of a portable nested data-parallel language. *Principles and Practice of Parallel Programming (PPoPP)*, pages 102–101, 1993.

[BK96] G.H. Botorog and H. Kuchen. Skil: An imperative language with algorithmic skeletons for efficient distributed programming. In *Proceedings of the Fifth International Symposium on High Performance Distributed Computing (HPDC-5)*, pages 243–252. IEEE Computer Society Press, 1996.

[BNOW94] A. Birrell, G. Nelson, S. Owicki, and E. Wobber. Network objects. Technical Report Digital SRC 115, DEC, Palo Alto, USA, 1994.

[Col89] M. Cole. *Algorithmic Skeletons: Structured Management of Parallel Computation*. MIT Press, 1989.

[DGTY95] J. Darlington, Y. Guo, H.W. To, and J. Yang. Functional skeletons for parallal coordination. In *Proceedings of Europar 95*, 1995.

[Kes94] M.H.G Kesseler. Uniqueness and lazy graph copying - copyright for the unique. In *Proceedings of the 6th International Workshop on the Implementation of Functional Languages*, Norwich, UK, 1994. University of East Anglia.

[Kes95] M.H.G Kesseler. Constructing skeletons in clean - the bare bones. In *Proceedings of High Performance Functional Computing (HPFC '95)*, pages 182–192, Denver, Colorado, 1995. Lawrence Livermore National Laboratory. CONF-9504126.

[KG91] H. Kuchen and G. Geiler. Distributed applicative arrays. Technical Report AIB 91-5, RWTH Aachen, 1991.

[KGGK94] V. Kumar, A. Grama, A. Gupta, and G. Karypis. *Introduction to Parallel Computing, Design and Analysis of Algorithms*. The Benjamin/Cummings Publishing Company, Inc., California, 1994.

[PvE93] M.J. Plasmeijer and M.C.J.D. van Eekelen. *Functional Programming and Parallel Graph Rewriting*. Addison-Wesley Publishers Ltd., 1993.

[Ser96] P.R. Serrarens. A clean conjugate gradient algorithm. In *Proceedings of the 8th International Workshop on Implementation of Functional Languages*, pages 367–376, Kiel, Germany, September 1996.

[vG96] J.H.G. van Groningen. The implementation and efficiency of arrays in Clean 1.1. In *Proceedings of the 8th International Workshop on Implementation of Functional Languages*, pages 131–154, Kiel, Germany, September 1996.

Design and Implementation of Parallel TRAM

Kazuhiro Ogata, Masaru Kondo, Shigenori Ioroi and Kokichi Futatsugi

JAIST, JAPAN ({ogata, m-kondo, ioroi, kokichi}@jaist.ac.jp)

1 Introduction

Algebraic specifications, introduced in the mid 70's as a method of modeling and specifying abstract data types, has been widely attracting attention since they have exact semantics and the ability to verify and reason about specifications of software systems. In addition, since specifications written in algebraic specification languages can be executed on stock hardware, the specification languages can be used as tools for rapid prototyping.

TRAM[7] is an abstract machine for order-sorted conditional term rewriting systems (OSCTRSs). The OSCTRSs [5] can serve as a general computation model for advanced algebraic specification languages such as *OBJ*[6] and *CafeOBJ*[2]. TRAM adopts the *E-strategy*[6] as its reduction strategy. Parallel TRAM is a parallel variant of TRAM that is designed to be executed on shared-memory multiprocessors. In Parallel TRAM, parallelism directives are specified by using the *Parallel E-strategy* that is an extension of the E-strategy. The Parallel E-strategy may control parallelism suitably by combining conditions. In this paper, we describe the design and implementation of Parallel TRAM and assess the current implementation on OMRON LUNA-88K^2.

Up to the present, several researches on designing rewrite engines for algebraic specification languages on parallel architectures have been done [4, 8]. Massively parallel computers have been mainly focused as the target architectures so that much faster rewritings by far could be achieved. But, almost all the rewrite engines have been gone no further than having been designed. Shared-memory multiprocessors have been chosen as our target architecture since we think the multiprocessors will undoubtedly become standard for the future workstations.

2 TRAM: Term Rewriting Abstract Machine

TRAM Programs. TRAM programs are rewrite (equational) programs that are similar to OBJ's modules. For example, the program defining a function that returns the nth element of (infinite) natural numbers' lists is as follows:

sorts: *Zero NzNat Nat List* .
order: *Zero < Nat   NzNat < Nat* .
ops: *0 : −> Zero   s : Nat −> NzNat*
     cons : Nat List −> List { strat: (1 0) }
     inf : Nat −> List   nth : Nat List −> Nat .
vars: *X Y : Nat L : List* .

rules: $inf(X) \rightarrow cons(X, inf(s(X)))$
$nth(0, cons(X, L)) \rightarrow X$
$nth(s(X), cons(Y, L)) \rightarrow nth(X, L)$.

TRAM adopts the *E-strategy* that lets each operation have its own *local strategy*. The local strategies indicate the order of rewritings of terms whose top operations have the strategies. The order is specified by using lists of numbers ranging from zero to the number of the arguments. Non-zero number n and zero in the lists are intended to reduce (evaluate) nth arguments of the terms and the terms themselves to a variant of normal forms respectively. We call the variant of normal forms *E-normal forms* (*ENFs*). Arguments whose numbers are not in the lists might or might not be evaluated lazily. The operation *cons* has the local strategy (1 0) that indicates a term whose top operation is *cons* is tried to be rewritten to another after evaluating the first argument to ENF when the term is evaluated. If the term is rewritten to another, the new one will be evaluated according to the local strategy of its top operation. The second argument might or might not be evaluated lazily. The eager local strategy (1 2...0) is attached to each operation to which explicit local strategies are not specified. Figure 1 shows the reduction sequence for $nth(s(0), inf(0))$ using the above program.

TRAM Architecture. TRAM consists of six regions (DNET, CR, CODE, SL, STACK and VAR) and three processing units (the rule compiler, the term compiler and the TRAM interpreter). DNET is the region for *discrimination nets* [3] encoded from the lefthand sides of rewrite rules. The righthand sides of rewrite rules (RHSs) are compiled and allocated on CR. Matching programs compiled from subject terms are allocated on CODE. SL contains strategy lists for subject terms. STACK is the working place for pattern matching. VAR contains substitutions.

In TRAM, subject terms are compiled into sequences of abstract instructions (*matching programs*) that are self modifying programs.

Definition 1. The matching program L_T for a term T whose top operation is f of arity n is as follows:

L_T: match_sym idx_f
$\qquad L_1$ // idx_f is the index for f.
$\qquad \vdots$ // $L_i(i = 1, \ldots, n)$ is the label of
$\qquad$ // the ith argument's matching program.
$\qquad L_n$

Figure 1 shows some terms and the corresponding matching programs. All applicable rewrite rules for T are gained by executing the program. The program is called by jumping the label L_T after pushing a continuation (a return address) onto STACK. match_sym tests whether f is in the root node of (a sub-tree of) DNET. If f is in the root, the continuations (the arguments) $L_n, \ldots, L_2$ are pushed onto STACK and the control is transferred to L_1. Backtracking is used so as to find all applicable rewrite rules for the term. The method for backtracking used in WAM [1] for Prolog is adopted.

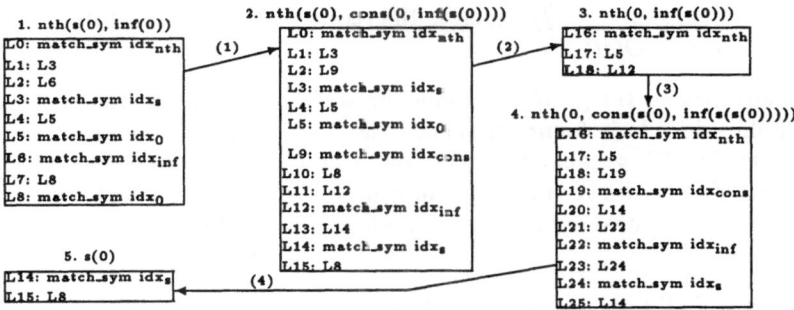

Fig. 1. Reduction sequence for $nth(s(0), inf(0))$

Matching program templates are compiled from RHSs and are instantiated when the corresponding rules are used. The matching program at L9 through L15 in Fig. 1 is the instantiated one of the matching program template of $inf(X)$ $->$ $cons(X, inf(X))$ that is used in the rewriting (1) in Fig. 1.

Rewriting Machinery. The implementation of the E-strategy in TRAM is based on *strategy lists*. A strategy list for a term t is basically a sequence of all eager positions (reachable nodes) in t. The order of the strategy list corresponds to the order of the evaluation of t. Elements of strategy lists in TRAM are triples $\langle label, pslot, skip \rangle$ of matching programs' labels, parent slots holding the labels, and the number of elements to be skipped. The last element of the strategy lists is the BINGO triple $\langle BINGO, subject, \_ \rangle$. BINGO is the label of the instruction bingo that is executed when an ENF is got. The second of the BINGO triple is the parent slot of subject terms. The TRAM interpreter executes the matching program to reduce a subject term according to the strategy list until the BINGO triple. The strategy list for $nth(s(0), inf(0))$ in Fig. 1 is $[\langle L5, L4, 0 \rangle, \langle L3, L1, 0 \rangle, \langle L8, L7, 0 \rangle, \langle L6, L2, 0 \rangle, \langle L0, RESULT, 0 \rangle, \langle BINGO, L0, \_ \rangle]$.

When an applicable rule is got by executing the matching program for a subterm of a term, the strategy list template of the RHS is instantiated and the instantiated list is appended to the remaining strategy list of the term so that the strategy list for the new term is gained after replacing the subterm (redex) with the corresponding contractum. $[\langle L9, L2, 0 \rangle]$ is the instantiated one for the strategy list template of $inf(X) -> cons(X, inf(s(X)))$ that is used in the rewriting (1) in Fig. 1. The strategy list for $nth(s(0), cons(0, inf(s(0))))$ is $[\langle L9, L2, 0 \rangle, \langle L0, RESULT, 0 \rangle, \langle BINGO, L0, \_ \rangle]$ after the rewriting (1) is done.

The TRAM interpreter executes the following instruction sequence when it begins to interpret the matching programs to reduce terms:

```
          init            // initializes TRAM's registers
LOOP: next                // pops a label from SL and puts it at L_Dummy
          jump L_Dummy
          go_ahead
          select          // selects one among the applicable rules
```

```
rewrite        // replaces the redex with its contractum
jump LOOP
```

next also pushes the go_ahead's label onto STACK. go_ahead is executed when an applicable rule is found. If there is no applicable rule, the control is transferred to *LOOP*. go_ahead triggers off backtracking if there is a choice point frame [1]. Otherwise it transfers the control to select. rewrite also appends the instantiated strategy list of the used rule to one for the subject term.

3 Parallelization for TRAM

The Parallel E-strategy. Since rewrite programs have parallelism inherently, explicit parallelization directives are not necessary for executing them in parallel [8]. Generally speaking, however, we can gain better performance of a program when its restricted subtasks with sufficiently large amount of work are only assigned to parallel processes than when its all subtasks are assigned to parallel processes. The ability to control parallelism is very desirable for this reason. Especially, it is indispensable to control parallelism of rewrite programs so that they are executed efficiently in parallel on a multiprocessor with a small number of processors.

In Parallel TRAM, we adopt the *Parallel E-strategy*, that is an extension of the E-strategy and is a subset of the *Concurrent E-strategy* [4], in order to control parallelism of rewrite programs. The Parallel E-strategy lets each operation have its own *parallel local strategy*. A parallel local strategy for an operation f of arity n is specified by using a list defined by the following extended BNF notation:

Definition 2.

⟨ParallelLocalStrategy⟩ ::= ( ) | (⟨SerialElem⟩* 0)
⟨SerialElem⟩ ::= 0 | ⟨ArgNum⟩ | ⟨ParallelElem⟩
⟨ArgNum⟩ ::= 1 | 2 | ... | n
⟨ParallelElem⟩ ::= { ⟨ArgNum⟩+ }

⟨ParallelElem⟩ specifies some arguments of a term whose top operation is f that are reduced in parallel. Each element of ⟨ParallelLocalStrategy⟩ are evaluated in sequence from left. For example, a TRAM program computing Fibonacci numbers in parallel can be defined as follows:

sorts: *Nat* .
order: .
ops: *0 : -> Nat   s : Nat -> Nat*
    padd : Nat Nat -> Nat { strat: ({1 2} 0) }
    pfib : Nat -> Nat .
vars: *X Y : Nat* .
rules: *padd(X, 0) -> X   padd(X, s(Y)) -> s(padd(X, Y))*
    pfib(0) -> 0   pfib(s(0)) -> s(0)
    pfib(s(s(X))) -> padd(pfib(s(X)), pfib(X)) .

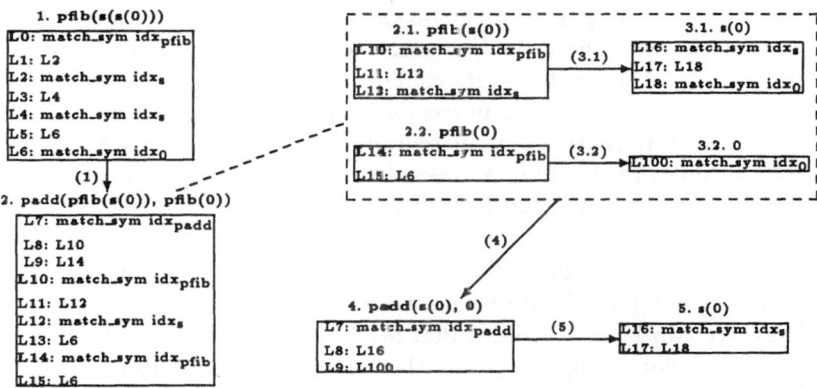

Fig. 2. Parallel reduction sequence for $pfib(s(s(0)))$

The operation *padd* has the parallel local strategy $(\{1\ 2\}\ 0)$ that indicates a term whose top operation is *padd* is tried to be rewritten to another after evaluating the first and second arguments to ENFs in parallel when the term is evaluated. Figure 2 shows the parallel reduction sequence for $pfib(s(s(0)))$. The two rewritings enclosed with the dash frame in Fig. 2 are done in parallel.

The Parallel E-strategy may control parallelism more suitably by combining conditions. Instead of $pfib(s(s(X)))\ ->\ padd(pfib(s(X)),\ pfib(X))$, the following rules can be defined for gaining a more efficient program:

$$pfib(s(s(X)))\ ->\ padd(pfib(s(X)),\ pfib(X))\ \text{ if } threshold(X) = false$$
$$pfib(s(s(X)))\ ->\ add(fib(s(X)),\ fib(X))\ \text{ if } threshold(X) = true$$

threshold returns (is reduced to) *false* if its argument is larger than a number. Otherwise it returns *true*. *add* is a sequential addition and *fib* computes Fibonacci numbers in sequence by using *add*.

Parallel TRAM Architecture. In Parallel TRAM, each processor has its own TRAM interpreter and four mutable regions. Two stable regions are shared with all processors. One of the processors is also an interface processor that also plays a role of the user interface by using the rule compiler and the term compiler. A strategy list represents a process queue. A chunk of consecutive elements of the strategy list represents a process. Each process has two possible states: *active* and *pending*. Each process queue contains zero or more pending processes and zero or one active process. If there is an active process in a process queue, it is at the top of the queue. A processor whose process queue contains an active process is in *busy*. Otherwise it is in *idle*.

Parallel Rewritings. It is not necessary to change matching programs even though terms are reduced in parallel. Strategy lists control parallel rewritings by holding labels at which new parallel instructions are stored. The new parallel instructions are fork, join, exit, sleep and nop. fork creates a new process reducing a subterm in parallel and allocates the processor to it if there is an idle processor. Otherwise the caller processor reduces the new process, or the new process is *in-lined*. join delays its caller process until all of its child processes terminate and makes the caller processor idle. exit terminates its caller process and reports the termination to its parent process. After executing exit, the caller processor resumes the pending process at the top of its process queue if there are pending ones in the queue. Otherwise the processor becomes idle by executing sleep. sleep makes its caller processor idle. nop does nothing. fork creating an empty process may appear in a strategy list by instantiating a strategy list template. nop is replaced with such a wasteful fork.

There are two kinds of triples that are elements of strategy lists in TRAM. One is for matching programs, the other for bingo. In addition to these ones, Parallel TRAM has five kinds of triples as elements of strategy lists. These new ones correspond to the new five parallel instructions: $\langle FORK, SIZE, A_J \rangle$, $\langle JOIN, PNUM, \_ \rangle$, $\langle EXIT, A_{PJ}, PID \rangle$, $\langle SLEEP, \_, \_ \rangle$ and $\langle NOP, \_, \_ \rangle$. FORK is the label of fork. $SIZE$ is the size of the forked process. A_J points to the JOIN triple corresponding to the FORK triple. JOIN is the label of join. $PNUM$ is a number of the caller's child processes that do not finish their work. EXIT is the label of exit. A_{PJ} points to the JOIN triple of the caller's parent process. PID is the processor ID of the caller's parent process. SLEEP is the label of sleep. NOP is the label of nop. For example, the strategy list for $padd(pfib(s(0)), pfib(0))$ after rewriting (1) in Fig. 2 is $[\langle FORK, 2, A_J \rangle, \langle L12, L11, 0 \rangle, \langle L10, L8, 0 \rangle, \langle L14, L9, 0 \rangle, \langle JOIN, 1, \_ \rangle, \langle L7, RESULT, 0 \rangle, \langle BINGO, L7, \_ \rangle]$. Suppose that there are two processors P0 and P1 that are in active and in idle respectively. After rewriting (1), P0 executes fork, and a new process is created and is allocated to P1. Then, P1's strategy list is $[\langle L12, L11, 0 \rangle, \langle L10, L8, 0 \rangle, \langle EXIT, A_J, P0 \rangle, \langle SLEEP, \_, \_ \rangle]$. The two processors reduces terms in parallel soon after.

4 Implementation of Parallel TRAM on Multiprocessor

Parallel TRAM has been implemented on OMRON LUNA-88K^2 in C. LUNA-88K^2 carries four MC88100 processors and adopts Mach 2.5 as its OS.

Processors are represented by using C structures in which four mutable regions and the related registers such as the instruction pointer are packed. Each structure also contains states of processors: $BUSY$ and $IDLE$. Parallel execution is realized by giving each processor a Mach thread. Pointers to the structures are passed to C functions in which four mutable regions or the related registers are accessed so that processors (threads) can access them as quickly as possible. Idle processors are managed by using a processor idle queue $IdleQueue$. When fork is executed, its caller processor tries to get one of idle processors from IdleQueue and to allocate it to a newly created process. If IdleQueue is empty, the new process is in-lined.

Parallel TRAM adopts a sequential stop-and-copy algorithm as the garbage collection (GC). A processor detecting a necessity of a GC performs the GC after all other processor pause. In Parallel TRAM, one global creation space (where

Table 1. Experimental results in computing the 24th Fibonacci number

system	time (s)	rewritings	r/s	GCs	created processes	in-lined processes	speed
tram	13.63	514105	37719	6	–	–	base
ptram1	14.71	527377	35852	6	0	376	0.93
ptram2	8.84	527377	59658	6	8	368	1.54
ptram3	7.55	527377	69851	7	39	337	1.81
ptram4	6.04	527377	87314	6	20	356	2.26

ptram$i$ means the Parallel TRAM system running on i processors.
The threshold is 10 for parallel computations of Fibonacci numbers.

new matching programs are allocated) is divided into multiple chunks for the purpose of efficient usage of storage and each processor gets one chunk from the global space when necessary. A GC is triggered when there is no space left in the global space. In addition to *BUSY* and *IDLE*, *GC* is used as a processor state for a GC. After a GC, the GC processor resumes the processors pausing for the GC. It is necessary to distinguish processors pausing for the GC from idle ones at the moment. The state *GC* is given to processors that are the second or later to detect the necessity of the GC.

5 Performance of Parallel TRAM

Table 1 shows the experimental results in computing the 24th Fibonacci number in parallel with the Parallel TRAM system on LUNA-88K[2]. For comparison with TRAM, the results of the TRAM system are shown. Due to limitations of space, the overhead of the current implementation is only discussed here.

We calculate an ideal improvement in speed when the 24th number is computed in parallel on four processors. When the 24th number is computed in parallel, two processors add the computations of the 22nd and 21st numbers, and of the 21st and 20th numbers in parallel after four processors compute the 22nd, 21st, 21st and 20th numbers in parallel. Then, the two sums are added so as to get the 24th number. The computations of the 22nd, 21st, 21st and 20th numbers involve 186579, 112286, 112286 and 67596 rewritings, respectively. If these rewritings are fairly done in parallel with four processors, it takes a time in proportion to 119687 rewritings (the mean of the four rewritings) to compute the four numbers. The additions of the computations of the 22nd and 21st, and of the 21st and 20th numbers involve 10948 and 6697 rewritings, respectively. Since these two additions are done independently with two processors, it takes a time in proportion to 130635 rewritings (119687 + 10948) to compute the 23rd and 22nd numbers. It takes a time in proportion to 148348 rewritings (130635 + 17713) to compute the 24th number in parallel with four processors since the addition of the computations of the 23rd and 22nd numbers involve 17713 rewritings. Hence, the ideal improvement in speed is 3.47 (514105/148348) when the 24th number is computed in parallel on four processors.

Since the actual improvement in speed is 2.26, there are some considerable overheads in the current implementation. The overheads may depend on the sequential GC, the process creation, the parallelization of TRAM and the architectural characteristic of LUNA-88K^2 such as shared bus. The process creation and the parallelization of TRAM may slightly contribute to the overhead judging from the experimental result of ptram1 in Table 1. It is currently unclear that how much the architectural characteristic of LUNA-88K^2 affects the overhead. We should implement Parallel TRAM on another multiprocessor so as to assess that thing. The computation of the 24th Fibonacci number was done in parallel with four processors on the Parallel TRAM system with enough size of the creation space to involve no GCs so that the overhead of the sequential GC was confirmed. It took 4.90 seconds. The overhead caused by the sequential GC is about 19% in the Parallel TRAM system while it is about 9% in the TRAM system. One of the main source of the overhead in the current implementation of the Parallel TRAM system is the sequential GC.

6 Conclusion and Future Work

We have described the design and implementation of Parallel TRAM where parallelism directives are specified by using the Parallel E-strategy. The Parallel TRAM system on LUNA-88K^2 was found to be about twice times faster than the TRAM system on the same workstation. Parallel TRAM will be implemented on other multiprocessors so that Parallel TRAM and its implementation technique will be improved and be made secure. One of the points that should be improved in the current implementation is to adopt a parallel or on-the-fly garbage collection.

References

1. Aït-Kaci, H.: Warren's Abstract Machine. A Tutorial Reconstruction. The MIT Press. 1991
2. CafeOBJ home page: http://ldl-www.jaist.ac.jp:8080/cafeobj
3. Christian, J.: Flatterms, Discrimination Nets, and Fast Term Rewriting. *Journal of Automated Reasoning.* **10** (1993) 95–113
4. Goguen, J., Kirchner, C. and Meseguer, J.: Concurrent Term Rewriting as a Model of Computation. Proc. of the Workshop on Graph Reduction. LNCS **279** Springer-Verlag. (1986) 53–93
5. Kirchner, C., Kirchner, H. and Meseguer, J.: Operational Semantics of OBJ-3. Proc. of the 15th International Colloquium on Automata, Languages and Programming. LNCS **317** Springer-Verlag. (1988) 287–301
6. Futatsugi, K., Goguen, J. A., Jouannaud, J. P. and Meseguer, J.: Principles of OBJ2. Conference Record of the Twelfth Annual ACM Symposium on Principles of Programming Languages. (1984) 52–66
7. Ogata, K., Ohhara, K. and Futatsugi, K.: TRAM: An Abstract Machine for Order-Sorted Conditional Term Rewriting Systems. Proc. of the 8th International Conference on Rewriting Techniques and Applications. (1997) (to appear)
8. Viry, P. and Kirchner, C.: Implementing Parallel Rewriting. Proc. of the International Workshop on Programming Language Implementation and Logic Programming. LNCS **456** Springer-Verlag. (1990) 1–15

Changing the Distribution Depth During a Parallel Tree Search

Nicolas Prcovic

Groupe Contraintes du CERMICS-INRIA Sophia Antipolis
2004, Route des Lucioles, B.P. 93
06902 Sophia Antipolis Cedex, France

Abstract. We present a new parallel tree search method for finding one solution to a constraint satisfaction problem. It consists in dynamically choosing the depth of the subtrees to distribute. This choice is based on a criterion which takes into account a heuristic valuation of the size of the subtree to parallelize in order to estimate the maximum size of the search space. We show how this method is likely to increase the speedup.

1 Introduction

This paper deals with the depth-first search (DFS) managed by the Backtrack algorithm when only one solution is needed. When all the solutions of a problem have to be found by expanding the whole search tree, satisfactory performances have been achieved: the speedup can be close to linearity, that is, the resolution time is almost divided by the number of engaged processes. It is established in theory [KGR94] and verified in practice [FM87]. However, an additional difficulty has to be faced when only one or the best solution is expected: some processes running in parallel may explore a part of the search tree which would not have been examined during a sequential search, and this may produce useless work overheads and then an important loss of efficiency. We present a parallel method which intends to minimize these risks while keeping the interprocess communication overheads as low as possible.

2 Distributed Tree Search Algorithms

This paper focuses on distributed memory parallelism. Several processes run in parallel and communicate to each other by sending messages. The goal of parallelization is to increase the performances linearly, that is, dividing the sequential time by the number of engaged processes. To approach this goal, we have to minimize the overheads due to additional operations brought by distribution: (1) the number of interprocess messages and (2) the idle-time of a process (due to a bad sharing of the work). We define p as the number of processes.

Now, we quickly recall the general properties of parallel depth-first search (PDFS) algorithms dedicated to the search for all solutions. As to parallelize the expansion of the whole tree, the root nodes of the highest subtrees are allocated to the processes. Each process runs a DFS of his own subtree(s). As soon as it terminates, it asks another process for one or several new root nodes. In [KGR94] is proven that we can make the communication overheads only grow in a polynomial way while guaranteeing an equitable work sharing. This confirms previous experimental results that show a speedup close to linearity for problems solved by this kind of algorithms [FM87]. Applied to CSPs, as the solving time is often exponential in the number of variables, this implies that these overheads become negligible if the problem is big enough. Hence, the speedup, defined by $S = \frac{T_{seq}}{T_{par}+T_{com}}$ (where T_{com} is the overall time spent by the communications, and T_{seq} and T_{par} the times consumed by the sequential and parallel searches), is close to p. We assume in this paper that we are only dealing with CSPs whose size is such that the existing PDFS algorithms achieve a linear speedup.

3   Changing the Distribution Depth

When only one solution is needed, if the same distribution method of the highest subtrees is kept, the linearity of the speedup can no longer be expected. Actually, one may have no speedup at all or a superlinear one. It depends on the location of the solution found in the tree. If we manage to distribute the nodes in the same order that they would be found by a sequential algorithm, then in the worst case, there will be no speedup but also no deceleration (if we neglect the communication cost) [LW86]. Nevertheless, we have not $S \approx p$ anymore but only $S \in [1; +\infty[$. Despite its variability, the PDFS algorithm has been used for several reasons: communication overheads are small and the speedup is linear on the average when the solution density is uniform in the tree. It is even superlinear in some cases [RK93]. Still, when heuristics order the search to reach one solution sooner, the solution density is not uniform anymore so the speedup may be sublinear or superlinear on the average. That is why another method is also used when a PDFS generate too much nodes. It consists in partially expanding the tree until nodes are met at a fixed depth before applying the PDFS (started from each discovered node). We shall refer to it as the Fixed-depth PDFS method (F-PDFS). The number of generated nodes gets lower and lower when the distribution depth increases. However, the number of messages grows exponentially with this depth and the communication overheads loose their polynomial behavior.

3.1   The V-PDFS Algorithm

Instead of choosing a fixed depth, we make it vary all along the search according to a criterion which takes into account a heuristic valuation of the size of the subtree to parallelize and thus estimate the maximum size of the explored search space. We call this the Variable-depth PDFS method (V-PDFS). We have conceived a parallel algorithm composed of two alternatively running phases: (1) sequential DFS *until the kth node visited at a depth is met*, where k is a constant to choose, or until the maximum depth M is reached. (2) search for one solution in the subtree starting from the selected node thanks to a PDFS algorithm like those introduced in Section 2.

Now, we are going to justify the relevance of our distribution method. All the following formulas assume we are only dealing with one-solution problems. Let $F = \frac{N_{seq}}{N_{par}}$ be the efficiency of the exploration of the search space, that is, the quotient between the number of nodes generated by a sequential DFS algorithm and the number of nodes generated by a PDFS algorithm before finding the solution. Let X be a subtree we intend to parallelize, N^- be the number of the nodes already generated in the search tree before to meet the root of X and N^+ be the number of nodes that must be generated before to reach the solution in X (if any). Thus, we can express F as : $F = \frac{N_{seq}}{N_{par}} = \frac{N^- + N_{seq}^+}{N^- + N_{par}^+}$.

Let ν be the number of nodes generated by any of the p processes that expands X before one of them finds the solution. We have $N_{par}^+ = p\nu$ and $F = \frac{N^- + N_{seq}^+}{N^- + p\nu}$. We have no way of knowing where the solution is located so we can only calculate the worst case and give a lower bound. We know that $N_{seq}^+ \geq \nu$ when the PDFS algorithm distributes the subtrees in the same order it examines them in sequential so : $F \geq \frac{N^- + \nu}{N^- + p\nu}$. Since $\frac{\partial}{\partial \nu}\left(\frac{N^- + \nu}{N^- + p\nu}\right) = \frac{(1-p)N^-}{(N^- + p\nu)^2} \leq 0$, ν has to be maximized as to minimize F, so finally : $F_{min} = \frac{N^- + N/p}{N^- + N}$ where N stands for the total number of nodes in X. The distribution criterion, which lays on an efficiency threshold f, is then : $\frac{N^- + N/p}{N^- + N} \geq f$ or $N \leq \frac{1-f}{f - \frac{1}{p}} N^-$.

If we knew the value of N, we could determine if the criterion is filled for X and if running a PDFS from its root would guarantee that $F \geq f$. If the criterion is not filled then it is necessary to go deeper into the tree to decrease N. However, we are obliged to fix a maximum depth of distribution M beyond which a PDFS becomes too inefficient. Otherwise, T_{com} would be too high to be negligible compared to T_{seq} and T_{par} and the criterion formula would not be sound anymore.

In so far as we have no information about the search tree structure, we propose, when searches with a static variable ordering are done, to make the distribution as if all the subtrees located at the same depth had the same number of nodes. This model, even if it is likely to be far from reality, is the one implicitly chosen when distribution is done at a fixed depth: we hope that, on the average, the size of the subtrees is such that we can distribute them at this depth but we do not take the variability of that size into account. With this model of tree, we have the relation $N^- = kN + N_0$, where k is the number of already explored subtrees at the same depth as X, and N_0 is the number of nodes sequentially examined before to reach X. When the maximum depth M is small, N_0 is negligible and $N^- \approx kN$. Introducing this expression in the criterion formula, we obtain : $k \geq \frac{f - \frac{1}{p}}{1 - f}$.

We can see that k is constant and independent of the depth. The criterion for distribution is simplified to a counting of generated nodes at each depth as to decide to parallelize or not. We can notice that if we set f to 100% then we ensure that k is infinite, so we fall into a F-PDFS. Thus, V-PDFS can be seen as a generalization of F-PDFS when adding the f parameter.

3.2 Experimental Results

We have chosen to run tests on randomly generated CSPs following the model described in [Pro94] on a network of 8 SUN stations. All tests have been made on two hundred instances all containing at least one solution. We made M and k vary. We recorded the number of generated nodes and the number of interprocess messages. As expected, we found that the search efficiency and the communication cost were increasing functions of k and M. From these results we can show that our distribution method can increase the speedup. As a matter of fact, for the kind of problems which are likely to produce a sublinear speedup on the average, the quality of speedup depends on the compromise made between F and the communication cost. When T_{par} decreases, T_{com} increases (and vice-versa), so we have to experimentally find the best pair of values which minimizes their sum, thanks to tests on a representative sample of the kind of problems we want to solve. It is clear that, using a F-PDFS algorithm, we have a poor choice of value combinations whereas a V-PDFS let us choose a much more accurate value combination which minimizes $T_{par} + T_{com}$ and then makes the speedup better.

4 Conclusion

We have presented the V-PDFS algorithm, which allows us to dynamically select the most adequate depth to distribute the nodes. By adding a new parameter, we have conceived a new means to refine the distribution and to enhance the resulting speedup of problems whose parallelization is inefficient with usual methods.

References

[FM87]   R. Finkel and U. Manber. DIB—A Distributed Implementation of Backtracking. *ACM Transaction on Programming Languages and Systems*, 9(2):235–256, 1987.

[KGR94] V. Kumar, A. Grama, and V. Rao. Scalable load balancing techniques for parallel computers. *Journal of Parallel and Distributed Computing*, 22(1), 1994.

[LW86]   G.-J. Li and B.W. Wah. How Good are Parallel and Ordered Depth-First Searches. In *Proceedings of the 1986 International Conference on Parallel Processing*, pages 992–999. IEEE Computer Society Press, 1986.

[Pro94] P. Prosser. Binary constraint satifaction problems : Some are harder than others. In *11th European Conference on Artificial Intelligence*, pages 95–99, 1994.

[RK93]   V. N. Rao and V. Kumar. On the Efficiency of Parallel Backtracking. In *IEEE Transactions on Parallel and Distributed Systems*, volume 4:4, pages 427–437, 1993.

Abstract And-Parallel Machines*

Nachum Dershowitz and Naomi Lindenstrauss

Department of Computer Science, The Hebrew University, Jerusalem 91904, Israel

The deterministic Turing machine, though abstract, can still be seen as a
model of a realistic computer. The same cannot be said for the nondeterministic
Turing machine as a model of parallel computing. We introduce several abstract machines with fine-grained parallelism—the *and-parallel Turing machine*,
the stronger *parallel rewriting machine*, and extensions of both with an *interrupt capability*. These machines are very powerful: the parallel rewriting machine
can compute the permanent in polynomial time and the and-parallel machine
with interrupt can simulate nondeterministic and alternating Turing machines
of polynomial time complexity in polynomial time. All the same, they may be
viewed as realistic models if time and space are suitably restricted.

And-Parallel Turing Machines. One way of viewing the nondeterministic
Turing machine is to say that at each stage, when confronted with k choices, it
splits into k replicas of itself, each of which makes one of the choices. The input
string is accepted if one of these machines enters an accepting state. This kind of
parallelism can be termed "or-parallelism". However, the nondeterministic Turing machine does not model real parallel machines in which different processors
communicate with each other. Also, when faced with an input of size n that
must be read, it will need at least n units of time. But by then the machine may
have arrived at an exponential number of possibilities, an unrealistic scenario.

Our *and-parallel machine* works with an unbounded binary tree. (We take
a binary tree for simplicity—any bounded arity will do.) Instead of the square
scanned by the "head" (representing the memory cell dealt with at the moment
by a sequential computer), the tree has a "frontier" of nodes, representing the
memory cells dealt with at the moment by the processors comprising the parallel
computer. (By "frontier" we mean a subset of the binary tree such that none
of its nodes is an ancestor of another, while every node in the binary tree is
either an offspring or an ancestor of a node in the frontier.) In the beginning
the frontier consists of the root node. The input is a finite binary tree. Each
node in the frontier is in one of a finite number of states, and can be thought
of as being taken care of by a different processor. Given a binary tree with
frontier, the transition function δ acts on its frontier nodes and transforms it
into another binary tree with frontier. It may change the content of the nodes
of the frontier in the following way: if N is a frontier node in state q with
symbol t, δ can change the symbol at N. According to q and t, N may remain
a frontier node, although its state may change. It can also be replaced in the
frontier by its two children, who get states determined by δ. If q and t are
suitable, and N's sibling N' also has suitable symbol and state, then both N

---

* The full version of this extended abstract can be found on the World Wide Web at
http://www.cs.huji.ac.il/~naomil.

and N' are removed from the frontier and their parent enters the frontier with a state determined by δ according to the states and symbols of its children. This step models communication between different processors. If N is ready to be replaced by its parent and N' is not (see the following figure), it waits for N'. This models suspension and synchronization. If N has a blank symbol, then it may be overwritten with a nonblank, and have two children appended. At each stage, δ performs all the actions it can perform on the nodes of the frontier according to the above rules. We use deterministic rules, so the machine is deterministic. If there is a frontier node in a non-accepting state on which δ cannot act, the machine stops with failure. If a node reaches an accepting state it will remain in it. If all the nodes of the frontier are in an accepting state the machine accepts the input.

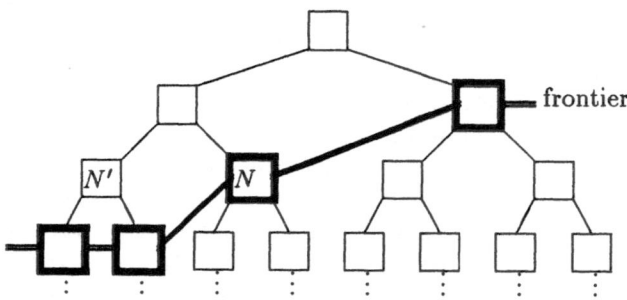

Throughout this paper, we make the following assumptions:

Assumption 1: *An unlimited number of processors is available.* Of course there is a real-life limit to the number of processors, but we can assume that this is transparent to the user (just as physical limitations on the size of memory are ignored by the deterministic Turing machine model). If the depth of the tree is $O(\log n)$, where n is the input size, then the number of processors will be polynomial in n (hence problems that can be solved on such a machine in polylogarithmic time and space of logarithmic depth are in NC).

Assumption 2: *The input should be given in a form suitable for parallel processing, that is, lists have to be replaced by balanced trees.* If the depth of the input is not logarithmic in its size, just looking at it will take too much time.

Assumption 3: *The memory on which the processors work is organized as a tree and not as a directed acyclic graph (dag).* Clearly this simplifies matters. Though it may seem that a dag is more economical, because it avoids having the same computation performed several times, on the average not that much is saved. Under a variety of probabilistic models, a tree of size n has a maximally compacted form as a dag (where all common subtrees are represented only once) of expected size (asymptotically) $cn/\sqrt{\log n}$, where c is a constant depending on the type of trees compacted and the statistical model. Though this represents a drastic savings in storage, what is important in our case is the logarithm of the size, for which the savings is only marginal ($\log \log n$).

Assumption 4: *On each branch of the memory tree there is at most one node* *on which a processor works at any given time.* Again this simplifies matters, since it will never happen that two processors will attempt to work on the same node, and is enough for good performance.

Parallel Rewriting Machines. We would like to have a parallel machine that is easy to program yet stronger than the and-parallel machine, for instance one that can sort in polylogarithmic time. Assume that the input to the machine is given as a tree and that its behavior is determined by "flat" conditional rewrite rules of the form $C \mid L \longrightarrow R$ where L is a tree "pattern" containing variables representing subtrees, R is a replacement pattern for L, and C is some simple property of the tree that can be checked locally at the location in the tree where L appears. The computational paradigm associated with rewriting has the full power of Turing machines, while lending itself naturally to parallel execution, since different subtrees can be rewritten in parallel. The advantage of this formalism is that machine operations are very straightforward (almost as simple as for a Turing machine), and the machine gets the problem in a form that preserves its "meaning". A tree to which no rewrite rule can be applied is said to be in *normal* form, and computation essentially amounts to bringing a tree to its normal form. Of course there is no guarantee that the rewriting process will terminate, and even if it does, that the normal form will be unique. Different strategies for rewriting may be used (under certain well-defined conditions, the order does not affect the final result) and in a parallel environment several steps of rewriting may be performed simultaneously.

The *parallel rewriting machine* implements a top-down strategy by first trying to rewrite a tree by a rule that applies at its root, and only if this is not possible it tries to rewrite the children of the root, and so on. After the children of a node have been rewritten to normal form, rewriting continues at the parent node. The machine has an unbounded tree-tape available. In the beginning this tree consists of the input. At each stage there is a frontier of nodes at which processing is taking place. These nodes can be in any of five states: *up, down, going-up, going-down, stopped.* In the beginning the frontier consists of the root node in the state *going-down.* As long as there is a rule which can be applied at the root this rule is applied, possibly causing some local change in the form of the tree. If no rule can be applied the node goes into state *down.* A node in state *down* is replaced in the frontier by its children, each in state *going-down.* Now the machine tries to apply rules at the nodes corresponding to the children. Thus the processing continues until it arrives at nodes that cannot be rewritten and have no children. Such nodes go into state *up.* If all the children of a node are in state *up*, they are replaced in the frontier by their parent node in state *going-up.* If a rule can be applied to the node in state *going-up* it is applied and the node goes into state *going-down.* Otherwise the node goes into state *up.* If the root node is in state *up* it goes into state *stopped.* In this case the computation terminates and the tree holds the normal form of the original input.

The essential advantage of this machine over the previous one is its local pointer capability: it can graft new nodes within the tree and can replace the children of a node by a subset of them in any order and possibly with repetitions. (The last property bears similarity with top-down tree transducers, which traverse a tree from top to bottom, rewriting it.)

When the rewrite rules are linear (that is, no variable appears more than once in either L or R) each transition takes constant time, since applying a rewrite rule consists of checking a simple condition and performing a local change in the tree. Otherwise:

Theorem. *In the case of nonlinear rules the time is bounded by the number of machine transitions multiplied by a constant times the depth of the tree.*

When there are rewrite rules in which the same variable appears more than once in L, we must replace them by rules where this does not happen and instead check for equality. With our rewriting machine, it is not difficult to see that checking equality of subtrees can be done in time which is of the order of magnitude of the depth of the trees.

When the same variable appears more than once on the right hand side this means that the tree it represents has to be copied. One can use the rewriting machine to copy trees in time that is of the order of magnitude of their depth. Without loss of generality assume that the tree is binary with inner nodes of the form $t(Symbol, Left, Right)$ and its leaves are constant symbols. The normal form of $makecopy(T)$ is $copied(T', T'')$ where T' and T'' are copies of T. The rules are

$$
\begin{aligned}
&makecopy(T) \longrightarrow top(double(T)) \\
&double(t(S, Left, Right)) \longrightarrow dt(S, S, double(Left), double(Right)) \\
&leaf(A) \mid double(A) \longrightarrow build(A, A) \\
&dt(S', S'', build(X, X'), build(Y, Y')) \longrightarrow build(t(S', X, Y), t(S'', X', Y')) \\
&top(build(T', T'')) \longrightarrow copied(T', T'')
\end{aligned}
$$

(Notice that while S and A appear twice in the second and third rules they stand for function and constant symbols, so the rules can be thought of as schemas for linear rules.) Basically there is one movement down where things are doubled, and then one movement up where things are appropriately shuffled. The time needed to copy a term is proportional to its depth.

The class $\#P$ comprises those functions that can be computed by *counting Turing machines* of polynomial time complexity. Computing the permanent is an $\#P$-complete problem, yet the parallel rewriting machine can compute the permanent in polynomial time. Of course using our machine in this case is not realistic. However, if the depth of the space used is logarithmic, only a polynomial number of processors will be required. All the following examples are of this kind:

1. **Parallel Bits Algorithm:** adding two numbers in time logarithmic in their length.
2. **Parallel Merging and Sorting:** in time polylogarithmic in length of list.
3. **Evaluating Algebraic Expressions:** in average time $O(\sqrt{n})$ for size n.
4. **Scalar Product:** in time logarithmic in length of vectors.
5. **Matrix Multiplication:** matrices of size $n \times n$ in time $O(\log^2 n)$.
6. **Reachability:** for a graph of n vertices in time $O(\log^3 n)$.

In each, the machine, given a straightforward recursive description of the problem in the form of rewrite rules, works out an efficient implementation of the rewrite algorithm.

Parallel Machines with Interrupt. Once a node in the tree representing the memory of either the and-parallel machine or the rewriting machine has launched its children, it must wait for their answers—it must wait for both its children to become frontier nodes in a suitable state before it can become a frontier node again. There may, however, be cases when an answer from one child is enough (for example, when evaluating a Boolean expression of the form $E \vee F$, we know that if E evaluates to *true* so does the whole expression, irrespective of the value of F). It even may happen that the answer of one child is sufficient while the other child embarks on an infinite computation. We therefore consider more powerful machines, in which the frontier may move to the parent even if only one child is in a suitable state. This means that instead of having all processors work according to a uniform program, there may be cases in which one informed processor can cause other processors to terminate. Suppose that we are evaluating $E \vee F$ where E is small and F is large. In that case the more powerful machine may be faster, because if it discovers that E evaluates to *true* it can abandon evaluation of F. It is clear that an and-parallel machine evaluating a Boolean expression will take time proportional to the depth of the expression, because the frontier moves once from the root to the leaves of the input and once back to the root. If we consider all different Boolean expressions of the same length n to have the same probability, we can show that the and-parallel machine can evaluate them in $\Theta(\sqrt{n})$ time, and, for the interrupt machine, in asymptotically constant time.

Theorem. *The nondeterministic (and alternating) Turing machine can be simulated by an interrupt parallel machine. If a problem can be solved in time polynomial in the size of the input with the nondeterministic (alternating) Turing machine, the same is true for the interrupt parallel machine.*

Theorem. *Nondeterministic versions of our interrupt machines are not stronger than the deterministic ones.*

The machines described here are very powerful. However, if we restrict ourselves to polylogarithmic time and space of logarithmic depth, as in all our examples, they can be seen as models of realistic machines. It is also interesting to observe that the and-parallel machine gained much power by adding a pointer capability and an interrupt possibility—two things that can be easily implemented. Because of the tree structure of the memory and the way the computation proceeds it can never happen that two processors try to write to the same node (a problem that arises in the PRAM). Using rewrite rules makes the machine very easy to program—rules simply cannot be applied as long as their arguments have not reached the right form.

Workshop 20

Real-Time Systems and Constraints

Workshop 20:
Real-Time Systems and Constraints

Gérard Berry

Goal of the Workshop

For the first time, Europar includes a workshop on real-time systems and constraints. Real-time systems are nowadays ubiquitous, with applications ranging from small domestic appliances to full airplane control. Real-time systems call for specific programming, implementation, and validation technologies, which are the subject of the workshop.

Classical languages are somewhat unsufficient for real-time systems, because of their natural concurrent and reactive structure. Therefore, new programming paradigms and languages must be developed. Unlike classical programs, real-time ones are often implemented using mixed software / hardware architectures, which calls for new compiling and synthesis techniques (codesign). The performance of generated codes and circuits must be predictable, which calls for new dynamic and static performance measurement tools. The underlying operating systems and scheduling strategies must themselves obey response-time constraints, unlike conventional ones. Finally, since real-time systems often operate in safety-critical environments, verification of program properties is a must. Verification should not limited to logical correctness issues since the timeliness of a reaction may be as important as its correctness.

Presentation of the Papers

The workshop's ten papers address the issues above. We briefly summarize them below.

Three papers deal with design and language issues. The paper "An ML-like Module System for the Synchronous Language Signal" by D. Nowak, J.-P. Talpin, T. Gautier, and P. Le Guernic extends the Signal data-flow synchronous languages with a module system inspired by that of the functional language ML, with polymorphism and type inference. Module interfaces are not limited to types but also deal with the clock calculus on which Signal is based. The paper "Reactive Real-Time Programming with Distributed Agents" by G. Schrott introduces a new way of designing systems using the notion of a contract requested by an agent and executed by another agent after a bidding step. In that step, various agents can propose realizing the contract at some cost. The paper "Synchronous Thread Management in a Distributed Operating System's Micro-Kernel" by O. Potonnié and J-B. Stefani links languages and deterministic real-time operating systems. Deterministic process management algorithms are programmed in the Esterel synchronous language and verification tools are used for checking correctness properties.

Two papers deal with scheduling. The paper "Analyzing Schedulability of Astral Specifications using extended Timed Automata" by K. Brink, J. van Katwijk, R.F. Lutje Spelberg, and W.J. Toetenel uses timed automata based model-checking techniques to analyse task schedulability. The problems are described in a requirements specification language and translated into timed automata extended by urgent transitions that can be formally analyzed. The paper "Schedulers for Age Constraint Tasks and their Performance Evaluation" by W. Albrecht and R. Wisser deals with age-constrained tasks that have to be executed at a certain rate but not necessarily in a periodic way. The authors present new scheduling techniques that perform better than the usual periodic scheduling ones and present experiments.

Three papers deal with generated code run-time performance. The paper "A Methodology for Compilation of High-Integrity Real-Time Programs" by K. Lermer and C. Fidge presents a methodology called Topaz in which timeliness requirements are provided by adding additional deadlines annotations to source code written in a conventional language. The annotations are carried through the compilation phase and timing verification is done on the object code. The paper "Cinderella: A Retargetable Environment for Performance Analysis of Real-Time Software" by Yau-Tsun Steven Li, S. Malik, and A. Wolfe is devoted to best- and worst-case analysis of generated code performance on modern microprocessors. It present path analysis techniques coupled with cache and pipeline analysis techniques. Actual experimental results are given. The paper "Deriving Annotations for Tight Calculation of Execution Times" by A. Ermedahl and J. Gustafsson tackles the annotation problem for code speed analysis. Often, speed analysis is based on possibly incorrect user-provided annotations about false paths. The authors use abstract interpretation techniques to automatically derive annotations, eliminate false paths, and get tight bounds.

The paper "Task-System Analysis using Slope-Parametric Hybrid Automata" by A. Burgueño and V. Rusu extends the classical hybrid automata techniques to parametric hybrid automata where the slopes of state variables can be parameters. To find which parameters can make a property satisfied, the authors use parametric polyhedra analysis techniques that are implemented using Maple and Prolog IV.

Finally, the paper "Designing an Embedded Hard Real-Time System: A Case Study" by M. Colnaric and W. Halang is an attempt to globally master determinism and predictability through all the layers from programming to implementation. Of course, this cannot be done in an abstract way and the authors present an architecture case study.

Designing an Embedded Hard Real-Time System: A Case Study

Matjaž Colnarič[1], C. T. Cheung[2] and Wolfgang A. Halang[3]

[1] University of Maribor, Slovenia, colnaric@uni-mb.si
[2] Hong Kong Polytechnic University, csronnie@comp.polyu.edu.hk
[3] FernUniversität Hagen, Germany, wolfgang.halang@fernuni-hagen.de

Abstract. In this paper a description of a consistent design of an embedded hard real-time control system is given. To provide for the overall predictability of tasks' temporal behaviour, which is the ultimate requirement in such systems, all influencing factors are taken into account in a holistic manner: system and hardware architecture, operating system issues, programming language and application design methodology. Based on the resulting guidelines, a consistent prototype was implemented.

1 Introduction

Instead of computer speed, which cannot guarantee that specified timing requirements will be met, almost a decade ago a different ultimate objective in designing consistent systems for embedded hard real-time applications was generally accepted: predictability of temporal behaviour. While, for the systems usually employed in process control, testing of conformance to functional specifications is well established, temporal circumstances, being an equally important design aspect, are seldom verified consistently. Almost never it is proven at design time that such a system will meet its temporal requirements in every situation that it may encounter.

Although the domain of real-time systems substantially gained interest in recent years, the results of fundamental research are not broadly used in hard real-time applications, yet. One of the reasons for this situation is that most studies consider selected topics only, and assume that other system constituents behave predictably. This assumption makes it very hard for the application designers to set up consistent systems from different components. For that reason we initiated a project systematically addressing all crucial layers to provide practically usable design guidelines and techniques for hard real-time control systems. A consistent prototype implementation of an embedded hard real-time system was designed and will be briefly sketched in this paper. For those who are interested in the details, thorough reference to our detailed publications on specific topics resulting from this project is given, where also comments on, and references to, related work of other research groups can be found.

2 Concept of the Experimental Hardware Platform

Using static scheduling algorithms, severe problems regarding the non-determinism of temporal task execution behaviour could be avoided. However, following the very nature of real-time applications, it is necessary to provide for dynamic task scheduling. The algorithms which guarantee that, once successfully scheduled, all tasks will meet their deadlines, are referred to as feasible. In the literature, several such algorithms have been reported. For our purpose, the earliest-deadline-first scheduling algorithm, which was shown to be feasible for scheduling tasks on single processor systems, was chosen and implemented in the kernel of an operating system [3].

For process control applications, where process interfaces are usually physically hard-wired to sensors and actuators establishing the contact to the environment, it is natural to implement either single processor systems or dedicated asymmetrical multiprocessors acting and being programmed as separate units. Thus, the earliest-deadline-first scheduling policy can be employed without causing any restrictions.

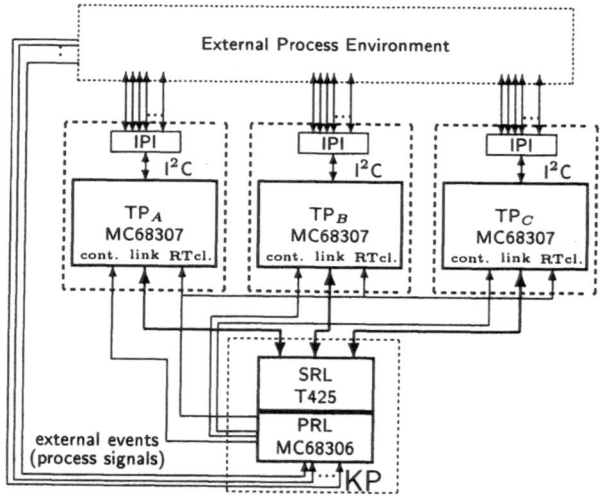

Fig. 1. Scheme of an experimental hardware platform

In classical computer architectures the operating systems are running on the same processor(s) as the application software. In response to any occurring event, the context is switched, system services are performed, and schedules are determined. Although it is rather likely that the same process will be resumed, quite a bit of computer capacity is wasted by superfluous overhead. This suggests to employ a second, parallel processor, to carry out the operating system services. Such an asymmetrical architecture turns out to be advantageous, since, by dedicating a special-purpose, multi-layer processor to the real-time operating system kernel, the user task processor(s) are relieved from any administrative overhead. The kernel processor and the task processors are connected point-to-point by serial links, thus avoiding collisions on common communication media as a possible source of non-determinism. This concept was in detail elaborated in [7] and [2]. The implementation is shown in Figure 1.

The **kernel processor** (KP) is responsible for all operating system services. It maintains the real-time clock, and observes and handles all events related to it, to the external signals and to the accesses to the common variables and synchronisers, each of these conditions invoking assigned tasks into the ready state. It performs earliest-deadline-first scheduling on them and offers any other necessary system services.

External processes are controlled by tasks running in the **task processors** (TP) without being interrupted by the operating system functions. A running task is only pre-empted if, after re-scheduling caused by newly invoked tasks as a consequence of an event, it is absolutely necessary to execute one of the arriving tasks immediately in order to allow for all tasks to meet their deadlines. Each task processor supports a peripheral serial I^2C bus. To these buses intelligent peripheral interfaces are connected, enhancing fault tolerance by adding certain intelligence to them enabling reasonable reactions in exceptional situations, and increasing system performance by providing higher level I/O services.

3 Concept of a Real-Time Programming Tool

To program applications for the above hardware platform a tool is being constructed [8], into which also a specific exception handling mechanism [4] is integrated. Its ultimate objective is to produce the best possible program code for embedded hard real-time applications, and realistic upper bound estimations of the code's execution time. In the tool two parts are closely interleaved: a compiler for an adapted standard real-time programming language, and a program execution time analyser. The latter is providing the necessary information for a schedulability analyser currently being beyond the scope of our research.

Being aware of the unpopularity of defining new languages and writing "own" compilers, we had several arguments for doing so. One is that commercially available and widely used high-level programming languages and compilers still do not meet all the needs of hard real-time application programming. Another argument in favour was the co-design of the particular experimental hardware architecture, the corresponding operating system, and the application development tool rendering it impossible to use any of the existing compilers. Finally, experience with run-time analysis of high-level source code programs demonstrated that access to the internal structures of a compiler is required in order to gain reasonably realistic results.

The language introduced was called miniPEARL [8]. It is a simplified version of PEARL [6], a standard language for programming real-time applications, which, however, may produce temporally unpredictable code for several reasons. To eliminate these problems, PEARL's syntax was modified. Further, to support efficient mapping onto typical target architectures certain features were removed.

Finally, it was enhanced by some constructs specific to real-time systems, as proposed by Halang and Stoyenko [7].

The main differences between PEARL and miniPEARL are:

- no GOTOs (LOOP and REPEAT instead);
- pointers and recursion renounced;
- number of loop iterations strictly bounded;
- signals not directly supported;
- temporally bounded statement execution, including synchronisation mechanisms and process I/O operations;
- possibility to explicitly assert execution time of software blocks;
- PEARL's DATIONs not implemented;
- improved task activation scheme and scheduler support.

To allow for schedulability analysis, precise execution times of application tasks must be known in advance. In our tool, two methods for the estimation of program run-times are supported:

1. *Analysis of executable code.* In this method, an automatic analyser is used to estimate execution times. Source code is transformed into an intermediate form (modified syntax tree) prior to the generation of executable code. Each element of this form is associated with a macro block that is used for two purposes. The first one is to generate code, and the second one to obtain its execution time. Since the execution time of a block can be data-dependent, as much information as possible about operands should be passed to it. An operand can be a register, a constant, a local or a global variable. When a macro is expanded, the sum of times needed for accessing these operands is added to the basic execution time of the macro.

2. *Direct measurement of executable code.* This method can be used when a more precise value for an execution time than estimated is desired. To achieve this, object code is executed on the target system and the execution time is recorded. Direct implementation of this method has some obvious disadvantages. First, the complete target system must be implemented; thus, co-design of the hardware and the software of an application is not possible. Further, through recording, only average execution times can be obtained. For a usable analysis, however, worst-case execution times are needed, and it is usually difficult to create a test scenario leading to a worst-case situation. Finally, the impact of the delays caused by the input/output devices is dependent on the run-time circumstances.

By our approach, these disadvantages are eliminated. Only a task processor or its equivalent must be implemented. The longest path through a task is determined by the compiler, and "pilot code" is generated running only along that path. From a set of alternative constructs (IF and CASE statements, for example), the longest one is statically routed. All time-guarded commands, system calls and input/output variable accesses are replaced by corresponding delays. This pilot code is then executed on the hardware platform.

To practically and adequately support the design of embedded real-time applications, it was considered to include new, and more explicitly emphasise the existing, features in miniPEARL to enhance its suitability also for purposes of hardware and software system specification [5]. Instead of using strictly formal specifications, systems can be described in a simple and straightforward manner with a terminology which is close to application programmers and their way of thinking. Such descriptions are mixtures of clauses in syntactically correct formal notation and natural language inserts. Employing appropriate artificial intelligence methods [1], specifications are then gradually refined until programs in the real-time programming language miniPEARL are obtained.

4 Conclusion

To provide for application layer predictability of an embedded real-time system, the main objectives pursued in its design as presented here were determinism and predictability of each of its layers. The alternatives to design such a system to an as large as possible extent of consistency with the guidelines set, using commercial off-the-shelf components and other easily available means, were verified and validated. Applications designed this way fulfill the requirements of hard real-time systems, viz., timeliness, simultaneity, predictability, and dependability, better than conventionally constructed ones.

References

1. C. T. Cheung. Transforming Mixed Formal and Natural Language Specifications into High Level Language Code. Internal Report, Department of Computing, Hong Kong Polytechnic University, 1997.
2. Matjaž Colnarič and Wolfgang A. Halang. Architectural support for predictability in hard real-time systems. *Control Engineering Practice*, 1(1):51–59, February 1993. ISSN 0967-0661.
3. Matjaž Colnarič, Wolfgang A. Halang, and Ronald M. Tol. Hardware supported hard real-time operating system kernel. *Microprocessors and Microsystems*, 18(10):579–591, December 1994. ISSN 0141-9331.
4. Matjaž Colnarič, Domen Verber, and Wolfgang A. Halang. Supporting high integrity and behavioural predictability of hard real-time systems. *Informatica, Special Issue on Parallel and Distributed Real-Time Systems*, 19(1):59–69, February 1995. ISSN 0350-5596.
5. Matjaž Colnarič, Domen Verber, and Wolfgang A. Halang. A Real-Time Programming Language as a Means to Express Specifications. *Control Engineering Practice*, 5(7), July 1997.
6. *DIN 66 253: Programming Language PEARL, Part 1: Basic PEARL*. Beuth Verlag, Berlin, 1981.
7. Wolfgang. A. Halang and Alexander D. Stoyenko. *Constructing Predictable Real Time Systems*. Kluwer Academic Publishers, Boston–Dordrecht–London, 1991.
8. Domen Verber, Matjaž Colnarič, and Wolfgang A. Halang. Programming and time analysis of hard real-time applications. *Control Engineering Practice*, 4(10):1427–1434, October 1996. ISSN 0967-0661.

Reactive Real–Time Programming with Distributed Agents

Gerhard Schrott

Technische Universität München
Institut für Informatik
D-80290 Munich, Germany

Abstract. The proposed reactive real–time programming system is a new approach to implement complex distributed heterogeneous real-time applications. It is based on the notion of distributed multi–agent systems. The whole control task is decomposed top down into small execution units, called agents which communicate by sending and executing contracts and are specified in a hardware independent language based on states and guarded commands. At compile time the agents are distributed to specified targets. PC's, micro-controllers, programmable logic controllers and even programmable logic devices are supported. The system automatically translates each agent to the particular code and realizes the communication including a bidding protocol between the agents either on the same processor or within a network. Due to a strictly cyclic processing of the agents exact response times can be guaranteed. Zero delay agents can be implemented in hardware.

1 Introduction

Reactive programming of distributed systems connected by a field–bus is nowadays imperative in process control. Networks comprising PC's and/or micro-controllers are available at low prices and will replace the "one for all" computer solution based on conventional real-time operating systems and will solve its inherent problems: Micro-controllers dedicated to time–critical applications guarantee the needed reactiveness; the complexity of control applications is mastered by a natural distribution of tasks; modern programming paradigms are used to improve the flexibility and fault tolerance of the system.

1.1 Multi–agent systems

Agents [4] are small autonomous software units which are able to perceive, to plan, to communicate with each other, to decide and to act. Multi–agent systems (e.g. [1],[5]) are in many respects similar to distributed real–time systems. Tasks correspond to agents, messages to contracts between agents, perception and action are equivalent to the input/output of the technical process and planning and deciding is implicitly programmed in intelligent autonomous applications. It is obvious that the paradigm of an agent can easily be transfered to distributed real–time systems and may give new impulses to the programming of distributed process control.

1.2 Cyclic predictable scheduling

Cyclic scheduling was already used in the earliest real–time systems. It is still applied to safety critical applications. Lawson [3] stresses that only cyclic systems can be proved in their time behavior.

1.3 Synchronous programming

Synchronous languages as ESTEREL, LUSTRE or SIGNAL ([2]) provide statements which allow a program to be considered as instantaneously reacting to external events. Their behavior is also fully deterministic, however their computational power is only equivalent to finite automaton. They create reactive kernels which need additional layers for physical input/output, for data management and for communication between different kernels.

1.4 Real–time programming with distributed agents

Most available operating systems are based on the execution of parallel tasks causing the known problems of scheduling and interrupt response times. In this paper a new approach for programming distributed heterogeneous real-time applications is proposed based on the notion of distributed multi-agent systems. The communication between these agents is transparent to the programmer. The system guarantees response times by a strictly cyclic and therefore predictable scheduling. Using a restricted subset of the system specification even the hard real-time requirements of synchronous programming are met: the resulting code is the definition of a finite automaton specified in VHDL.

2 MAD–RTS Specification

The proposed reactive multi–agent distributed real-time system (MAD–RTS, [6]) combines specification and coding of the control programs in a top down approach. It supports the definition of small agents which communicate with each other by sending contracts to start activities of other agents. If more than one agent can perform the needed activity, bids are sent and the 'cheapest' agent will get the contract. Each agent has a set of defined states and executes the guarded actions of this state. It interacts via generic sub-agents with physical input/output channels, timers and other specific hardware. As in object oriented programming the agent executes the contract like a method without showing the real implementation.

A complete MAD program starts with the declaration of the available target processors and is followed by a number of agents (see examples in chapter 4). The complete definition of an agent in MAD is divided into the following parts:

– Declaration of the target microprocessor, the agent will be executed on

– Instantiation of sub-agents used by the agent to interface to the hardware and to special functions

- Declaration of bids for contracts
- Declaration of contracts, the agent will accept
- Action part

2.1 Target definition

The target processor chosen within the network to execute the agent is defined after the keyword **target**. More than one agent may of course run on one processor. The final distribution is determined both by the physical input/output channels used by the agent and connected to the processor, and by the load on one processor resp. the response time needed by the agent. This assignment can be changed at any time; only recompiling is necessary to get a new running system.

2.2 Sub-agents

At lowest level generic sub-agents are defined to realize the interface to the technical process (e.g. digital and analog input/output, timer, stepper motor, pid-controller). They are instantiated in the **decls** definition of a MAD–program with the actual i/o-address and mnemonic identifiers to enhance self documentation of the program. Consequently, local variables and arrays are also declared as sub-agents. Sub-agents written in C may be added to the libraries.

2.3 Contracts and bidding

The only interface between agents is the contract protocol. In the **contracts** definition every contract which can be called by other agents is listed. A contract transfers parameters and causes the execution of actions or in most cases a state transition in the agents action part. Included in the communication system is a contract net bidding protocol. If more than one agent can execute a contract each agent sends its cost to perform the activity of the contract, computed by a cost function supported in the **bids** definition. The runtime system chooses the 'cheapest' agent and sends the contract to it.
Sending a contract only starts the activity of another agent but does not wait for its completion. Therefore, deadlocks in contract calls cannot occur. It is the duty of the programmer to avoid cyclic dependencies of calls to contracts e.g. by using strictly hierarchical dependencies of contracts.

2.4 States and actions

The action part is subdivided into a set of states. Control tasks usually change between different states of operation, e.g. at lowest level 'on' or 'off', at higher level 'open', 'closed' or 'error'. One state is active and the actions comprising it are executed. Each action is bound by a condition (guard) and only if it is true the corresponding statements are executed. At lower level these conditions will be signals from the controlled process, at higher level it may be timers or conditional expressions on variables. There is also a special condition once to ensure that the following statements are executed only once when entering the corresponding state. All other conditions in one state are tested cyclically and all agents fixed to one processor are executed one after the other to guarantee an exactly predictable time behavior. Special agents may be defined for time consuming computations which get a specified time slice per cycle. Two distinguished states are obligatory on each agent: At start up the agent goes into the initial state, in case of emergency the shutdown state is entered.

3 Implementation

The programming system MAD–RTS is hosted on a PC with MS-DOS. It contains the compiler for MAD and code generators and run time systems for different targets. The compiler is written in the object-oriented language C++. The syntax of MAD–RTS is defined in YACC, for lexical analysis the tool 'FLEX', for syntactical analysis the tool 'BISON' is used. Therefore, the compiler can easily be extended by additional language features as well as additional code generators and sub-agents. The compiler uses the contract specification to generate automatically the code for the transmission of contracts between agents on the same or on different targets. A small run time kernel is added which executes sequentially all agents on one target and realizes the physical communication in the network.

Code generators are available for Intel80x86, MC68HC11 and programmable logic controllers; the communication link is implemented for RS 232 serial link and the CAN–field–bus. If only digital input/output, no numeric expressions and no parameter passing to contracts are used, a further code generator is available which translates an agent program into a table of a finite automaton [7] which can easily be transcribed into VHDL and put into the hardware of a programmable logic device (PLD).

Figure 1 shows the generation of a MAD–RTS application, a small plant with one robot, a machine tool and a conveyer belt. At source level the target of each agent is defined. The agents are compiled into intermediate code which is translated according to the target definition to the dedicated hardware and linked with the run time kernel and the needed libraries. The resulting code is then loaded into the target or burned into a PLD. The resulting distributed multi-agent system is now ready to run; the agents communicate via CAN-Bus or RS232.

No explicit programming of this communication between agents is necessary. Each agent can be shifted to other hardware with only minor changes in the definition part of the agent. Likewise, the communication links are automatically altered without the need for changes in the application program. A contract net protocol with simple bidding is embedded into the runtime system kernel. Specification and coding of the control programs are closely related. The system engineer can design the control application at first according to problem ori-

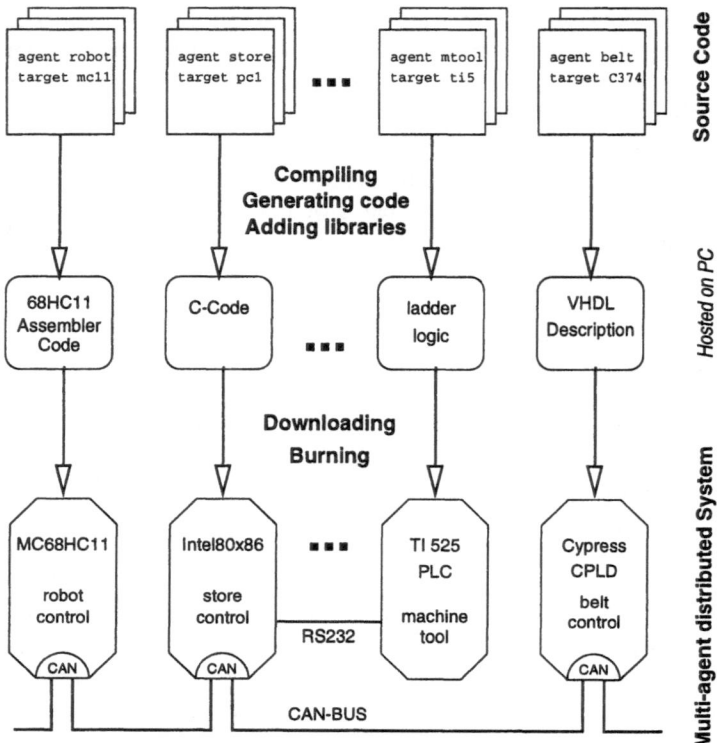

Fig. 1. Generation of a MAD–RTS application

ented criteria and afterwards decide about the optimal hardware structure for his application.

4 Programming examples

To test the MAD–System on a real platform we use a FischerTechnik model of a twin elevator with four floors. The model has the complete functionality of a twin elevator with all keys, switches, lamps, displays and motors. It is controlled by one microprocessors MC68HC11 on each floor, one in each cage and one at each motor platform, all connected via the CAN–field–bus (see figure 2).

4.1 Door agent

The implementation of the door agent shows an example of a MAD program. Every agent may send the contract open in order to induce the door agent to open the elevator door, wait for 10 seconds and close it again. If the light barrier is interrupted during closing, the door is opened again. If the door is closed, the contract start is sent to agent elevator1.

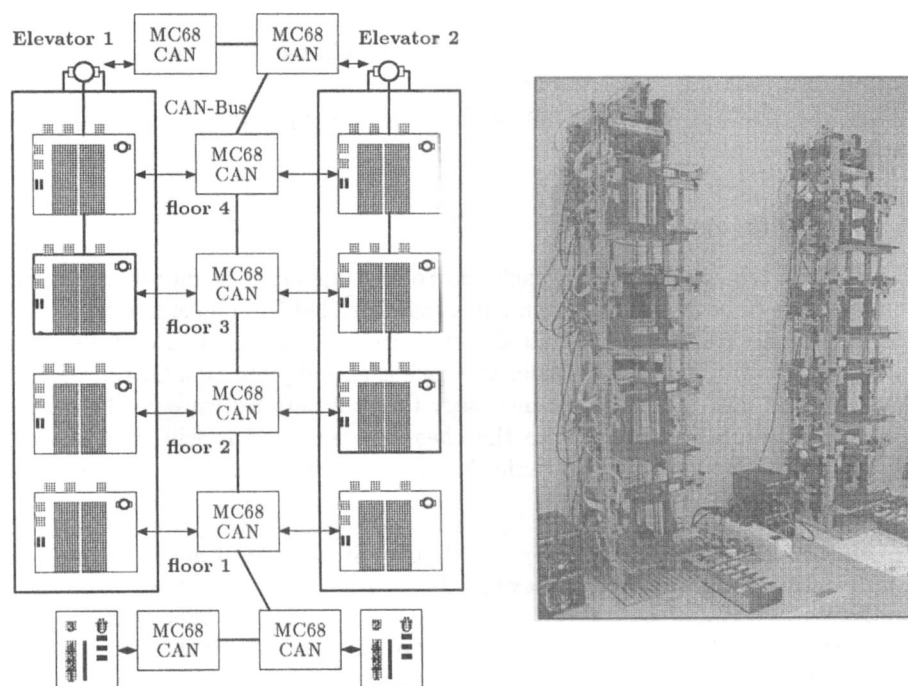

Fig. 2. Microprocessor network for twin elevator and a View of the model

```
targetdecl mc68hc11 mcfloor2; % target definition
agent door2 target mcfloor2;  % door agent elevator1, 2nd floor
decls                         % declaration of subagents
  DigOut motor(out1,on,off);
  DigOut direction(out2,open,close);
  DigIn  open_key(in1,on,off);
  DigIn  closed_key(in2,on,off);
  DigIn  light_barrier(in7,interrupted,ok);
  Timer  delay;
contracts                     % accepted contract
  open do newstate opening;
states                        % start of action part
  closed/shutdown:
    once          => motor.off;
  opening:
    once          => {direction.open; motor.on;}
    open_key.on   =>  newstate waiting;
  waiting:
    once          => {motor.off; delay(10000);}
    delay.tout    =>  newstate closing;
  closing/initial:
```

```
once             => {direction.close; motor.on;}
closed_key.on => {newstate closed; elevator1.start;}
light_barrier.interrupted
                 => {motor.off; newstate opening;}
endagent;
```

4.2 Bidding agent

There exist two similar agents each controlling one elevator motor. A contract may be sent to both elevator agents if a cage is called from a certain floor. Both elevator agents compute their actual cost to go to this floor based on their actual state and the difference to the floor the cage is actually located. Of course, more sophisticated cost functions are necessary to get a better strategy. The runtime system will send the contract to the cheaper agent and it will send the cage to this floor. The following example shows the relevant parts for bidding in the definition of the contract:

```
targetdecl mc68hc11 mcelev1;    % target definition
agent elevator1 target mcelev1; % motor agent elevator1
decls                           % declaration of subagents
  varinteger aktfloor;          % for a local variable
  arraybool(4) stop;            % and a local array
bids                            % cost function for contract get
  get (integer floor)
    cost ((instate busy)*4 + abs(floor-aktfloor));
contracts                       % accepted contract
  get (integer floor)
    do stop(floor).set(true);
states                          % start of action part
    .
    busy:                       % motor busy
    .
endagent;
```

An agent controlling the keys at the second floor will send the following contract to call one of both elevators:

```
elevator1 | elevator2.get(2)
```

5 Results

Due to the strictly cyclic processing you get the advantages of a distributed multi–agent programming system combined with the exact preview of the maximal delay time until the acceptance of a new signal. For hard real-time requirements a 'hardware agent' with zero delay can be created using the VHDL code generator. The strict cyclic approach produces a minimum of overhead and results in fast reaction times; on the micro-controller MC68HC11 (8 MHz) the periodic execution time for the model elevator is between 3 to 10 msec, including the transfer on the CAN-Bus. The produced code is very small, max. 20 KBytes for one controller.

The complex program can be easily tested: A monitor shows the actual states of each agent and the contracts sent between the agents. Reaction on errors can be embedded into the guarded actions and fault tolerance can be realized by distributing tasks on different hardware. Our experience showed that even severe errors in one agent don't lead to a total breakdown of the controlled process.

6 Conclusions

Reactive real–time programming with distributed agents covers a wide range of distributed real-time applications, supporting conventional PC based applications, micro-controllers and programmable logic controllers, down to hardware/software codesign. It introduces the notion of multi–agent systems, the security of state driven design and the flexibility of distributed multi–processing. Because of its strictly cyclic execution, exact response times can be guaranteed. Problems with priorities, task scheduling and interrupts don't occur. A redesign of the system is planned embedding the MAD-RTS into the object oriented language JAVA. This promises more flexibility, portability and a wider acceptance.

References

1. S. Hahndel and P. Levi: A Distributed Task Planning Method for Autonomous Agents in a FMS. *Proc. IEEE/RSJ/GI Int. Conf. on Intelligent Robots and Systems (IROS '94) Sept. 12-16, 1994, München, Germany.* Los Alamitos, CA: IEEE Computer Society Press, pp. 1285–1292, 1994..
2. Halbwachs, N.: Synchronous Programming of Reactive Systems. Kluwer Academic Publishers, Dordrecht 1993.
3. H.W. Lawson: Parallel Processing in Industrial Real-Time Applications. Prentice Hall, Englewood Cliffs, NJ., 514 p., 1992
4. J. P. Müller: The design of intelligent agents. Lecture notes in computer science; 1177 : Lecture notes in artificial intelligence, Springer, Berlin, 227 p., 1996.
5. G. Schrott: An Experimental Environment for Task-Level Programming of Robots. *Proceedings of the 2nd Int. Symposium on Experimental Robotics,* Toulouse, Juni 25-27, 1991. R. Chatila (Ed.), Lect. Notes in Control and Information Sciences 190, Springer, Berlin, pp. 196-206, 1992.
6. G. Schrott: A Multi-Agent Distributed Real-Time System for a Microprocessor Field-Bus Network. *Proc. of 7th Euromicro Workshop on Real-Time Systems, Juni 14–16, 1995, Odense, Denmark.* IEEE Computer Society Press, Los Alamitos, California, pp. 302-307, 1995.
7. G. Schrott and T. Tempelmeier: Putting Hardware-Software Codesign into Practice. *Proc. of 22nd IFAC/IFIP Workshop on Real-Time Programming, Sept 15-17, 1997, Lyon, France,* to be published.

An ML-Like Module System
for the Synchronous Language SIGNAL

David Nowak, Jean-Pierre Talpin, Thierry Gautier and Paul Le Guernic

IRISA (INRIA-Rennes & CNRS) Campus de Beaulieu, F-35000 Rennes

Abstract. Synchronous languages, such as SIGNAL, are best suited for the design of dependable real-time systems. Synchronous languages enable a very high-level specification and an extremely modular implementation of complex systems by structurally decomposing them into elementary synchronous processes. Separate compilation in reactive languages is however made a difficult issue by global safety requirements. To enable separate compilation of the functional components of reactive systems while preserving their global integrity, we introduce a module system for SIGNAL. Just as data-types describe the invariants of program modules in functional languages, temporal and data-flow invariants interface SIGNAL processes to their environment. In conventional languages, typing is the medium allowing the separate compilation of functions in a program. In SIGNAL, the notion of conditional data-flow graph can similarly be used for separately compiling reactive processes and for assembling them in complex systems. Following this principle, we present the first design and implementation of a polymorphic type system and of a module system for the synchronous language SIGNAL.

1 Introduction

A reactive system is a computer system which continuously *reacts* to its environment. Many industrial systems are reactive in nature: process control systems, monitoring systems, signal processing systems, communication protocols. Reactive systems are commonly characterized by critical requirements such as fast reaction time or bounded memory usage. Classical design tools for implementing reactive systems, such as real-time operating system or general-purpose concurrent languages (e.g. ADA), neither provide a global and formal view of the system (separated into tasks or services) nor preserve its determinism. Synchronous languages, such as SIGNAL [2], LUSTRE [4] or ESTEREL [3], STATEMATE [5], are specifically designed to ease the development of reactive systems by providing both a formal view and a logical notion of concurrency preserving determinism: *synchronous* concurrency, where operations and communications are instantaneous. In a synchronous language, concurrency is meant as a way to logically decompose the description of a system into a set of elementary communicating processes. Interaction between concurrent components within the program is conceptually performed by broadcasting events. In practice, a synchronous program is usually translated into a circuit or into a monolythic automaton. The hypothesis of synchrony is translated into the requirement that the program reacts rapidly to its environment. As a result, synchronous languages allow a very high-level specification and an extremely modular implementation of complex systems by structurally decomposing their functional components into elementary processes. Although modularity is a key advantages of synchronous languages, separate compilation is made a difficult issue by the requirements of proving global safety properties of the system. To enable separate compilation of the functional components of reactive systems while preserving their global integrity, we introduce a module system for SIGNAL. Just as data-types describe the invariants of program modules in functional languages, temporal and data-flow invariants interface SIGNAL processes to their environment. In conventional languages, typing is the medium enabling the separate compilation of the functions of a program. In SIGNAL, conditional data-flow graphs can similarly be used for separately compiling reactive processes.

2 An Overview Of SIGNAL

SIGNAL is an equational synchronous programming language: a SIGNAL program is modularly organized into processes consisting of simultaneous equations on signals. In SIGNAL, an equation is an elementary and instantaneous operation on input signals which defines an output. A signal is a sequence of values defined over a totally ordered set of instants.

Syntax The syntax of SIGNAL is defined in the figure 1. An expression e is either a value v (event, boolean, integer, real, etc), a signal x, the reference $x\$1$ to the previous value of a signal x, the down-sampling x when y of the signal x when y is present and true, the deterministic merge x default y of two signals x and y, or a synchronous operation $o(x_{1..n})$ (e.g. boolean, numerical operations). A process p is either an equation $x := e$, the synchronous composition of two processes, or a call $y_{1..m} := f(x_{1..n})$ to a process f with input sequence $x_{1..n}$ and output sequence $y_{1..m}$. A declaration p/x or d defines either a local signal x or a process f of inputs $x_{1..n}$, outputs $y_{1..m}$ and body p.

v	value	$d ::= \textbf{process } f(?\ x_{1..n}\ !\ y_{1..m})\ p$	definition
x, y, z	signal	$e ::= v \mid x \mid x\$1 \text{ init } v \mid x \text{ when } y \mid x \text{ default } y \mid o(x_{1..n})$	expression
f	process	$p ::= (x := e) \mid (p \mid p') \mid y_{1..m} := f(x_{1..n}) \mid p/x \mid p \textbf{ where } d$	process

Fig. 1. Syntax of SIGNAL

Dynamic Semantics The dynamic semantics of SIGNAL, written $p \xrightarrow{m} p'$, presented in [2], outlined in the figure 2, describes how a process p evolves over time. A transition in the dynamic semantics defines an instant. Each instant is characterized by an instantaneous transition from a state p to a state p' and by a set m of simultaneous events (written $x(v)$ for x present with v or $x(\bot)$ for x absent).

$$\frac{p \xrightarrow{m} p'' \quad p' \xrightarrow{m'} p''' \quad m \cap m'}{p \mid p' \xrightarrow{m \cup m'} p'' \mid p'''} \qquad \frac{p \xrightarrow{m \cup x(v)} p'}{p/x \xrightarrow{m} p'/x}$$

$x := v$	$\xrightarrow{x(v)}$	$x := v$	$x := y$ when z	$\xrightarrow{x(\bot)\ y(\bot)}$ $x := y$ when z
$x := o(y, z)$	$\xrightarrow{x(\bot)\ y(\bot)\ z(\bot)}$	$x := o(y, z)$	$x := y$ when z	$\xrightarrow{x(v)\ y(v)\ z(t)}$ $x := y$ when z
$x := o(y, z)$	$\xrightarrow{x(o(v,v'))\ y(v)\ z(v')}$	$x := o(y, z)$	$x := y$ when z	$\xrightarrow{x(\bot)\ y(v)\ z(f)}$ $x := y$ when z
$x := y\$1 \text{ init } v$	$\xrightarrow{x(\bot)\ y(\bot)}$	$x := y\$1 \text{ init } v$	$x := y$ default z	$\xrightarrow{x(v)\ y(v)}$ $x := y$ default z
$x := y\$1 \text{ init } v$	$\xrightarrow{x(v)\ y(v')}$	$x := y\$1 \text{ init } v'$	$x := y$ default z	$\xrightarrow{x(v)\ y(\bot)\ z(v)}$ $x := y$ default z
$x := y$ when z	$\xrightarrow{x(\bot)\ z(\bot)}$	$x := y$ when z	$x := y$ default z	$\xrightarrow{x(\bot)\ y(\bot)\ z(\bot)}$ $x := y$ default z

Fig. 2. Dynamic semantics of SIGNAL

Parallel composition $p \mid p'$ synchronizes the events m and m' produced by p and p'. The relation $m \cap m'$ is defined iff. $m(x) = m'(x)$ for all x in $dom(m) \cap dom(m'))$. A synchronous operation $x := o(y, z)$ instantaneously computes the value of o by y and z and outputs it to the signal x. A delay $x := y\$1$ init v stores the value v' of y and outputs the previous value v to x. A merge $x := y$ default z outputs the value of y to x when y is present (z is not needed) and the value of z otherwise (y is absent). A sampling $x := y$ when z outputs the value of y to x when z is present and true. When the inputs of an equation are absent, a transition takes place but no value is given to its output. The dynamic semantics of a definition p where process $f(? x_{1..n} ! x_{n+1..m}) \, p'$ is equivalent to that of p where every expression $y_{n+1..m} := f(y_{1..n})$ is executed as $p'[y_{1..m}/x_{1..m}]$.

Compilation The compilation of SIGNAL requires the static resolution of boolean equations where signals are abstracted by clocks. The SIGNAL compiler ensures the respect of the synchronization constraints expressed in programs by computing temporal relations between signal clocks and analyzing the conditional data dependencies between signals. The control model of a SIGNAL program is represented by a set of temporal relations $\hat{x} = c$ between signal clocks $\hat{x}$ and expression clocks c. The clock $\hat{x}$ denotes the instants when x is present. The clock $[x]$ denotes the presence of a boolean signal x with the value true. The clock $\hat{x} \backslash \hat{y}$ denotes the instants where x is present and y absent. The formula $c \wedge c'$ and $c \vee c'$ denote the conjunction and disjunction of the instants c and c'. The data-flow model of a program is represented by a graph composed of arrows $x \xrightarrow{c} y$. An arrow $x \xrightarrow{c} y$ denotes a dependency from x to y at the clock c. References to local signals x in a graph g are bound by existential quantification $\exists x.g$. We write $fv(g)$ and $bv(g)$ for the free and bound signals of g. We write $\exists y.g = \exists x.(g[x/y])$ and $(\exists x.g') \cup g = \exists x.(g \cup g')$ iff. $x \notin fv(g) \cup bv(g)$. In most cases, references to bound signals in a graph g can be eliminated by determining its transitive closure on input/output signals[1].

$$c ::= \hat{x} \mid \hat{x}\backslash\hat{y} \mid [x] \mid c \wedge c \mid c \vee c \qquad g ::= \emptyset \mid g \cup (\hat{x} = c) \mid g \cup (x \xrightarrow{c} y) \mid \exists x.g \mid g \cup g'$$

Fig. 3. Conditional data-flow graphs g

Clock Calculus The clock calculus of SIGNAL, defined by the proof system $G \vdash p \Rightarrow g$ of the figure 4, determines the conditional data-flow graph g of programs[2].

$$G \vdash x := y\$1 \Rightarrow (\hat{x} = \hat{y}) \qquad\qquad G \vdash x := y \text{ when } z \Rightarrow (y \xrightarrow{[z]} x, \hat{x} = \hat{y} \wedge [z])$$

$$G \vdash x := o(y_{1..n}) \Rightarrow \begin{pmatrix} y_i \xrightarrow{\hat{y}_i} x \\ \hat{x} = \hat{y}_i \end{pmatrix}_{i=1..n} \qquad G \vdash x := y \text{ default } z \Rightarrow \begin{pmatrix} y \xrightarrow{\hat{y}} x, z \xrightarrow{\hat{z}\backslash\hat{y}} x \\ \hat{x} = \hat{y} \vee \hat{z} \end{pmatrix}$$

$$\frac{G \vdash p \Rightarrow g \quad G \vdash p' \Rightarrow g'}{G \vdash p \mid p' \Rightarrow g \cup g'} \qquad\qquad \frac{G \vdash p \Rightarrow g}{G \vdash p/x \Rightarrow \exists x.g}$$

$$\frac{f : \forall x_{1..m}.g \in G}{G \vdash y_{n+1..m} := f(y_{1..n}) \Rightarrow g[y_{1..m}/x_{1..m}]} \quad \frac{G \vdash p' \Rightarrow g' \quad G \cup \{f : \forall x_{1..m}.g'\} \vdash p \Rightarrow g}{G \vdash p \text{ where process } f(? x_{1..n} ! x_{n+1..m}) \, p' \Rightarrow g}$$

Fig. 4. Clock calculus $G \vdash p \Rightarrow g$

---

[1] $g|_x = g\backslash g_x \backslash g^x \cup \{(y \xrightarrow{c \wedge c'} z) \mid (y \xrightarrow{c} x, x \xrightarrow{c'} z) \in g_x \times g^x\}$ is the closure of g w.r.t. x where $g^x = \{x \xrightarrow{c} y \in g\}$ and $g_x = \{y \xrightarrow{c} x \in g\}$

[2] The graph environment G associates process names f to process specifications $\forall x_{1..n}.g$

Using the graph g produced by the clock calculus, the SIGNAL compiler generates an optimal compile-time scheduling of the actions specified in the source program by hierarchizing temporal relations in g and by ruling the execution of the program using its master clock [1]. Using the graph g, causal dependencies in the source program can easily be detected as constrained boolean conditions on signals (e.g. $[x] = \hat{y}$) or cyclic data dependencies (e.g. $x \xrightarrow{c} x$). This allows proving the absence of dead-locks in the program.

Example As the conditional data-flow graph of a SIGNAL program is the essential medium for checking its safety and compiling it, separate compilation is made a difficult issue by the requirements of proving global safety properties. To illustrate this issue, let us consider a typical situation raised in the following process definition. Let copy be the process which assigns the value of its input signals a and b to its output signals x and y: (x:=a | y:=b). To compile it, the SIGNAL compiler has the choice between scheduling either if present(a) x:=a; if present(b) y:=b or if present(b) y:=b; if present(a) x:=a. However, the appropriate choice depends on the way the process is invoked. For instance, if we write (u,v) := copy(w,u), only the first scheduling (i.e. if present(w) {u:=w; v:=u}) is correct (the second dead-locks). Fortunately, this problem can be solved by determining the data dependencies $a \rightarrow x$ and $b \rightarrow y$ between (a,b) and (x,y) where copy is defined and the actual data dependencies $w \rightarrow u$ and $u \rightarrow v$ between u, v and w where copy is used. In a situation of separate compilation that information is however not directly accessible by the textual definition of the process copy. In this case, the explicit declaration of the temporal and data-flow relations between signals appears to be necessary for separate compilation.

3 Polymorphic Type Inference in SIGNAL

In this section, we propose a polymorphic type system for SIGNAL. The type system complements the clock calculus of SIGNAL with the static determination of signal types. Just as the clock calculus allows the detection of non-reactive or dead-lockingSIGNAL programs, the type inference system allows the detection of erroneous access to data-structures. For instance, consider the definition of the generic identity function id applied to either integers or booleans below. The SIGNAL compiler [6] handles this specification by textually substituting the definition of the generic process id everywhere the process is used. This results in constructing a program in elementary syntax: "a:=true | b:=1". Type checking is then performed on the elementary syntax. However, if the process id is separately compiled, this translation is not possible. The introduction of type polymorphism is the solution to this issue. Just as the conditional data-flow graph $x \xrightarrow{a} y$ s.t. $\hat{x} = \hat{y}$ generically represents the behavior of id, the polymorphic type $\forall \alpha. \alpha \rightarrow \alpha$ characterizes the type of all its possible instances.

```
(a := id(true) | b := id(1)) where process id (? x ! y) (y := x)
```

Type System The type system of SIGNAL has a rich structure to handle the variety of data structures and formats of real-time systems. Some predefined data-types are given in the figure 5 in order to address an important issue: subtyping. The smallest signal type t is the event and denotes a boolean which is always true when present. A type variable α denotes a generic signal type.

The types t of the signals manipulated by a process are subject to simple[3] subtyping constraints denoted by a set C. The subtyping relation, written "$t \leq t'$", reads "the signal type t is a subtype of t'" and is defined as follows.

$$t ::= \text{event} \mid \text{boolean} \mid \text{short} \mid \text{integer} \mid \text{long} \mid \text{real} \mid \text{double} \mid \alpha \qquad \text{signal types}$$

$$\tau ::= t_{1..n} \to t'_{1..m} \quad \text{process types} \qquad A ::= \emptyset \mid A \cup \{x{:}t\} \mid A \cup \{f{:}\sigma\} \quad \text{assumptions}$$
$$\sigma ::= \tau/C \mid \forall\alpha.\sigma \quad \text{generic types} \qquad C ::= \emptyset \mid C \cup \{t \leq t\} \qquad \text{constraints}$$

Fig. 5. Type system

$$C \supset \text{short} \leq \text{integer} \qquad C \supset \text{integer} \leq \text{long} \qquad C \supset \text{real} \leq \text{double} \qquad C \supset t \leq t$$

$$C \cup \{t \leq t'\} \supset t \leq t' \qquad \frac{C \supset t \leq t' \quad C \supset t' \leq t''}{C \supset t \leq t''} \qquad \frac{C \supset t \leq t' \quad C \supset t' \leq t}{C \supset t = t'}$$

Fig. 6. Subtyping relation

Type Polymorphism The type τ of a process is a map $t_{1..n} \to t'_{1..m}$ from inputs types $t_{1..n}$ to output types $t'_{1..m}$. A polymorphic process type σ is a process type τ universally quantified over the set of its generic signal type variables α (eventually constrained by a set C of subtyping relations). Assumptions A in the type inference system declare signal types "$x{:}t$" and process types "$f{:}\sigma$". The generalization and the instantiation of process types is defined in the figure 7. The function fv gives the set of free signal variables α of a term (a type t, a polymorphic type σ or an environment A). We write $\overline{C}$ for the transitive closure of C for the relation $\leq$ (i.e. $t \leq t'' \in \overline{C}$ for all $(t \leq t'), (t' \leq t'') \in C$). We write $\overline{C}_{\alpha_{1..n}}$ for the restriction of $\overline{C}$ to all constraints referencing $\alpha_{1..n}$.

$$gen(A, C, \tau) = \forall\alpha_{1..n}.(\tau/\overline{C}_{\alpha_{1..n}}) \text{ iff. } \alpha_{1..n} = (fv(\tau) \cup fv(C)) \backslash fv(A)$$
$$\tau/C \preceq \forall\alpha_{1..n}.(\tau'/C') \text{ iff. } \tau = \tau'[t_{1..n}/\alpha_{1..n}] \text{ and } C \supset C'[t_{1..n}/\alpha_{1..n}]$$

Fig. 7. Type generalization and instantiation

Type Inference System The type inference system of SIGNAL is defined in the figure 8 by the sequent $A, C \vdash p$ which reads "p is well-typed with the assumptions A and constraints C". The auxiliary sequent $A, C \vdash e{:}t$ checks that the expression e has signal type t under the assumptions A and constraints C.

$$\frac{A(x) = t}{A, C \vdash x{:}t} \qquad \frac{\mathcal{R}(v) = t \quad A, C \vdash x{:}t}{A, C \vdash x\$1 \text{ init } v{:}t} \qquad \frac{A, C \vdash e{:}t \quad C \supset t \leq t'}{A, C \vdash e{:}t'} \qquad \frac{A, C \vdash x{:}t \quad A, C \vdash y{:}t}{A, C \vdash x \text{ default } y{:}t}$$

$$\frac{\mathcal{R}(v) = t}{A, C \vdash v{:}t} \qquad \frac{A, C \vdash x{:}t \quad A, C \vdash y{:}\text{boolean}}{A, C \vdash x \text{ when } y{:}t} \qquad \frac{A, C \vdash x{:}t \quad A, C \vdash e{:}t}{A, C \vdash x := e} \qquad \frac{A \cup \{x{:}t\}, C \vdash p}{A, C \vdash p/x}$$

$$\frac{A, C \vdash p \quad A, C \vdash p'}{A, C \vdash p \mid p'} \qquad \frac{(A, C \vdash x_i{:}t_i)_{i=1..m} \quad t_{1..n} \to t_{n+1..m}/C \preceq A(f)}{A, C \vdash x_{n+1..m} := f(x_{1..n})}$$

$$\frac{A \cup \{x_i{:}t_i, i = 1..m\}, C \vdash p' \quad A \cup \{f{:}gen(A, C, t_{1..n} \to t_{n+1..m})\}, C \vdash p}{A, C \vdash p \text{ where process } f(?\ x_{1..n}\ !\ x_{n+1..m})\ p'}$$

Fig. 8. Type inference system

---

[3] Note that signal types, and only signal types t, are referenced in subtyping constraint sets C. Hence, each constraint references either ground types (e.g. event) or type variables α.

The function $\mathcal{R}$ relates each constant to its minimal predefined type. The type inference system proceeds by structurally scanning expressions e and processes p to determine output signal types t and subtyping constraints C. All rules but the rule of subtyping apply on a specific syntactic category. Every expression rule $A, C \vdash e : t$ explicits a type containment constraint. For instance, the rule for a merge statement "x default y" says that the type of both x and y must match t for the typing of the expression to be preserved. The rule of subtyping is however applicable either to x or to y in order to eventually coerce types to a common upper bound. The rules for processes p, written $A, C \vdash p$ differ in that they just check that the process p agrees with a set of assumptions A and a set of constraints C i.e. that A and C are sufficient to give a type to every expression in p. Process definitions and applications make use of type polymorphism in order to represent and use all the possible types of a process. A type reconstruction algorithm can easily be derived from the inference system of the figure 8 by composing every expression rule $A, C \vdash e : t$ with the subtyping rule and by introducing a fresh signal type α wherever a type t is introduced or referenced in the expression rules. In addition, some simplifications rules for constraints sets can be implemented by removing trivial constraints (e.g. constraints between ground types or constraints on unreferenced type variables) and by substituting constraints between type variables α and minima (e.g. $\alpha \leq$ event) or maxima (e.g. boolean $\leq \alpha$). The definition and proof of a type reconstruction algorithm is presented in [7].

Correctness of the Inference System The correctness of the type inference system with respect to the dynamic semantics is proven by showing that typing is an invariant of process execution (theorem 1). We say that an event map m is consistent with A and C, written $A, C \vdash m$, iff. $A, C \vdash m(x) : A(x)$ for all $x \in dom(m)$. We write $out(p)$ (resp. $in(p) = fv(p) \setminus out(p)$) for the input/output signals of p.

Theorem 1. *If $A, C \vdash p$, $A, C \vdash m_{in(p)}$ and $p \xrightarrow{m} p'$ then $A, C \vdash p'$ and $A, C \vdash m_{out(p)}$.*

4 A Module System for SIGNAL

In the previous sections, we did present the clock calculus and the polymorphic type inference system of SIGNAL. Just as data-types describe the invariants of program modules in functional languages, temporal and data-flow invariants interface SIGNAL processes to their environment. In conventional languages, typing is the medium enabling the separate compilation of the functions of a program. In SIGNAL, conditional data-flow graphs can similarly be used as a medium for separately compiling reactive processes. The module system presented in this section put this principles to work by defining a notion of specification and of implementation for all components of a program. In addition, modules can be parameterized by other modules, in a way which was first introduced in STANDARD ML [8, 9] and CAML [10]. The integration of the principles of type polymorphism and modules to SIGNAL form a system which considerably augments the expressive power of SIGNAL and generalizes its underlying programming methodology. We present a generalized notion of signature subtyping which allows characterizing the refinement of modular components of reactive systems. This notion of refinement makes use of the notion of binary decision diagrams in order to prove that the invariants expressed in the signature of a module are more general than the invariants inferred from its implementation.

Module Language The syntax of the language which implements the module system of SIGNAL is defined in the figure 9. It consists of signatures S, modules M and functors F. The figure 9 defines the syntax of programs P as a sequence of signatures S, functors F and module declarations M. A signature s gives the specification or interface of a module in terms of type declaration and process specification. A process specification "process $f(? x_1:t_1..!..x_n:t_n)/C$ spec g" consists of the declaration of its input/output signal types $t_{1..n}$ and of its temporal and data-flow invariants modeled by the conditional data-flow graph g. A functor declaration is a module object parameterized over a sequence of modules. A module declaration gives the implementation of a program component.

$$
\begin{aligned}
P ::= &\ \text{signature } S = s \text{ end} \\
&|\ \text{functor } F(M_{1..n}:S_{1..n}) = m \text{ end} \\
&|\ \text{module } M = F(M_{1..n}) \\
&|\ \text{module } M = m \text{ end} \\
&|\ P; P'
\end{aligned}
\qquad
\begin{aligned}
s ::= &\ s; s' \\
&|\ \text{type } T \\
&|\ \text{type } T = t \\
&|\ \text{process } f(? x_1:t_1..!..x_n:t_n)/C \text{ spec } g \\
m ::= &\ \text{type } T = t \mid d \mid m; m'
\end{aligned}
$$

Fig. 9. Syntax of the module system

Subtyping of Signatures The key principle of a module system is that the interface or signature of a module should contain enough information to allow separate verification and implementation of other modules using it. The notion of signature matching and subtyping is central for checking the correctness of this information flow. In conventional languages, such as STANDARD ML [8, 9], this amounts to defining a relation $s \leq s'$ between signatures which specifies the agreement protocol between the interface of a module and its implementation.

$$
A \vdash \text{type } T \leq \text{type } T \qquad A \vdash \text{type } T = t \leq \text{type } T \qquad \frac{A \vdash t \simeq t'}{A \vdash \text{type } T = t \leq \text{type } T = t'}
$$

$$
\frac{A(T) = t}{A \vdash T \simeq t} \qquad \frac{t'_{1..n} \to t'_{n+1..n'}/C' \preceq gen(A, C, t_{1..n} \to t_{n+1..n'}) \quad g \leq g'[x_{1..n'}/y_{1..n'}]}{\begin{array}{c} A \vdash \text{process } f(x_{1..n}:t_{1..n} \,!\, x_{n+1..n'}:t_{n+1..n'})/C \text{ spec } g \\ \leq \text{process } f(? y_{1..n}:t'_{1..n} \,!\, y_{n+1..n'}:t'_{n+1..n'})/C' \text{ spec } g' \end{array}}
$$

Fig. 10. Subtyping of signatures

This relation is defined in the figure 10. Unlike for conventional modules systems, it does not reduce to checking the containment of type definitions and of process types declarations. Let us consider, as in the above figure, a process signature of declared data-flow graph g' and d be its implementation of data-flow graph g. Checking that the temporal and data-flow information propagated in the signature (i.e. g') contains that inferred in the implementation of the process (i.e. g) amounts to verifying the relation written $g \leq g'$.

Definition 2. For all g, $\vec{g} = [(x \xrightarrow{c} y) \in g]$ and $\hat{g} = [(\hat{x} = c) \in g]$. For all g and g', $g \leq g'$ iff. $\vec{g} \subseteq \vec{g'}$ and $\hat{g'} \Rightarrow \hat{g}$.

The relation $g \leq g'$ guaranties the safety of the interface of a module with respect to its implementation. The interface must declare any interaction performed in the implementation of a module in order to verify the safety of the use of the module.

The refinement of $\vec{g}$ by a bigger $\vec{g'}$ ensures that whenever $\vec{g'}$ is safe (it does not contain cyclic data dependencies $x \xrightarrow{c} x$ with non-trivial clock c) then so is $\vec{g}$. The refinement of $\hat{g}$ by $\hat{g'}$ means that every clock equation in $\hat{g}$ can be deducted from $\hat{g'}$. Hence, a trivial cycle $x \xrightarrow{c} x$ in g (i.e. s.t. $\neg c$) cannot be non-trivial in g'. Furthermore, a clock equation in g' cannot induce a causal dependency of the form $[y] = c$ in g.

Refinement Checking Checking satisfaction of the relation $g \leq g'$ amounts to solving the problem $\hat{g'} \Rightarrow \hat{g}$. As $\hat{g'}$ and $\hat{g}$ are systems of boolean equations, this problem can be expressed on $Z/2Z$ [1] as $\hat{g} : (P_i(\vec{x}) = 0)_{i \in I}$ and $\hat{g'} : (P'_j(\vec{x}) = 0)_{j \in J}$ where the formula P_i and P'_j (of index I and J) are polynomials on $Z/2Z$ and where $\vec{x}$ is the vector of signals defined in the system of equations. Now, let us define the boolean operator $\oplus$ as $a \oplus b = a + b + a.b$. A property of $\oplus$ is that $P_1(\vec{x}) = 0$ and $P_2(\vec{x}) = 0$ iff. $(P_1 \oplus P_2)(\vec{x}) = 0$. Using $\oplus$, the system of equations reduces to $\hat{g} : P(\vec{x}) = 0$ and $\hat{g'} : P'(\vec{x}) = 0$ where $P = \oplus_{i \in I} P_i$ et $P' = \oplus_{j \in J} P_j$. Let V and V' be the solutions of P and P', $V \subseteq V'$ iff. $\forall \vec{x} .(1 - P(\vec{x})).P'(\vec{x}) = 0$, which amounts to proving a tautology.

Signature Checking Having formally defined a notion of signature matching in our module system allows us now to define signature checking by the sequent $A \vdash P : A'$ in the figure 11.

$$A \vdash \text{signature } S = s \text{ end}: A \cup \{S : s\} \qquad A \cup \{M : s\} \vdash M : s \qquad A \cup \{S : s\} \vdash S : s$$

$$\frac{A(F) = (M'_{1..n} : s_{1..n}) \to s \quad (A \vdash M_i : s_i)_{i=1..n}}{A \vdash \text{module } M = F(M_{1..n}) : A \cup \{M : s[M'_{1..n}/M_{1..n}]\}} \qquad \frac{A \vdash P : A' \quad A' \vdash P' : A''}{A \vdash P; P' : A''}$$

$$\frac{A \vdash m : s}{A \vdash \text{module } M = m \text{ end}: A \cup \{M : s\}} \qquad \frac{A \cup \{T : t\} \vdash m : s}{A \vdash \text{type } T = t; m : \text{type } T = t; s}$$

$$\frac{(A \vdash S_i : s_i)_{i=1..n} \quad A \cup \{M_{1..n} : s_{1..n}\} \vdash m : s}{A \vdash \text{functor } F(M_{1..n} : S_{1..n}) = m \text{ end}: A \cup \{F : (M_{1..n} : s_{1..n}) \to s\}} \qquad \frac{A \vdash m : s' \quad A \vdash s' \leq s}{A \vdash m : s}$$

$$\frac{A \cup \{x_i : t_i\}_{i=1..n'}, C, G \vdash p \quad A, G \vdash p' \Rightarrow g}{A \cup \{f : gen(A, C, t_{1..n} \to t_{n+..n'})\}, G \cup \{f : \forall x_{1..n'}.g\} \vdash m : s}{A \vdash \text{process } f(? x_{1..n} ! x_{n+1..n'}) \ p; m : \text{process } f(? x_{1..n} : t_{1..n} ! x_{n+1..n'} : t_{n+1..n'})/C \text{ spec } g; s}$$

Fig. 11. Signature checking

$$\frac{A \vdash M : s; \text{type } T = t; s'}{A \vdash M.T \simeq t[M.T'/T']\text{type } T' = t' \in s} \qquad \frac{\text{process } \tilde{f}(? x_1 : t_1.. ! ..x'_n : t'_n)/C \text{ spec } g \in A(M)}{A, G \vdash y_{n+1..n'} := M.f(y_{1..n}) \Rightarrow g[y_{1..n'}/x_{1..n'}]}$$

$$\frac{A, C \vdash e : t \quad A \vdash t \simeq t'}{A, C \vdash e : t'} \qquad \frac{\text{process } f(? x_{1..n} : t_{1..n} ! x_{n+1..n'} : t_{n+1..n'})/C' \text{ spec } g \in A(M)}{A \cup \{M.f : gen(A, C \cup C', t_{1..n} \to t_{n+1..n'})\}, C \vdash y_{n+1..n'} := M.f(y_{1..n})}{A, C \vdash y_{n+1..n'} := M.f(y_{1..n})}$$

Fig. 12. Modular clock and type inference

The principle of the signature checker is to verify the correctness of a signature declaration w.r.t. the information inferred from its implementation, and then checking that it is correctly used elsewhere. The module checker sequentially analyses the signatures and modules declared in a program P, verifies that declarations in A match uses in P and then returns a set of assumptions A' augmented with the declarations made in the processed component P. The definition of the module checker centralizes the verification of the program by using the type inference system $A, C \vdash p$ to check the implementation of programs, and of the clock calculus $A, G \vdash p \Rightarrow g$ in order to verify their safety. In order to be usable in the module checker, the type and clock inference systems (figures 4 and 8) need to understand type declarations and references to processes made in separate modules (figure 12).

Correctness of the Module System In a conventional module system, the type interface of a program module should contains enough information to allow its separate verification and compilation. This requirement is usually implemented using a notion of signature matching, which checks the correctness of the type-information flow. Conventionally, implementing signature matching amounts to defining a relation $s \leq s'$ between module signatures. In our system, such a relation is extended to checking the containment of temporal and data-flow information using the relation $g \leq g'$. As a result, previous formal propositions for establishing the correctness of module systems [9, 10] hold. The correctness of our module system could be stated by showing that the decomposition of a system of modules P in a context A using an appropriate sequent translation $A \vdash P \Rightarrow A', C \vdash p$ produces the appropriate context (A', C) to check the corresponding program p well-typed.

5 Implementation

Since its first implementation, more than a decade ago, the SIGNAL compiler has been in constant evolution to integrate new principles of programming, compilation and verification. The most recent version of the compiler, SIGNAL V4 [6], implements a common format called DC+ which is the result of the SYNCHRON initiative. The type inferencer described in this paper has been integrated in SIGNAL V4. Detailed proofs and examples are available in [7]. The module system is currently implemented as a separate "*module processor*", to maximize the flexibility of its use. An important point on the compilation of SIGNAL is that, although it is best to perform type inference at an early stage of the compilation process, type information in the SIGNAL compiler is not needed until the very last step of the compilation process: code generation (usually C or VHDL). Our module processor has been implemented in such a way to allow the separate verification of program modules and functors "$M \Rightarrow g$". From the implementation of a functor F or module M, the module checker can produce a specification g which can then be checked by the SIGNAL compiler for safety. The separate compilation of a hierarchy of modules "$P \Rightarrow p$" is also supported. Since code generation is the very last step of the compilation for SIGNAL programs where type information is needed, all earlier verification and optimizations can be performed on the intermediate representation of a SIGNAL program while preserving the information declared in its signature. Experiments show that extending our module system with a notion of higher-order modules (i.e. by parameterizing functors with functors as in [10], for instance), would not be technically a problem. However, from a practical point of view, it would incur the introduction of a notion of *contravariant refinement*, which would certainly be very difficult to understand and to use. From a methodological point of view, such a feature would also not provide any significant improvement over the present system. Another direction for improvements in our module system is the representation of temporal and data-flow invariants of programs using conditional data-flow graph. We have seen that the interaction on local signals is represented using existential quantification. A general method for eliminating such quantifications would either be to introduce clock inequations or use reactive types, as advocated in [12].

6 Conclusion

We have presented the design and implementation of a polymorphic type inferencer and of a module system for the synchronous language SIGNAL. This system is inspired from principles introduced in previous implementations of functional languages such as STANDARD ML [8, 9] or CAML [10]. Our main contribution is to show that, just as data-types describe the invariants of program modules in functional languages, temporal and data-flow invariants interface SIGNAL processes to their environment: the notion of signature matching employed in conventional module systems naturally extends to a notion of signature refinement in synchronous languages.

References

1. T. P. Amagbegnon, L. Besnard and P. Le Guernic. Implementation of the data-flow synchronous language SIGNAL. In *Proceedings of the 1995's ACM Conference on Programming Language Design and Implementation*, p. 163–173. ACM, 1995.
2. A. Benveniste, P. Le Guernic and C. Jacquemot. Synchronous programming with events and relations: the SIGNAL language and its semantics. In *Science of Computer Programming*, v. 16, p. 103–149, 1991.
3. G. Berry and G. Gonthier. The Esterel synchronous programming language: design, semantics, implementation. In *Science of Computer Programming*, v. 19, p. 87–152, 1992.
4. N. Halbwachs, P. Caspi, P. Raymond and D. Pilaud. The synchronous data-flow programming language Lustre. In *Proceedings of the IEEE*, v. 79(9). IEEE Press, 1991.
5. D. Harel and A. Naamad. *The STATEMATE semantics of STATECHARTS*. I-Logix, 1995.
6. Thierry Gautier, Paul Le Guernic, and François Dupont. SIGNAL v4 : manuel de référence. Technical report n. 832. IRISA, 1994.
7. D. Nowak, Talpin, J.-P., and Gautier, T. Un système de modules avancé pour SIGNAL. Rapport de recherche n. ??, 1997. (To appear, available from nowak@irisa.fr).
8. R. Milner, M. Tofte, R. Harper. *The definition of STANDARD ML*. MIT Press, 1990.
9. M. Tofte. Principal signatures for higher-order program modules. In *Proceedings of the 1992's ACM Symposium Principles of Programming Languages*, p. 189-199. ACM, 1992.
10. X. Leroy. Applicative functors and fully transparent higher-order modules. In *Proceedings of the 1995's ACM Symposium Principles of Programming Languages*. ACM, 1995.
11. O. Maffeïs and P. Le Guernic. Distributed implementation of SIGNAL: scheduling and graph clustering. In *Symposium on Formal Techniques in Real-Time and Fault-Tolerant Systems*. Lecture Notes in Computer Science n. 863. Springer Verlag, 1994.
12. J.-P. Talpin. Reactive Types. In *Conference on the Theory and Practice of Software Development (TAPSOFT'97)*. Lecture Notes in Computer Science. Springer Verlag, 1997.

Synchronous Thread Management in a Distributed Operating System's Micro Kernel

Olivier Potonniée[1], Jean-Bernard Stefani[2]

[1] Alcatel Alsthom Recherche, Route de Nozay, 91460 Marcoussis, France,
[2] France Telecom/CNET, 38-40 av. du Gal Leclerc, 92794 Issy Moulineaux Cedex 9, France

Abstract. This paper describes an experiment in programming part of an operating system kernel using the Esterel synchronous programming language. Using a synchronous programming language allows the construction of provable, deterministic reactive systems. The paper describes and analyzes the small executive realized and the formal verification of some of its properties. It also presents how multiple interconnected instances of this executive can be synchronized, yielding a distributed real-time platform operating under a sparse-time model.

Key Words : Synchronous programming, distributed systems, thread management, real-time systems, deterministic systems

1 Introduction

Synchronous languages [2] such as Signal [3], Lustre [4], Esterel [5] or Argos [6], provide a means to program real time systems having a formal description of their (deterministic) behavior. One potential interesting area of application is the construction of a real time operating system kernel.

Current operating systems are built using standard languages, which make it difficult to formerly verify their behavior. In this paper, for the first time to best of our knowledge, we use a synchronous programming language, Esterel, to develop a small real time executive corresponding to the thread management function of the Chorus micro-kernel. We used the Mauto tool [7] to verify properties of our implementation through bisimulation reductions.

Based on our synchronous executive, we further implemented a ditributed computational model, originally proposed in the Saturn project [8]. The resulting distributed platform realizes a sparse time execution model as proposed by H. Kopetz [9], where events can only occur in specified intervals of time, thus simplifying the synchronization of distributed applications.

2 Overview of the prototype

We did not aim to reinvent thread management, but only to investigate a new way to realize it. We thus reproduced that of the Chorus [10] micro-kernel.

Thread management in Chorus comes with a specific scheduler which is preemptive and piority driven : threads having a priority higher than a given threshold can only be preempted by higher priority threads, while the others may share time with equal priority threads.

The analysis of an existing thread management is not an easy work : the documentation presents isolated possible states of a thread, but it misses information on the combination of those states. The kernel sources are usually the only way to get precise information on specific behavior. This is due to the large numbers of combined states that a thread can take (hundreds), that can't be documented and do not directly appear in classical programming languages. Esterel, being state oriented, provides a much simpler description of those states.

We decomposed Chorus thread management into four different components, corresponding to the four basic functions involved in thread management, namely interrupt management, handling of time, thread management proper, and scheduling. Each of these components corresponds in our prototype to one or more Esterel module(s). The structure of the prototype, together with the flow of communication between the different components, is shown in figure 1.

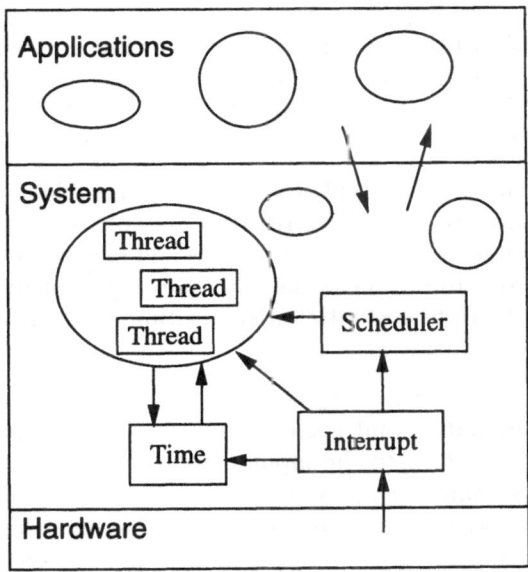

Fig. 1. Architecture of the synchronous process management

- **Interrupt**

  This component receives interrupt signals from the processor, and redirects them to the involved components. It contains two Esterel modules : one launches handlers to treat hardware interrupts, the other redirects software interrupts (traps) and exceptions to the concerned threads.

- **Time**
  The Time component handles the physical clock interrupt, and timers.
- **Scheduler**
  This component allocates the processor to the threads. It is divided in a generic module, and an interchangeable specific one, that implements the Chorus scheduling policy.
- **Thread**
  This is the largest component. It manages all possible states that a thread can take. Whereas each of the previous components was unique in the system, there is one instance of Thread per thread in the system.
  The number of states managed by this component being very large, it is decomposed in five modules : three of them handle states concerning different kinds of blocked states, and communicate a sub-state to a fourth module, that deduces if the thread is ready or not to execute. This information is transmitted to the scheduler. The fifth module handles transitions to and from system mode.

3 Implementation and verification

The compilation of an Esterel module produces an automaton coded in a set of C++ classes [12]. We call an instance of the automaton a *reactive object*. Ideally, all modules should be compiled together, providing a single object. However, this is not possible for two reasons :

- The resulting automaton would be so large that it would not even be possible to generate it due to a combinatorial explosion of states [3].
- Esterel does not allow dynamicity (which would require dynamic reconstruction of the automaton !). Thus the Thread component has to be compiled separately, so it can be instantiated at run time when new threads are created.

The interface of a reactive object provides methods to set input signals awaited by the automaton, and get output signals that it produces. An output signal can be connected to the input signal of another reactive object, ensuring synchronous communications between separately compiled modules. This enables to chain reactive objects, provided that they only have unidirectional communication, so that an order can be decided for the execution of the reactions [4]. We call this composition method *synchronous sequencing*. Our prototype has five reactive objects, linked by synchronous sequencing.

To execute those objects, it is necessary to have an execution machine, that will activate each object in the sequencing order, ensuring the synchronous semantic. This leads to the definition of a *synchronous execution machine*, which

---

[3] However, the new version of the Esterel compiler which was not available to us at the time, resolves this problem, by using a new coding structure for the automaton.

[4] Sequencing allows to have loop in the communication graph by introducing a *delay* object [12].

performs signal communications inside and at the interface of the thread management, and guarantees atomicity of the reactions. This execution machine being written in classical programming language, it has to be minimum, so that it does not contain unproven behavior. We adopted the simplest policy to start objects' activation : a reaction sequence is executed as soon as there is one input signal, unless there is already one running. There are two concurrent sources of events in an operating system : threads, invoking system operations or raising exceptions, and hardware interrupts. A lock mechanism is used to ensure atomicity of the reaction sequence when concurrent events occur simultaneously.

To certify a program, a formal verification, must be used. This is done with the Mauto tool [11], that uses a mathematical representation of the automaton, as a set of labeled transitions. To verify that a situation cannot occur, we define the abstract actions specifying it. Those actions are described by a sequence of combinations of input and output signals. Each combination can contain AND, OR and NOT boolean operators, allowing complex situations to be expressed. Given the automaton and one abstract action, Mauto certifies, using bisimulation reductions, that the transition system does not hold the undesired action.

Example 1. The property *"the Run signal cannot be emitted by the scheduler while it is in interrupted mode"* is verified by proving that the following abstract action is unreachable by the scheduler automaton :

   true* : /EnterInterrupt? : (not /ExitInterrupt?)* : (not /ExitInterrupt?) and /Run!

For each Esterel module of our executive, its important properties are checked with this tool. However, the use of the sequencing composition instead of the synchronous parallel operator prohibits verifications of behavior implicating several modules.

4 A sparse time distributed platform

Based on our executive, we have implemented a distributed execution platform which realizes the sparse time execution model proposed by Kopetz [9]. We detail it in this section.

4.1 Different distributed execution models for reactive objects

The execution of multiple reactive objects in a distributed environment may follow different schemes :

- **Asynchronous** ; no guarantees on module execution time and communication delays.
  Two executions of the same system may lead to different results, depending on arbitrary parameters. It is an indeterministic system.

- **Weak synchronous**

  A logical time may be defined, that is shared by all the objects of the system. If the reaction of each object occurs at specified instants of that logical time, and if the communications between objects have a fixed length in that logical time, then the system is deterministic. This can be accomplished, for example, by a master clock, broadcasting ticks to all reactive objects, and waiting for their acknowledgment to advance to the next instant. Each object starts a reaction when it receives a tick, and emits its acknowledgment when it is sure to have received all external signals for the next instant.

  This is called the weak synchronous model in the literature, as first introduced by Milner in [13], and implemented in the Saturn project [8].

- **Timed weak synchronous**

  The logical time is anchored in the physical time. That is each instant of the logical time corresponds to a defined granularity of the physical time. The system is then time deterministic, i.e. its result is deterministic in value and time. A distributed system providing this determinism considerably eases the development of real-time distributed applications, that necessitate time guarantees. We present in the next subsection an efficient implementation of such a synchronization.

4.2 Offering Sparse Time to distributed applications

H. Kopetz, in [9], presented an execution model restricting event occurrences in specific points of a distributed synchronized time. The set of those points constitutes a sparse timebase. This model, when synchronized points are properly chosen, provides temporal order on events : an event E_1 is either earlier, simultaneous or later than an event E_2.

Our reactive process management, timely synchronized, constructs such a sparse timebase for the applications, realizing temporal order determinism of their events.

To determine the synchronization points, we rely on a classical clock synchronization mechanism, with a known precision p.

$$p = MAX((\forall i, (\forall k, (\forall l, \quad \|z(k_i) - z(l_i)\|)) \tag{1}$$

where $z(e)$ is the time-stamp, on a reference clock, of the event e, and k_i and l_i are the ith ticks of local clocks k and l.

We build a global timebase, with a granularity g_g superior to the precision p. Between two ticks of this global time, there is an interval K within which all clocks have the same value.

$$K = g_g - p \quad (g_g > p) \tag{2}$$

Those K intervals are the synchronization points, defining a sparse timebase. If all events of our system occur during one of the K intervals, then they will be time-stamped with the same value on every machine.

We suppose the existence of a maximum event communication delay d_{max}. If an event e is emitted from k at $z(k_i)$, we are sure that all recipients have received it at $z(k_i) + d_{max}$.

A receiver time-stamps an event, with the global timebase, as soon as it receives it. For this value to be identical on all receivers, the emission must respect two constraints (represented in figure 2) :

- An event must not be emitted before $g(i) + p$, so that its receiver won't get it too soon (imprecision constraint)
- An event must not be emitted after $g(i + 1) - p - d_{max}$, so that its receiver won't get it too late (transit delay constraint)

Those two constraints define the interval K :

$$K = g_g - 2.p - d_{max} \qquad (3)$$

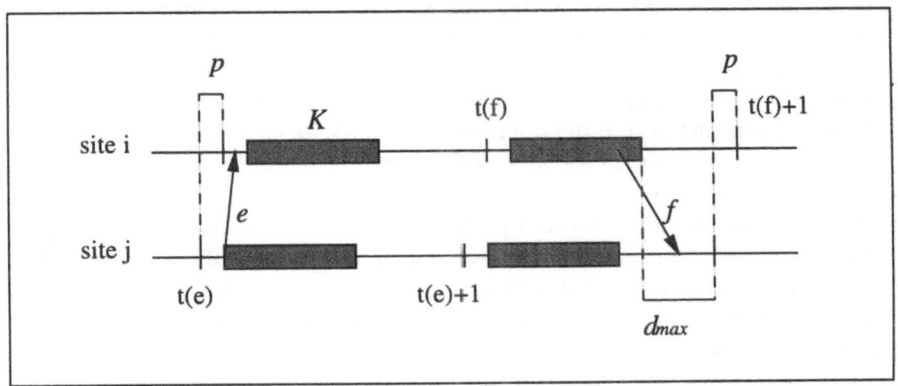

Fig. 2. Restricting emissions to K ensures uniform time-stamps

p and d_{max} values are fixed by the environment. g_g has thus to be determined to offer a sufficient K length to allow event processing and emissions.

In our implementation, all kernel instances start their reactions at the beginning of g_g, and must complete it and emit its resulting signals at least at time $g_g + p + K$. K value has hence to be superior or equal to the worst case execution time of this treatment, minus p. After their reaction, kernels receive signals emitted from other kernels. They will be taken into account in the next reaction, starting at next g_g interval. All kernels will then have the same set of message signals.

To prototype this model, we isolated the triggering mechanism of the synchronous kernel, making it replaceable. We implemented three triggers, realizing respectively the asynchronous, weak synchronous, and timed weak synchronous models. In the first one, kernels react as soon as they detect an input signal.

In the second, an external sequencer broadcast stimulation to start reactions, and collects acknowledgment of their completion. The last one implements the model presented above by relying on a physical clock synchronisation protocol.

5 Results and conclusion

Our implementation revealed that it was indeed possible to use synchronous programming to realize real time operating system kernels, gaining the inherent benefits : determinism and formal descriptions.

Our distributed implementation also showed how we could realize a distributed timed synchronous model, preserving the benefit of determinism and time determinacy in a distributed setting.

However, the following difficulties must be noted :

1. We had difficulties in using Esterel synchronous parallel composition, essentially due to the inability to dynamically create reactive objects. To circumvent Esterel limitations, we had to rely on an ad-hoc synchronous sequencing operator, implementing a weaker form of composition. Note that the recent reactive objects model proposed in [15] alleviates the problem by providing a new synchronous parallel operator which allows dynamic creation of reactive objects, at the expense of a slightly weaker semantic for interrupts (trap constructs in Esterel).
2. The implementation exhibited poor performances, compared to the original Chorus kernel. The sources of these inefficiencies lie mainly in the way events are implemented in the C++ library. The document [14] provides a detailed analysis of the sources of inefficiencies, and suggests ways to remove them.
3. Verification tools at our disposal did not allow us to verify global properties of the executive, because they did not take into account the weak parallel operator used in the implementation. Note however that these tools could still be used if we introduced bounds on the maximum number of threads allowed.

These difficulties, in turn, suggest different avenues for further research.

References

1. Ferrari, D.: Client requirements for real time communication services. Research Reports ICSI TR-90-007, Berkeley, California, USA, 1990
2. Benveniste, A., Berry, G.: The Synchronous Approach to reactive and real-time systems. IEEE, 1991
3. le Guernic, P., Gautier, T., le Borgne, M., le Maire, C.: Programming Real-Time Applications with Signal. Proceedings of the IEEE, 1991
4. Halbwachs, N., Caspi, P., Raymond, P., Pilaud, D.: The Synchronous Data Flow Programming Language Lustre. Proceedings of the IEEE, 1991

5. Berry, G., Gonthier, G.: The Esterel synchronous language : Design, Semantic, Implementation. Journal of Science Of Computer Programming, Vol 19, Num 2., pp87-152, 1992
6. Jourdan, M., Maraninchim, F., Olivero, A.: Verifying qualitative real-time properties of synchronous programs. International Conference on Computer Aided Verification, Elounda, 1993, LNCS697
7. Lecompte, V.: Vérification automatique de programmes Esterel. Ph.D thesis from Paris VII University, 1989
8. Boniol, F., Adelantado, M.: Programming distributed reactive communicating distributed reactive automata : the weak synchronous paradigm. International Conference on Decentralized and Distributed Systems, Spain, 1993
9. Kopetz, H.: Sparse time versus Dense time in distributed real time systems. IEEE Symp. On Distributed Systems, 1992
10. Rozier, M., Abrossimov, V., Armand, F., Boule, I., Gien, M., Guillemont, M., Herrmann, F., Kaiser, C., Langlois, S., Lanard, P., Neuhauser, W.: CHORUS distributed operating systems. Computing Systems 1(4), pp 305-367, 1988
11. Vergamini, D.: Auto/Mauto User Manual, version 2-3. INRIA CERICS, 1992
12. Boulanger, F.: Intégration de Modules Synchrones dans la Progammation par Objets. Ph.D thesis from Orsay University, 1994
13. Milner, R.: Calculi for synchrony and asynchrony. Theoretical Computer Science, 25(3), 1983
14. Potonniée, O.: Etude et prototypage en Esterel de la gestion de processus d'un micro-noyau de système d'exploitation réparti avec garantie de service. Ph.D thesis from Paris VI University, 1996
15. F. Boussinot, G. Doumenc, J.B. Stefani: Reactive Objects. Research Report 2664, INRIA, France, October 1995.

Task-System Analysis Using Slope-Parametric Hybrid Automata

Augusto Burgueño[1] * and Vlad Rusu[2]

[1] ONERA-CERT, Département d'Informatique,
2 av. E. Belin, BP4025, 31055 Toulouse Cedex 4, France.
a.burgueno@acm.org
[2] IRCyN (UMR CNRS N. 6597), Ecole Centrale de Nantes, Université de Nantes),
1 rue de la Noë, BP92101, 44321 Nantes Cedex 3, France.
Vlad.Rusu@lan10.ec-nantes.fr

Abstract. Slope-parametric hybrid automata (SPHA) are hybrid automata whose variables can have parametric slopes. SPHA are useful, in particular, for modeling task-control systems in which the task speeds can be adjusted for meeting some safety requirement. In this paper, we present an example of parametric analysis for a simple task system. We introduce a prototype verification tool that fully automates the analysis.

Keywords: real-time systems, hybrid automata, parametric polyhedra.

1 Introduction

The verification of real-time properties is nowadays a well-known problem, and its most successful resolution techniques [ACH+95] have been automated and applied to real-size systems [DY95, HH94, BGK+96]. It consists, clasically, in verifying a given (timed) property on a given model of the system, and thus obtaining a binary answer: 'the system satisfies/does not satisfy the property'. However, many problems arising in the field of verification are *parametric*. Indeed, for a system designer it is often more important to obtain quantitative information such as: (α) 'for protocol safety, the messages should arrive at destination in no more than 1 second' or (β) 'for the task system to operate correctly, task 1 should run at least 2 times faster than task 2'.

Parametric analysis is the subject of some study and application, although mainly dealing with (α)-like analysis: finding the possible values of *delays* for some property to be satisfied. In a short paper [BBRR97] we presented a first approach to (β)-like analysis: the parameters to be computed are *speeds* rather than delays.

---

* Partially supported by Research Grant of the Spanish Ministry of Education and Culture. This research was carried out in part while the author was visiting the IRCyN (formerly LAN).

In this paper, we present an application of this approach. The goal is to have a situation in which the correct parameter values are hard to find by classical, 'try and fail until success' verification, since the correct values lie in a broad interval. We find these values automatically, using a prototype tool that we have implemented using MAPLE V and Prolog IV.

Related work. Parametric analysis of hybrid and timed automata has also been treated from other view points: [AHV93, HH94] focus on delays, that is, parameters appear on guards; [CY91] follows a similar approach; [Wan96] defines Parametric TCTL and redefines the classical model checking algorithm; [KS97] and [HLM97] deal with the somehow similar problem of controller synthesis; finally, the only other work to our knowledge combining constraint solving and model checking is [CABN97].

2 Slope-Parametric Hybrid Automata

Slope-Parametric Hybrid Automata (SPHA) are a generalization of Multirate Automata [ACH+95] in which the rates (slopes) of variables can be parameters. Let $\mathbb{R}$ be the set of real numbers.

Syntax. A SPHA is a tuple $(\mathcal{L}, \mathcal{E}, \mathcal{V}, \mathcal{K}, invar, diff, guard, reset)$ where

- $\mathcal{L}$ is a finite set of *vertices*
- $\mathcal{E} \subseteq \mathcal{L} \times \mathcal{L}$ is a finite set of *edges*
- $\mathcal{V} = \{x_1, .., x_n\}$ is a finite set of *variables*
- $\mathcal{K} = \{k_1, .., k_m\}$ is a finite set of *parameters*
- *invar* is a function that associates to each vertex an *invariant* i.e. a predicate of the form $(\bigwedge_i x_i \sim c_i)$, where $\sim \in \{<, \leq, >, \geq\}$ and $c_i \in \mathbb{R}$
- *diff* is a function that associates to each vertex, a *parametric differential law* for each variable, i.e an expression of the form $dx/dt = \sum_{j=1}^m a_j \cdot k_j + b$, where $a_j, b \in \mathbb{R}$ and $k_j \in \mathcal{K}$ are parameters
- *guard* is a function that associates to each edge a *guard* i.e. a predicate of the form $(\bigwedge_i x_i \sim c_i)$, where $\sim \in \{<, \leq, >, \geq\}$ and $c_i \in \mathbb{R}$
- *reset* is a function that associates to each edge a *reset expression* i.e. an expression of the form $(\bigwedge_i x_i := 0)$.

The SPHA in figure 1 has five vertices (L_1, L_2, L_3, L_4, L_5), three variables (a_1, a_2, t) and one parameter K. Some evolution laws are parametric and some are constant. For example, at vertex L_2 the evolution laws of a_1 and a_2 are parametric ($\dot{a}_1 = K, \dot{a}_2 = 1 - K$) while the evolution law for t is constant ($\dot{t} = 1$). The edges are labeled with guards ($a_2 = 40$ for the edge from L_2 to L_4) and variable resets ($a_2 := 0$ for the same edge). The invariant for each vertex is also shown (e.g. $t \leq 100 \wedge a_1 \leq 30 \wedge a_2 \leq 40$ for vertex L_2).

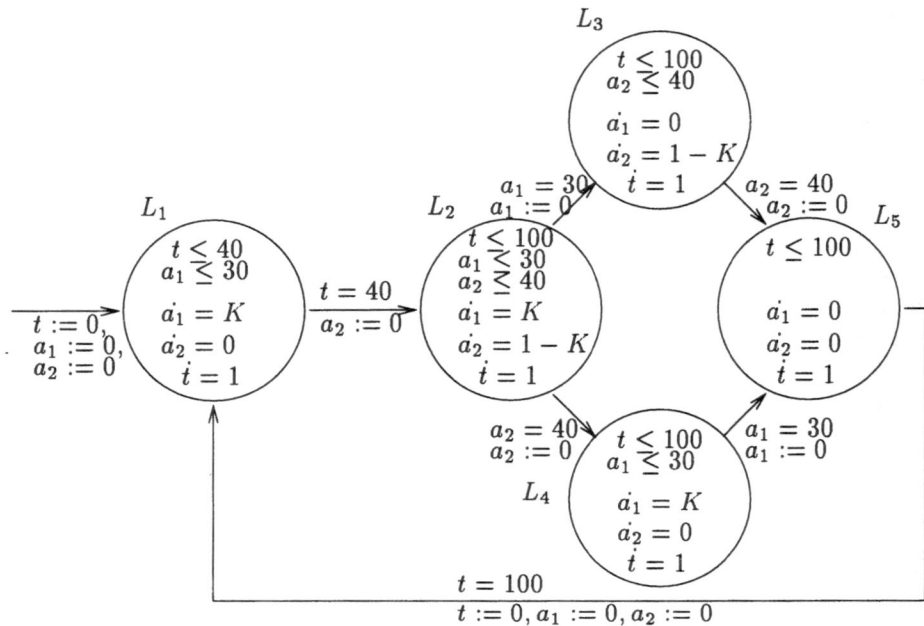

Figure 1. Example of slope-parametric hybrid automaton

Semantics. A *state* of a SPHA is defined, as in the case of Multirate Automata [ACH+95], by a couple (L, v) where $L \in \mathcal{L}$ is a vertex and v is a variable valuation, assigning a value $v(x)$ to each variable $x \in \mathcal{V}$. But in SPHA, variables can have *parametric* values i.e. a variable's value can be not only a real number, but also an expression on the parameters, as for example $v(x) = K^2 - 3 \cdot K + 1$.

A *run* of a SPHA consists in a sequence of states, obtained by letting the variables continuously evolve by their differential laws in some vertex, such that the vertex invariant is continuously satisfied; in crossing some outgoing edge, when the edge's guard is satisfied; and finally resetting the corresponding variables. The process is pursued in the newly reached vertex. For instance, in the SPHA represented in figure 1, consider a run starting at vertex L_1 with variables $t, a_1, a_2 = 0$. Then, evolution at vertex L_1 is given by $(\dot{a}_1, \dot{a}_2, \dot{t}) = (K, 0, 1)$, while the invariant $t \leq 40 \wedge a \leq 30$ holds. Vertex L_1 will be left, by crossing the edge from L_1 to L_2, when $t = 40$. But a_1, whose value is $K \cdot 40$ at the crossing moment, cannot exceed 30 (as stated by the vertex L_1 invariant); thus we have a condition on the parameter, in order to continue the run: $K \leq 3/4$.

The previous remark showed that *any run of a SPHA defines a sequence of conditions on the parameters*. Indeed, any edge guard must be satisfied by the variable values when that edge is crossed; since these values are expressions·on the parameters, the guard satisfaction translates to a condition on the parameters. Thus, the analysis of SPHA focuses not only on the existence of a run between states (like in the case of plain hybrid automata), but also on computing and solving the associated sequence of conditions on the parameters. This is the goal of parametric analysis.

Parametric Analysis. The problem of parametric analysis is formulated as follows: given a *slope-parametric hybrid automaton*, a *temporal logic formula expressing a reachability property* and a *set of intervals in* $\mathbb{R}$ (one for each parameter), find the relations among parameters (in the corresponding intervals) such that the formula is true. For instance, in the automaton of figure 1, find values for parameter $K \in]0, 1[$, for the TCTL formula $\varphi = \{(L_1 \wedge a_1 = 0 \wedge a_2 = 0 \wedge t = 0) \Rightarrow \neg[(\neg L_5)\exists \mathcal{U}_{=100}(\neg L_5)]\}$ to be true. This formula means that, starting from the initial set of states defined by vertex L_1 with $a_1 = 0, a_2 = 0, t = 0$, it is not possible to avoid vertex L_5 for a period of 100 units of time. In other words, vertex L_5 is inevitably reached within 100 time units.

3 Example

Consider the problem of assigning CPU time to processes that share a single processor. There are several ways to cope with this problem. One of them is to allocate resources statically, such that every time a process must execute, it will have its *time slice* reserved to do it. This is an approach followed in reactive programming, when processes must be launched as a reaction to an observed event: if the process has to compete at run time for resources, it may fail to get them and, therefore, it may fail its goal; if, on the contrary, resources are reserved in advance, it will always be able to run. In our example, processes a_1 and a_2 will share a CPU in fixed proportions: a percentage of time of $100 * K$ for a_1 and of $100 * (1 - K)$ for a_2.

Apart from these allocation considerations, there are other facts in the example that interest us. In particular, processes a_1 and a_2 must both be executed within a period of 100 time units. Total execution times are 30 time units for a_1 and 40 time units for a_2 (measured on the processor when each process takes all resources). Process a_2 must wait 40 time units from the begining of the period $(t = 0)$ to be ready to execute, therefore, during the first 40 time units $(t \in [0, 40])$ only a_1 runs. From that instant on $(t \in]40, 100])$ a_1 and a_2 run concurrently (once one of them has finished, the other will continue alone until completion). As we said before, the processor allocation is fixed for each process disregarding whether they run alone or concurrently. We assume that these processes don't have any interaction and the waiting times (for I/O, for example) are null.

The slope-parametric hybrid automaton of figure 1 represents this schema. The expression $\dot{a_1} = K$ indicates that process a_1 executes using $100 * K$ per cent of CPU time. Similarly, the expression $\dot{a_2} = 1 - K$ indicates that process a_2 executes using $100 * (1 - K)$ per cent of CPU time. The expression $\dot{a_1} = 0$ says that the process a_1 does not execute at all. The fact of having a parameter (K) in our automaton allows us to define parametric analysis problems. One could be: determine the possible values of $K \in]0, 1[$ such that processes a_1 and

a_2 are both executed, once every 100 seconds. On the SPHA of figure 1, this translates to the property (already stated in section 2) that, starting from the initial situation, vertex L_5 is inevitably reached within 100 time units. A non trivial hand calculation shows that, for the above property to hold, K must lie in a broad interval: $K \in]0.3, \frac{1}{3}[$. We recall how to automate this calculus [BBRR97].

4 Computing parameter values

Parametric analysis comes to operating with parametric polyhedra. The latter are described by sets of linear equations and inequations whose coefficients can be either *constants* or *symbolic expressions on the parameters*; for instance, $0 \leq a_1 \leq 30 \wedge 0 \leq a_2 \leq 40 \wedge 40 \leq t \wedge a_1 - K \cdot t = 0$, where K is a parameter. The operations involved in parametric analysis are *extension, restriction* and *projection*, that incrementally generate conditions on the parameters, at each passage from a vertex to the next one.

Example. Consider the vertex L_2 of the automaton in figure 1 and an initial parametric polyhedron $0 \leq a_1 \leq 30 \wedge 0 \leq a_2 \leq 40 \wedge 40 \leq t \wedge a_1 - K \cdot t = 0$. We want to find the conditions on the parameter K, for the control to go from L_2 to L_4. For this we *extend* the initial polyhedron in the parametric direction given by $\dot{a_1} = K, \dot{a_2} = 1 - K$ and $\dot{t} = 1$; as we shall see, this extension has a finite number of different forms, under different conditions on K. Then, we *restrict* the values of parameter K such that the previous extension intersects the guard $a_2 = 40$ and the invariant $t \leq 100 \wedge a_1 \leq 30 \wedge a_2 \leq 40$: the intersection polyhedron is non-empty iff parameter K satisfies some further conditions. Once we have obtained these conditions, we continue at vertex L_4 by *projecting* the intersection polyhedron on plane $a_2 = 0$ (this corresponds to reinitializing a_2 on the transition). We should of course, apply the same sequence of operations in L_4 to find the conditions on K to cross the edge from L_4 to L_5.

4.1 Extension

Extension means: given a parametric polyhedron $P = \bigwedge_{j=1}^{m} \left(\Sigma_{i=1}^{n} a_{i,j} \cdot x_i \succ b_j \right)$ where $a_{i,j}, b_j$ can be constants or symbolic expressions and $\succ \in \{>, \geq\}$, find the polyhedron: $\overrightarrow{P} = \exists \tau \geq 0. \left[\bigwedge_{j=1}^{m} \left(\Sigma_{i=1}^{n} a_{i,j} \cdot (x_i - k_i \cdot \tau) \succ b_j \right) \right]$ where each k_i is the slope for variable x_i (constant or symbolic expressions on parameters). This is *forward continuous simulation* [ACH+95] except that we consider parametric polyhedra and directions. This imposes to consider several cases for eliminating the $\exists$ quantifier in the equivalent expression $\overrightarrow{P} = \exists \tau \geq 0. \left[\bigwedge_{j=1}^{m} \left(\Sigma_{i=1}^{n} a_{i,j} \cdot x_i - \tau \cdot \left(\Sigma_{i=1}^{n} a_{i,j} \cdot k_i \right) \succ b_j \right) \right]$.

The cases to consider are, for each sum-of-products $\Sigma_{i=1}^{n} a_{i,j} \cdot k_i$, the possibility that it is negative, positive or 0. As in [AHH93] we eliminate the existential quantifier by dropping the inequations that correspond to negative sums-of-products ($\Sigma_{i=1}^{n} a_{i,j} \cdot k_i < 0$), keeping the inequations that correspond to positive or zero sums-of-products ($\Sigma_{i=1}^{n} a_{i,j} \cdot k_i \geq 0$), and linearly combining pairs of inequations that correspond to one negative and one positive sum-of-product. The point here is that the sums-of-products are symbolic, so in general we will not be able to tell at sight the sign of the expressions $\Sigma_{i=1}^{n} a_{i,j} \cdot k_i$. We must then consider all the possible cases for these signs, so the extended polyhedron $\vec{P}$ has at most 3^m different forms, following the possible signs of the m expressions $\Sigma_{i=1}^{n} a_{i,j} \cdot k_i$.

For the example of figure 1, vertex L_2, the extension of the polyhedron $P = 0 \leq a_1 \leq 30 \wedge 0 \leq a_2 \leq 40 \wedge 40 \leq t \wedge a_1 - K \cdot t = 0$ by $\dot{a}_1 = K, \dot{a}_2 = 1 - K, \dot{t} = 1$ would give three cases, depending on the sign of expression $1 - K$. For instance, when $1 - K > 0$, the extended polyhedron (as automatically generated by our verification tool described in section 5) is:

$$\vec{P_1} = [0 \leq a_1 \wedge 0 \leq a_2 \wedge 40 \leq t \wedge 0 = a_1 - K \cdot t \wedge 30 \cdot K - 30 \leq$$
$$(K - 1) \cdot a_1 + K \cdot a_2 \wedge -30 \leq K \cdot t - a_1 \wedge 40 \cdot K - 40 = a_2 + (K - 1) \cdot t \wedge 0 \leq$$
$$(1 - K) \cdot a_1 - K \cdot a_2 \wedge -40 \cdot K \leq a_1 - K \cdot t \wedge 40 \cdot K - 30 \leq K \cdot t - a_1 \wedge 0 \leq (1 - K) \cdot t - a_2].$$

The extended polyhedra have different forms when $1 - K = 0$ and $1 - K < 0$. In practice, we do not have to generate all 3^m cases, since most of them lead to unsatisfiable conditions on the parameters (for instance, in the above example, the cases $1 - K = 0$ and $1 - K < 0$ are eliminated from start since we are looking for parameters in the interval $]0, 1[$ cf. section 3). We will come back to this point in section 5.

Note also that the parametric slopes have become coefficients in the inequations of the extended polyhedron e.g. $40 \cdot K - 30 \leq K \cdot t - a_1$.

4.2 Restriction

Restriction is: given a parametric polyhedron $P = \bigwedge_{j=1}^{m} (\Sigma_{i=1}^{n} a_{i,j} \cdot x_i \succ b_j)$ (remember that $a_{i,j}, b_j$ are constants or symbolic expressions on the parameters) find the possible values of parameters such that P is non-empty.

For the parametric polyhedron $\vec{P_1}$ obtained in the previous step, we compute its intersection with the guard $a_2 = 40$ and the invariant $t \leq 100 \wedge a_1 \leq 30 \wedge a_2 \leq 40$, and find the conditions on K such that this intersection is not empty. We obtain several cases from which only one is satisfiable (with respect to the values of K):

- Condition: $0 < K \wedge K < 1 \wedge (K - 1) \cdot K < 0 \wedge K^2 \cdot (K - 1) < 0$
- Intersection: the same as for extension plus three additional inequations ($\vec{P_1} \wedge t \leq 100 \wedge a_1 \leq 30 \wedge a_2 = 40$).

The general case can be treated as follows: the polyhedron P is non-empty iff the expression $\exists x_1 . \exists x_2 \ldots \exists x_n . P$ is 'true'. The formal elimination of all variables in the previous expression generates a symbolic condition on the parameters, *that precisely constitutes the condition for P to be non-empty.* Variables $x_1, .., x_n$ can be eliminated one by one by successively applying the Fourier-Motzkin elimination algorithm (see for instance chapter 1 of [Zie95]) that we now describe.

The Fourier-Motzkin elimination algorithm. This algorithm computes, given a system of linear inequations $P = \bigwedge_{j=1}^{m} (\sum_{i=1}^{n} a_{i,j} x_i \succ b_j)$, the system obtained by eliminating a variable say x_k : $P \updownarrow_k = \exists x_k . \bigwedge_{j=1}^{m} (\sum_{i=1}^{n} a_{i,j} \cdot x_i \succ b_j)$. The idea is to consider the possible signs of the coefficients $a_{k,j}$: as in the case of extension, eliminating variable x_k leads to at most 3^m possible forms of the result, depending on the signs of the m coefficients $a_{k,j}$. For a given combination of signs, denote $J_>$ (respectively, $J_=$, $J_<$) the subsets of indices of $\{1, \ldots, m\}$ such that $a_{k,j} > 0$ (respectively $= 0$, < 0). Then, $P \updownarrow_k$ is obtained by keeping the inequations indexed by $J_=$, by eliminating the inequations indexed by $J_<$ and $J_>$ (i.e. keep only the inequations where x_k does not occur), and by linearly combining pairs of inequations (one indexed by some $j_> \in J_>$, the other indexed by some $j_< \in J_<$) to eliminate variable x_k. For all $j_> \in J_>$ and $j_< \in J_<$, generate the following linear combination: $a_{k,j_>} [\sum_{i=1}^{n} a_{i,j_<} x_i \succ b_{j_<}] - a_{k,j_<} [\sum_{i=1}^{n} a_{i,j_>} x_i \succ b_{j_>}]$ which is equivalent to $\sum_{i=1}^{n} [a_{k,j_>} a_{i,j_<} - a_{k,j_<} a_{i,j_>}] x_i \succ [a_{k,j_>} b_{j_<} - a_{k,j_<} b_{j_>}]$. In this last inequation the coefficient of x_k is 0 so variable x_k has been eliminated.

4.3 Projection

Projection on the $x_k = 0$ plane means: given a parametric polyhedron $P = \bigwedge_{j=1}^{m} (\sum_{i=1}^{n} a_{i,j} \cdot x_i \succ b_j)$, find the parametric polyhedron $P|_{x_k=0}$ obtained by projecting P on the plane $x_k = 0$. Consider the polyhedron $P = \vec{P_1} \cap \{t \leq 100 \wedge a_1 \leq 30 \wedge a_2 = 40\}$ of the previous example. The projection of P on $a_2 = 0$ for conditions $0 < K \wedge K < 1 \wedge (K-1) \cdot K < 0 \wedge K^2 \cdot (K-1) < 0$ is

$$
\begin{aligned}
P|_{a_2=0} = & 0 \leq a_1 \wedge 0 = a_2 \wedge a_1 \leq 30 \wedge 40 \leq t \wedge a_1 - K \cdot t = 0 \wedge \\
& t \leq 100 \wedge -40 \cdot K \leq a_1 - K \cdot t \wedge \\
& -30 \leq K \cdot t - a_1 \wedge 40 \cdot K - 30 \leq K \cdot t - a_1 \wedge \\
& 30 \cdot K - 30 \leq (K-1) \cdot a_1 - (K-1) \cdot K \cdot t \wedge \\
& -10 \cdot K - 30 \leq (K-1) \cdot a_1 \wedge \\
& 40 \cdot (K-1) \cdot K \leq (1-K) \cdot a_1 + (K-1) \cdot K \cdot t \wedge \\
& 30 \cdot K - 30 \leq (K-1) \cdot a_1 \wedge \\
& 70 \cdot K - 40 \cdot K^2 - 30 \leq (K-1) \cdot a_1 - (K-1) \cdot K \cdot t \wedge \\
& 0 \leq (1-K) \cdot t \wedge 40 - 40 \cdot K \leq (1-K) \cdot t \wedge \\
& 0 \leq (1-K) \cdot a_1 \wedge 40 \leq (1-K) \cdot t \wedge \\
& 40 \cdot K - 40 \leq (K-1) \cdot t \wedge 40 \cdot K \leq (1-K) \cdot a_1 \wedge \\
& 80 - 40 \cdot K \leq (1-K) \cdot t \wedge 40 \cdot K - 80 \leq (K-1) \cdot t
\end{aligned}
$$

In general, to obtain the projected parametric polyhedron we use the Fourier-Motzkin elimination algorithm and the identity $P|_{x_k=0} = \exists x_k . P \wedge (x_k = 0)$.

4.4   The operations at work

To obtain the conditions on the parameters for a reachability formula to be true, combine the three operations as follows. First, choose a vertex path that links a vertex from the initial region to a vertex from the final region (as defined by the reachability formula). Then, iterate the three operations on that vertex path, to generate a *tree* whose nodes are pairs (vertex, parametric polyhedron), and whose edges are labeled by symbolic conditions on the parameters.

Starting from the pair (initial vertex, polyhedron defined by the initial values of variables) as the root, apply the *extension* procedure to the initial polyhedron to generate all the possible extended polyhedra, and associate a node (successor of the root) to each one. The branch leading to a node is labeled with the condition under which the node's extended polyhedron was obtained. Likewise, for each new node, generate its successors by applying the *restriction* procedure, and label the new branches correspondingly. For the lastly obtained successors, continue with the *projection* procedure. Iterate the sequence of procedures in this order until the final vertex is reached, and terminate by a restriction to intersect the final region.

At this point, any sequence of branches of the constructed tree, from the root to a leaf, defines a sufficient condition on the parameters, for the final region to be reachable from the initial one (it is the conjunction of the conditions on all branches). The *disjunction* of these sufficient conditions, for all the sequences of branches in the tree (from the root to a leaf), constitute the necessary and sufficient condition for the final region to be reachable from the initial one, just by the particular vertex path *in the automaton* that was initially chosen.

But, in general, there are an infinity of vertex paths in the automaton, from a formula's initial vertex to a final one; this means that in order to generate the necessary and sufficient conditions on the parameters for the reachability to hold, we might need to iterate the above operations on an infinite number of vertex paths, making our problem undecidable. However, we have noted in [BBRR97] that, for a restricted class of SPHA and formulas, the problem remains decidable: these are *uniformly low-bounded SPHA* and *time-bounded reachability formulas*.

Uniformly low-bounded SPHA are characterized by the fact that any cyclic run has a duration at least equal to some strictly positive ϵ, which does not depend on the run or the parameter values. For instance, the 2 cycles of the automaton in figure 1: $L_1 - L_2 - L_3 - L_5 - L_1$ and $L_1 - L_2 - L_4 - L_5 - L_1$, always have a duration 100 whatever the parameters, so this SPHA falls into the uniformly low-bounded case. Now for such SPHA and for time-bounded reachability formulas, it is necessary to propagate extension, restriction, and projection, only on a finite number of vertex paths [BBRR97].

Furthermore, some conditions on the parameters might be unsatisfiable. Therefore, every time a condition is generated it is important to test if it is satisfiable; this will prevent us from analyzing branches that would lead us nowhere. As it will be described in section 5, there exist automatic means to do it.

4.5 Solving our example problem

Consider the above automaton and the time-bounded reachability formula $\varphi = \{(L_1 \wedge a_1 = 0 \wedge a_2 = 0 \wedge t = 0) \Rightarrow \neg[(\neg L_5)\exists\mathcal{U}_{=100}(\neg L_5)]\}$. This means that, starting from the initial state, vertex L_5 is inevitably reached within 100 time units. Alternatively, this can be written as: $\neg\{(L_1 \wedge a_1 = 0 \wedge a_2 = 0 \wedge t = 0) \wedge [(L_1 \vee L_2 \vee L_3 \vee L_4)\exists\mathcal{U}_{=100}(L_1 \vee L_2 \vee L_3 \vee L_4)]\}$ meaning that, starting from the initial region $(L_1 \wedge a_1 = 0 \wedge a_2 = 0 \wedge t = 0)$, it is not possible to stay within vertices L_1, L_2, L_3 and L_4 for 100 time units.

In order to solve our initial parametric problem, we shall find values for K such that (α): *it is possible to stay vertices L_1, L_2, L_3 and L_4 for 100 time units*, and then take the complement for K (in $]0,1[$) .

To solve (α) we have to iterate the extension, restriction and projection operations on two vertex paths $(L_1 - L_2 - L_3$ and $L_1 - L_2 - L_4)$. At each vertex L_i ($i \in \{1, 2, 3, 4\}$), we find a set $\mathcal{P}_i$ of parametric polyhedra with their associated conditions on K; they constitute the result of propagating the initial region up to that vertex. In order to see if it is possible to be at vertex L_i at time $= 100$, we *restrict* each polyhedron in $\mathcal{P}_i$ with $t = 100$, that is, we find the conditions on parameter K such that their intersection with $t = 100$ is non-empty.

For vertices L_1 and L_2, these final conditions are unsatisfiable. For vertex L_4, the condition is $K \leq 0.3$. For vertex L_3, it is $K \in [1/3, 3/4]$. The complement of these conditions in interval $]0, 1[$ is: $K \in]0.3, 1/3[\cup]3/4, 1[$. But for $K > 3/4$, the system is so-called *Zeno* (cf. section 4.6). Thus, the answer to our parametric problem is: $K \in]0.3, 1/3[$.

4.6 Non-Zenoness and backwards reachability

Consider again the SPHA of figure 1. For some values of the parameter $K \in]0, 1[$, the automaton is affected by so-called *Zeno* behaviours [HNSY94]. We have seen an example of such behaviour in section 2: if $K > 3/4$, the run starting from initial state $L_1 \wedge a_1 = 0 \wedge a_2 = 0 \wedge t = 0$ is unable to leave vertex L_1 but it is also unable to stay in L_1 forever, since L_1's invariant eventually becomes false (after at most 40 time units). The system is so-called *Zeno* [HNSY94], meaning that *values $K > 3/4$ are intrinsically bad* for the system, whatever further properties we might want it to satisfy (e.g. tasks terminating within 100 time units). So a new problem of parametric analysis is to find the possible values of parameters of a SPHA, for the SPHA to be non-Zeno. To solve this problem, we use an existing technique [HNSY94]: a hybrid automaton is non-Zeno iff from any state of the automaton, it is possible to have a run of duration 1. In our parametric context, checking non-Zenoness means *finding the possible values of the parameters, such that from any state of the automaton, it is possible have a run of duration 1.*

Parametric backward analysis. The existence of a run of duration 1 from each state cannot be checked directly by the *forward* parametric analysis as presented in section 4, because one would have to perform the analysis starting from all the (continuously infinite number) states of the SPHA. Instead, the analysis should proceed *backwards*. For example, consider the SPHA in figure 1; to measure the global time, we enrich the SPHA with a new variable z behaving like a *clock* (constant slope 1 at all vertices), which is never reset[1]. Let us start indicating how to compute, for instance, the states from which it is possible to reach vertex L_2 in 1 time unit. For this, we first consider all the states of the automaton at vertex L_2 and at instant 1, given by L_2's invariant: $P_0 = t \leq 100 \wedge a_1 \leq 30 \wedge a_2 \leq 40 \wedge z = 1$. Then, we compute the set of states P_1 from which it is possible to reach some state in P_0 by *continuously remaining in vertex L_2*: this can be done by the *backwards extension* of P_0. Next, we compute the states P_2 from which it is possible to reach some state in P_1 by some *discrete edge crossing* i.e. by crossing the edge from L_1 to L_2: this is done by *backwards projection* on the edges's *reset* $(a_2 := 0)$ followed by *restriction* to the edges's *guard* $(a_2 = 40)$.

Restriction has been defined in section 4.2, and backwards extension/projection are similar to the restriction/projection described in sections 4.1 and 4.3. For instance, backwards extension of polyhedron $P_0 = t \leq 100 \wedge a_1 \leq 30 \wedge a_2 \leq 40 \wedge z = 1$ in direction $(\dot{a}_1, \dot{a}_2, \dot{t}, \dot{z}) = (K, 1 - K, 1, 1)$ coincides with the forward extension of the polyhedron in the *opposite* direction $(-K, K-1, -1, -1)$; and backwards projection of polyhedron P_1 following the reset $(a_2 := 0)$ is just $\exists a_2.P_1$ i.e. the Fourier-Motzkin elimination defined in section 4.2.

These operations should be iterated backwards on all the vertex paths of the SPHA, and at each step i, by restricting the obtained set of parametric polyhedra $\mathcal{P}_i$ with condition $z = 0$, we obtain states from which it is possible to have a run of duration 1. The iteration should continue until for the lastly obtained set of states, *the restriction to $z = 0$ is empty*. This last condition becomes true after a finite number of iterations, if SPHA has uniformly low-bounded cycles [BBRR97]. For example, in the hybrid automaton of figure 1, the above operations should be iterated only once on each cycle, because each cycle is guaranteed to last more than 1 time unit.

Thus, after a finite number of steps, the algorithm terminates, and one obtains the whole set of states $\mathcal{P}$ from which it is possible to have a run of duration 1, under the form of a set of parametric polyhedra. By imposing the condition that $\mathcal{P}$ contains *all* the states of the automaton, on obtains the necessary and sufficient condition on the parameters, for the SPHA to be non-Zeno.

5 The tool

We have implemented the *extension*, *restriction* and *projection* procedures in MAPLE V, a tool for symbolic computation. As we have seen, these operations generate conditions on the parameters that, in general, take the form of systems of non-linear inequalities whose unkowns are the variables of the hybrid automaton and the parameters. Such systems of inequalities are not solvable by the classical numerical methods but, instead, *propagation* methods [Sac87, Bro81]

---

[1] in section 4.5, we used directly t as a clock to measure global time, since it was never reset on the paths that interested us.

must be used to calculate a "rectangular envelope" of the solution. This envelope defines upper and lower bounds on the unknowns and its meaning is as follows: (α) if the envelope is empty then the system of inequalities has no solution, and (β) if the envelope is non-empty and if the system of inequalities has a solution, then the envelope covers the solution. Note that, as we are interested in those values of parameters that make an unsafe state reachable, in order to discard them, an over approximation is always valid (as long as its complement is not empty). We use in our prototype the propagation features of Prolog IV.

Furthermore, we have implemented the following optimizations:

- **Consistent case generation.** Considering all possible combinations of signs of sum-of-products, in the extension procedure, and of coeficients, in Fourier-Motzkin, may lead to generation of unsatisfiable cases as $K \land -K \land K \neq 0$. The prototype is aware of this, and does not generate unsatisfiable cases.
- **On the fly branch pruning.** The conditions associated with a case may be unsatisfiable in conjunction with the corresponding parametric polyhedron, leading to a useless branch in the tree. The tool prunes all such branches.
- **Alternative restriction procedure.** The aim of the restriction procedure is to generate conditions on the parameters for a given parametric polyhedron to be non-empty. This is a very expensive procedure whose result can be approximated as follows: approximate the conditions on the parameters by the envelope obtained *by propagating* the system of inequalities defined by the parametric polyhedron.

Table 1 shows the execution times for the example of section 4.5 measured on a Sun SPARCclassic. Times are in seconds.

Path	Tree generation (Maple)	Branch pruning (Prolog)	Envelope calculation (Prolog)	Total
$L_1 - L_2 - L_3$	3.750	0.350	0.060	4.16
$L_1 - L_2 - L_4$	3.333	0.380	0.060	3.773

Table 1. Execution times

The maximum number of cases generated is 9. This extremely low number of cases and the low execution times are due to the application of the three optimizations described above.

6 Conclusion

In this paper we have presented an example of system analysis using the slope-parametric hybrid automata presented in [BBRR97]. The method focuses on computing slopes of variables (rather than delays) for some safety requirement to be respected, and it may be seen as an extension of the polyhedra-based symbolic analysis [ACH+95] of hybrid automata. The main operation is the

Fourier-Motzkin's algorithm to deal with parametric polyhedra, which imposes, in theory, to consider a large number of cases. In practice, however, many of these cases are either redundant or unsatisfiable and can be discarded, making the problem computationally tractable. We have presented a prototype verification tool which fully automates the analysis, and shown its applicability to a simple example.

Acknowledgments. We thank Gérard Verfaillie for suggesting us to use Prolog IV, and Olivier Roux and Frédéric Boniol for following this work.

References

ACH+95. R. Alur, C. Courcoubetis, N. Halbwachs, T. A. Henzinger, P-H. Ho, X. Nicollin, A. Olivero, J. Sifakis, and S. Yovine. The algorithmic analysis of hybrid systems. *TCS*, 138:3–34, 1995.

AHH93. R. Alur, T. A. Henzinger, and P.-H. Ho. Automatic symbolic verification of embedded systems. In *Proc. IEEE RTSS'93*, pages 2–11, 1993.

AHV93. R. Alur, T. A. Henzinger, and M. Y. Vardi. Parametric real-time reasoning. In *Proc. ACM STOC'93*, pages 592–601, 1993.

BBRR97. F. Boniol, A. Burgueño, O. Roux, and V. Rusu. Analysis of slope-parametric hybrid automata. In *Proc. HART'97*, volume 1201 of *LNCS*, pages 75–80, 1997.

BGK+96. J. Bengtsson, D. Griffioen, K. Kristoffersen, K. G. Larsen, F. Larsson, P. Pettersson, and W. Yi. Verification of an audio protocol with bus collision using UPPAAL. In *Proc. CAV'96*, volume 1102 of *LNCS*, 1996.

Bro81. R. A. Brooks. Symbolic reasoning among 3-D models and 2-D images. *Art. Int.*, 17:285–348, 1981.

CABN97. W. Chan, R. Anderson, P. Beame, and D. Notkin. Combining constraint solving and symbolic model checking for a class of systems with non-linear constraints. In *Proc. CAV'97*, 1997.

CY91. C. Courcoubetis and M. Yannakakis. Minimum and maximun delay problems in real-time systems. In *Proc. CAV'91*, volume 575 of *LNCS*, pages 399–409, 1991.

DY95. C. Daws and S. Yovine. Two examples of verification of multirate timed automata with Kronos. In *Proc. IEEE RTSS'95*, 1995.

HH94. T. A. Henzinger and P.-H. Ho. HyTech: the Cornell HYbrid TECHnology tool. In *Hybrid Systems II*, volume 999 of *LNCS*, pages 265–294, 1994.

HLM97. M. Heymann, F. Lin, and G. Meyer. Control synthesis for a class of hybrid systems subject to configuration-based constraints. In *Proc. HART'97*, volume 1201 of *LNCS*, pages 376–390, 1997.

HNSY94. T. A. Henzinger, X. Nicollin, J. Sifakis, and S. Yovine. Symbolic model checking for real-time systems. *Inf. and Comp.*, 111(2):193–244, 1994.

KS97. D. Kapur and R. K. Shyamasundar. Synthesizing controllers for hybrid systems. In *Proc. HART'97*, volume 1201 of *LNCS*, pages 361–375, 1997.

Sac87. E. Sacks. Hierarchical reasoning about inequalities. In *Proc. AAAI'87*, volume 2, pages 649–654, 1987.

Wan96. F. Wang. Parametric timing analysis for real-time systems. *Inf. and Comp.*, 130(2):131–150, 1996.

Zie95. G. M. Ziegler. *Lectures on Polytopes*, volume 152 of *Graduate Texts in Mathematics*. Springer-Verlag, 1995.

A Methodology for Compilation
of High-Integrity Real-Time Programs

Karl Lermer   Colin Fidge

Software Verification Research Centre
School of Information Technology
The University of Queensland

Abstract. A practical methodology for compilation of trustworthy real-time programs is introduced. It combines new program development and timing analysis techniques with traditional compilation and assembly technologies.

1   Introduction

Programming real-time systems in a high-level language is difficult because it is the machine code generated by the compiler and assembler, not the high-level source program, that ultimately determines timing correctness. Contemporary compilers make no attempt to generate code with *predictable* timing characteristics [8], undermining their value for real-time applications. Consequently, safety-critical real-time programs are often written in assembly language, forsaking the well-established productivity benefits of high-level language programming.

The *Topaz* project is currently applying formal methods to compilation of trustworthy real-time programs. Topaz comprises

- a real-time refinement theory for formally translating high-level programming language 'specifications' to time-verified machine code 'implementations', and
- a practical methodology for instantiating this theory in existing, or planned, programming environments.

In this article we introduce the second of these two aspects, the Topaz methodology, via a small example.

2   The Topaz methodology

As shown in Figure 1, the Topaz methodology extends traditional high-level language (HLL) program compilation. The *program development*, *compilation* and *assembly* phases are extensions of their traditional (untimed) counterparts. The *timing path analysis* and *timing verification* phases are new. Below we illustrate key aspects of the methodology using a small "transmitter" example [1].

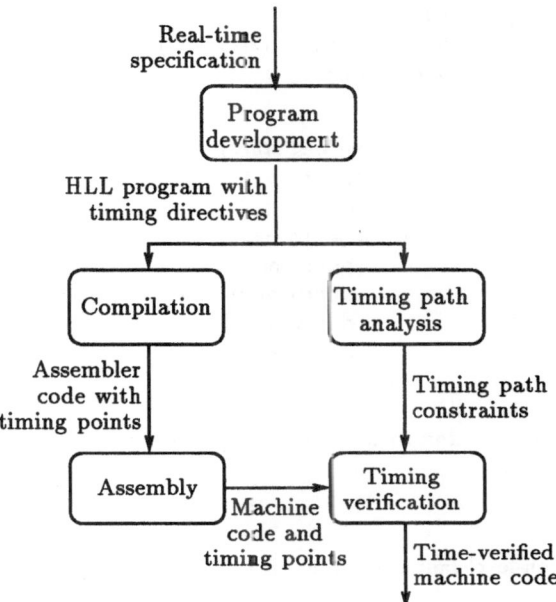

Fig. 1. Overview of the Topaz real-time compilation strategy.

2.1 Program development

Figure 2 shows the time-annotated HLL program fragment for our case study, in guarded command language notation. Hayes and Utting explain in detail how this program can be formally developed using stepwise refinement [1].

Its functional behaviour is straightforward. It assumes the existence of an array *msg*, of length *size*, containing a message to be transmitted. Special variable *out* is a memory-mapped output location. The program fragment of interest, between labels A and D, simply writes each character in *msg* to *out*.

The required timing behaviour is expressed through the directives labelled A, B, C and D. Statement A is an *assumption* [6] that we may expect to be true when this program segment starts executing. Let *start* be the absolute time by which the first character must be available, and *early* be the minimum duration before *start* at which the program fragment begins. Statement A tells us that initially the current time, denoted by imaginary specification variable **now**, is expected to be at least *early* time units before *start*. No executable code is generated for assumptions [6], but we may use them when analysing the program.

Directive B is a so-called real-time *coercion* [1]. Let duration *chsep* be the required separation of characters transmitted. The **deadline** directive occurs immediately after transmission of the nth character, and tells us that we must reach this point no later than n times *chsep* units from time *start*. This de-

```
|[ var size : nat;
      msg : array(0 .. (size − 1)) of char;
      out : char •
            ⋮
A:    {now + early ⩽ start};
      |[ var n : nat •
          n := 0;
          do n ≠ size→
              out := msg(n);
B:            deadline start + chsep ∗ n;
C:            delay until start + chsep ∗ n + chdef;
              n := n + 1
          od
      ]| ;
D:    deadline start + chsep ∗ size
   ]|
```

Fig. 2. High-level language program annotated with timing directives [1].

fines the *initial* availability of the nth output character, since it constrains the preceding assignment statement to finish *before* this time.

Statement C is a conventional **delay until** statement of the form found in a programming language like Ada 95 [2]. Let *chdef* be the minimum duration for which each output character must be stably defined. The delay statement tells us that we may pass this point no earlier than n times *chsep* plus *chdef* units from time *start*. This defines the *final* availability of the nth output character, since it prevents the program from performing any actions that may change *out* until *after* this time.

Finally, statement D is another coercion, this time placing an overall timing constraint on the whole code segment to finish within a time proportional to the number of characters in *msg* from *start*.

2.2 Timing path analysis

Hayes and Utting explain how the timing constraints expressed in Figure 2 can be simplified using control flow analysis [1]. All control paths between significant pairs of timing directives can be statically identified, and an overall worst-case execution time constraint for each such path defined. This reduces the timing requirements to simple execution time bounds on 'straight-line' code segments.

For the example in Figure 2, Hayes and Utting identify the following worst-case execution time constraints.

$$WCET(A\leadsto D) = early \qquad\qquad WCET(A\leadsto B) = early$$
$$WCET(C\leadsto B) = chsep - chdef \qquad WCET(C\leadsto D) = chsep - chdef$$

Path $A \rightsquigarrow D$ is that followed when the loop is not entered at all, i.e., when *size* is 0. In this case the expression at point D simplifies to '*start*', so the time available to get from A to D may be as low as '*early*'.

Path $A \rightsquigarrow B$ is the time taken to transmit the first character, including the loop initialisation activities. In this case n is 0 at point B, so again as few as *early* time units may be available to reach this point from A.

Path $C \rightsquigarrow B$ is the time taken to iterate, after the end of the stable period for the nth character at time $start + chsep * n + chdef$, up to the beginning of the stable period for the $(n + 1)^{\text{th}}$ character , and includes the time required to evaluate the (true) loop guard. In this case n is incremented in going from C to B, so '$n + 1$' must be substituted for 'n' in the expression at point B. Simplification then yields the overall constraint shown above.

Path $C \rightsquigarrow D$ is the time required to exit the program after the last character has been transmitted at time $start + chsep * (size - 1) + chdef$, and includes the overhead of evaluating the (false) guard for the last time. In this case it is known at point C that n equals $size - 1$.

The final machine code program must be proven to satisfy these worst-case path constraints. However, it is not necessary to check the timing behaviour of 'best-case' paths, such as $B \rightsquigarrow C$, because their correctness is guaranteed by correct implementation of the **delay until** statement.

2.3 Compilation

Figure 3 shows a MIPS R3000 [3] assembler code program developed from the HLL program in Figure 2. Here *size* and *msg* are symbolic constants representing local addresses for these variables; *body*, *delay* and *test* are symbolic instruction addresses; *chsep* and *dt* are compile-time constants, where *dt* equals $start + chdef$; and *clock* and *out* are hardware-dependent global addresses used to read the current absolute time and write to the output location, respectively (we assume an on-board synchronous clock that can be read with a normal load instruction, and a memory-mapped output location). Variables \$t0 to \$t7 are symbolic register names [3].

Keep in mind that, due to the way instructions overlap in a RISC pipeline, the instruction following a branch is *always* executed, whether the branch is taken or not [3]. Also the result of a load from memory that takes more than one cycle is not immediately available to the following instruction [3].

Timing points corresponding to the labelled directives in Figure 2 are prominently marked. A, B and D mark the points where upper timing bounds occurred. No code is generated for these constructs. C^α and C^ω mark the beginning and end of the **delay until** statement, respectively. The compiler generates several instructions to implement the lower timing bound required by HLL statement C. Since this statement is no longer atomic we separately mark both its beginning and end. The functional code appears in paths $A \rightsquigarrow B$ and $C^\omega \rightsquigarrow D$.

The **delay until** implementation in path $C^\alpha \rightsquigarrow C^\omega$ is a busy wait on label *delay*. At each iteration it reads the clock value, and compares it with the time at which the delay expires.

```
                  —— A ——
          la $t0, msg          t0 := msg base address
          lb $t1, size         t1 := size (length of msg)
          li $t2, 0            t2 := 0 (loop counter n)
          lb $t3, dt           t3 := first delay time (start + chdef)
          lb $t4, chsep        t4 := constant chsep
          j test               goto loop test
          nop                  jump delay slot
  body :  lb $t6, ($t5)        t6 := msg(n)
          nop                  load delay slot
          sb $t6, out          output location := msg(n)
                  —— B ——
                  —— Cᵅ ——
  delay : lb $t7, clock        t7 := current time
          nop                  load delay slot
          sub $t7, $t3, $t7    t7 := delay time − current time
          bgtz $t7, delay      busy wait if current time < delay time
          nop                  branch delay slot
          add $t3, $t3, $t4    next delay time := delay time + chsep
                  —— Cᵚ ——
          add $t2, $t2, 1      increment n
   test : bne $t1, $t2, body   goto loop body if n ≠ size
          add $t5, $t2, $t0    address of msg(n) := msg base + n
                  —— D ——
```

Fig. 3. MIPS R3000 assembler code for the high-level program in Figure 2.

2.4 Assembly

Figure 4 shows binary machine code generated for our program by the SPIM assembler [4]. The first column contains instruction addresses, and the second the instructions themselves. A mnemonic version is shown on the right.

Symbolic constants, addresses and register names have been substituted with particular values. More significantly, however, some assembler instructions have generated several machine code instructions. For instance, the first assembler 'load byte' from Figure 3 puts the value of local variable *size* in register '$t1'. The equivalent machine code consists of two instructions: firstly, a 'load upper immediate' stores data area base pointer 4097 in machine register number 1; secondly, a machine-level load byte instruction stores the contents of memory location 4097 plus offset 5 in register number 9 (the register number corresponding to name '$t1' [3]). It is for this reason that we can undertake accurate timing verification only on the final *machine code* program.

To support the timing verification phase some way of associating timing points with particular instructions is needed. The assembler must generate a table associating timing point *A* with address 0x00400020, point *B* with address 0x00400054, and so on.

```
                          —— A ——
[0x00400020] 0x3c081001    lui $8, 4097
[0x00400024] 0x3c011001    lui $1, 4097
[0x00400028] 0x80290005    lb $9, 5($1)
[0x0040002c] 0x340a0000    ori $10, $0, 0
[0x00400030] 0x3c011001    lui $1, 4097
[0x00400034] 0x802b0006    lb $11, 6($1)
[0x00400038] 0x3c011001    lui $1, 4097
[0x0040003c] 0x802c0007    lb $12, 7($1)
[0x00400040] 0x0810001c    j 0x00400070
[0x00400044] 0x00000025    or $0, $0, $0
[0x00400048] 0x81ae0000    lb $14, 0($13)
[0x0040004c] 0x00000025    or $0, $0, $0
[0x00400050] 0xa38e8000    sb $14, −32768($28)
                          —— B ——
                          —— Cᵅ ——
[0x00400054] 0x838f8001    lb $15, −32767($28)
[0x00400058] 0x00000025    or $0, $0, $0
[0x0040005c] 0x016f7822    sub $15, $11, $15
[0x00400060] 0x1de0fffd    bgtz $15, −12
[0x00400064] 0x00000025    or $0, $0, $0
[0x00400068] 0x016c5820    add $11, $11, $12
                          —— Cᵂ ——
[0x0040006c] 0x214a0001    addi $10, $10, 1
[0x00400070] 0x152afff6    bne $9, $10, −40
[0x00400074] 0x01486820    add $13, $10, $8
                          —— D ——
```

Fig. 4. MIPS R3000 machine code for the assembler program in Figure 3.

2.5 Timing verification

The timing verification phase checks the timing behaviour of machine code paths against the timing constraints derived from the HLL timing directives. For the purposes of this paper we used the SPIM S20 simulator [4] to verify the timing behaviour of the machine code program in Figure 4.

For clarity below, assume there are no caching overheads (although the SPIM tool is capable of simulating such overheads [7]). We can then treat all machine code instructions in Figure 4 as taking exactly one cycle. It is then merely necessary to count all instructions a path contains. Let $numi$ be the number of instructions for each path in Figure 4:

$$numi(A \rightsquigarrow D) = 12 \qquad\qquad numi(A \rightsquigarrow B) = 15$$
$$numi(C^\omega \rightsquigarrow B) = 6 \qquad\qquad numi(C^\omega \rightsquigarrow D) = 3$$

We must then compare the number of instructions in each machine-level path with the WCET timing constraints calculated during the HLL timing analysis phase (Section 2.2). This tells us that we must satisfy the following four equations. Let constant t be the elapsed 'real' time per instruction, calculated as the number of cycles per instruction multiplied by the machine-specific elapsed time per processor cycle [3]. The '*lateness*' value is explained below.

$$numi(A{\rightsquigarrow}D) * t \leqslant early$$
$$numi(A{\rightsquigarrow}B) * t \leqslant early$$
$$(numi(C^{\omega}{\rightsquigarrow}B) + lateness(C)) * t \leqslant chsep - chdef$$
$$(numi(C^{\omega}{\rightsquigarrow}D) + lateness(C)) * t \leqslant chsep - chdef$$

We are not required to prove any timing properties of path $B{\rightsquigarrow}C^{\alpha}$, which ends with a **delay until** statement. Nevertheless, the possible 'lateness' of this path must be considered in analysing the *following* paths $C{\rightsquigarrow}B$ and $C{\rightsquigarrow}D$. This is because any practical implementation of the HLL statement '**delay until** E' cannot guarantee to finish at *exactly* time E. Thus *lateness* above represents the number of instructions by which a path ending in a **delay until** statement may exceed its specified finishing time. For the implementation of statement C above we can calculate this as follows.

Firstly, if *chdef* is very small, then the delay time may pass while we are still executing the first few instructions following point C^{α}. In the worst case *every* instruction executed between C^{α} and C^{ω} contributes to an overrun.

$$lateness_{passed} = 6 - \left\lfloor \frac{chdef}{t} \right\rfloor$$

It takes 6 cycles to get from C^{α} to C^{ω} when the '*delay*' loop in Figure 3 does not iterate. However, we subtract the number of *whole* processor cycles that can be completed before duration *chdef* expires, because these do not contribute to lateness.

Secondly, when *chdef* is large, the **delay until** implementation in Figure 3 may overrun by going around the '*delay*' loop too often. The worst case is where the delay period expires fractionally *after* reading the time from the hardware clock. In this situation it may take up to two entire iterations beyond the desired finishing time to reach point C^{ω}.

$$lateness_{loops} = 11$$

Thus, since the actual value of *chdef* is unknown, the most we can say about lateness is as follows.

$$lateness(C) = max\{lateness_{passed}, lateness_{loops}\} = 11$$

Substituting these calculated execution times into the WCET constraints, and simplifying, yields the following inequalities.

$$15 * t \leqslant early \qquad\qquad 17 * t \leqslant chsep - chdef$$

If the programmer-supplied values for durations *early*, *chsep* and *chdef*, together with the machine-specific constant t, satisfy these equations then the program in Figure 4 can be considered to conform with the programmer's timing requirements.

3 Conclusion

The Topaz methodology for development of verified real-time machine code integrates recent advances in real-time program specification and timing analysis into traditional compiler technology. At the time of writing we are furthering the approach through the development of a formal model for representing program compilation [5].

Acknowledgements We are indebted to Ian Hayes, Mark Utting, Peter Kearney and Steve Grundon for inspiring many of the ideas used in this paper. This work is funded by Australian Research Council grant A49600176.

References

1. I. J. Hayes and M. Utting. Coercing real-time refinement: A transmitter. In D. J. Duke and A. S. Evans, editors, *BCS-FACS Northern Formal Methods Workshop*, Electronic Workshops in Computing. Springer-Verlag, 1997. http://www.springer.co.uk/ewic/workshops/NFM96/.
2. ISO. *Ada Reference Manual: Language and Standard Libraries*, 6.0 edition, December 1994. International Standard ISO/IEC 8652:1995.
3. G. Kane and J. Heinrich. *MIPS RISC Architecture*. Prentice-Hall, 1992.
4. J. R. Larus. Assemblers, linkers and the SPIM simulator. In J. L. Hennessy and D. A. Patterson, editors, *Computer Organization and Design—The Hardware / Software Interface*. Morgan Kaufmann, 1994.
5. K. Lermer and C. J. Fidge. Compilation as refinement. In *Proc. Formal Methods Pacific '97*, Wellington, New Zealand, July 1997. To appear.
6. C. Morgan. *Programming from Specifications*. Prentice-Hall, second edition, 1994.
7. A. Rogers and S. Rosenberg. *Cycle Level SPIM*. Department of Computer Science, Princeton University, July 1993.
8. K. G. Shin and P. Ramanathan. Real-time computing: A new discipline of computer science and engineering. *Proceedings of the IEEE*, 82(1):6–24, January 1994.

Schedulers for Age Constraint Tasks and their Performance Evaluation

Wolfgang Albrecht and Ralf Wisser

Universität Koblenz-Landau, Institut für Informatik,
{wal|wisser}@informatik.uni-koblenz.de

Abstract. Static hard real-time scheduling has to construct a schedule off-line by interleaving all required computations in a way which meets the application's timing requirements. We focus on the special case where real-time requirements impose so-called age constraints on tasks. An age constraint bounds the maximum distance between the repetitive executions of a task. For schedule construction such tasks are usually transformed into periodic ones, thereby imposing unnecessary load and thus causing unnecessary rejections of task sets. We present and compare new scheduling schemes better oriented towards the tasks' age constraints.

1 Introduction

In hard real-time systems a timing failure may lead to some intolerable catastrophic behavior. Thus a guarantee is needed that no timing failure will occur at runtime. We will construct a so called static schedule for the tasks carrying out the required computations. Such a schedule represents all reservations over time for individual tasks on the processor and is constructed off-line. If successful, the schedule will be executed cyclic at run-time.

Depending on the application there may be different kinds of real-time requirements, imposing different kinds of timing constraints on the placement of the associated tasks in the schedule. We focus on hard real-time static scheduling for tasks where a so-called age constraint is imposed on each task. An age constraint forces a task τ_i to be repeatedly executed. This has to be done in such a manner that the distance between two consecutive executions of the same task is bound by the age constraint A_i. To be exact, A_i bounds the maximum distance allowed between the beginning of a task's execution and the end of the task's consecutive execution. We now give some examples of what kind of hard real-time requirements will impose age constraints on the associated tasks.

The first example is about how to guarantee responsiveness. Looking at Fig. 1 assume a task τ_i to be responsible for computing some control instruction if a stimulus E_i occurs at some unforeseeable time. The instructions will be carried out as soon as the task's next starting execution, called τ_i^{k+1}, has finished. Thus an age constraint placement of τ_i will bound the response time – regardless of when E_i occurs. As another example, assume task τ_i is responsible for refreshing a data value (may be sensor readings or derived information) representing state information about the physical world. Assume the time the data value will be requested is unforeseeable and the data value(s) are up-to-date when used as input by τ_i starting execution. Looking again at Fig. 1, the freshness of the computed data value – requested for example at R_i – depends on the distance from the beginning of the last already finished execution τ_i^k of τ_i. Thus the maximum 'age', a state information is allowed to have, can be bound by an age constraint A_i.

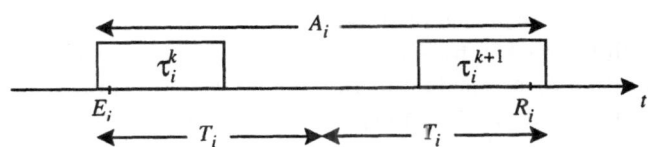

Fig. 1. Requirements imposing an age constraint on task τ_i^k

As in [Alb96] we have shown that preemptive[1] scheduling of a set of age constraint tasks is NP-hard, we can not suspect an efficient optimal solution, but only heuristic ones of varying quality. The problem of hard real-time scheduling subject to age constraint tasks has been examined in [Mok83, Kop86, ABR$^+$92, SL92]. The first two doing (explicit) static scheduling, the others focusing on (implicit) dynamic scheduling, assigning priorities to tasks before runtime. The scheme presented in the seminal work [Mok83] for scheduling age constraint tasks, i.e. transforming them into periodic tasks, has been adopted by all latter ones. In [ABR$^+$92] the usage of deadline monotonic scheduling (DMS) for age constraint tasks was proposed. This was done for dynamic scheduling, but clearly any dynamic scheduling discipline can be used for static schedule construction as well. In [Alb96] we have shown analytically that the basic version of our new scheduler $\mathcal{A}(\delta)$ will outperform a reasonable scheduling scheme based on DMS.

We will show later that transforming a task's age constraint into a periodicity constraint forces an unnecessary increase of utilization in most cases. That is way we will try in our first approach to avoid some of the unnecessarily imposed load by making some changes in the standard periodic tasks' scheduling scheme. Our second approach will prevent the transformation into periodic tasks at all; the placement of tasks is done with respect to the tasks' actual age constraints only.

The rest of this paper is organized as follows. Section 2 describes the underlaying model. In Sec. 3 we revisit the standard, periodic approach and show how to improve it. Section 4 will present our scheduling scheme that prevents the transformation into periodic tasks. In Sec. 5 the schedulers' performance evaluation is presented. Concluding remarks are offered in Sec. 6.

---

[1] A corresponding reduction in [Mok83] assumes tasks may not be preempted.

2 Model

An age constraint task $\tau_i = (C_i, A_i)$ is defined by its required maximum execution time C_i and its age constraint A_i. We force $C_i \leq A_i/2$ to hold. All parameters will be integers. Tasks are assumed to be preeemptible; but a task doesn't preempt itself before it's computation is finished. There are no exclusion or precedence constraints between different tasks. The repeated executions of a task τ_i is defined by its sequence $\tau_i^1, \tau_i^2, \tau_i^3, \ldots$ of jobs. A schedule for a set of age constraint tasks $\Theta = \{\tau_1, \ldots, \tau_n\}$ is a function $s : I\!N \to \{0, 1, \ldots, n\}$, where $I\!N$ corresponds to the sequence of timeslices to be divided among the tasks. A free timeslice will be marked by zero. For sake of brevity we will not discuss the cyclic representation of a schedule in this paper. With respect to a schedule s for each job τ_i^k let us mark b_i^k the beginning of the first and e_i^k the end of the last timeslice reserved for this job.

An age constraint A_i of a task τ_i is *satisfied* with respect to schedule s if for any two consecutive jobs τ_i^{k-1}, τ_i^k ($k \in I\!N$): $e_i^k - b_i^{k-1} \leq A_i$ is true and the intra task precedence relationship $e_i^{k-1} < b_i^k$ ($k > 1$) is met. For initialization we have chosen[2] a task's first deadline to be $d_i^1 = A_i - C_i$. A schedule s is *feasible* for a set of age constraint tasks Θ if all tasks $\tau_i \in \Theta$ satisfy their age constraint with respect to s. We will call $\delta_i = A_i - C_i$ the density of τ_i.

In the standard classical periodic process model a task $\tau_i = (C_i, T_i)$ is defined by its period T_i instead of an age constraint. A periodicity constraint T_i of a task τ_i is *satisfied* in a schedule s if for each job τ_i^k ($k \in I\!N$) there are C_i timeslices reserved between its ready time $r_i^k = (k-1)T_i$ and its deadline $d_i^k = kT_i$. We will call the interval $I = [r_i^k, d_i^k)$ the period-frame of τ_i^k. A schedule s is *feasible* for a set of periodic tasks Θ if all tasks $\tau_i \in \Theta$ satisfy there periodicity constraint with respect to s. If there is a periodic task $\tau_{i'} = (C_{i'}, T_{i'})$ such that any schedule satisfying its periodic constraint $T_{i'}$ will also satisfy the age constraint A_i of some task $\tau_i = (C_i, A_i)$ then we will say that τ_i can be *realized* by $\tau_{i'}$.

3 Realizing Age Constraints using the (Weakened) Periodic Task Model and EDF

The best results scheduling periodic tasks on a single processor are achieved using earliest deadline first scheduling (EDF) [LL73]. The strategy is: At any moment of time select – out of all ready jobs – the one with nearest deadline for execution. A necessary and sufficient condition for feasible scheduling a periodic task set Θ is given by $\sum_{i:\tau_i \in \Theta} C_i/T_i \leq 1$. In [Mok83] it is proposed that an age constraint task $\tau_i = (C_i, A_i)$ can be realized by a corresponding periodic one $\tau_{i'} = (C_i, T_{i'})$ setting its period $T_{i'} = A_i/2$, then using EDF for schedule construction. This method reflects the fact that in one period a job τ_i^k may be scheduled just at the beginning of that period-frame, while there may be other

---

[2] Our implementations [Wis96] start with $d_{1 \leq k \leq n}^1 = \sum_{i:\tau_i \in \Theta} C_i$, but any cycle found will guarantee the model's assumption.

jobs causing τ_i^{k+1} to finish not earlier than at the end of its period-frame (see Fig. 1). Instead of the optimum possible load of $C_i/(A_i - C_i)$ we will then have as much as $2C_i/A_i$ imposed by τ_i. Because the worst possible placement of jobs at the beginning of a period may only happen in rare cases, we can decrease the imposed load in the following way.

Assume a job τ_i^k (just integrated into the schedule) has a beginning time b_i^k that lay after it's ready time r_i^k by some amount $shift_i^k = b_i^k - r_i^k$. In order to satisfy the task's age constraint, it is no longer necessary to meet the deadline $d_i^{k+1} = (k+1)T_i$; instead the later $d_i^{k+1} = b_i^k + A_i$ will be sufficient. Thus we shift τ_i^{k+1}'s period-frame (it's ready time and deadline) by $shift_i^k$. If the scheduler can place τ_i^{k+1} somewhere in this shifted period-frame the age constraint is satisfied. Note, the gap in the sequence of period-frames spanned by τ_i leads to less load imposed by τ_i and that the probability that the scheduler can reserve the required C_i timeslices is the same if we shift or don't, at least in average case. We call this scheme ready time adjustment and if EDF is using it, we will call the scheduler EDF-RTA. The scheme can further be supported by manipulate some tasks' reservations. The goal is to delay a job's beginning time b_i^k as much as possible, but keep in mind not to harm other tasks' placements.

The first kind of situation where this can be done is if there are x free timeslices between the actual finishing time e_i^k of some job τ_i^k and it's deadline d_i^k. We can move the reservations of τ_i^k towards its deadline using the free timeslices, thereby getting the modified beginning $b_i^{k'} = b_i^k + x$. The second kind of scheduling situations where the beginning of a task τ_i^k can be delayed is if τ_i^k preempts a job τ_j^l of an (or some) other task(s) τ_j. We can move the reservations of τ_i^k towards its deadline by exchanging τ_i^k's timeslices with those of τ_j^l. There is no harm for τ_j, because b_j^l will not be changed – but we will get a larger $shift_i^k$ for τ_i^k. In order to do not harm another job's placement, one has to perform the both rearrangement as late as possible. That is when the task's consecutive job τ_i^{k+1} – the one that will have the shifted period-frame – would like to be started. We will call the modified scheduler realizing the both optimization EDF-RTA-Opt.

4 Realizing Age Constraints using Fixed Priority Scheduling

Our static schedule construction scheme called "age constraint fixed priority scheduling, according to tasks' density δ" ($\mathcal{A}(\delta)$, for short) works very much like standard fixed priority scheduling, such as rate monotonic scheduling (RMS). Remember, RMS first assigns priorities to tasks according to their period: the smaller the period the higher the priority (ties being broken arbitrarily). Each task's jobs will have the priority of the task. Then at any instant of time the job with highest priority – out of all ready jobs – is select for execution.

For $\mathcal{A}(\delta)$ we first assign priorities to each age constraint task with respect to its density, leading to a priority sorted task set $\Theta = \{\tau_1, \ldots, \tau_n\}$ of age constraint tasks $\tau_i = (C_i, A_i)$, $(1 \le i \le n)$. Tasks are labelled with respect to their priority,

τ_1 being the one with highest priority (i.e. smallest density). Constructing the schedule proceeds from the task with highest priority to the lower ones by reserving timeslices for their sequences of jobs $\tau_i^1, \tau_i^2, \ldots, (1 \leq i \leq n)^3$. The difference between ours and standard fixed priority scheduling is now in the reservation scheme. For each job τ_i^k actually to be integrated into the schedule, its placement will not be adjusted to some ready time but according to its deadline d_i^k. This deadline will be set depending on the job's predecessor's start time b_i^{k-1} and the age constraint A_i, that is $d_i^k = b_i^{k-1} + A_i$. Having got d_i^k, we then try to reserve free timeslices for τ_i^k going back in time from d_i^k until either C_i free timeslices could have been reserved or until a timeslice is reached which has been reserved for the job's predecessor τ_i^{k-1}. In the latter case the scheduler stops with no feasible schedule being constructed. Otherwise we go one until all tasks are integrated into the schedule[4].

Orienting the reservation scheme according to the deadline directly there is no more need for the concept of ready time. That is: no more need to guess when the time is right to begin to compete for the processor in order to be finished until deadline. The scheduler tries to place the consecutive jobs as far as possible into the future, thus trying to minimize the imposed load and in this way making it more amenable that lower priority tasks can satisfy their age constraints.

The reservation of timeslices for some job τ_i^k can further be supported by an optimization. If τ_i^k is looking out for timeslices to be reserved, it can sometimes get an already reserved timeslices of another (higher priority) job τ_j^l without any harm to that task. This is possible if τ_j^l can reserve another timeslice in the interval $I_j^{l+1} = [e_j^{l+1} - A_j, t_1)$ instead, with t_1 being the position of the timeslice under consideration to be given to τ_i^k. (In order to decide this τ_j^l's consecutive job τ_j^{l+1} must have been placed into the schedule already.) There may be such a timeslice if $e_j^{l+1} \neq d_j^{l+1}$, that is if τ_j^l's successor τ_j^{l+1} couldn't have been placed just before it's deadline $d_j^{l+1} = b_j^l + A_j$, because of some higher priority task laying there. Then – looking into the past – τ_j^l can start as early as $e_j^{l+1} - A_j$ and still satisfy it's age constraint.

We call this optimization backpropagation. The scheduler, called $\mathcal{A}(\delta)$-BP, will perform this strategy in a recursive manner for each timeslice in interval I_j^{l+1} in the following way: If τ_j^l is looking for an alternative timeslice because he will give one to τ_i^k, then τ_j^l can use the backpropagation technic itself in order to find another, third job that can give one of it's timeslices to τ_j^l etc. One may benefit from marking a job if it can not give one of it's timeslices for backpropagation and never ask him again – but as we don't bound the deep of this kind of recursion we have to expect exponential complexity with respect to schedule length as worst case behavior for this optimization technic. Thus we have to be cautious whether such an expensive optimization can produce acceptable results in performance evaluation.

---

[3] An implementation in fact doesn't have to place all jobs of all higher priority tasks before a lower priority job, but only those that might compete for some timeslices.

[4] In fact one will proceed until some cycle is found in the schedule.

5 Experimental results

The schedulers have been implemented [Wis96] in order to compare them experimentally. We will measure the probability of success P for feasible scheduling of task sets by slightly increasing the load. This is done with respect to the load that would have been imposed if we had used the conventional approach, setting tasks' periods to the half their age constraints. We call this load the compare utilization $U_{A/2} = \sum_{i:\tau_i \in \Theta} 2C_i/A_i$. Synthetic work load has been constructed for individual compare utilization values, ranging from $U_{A/2} = 0.9$ up to $U_{A/2} = 1.8$, passing through in steps of wide 0.02. For each step we have generated 100 tasks sets. The generation of a single task set consisting of 20 tasks works as follows.

Generate a pseudo-random number with a normal distribution for each task's age constraint and for it's share of the compare utilization as well. Having these two one can compute the corresponding execution time for each task. We assign each task a minimum execution time of one timeslice. Age constraints and execution times are represented as integers in the range of $10, \ldots, 1000$ and $1, \ldots, 1000$, respectively.

The machine used to run the tests was a work station Sun Sparc (75 MHz SPARCstation 20 MP). Because the length of the schedule to be constructed can in general make the problem intractable, we have limited the computation time for each run. Luckily the average case behavior is drastically better. The cycle- (and thus schedule-) lengths are less then 2 times the maximum age constraint in most of the (relevant) cases (Fig. 2 (b)). As the complexity of all schedulers except for $\mathcal{A}(\delta)$-BP is bound polynomial if the measure is with respect to the constructed schedule length, we found that there is no more need to bound the computation time (some seconds) in practice. This is different for $\mathcal{A}(\delta)$-BP. We have set a 5 minutes timeout for the schedule construction of an individual task set. The timeout has expired only for task sets with a compare utilization exceeding 1.6. We will see, that because of the corresponding success-ratio this is beyond practically interesting values anyway.

Figure 2 (a) shows the success-ratio of the schedulers with respect to the increasing compare utilization. $\mathcal{A}(\delta)$ has always found a feasible schedule for task sets imposing compare utilizations up to 1.45 – far above the 100% utilization bound the standard approach can reach. In order to show an upper bound for feasible scheduling we have included a curve representing the density utilization U_δ, measured with respect to the tasks' density $\delta_i = A_i - C_i$. That is, for each step of compare utilization we draw the mean value over all test sets Θ according to $U_\delta^\Theta = \sum_{i:\tau_i \in \Theta} C_i/\delta_i$. There will we no feasible schedule for an age constraint task set Θ with $U_\delta^\Theta > 1$. Note the obscuring fact that EDF has constructed feasible schedules for task sets exceeding the 100% utilization bound! This is because we don't follow standard period EDF scheduling when looking for a cycle. We have repeated the test for task set sizes 5,10 and 30 with similar results[5].

---

[5] Unexpectedly, the chance to find feasible schedules increases as the number of tasks increases – but this is only if the measure is with respect to the task sets' compare utilization $U_{A/2}$. If we underlay density utilization U_δ this effect vanishes.

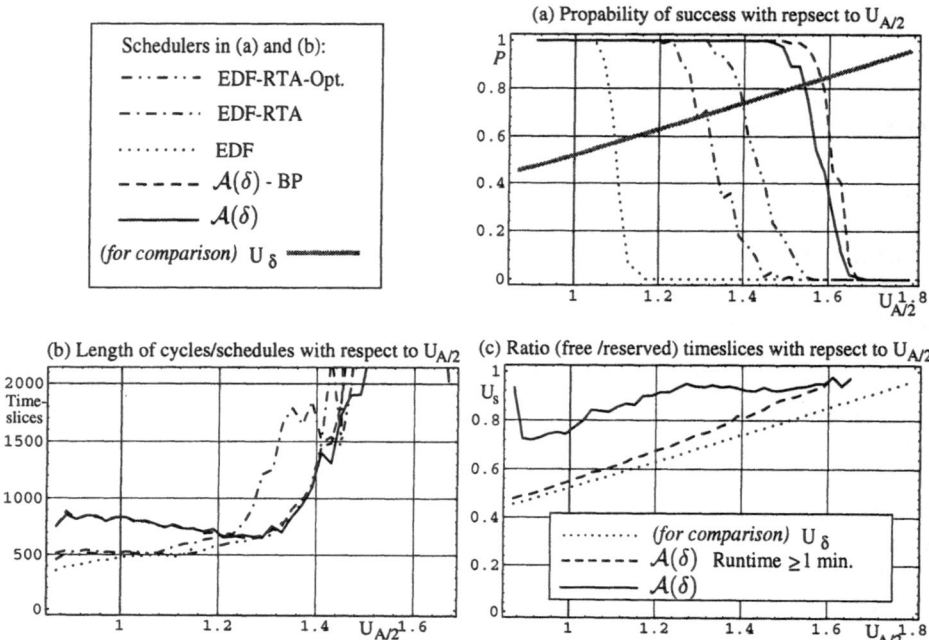

Fig. 2. Performance evaluation of the schedulers

Looking at Fig. 2 (b), the achieved cycle length seems to be mostly independent from an actual scheduler used for schedule construction; all schedulers' (except EDF-RTA) cycle length leaps up at $U_{A/2} \approx 1,4$. The strategy for finding cycles (not presented in this paper) used by all schedulers seems to dominate the influence of the schedulers themselves.

Up to now the schedulers proceeded, by constructing longer and longer schedules, until a first possible (with respect to the age constraints) cycle is found. Note – in opposite to the scheduling of periodic tasks – one may benefit proceeding further after the first possible point to close a cycle is found. There may come other points to close a cycle. In order to run some soft or non real-time tasks one may be interested in generating schedules with as many as possible timeslices left free. In Fig. 2 (c) the resulting fraction of free timeslices to schedule length U_s (the mean value of the corresponding test sets) is shown if we let the schedulers proceed at least one minute and then select the schedule (that is: the cycle) imposing minimum load. For comparison, the mean value to find the first cycle was less then 10 seconds. As the curve for the one-minute-version of $\mathcal{A}(\delta)$ is very close to the task sets' mean density, there is no need to increase the generation time further.

6 Conclusion

We have presented new heuristic scheduling schemes for the NP-hard problem of scheduling age constraint tasks. The first kind is based on earliest deadline first, the other on fixed priority scheduling. They impose less load because they are oriented on the real age constraints instead of on some derived periodicity constraints when constructing the schedule and in this way making it more likely to find a feasible schedule. Performance evaluation has confirmed this and shows that our schedulers can construct feasible schedules for a huge class of task sets that would have been rejected using the standard approach.

A pleasant by-product of leaving the periodic task model is the possibility to produce feasible schedules of very short cycle length. As in standard periodic scheduling only the smallest common multiple of all periods bounds the length [LM80], one usually has to shorten some periods in order to make them harmonic. But then we have a trade-off with respect to the increasing load imposed by smaller periods, thus making it less likely to find a feasible schedule at all.

In order to solve realistic problems future work will have to integrate tasks with real periodicity and synchronization requirements. The new schedulers' better capability in finding a feasible schedule if hard real-time requirements are given in form of age constraints may be of particular interest for distributed real-time systems. In order to maintain the state of the physical world with sufficient actuality one may consider using our scheduling schemes for a shared communication medium instead for the processor.

References

[ABR+92] N. Audsley, A. Burns, M. Richardson, K. Tindell, and A. Wellings. Absolute and relative temporal constraints in hard real-time databases. In *Proc. of IEEE Euromicro Workshop on Real Time Systems*, Feb. 1992.

[Alb96] W. Albrecht. Echtzeitplanung für Alters- oder Reaktionszeitanforderungen. Technical report, Institut für Informatik, Universität Koblenz, Dez. 1996.

[Kop86] H. Kopetz. Design of real-time systems. In R. Güth, editor, *Computer Systems for Process Control*. Plenum Press, New York, 1986.

[LL73] C. L. Liu and J. W. Layland. Scheduling algorithms for multiprogramming in a hard-real-time environment. *Journal of the ACM*, 20(1):46–61, Jan. 73.

[LM80] J. Y-T. Leung and M. L. Merrill. A note on preemptive scheduling of periodic real-time tasks. *Information Processing Letters*, 11(3), Nov. 1980.

[Mok83] A. K. Mok. *Fundamental design problems of distributed systems for the hard-real-time environment*. PhD thesis, Massachusetts Institute of Technology, 1983.

[SL92] X. Song and J. W. S. Liu. Maintaining temporal consistency: Pessimistic vs. optimistic concurrency control. In *Proc. of the IEEE Symposium on Computer-Aided Control System Design*, Napa, California, Mar. 1992.

[Wis96] R. Wisser. Verplanung von Prozessen mit Altersanforderungen. Studienarbeit, Institut für Informatik, Universität Koblenz, Okt. 1996.

Analyzing Schedulability of Astral Specifications using Extended Timed Automata

K. Brink, J. van Katwijk, R.F. Lutje Spelberg, W.J. Toetenel

Delft University of Technology
Department of Technical Mathematics and Computer Science
Zuidplantsoen 4, 2628 BZ Delft, The Netherlands
{k.brink, j.vankatwijk, r.f.lutjespelberg, w.j.toetenel}@twi.tudelft.nl

Abstract. This paper reports our experiences with using an extension of timed automata [1] for schedulability analysis of prototype implementations. The approach builds upon requirements specifications constructed using the formal real-time specification language Astral [7]. Astral specifications are translated into extended timed automata. The resulting automata are augmented with implementation details like assignment of processes to processors, priorities, worst-case execution times of operations, and scheduling policies. Schedulability analysis is then performed by (automated) formal verification of the extended automaton.

1   Introduction

Our research is aimed at constructing a software development framework for real-time embedded control software which is based upon the use of the formal specification language Astral [9, 7]. The choice for Astral was motivated by its applicability in requirements specification [3] and its possibilities for verification [6]. Research has addressed implementation of real-time Astral specifications and resulted in a mapping of Astral primitives upon Ada95 language constructs [4, 5]. It proved to be very difficult to assess the correctness of the implementation's timing behavior. Rate-Monotonic Analysis (RMA) showed to be applicable to a limited extent only [5].

An alternative approach to schedulability analysis was proposed by Corbett [8]. He showed that abstract models of Ada95 programs could be constructed and subsequently analyzed using hybrid automata. Our approach starts from the opposite direction, by augmenting Astral specifications with implementation details concerning timing behavior. An overview of our approach is given in Fig. 1. We defined an extension of timed automata [1] called urgent extended timed automata (UXTA). Requirements analysis of Astral specifications can be done by translating Astral specifications into these automata (referred to as *abstract automata*) and subsequently applying model checking techniques.

To be able to perform schedulability analysis, a number of details of the intended implementation have to be known. These details, called an *implementation scheme*, could be extracted from a (partial) implementation, or based on experience or design decisions. An implementation scheme describes four aspects

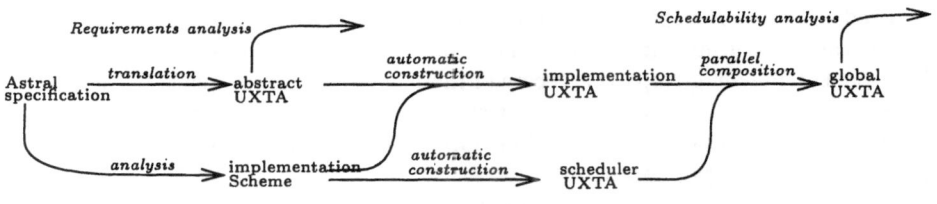

Fig. 1. Overview of the approach

of an Ada95 implementation of an Astral specification: (i) a mapping from processes to processors, (ii) a mapping from processes to priorities, (iii) worst case execution times for actions, and (iv) a scheduling policy. The work described here is focused on (timing) analysis of *prototype implementations* of Astral specifications. To that end the abstract automata resulting from the translation are refined by taking into account the worst-case execution times and the mapping of processes to physical processors. We will refer to these refined automata as *implementation automata*. A separate automaton is constructed which encompasses the scheduling policy as described in the implementation scheme. Then the composition of this *scheduling automaton* and the implementation automaton is taken and used as input for our schedulability analysis. To show the feasibility of the approach, we defined transformations from UXTA's into timed automata variants for which verification tools exist. The resulting automata are verified using the model-checking tools HyTech[10] and Uppaal [11].

Section 2 discusses the specification language Astral and illustrates its use with fragments of an example specification. Section 3 defines urgent extended timed automata, and discusses the translation from Astral specifications. Section 4 shows how the resulting UXTAs are refined using an implementation scheme and briefly discusses the final verification step. Section 5 presents conclusions and suggestions for future work.

2 Real-Time System Specification in Astral

An Astral system specification consists of a global specification and a set of process-type specifications. Process behavior is specified operationally, i.e. by defining state and a set of transitions. Furthermore Astral offers constructs for the specification of constraints over the system's behavior.

To illustrate the use of Astral an example specification of a packet-maker system [9] is discussed. The packet-maker system collects messages from input channels, combining them into packets and sending these packets out. The system contains N receivers and a single packet maker. Upon receipt of a data message from its input channel, a receiver stores the data message in its buffer and marks the input channel as active. Whenever *Input_Tout* time units have elapsed since the receipt of a data message, the receiver stores a special message, called the *Closed* message in its buffer and marks its input channel as inactive. The packet maker groups the received messages into larger packets of K messages and delivers them to the output channel. An incomplete packet is delivered in case all receivers have marked their input channel as inactive and at least one message has been processed.

The packet maker specification defines two kinds of processes: input processes which receive messages from an input channel and a packet maker process which assembles the messages into packets. The `global specification` part contains a process declaration part in which a (fixed) number of processes of the system are instantiated. A process-type specification defines the behaviour of a process, by specifying a *state definition*, *initial states* and a set of *transitions*. Each transition is defined by a duration, a pre-condition defining the set of states in which the transition is enabled and a post-condition defining the effect of the transition. The following excerpt from the `Packet_Maker` process type illustrates how state variables, initial states and transition are specified:

```
variable
  No_Of_Msg    : integer;
  Previous     : array [1..N] of Time;
initial
  No_Of_Msg = 0 & forall i:Receiver_ID; (Previous[i] = 0)

transition Deliver                Del_Dur
pre
  No_Of_Msg = K ∨ { No_Of_Msg > 0 &
  forall R_id:Receiver_ID; (Receiver[R_id].Msg = Closed) }
post
  No_Of_Msg = 0
```

Each process is assumed to execute on its own processor and transitions are executed as soon as they are enabled unless another transition of the same process is already executing. If more than one transition of the same process is enabled, one is chosen non-deterministically. Start and end of a transition execution are assumed to be atomic. The amount of time between start- and end time of a transition execution is defined by a (non-null) transition duration.

In Astral time is dense and the existence of a global clock is assumed. Every process can refer to the current time through the variable *now*. Two standard functions Start(<transname>) and End(<transname>) return the time instant in the past at which the execution of the transition named <transname> started (resp. ended) for the last time. Processes communicate through shared variables; Processes can read variables that are exported by other processes by importing them. Transitions can also be exported and allows other processes to refer to their latest start- and end-times through the Start()- and End()-construct.

The fragment below shows the specification of the transitions of the receiver process type. A more comprehensive overview of Astral can be found in [7].

```
transition New_Info                N_I_Dur
```

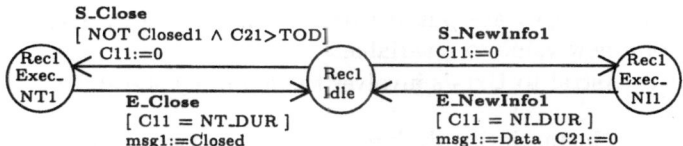

Fig. 2. UXTA of receiver process

```
pre
  now - Start(New_Info) >= NI_delay
post
  Msg = Data & not Channel_Closed & NI_delay ISIN [0,Max_Del_Ni]

transition Notify_Timeout                    N_T_Dur
pre
  now - Start(New_Info) >= Input_Tout & not Channel_Closed
post
  Msg = Closed & Channel_Closed
```

3 Translation into Urgent Extended Automata

Timed automata [1] extend automata through the introduction of clocks that change uniformly with time. The clocks are used to constrain the execution of transitions and can be reset by any transition. UXTA's extend traditional timed automata in two ways. Firstly, they allow the modeling of real valued variables. Secondly, transitions represent *urgent* actions meaning that an enabled transition must be taken as soon as possible. This matches closely to the maximal progress semantics of Astral. Urgent transitions are also present in other timed automaton variants [10, 11] but they only allow urgency on unguarded edges. Our general form of urgent transitions allows the modeling of transitions triggering on data conditions: a transition of an automaton changing the value a data element may trigger the execution of a transition of another component automaton.

An UXTA is a tuple $A = (Loc, Clk, Var, Lab, Edg)$, where Loc is a set of locations, Clk is a set of clocks, Var is a set of real-valued variables, Lab is a set of transition labels, and Edg is a set of edges. An edge $(l, a, \mu, l') \in Edg$ is defined by a source location and a destination location $l, l' \in Loc$, a label $a \in Lab$, and a transition relation $\mu \subseteq (V_v \times V_c) \times (V_v \times V_c)$. V_v denotes the set of variable valuations, and V_c the set of clock valuations. The transition relation incorporates both the guard of a transition, as well as the associated update of clocks and data. The semantics is defined similarly to that of timed automata, and is not given here.

Figure 2 shows an example UXTA representing a model of the receiver of the packet maker system. In this figure, transition labels are typeset in bold-face, conditions on the edges are put between square brackets, and ':=' expresses assignment of a new value to a variable.

Translating Astral to Uxta's involves the following steps:

- Each process instantiation declared in the global specification is mapped to a corresponding UXTA.
- Expressions referring to absolute time instants are converted to expressions that refer to clocks. This is discussed in more detail below.
- Data elements are either: left untouched, expanded (if finite), or abstracted away. When abstracting from variables, all constraints referring to them are set to *True*, which causes additional behaviors to be added to the system. Then our verification procedure becomes partial; a negative result does then not necessarily imply incorrectness.
- Finally, the composition of the resulting UXTA's is taken.

The mapping of process instances results in UXTA's with one *Idle* location and a set of locations *Exec_i*, one for each transition i. There is an edge from the *Idle* location to each *Exec_i* location representing the start of the corresponding transition. From each *Exec_i* location there is an edge leading back to the *Idle* location. With each process a clock is associated which enforces the duration of Astral transitions. The condition associated with an edge to an *Exec_i* location is equal to the precondition of the corresponding Astral transition and a reset of the clock variable which is associated with the start time of this transition. The condition associated with the edge from the the *Exec_i* to the *Idle* location equals the postcondition of the corresponding Astral transition and an additional clock condition which ensures that the amount of time passed in the *Exec_i* location equals the duration of the corresponding Astral transition.

In Astral, there are four ways to refer to time instants: the Start() and End() constructs, the variable *now*, and variables of type *Time*. The latter are variables that can be assigned the current value of *now*. When expressing behavior in time, one is usually interested in relative time constraints. This means that the four Astral primitives that refer to absolute time instants are in most cases used to specify minimum or maximum durations. In timed automata, relative time constraints can easily be expressed by means of clocks. The translation of a specification which refers to absolute time instants into a specification in which the current time is left implicit, is done by rewriting the terms in which absolute time references occur. For this purpose, additional clocks are introduced to register the amount of time that has elapsed since a relevant event has occurred.

When converting an expression containing a Start(T) primitive, a clock C_T is introduced which is reset on the UXTA edge that is associated with the start of T. Then at any moment the equation Start(T) = now − C_T holds. Similar conversions can be made for End(T) primitives. When dealing with relative time constraints, this rewriting will result in expressions in which the now variable is always factored out. As an example, consider the rewriting of the pre-condition of the New_Info transition now − Start(New_Info) ≥ NI_delay to $C_{New\_Info} \geq$ NI_delay. An analogous process is used to replace state variables of type *Time* (used to register the time instant of a specific event). Figure 2 presents the automaton that results from translating a receiver process from the packet maker specification. For a more detailed discussion and a formal definition, see [2].

4 Schedulability Analysis

To be able to apply schedulability analysis, the automata resulting from the above described translation are refined to include the consumption of resources. The duration of Astral transitions are interpreted as upper bounds on the time between enabling and completion of the transition. When applying schedulability analysis, one also takes in account the actual execution times. Goal of our analysis is to determine, given worst-case bounds on execution times, whether or not all transitions will always execute within their transition durations.

Refinement of the UXTA. Every $Exec_i$ location is split in three locations, which represents the fact that the corresponding Astral transition (1) is waiting to be executed, (2) is executing, or (3) has completed. An additional clock is introduced for every process to keep track of the execution times. Figure 3 shows how the refinement works out for both non-preemptive and preemptive scheduling. The leftmost figure shows an automaton fragment associated with some transition y of process x. $C1x$ is the clock that enforces the Astral transition time. The middle figure shows how the $Exec$ location is refined in case of non-preemptive scheduling. The $Exec\_i$ location is replaced by three locations $Wait\_i$, $Running\_i$, and $Ready\_i$. The G_x transition marks the start of the actual execution, Y_x its completion. The clock $C0x$ enforces enforces a delay equal to the execution time of the transition. The latter is represented by the constant RT_y. The rightmost automaton fragment show the refinement for preemptive scheduling. In this case an additional preemption transition P_x is incorporated. Note that G_x and P_x synchronize with the scheduling process.

Specification of the Scheduling Policy. The next step is to model the scheduling of transitions according to some scheduling policy. Figure 4 shows a pair of extended automata that models a non-preemptive priority-driven scheduler for three processes with different priority levels. The left automaton takes care of the start actions (S_i). An S_i action, which represents the fact that an action from process i becomes enabled, causes the req_j counter to be increased by one. j denotes the priority level of process i. In this particular case we chose priority levels to match process identifiers. The rightmost automaton selects a transition with the highest priority. The complete system is now constructed by taking the composition of the implementation UXTA and the automaton defining the scheduling policy. Other scheduling policies, like for example preemptive scheduling, can equally simple be defined by means of automata.

Verification. The techniques discussed above provide us with an interpretation of Astral specifications that incorporates the binding of Astral processes to resources. Goal is to verify whether or not this system is a correct refinement of the

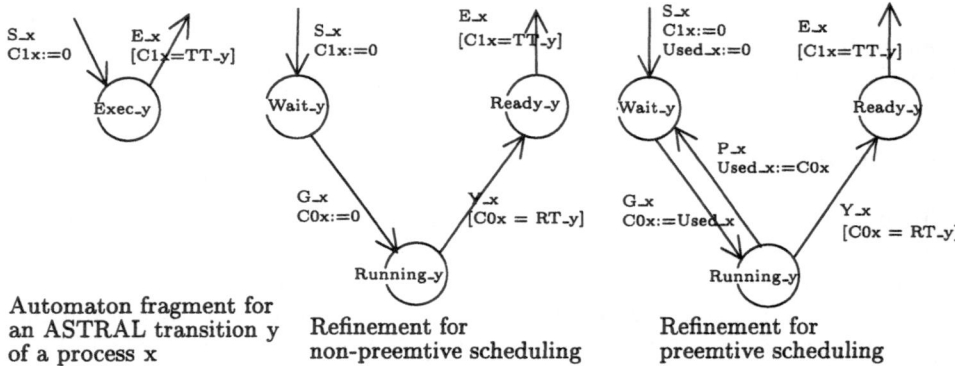

Automaton fragment for
an ASTRAL transition y
of a process x

Refinement for
non-preemtive scheduling

Refinement for
preemtive scheduling

Fig. 3. Refinement of transitions

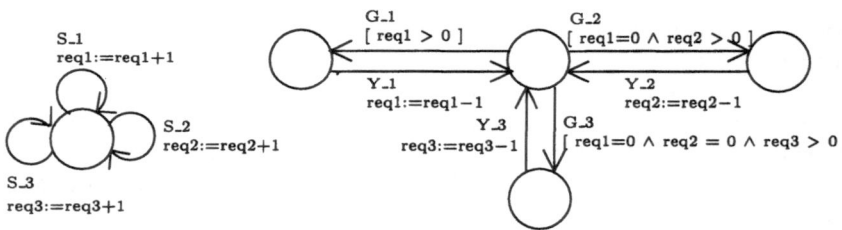

Fig. 4. Non-preemptive priority-driven scheduler

abstract automaton. Obviously, this is the case if every transition always completes within its deadline. This means that for any transition y of a process x, the *Ready_y* location is reached no later than the moment $C1x$ becomes TT_y. If an action does not complete within its deadline it will cause its process automaton to block in the *Ready_y* state. Since we do not have a verification tool for our type of automata, we used the symbolic model checking tools HyTech [10] and Uppaal [11] to show the feasibility of our approach. Since the timed automaton variants that are employed by these model checkers do not allow our general form of urgency, transformations from our timed automaton extension had to be defined. The page limit does not allow us to discuss this transformation. Also we will not discuss the details of verification. In short, when using HyTech we applied a *non-zenoness* analysis, while with Uppaal we performed an analysis by introducing *error states*. As a first experiment we applied our approach to a packet maker system with one receiver and a non-preemptive scheduling policy. With HyTech, we were able to verify this system, both through non-zeno analysis, and by introducing error states. For now HyTech seems to fail on larger systems. With Uppaal on the other hand, we were able to verify a packet maker system with two receivers, using error states.

5 Conclusion

The main advantage of the approach we experimented on, is its generality. Approaches like RMA are focused on very specific types of systems, and are not generally applicable. The limitations of our approach on the other hand, lie in the size of the systems that can be analyzed. We applied the discussed techniques to a small example and have shown that for this example schedulability analysis is possible. It is clear that when taking up larger systems, one will at some point be faced with the state space explosion that is inherent to model checking. When judging the scalability of our approach, one must keep in mind that a considerable amount of complexity is introduced when UXTA models are translated into hybrid automata. A verification tool that allows general urgent transitions would avoid this redundancy. Furthermore, it seems that current model checking tools have not yet reached the full potential of the model checking approach.

References

1. R. Alur and D. Dill. The theory of timed automata. In *Proceedings Real-Time: Theory and Practice*, volume 600 of *LNCS*, pages 45–73. Springer-Verlag, 1991.
2. K. Brink. *Interfacing Control and Software Engineering: A Formal Approach*. PhD thesis, Delft University of Technology, 1997.
3. K. Brink, L. Bun, J. van Katwijk, and W.J. Toetenel. Hybrid Specification of Control Systems. In *Proceedings of ICECCS'95*, pages 149–152, 1995.
4. K. Brink, J. van Katwijk, and W.J. Toetenel. Ada95 as Implementation Vehicle for Formal Specifications. In *Proceedings of RTCSA'96*, pages 98–105. IEEE Computer Society Press, 1996.
5. K. Brink, J. van Katwijk, and W.J. Toetenel. Implementing Distributed Real-Time Specifications in Ada95. In *Proceedings of the 2nd annual conference of the advanced school for computing and imaging*, pages 84–89, 1996.
6. L. Bun and J. van Katwijk. The ASTRAL Specification of the Railroad Controller. Technical Report 95-104, Faculty of Technical Mathematics and Informatics, Delft University of Technology, 1995.
7. A. Coen-Porisini, C. Ghezzi, and R.A. Kemmerer. Specification of Realtime Systems Using ASTRAL. Technical report, TRCS96-30, Faculty of Computer Science, University of California, Santa Barbara, 1996.
8. J.C. Corbett. Timing analysis of ada tasking programs. *IEEE Transactions on Software Engineering*, pages 461–483, July 1996.
9. C. Ghezzi and R.A. Kemmerer. Executing formal specifications: the ASTRAL to TRIO translation approach. In *Proceedings TAV4*, pages 112–119, 1991.
10. T. A. Henzinger and P. H. Ho. A user guide to HyTech. In *Proceedings of TACAS'95*, volume 1019 of *LNCS*, pages 41–71. Springer-Verlag, 1995.
11. K. G. Larsen, P. Petterson, and W. Yi. Model checking for real-time systems. In *Proceedings of FOCS'95*, volume 965 of *LNCS*, pages 62–88. Springer-Verlag, 1995.

Deriving Annotations for Tight Calculation of Execution Time[*]

Andreas Ermedahl[1] and Jan Gustafsson[2]

[1] Dept. of Computer Systems, Uppsala University, Sweden, ebbe@docs.uu.se
[2] Dept. of Computer Engineering, Mälardalens högskola, Sweden, jgn@mdh.se

Abstract. A number of methods have been presented to calculate the worst case execution time (WCET) of real-time programs. However, to properly handle semantic dependencies, which in most cases is needed to reduce overestimation, all these methods require extra semantic information to be given by the programmer (manual annotations for paths, loops and recursion depth). To manually derive these annotations is often difficult and the process is error-prone. In this paper we present a new method to automatically derive safe and tight annotations for paths and loops. We illustrate our method by giving some examples and by presenting a prototype tool, implementing the method for a subset of C.

1 Introduction

Real-time systems are systems in which the correctness depends not only on the results of computations, but also on the time at which the result is produced. To be able to guarantee the deadlines of a real-time system, the software execution time is needed before run-time.

The execution time of most programs depends on the input data and the system state. For programs with some complexity, it is intractable to find the data and the state which causes the *actual* worst case execution time ($WCET_A$). This approach is therefore not a feasible method. Instead, *static analysis*, which from the source code derives the *calculated* worst case execution time ($WCET_C$), has been proposed by many researchers. The calculation must be *safe* (i.e., $WCET_C \geq WCET_A$), yet as *tight* as possible, to avoid waste of resources.

To achieve a tight $WCET_C$, more information about the program behaviour, than what is contained in the flow graph, is needed. *False paths* (non-executable paths, i.e., paths that never can be taken) must be identified and excluded from the calculation. Maximum number of *iterations* in loops and maximum *depth of recursion calls* must be given, because to calculate them is, in the general case, equivalent to the well-known halting problem.

Existing methods require this information to be given as *manual annotations*. However, a fundamental problem with this approach, beside being difficult and

---

[*] This work is performed within the ART-project within the Advanced Software Technology (ASTEC) competence centre, and is supported by the Swedish board for technical development (NUTEK), IAR Systems AB, Mecel AB and Uppsala University.

time-consuming for the programmer, is that the annotations may be *incorrect*. The WCET$_C$ may be untight or, worse, unsafe.

In this paper we present a static analysis method which *automatically* derives safe and tight annotations from the program semantics. It can be seen as a first phase in a tight WCET$_C$ calculation. These annotations can be used by the following phase, an object code analysis, which also considers modern hardware architectures. If simple hardware is used, this next phase may be unnecessary, and the WCET$_C$ calculated by our method can be used.

The remainder of the paper is organised as follows: The next section illustrates how our approach relates to other methods. In section 3, we introduce our new method. In section 4 we present our tool and an illustrative example. Finally, section 5 gives some results, conclusions and ideas for future work.

2 Related work

Timing analysis of software has been an important area in real-time research during the last 8 - 10 years. The issues this research has dealt with are:

1. Mapping the high-level source code constructions to the corresponding object code instructions by *static analysis*. Compiler optimisations will make this more difficult [19].
2. Deriving *dynamic* properties of programs, i.e., how many times each instruction will be executed in the worst case. Information of *false paths*, maximum number of *iterations* in loops and maximum *depth of recursion calls* are normally given as manual annotations. However, *symbolic execution methods* has been proven to find some false paths automatically [1,4].
3. Calculating the WCET$_C$ using the static and dynamic information above. *Timing schema* [15,16] analysis calculates WCET$_C$ by recursively adding the times for the constructs in the program. Another method is *Integer Linear Programming* (ILP) [9,17] analysis, which is used in most recent research. It transforms the program to a flow graph where each edge corresponds to a basic block. For each edge, the worst execution time is calculated from the instructions in the basic block. The total execution time is represented as a linear expression, and the WCET$_C$ is calculated by finding the maximum of the expression with linear programming optimisation.
4. Modern hardware with cache and pipeline is analysed in extensions to TS [11], ILP [10] or using constraint techniques [13]. This analysis must take into consideration the current cache and pipeline contents before each instruction execution. To get a tight calculation, detailed information on dynamic program behaviour, especially nested loops, is needed.

As we can see, all published methods rely more or less on manual annotations to work and to give a small overestimation. But to give these annotations is extra work for a stressed programmer, and it is certainly easy to find very simple examples where, e.g., the maximum number of iterations in a loop is very hard to calculate (see Fig. 4(a) for an example). And what happens if the manual annotation is incorrect? One possible effect is that a program could be given a too small time slot in a schedule, ending up in a missed deadline, with possible catastrophic consequences.

Park discusses this problem in [14], proposing a method that verifies the correctness of the annotations. But why not let the analysis method try to find the annotations automatically, as they are inherent in the semantics of the program?

3   Deriving Annotations

In our analysis of a program we will use data flow analysis to find out values of variables at different points in the program. Using this information we can automatically derive path and loop annotations.

Control points, $c_0, c_1, \cdots, c_m$, are introduced at points in the program where the value of a variable may change (after assignments), or when we can constrain the possible values of variables (after conditions). For an example, see Fig. 1(a).

With each control point, c_i, we associate an *environment*, σ_i^h. An environment holds all combinations of variables values that are possible at the control point in a program execution. $\sigma_i^h = \{a_1 \mapsto v_1, \cdots, a_m \mapsto v_m\}$ denotes an environment where the variables $a_1, \cdots, a_m$ have been assigned the values $v_1, \cdots, v_m$, respectively. We will by the index h separate between different passings of the same control point, e.g., in loops and continuations after selection statements (h will in the sequel only be included when necessary).

$[c_0]$	σ_0	$\sigma_0 = \{a \mapsto 1, b \mapsto 4\}$	$\sigma_0 = \{a \mapsto 1..3, b \mapsto 2..4 \lor 7\}$
$\mathtt{a = 5}; [c_1]$	$\sigma_1 = [\![\mathtt{a = 5}]\!]\sigma_0$	$\sigma_1 = \{a \mapsto 5, b \mapsto 4\}$	$\sigma_1 = \{a \mapsto 5, b \mapsto 2..4 \lor 7\}$
$\mathtt{b = b - 2}; [c_2]$	$\sigma_2 = [\![\mathtt{b = b - 2}]\!]\sigma_1$	$\sigma_2 = \{a \mapsto 5, b \mapsto 2\}$	$\sigma_2 = \{a \mapsto 5, b \mapsto 0..2 \lor 5\}$
$\mathtt{a = a + b}; [c_3]$	$\sigma_3 = [\![\mathtt{a = a + b}]\!]\sigma_2$	$\sigma_3 = \{a \mapsto 7, b \mapsto 2\}$	$\sigma_3 = \{a \mapsto 5..7 \lor 10, b \mapsto 0..2 \lor 5\}$
(a)	(b)	(c)	(d)

Fig. 1. The statements (with inserted control points) in (a) gives the semantic rules in (b) and the concrete and abstract evaluation in (c) respectively (d).

3.1   Concrete and abstract semantics

The *concrete semantics* (meaning) of a program is defined as the environments that can be generated by the program description [12]. We will use a semantic function, $[\![\cdot]\!]$, that takes an environment, σ, and a *rule* on how to modify (or constrain) the environment and return a modified environment: $\sigma' = [\![rule]\!]\sigma$. Depending on the nature of the rule, i.e., if it is a statement or a condition, we will further subdivide $[\![\cdot]\!]$ into $\mathcal{S}[\![\cdot]\!]$ and $\mathcal{C}[\![\cdot]\!]$[1]. See Fig. 1(b) for the semantic rules that corresponds to the statements in 1(a). If each variable in the initial environment, σ_0, is assigned to a single value, like $\sigma_0 = \{a \mapsto 1, b \mapsto 4\}$, the evaluation of the equations will correspond to a "normal" execution of the program, see Fig. 1(c).

---

[1] $\mathcal{S}$ stands for *statement* and $\mathcal{C}$ for *condition* rule.

For our analysis, we will define an *abstract environment* where variables can be assigned to several values. For each concrete semantic rule, in the programming language, a corresponding abstract rule is defined. For example, our abstract version of the '+' operator handles sets of values. An abstract evaluation can be seen in Fig. 1(d). Note that the abstract evaluation corresponds to a set of concrete evaluations and that each concrete evaluation corresponds to a possible execution.

In this paper, and in our tool (see section 4), we will represent abstract values with *split integer intervals*. For example, b $\mapsto$ 2..4 $\vee$ 7 means that b:s value is either 7 or between 2 and 4. This representation has some drawbacks, e.g., it does not express conditions between variables in an environment, on the other hand it is simple to manipulate and allows efficient implementations.

The domain of the environments is in the general case a partially ordered set (poset), $\langle \mathcal{D}, \subseteq, \bot, \cup, \cap \rangle$, The set $\mathcal{D}$ contains possible combinations of value tuples for variables. The bottom element, $\bot$, means that one or several of the variables within the environment can not have any value at all. $\sigma_1 \subseteq \sigma_2$ is true iff at least all tuples of variable value combinations that exists in σ_1 also exists in σ_2. $\sigma_1 \cup \sigma_2$ creates a new environment that holds exactly all tuples of variable value combinations in *both* σ_1 and σ_2.

The cost, in terms of time and memory, to express an exact environment, i.e., all possible tuples of variables value combinations, is often to expensive. The representation can then be simplified by a safe abstraction, e.g., $a \mapsto 1 \vee 3 \vee .. \vee 99$ can be safely approximated to be in the interval 1..99. The approximation must be safe (possible values must not be removed), tight (as few extra values as possible), and efficient (in terms of time and memory). We will face a trade-off between cost of computation and quality of results. For the above reason we will also use an approximative meet operator $\cup'$ that applied on σ_1 and σ_2 creates a new environment that holds *at least* all tuples of variable value combinations in σ_1 and σ_2. Abstract interpretation techniques [5] can be used to define a correct relation between the abstract and the concrete domains.

3.2 Finding false paths

We will use a sequence of if-statements to illustrate how dependencies between different program parts can be found.

A condition can be seen as a constraint to be applied on the variables in a given environment. Fig. 2(c) shows how the start environment, $\sigma_0 = \{a \mapsto 0..20\}$, will be constrained[2] by the conditions in the two if-statements in Fig. 2(b). The evaluation in Fig. 2(c) has the disadvantage that it does not take into account the dependencies between the if-statements (a > 10 implies a > 5). Thus, it will not detect that $S_1 \to S_4$ is a false path. Our solution is to continue the analysis from each of the two environments that are generated after an if-statement, giving the semantic rule in Fig. 2(a).

---

[2] We are assuming that a will not be changed in any of the statements $S_1 \ldots S_4$.

$\mathcal{S}[\![\texttt{if}(C)\ S_1\ \texttt{else}\ S_2]\!]\sigma = \{\sigma_2, \sigma_4\}$
  where
  $\sigma_1 = \mathcal{C}[\![C]\!]\sigma$
  $\sigma_2 = \mathcal{S}[\![S_1]\!]\sigma_1$
  $\sigma_3 = \mathcal{C}[\![\neg C]\!]\sigma$
  $\sigma_4 = \mathcal{S}[\![S_2]\!]\sigma_3$

(a)

$[c_0]$
if$(a > 10)\ [c_1]\ S_1\ [c_2]$
else $[c_3]\ S_2\ [c_4]$
$[c_5]$
if$(a > 5)\ [c_6]\ S_3\ [c_7]$
else $[c_8]\ S_4\ [c_9]$
$[c_{10}]$

(b)

$\sigma_0 = \{a \mapsto 0..20\}$
$\sigma_1 = \sigma_2 = \{a \mapsto 11..20\}$
$\sigma_3 = \sigma_4 = \{a \mapsto 0..10\}$
$\sigma_5 = \{a \mapsto 0..20\}$
$\sigma_6 = \sigma_7 = \{a \mapsto 6..20\}$
$\sigma_8 = \sigma_9 = \{a \mapsto 0..5\}$
$\sigma_{10} = \{a \mapsto 0..20\}$

(c)

$\sigma_0 = \{a \mapsto 0..20\}$
$\sigma_1 = \sigma_2 = \{a \mapsto 11..20\}$
$\sigma_3 = \sigma_4 = \{a \mapsto 0..10\}$

$\sigma_5^{c2} = \{a \mapsto 11..20\}$
$\sigma_6^{c2} = \sigma_7^{c2} = \{a \mapsto 11..20\}$
$\sigma_8^{c2} = \sigma_9^{c2} = \{a \mapsto \bot\}$

$\sigma_5^{c4} = \{a \mapsto 0..10\}$
$\sigma_6^{c4} = \sigma_8^{c4} = \{a \mapsto 6..10\}$
$\sigma_8^{c4} = \sigma_9^{c4} = \{a \mapsto 0..5\}$

$\sigma_{10}^{c2,c7} = \{a \mapsto 11..20\}$ $\sigma_{10}^{c2,c9} = \{a \mapsto \bot\}$ $\sigma_{10}^{c4,c7} = \{a \mapsto 6..10\}$ $\sigma_{10}^{c4,c9} = \{a \mapsto 0..5\}$

(d)

Fig. 2. The semantic rule for an if-statement in (a) gives for the program in (b) the evaluation in (d) instead of the one in (c).

As seen in Fig. 2(d) each if-statement will now generate *two* different environments. It can now be seen that the path S1 $\to$ S4 is a false path.

3.3 Finding the number of iterations in loops

We will use a while-statement to illustrate how our analysis method will work for loops. The core idea is to transform them into if-statements, giving the semantic rule in Fig. 3(a). The if-statement yields two environments each time it is analysed (as before):
1. The environment in which the loop shall be executed again, σ_{true}.
2. The environment in which the loop terminates, σ_{false}.

$\mathcal{S}[\![\texttt{while}(C)\ S]\!]\sigma =$
  $\mathcal{S}[\![\texttt{if}(C)\{S; \texttt{while}(C)\ S\}]\!]\sigma$

(a)

while (a < 9) { $[c_{true}]$
    a = a + 2;
} $[c_{false}]$

(b)

Iter	σ_{true}	σ_{false}	
0	$\{a \mapsto 0..2 \vee 5\}$	$\{a \mapsto \bot\}$	
1	$\{a \mapsto 2..4 \vee 7\}$	$\{a \mapsto \bot\}$	
2	$\{a \mapsto 4..6\}$	$\{a \mapsto 9\}$	min = 2
3	$\{a \mapsto 6..8\}$	$\{a \mapsto \bot\}$	
4	$\{a \mapsto 8\}$	$\{a \mapsto 9..10\}$	
5	$\{a \mapsto \bot\}$	$\{a \mapsto 10\}$	max = 5

(c)

Fig. 3. The semantic rule for loop-evaluation in (a) gives for the program in (b) the table in (c).

A loop is "rolled out" until it cannot execute again, or until the time budget is exceeded (see section 3.5). For example, with the start environment $\sigma = \{a \mapsto 0..2 \lor 5\}$, the code in Fig. 3(a) will generate the table in Fig. 3(b). The analysis shows that the loop will iterate at least two times (since 2 iterations is needed to set $\sigma_{false} \neq \perp$) and at most 5 times (since after 5 iterations $\sigma_{true} = \perp$, which means that we cannot enter the loop again). The analysis also shows that a always will be in the interval 9..10 after the loop.

3.4 Merging environments

Environments will, for several reasons, be merged (using the $\cup$ or $\cup'$ operations) at certain points during the analysis. In our current tool the chosen merge points are the ends of loops, functions and programs. The reasons for merging are:

1. Many evaluations from σ_{false}- environments will be redundant. For example, $\sigma_{false}^{h_i} \subseteq \sigma_{false}^{h_j}$, means that $\sigma_{false}^{h_i}$ is redundant since the evaluation from $\sigma_{false}^{h_j}$ will include all possible executions that could result from $\sigma_{false}^{h_i}$ [3].
2. To reduce the number of continuous evaluations. For example, a loop body with n if-statements will generate 2^n environments for each iteration. Merging of (possible) non-redundant environments will reduce the computational cost. However, overestimation may occur.
3. The goal for the analysis of a program is to generate annotations for the corresponding flow graph. Several methods for low level cache- and pipeline-analysis demands this [8,10,11,13]. The annotations must then be true for *all* iterations of each loop.

3.5 Introducing time

The analysis described so far will often not terminate if the program does not terminate. To terminate our analysis, we will use the fact that a real-time program must complete its task within a given deadline. A program is given a time budget, T_{budget}, which should be a realistic upper time limit for the program on a given hardware. The time budget may be calculated during the design phase and can be seen as part of the specification of the program. The T_{budget} is the only manual "annotation" needed by our method[4].

Each statement or program block has a minimum and a maximum execution time, t_{minc} and t_{maxc}. For each analysed statement the corresponding time interval will be added to a accumulated time. The time for the longest path, $T_{minc}..T_{maxc}$, will be compared to the time budget during the analysis. Three cases can be identified:

---

[3] In Fig. 3(b) the continuing analysis from σ_{false} in both the second and fifth iteration can be included in the continuing analysis from σ_{false} in the fourth iteration.

[4] Note that our time annotation is different from the annotations of other methods. Erroneous path or loop annotations may lead to wrong $WCET_C$, but an erroneous T_{budget} may in the worst case only lead to too early ending of the analysis.

1. $T_{minc} < T_{maxc} < T_{budget}$: In this case we can guarantee that the program will not exceed its time budget.
2. $T_{minc} < T_{budget} < T_{maxc}$: There is now a risk that the program does not terminate within the time budget. Our analysis tool (see section 4) will stop and generate a warning message. If we suspect that the time budget is too narrow, we may extend it and continue further.
3. $T_{budget} < T_{minc} < T_{maxc}$: In this case we know that the program will exceed its time budget. The analysis will normally not reach this point.

Thus, our method calculates $WCET_C$ for the program. However, note that the main reason for this calculation is not to get the $WCET_C$ for the program, but to make sure that the analysis terminates.

4 Implementation and Example

To test the described ideas a prototype tool has been implemented in the programming language Erlang [2]. The tool uses a split integer interval representation of environments. So far only a subset of C is handled, including integer variables and the standard arithmetical operations (+, -, *, /), declarations, assignments, selection statements (if- and if-else- statements) and loop-constructs (while- and for-statements). Still, this simple language serves to illustrate our ideas. To add more types (e.g., floats), more complicated constructions, (e.g., arrays and structs) will be relatively simple. Functions calls, dynamic memory and pointers demands a much more complicated analysis.

There is also an option in our tool to annotate the code manually with possible input values. This can be used by the programmer to, for instance, study the program behaviour for different inputs.

4.1 Example

The information retrieved from the analysis of the program in Fig. 4(a) is presented in Fig. 4(b) and (c). We are assuming that both a and b are within the interval 1..30 at the beginning of the program, that is: $\sigma_0 = \{a \mapsto 1..30, b \mapsto 1..30\}$. The values presented in Fig. 4(d) and (e) has been extracted by running the program for all its possible combinations of input values, in this case: $30 * 30 = 900$ executions[5]. For a program with large number of arguments, with varying input values, this is not a feasible option. Our analysis tool extracted, among other things, the following information:

- A safe estimation of the minimum and maximum number of iterations in the outer loop, Fig. 4(a).
- A safe estimation of the minimum and maximum number of iterations in the inner loop for each iteration of the outer loop, Fig. 4(c).
- The fact that the a = a + 10; statement never will be executed and therefore is dead code.

---

[5] Without any abstractions the analysis would derive these values as well.

- A safe estimation of the possible values for a and b within the program, (1..41 and −9..87 respectively).

An interesting comparison can be made between the actual maximum number of iterations in the inner loop: 13, (given by $\sigma_0 = \{a \mapsto 1, b \mapsto 7\}$), the number derived without abstractions: 40, the number derived with our tool: 66, and a coarse manual annotation given by a programmer[6], which could be 300. The reason for the difference, between the actual maximum number of iterations in the inner loop and the values given by the analysis tool, is that the analysis result includes *all* possible executions and reduces the computational cost by using abstractions. When the program iterates very differently depending on the input values our analysis results will of course deteriorate.

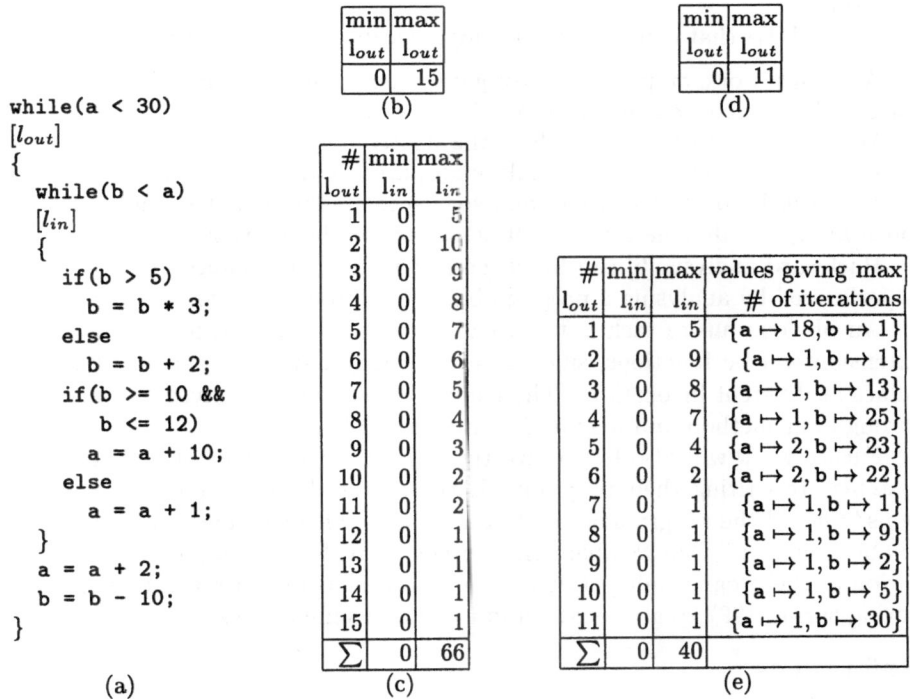

```
while(a < 30)
[l_out]
{
    while(b < a)
    [l_in]
    {
        if(b > 5)
            b = b * 3;
        else
            b = b + 2;
        if(b >= 10 &&
            b <= 12)
            a = a + 10;
        else
            a = a + 1;
    }
    a = a + 2;
    b = b - 10;
}
```

(a)

(b)

min l_{out}	max l_{out}
0	15

(d)

min l_{out}	max l_{out}
0	11

(c)

# l_{out}	min l_{in}	max l_{in}
1	0	5
2	0	10
3	0	9
4	0	8
5	0	7
6	0	6
7	0	5
8	0	4
9	0	3
10	0	2
11	0	2
12	0	1
13	0	1
14	0	1
15	0	1
$\sum$	0	66

(e)

# l_{out}	min l_{in}	max l_{in}	values giving max # of iterations
1	0	5	$\{a \mapsto 18, b \mapsto 1\}$
2	0	9	$\{a \mapsto 1, b \mapsto 1\}$
3	0	8	$\{a \mapsto 1, b \mapsto 13\}$
4	0	7	$\{a \mapsto 1, b \mapsto 25\}$
5	0	4	$\{a \mapsto 2, b \mapsto 23\}$
6	0	2	$\{a \mapsto 2, b \mapsto 22\}$
7	0	1	$\{a \mapsto 1, b \mapsto 1\}$
8	0	1	$\{a \mapsto 1, b \mapsto 9\}$
9	0	1	$\{a \mapsto 1, b \mapsto 2\}$
10	0	1	$\{a \mapsto 1, b \mapsto 5\}$
11	0	1	$\{a \mapsto 1, b \mapsto 30\}$
$\sum$	0	40	

Fig. 4. For the program in (a) our tool gives the estimated min and max iterations in the outer (b) respectively inner (c) loop. The actual values are those in (d) and (e).

---

[6] A programmer may see that a increases with at least 2 every outer iteration, giving $30/2 = 15$ iterations in the outer loop. He may also note that b increases with at least 2 for each iteration in the inner loop. As b also is decreased with 10 for each iteration in the outer loop, the maximum number of iterations in the inner loop will be: $(30 + 10)/2 = 20$ times. This gives a total of $15 * 20 = 300$ iterations.

5 Conclusions, results and future work

We have presented a static analysis method which automatically derives safe and tight annotations from the semantics of the source-code program. Normally, these annotations are given manually, but to derive them is often difficult and error-prone. The analysis shall be seen as a first phase in a tight worst case execution time (WCET$_C$) calculation. The derived annotations can be used by the following phase, an object (micro- or assembler) code analysis, which also considers modern hardware architectures.

A short summary of the information derivable with our method are:

- Information of false paths and dead code within programs (section 3.2).
- Safe estimations of maximum and minimum of iterations both for single and nested loops (section 3.3).
- Possible values for all variables in each point[7], program block or entire program[8] [9].
- A WCET$_C$ that can be used on simple hardware architectures (section 3.5).

As future work we plan to investigate other forms of environment representation. General constraint techniques is one of the candidates [18].

We also plan to investigate how the degree of merging affects the analysis result. We can in one extreme analyse all paths, without merging, but such an analysis will be both very time- and space-consuming. On the other hand, too much merging will generate a lot of pessimism in the analysis.

Backward analysis [6] can be of interest to further enhance our analysis. It is performed by analysing a program backwards from the goal environments.

An obvious future task is to extend the analysed language. An interesting extension will be functions, since a function may have different possible input values at different invocations. These input values can be derived automatically through our method and may lead to a tighter WCET$_C$[10].

Future work will also be to investigate *how* programmers of hard real-time systems are writing their programs. Is there a need for complicated constructions? Should the programmer be forced to write his programs in a certain way, to allow analysis? Can we abandon recursion in real-time programs? An investigation of the programming style used in real-time companies will be performed during spring 1997 to give answers to these and similar questions [7].

References

1. P. Altenbernd. On the false path problem in hard real-time programs. In *Proceedings of the Eight Euromicro Workshop on Real-Time Systems*, pages 102–107, June 1996.

[7] Derived by merging all n environment generated at the i:th control point: $\bigcup_{j=1..n} \sigma_i^{hj}$.

[8] Derived by merging all environments generated at all different control points.

[9] This information can be used for compiler optimisations (e.g., reduction of size of variables) and program verification (e.g., index checking) [3].

[10] The only other method we know that considers input values for WCET$_C$ calculation is [4], but it relies on manual annotations.

2. J. Armstrong, R. Virding, C. Wikström, and M. Williams. *Concurrent programming in Erlang.* Prentice Hall, 2 edition, 1996. ISBN 0-13-508301-X.
3. F. Bourdoncle. Abstract debugging of high-order imperative languages. In *Proceedings of SIGPLAN'93 Conference on Programming Language design and Implementation*, pages 46–55, 1993.
4. R. Chapman, A. Burns, and A. Wellings. Integrated program proof and worst-case timing analysis of SPARK Ada. In *ACM Sigplan Workshop on Language, Compiler and Tool Support for Real-Time Systems*, June 1994.
5. P. Cousot and R. Cousot. Abstract interpretation: A unified model for static analysis of programs by construction or approximation of fixpoints. In *4th ACM Symp. on Principles of Programming Languages*, pages 238–252, 1977.
6. P. Cousot and R. Cousot. Comparing the Galois connection and widening/narrowing approaches to abstract interpretation. In *Programming Language Implementation and Logic Programming, Proceedings of the Fourth International Symposium, PLILP'92*, volume Lecture Notes in Computer Science 631, pages 269–295, Aug 1992.
7. A. Ermedahl and J. Gustavsson. Real-time industry inquiry of execution time analysis tools. Technical report, Department of Computer Systems, Uppsala University, Sweden, 1997. To be published.
8. M. Harmon, T. Baker, and D. Whalley. A retargetable tecnique for predicting execution time of code segments. *The Journal of Real-Time Systems*, 7, 1994.
9. Y.-T. Li and S. Malik. Performance analysis of embedded software using implicit path enumeration. In *ACM Workshop on Lang., Comp. and Tools for RTS*, May 1995.
10. Y.-T. Li, S. Malik, and A. Wolfe. Cache modeling for real-time software: Beyond direct mapped instruction caches. In *17th IEEE Real-Time Systems Symposium, RTSS'96*, pages 254 – 263, 1996.
11. S. Lim, Y. Bae, G. Jang, B.-D. Rhee, S. Min, C. Park, H. Shin, K.Park, S.-M. Moon, and C. Kim. An accurate worst case timing analysis for risc processors. *IEEE Trans. on Software Engineering*, 21(7):593 – 604, July 1995.
12. H. R. Nielson and F. Nielson. *Semantics with Applications.* John Wiley & Sons, 1992.
13. G. Ottosson and M. Sjödin. Worst-case execution time analysis for modern hardware architectures. In *Proc. SIGPLAN 1997 Workshop on Languages, Compilers and Tools for Real-Time Systems*, June 1997. To appear.
14. C. Park. Predicting program execution times by analyzing static and dynamic program paths. *The Journal of Real-Time System*, 5:31–62, 1993.
15. C. Park and A. Shaw. Experiments with a program timing tool based on a source-level timing schema. *Proceeding of 11th IEEE Real-Time Systems Symposium*, pages 72–81, Dec 1990.
16. P. Puschner and C. Koza. Calculating the maximum execution time of real-time programs. *The Journal of Real-Time Systems*, 1(2):159–176, Sep 1989.
17. P. Puschner and A. Schedl. Computing maximum task execution times with linear programming techniques. Technical report, Report, Techn. Univ., Inst. für Technische Informatik, Vienna, April 1995.
18. E. Tsang. *Foundations of Constraint Satisfaction.* Academic Press, 1993.
19. A. Vrchoticky. *The Basis for Static Execution Time Prediction.* PhD thesis, Institut für Technische Informatik, Technische Universität Wien, Austria, April 1994.

Cinderella: A Retargetable Environment for Performance Analysis of Real-Time Software

Yau-Tsun Steven Li[1], Sharad Malik[2], Andrew Wolfe[2]

[1] Hewlett-Packard Company, 1501 Page Mill Rd, MS 6U-J, Palo Alto, CA 94304, USA
[2] Dept of Electrical Engineering, Princeton University,
Princeton, NJ 08544, USA.

Abstract. Real-time systems are characterized by the presence of timing constraints that a task must be completed within a given deadline. In this paper, we present a complete environment for determining best-case and worst-case execution time of a program when running on a given hardware. Our analysis technique is unique in that it allows user to annotate complex program path information and at the same time, models cache memory and pipeline accurately. This results in tight estimations even for complicated programs running on modern hardware. The technique has been implemented on a timing analysis tool — `cinderella`[3], which provides retargetable back-ends for analyzing programs written in different languages and executed on different hardware. We present some experimental results of using this tool.

1 Introduction

The execution time of a program running on a given system may vary significantly according to different input data and initial system states. In many cases it is essential to determine the extreme case (best case or worst case) execution time of a program. This information is needed in many real-time operating systems for task scheduling. It is also needed in the hardware/software partitioning step in embedded system designs.

The *actual* extreme case execution time of a program cannot be determined unless all feasible input data and system states are simulated. Since a large number of simulations is required, this method is impractical. Instead, our objective is to determine a *tight bound* on all feasible execution times of a program. This bound is denoted as the **estimated bound** of the program. We tackle the problem of determining tight estimated bound by dividing it into two smaller ones:

Program path analysis This analyzes the structure of the program statically and determines the set of paths that corresponds to the extreme case program execution time. Since the number of program's feasible paths is in general exponential in its size, no path enumeration is allowed in this analysis. Also, many of the statically feasible program paths are never executed in practice. A mechanism for the programmer to mark infeasible paths is essential in tightening the estimation.

Microarchitecture modeling This determines the extreme case execution times of known sequences of instructions and passes them to program path analysis. The presence of modern microarchitecture features, such as pipelines and caches, varies the instruction execution time significantly. The instruction execution time is no longer constant and it depends on the execution trace. Hence, this analysis interferes with program path analysis.

---

[3] In recognition of her hard real-time constraint — she had to be back home at the stroke of midnight!

Both problems are *equally important* in determining tight estimated bound of the program. In solving the above problems, we also need to consider the retargetability issues so that the solution can be applied to a wide range of programs running on different hardware. In the following, we will describe our analysis in determining the estimated worst case execution time (WCET) of the program. The analysis for the best case execution time is similar.

2 Related Work

Early researchers [11, 15, 17, 18, 20] in this area adopted simple microarchitecture modeling where instruction execution times are assumed to be constant and independent of each others. They focused on program path analysis and proposed different techniques to eliminate false program paths. In particular, Park [17] recognized the use of regular expressions to annotate various path information. However, the analysis of regular expression is complicated and some pessimism approximations are used.

More recently, many techniques have been proposed to model pipelines [4, 7, 13, 16, 21] and caches [2, 3, 9, 13, 14, 19] with various success. In modeling these microarchitecture features, the importance of path annotation are neglected. As a result, these techniques could only handle simple programs with fixed loop bounds (e.g. matrix multiplication routines). For more complicated programs, like sorting routines, they generated loose estimations.

Most researchers claimed their methods to be retargetable. However, there are no timing analysis tools that actually implement the retargetable framework. Only a few retargetable tools exist in pipeline modeling [5, 16].

3 Program Path Analysis

In this analysis, we assume that instruction execution times are all constant. This assumption will be removed in Sect. 4. Our analysis technique uses the counting approach to compute the estimated WCET. The method converts the problem of solving the estimated WCET into a set of **integer linear programming** (ILP) problems in which the estimated WCET, and the worst case execution counts of the instructions are solved for.

For each basic block [1] B_i in the program, we let variable x_i be its execution count and constant c_i be its single execution time. For program with N basic block, the total execution time is:

$$\text{Total execution time} = \sum_{i=1}^{N} c_i x_i. \tag{1}$$

The possible values of x_i's are constrained by the program structure and the program input data. This constraints are represented by a set of linear constraints. The linear constraints are divided into two parts: (i) **structural constraints**, which are derived automatically from the program's control flow graph (CFG) [1], and (ii) **functionality constraints**, which are provided by the user to specify loop bounds and other path information. Fig. 1 shows a simple code fragment and its CFG. Each edge in the CFG is labeled with a variable d_i which serves both as a label for that edge and as a count of the the number of times that the program control passes through that edge. Analysis of the CFG is equivalent to a standard network-flow problem. Structural constraints can be derived from the CFG from the fact that, for each node B_i, its execution count is equal to the number of times that the control enters the node (inflow), and is also equal to the number of times that the control exits the node (outflow):

$$x_i = \sum d\_inflow = \sum d\_outflow \tag{2}$$

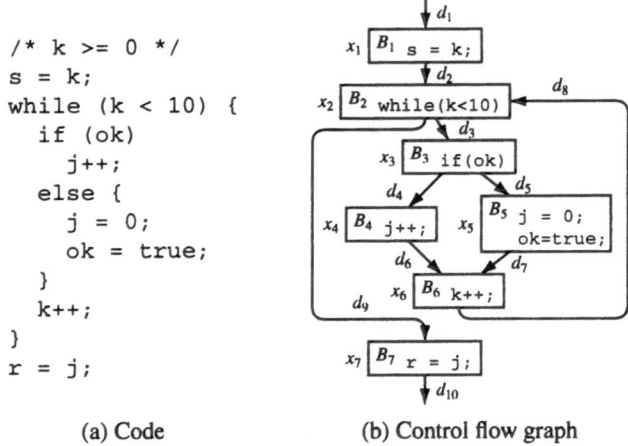

```
/* k >= 0 */
s = k;
while (k < 10) {
    if (ok)
        j++;
    else {
        j = 0;
        ok = true;
    }
    k++;
}
r = j;
```

(a) Code   (b) Control flow graph

Fig. 1. An example showing how the structural and functionality constraints are constructed.

The loop bound information must be provided by the user using functionality constraints. Otherwise, the estimated WCET is unbounded. In this example, since k is positive before it enters the loop, the loop body will be executed at most 10 times each time the loop is entered. This information can be represented by the functionality constraints: $0x_1 \leq x_3 \leq 10x_1$.

Additional path information can also be described by functionality constraints. We have been able to show that the functionality constraints are more powerful than Park's IDL [17] in describing path information [12]. As a simple example, the else statement (B_5) can be executed at most once inside the loop. This information can be specified as: $x_5 \leq 1x_1$.

4 Microarchitecture Modeling

Microarchitecture modeling models CPU pipeline and cache memory — two dominant microarchitecture features that affect the execution time of an instruction. In modeling these features, our analysis technique is unique in that the program path analysis model described in previous section is retained.

4.1 Pipeline Modeling

The pipeline modeling is relatively easy and straightforward. We model the pipeline within each basic block and add up the execution time each instruction spent in the execution stage of the pipeline [8]. In determining the estimated WCET of the basic block, we assume the pipeline is flushed at the end of the basic block. This model is used by many researchers [4, 16, 21]. Our experiments in modeling the pipeline of Intel i960KB processor showed that it is accurate. Detailed results will be given in Sect. 6.

4.2 Cache Modeling

Cache is much harder to model than pipeline. In a pipeline, the instruction execution time depends only on a few of its preceding instructions. But with the presence of cache memory, it depends on all instructions that are mapped to the same cache set. A *global* analysis is required for accurate cache modeling. If not properly modeled, the cache analysis will introduce more pessimism than the pipeline analysis. Direct mapped instruction cache analysis will be described first. This is followed by set associative instruction cache analysis.

The goal of cache modeling is to determine for each instruction, the number of fetches that result in cache hits/misses. We first partition each basic block into smaller units (called *l-blocks*) that are aligned with the instruction cache line. Suppose a basic block B_i is partitioned into n_i l-blocks, they are denoted as $B_{i.1}, B_{i.2}, \ldots, B_{i.n_i}$. Since the cache controller always fetches a line of code whenever this is a miss, an l-block $B_{i.j}$ is either in the i-cache completely, or not in it at all. These two cases correspond to two possible execution times of the l-block, which are represented by constants $c_{i.j}^{hit}$ and $c_{i.j}^{miss}$ respectively. We let $x_{i.j}^{hit}$ and $x_{i.j}^{miss}$ be integer variables that represent an l-block $B_{i.j}$'s hit and miss counts. Given these variables, the total execution time of the program can be refined as:

$$\text{Total execution time} = \sum_{i=1}^{N} \sum_{j=1}^{n_i} (c_{i.j}^{hit} x_{i.j}^{hit} + c_{i.j}^{miss} x_{i.j}^{miss}). \tag{3}$$

Since l-block $B_{i.j}$ is inside the basic block B_i, its total execution count is equal to x_i. Hence

$$x_i = x_{i.j}^{hit} + x_{i.j}^{miss}, \qquad j = 1, 2, \ldots, n_i \tag{4}$$

Eq. (4) links the new cost function (3) with the structural constraints and the functionality constraints, both of which remain unchanged. In addition, the cache activities can now be described in terms of the new variables $x_{i.j}^{hit}$'s and $x_{i.j}^{miss}$'s.

For any two l-blocks mapped to the same cache set, we say that they *conflict* with each other if their address tags [8] are different. Otherwise, they are called *non-conflicting* l-blocks.

Direct Mapped Instruction Cache Analysis Consider a simple case. For each cache set, if there is only one l-block $B_{k.l}$ mapped to it, then only its first execution *may* result in a cache miss, therefore,

$$x_{k.l}^{miss} \le 1. \tag{5}$$

When a cache set contains two or more conflicting l-blocks, the hit/miss counts of all the l-blocks mapped to this set will be affected by the execution sequence of these l-blocks. In this case, a **cache conflict graph** (CCG) [12] is constructed. A CCG captures the control flow of a set of l-blocks mapped to the same cache set. Each edge of CCG is labeled with a p variable to count the number of times that the control passes through that edge. Suppose that in Fig. 1(b), basic blocks B_1, B_4 and B_5 are partitioned into l-blocks and l-blocks $B_{1.1}$, $B_{4.1}$ and $B_{5.1}$ are mapped to the same cache set and they conflict with each other, the CCG is shown in Fig. 2(a). The control flow from one l-block $B_{i.j}$ to the other $B_{k.l}$ is represented by a variable $p_{(i.j, k.l)}$.

A set linear constraints can be derived from the CCG to link with the structural and functionality constraints, and also to describe cache hit/miss constraints. At each node $B_{i.j}$, the sum of control flow going into the node must be equal to the sum of control flow leaving the node, and it must also be equal to the total execution count of l-block $B_{i.j}$. Therefore, two constraints are constructed at each node $B_{i.j}$:

$$x_i = \sum p\_inflow = \sum p\_outflow \tag{6}$$

Due to the existence of x_i's, this set of constraints is linked to the structural and functionality constraints.

All self loops in CCG indicate cache hits. Therefore,

$$x_{i.j}^{hit} = p_{(i.j, i.j)} \tag{7}$$

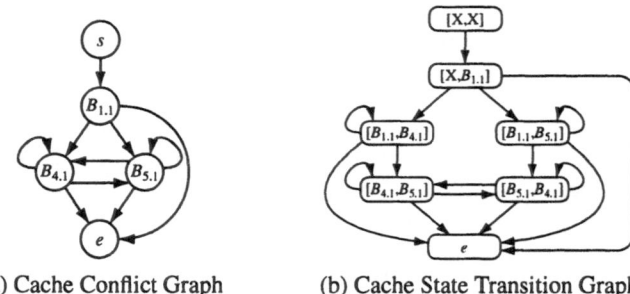

(a) Cache Conflict Graph   (b) Cache State Transition Graph

Fig. 2. The CCG captures control flow of l-blocks mapped to the same cache set. The CSTG, shown in 2-way set associative i-cache, represents all feasible cache states and their transitions due to the flow of these l-blocks. In this example, the graphs are constructed when l-blocks $B_{1.1}$, $B_{4.1}$ and $B_{5.1}$ in Fig. 1 conflict with each other.

The above linear constraints are the **cache constraints** for direct mapped i-cache. These constraints, together with (4), the structural constraints and the functionality constraints, are passed to the ILP solver with the goal of maximizing the cost function (3). Because of the cache information, a tighter estimated WCET will be returned. The CCGs are network flow graphs and thus the cache constraints are typically solved rapidly by the ILP solver. For programs with function calls, the functions are treated as if they are inlined [12].

Set Associative Instruction Cache Analysis The modeling of set associative i-cache is very similar to that of direct mapped i-cache. Only the direct mapped i-cache constraints (5)–(7) are replaced by a set of new ones.

A **cache state transition graph** (CSTG) [12] is constructed to model all feasible cache state transitions of a cache set. Each node of the graph contains a state $[B_{i.j}, B_{m.n}]$ showing two l-blocks in the least and most recently used entries. For 2-way set associative i-cache with least recently used (LRU) replacement policy, a CSTG is shown in Fig. 2(b). A transition from state $[B_{i.j}, B_{k.l}]$ to $[B_{k.l}, B_{m.n}]$ represents an execution of l-block $B_{m.n}$. Similar to the CCG, self loops in CSTG represent cache hits. In addition, transition from $[B_{i.j}, B_{m.n}]$ to $[B_{m.n}, B_{i.j}]$ also results in cache hits. A p-variable is associated with each edge of the graph to represent the number of transition. Based on x's and p's, a new set of flow equations and cache hit linear constraints can be generated.

5 Implementation and Retargetability Issues

The above analysis technique has been implemented in our timing analysis tool cinderella. The tool features several retargetable back-ends (Fig. 3(a)) so that it can be easily ported to model different hardware, as well as programs written in different source language. The core performs program path analysis and cache analysis. The *object file handler* reads the executable file of the program directly and utilizes debugging information to map binary code to source code so that path information can be entered at source level. This approach allows programs written in different languages and compiled by different compilers to be analyzed. The *instruction set handler* decodes the binary code and passes information to the core for building control flow graph and also to the *machine handler*, which models instruction pipelines and provides instruction timings and cache configurations to the core. The separation of instruction set decoding and machine

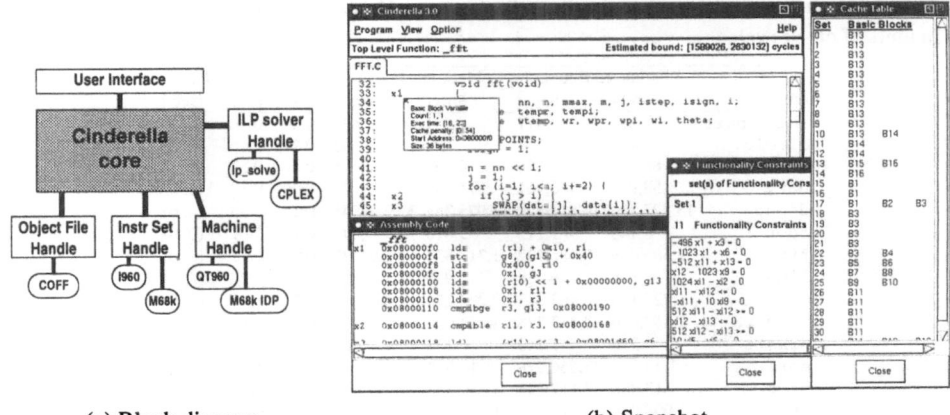

(a) Block diagram (b) Snapshot

Fig. 3. Cinderella's retargetable back-ends and its graphical user interface.

Table 1. Set of benchmark examples, their descriptions, source file line sizes and Intel i960KB binary code sizes.

Program	Description	Lines	Bytes
check_data	Check if any element in an array is negative, from Park [17]	23	88
circle	Circle drawing routine, from Gupta [6]	100	1,588
des	Data Encryption Standard	192	1,852
dhry	Dhrystone benchmark	761	1,360
djpeg	Decompression of 128×96 color JPEG image	857	5,408
fdct	JPEG forward discrete cosine transform	300	996
fft	1024-point Fast Fourier Transform	57	500
line	Line drawing routine, from Gupta [6]	165	1,556
matcnt	Summation of 2 100×100 matrices, from Arnold [2]	85	460
matcnt2	Matcnt with inlined functions	73	400
piksrt	Insertion sort of 10 elements	19	104
sort	Bubble sort of 500 elements, from Arnold [2]	41	152
sort2	Sort with inlined functions	30	148
stats	Calculate the sum, mean and variance of two 1,000-element arrays, from Arnold [2]	100	656
stats2	Stats with inlined functions	90	596
whetstone	Whetstone benchmark	196	2,760

timing model allows one to model a family of processors easily. Currently, we have implemented modules for modeling programs running on Motorola M68000 and Intel i960KB processors. The tool also features a user-friendly graphical interface (Fig. 3(b)). It can be downloaded from its WWW home page (http://www.ee.princeton.edu/~yauli/cinderella/).

6 Experimental Results

We have analyzed a large set of programs to validate our analysis technique. The programs are shown in Table 1. Some of them comes from other researchers. Others are much more complicated software benchmarks and application programs.

Table 2. Benchmark program analysis results. Estimated bounds and measure bounds are shown in units of clock cycles.

Program	Estimated Bound lower	Estimated Bound upper	Measured bound lower	Measured bound upper	Pessimism[a] lower	Pessimism[a] upper	CPU Time (sec.)
check_data	34	471	34	430	0.00	0.10	(0, 0)
circle	465	15,364	585	14,483	0.21	0.06	(0, 0)
des	86,570	369,840	111,468	243,676	0.22	0.52	(6, 4)
dhry	458,966	756,961	575,492	575,622	0.20	0.32	(2, 0)
djpeg	13,225,736	70,414,320	14,975,268	35,636,948	0.12	0.98	(4, 6)
fdct	6,145	9,115	7,616	9,048	0.19	0.01	(0, 0)
fft	1,589,026	2,630,132	1,719,832	2,204,472	0.08	0.19	(0, 0)
line	578	6,088	929	4,836	0.38	0.26	(0, 0)
matcnt	1,722,105	5,463,383	2,202,276	2,202,698	0.22	1.48	(0, 0)
matcnt2	1,482,086	2,113,328	1,862,007	1,862,333	0.20	0.13	(0, 0)
piksrt	236	1,740	337	1,705	0.30	0.02	(0, 0)
sort	13,965	27,866,978	16,942	9,991,172	0.18	1.79	(0, 0)
sort2	13,965	7,117,043	16,507	6,747,664	0.15	0.05	(0, 0)
stats	1,008,085	2,213,764	1,158,142	1,158,469	0.13	0.91	(0, 0)
stats2	894,017	1,235,696	1,060,118	1,060,380	0.16	0.17	(0, 0)
whetstone	5,970,554	10,546,246	6,935,612	6,935,668	0.14	0.52	(0, 0)

[a] Pessimism is calculated as: $lower = \frac{Mea.\ lower - Est.\ lower}{Mea.\ lower}$, $upper = \frac{Est.\ upper - Mea.\ upper}{Mea.\ upper}$.

We studied each program carefully, determined its loop bounds and path information, and used cinderella to compute the estimated bound. To validate this estimation, we determined each program's extreme case input data sets and then used logic analyzer to measure the execution time of the program when running on an Intel QT960 evaluation board [10] containing a 20 MHz i960KB processor. The processor has an on-chip 512 bytes direct-mapped i-cache and a 4-stage execution pipeline.

Table 2 shows the results of analysis. All estimated bounds bound their corresponding measured bounds. For most programs, the estimated bound is very close to the measured bound. We have also conducted other experiments to validate program path analysis and cache analysis separately [12]. These experiments indicate that when given enough path information, our analysis technique is very accurate. A few programs have looser estimation. For programs des and djpeg, this is because the extreme case input data set could not be determined, random input data sets were used and as a result, the measured bound might not be close to the actual bound of the program. For programs matcnt, sort and stats, the reason for loose estimation is that i960KB processor features 4 register windows, which were not modeled in our tool. Conservation assumptions were used in modeling the execution times of call and return instructions and large pessimism will occur for programs with lots of small function calls. We inlined the frequently called functions in these programs. The results are programs matcnt2, sort2 and stats2. Their estimated bounds are much tighter. The ILP problems were solved by a commercial ILP solver CPLEX. They were solved efficiently on a Silicon Graphics Indigo2 workstation containing a 150 MHz MIPS R4400 processor with 256 MB main memory.

7 Concluding Remarks

In this paper, we have described an efficient and powerful method based on integer linear programming to determine the execution time bounds of real-time programs. When compared with other existing methods, ours offers more powerful path annotation mechanism and more accurate cache modeling. Even better, the ILP formulation allows both to be applied simultaneously.

This results in very tight estimations even for complicated programs. Our implementation is more complete than others. `Cinderella` features a user-friendly graphical interface and retargetable back-ends. We have conducted extensive experiments to validate that our tool is capable of analyzing large and complicated programs accurately.

References

1. A. Aho, R. Sethi, and J. Ullman. *Compilers Principles, Techniques, and Tools*. Addison-Wesley, 1986.
2. R. Arnold, F. Mueller, D. Whalley, and M. Harmon. Bounding worst-case instruction cache performance. In *Proc. of the 15th IEEE Real-Time Systems Symposium*, Dec 1994.
3. S. Basumallick and K. Nilsen. Cache issues in real-time systems. In *Proc. of ACM PLDI Workshop on Language, Compiler, and Tool Support for Real-Time Systems*, Jun 1994.
4. S. Bharrat and K. Jeffay. Predicting worst case execution times on a pipelined RISC processor. Technical report, Dept of Computer Science, University of North Carolina at Chapel Hill, Apr 1994. TR94-072.
5. D. Bradlee. *Retargetable Instruction Scheduling for Pipelined Processors*. PhD thesis, University of Washington, 1991.
6. R. Gupta. *Co-Synthesis of Hardware and Software for Digital Embedded Systems*. PhD thesis, Stanford University, Dec 1993.
7. C. Healy, D. Whalley, and M. Harmon. Integrating the timing analysis of pipelining and instruction caching. In *Proc. of 16th IEEE Real-Time Systems Symposium*, Dec 1995.
8. J. Hennessy and D. Patterson. *Computer Architecture: A Quantitative Approach, 2nd Ed*. Morgan Kaufmann Publishers, Inc., 1996. ISBN 1-55860-329-8.
9. Y. Hur, Y.-H. Bae, S.-S. Lim, S.-K. Kim, B.-D. Rhee, S.-L. Min, C.-Y. Park, M. Lee, H. Shin, and C.-S. Kim. Worst case timing analysis of RISC processors: R3000/R3010 case study. In *Proc. of 16th IEEE Real-Time Systems Symposium*, Dec 1995.
10. Intel Corp. *QT960 User Manual*, 1990. Order Number 270875-001.
11. E. Kligerman and A. Stoyenko. Real-time Euclid: A language for reliable real-time systems. *IEEE Trans. on Software Engineering*, Sep 1986.
12. Y.-T. Li. *Performance Analysis of Real-Time Embedded Software*. PhD thesis, Princeton University, 1997.
13. S.-S. Lim, Y.-H. Bae, G.-T. Jang, B.-D. Rhee, S.-L. Min, C.-Y. Park, H. Shin, K. Park, and C.-S. Kim. An accurate worst case timing analysis technique for RISC processors. In *Proc. of the 15th IEEE Real-Time Systems Symposium*, Dec 1994.
14. J.-C. Liu and H.-J. Lee. Deterministic upperbounds of the worst-case execution times of cached programs. In *Proc. of the 15th IEEE Real-Time Systems Symposium*, Dec 1994.
15. A. Mok, P. Amerasinghe, M. Chen, and K. Tantisirivat. Evaluating tight execution time bounds of programs by annotations. In *Proc. of the 6th IEEE Workshop on Real-Time Operating Systems and Software*, May 1989.
16. K. Narasimhan and K. Nilsen. Portable execution time analysis for RISC processors. In *Proc. of ACM PLDI Workshop on Language, Compiler, and Tool Support for Real-Time Systems*, Jun 1994.
17. C.-Y. Park. *Predicting Deterministic Execution Times of Real-Time Programs*. PhD thesis, University of Washington, Aug 1992.
18. P. Puschner and Ch. Koza. Calculating the maximum execution time of real-time programs. *The Journal of Real-Time Systems*, Sep 1989.
19. J. Rawat. Static analysis of cache performance for real-time programming. Master's thesis, Iowa State University of Science and Technology, Nov 1993. TR93-19.
20. A. Shaw. Reasoning about time in higher-level language software. *IEEE Trans. on Software Engineering*, Jul 1989.
21. N. Zhang, A. Burns, and M. Nicholson. Pipelined processors and worst-case execution times. *Journal of Real-Time Systems*, Oct 1993.

Esprit Workshop

Esprit Projects on High-Performance Computing and Networking

Sergei Gorlatch

University of Passau, D–94030 Passau, Germany.
email: gorlatch@fmi.uni-passau.de
http://www.uni-passau.de/~gorlatch

The Esprit Workshop at Euro-Par'97 is a new event in the series of the Euro-Par conferences. Our decision to invite the Esprit participants to present their research and development projects, carried out with the support of the European Commission, has been met with a good response. Each of the eleven submissions received was refereed by at least three experts, after which nine submissions were accepted for presentation at the conference.

The importance of high-performance computing and networking (HPCN) is recognized all over the world. In the European Union, it has lead to special HPCN programmes. The main objective of these programmes is to exploit the opportunities in the information technology to support the competitiveness of the European industries. The Esprit projects address the great expansion of application areas in which parallel and distributed computing become important or even necessary to solve complex and time-consuming problems. The current trend is that the growth of parallel computing power is realized not only by increasing the number of processors in a parallel computer, but also by combining the power of several geographically distributed computers by coupling them via communication networks.

Whereas the early European programmes put a strong emphasis on hardware and software technology, recent programmes are much more user-driven, with an emphasis on applications. Most projects are organized vertically, i.e., they involve vendors, implementors of software and application users, with users playing a key rather than a marginal rôle.

The Esprit workshop exhibits much more diversity in the topics than the traditional, more focussed workshops of the conference. Another distinguishing feature is a broad international cooperative effort behind each of the presented projects. The papers of the workshop report on using parallel and distributed systems in diverse areas of science and industry. The authors formulate also the challenges for the future and their vision of meeting these challenges.

To fit into the conference format, the workshop has been divided into two sessions: one session is on the projects that address practical problems of parallel and distributed computing and the other is on the HPCN application projects.

The first session starts with the presentation of projects PHASE and MICA devoted to metacomputing – the coordinated use of geographically distributed high-performance computers. The projects implement application-specific meta-computer environments: a pharmaceutical WWW-based application server and a model for CFD applications. The topic of the Internet is addressed also by the next presented project, FRONTIER. Its primary objective is to develop a par-allelizable hybrid technology for collaborative design optimization, which would enable the use of the Web technology in large joint projects. The PINEAPL project is a cooperative effort to produce a general-purpose library of parallel numerical software, suitable for a wide range of computationally intensive indus-trial applications. Coordinated by an industrial partner (NAG Ltd.), the project pays special attention to the reusability of the developed software and the tech-nology transfer. The next project, RAINS, aims at neural network applications for industry and medicine. The objective is to enable the use of comparatively in-expensive clusters of workstations instead of dedicated neurocomputers. Finally, the PARSAR project describes the porting of a data- and computation-intensive application – obtaining high-resolution images of the Earth's surface – to a clus-ter of UNIX workstations under PVM.

The second session starts with the OCEANS project, whose goal is to in-vestigate and develop state-of-the-art compilation techniques to allow high per-formance implementations of embedded applications. As a case study, compiler-performed optimizations on multimedia applications are considered. The next, SEEDS, develops a general-purpose toolkit for the simulation of distributed traf-fic control systems on a cluster of workstations. As a case study, the ground traffic control in airports is considered. The area of embedded systems with their high requirements on fault-tolerance is addressed by the EFTOS project. Through the development of a generic fault tolerance framework, the burden of ad hoc programming of corresponding mechanisms is removed from the application de-velopers. The final presentation of the session is on project STAMPAR, which demonstrates the use of explicit dynamic programming for the solution of indus-trial sheet stamping problems.

The projects presented at the Esprit workshop provide a snapshot of the current European R&D activities in the area of parallelism and, thereby, make a valuable contribution to the exchange of ideas and experience from academia and industry at Euro-Par'97.

PHASE and MICA:
Application Specific Metacomputing*

Jörn Gehring, Alexander Reinefeld, Anke Weber

Paderborn Center for Parallel Computing (PC²)
Fürstenallee 11, D-33095 Paderborn, Germany
email: {joern, ar, weber}@uni- paderborn.de

Abstract. Metacomputing refers to the coordinated use of a pool of geographically distributed high-performance computers. The EU projects PHASE and MICA, presented in this paper, are settled in the field of application specific metacomputing combining high-performance computing expertise and industrial requirements.

1   Introduction

Distributed high-performance computing – so-called metacomputing – refers to the coordinated use of a pool of geographically distributed high-performance computers. The advantages of metacomputing are obvious: Metacomputers provide true supercomputing power at little extra cost, they allow better utilization of the available high-performance computers, and they can be easily upgraded to include the latest technology. It seems, however, that up to now no system has been built that rightfully deserves the name *metacomputer* in the above sense. The following two types of metacomputing can be distinguished: supercomputer metacomputing and workstation metacomputing

Supercomputer metacomputing is made up by a collection of parallel high-performance computers that are used in a coordinated way. The rationale is that it is easier to harness a few HPC systems than several hundreds or thousands of (unreliable) workstations. A first attempt in this direction is the Paderborn MOL project (Metacomputer Online) [3]. MOL is based on the design of versatile interfaces combining existing, well-proven MPP tools to a homogeneous environment.

In *workstation cluster metacomputing*, a collection of geographically distributed workstation clusters is used as a single computational resource. Examples of such initiatives can be found at the Dutch Polder [7] and the German Hypercomputing [8].

Many obstacles in building a metacomputer can be overcome, if we restrict its capabilities to what is needed by industrial end-users. These people want to use the computing power available in the Internet for getting problems solved with commercial applications. The projects MICA and PHASE both address the needs of these users by implementing an *Application Centered Metacomputer*. In

---

* This work has been supported by the EU, ESPRIT projects 23486 and 20966

application centered metacomputing, users may execute a number of predefined applications on a set of WAN-connected supercomputers. They specify job requirements (e.g., expected response time, data requirements, cost) and submit problems to an automized load-balancing scheme that distributes the parallel tasks to the best suited hardware platforms. Within the EU-funded project MICA, it was proved that the prototype of such a metacomputer can be built in less than two years. This metacomputer prototype is now being enhanced within the PHASE project.

Section 2 gives a more detailed view on MICA while PHASE will be discussed in Section 3. An overview of the current status of both projects is given in Section 4.

2 MICA – A Model for Industrial CFD Applications

The MICA project puts together existing technology in order to implement a powerful and easy-to-use CFD[2] -engine. By having the frontend of this engine running on low-cost personal computers and the backend on high-end supercomputers, MICA delivers the power of high-performance-computing onto the desks of small and medium sized industrial companies.

The system consists of three main components: The first component is the PHOENICS CFD code which runs on personal computers as well as on massively parallel supercomputers. The capabilities of this code have been improved significantly within the project. The second component is a graphical virtual environment editor which enables non-CFD-experts to set up complex problem descriptions and to examine the results within a powerful 3-D-viewer. These two components are connected by MICA-Net, which receives problem and description files from end-users and takes responsibility for providing the results as soon as possible and for the best achievable price.

Fig. 1 depicts the system architecture of MICA-Net. On the right there are the various users who may be distributed all over Europe. The users submit CFD-problems to the virtual entry point of MICA-Net. A virtual entry point is given by the collaboration of a set of MICA-servers. Each participating HPC center operates its server autonomously. Servers can be signed on or removed from the MICA-Net at any time. A user submits a problem to the nearest server which will then contact all other servers to find out where the best suited machine for this problem is located. For this decision the following parameters are taken into account: current load of all machines, network throughput, problem size and type, desired response time, and maximum cost.

In order to keep MICA open to new platforms, the hardware and its management system are encapsulated by *Load-Monitors* and *Job-Managers*. The Load-Monitors provide abstract load information to the MICA-servers while the Job-Managers supply routines for starting and controlling parallel jobs. Within the project, these modules have been developed for Unix, CCS [2], and CODINE [5].

---

[2] Computational Fluid Dynamics

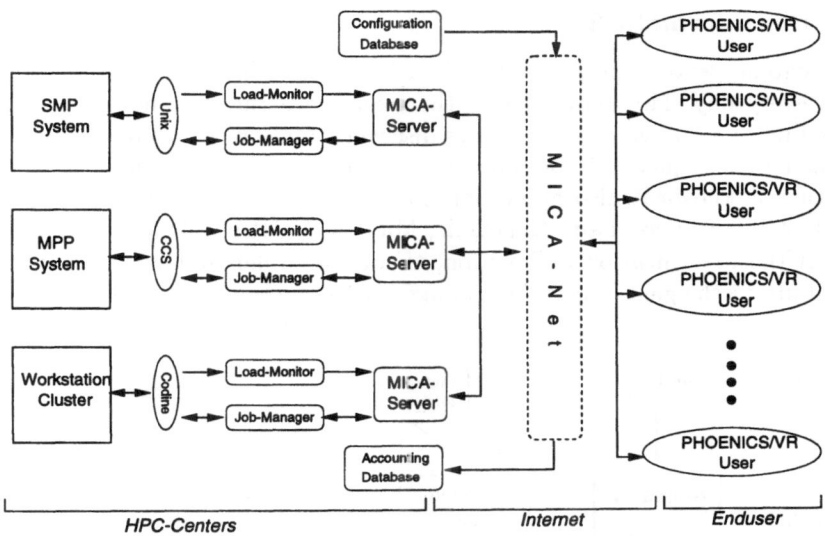

Fig. 1. System architecture of the MICA network

The system is currently limited to one application only and it does not yet support applications that are distributed among several centers. We consider it as a first step towards a metacomputer which brings the huge computing power of the Internet onto the desks of industrial end-users. This is especially true as the MICA-Net services can be seamlessly integrated into the user interfaces of the applications. Thus, it was decided to extend the MICA system within the PHASE project towards a wider range of applications and towards jobs of interacting applications.

3 PHASE –
A Distributed Pharmaceutical Applications Server

Drug development in the pharmaceutical industry is a long drawn-out process that starts with the discovery of drug targets and ends in years of clinical testing. The total process can take more than a decade and costs hundreds of millions of ECUs. High-performance computing can make a major contribution to improve both speed and scope of the drug target discovery process. Work that, due to the lack of adequate computing facilities, would have been submitted in batch mode (e.g., profile searches in genome sequence databases) now becomes nearly interactive. This allows the researcher to approach a problem from many different angles. It improves iteratively the detection of similarities leading to the specification of new drug targets and the automatic analysis of large data volumes, e.g. entire genomes. In addition, simulation jobs now become realistic for the first time, e.g., the simulation of protein molecular complexes using distance geometry approaches ([4], [1]).

1324

3.1 Bio-Informatic Tools

The ongoing genome projects identify a large number of partial and complete genes every day. The knowledge of the sequence alone does not give any clue on its function and is useless for drug discovery. Only when a function has been assigned to a sequence it becomes a potential drug target, and once the 3D structure is known it also allows for small molecule drug design, the ultimate target of the process chain. GeneQuiz, MaxHom, DRAGON and MSAP [6] all help in the identification of function, though at different levels. All four also contribute to the generation of meaningful 3D models of potential drug targets.

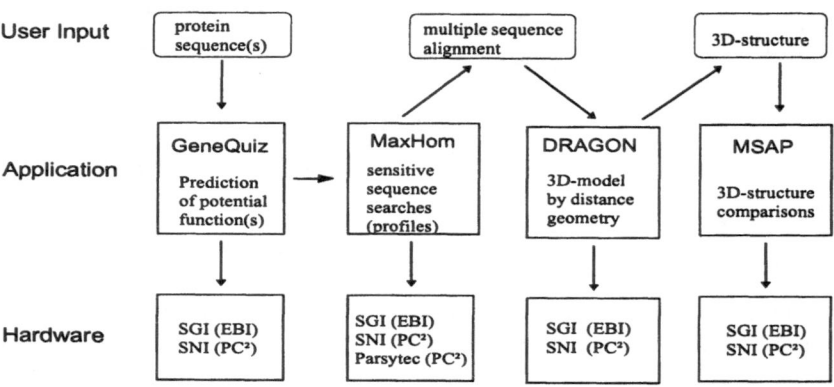

Fig. 2. Computational contribution to the drug target discovery process

Fig. 2 illustrates how the four PHASE applications build on each other to cover the complete drug target discovery process. All applications have been or will be efficiently parallelized using the portable PVM message passing library as a hardware independent programming model. MaxHom, the code with the highest computing demand has been benchmarked on the largest MPP system, a 64 processor system with more than 90% efficiency. The other codes run efficiently on workstation clusters with a moderate number of powerful computing nodes, e.g. 8 to 16 processors.

Currently, the four applications are driven by either a command-line interface or by a graphical user interface, both facilitating the integration into the server interface.

3.2 Distributed Applications Server

The distributed application server provides a central user access point to the described bio-informatic applications. The server accepts requests from the end-users via a graphical user interface and submits them to the best suited hardware system. It builds upon the MICA-Net architecture (see Fig. 1) and consists of the following main components:

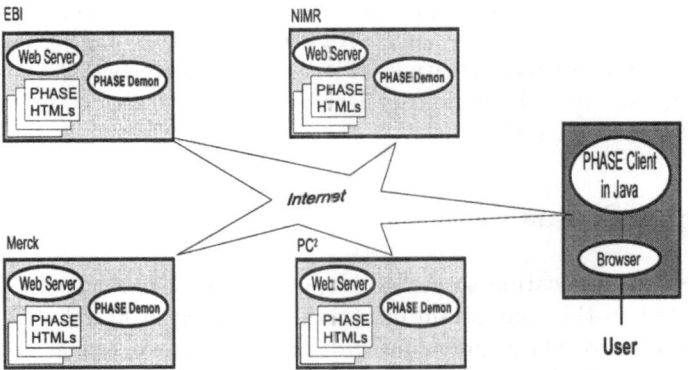

Fig. 3. Integration of PHASE-Net into the WWW

- A *Graphical User Interface* based on HTML and Java provides a user-friendly, portable entry point for incoming requests. HTML and Java are standardized WWW languages that run on virtually any hardware platform with standard web browsers.
- The *PHASE Access Layer* collects all incoming requests, checks authorization and registers them in the Accounting Database.
- A *Configuration Database* maintains a list of resources (programs and HPC systems) that are currently accessible by the PHASE server.
- A *Request Analyzer* scans incoming requests and determines their hard and software requirements.
- A *Dynamic Job Distributor* forwards the requests to the best suited platform. Its decisions are based on the system load factors and the current network throughput.
- *Distributed Load Monitor Daemons* provide information on the current system and network performance.
- *Cluster management software* like CODINE or the Computing Center Software CCS [2] will be used as an access point to the distributed resources.

Fig. 3 illustrates the planned Internet architecture for PHASE. The WWW user interface builds on GeneQuiz and Predict Protein access points, which are well-established WWW interfaces in the bio-informatics community.

3.3 Security and Confidentiality Aspects

Security issues are concerned with respect to industrial requirements and the repeated misuse of the Internet by intruders. MICA concentrates on data encryption techniques. Extending these concepts, the Internet version of PHASE is based on Java applets [9]. Applets loaded over the network are untrusted by the Web browser. Untrusted applets are run in a restricted environment and the Web browser carefully controls, what an applet is allowed to do, e.g., deleting files on the local host, sending out emails, list directory contents, open socket connections to remote hosts, and many more. Additionally, a two-stage security system will be provided: a normal security level with standard Kerberos/Netscape security mechanisms, and a high-security level with email access to trusted servers guarded by UNIX fire walls.

4   Current Status

The distributed application server developed in MICA is ready for use. It has been presented to the consortium in a review meeting in March 1997. Current work concentrates on the seamless integration of human experts as an additional resource. The distributed application server will be adopted and extended to the specific needs of PHASE within the next months. As PHASE has started in February 1997, conceptual work regarding the WWW based user interface and security issues is going on.

References

1. Casari, G., Andrade, M.A., Bork, P., Daruvar, A., Ouzounis, C., Schneider, R., Tamanes, J., Valencia, A., Sander, C.: Challenging times for bio-informatics. Nature (1995), Vol 376, 24 August 1995.
2. Ramme, F., Römke, T., Kremer, K: A Distributed Computing Center Software for the Efficient Use of Parallel Computer Systems. Springer-LNCS 797, 1994, 129–136.
3. Reinefeld, A., Baraglia, R., Decker, T., Gehring, J., Laforenza, D., Ramme, F., Römke, T., Simon, J.: The MOL Project: An Open, Extensible Metacomputer. Heterogenous computing workshop HCW'97 at IPPS'97, April 1-5, Geneva.
4. Sander, C., Schneider, R., Stouten, P.: The Human Genome and High Performance Computing in Molecular Biology. Supercomputer 1992.
5. GENIAS Homepage: http://www.genias.de/
6. PC^2 project page: http://www.uni- paderborn.de/pc2/projects/
7. POLDER Homepage: http://www.wins.uva.nl/projects/polder
8. Hypercomputing Homepage:
   http://www.tec.informatik.uni-rostock.de/hypercomp
9. Flanagan, D.: Java in a Nutshell. O'Reilly & Associates, 1996.

Acknowledgements

This work has been done in a stimulating environment. Many thanks to our MICA and PHASE project partners, the EU project officer Massimo Luciolli and the EU reviewers.

FRONTIER: Use of HPCN Technologies[*]

J.M.R.Shaw, D.C.Spicer

British Aerospace Defence Ltd, Preston, PR4 1AX, Lancashire, U.K.
email: John.Shaw@bae.co.uk, David.Spicer@dial.pipex.com

Abstract. Business trends are leading companies to work collaboratively both internally and externally. Data exchange technology has been driven by this trend, and the growth of World Wide Web technology and usage, supported by HPCN capabilities, is opening further possibilities for organisations to carry out product design and optimisation cooperatively. The FRONTIER project seeks to demonstrate the viability of collaborative design optimisation by putting into place a system whose architecture and technology addresses these needs.

1 Introduction

The FRONTIER project is concerned with *design optimisation*. It addresses directly the problems of tradeoff of *multiple conflicting objectives*, and includes 6 case studies from a range of engineering sectors.

The current industrial trend is towards Integrated Product Definition, involving collaborative consideration of all aspects of design, rather than via traditional tasks based on single-function responsibilities. This applies to inhouse design and manufacturing teams; and also to equivalent close working arrangements between collaborators, partners and customers. In I.T. terms, this places emphasis on systems integration, data sharing, and distributed operations.

In practice, an engineer's aim in optimising a design is to be able to consider as wide a range of design cases as possible. Firstly, to identify the best concept; and then to refine it sufficiently, within the time available. HPCN technologies are essential to support this. However, the successful industrial exploitation of HPCN's potential requires more than parallelised solvers and a respectable HPC platform to run them on. In order for design analysis systems to play their full part, they need to be able to operate responsively within the design and manufacturing framework, and integrate in particular with CAD and product definition systems.

---

[*] This work has been supported by the EU, ESPRIT project 20082

2 FRONTIER Technology

In FRONTIER, we want to minimise n objective functions $F_1(\tilde{X})$, $F_2(\tilde{X})$, $F_3(\tilde{X})$, $F_n(\tilde{X})$ simultaneously with respect to the design variables $\tilde{X}$, subject to a number, m, of constraints $c_i(\tilde{X}) \geq 0$; $i = 1$, m. In practice, these objectives are in conflict, so they need to be traded off against each other.

Tradeoffs are viewed by considering the performance of design options plotted in *objective space*. The region of feasible designs is limited by a notional tradeoff boundary, the *Pareto frontier*. This is illustrated in Figure 1. To identify the boundary, several approaches are possible. They include: *explicit multiobjective search* [1], conversion of multiple objectives into a single objective via suitable decision support, and fixing objectives as constraints. FRONTIER caters for all of these. A repertoire of probabilistic *Genetic Algorithms* [2] and hill climbing algorithms is being provided to address search strategies, as well as allowing full user participation if required.

In optimisation, robust *global search* has traditionally been a major problem. FRONTIER incorporates genetic algorithms to address this. G.A.'s lend themselves naturally to generalisation to cover the case of more than one objective function. Also, natural parallelism is available when using a generational-based G.A. The G.A. technology being used in FRONTIER, which includes new inheritance techniques and automatic parameter setting, is currently under development by Trieste University.

It is also crucial to provide efficient hill-climbing techniques to address single criterion *constrained optimisation*. Within the FRONTIER project this will be provided by Quasi-Newton, Sequential Quadratic Programming and gradient-free Multi-Directional Search (MDS) algorithms. Parallel versions of these are being developed by Manchester University within the PINEAPL project [3].

The aim is to combine the robustness of the G.A. approach with the superior convergence performance of hill climbers in a *hybrid optimisation algorithm* which switches between the two approaches when the region of the global minimum can confidently be identified.

Optimisation involves many design analyses, particularly where tradeoff boundaries need to be explored. *Decision support tools* are being provided, to enable the user to input tradeoff preference information as the optimisation proceeds. This allows several disparate objective functions to be combined into a single objective, expressed as a 'utility function'. The interactive coupling of these *'Multi Criteria Decision Making'* [4] tools within the main optimisation allows attention to be focussed early on the part of the boundary which is of most interest. This focussing operation is important in limiting computing resource demands, because it avoids making a large expenditure on solutions around the entire tradeoff boundary before facing up to the tradeoff decision. These tools are being developed in FRONTIER by Newcastle University.

3 Architectural Approach

The FRONTIER conceptual architecture is shown in Figure 2. Optimisation involves proposing designs to consider, by creating the design variable sets which define them (Optimisation layer); then translating, where necessary, these design variable parameters into real geometry (Parametrisation layer); and evaluating the designs (Design Evaluation layer). A *heterogeneous* collection of hardware platforms and operating systems will typically be involved, in dispersed geographical locations. *Design evaluation* will be carried out using the appropriate HPC platforms for the analysis codes concerned. *Geometry parametrisation* is likely to require repeated use of CAD systems (CATIA, Euclid, I-DEAS, ..) running within the main organisational CAD platform setup. Pre and postprocessing operations are needed on the user's desktop machine.

It is thus apparent that the FRONTIER architecture needs to be distributed across hardware and operating systems, in particular across Unix and Windows-NT. Operations and results may also need to be shared between design partners. The infrastructure thus needs environment and communication protocol independence. The emergence of *CORBA* object request brokers [5] and the Java programming language [6], and their ability to interact relatively easily with Microsoft tools and standards makes it possible for FRONTIER to realise the distributed system needed.

Design optimisation is a major HPCN customer area of the future. However, extending operations beyond the boundaries of a single enterprise, in line with the collaborative needs mentioned earlier, will involve many contractors. Interactive design will then need *web technology* combined with HPCN facilities. This subject area is debated in [7], motivated in part by the needs of the U.S. 'Affordable Systems Optimisation Project'[8]. The needs are architecturally complex and multileveled. A goal of FRONTIER is to expose the architectural issues and adopt the appropriate strategic technologies.

4 HPCN Content

Although the FRONTIER system will make major use of HPCN resources, FRONTIER is not primarily a code parallelisation project. Rather, it can be viewed as a *systems integration project*, in which advantage is taken of the increasing number of codes already parallelised.

These include: STAR-CD, FIDAP, KIVA II and SAUNA. In addition, one 'in-house' solver, 'HISSS-D' from Daimler-Benz Aerospace, is being parallelised by Bergen University.

FRONTIER project trial cases involve optimisations of *aircraft wings, marine pumps, heat exchangers, electrodomestic equipment*, and *diesel engine design*. Analysis codes used are shown in Table 1, with single-analysis serial code run-times on platforms listed. Some of the codes, particularly structural optimisation codes, include their own 'private' sub-optimisations using additional design variables owned by them. Others are pure analysis codes. Computational affordability is an important element in the trials. However, in all cases significant optimisation can be done. As an example, trials conducted on the design of a single stage of a heat exchanger using FIDAP have been run using 512 evaluations in an initial G.A. computation followed by refinement of the best solution using a hill climber. This has produced a design weight reduction of around 20%.

The platforms to be utilised in FRONTIER include workstation clusters and MPP machines. The system must allow for very flexible use of available processors, reconciling the needs of individual parallel codes with parallel design evaluations required by G.A. operations. Both generational and steady-state G.A.'s are provided, and a 'pool of tasks' approach is adopted to ensure that processors are not idle whilst individual evaluators are converging at different rates.

File transfer times between platforms are also a potential bottleneck. Each design evaluation needs a *geometry transfer*, which could take several hours, and will also produce extensive results files even after reduction and compression. Thus the overlapping of these operations with computation is important.

References

[1] P. Hajela, and C.Y. Lin: Genetic search strategies in multicriterion optimal de sign. *Structural Optimisation*. 4, pp 99-107. Springer-Verlag, 1992.

[2] L. Davis: *Handbook of Genetic Algorithms*. Van Nostrand Reinhold, 1991.

[3] ESPRIT Project 20018 'PINEAPL' home page. http://www.nag.co.uk/projects/PINEAPL.html, 1996.

[4] P. Sen, and J.-B. Yang: Multiple-criteria Decision-making in Design Selection and Synthesis. *Journal of Engineering Design*, Vol. 6, No. 3, 1995

[5] Object Management Group (OMG) home page. http://www.omg.org/, 1996.

[6] Sun Microsystems. Java homepage. http://java.sun.com/, 1996.

[7] H.W. Yau, A. Leung, W. Furmanski, and G.C. Fox: Exploration of Emerging HPCN Technologies for Web-Based Distributed Computing. *Lecture Notes in Computer Science*. 1067, pp 869-874. Springer-Verlag, 1996.

[8] G.C. Fox: 'Implementation of Affordable Systems Optimization Process in the NII'. http://www.npac.syr.edu/users/gcf/ASOPSept95A/fullhtml.html, 1995.

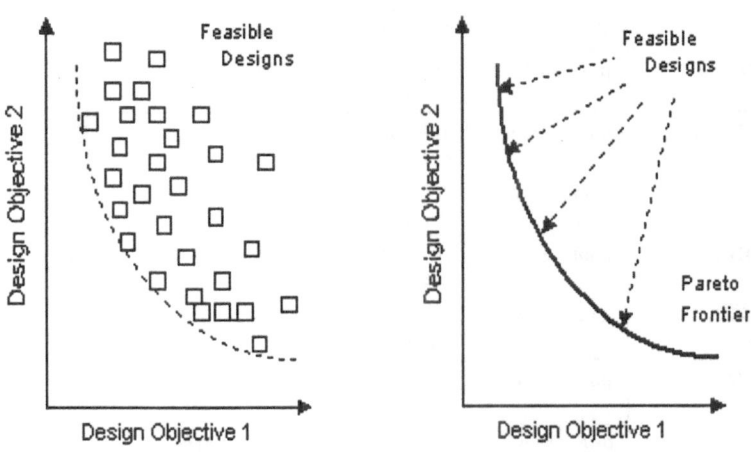

Figure 1

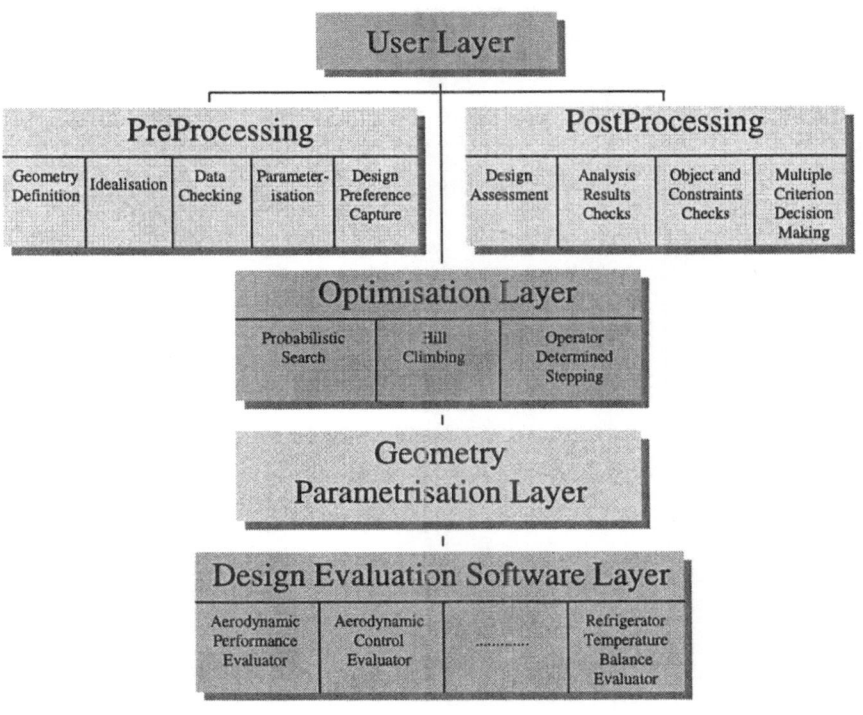

Figure 2

Code	Type of code	Run time	Platform	Intensity
FIDAP	pure analysis	500 sec (one analysis)	Cray J/916	medium
HISSS-D	anal. & opt.	100 min (with optimisation)	AIX 370 AT	high
LAGRANGE	anal. & opt.	15 min (one analysis)	HP-715	medium
CODAS	anal. & opt.	10 min (one analysis)	Cray YMP	medium
STARS	anal. & opt.	3 min (one analysis)	HP9000	low
KIVA-II	pure analysis	6-8 hours	IBM RS6000	high
MODEL	pure analysis	0.5 min	IBM W/S	very low
STAR-CD	pure analysis	3 hours 6 min	SGI Power Challenge 1CPU	high

Table 1

PINEAPL: A European Project to Develop a Parallel Numerical Library for Industrial Applications

Daniela di Serafino[1], Lucia Maddalena[1] and Almerico Murli[1]

Center for Research on Parallel Computing and Supercomputers (CPS) - CNR
Via Cintia, Monte S. Angelo, 80126 Naples, Italy

Abstract. PINEAPL (Parallel Industrial NumErical Applications and Portable Libraries) is an ESPRIT project, in the area of High Performance Computing and Networking, with partners from industral, academic and research organizations. The aim of the PINEAPL project is to produce a library of parallel numerical software that is relevant to a wide range of industrial applications. Therefore, the choice of the areas of numerical software is driven by the applications provided by the industrial partners. In this paper we describe the structure and the strategies of the project and report on progress so far. Moreover, as an example, we present an application provided by an industrial partner and the parallel routine that has been developed taking into account the requirements of the application.

1 Introduction

PINEAPL (**P**arallel **I**ndustrial **N**um**E**rical **A**pplications and **P**ortable **L**ibraries) is a three years IV Framework ESPRIT project in the area of High Performance Computing and Networking, which started on January 1996. The main goal of the project is *to produce a general purpose library of parallel numerical software suitable for a wide range of computationally intensive industrial applications and portable and efficient across a wide range of high performance machines.*

Numerical software libraries supply *building blocks* for the solution of mathematical problems common to different applications. The availability of reliable, accurate and efficient numerical software allows users to solve application problems without dealing with details related to numerical algorithms and their implementation. This enables non-expert users to exploit the experience of numerical analysts and mathematical software developers, thus reducing their efforts and improving the quality of the results. Presently, there are more than 3000 industrial and academic organisations using general purpose numerical libraries in Europe alone. On the other hand, the widespread and effective use of High Performance Computing resources is inhibited by the lack of software libraries that are suitable for such advanced environments. The demand for these libraries is increasing as more end-users take advantage of networks and clusters of workstations and distributed memory computers. In this context, the PINEAPL project aims at reducing this gap, enabling users to take advantage of the increasing computing power and memory offered by multiple processors.

The project relies on a close collaboration between *academic, research* and *industrial organizations*, which are the partners of the PINEAPL Consortium. The industrial organizations, that make use of numerically intensive application codes for design and simulation, are the *end-users* of the project. They take advantage of the opportunity of introducing new mathematical models and techniques into existing application codes and of enhancing their performance by the incorporation of parallel library routines. The end-users' application codes have been chosen to represent a varied cross-section of industrial problems and the PINEAPL library coverage has been primarily determined by the requirements of these applications. The research and academic organizations, with well-known experience in the development of sequential and parallel mathematical software, act in the project as *numerical software suppliers* and *application parallelization experts*. Their expertise is used for the development of parallel numerical software of interest for the industrial applications. Each industrial partner has one or more associated parallelization experts, as it is shown in Table 1.

Table 1. PINEAPL Consortium

Project Coordinator: Numerical Algorithms Group (NAG), UK	
Industrial Partners	**Related Parallelization Experts**
British Aerospace (BAe), UK	University of Manchester (ManU), UK
Piaggio Veicoli Europei, Italy	IBM SEMEA, Italy; CPS, Italy
Thomson-CSF, France	CERFACS, France; CPS, Italy
Danish Hydraulic Institute (DHI), Denmark	Math-Tech, Denmark

Contacts with external collaborators and other related EC funded projects (EUROPORT) and user groups, such as the NAG User's Association, give further input to the software design and specification, to ensure that the needs of a larger industrial audience are met.

More details on the project can be found in the PINEAPL Web page located at the URL http://www.nag.co.uk/projects/PINEAPL.html.

2 Project Description and Progress

The choice of the areas of numerical software for the PINEAPL library has been driven by the applications provided by the industrial partners, summarized in Table 2. A first stage of the project has been devoted to a careful analysis of these applications and has led to the identification of the main areas to be covered by the numerical library. Some more areas have been added, that are likely

1335

Table 2. PINEAPL applications

Applications	End-users
Wing design: maximum lift/minimum drag	BAe
Electro-magnetic wave reflection of ducts	
Estuarine and coastal hydraulics and oceanography, and environmental simulation	DHI
Oil reservoir simulation	Math-Tech
Engine simulation: chemically reactive flows with sprays	Piaggio
Design of nanometric recording devices	Thomson-CSF
Simulation of beam propagation in rod lasers	

to be of use in a wide range of computationally intensive applications outside the Consortium. The parallel library includes also dynamic load balancing and support/utility routines, that aid the development and the use of parallel software. All these areas are summarized in Table 3.

Table 3. Areas covered by the PINEAPL library

Areas	Routines	SW suppliers
Dense Linear Algebra	Banded solvers	NAG
	Condition number estimators	
	Eigensolvers	NAG, CERFACS
Sparse Linear Algebra	Krylov iterative solvers	NAG
	Preconditioners	
Non-linear Optimization	Constrained	ManU
	Unconstrained	
Discrete Fourier Transform	1-D, 2-D, 3-D FFT	CPS
Partial Differential Equations	2-D, 3-D Fast Poisson Solvers	CPS
	3-D Multigrid for Helmholtz equation	
	Mesh partitioning	NAG
Dynamic Load Balancing		ManU
Support/Utility	Basic sparse matrix operations	NAG
	Input/output	
	Data distribution	NAG, ManU

The library software is written in Fortran 77 and uses popular *de facto* standard message-passing libraries, mainly the BLACS [3], and MPI [6] and PVM [4] where the BLACS do not have the required functionality. The library is targeted primarily at distributed memory machines, ranging from low cost networks of workstations to high-end distributed memory parallel computers, but can be used also on shared memory machines provided that implementations of MPI or PVM are available. The driving criteria in the development of the library routines are ease of use, portability, performance, flexibility and reliability [5].

Specification documents have been produced for all the parallel library routines. Most of these routines are currently under development, and some of them have been already completed in the areas of Dense and Sparse Linear Algebra, Discrete Fourier Transforms, Optimization and Support/Utility.

Since the application codes provided by the end-users act as a testbed for most of the library routines, initial working implementations of these codes on the target architectures have been produced, taking into account the library software specifications. Moreover, benchmark problems have been prepared for future testing of the parallel versions of the codes that will be based on the library routines.

To test the quality of numerical algorithms and software developed within the PINEAPL project, the tool PRECISE (PRecision Estimation and Control in Scientific and Engineering computing) is being used [2]. PRECISE is a set of tools to perform numerical experiments exploring the accuracy and stability of computational schemes, and consists of two modules, one performing statistical backward error analysis and the other performing sensitivity analysis. At the beginning of the PINEAPL project it was in a Matlab prototype form; part of the work in the project is devoted to bring it to a mature and commercial quality. Most of the computational routines required for a PRECISE analysis have been converted to Fortran 77; the core PRECISE routines have been completed, and further work is devoted to allowing the user or developer of parallel algorithms to easily call PRECISE routines from the parallel applications. Moreover, PRECISE has been applied to numerical problems obtained from DHI, Piaggio and Thomson-CSF applications.

The project progress can also be found in the first project Annual Report, available on the WEB page.

3 Example: a 2-D FFT Routine for Simulation of Beam Propagation in Rod Lasers

In this section we give a short presentation of a pair application / library routine, where CPS is involved both as application parallelization expert and numerical software supplier.

3.1 Application and Numerical Solution Method

The application presented here is the modelling of the propagation of beams in diodes pumped rod lasers (coupled with thermal effects), provided by Thomson-CSF. The modelling is mainly devoted to describe the evolution of an optical beam inside the rod laser, in order to estimate the deformation of the plane wave, the resulting quality and amplification of the beam, and thus the energy efficiency of the whole component.

Basically, the beam propagation is modelled by a Helmholtz equation:

$$\Delta \Phi + k_0^2 n^2 \Phi = 0,$$

where $\Phi = \Phi(x, y, z)$ is the (polarized) electromagnetic field, n is the index of the medium, and k_0 is the wave number $2\pi/\lambda$, where λ is the wavelength. Φ and n have complex values. The separation of the fast variation of the field along the direction of propagation, say z, leads to an equation of the form:

$$\Delta \Psi + 2i\, k_z \frac{\partial \Psi}{\partial z} + k_0^2 (n^2 - n_z^2)\Psi = 0. \tag{1}$$

The solution of (1) is performed using a method in the class of *Beam Propagation Methods* (BPMs) [8], that are methods widely used in integrated optics. The version of BPM used here is based on a splitting of (1) into two equations, modelling the propagation and the "lens" diffraction, which are solved in turn, step by step, along the z-direction:

$$\Delta_t \Psi + \frac{\partial^2 \Psi}{\partial z^2} + 2i\, k_z \frac{\partial \Psi}{\partial z} = 0, \quad 2i\, k_z \frac{\partial \Psi}{\partial z} + k_0^2 (n^2 - n_z^2)\Psi = 0.$$

(Δ_t is the Laplace operator in the plane of the transverse field, i.e. in the (x, y)-plane.) The propagation from z to $z + dz$ is computed by a transformation from the (x, y)-plane to the Fourier space, where the solution is given analytically. This transformation, and its inverse, are performed as 2-D Discrete Fourier Transforms (DFTs) on complex data obtained with a uniform grid discretisation with periodic boundary conditions. The DFTs are carried out using a 2-D Fast Fourier Transform (FFT) algorithm.

At each step along the z-direction, the most time consuming part of the algorithm is the execution of direct/inverse FFTs, which requires at least 70% of the total computation time. Therefore, a significant speedup can be achieved using a parallel 2-D FFT routine.

3.2   The 2-D FFT Routine

The 2-D FFT routine developed for the PINEAPL library computes the 2-D DFT of a bivariate sequence of complex data values $Z = (z_{j_1 j_2})$:

$$\hat{z}_{k_1 k_2} = \frac{1}{\sqrt{mn}} \sum_{j_1=0}^{m-1} \sum_{j_2=0}^{n-1} z_{j_1 j_2} \times e^{\pm 2\pi i \left(\frac{j_1 k_1}{m} + \frac{j_2 k_2}{n} \right)}, \quad \begin{array}{l} j_1, k_1 = 0, \ldots, m-1 \\ j_2, k_2 = 0, \ldots, n-1 \end{array}.$$

The 2-D DFT is obtained performing m 1-D transforms of length n, on the rows of the matrix Z, and n 1-D transforms of length m, on the rows of the transposed resulting matrix, i.e. on the columns of the resulting matrix. The FFT algorithm used to perform the 1-D transforms is the *Stockam self-sorting algorithm*, described in [7].

As all the routines of the PINEAPL Library, the FFT routine assumes that a 2-D grid of p logical processors is avalable. The parallel algorithm is based on a *row-block distribution* of the matrix Z, that is $M_b \simeq m/p$ consecutive rows of Z are allocated to processors on the 2-D grid row by row (i.e. in row major ordering of the grid), starting from the $\{0, 0\}$ processor. Each processor performs M_b 1-D transforms of length n, on the rows assigned to it, contributes to the transposition of the global resulting matrix, performs $N_b \simeq n/p$ 1-D transforms of length m, and finally, contributes to the transposition of the global final matrix. All the communication is performed in the transposition of the distributed matrices. More details are given in [1].

We note that the routine allows to avoid the second transposition of data, leading to a significant reduction of the execution time. This choice is recommended, for example, when a forward DFT followed by a backward DFT has to be computed, since the second DFT can be computed directly from a matrix in transposed order. Further time saving is allowed when several DFTs of the same length must be computed, since the exponential coefficients required by the FFT algorithm can be stored and reused.

Preliminary experiments have been carried out to test the parallel performance of the 2-D FFT routine, on an Intel iPSC/860 at CPS, with 8 nodes and 16 MB of memory per node. Each node has a peak performance of 40 Mflops in double precision; the peak bi-directional bandwidth among two nodes is 2.8 MB/s and the latency is 6.5 ms. The routine has been executed in double precision, using the NX (Intel native node operating system) version of the BLACS.

Figure 1 shows the parallel efficiency of the routine, with and without the second global transpose, for $p = 2, 4, 8$ and different values of $m = n$. Such values are radix-r, with $r = 2, 3, 5$ ($256 = 2^8$, $512 = 2^9$, $243 = 3^5$, $625 = 5^3$), and mixed-radix, also including large prime factors ($450 = 2 \times 3^2 \times 5^2$, $600 = 2^3 \times 3 \times 5^2$, $249 = 3 \times 83$, and $544 = 2^5 \times 17$). On one processor the (sequential) NAG Fortran Library routine which implements the Stockam self-sorting algorithm has been executed. As it was expected, avoiding the second global transposition of data improves the efficiency of the routine. Among the radix-r DFTs, the case $r = 3$ corresponds to the lowest efficiency; this can be explained since the sequential radix-3 FFT algorithm has a relatively lower operation count. The most efficient parallel executions are those corresponding to the mixed-radix case with large prime factors, since their sequential counterparts require a larger computation time. These results can be considered satisfactory, since communication has a significant weigth in parallel FFT algorithms, and typically accounts for a large part of the computational effort.

Acknowledgments. The authors wish to thank Thomson-CFS LCR for providing the description of their application and NAG Ltd for their helpful comments.

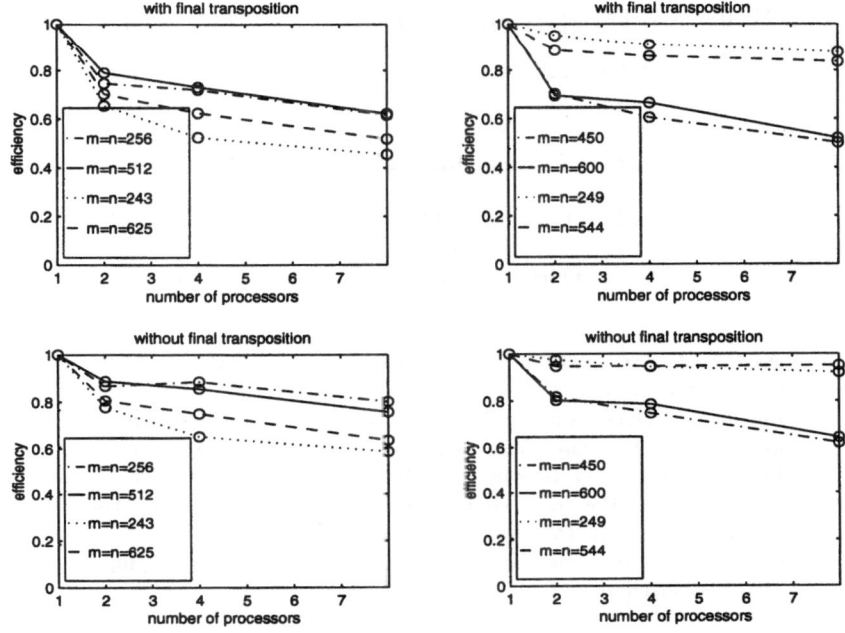

Fig. 1. Parallel efficiency of the 2-D FFT routine with and without the final global transposition, on 2, 4 and 8 processors

References

1. Carracciuolo, L., De Bono, I., De Cesare, M.L., di Serafino, D., Perla, F.: Development of a Parallel Two-dimensional Mixed-radix FFT Routine. Tech. Rep. TR-97-8, Center for Research on Parallel Computing and Supercomputers (CPS) - CNR, Naples, Italy (1997)
2. Chaitin-Chatelin, F., Fraysse, V.: Lectures on Finite Precision Computations. SIAM, Philadelphia (1996)
3. Dongarra, J.J., Whaley, R.C.: A Users' Guide to the BLACS v 1.0. LAPACK Working Note No. 94, Technical Report CS-95-281, Department of Computer Science, University of Tennessee, 107 Ayres Hall, Knoxville, TN (1995)
4. Geist, A., Beguelin, A., Dongarra, J.J., Jiang, W., Manchek, R., Sunderam, V.: PVM: Parallel Virtual Machine. A Users' Guide and Tutorial for Networked Parallel Computing. MIT Press, Cambridge, MA (1994)
5. Krommer, A., Derakhshan, M., Hammarling, S.: Solving PDE Problems On Parallel and Distributed Computer Systems Using the NAG Parallel Library. Communication at the Conference HPCN'97, April (1997)
6. Snir, M., Otto, S.W., Huss-Lederman, S., Walker, D.W., Dongarra, J.J.: MPI: The Complete Reference. MIT Press, Cambridge, MA (1996)
7. Temperton, C.: Self-sorting Mixed-radix Fast Fourier Transforms. J. Comp. Phys. **52** (1983) 1–23
8. Van Roey, J., Van der Donk, J., Lagasse, P.E.: J. Opt. Soc. Am. **71** (1981) 803

RAIN: Redundant Array of Inexpensive workstations for Neurocomputing

Davide Anguita, Marco Chirico, Anna Marina Scapolla, and Giancarlo Parodi

DIBE - Dept. of Biophysical and Electronic Eng.
University of Genova
Via Opera Pia 11a, 16145 Genova, ITALY.

Abstract. We describe here the project RAIN, started at the beginning of 1997 and aimed to demonstrate the use of High Performance Computing and Networking technologies in neural network applications for industry and medicine. The target architecture of the demonstrators will be a cluster of workstations: this choice will allow a low-cost implementation and, at the same time, will provide some insight on the use of this relatively new (at least outside the academic world) parallel architecture. Several communication networks will be tested ranging from 10 Mb/s Ethernet to 155 Mb/s ATM.

1 Introduction

The RAIN project acts within the framework of the European Commission Call for Tender III/96/55 (Esprit Programme), Contract No. 85387 for Lot 2 "Demonstration and assessment of HPCN in neural network applications for industry and medicine". The main objective is the realization of two demonstrators of neural applications with the support of High Performance Computing and Networking (HPCN) technologies. The project started in January 1997 and will develop during 18 months.

In this paper we will briefly describe the rationale behind the RAIN project and its development. This section will address the main issues of artificial neural networks and their simulation on high performance computing systems. In Section 2 we will discuss the project with greater detail.

1.1 Artificial neural networks (ANNs).

The target ANN considered in the RAIN project will be mainly the Multi Layer Perceptron (MLP) and its companion learning algorithm named Back–Propagation (BP) [10]. This network has been widely used to solve difficult problems where a mathematical model is hard to find.

The success of the MLP is in part derived from its ability to approximate any (reasonable) input/output relation, learning from a finite amount of input/output samples (patterns). This approximation capability can be applied both in pattern recognition and in regression problems [1]. In the first case the MLP can be used as a classifier, approximating a probability density function, in the second case it can directly approximate an optimal (following some criterion) control function. Despite the relatively recent rediscovery and the fact that some of the aspects of their functioning are still object of basic research, Artificial Neural Networks (ANNs) had rapidly become an effective tool for a variety of real-world applications. The reason for this success is mainly due to at least three factors:

- the multipurpose ability of ANNs: similar connectionist architectures can be used for a variety of different applications (from pattern recognition to image compression) virtually without any change and with a high reusability of the developed code;
- the highly successful outcome when applied to complex and highly-nonlinear real-world problems, especially when compared to traditional methods;
- the intrinsic massive parallelism of the computational paradigm that suggests an easy implementation on parallel architectures or dedicated VLSI hardware.

1.2 Artificial Neural Networks and Parallel Computing.

The computational requirement of a MLP is, in general, very large even when dealing with small problems; this means that, as the problem size grows, the computational requirements can become prohibitive. For example, in [7] it is stated that a current target for neurocomputing, in the field of speech understanding, is "... *evaluating a network with a billion connections a hundred times per second* ..." or, in other words, 100 GCPS (Giga Connections Per Second). This means that a performance in the range of TFLOPS is needed to obtain a reasonable learning speed on problems of this size.

It has been shown that the computational requirements of MLP, given a certain problem size, can be easily estimated for each learning step [4]. In table 1, an example of the requirements for real-world problems are summarised (the data bases are courtesy of Univ. of California - Irvine).

Table 1. Computational requirements of ANNs (10 learning steps/second).

Database	Problem size	Computational req. (MFLOPS)
Pima Indians Diabetes	8x8x1, 768	3.1
Breast Cancer 1	9x9x1, 699	3.4
Credit Screening	15x15x1, 690	8
Water Treatment Plant	38x38x1, 527	34
Thyroid	29x29x1, 2800	110
Low Resolution Spectr.	98x98x1, 531	210
Letter Recognition	16x64x1, 20000	2000

In Table 2, some implementations of MLP–BP are summarized (references can be found in [6]). The first column shows the target computer and the number of processors; the second column indicates the speed obtained by the implementation.

Table 2. Implementations of the backpropagation.

Computer	MCUPS
CNS–1 (128/1024)	22000/166000
Adapt. Sol. CNAPS (512)	2379
TMC CM–2 (64k)	350
TMC CM–5 (32+VU)	167/4.75
HNC SNAP (16/64)	80.4/302
TMC CM–5 (512)	76
FUJITSU VP-2400/10	60
Cray Y–MP (2)	40
TMC CM–2 (4k/64k)	2.5/40
Cray X–MP (4)	18
IBM 6000/550	17.6
Intel iPSC/860 (32)	11
Cray 2 (4)	10
PC486	0.47
MasPar MP–1	0.3

Scanning the table one can make surprising comparisons. For example, the fastest implementation on a large–grain supercomputer (FUJITSU VP–2400/10) outperforms a conventional workstation (IBM 6000/550) only by a factor of three. The MasPar figure has been reported in the table only as simple curiosity. It's obvious that this machine can perform orders of magnitude better, but this shows how the implementation of the backpropagation algorithm is not widely very well understood. Our group has developed efficient implementations of neural algorithms both for scalar and distributed computing architectures [5, 3] and is working to transfer this know-how in the project presented here.

2 The RAIN project.

The rationale behind RAIN is the following:

- only few dedicated machines for neurocomputing come from a non-academic environment and are available on the market (most notably: Adaptive Solution CNAPS and Siemens Nixdorf SYNAPSE);
- the performance gap between dedicated neurocomputers and conventional high-performance microprocessors is narrowing at a very fast rate (for example, peak performances of DEC Alpha and HP PA8000 CPUs now exceed

0.5 GFLOPS, much less than an order of magnitude from multiprocessor neurocomputers);
- software for dedicated hardware is much less portable than code developed for general purpose machines; furthermore the latter can be developed making use of standard routines (e.g. BLAS [2, 8, 9]) that provide virtually a hardware abstracting layer;
- costs are greatly reduced both on start-up and on the medium and long term, because general purpose workstations allow easy upgradability towards more powerful systems;
- one factor that fueled much skepticism about the feasibility of network-based high-performance computing was the limitations imposed by traditional LANs: for applications such as communication-intensive, coarse-grain high-performance applications, traditional networks simply cannot provide adequate performances). On the other hand the network bandwith (and global performance) can be expanded through the use of faster LANs as soon as the technology is wide-spread (from 10Mbit/s Ethernet to 100Mbit/s VG AnyLAN to 155 Mbit/s ATM).

Following the above-mentioned considerations, we proposed to the EC the development of the industrial and medical demonstrators using a Redundant Array of Inexpensive workstations for Neurocomputing (RAIN) system.

2.1 Phase I: Requirements Analysis.

The requirements analysis has to identify the requirements of neural network applications which may be satisfied by the use of HPCN technologies in industrial and medical applications. Phase I of the project will be completed at the end of month 6 (June 1997).

One of the means for collecting information will be a questionnaire sent to all main European organisations in the field. Direct interviews to the most interesting organisations will complete the information collection.

The answers to the questionnaires will provide informations on: the involved organisation (sector of activity, size, IT infrastructure, etc.); the applications requiring either the handling of large amount of data or asking for high performances, currently without a satisfactory solution; the expectations from the use of neural networks and HPCN technologies; the budget that each organisation could allocate to adopt a high performance platform to solve its problems. The analysis of the answers to the questionnaires will provide a ranked list of industrial and medical applications which may be satisfied by the use of neural networks and HPCN technologies. From the set of candidate organisations, the two organisations (industrial and medical, respectively) that will provide the real data for the demonstrators will be selected.

Several other ESPRIT projects are expected to contribute, at least indirectly, to the information collection:

- SIENA Esprit Project 9811
  SIENA (Survey of the European neural network market) aimed at getting an

overview of the commercial European neural network market. The SIENA survey has collected case studies of commercial successful neural network applications in many different sectors: chemical process technology, recognition of exploitable oil wells, modelling market dynamics, qualification of shock-tuning for automobiles, etc.
- NeuroNet Esprit "Network of Excellence"
NeuroNet is the European Neural Networks "Network of Excellence" set up by the European Commission's Esprit Division for Basic Research in 1994 (DIBE is a Managing Node of NeuroNet).
- ONNI-ADC Esprit project 9266
ONNI-ADC aims to develop an advanced defect classification system prototype based on a neural network system approach and OMI technologies.
- ANNIE Esprit Project 2092
ANNIE (Application of Neural Networks for Industry in Europe) aimed at researching which of several generic problem areas are best approached using neural networks.
- ELENA Basic Research Esprit project
ELENA (Enhanced Learning for Evolutive Neural Architectures) has made available a set of databases to be used for tests and benchmarks of neurocomputing simulators.
- Other EC projects
The requirements analysis will consider the results obtained by other R&D EC projects such as: NEUFODI (Esprit, No. 5433 and 7534), NEUROBOT (Esprit No. 8338), CONNY (Esprit No. 6715), NEUROQUACS (Esprit No. 7185), PSYCHO (Brite/Euram BRE20976).

2.2 Phase II: Implementation phase.

2.3 Prototype development.

The demonstrators will be implemented in a workstation cluster, selectable from the machines available at DIBE: Hewlett Packard HP 9000s, IBM RISC 6000s, PC Intel PentiumPro. The workstations will be connected with three different LAN types: 10 Mbit/s Ethernet LAN, 100 Mbit/s 100VG AnyLAN, 155 Mbit/s ATM LAN.

The structure of the demonstrator will be divided in layers: at the bottom layer, a low level library will implement the basic optimised vector/matrix operations that are essential for the large majority of ANN algorithms. The low level library will be portable and at the same time exploit the architecture and computing units of the different CPU families. This has been shown to be a feasible task, both in neurocomputing [5] and in basic linear algebra (BLAS [2])

A communication library will allow to exchange data between the computing nodes taking in account that communications must be performed through the LAN. This means that it must evaluate both the reliability and performance of the connection between the workstations. The communication library will make use of Parallel Virtual Machine (PVM).

The high level software layer will implement the ANN algorithms on the workstation cluster making use of the communication layer and the low level library for fast core operations.

At the top of the software structure, the user interface will allow the control of the functionality of the demonstrators. To fully exploit the demonstrator, even at remote sites, it will allow the control of the workstation cluster through low speed channels.

3 Concluding remarks.

At the time of this writing the project is in the first phase (requirements analysis). More information can be found at http://dibe.unige.it/RAIN along with public domain software regarding neurocomputing in general. Based on our experience, we believe that neural network applications will mostly benefit from reaserch in the field of parallel computing. We hope that the RAIN project will help to demonstrate this.

References

1. Special Issue on Artificial Neural Network Applications. Proc. of the IEEE **84** 1353–1576.
2. Anderson, E.C., Dongarra, J.: Perfomance of LAPACK: A Portable Library of Numerical Linear Algebra Routines. Proc. of the IEEE **81** (1993) 1094–1101
3. Anguita, D., DaCanal, A., DaCanal, W., Falcone, A., Scapolla, A.M.: On the distributed implementation of the back–propagation. Proc. of the Int. Conf. on Artificial Neural Networks, Sorrento, Italy (1994) 1376–1379
4. Anguita, D., Gomes, A.G.: Mixing floating- and fixed-point fromats for neural network learning on neuroprocessors. Microprocessing and Microprogramming **41** (1996) 757–769
5. Anguita, D., Parodi, G., Zunino, R.: An Efficient Implementation of BP on RISC-based Workstations. Neurocomputing **6** (1994) 57–65
6. Anguita, D., Passaggio, F., Zunino, R.: Learning in large neural networks. Proc. of HPCN Europe 1995, Lecture Notes in Computer Science **919** 269–274.
7. Asanović, K., Beck, J., Feldman, J., Morgan, N., Wawrzynek, J.: Designing a Connectionist Network Supercomputer. Int. J. of Neural Systems **4** (1993) 317–326
8. Corana, A., Rolando, C., Ridella, S.: A Highly Efficient Implementation of Back-propagation Algorithm on SIMD Computers. High Performance Computing, J.-L.Delhaye and E.Gelenbe (Eds.) (1989) 181–190
9. Corana, A., Rolando, C., Ridella, S.: Use of Level 3 BLAS Kernels in Neural Networks: The Back–propagation algorithm. Parallel Computing 89 (1990) 269–274
10. Rumelhart, D.E. and McClelland, J.L.: Parallel Distributed Processing. Vol. 1, MIT Press, 1986

PARSAR: Parallelisation of a Chirp Scaling Algorithm SAR Processor*

Antonio Martínez, Francisco Fraile

Remote Sensing Dep., INDRA Espacio
C/ Mar Egeo s/n. 28850-S.Fernando de Henares, SPAIN
e-mail: amar@mdr.indra-espacio.es

Jordi Mallorquí, Leonardo Nogueira, Jordi Gabaldá, Antoni Broquetas,
Antonio González (#)

Signal Theory & Communications Dep., U.Politecnica Catalunya
(#) Computer Architecture Dep., U.Politecnica Catalunya
C/Sor Eulalía de Anzizu s/n, Ed. D-3. 08071-Barcelona, SPAIN

Abstract. A parallel SAR processor is presented in this paper. The target configuration is a cluster of UNIX workstations, available in most user sites. This fact allows to obtain an increased computing performance without the need of dedicated hardware investment.

1 Introduction

Synthetic Aperture Radar, SAR, is a remote sensing instrument capable of obtaining high resolution images of the Earth surface [1]. The goal of SAR processing is to transform the SAR raw data, or SAR signal, into an image. The generation of SAR images involves both a great amount of data and a complex focusing algorithm. These reasons make attractive the application of HPCN technology.

The objective of PARSAR project is the porting of a sequential SAR Processor to a parallel architecture. The target configuration is a cluster of UNIX workstations, allowing to exploit the benefits of parallelisation without the need of hardware investment. This paper describes the parallelisation strategy of the processor, based on a multi-block approach using PVM as interface among processes. Some preliminary performance results are presented, along with the currently on-going activities.

2 Description of Data and Algorithm

The SAR data used in the project correspond to the standard full frame scene (100 x 100 km^2) of the ERS satellite [2]. The raw data is a matrix of 26800 lines each one consisting of 5616 samples or pixels. The pixels are complex numbers, coded in 1 + 1 bytes; the size of the raw data file is about 300 MB. The calculations are done in floating point, resulting in a processing matrix of 1.2 GB. The output SAR image has 25000 lines each one consisting in 4912 samples. The pixels are complex and coded in 2 + 2 bytes, resulting in about 500 MB.

* This work has been supported by the EU, PCI-II ESPRIT project 21037

The data volumes involved in SAR processing are high. This fact along with the increased use of SAR data and the advent of applications requiring near real time response (ie. oil pollution monitoring), make attractive the application of HPCN technology. An EUROPORT project is involved in SAR processing; its objective is the porting of a library of functions for the analysis, not the generation, of SAR images.

The focusing of SAR raw data is essentially a 2-D correlation of the input signal with the SAR Impulse Response Function [3]. Classical methods of SAR processing implement the signal compression in the frequency domain. The SAR processor used in the project is called Chirp Scaling Algorithm, CSA [4]. The CSA involves only FFT and multiplications. The structure of the CSA is relatively simple, consisting on a sequence of 1-D FFT and matrix element by element products.

3 Design of the Parallelisation

The parallelisation strategy that has been implemented is called Multi-Block Approach, MBA. It consists on dividing the input data into independent processing blocks; each block is fully processed in one host, resulting in a small piece of the final image. MBA can be seen as a coarse grained parallelisation strategy. PVM, Parallel Virtual Machine, is used to control the whole process.

The main advantage of MBA is the full independence of the different tasks split among the available hosts. Another interesting point of MBA is the minimization of both the number of I/O operations and the amount of data to be transferred across the cluster.

The implementation of the MBA parallelisation is relatively easy. Each host in the cluster has a CSA processor, very similar to the sequential one. So, improvements in the sequential code are readily portable to the parallel software. On the other hand, the drawback of MBA is the correlation efficiency, that strongly depends on the size of the processing data blocks and hence, on the available RAM in each host.

The flow char of the parallel SAR processor is shown in figure 1. There is a main process in charge of launching different tasks in the computers of the network and managing the execution of them. There are three types of tasks or slaves processes:
- cutter. This task is responsible of reading the raw data file and generating the different data blocks that will be sent to the nodes.
- child. This is a simplified version of the CSA processor, that reads a data block and produces a small piece of the final image (imagette).
- builder. The builder reads the different pieces of the image and assembles the final SAR image file.

Both the raw data blocks and the imagettes generated by the processors are stored in temporary files. The use of temporary files can be seen as a drawback; however, tests conducted without temporary files showed the importance of disk access conflicts when different child processes read from and write to the same large files. Furthermore, disk operations by the child processes are by direct access implying high inefficiency.

The parallel code starts by generating a set of data blocks that are assigned to the available hosts. The parent process continuously examine the status of the hosts in the network. When one host is free, the parent assign a new block. The priorities in the main process are: 1. Cutter process, to ensure that there are data blocks available for the child processes; 2. CSA processes and 3. Builder process.

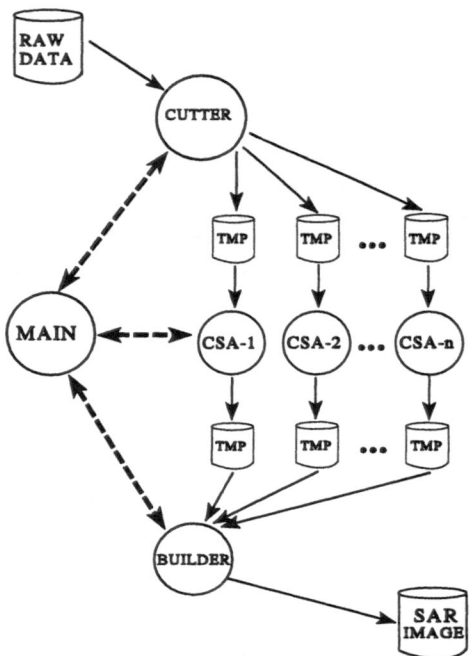

Fig 1. Flow char of the parallel processor

With this strategy, the disk bottleneck problems are alleviated by imposing that the different slave processes never access to the same file, and most disk operations can be carried out by using the more efficient sequential access. The number of hosts to be used in the "parallel machine" can be selected, as well as the tasks to be conducted by each node.

4 Preliminary Results

The classical parameters to estimate the performance of the parallel code, speed up factor and efficiency, are not readily portable for a heterogeneous cluster, as each node has its own processing time. We have used an alternative definition for these parameters that is intuitive and makes some sense for a heterogeneous set of computers:
- Efficiency: ratio of the number of standard products generated by the parallel code to the number of products generated by the sequential code running in all the computers of the cluster during the same time.
- Speed up: the efficiency times the number of computers in the cluster.

The results of the parallel processor running in different configurations are listed in the next tables, along with the characteristics of the workstations in the clusters and the processing time of the sequential processor in each workstation.

The first test used a block size of 32 MB (note that 2 of the computers in the cluster have 64 MB RAM). The main, cutter and builder processes were run in HP-720, so that this host is fully dedicated to data handling. The remaining two computers were in charge of SAR processing. The efficiency of the parallel code is 0.64, with a speed up factor of 1.93.

MODEL	Clock Rate	RAM	Proc. Time
HP-735	99 MHz	96 MB	150 min
HP-720	50 MHz	64 MB	320 min
HP-715	50 MHz	64 MB	320 min
CLUSTER	-	-	**120 min**

Table 1. Processing time in UPC cluster. Processing block 32 MB

The results of the tests at INDRA are presented in table 2; now, a block size of 64 MB was used; the slowest host executed the main, cutter and builder processes There are two test cases:
- Case 1: Only the 3 Sun computers were used. The processing time is 58 minutes, resulting in a efficiency of 0.68, and a speed up factor of 2.03.
- Case 2: All the computers are used. The processing time is 46 minutes, resulting in a efficiency of 0.56, and a speed up factor of 2.24.

MODEL	Clock Rate	RAM	Proc. Time
HP C160-L	160 MHz	128 MB	75 min
Sun Ultra 1	167 MHz	128 MB	75 min
Sun S-20	75 MHz	128 MB	135 min
Sun S-10	40 MHz	128 MB	210 min
3 Sun WSs	-	-	**58 min**
All hosts	-	-	**46 min**

Table 2. Processing time in INDRA cluster. Processing block 64 MB.

The results of the tests show that the parallel processor works relatively well in a cluster of three workstations. The efficiency figures obtained in the clusters at UPC and INDRA are equivalent. However, when including an additional host to the cluster, the efficiency decreases. This is due to the fact that the computers performing SAR processing are faster than the cutter, so that they have to wait for data blocks to process.

Work is currently on-going to upgrade and fine tune the performance of the parallel code. In particular, we may mention the following points:
- Optimization of the I/O operations, (cutter process). This should allow the use of more workstations in the cluster without loss of efficiency.
- Allowing the size of the processing block to be fitted for each host, so that computers with different RAM can be simultaneously used.

5 Conclusions

The first activities in the porting of a sequential SAR Processor to a parallel architecture have been performed. The parallel software is flexible and portable, so that it can be installed in most user sites. The preliminary results obtained with the parallel code are encouraging, and show a decrease in the processing time of the parallel code with respect to the sequential one. Good results were obtained with a cluster of three workstations. Additional work is on-going to enhance and fine-tune the code, so that the efficiency of the parallel code can be kept constant when adding more hosts to the cluster.

References

1. Elachi C.: Spaceborne Radar Remote Sensing. IEEE Press, 1988.
2. ESA.: ESA ERS-1 Product Specifications. ESA SP-1149, 1992.
3. Curlander J.C. & McDonough R.N.: SAR: Systems and Signal Processing. John Wiley & Sons, 1991.
4. Raney R.K., Runge H., Bamler R., Cumming I.G., Wong F.H.: Precision SAR Processing Using Chirp Scaling. IEEE Trans. Geosci. Remote Sensing (1994) Vol. 32 pp. 786-799.

OCEANS:
Optimizing Compilers for Embedded Applications*

Bas Aarts[1], Michel Barreteau[2], François Bodin[3], Peter Brinkhaus[4],
Zbigniew Chamski[3], Henri-Pierre Charles[2], Christine Eisenbeis[5], John Gurd[6],
Jan Hoogerbrugge[1], Ping Hu[5], William Jalby[2], Peter M. W. Knijnenburg[4],
Michael F. P. O'Boyle[7], Erven Rohou[3], Rizos Sakellariou[6], Henk Schepers[1],
André Seznec[3], Elena Stöhr[6], Marco Verhoeven[1], and Harry A. G. Wijshoff[4]

[1] Philips Research, Information and Software Technology, Prof. Holstlaan 4,
5656 AA Eindhoven, The Netherlands.
[2] Laboratoire PRiSM, Université de Versailles, 78035 Versailles, France.
[3] IRISA, Campus Universitaire de Beaulieu, 35042 Rennes, France.
[4] Department of Computer Science, Leiden University, P.O. Box 9512,
2300 RA Leiden, The Netherlands.
[5] INRIA, Rocquencourt, BP 105, 78153 Le Chesnay Cedex, France.
[6] Department of Computer Science, The University, Manchester M13 9PL, U.K.
[7] Department of Computer Science, The University, Edinburgh EH9 3JZ, U.K.

Abstract. This paper describes the recently funded ESPRIT project
OCEANS. Its aim is to investigate and develop advanced compiler infrastructure for embedded VLIW processors, such as the Philips TriMedia.
Such processors promise high performance at low unit cost. This paper
outlines the project's aims, presents the compiler infrastructure and its
application to a typical case study.

1 Introduction

Increasingly, general-purpose processors are used for embedded applications rather than customised hardware. As processor cost drops, it becomes more attractive to use one processor for several applications rather than designing specific
hardware. Multimedia based applications are typical of the growing uses of embedded systems, requiring cost-effective implementation and high performance.
Very Long Instruction Word (VLIW) processors are an attractive solution for
such applications as they provide potentially high performance, due to multiple
parallel functional units, and are relatively cheap to manufacture due to the
simple processor architecture. However, sophisticated optimizing compiler technology is necessary to exploit the fine-grain parallelism as assembly programming
of complex applications is not feasible. With current compiler technology, the
average number of operations per cycle in VLIW processors is only 2 to 2.5 [1].

---

* This research is supported by the ESPRIT IV reactive LTR project OCEANS, under
contract No. 22729.

The goal of the OCEANS project is to investigate and develop state-of-the-art compilation techniques to allow high performance implementations of embedded applications. In such applications, long compilation times can be afforded as each embedded processor will usually execute a limited number of applications throughout its lifetime. This makes it feasible to use more aggressive compilation techniques than previously considered. A brief presentation of the project is provided in this paper. In particular, the project objectives are highlighted in Section 2; Section 3 presents the structure of the compiler, that is, the front-end, the back-end, and their interaction. Finally, Section 4 examines the possible implications for the compiler when optimizing codes that are frequently used in embedded applications.

2 Project Objectives

Within the OCEANS project, we intend to meet the following objectives:

High-Level Optimizations Objectives: We aim to develop high-level restructuring transformations for the exploitation of VLIW processors. These transformations are primarily designed to enable successful later low-level exploitation of fine-grain parallelism. The strategy, or sequence of transformations, employed will be based on a cost model of the processor and will be guided by feedback from other stages.

Low-Level Optimizations Objectives: We intend to develop low-level restructuring techniques, concentrating on a highly retargetable object code scheduler that includes optimizing techniques suited for embedded applications and VLIW architectures. This is achieved through a multifunction testbed tool which can manipulate assembler code in order to implement low-level code restructuring as well as to provide the high-level code restructurer with information collected from the assembler code and from instruction profiling.

Integration Objectives: We intend to integrate the above into a prototype system based on iterative compilation, where a close interaction between the high and low-level exists, allowing better exploitation of available information. The validation and the evaluation of the efficiency of this approach will be carried out in close collaboration with the industrial validator, the compiler technology group at Philips Research. The main back-end target for this project is the Philips TriMedia (TM-1) processor [5]. The objective is to show that this approach yields more efficient code for this particular processor, eventually as optimal as hand-optimized code, while cutting down the code development time considerably.

An overall aim of this project is to achieve high retargetability of the code optimization process. The reason for this is that the cost of the development of compilers for embedded architectures must be amortized across variations of hardware implementations using the same instruction set architecture.

3 The OCEANS Compiler

The OCEANS compiler consists of a *front-end*, incorporating a high-level restructuring tool, MT1, and a *back-end*, incorporating a system for assembly

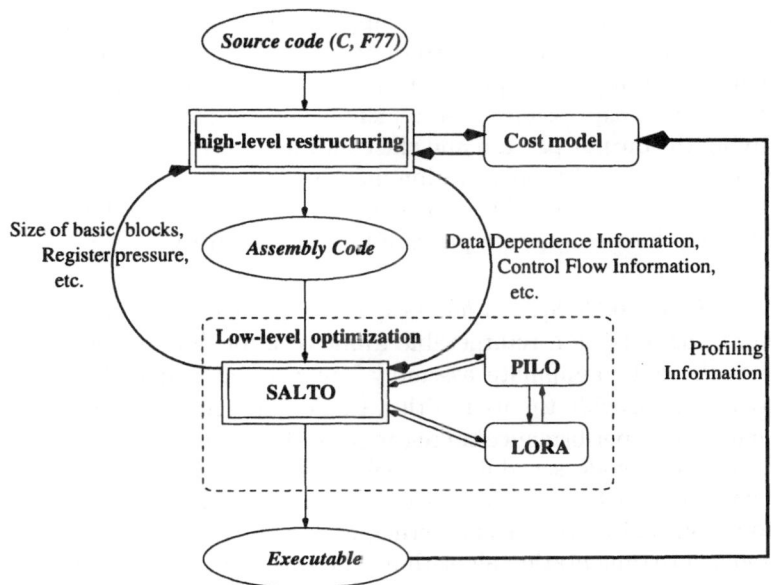

Fig. 1. The OCEANS compiler.

language transformation and optimization, Salto; the latter makes use of PiLo and LoRa, which are packages for software pipelining and loop register allocation, respectively. Their interaction is illustrated in Figure 1. It is intended that quantitative (e.g. number of instructions, slot occupancy, execution time) and qualitative information (e.g. pipelining failed due to register pressure) be used to guide the compilation process. By using an accurate cost model, the system can iterate until an acceptable level of performance is achieved.

The main back-end target of the OCEANS compiler is the Philips TriMedia (TM-1) processor [5], a state-of-the-art general purpose microprocessor, which has been enhanced to boost multimedia performance. At the heart of it, there is a 400 MB/s bus, which connects autonomous modules that include video-in, video-out, audio-in, audio-out, an MPEG variable length decoder, an image co-processor, a communications block and a VLIW processor. The VLIW processor includes a rich instruction set with many extensions for handling multimedia, and is capable of sustaining 5 RISC operations per clock cycle at 100 MHz. It contains 27 functional units which are pipelined ranging from 1 to 3 deep. The processor also includes 32 KB of instruction cache memory and 16KB of data cache memory.

The MT1 Restructuring Compiler: Over the past few years, a full Fortran 77 compiler, called MT1, has been developed at Leiden University [2]. An important aspect of the MT1 compiler is that it provides a facility for specifying program transformations that can be applied interactively. These are defined by an input pattern, an output pattern and a condition under which the transformation can be applied. Input patterns may contain meta-variables that are bound to program expressions or statements. These meta-variables may be used in specifying the output pattern and the condition. Moreover, functions that act directly on the internal representation may be defined and may be used in the output pattern and the condition. Conditions typically check for the existence of dependences between certain parts of the input pattern.

SALTO: A Retargetable System for Assembly Language Transformation and Optimization: Salto [7] is a retargetable framework for developing a whole spectrum of tools that manipulate assembly language programs. The objective of the system is to provide the user with a single environment that facilitates the implementation of performance tuning tools for low-level codes. This set of tools includes assembly code schedulers, profiling, and tracing tools. Salto is retargetable with respect to instruction sets and hardware details.

Salto consists of three parts; a kernel, a machine description file and an optimization or instrumentation algorithm. The kernel performs the parsing of the assembly code and of the machine description file, and the construction of the internal representation. The internal representation is then available via the user interface. The machine description file provides a model of hardware configuration and the complete description of the instruction set, including per-instruction resource reservation tables. The optimization or instrumentation algorithm is supplied by the user, via a user-supplied function `Salto_hook`.

The user interface of Salto is object-oriented and provides classes to represent a complete description of the control-flow graph of the program and a model of the target architecture.

PiLo and LoRa: PiLo and LoRa are packages for software pipelining and loop register allocation developed at INRIA Rocquencourt. PiLo has one heuristic mode based on the decomposed software pipelining algorithm [8], as well as one exact mode for code scheduling under register constraints based on an integer programming formulation. LoRa is a package that optimally allocates the loop variables into registers while controlling loop unrolling when necessary [6]. PiLo and Lora are connected to Salto via an interface describing architectural and dependency constraints among instructions.

Integration: In order to implement an iterative compilation process, MT1 constructs two output files for each input program. One consists of sequential TM-1 assembly code, to be scheduled by Salto, and the other consists of the result of program analysis by MT1, written in a format that can be read by Salto. This file includes high-level dependence information and information about loop structures. It also contains questions for Salto, such as "How does software pipelining perform on this loop?". After code scheduling by Salto and possibly profiling the resulting code, answers to these questions are used in the next compilation round and drive the selection of transformations and strategies to be applied.

4 A Case Study

In this section we present a case study of the optimizations a compiler can perform on multimedia applications. We consider four sets of benchmark programs, publically available on the Web. They consist of an MPEG2 encoder/decoder for converting uncompressed video frames into MPEG1/2 and *vice versa*, an MPEG1/2 player, an implementation of the CCITT G.711, G.721, and G.723 voice compression standards, and a very low bit-rate video encoder producing H.263 bitstreams. Profiling information has been obtained using gprof and further analysed using tcov. Although there are differences between the computationally most expensive functions of various programs, a regular pattern can be observed: typically, such functions contain double nested loops, having a rather small number of iterations, and embodying only a few statements mainly involving operations between array elements. For a more concrete example, consider the code fragment shown below taken from a function of the MPEG2 encoder in which approximately 67% of the program's execution time is spent.

```
for (j=0; j<h; j++)
{ for (i=0; i<16; i++)
  {   v = ( (unsigned int)(p1[i]+p1[i+1]+1)>>1 ) - p2[i];
      if (v >= 0)  s += v;  else  s -= v;
  }
  p1+= lx;  p2+= lx;
}
```

For typical RISC instruction sets, the innermost loop will require 14 instructions to execute one iteration. If we assume a latency of 3 cycles for load/store and a delay of 3 cycles for a jump, a naive sequential schedule would require 336 cycles to execute the entire i loop. Conversely, if we assume that a scheduler was able to utilise fully all five function units, without regard to resource and dependence constraints, and the latency of loads and jumps was masked, then the minimal execution time is 45 cycles. Thus, we have an upper and a lower bound on expected performance.

Modulo scheduling with an initiation interval $II = 5$ cycles generates code with prologue and epilogue costs of 5 cycles each, giving a total of 80 cycles for the inner loop. If more sophisticated scheduling with $II = 4$ is used, then the inner loop takes 68 cycles. The prologue and epilogue costs increase in this case to 6 and 10 cycles respectively. However, even this optimized schedule takes 50% longer to execute than our ideal lower bound. This is largely due to the prologue and epilogue overhead for small iteration counts. Unlike many scientific benchmarks, multimedia application codes are characterised by short inner loops, and therefore additional techniques are required to improve function unit utilisation.

Since scheduling with $II = 3$ is not possible due to dependence constraints, we cannot improve the performance by reducing II further and other techniques should be devised. Unroll and jam [3] is a technique which may be used to increase the size of inner loop bodies, thus reducing the loop overhead. In [4], a quantitative approach to the application of this technique is described showing improvements in most cases when applied to the Perfect Benchmarks. Unrolling the j loop once in this example and fusing (or jamming) the resulting two inner loops allows the new inner loop body to be scheduled with $II = 6$ and a prologue and an epilogue cost of 6 cycles each. Thus, two iterations of the outer loop take 96 cycles or 48 cycles for one iteration – just 6% longer than the lower bound. This has been achieved due to the greater freedom in scheduling more instructions and fewer jump instructions. This illustrates the importance of looking beyond simple pipelining of inner loops and applying high-level transformations when examining multimedia applications.

The above analysis, although encouraging, has not considered the impact of cache misses. Typical cache lines can hold 64 bytes or 16 words and therefore 2 new cache lines will be loaded on each iteration of the j loop. If we assume an 11 cycle delay per cache miss then the execution time for the inner loop will increase to 102, 90 or 70 cycles depending on the scheduling employed. In order that any gains from exploiting instruction level parallelism are not lost due to cache misses, it is necessary to prefetch the cache line towards the end of the execution of the inner loop. Thus, careful attention to prefetching is needed.

5 Conclusion

A brief overview of the OCEANS project has been presented. The main innovation of this project is the use of an iterative approach to compilation applying both high and low-level optimizations. These are guided by information gained from either level as well as previous compilation runs. A small example illustrated the need for both high-level transformations and low-level scheduling when optimizing typical multimedia codes, and indicated that VLIW architectures, given sufficient compiler support, are capable of delivering high performance.

References

1. G. Araujo *et al*. Challenges in Code Generation for Embedded Processors. In *Code Generation for Embedded Processors*. Kluwer Academic Publishers, pp. 49–64, 1995.
2. A. J. C. Bik, H. A. G. Wijshoff. MT1: A Prototype Restructuring Compiler. Technical Report 93-32, Department of Computer Science, Leiden University, Oct. 1993.
3. S. Carr, K. Kennedy. Improving the ratio of memory operation to floating-point operations in loops. *ACM ToPLaS*, 16(6), Nov. 1994, pp. 1768–1810.
4. S. Carr. Combining Optimizations for Cache and Instruction-Level Parallelism. *Proceedings of PACT'96*.
5. B. Case. Philips Hope to Displace DSPs with VLIW. *Microprocessor Report*, 8(16), 5 Dec. 1994, pp. 12–15. See also http://www.trimedia-philips.com/
6. C. Eisenbeis, S. Lelait, B. Marmol. The meeting graph: a new model for loop cyclic register allocation. *Proceedings of PACT'95*.
7. E. Rohou, F. Bodin, A. Seznec, G. Le Fol, F. Charot, F. Raimbault. SALTO: System for Assembly-Language Transformation and Optimization. Technical Report 1032, IRISA, June 1996. See also http://www.irisa.fr/caps
8. J. Wang, C. Eisenbeis, M. Jourdan, B. Su. Decomposed Software Pipelining: a New Perspective and a New Approach. *International Journal on Parallel Processing*, 22(3), 1994, pp. 357–379.

SEEDS — Simulation Environment for the Evaluation of Distributed Traffic Control Systems

Sebastiano Bottalico[1], Filippo de Stefani[1], Thomas Ludwig[2], and Günther Rackl[2]

[1] ALENIA
Via Tiburtina km 12.4, I-00131 Rome
email: sbot@lti.alenia.it
[2] Technische Universität München, Institut für Informatik
Lehrstuhl für Rechnertechnik und Rechnerorganisation
Arcisstr. 21, D-80333 München
email: {ludwig,rackl}@informatik.tu-muenchen.de

Abstract. The goal of the SEEDS project is to develop a simulation environment for the analysis and evaluation of distributed traffic control systems. It provides a general purpose tool kit for the simulation of applications like e.g. air and maritime traffic control systems. The environment will be composed of a cluster of workstations running the distributed simulator software. HPCN issues will be considered in order to set up a suitable HW/SW environment for the simulator.

1 Introduction

In the SEEDS project an open distributed interactive Simulation Environment (SE) for the analysis and evaluation of distributed traffic control systems is designed, set-up and tested. The SE provides a general purpose tool kit for the simulation of a wide range of industrial applications of strategic relevance, such as air and maritime traffic control systems.

In the frame of this project the application area chosen as case study is the ground traffic control in airports, that is the A-SMGCS (Advanced Surface Movement Guidance and Control System).

This system is devoted to support the safe, orderly and expeditious movement of aircrafts and vehicles on aerodromes under all circumstances and will be installed in airports or will replace/upgrade the existing ones to provide adequate capacity and safety in relation to the ever growing traffic density and bad weather conditions. The A-SMGCS has to be designed so as to facilitate its integration in existing systems in order that current investments are taking place in a cost-effective way.

Current research on air traffic management states that the limiting factors for the future of air traffic evolution are in principle the procedures for traffic management at airports, because new strategies are going to be applied for en-route traffic (ADS, free flight, etc.) which allow an increment of traffic efficiency.

Several international organisations, such as ICAO, are leading a standardisation effort for surface operations in airports through the FANS (Future Air Navigation System) Committee. With regard to Europe EU has started ECARDA (European Coherent Approach to Research and Technological Development for Air Traffic Management), in the frame of the Fourth Framework Programme; ECAC (European Civil Aviation Conference) has proposed the EATMS (European Air Traffic Management System) in the frame of the EATCHIP (European Air Traffic Control Harmonisation and Integration Programme), whose aim is the harmonisation among all the European air traffic services; EUROCONTROL is implementing EATCHIP and is going to put into effect EATMS.

The project is partially funded by the European Community in the Information Technology Programme: High Performance Computing and Networking task 6.2; the duration of the project is 30 months. The consortium, which is in charge of the project, is composed of Alenia (I), as co-ordinator, Sogitec (F), Rigel (B), as industrial partners, University of Siena (I) and Technical University of Munich (D), as associated partners, and Sicta (I) as final user-partner. Sicta is a consortium including as partner Enav (Italian Agency for Air Traffic Services). An European User Group (UG), composed of airport service and/or flight assistance Administrations, participates to all the phases of the project as associated partners of Sicta. Sogel working for Airport of Luxembourg and SEA (Società Esercizio Aeroportuale) of Milan Airport are members of the UG.

2 Simulation Environment

The SE is composed of powerful commercial workstations connected through a high performance LAN to cope with the application requirements. The use of HPCN is mandatory for this kind of application system because of the power capacity requested to simulate the distributed application environment where a team of human operators should co-operate. Heterogeneity of the HW and SW components of the SE should be faced; suitable programming paradigms have to be adopted to design and configure a portable, powerful, flexible and low-cost HPCN system in which should be possible to collect statistics and analyse performance, to generate 3-D high resolution scenarios and to organise distributed data-bases.

The basic functions supported by SE are traffic generator, actors' modelling and decision support tools. The traffic generator produces the scenario as seen by various actors using different set of sensors (eye, radar, camera, GPS, etc.) with the fidelity and accuracy of real world. Interactions which modify the predefined evolution of simulation are allowed. The actors' modelling simulates the behaviours of various actors at airport (pilots, controllers, drivers) and of the external world (airlines, airport managers, weather data, etc.): it is possible to mould new behaviours, to define new actors and to modify the stimuli of the external world . The decision support tools are introduced to help operators in the decision making process, which today relies only on the operator's mental capacity: many tasks, such as planning, re-planning, guidance can benefit of suggestions or decisions from automatic tools.

The SE does not reproduce a complete A-SMGCS but it is a friendly platform which can be used to introduce new values into traditional systems and is a support for various aspects of A-SMGCS design: change or redefinition of airport lay-out, definition of new procedures and their validation, simulation of new behaviours and interfaces. It is a validation environment for A-SMGCS defined by international organisations and it can be used for training purposes to allow traffic operators to learn new procedures and interfaces without interfere with the operativity of the real system.

In the future a complete A-SMGCS could be mapped onto the SE to test and validate a full system.

A test case suite is defined to verify the performance and to validate the SE at two typical airports: these test cases are potentially capable to represent the overall requirements, applications and operational realities.

3   SEEDS Functional Units

The SE is a distributed, modular, expandable, flexible and interactive HPCN architecture; it is composed of commercial off-the-shelf components and is open to be connected to other simulators.

The main functional units are:

- Scenario generation (SG): the sensors used at airports (radar, camera, eye, GPS, etc.) are modelled and the scenario is generated with the fidelity and accuracy of real world. The same scene can be viewed at the same time by different actors, using different set of sensors.
- Surveillance: a module to provide an accurate report and monitoring of all the movements within the aerodrome is produced to perform constraint checking in order to detect critical events and conflicts.
- Planning: tools to optimise taxi routes, traffic sequences and co-ordinate arrivals and departures are available to help operators in their routinely work. These tools are able to perform a re-planning to face critical situations or deviations.
- Guidance: this function provides all concerned operators with the methods of guiding aircraft and vehicles on the movement area from their current position to the intended destination in a fully automatic mode.
- Airport actors' modelling: the behaviours of the actors in an airport are simulated. They can interact with each other exchanging messages and commands with the possibility of modifying the evolution of the simulation session.
- Administration station: this subsystem configures the application and starts-up the simulation session, collecting performance measurements and statistics

Fig.1 shows a schema of the simulator environment as reported: it contains the main actors of the airport and the principal modules of the simulator.

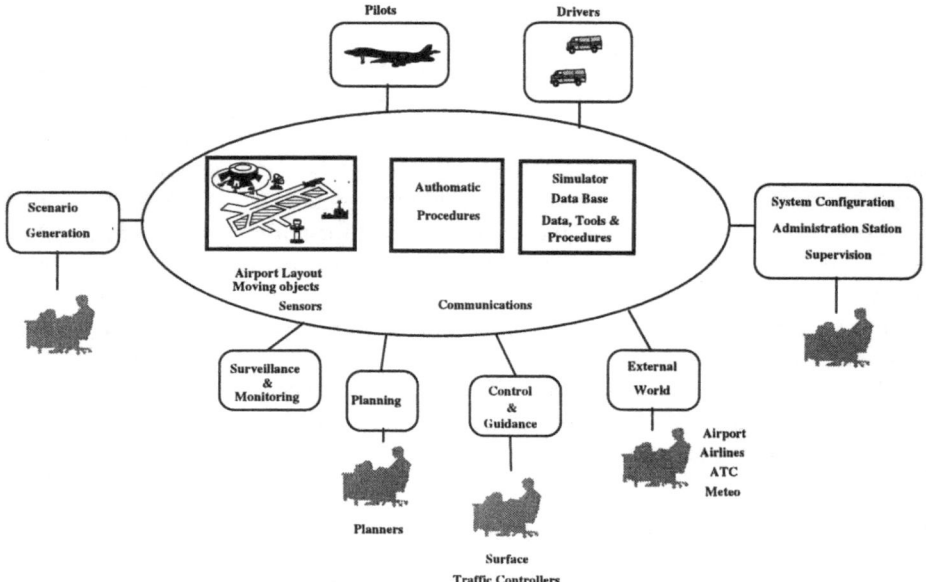

Fig. 1. Simulation environment: main actors and principal modules

4 Integration of HPCN Concepts

HPCN technology is needed to cope with the performance requirements of the application system. To realise a flexible and modular SE a network of powerful graphic workstations, on which the different actor models, scenario generation engine, and monitoring and supervision functions are mapped, is the most suitable architecture.

In defining such an SE architecture some requirements derived from industrial constraints should be taken into account: standard off-the-shelf platforms, that is commercial workstations with graphic features and largely diffused software tools, should be considered to obtain the following objective: to design and configure a portable, powerful, flexible and low-cost HPCN system in which it should be possible to collect statistics and analyse performance, to generate 3-D high resolution scenarios, and to organise a distributed data base. In particular the following HPCN related issues have to be investigated concerning their suitability for our project:

- environments for heterogeneous network computing: CORBA (Common Object Request Broker Architecture) and DCE (Distributed Computing Environment);
- distributed programming models like PVM (Parallel Virtual Machine) and MPI (Message Passing Interface);
- tools for debugging distributed programs;
- performance analysis tools for distributed applications;

- run time support tools for load and resource management;
- distributed data base systems;
- efficient network facilities such as FDDI, ATM, and Fast Ethernet.

In fact the composition of such a simulator environment has to be carefully planned as it comprises heterogeneous HW and SW components. A well balanced design of the SE should combine well known techniques from HPCN with standardised components suitable for commercial products. First a requirement analysis has to be performed to identify HW, SW, and performance related constraints of the SE. Especially performance issues like throughput, response time, and system load have to be considered carefully.

A HPCN oriented design of the SE will at first have to consider the degree of internal heterogeneity. Different workstation types with different operating systems will be used. An appropriate programming model for SW design and implementation has to be selected. High level programming paradigms like CORBA and DCE possibly handle problems arising from heterogeneity very well. However, ease of program development and performance might not fulfil the project specification. Medium level paradigms like PVM or MPI might be better suited for time critical applications. As a drawback commercial products can hardly be found here and programming is more difficult and error-prone. Possibly, a combination of high level programming platforms like CORBA with medium level programming models like PVM can provide a flexible environment for the management of the SE and fulfil the given performance requirements simultaneously.

Finally, the HW of the workstations and the interconnection network has to be selected such as to meet given requirements concerning e.g. graphics capabilties, computing power needed for planning tasks, and network throughput necessary for the scenario distribution.

All decisions will be made with respect to availability of HPCN program development and program maintenance tools. As said before, a large variety of commercial and public domain products for e.g. specification, debugging, performance analysis, or load and resource management, including tools from completed or ongoing Esprit projects has to be evaluated.

5 Conclusion and Future Work

The project started in January 1997 and has a duration of 30 months. Due to the schedule there are only few results to be reported yet. Up to now, specifications concerning SE functional requirements and A-SMGCS application requirements have been issued. Currently, these requirements are analysed in order to specify the hard- and software architecture of the SE subsequently. By December a deliverable is due that describes the architecture of the SE with respect to HPCN HW and SW. It lays the basis for the development of the simulator itself.

The main objectives of the SEEDS project are:

- to test scenario generation and interactivity between actors and SG;
- to evaluate the effectiveness of automatic supports (decision support tools, automatic agents, new HCI, etc.) to be used by different actors to reduce workload and to solve critical situations;
- to support analysis and design of new A-SMGCS;
- to validate new A-SMGCS (they can be mapped on SE);
- to experiment new procedures to be introduced at airports;
- to train operators to use new procedures, without interfere with the operativity of real system;
- to introduce HPCN methodologies in industrial applications;
- to evaluate HPCN tools and programming libraries;
- to propose high performance as well as low cost HPCN solutions for the software to be developed in order to occupy several market segments with our product.

The simulator being developed within the SEEDS project represents a prototype that shows how the listed objectives can be realised.

EFTOS: A Software Framework for More Dependable Embedded HPC Applications.

G. Deconinck[1], V. De Florio[1], R. Lauwereins[1], T. Varvarigou[2]

[1] K.U.Leuven, Dept. Elektrotechniek, Kard. Mercierlaan 94, B-3001 Leuven, Belgium
[2] N.T.U.A., Dept. Elect. Comp. Eng., I. Putechniou 9, GR-15733 Zographou, Greece

Abstract. Within the ESPRIT project EFTOS (Embedded Fault-Tolerant Supercomputing), a framework is developed to integrate fault tolerance flexibly and easily into distributed embedded HPC applications. This framework consists of a variety of reusable fault tolerance modules acting at different levels. The cost and performance overhead of generic Operating System and Hardware level fault tolerance mechanisms are avoided, while at the same time the burden of *ad hoc* fault tolerance programming is removed from the application developers. Integration of this functionality in real embedded applications validates this approach, and provides promising results.

Keywords. fault tolerance, embedded system, HPC

1   Introduction

Current industrial embedded applications require high computing performance that may be only provided by parallel computing systems. However they are prone to data-induced software faults and operate in industrial environments. Hence, these applications require fault tolerance to cope with the problems that will occur. Application developers tend to deal with these needs of availability and reliability of their applications by developing their own fault tolerance code as part of their application. This increases the cost of application development, maintenance and upgrade considerably.

Within the ESPRIT project EFTOS (Embedded Fault-Tolerant Supercomputing), several of these embedded applications have been investigated which require both high performance and fault tolerance [2]. We saw that different applications have many fault tolerance requirements in common, and incorporated similar ad hoc solutions. The goal of EFTOS is to streamline these requirements into a reusable fault tolerance framework that can be flexibly and easily integrated into the target applications. As such, application developers can enjoy the benefits of a variety of fault tolerance functions from which they can pick and choose what is necessary for their environment. The cost and performance overhead of generic Operating System and Hardware level fault tolerance mechanisms are avoided, while removing the burden of fault tolerance programming from the application developers at the same time.

Although the project aims at general embedded applications running on parallel systems, two representative applications have been analysed in more detail:

an image processing module in an automatic mail processing system [3] and a remote controller in a high voltage substation.

The distributed target system on which this framework is being implemented is a Parsytec *CC* System [1] which combines powerful processing nodes, I/O modules and routers into a parallel system (MIMD architecture). The parallel operating system, *EPX* (*Embedded Parallel extensions to uniX*), provides the functionality to operate the parallel environment via a message-passing API.

The EFTOS project is coordinated by Parsytec, who specialises in embedded HPC systems. AEG and ENEL represent the market of embedded applications. The knowledge on fault tolerance is supplied by K.U.Leuven, N.T.U.A. and D.L.R. The project started in April 1996 and runs for 2 years.

Several research projects investigated fault tolerance for embedded applications on distributed systems.

- The Delta-4 project (Esprit 818/2252) proposes a *generic architecture for dependable distributed computing*, based on multiplication of modules. Two variants of the architecture focus on portability and on performance of the approach. Both however rely on fail-silent hardware and on an atomic multicast protocol [6].
- The projects PDCS/PDCS2 (Esprit 3092/6362) consider *predictably dependable computing systems* aiming at making the design, development and production of dependable computing systems more predictable and cost effective (fault prevention, tolerance, removal and forecasting) [7]. Focus is on safety and security issues, as well as on quantitative assessments of dependability of the system. The MARS system (partly developed in this project) addresses embedded real time systems. It meets hard real time deadlines and tolerates interconnection and node faults, but requires dedicated hardware and a specific operating system (suited for time-triggered applications) [5].
- The FTMPS project (Esprit 6731) applies software fault tolerance solutions to number- crunching applications in massively parallel systems [4]. However, real-time aspects are not taken into account.

In fact, none of these projects considered standard hardware and operating systems for soft real time applications, to provide a powerful fault tolerance framework. It is clear however that there is an emerging need to provide such a framework to allow the application developer to incorporate *adaptable fault tolerance solutions into the embedded HPC applications*, based on standard hardware and software products. The availability of such a framework could decrease the current tendency of industrial application developers who frequently develop (similar) *ad hoc* solutions to improve the fault tolerance capabilities of their embedded applications.

2 Fault tolerance requirements

The errors that occur in systems in the scope of the EFTOS project can be summarised in broad terms in the following categories:

- software errors triggered by untested or unforeseen input values;
- hardware faults caused by electromagnetic interference;
- the propagation of errors through communication channels from one part of the system to the others;
- memory corruption by errors propagating from one process to another, as processes are running concurrently in the same memory space;
- the loss of a number of subsequent inputs due to a faulty input item;
- meeting time constraints and deadlines.

	detection	isolation and reconfiguration	recovery	associated mechanisms
thread level	watchdogs	termination of faulty thread	re-initialisation of a thread	exception handling
		re-mapping of threads	restoring saved state in application	checkpointing, atomic actions
memory level	memory access checking	freeing memory resources	restoring memory	stable memory
node level	I'm alive	disconnection of a faulty node	re-initialisation or fast rebooting of a node	
		reconfiguration of process graph / communication topology	integration of a recovered node into the system	
system level	monitoring system parameters		fast reboot fast restart	
message level	communication time-out	discard message	retransmission	reliable communication
	checking of message ordering		re-ordering	
link level	checking of communication channel/partner status before communication	termination of a faulty communication channel	re-initialisation of a communication channel	

Table 1. Fault tolerance requirements.

Based on this categorisation we have developed fault tolerance modules that deal with the different types of errors. Application developers were asked to prioritise these modules according to the relevance to the needs of their application, their impact on fault tolerance and the feasibility to integrate them into the target application. This lead to the following orthogonal classification of requirements, according to the *location* where fault tolerance is required (processing and networking modules), according to the *steps* to achieve fault tolerance (detection, isolation and recovery mechanisms). Table 1 shows the list of the set of fault tolerance functionality that our library contains and the corresponding classification. We also added associated mechanisms.

3 EFTOS framework with reusable software solutions

Based on these requirements, the EFTOS project is providing a framework for fault tolerance, elaborated on a Parsytec CC system, which acts at different levels. (This results in the architecture of figure 1.)

- At the lowest level, it contains *elements* for error detection (D tools) and for error recovery (R tools). These are parametrisable functions that can be used in stand-alone functionality, or in combination with the next levels to apply fault tolerance to processing or communication modules.

- At the middle level, the *DIR net* (detection, isolation and recovery network) combines these elements to tolerate faults. It ensures consistent decisions in the distributed system and serves as a backbone to pass information among the fault tolerance elements.
- At the highest level, these elements can be combined into mechanisms, such as to provide fault-tolerant communication or voting on results.

The **D tools** are meant to detect errors. They are dynamically started by the user during the execution of the application. As such, the programmer can supply them with the correct parameters. When an error is detected, the D tool passes the necessary information via a standardised interface to the DIR net. This includes the type of error that occurred, the location where the error is detected and the D tool that detected it. Examples of D tools that are being integrated include:

- watchdog timers for communication and computation, detecting if a message is delivered in time or if a task produces its results before a certain deadline;
- assertions that check invariant relationships in values of variables;
- support to detect wrong memory accesses, e.g. to unused or non-existing memory or write access to read-only areas;
- monitoring of system and environmental parameters to detect deviations from the normal behaviour;
- trap handlers to catch signals;
- application-specific D tools, etc.

The **R tools** provide recovery mechanisms. They are started by the DIR net and have to bring the application back into a consistent state. Possibly the system will have less functionality than before (graceful degradation). They include the following:

- restarting a single node or a set of nodes;
- disabling communication channels temporarily or permanently; thereafter they have to be reset, reinitialised, and brought back into a consistent state. New communication channels must be established if a communication channel was completely removed or if the reconfiguration steps require another application topology;

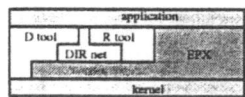

Fig. 1. EFTOS architecture. The application makes use of the message-passing environment EPX, running on top of the operating system kernel. The elements for detection (D tools) and recovery (R tools) interact with the application and are interconnected via the DIR net.

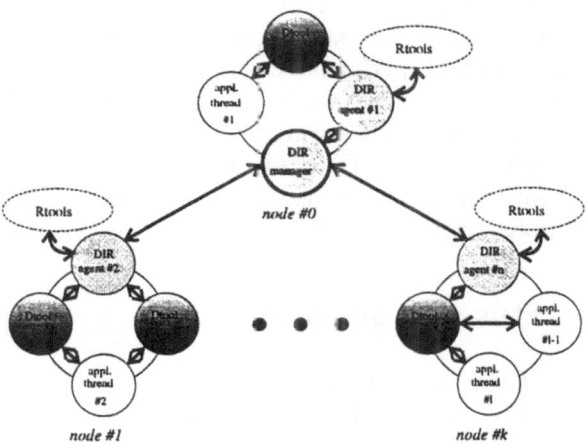

Fig. 2. Architecture of the DIR net.

- (virtually) disconnecting unusable threads or nodes from the rest of the application; releasing memory and other resources, that were assigned to threads or links that failed or had to be removed;
- validate output before releasing it, or selecting the correct answer from a set to mask faults;
- application-specific or user-specified recovery actions, etc.

The detection-isolation-recovery net, in short **DIR net**, is the backbone of the fault tolerance framework. It provides an interface to which all D tools and R tools should conform, while leaving enough room for flexibility within the tools. The DIR net allows distributed actions to take place. Therefore, it consists of a hierarchical network with a central manager and several distributed agents. This DIR net structure is shown in figure 2.

- The main module in the DIR net is the manager, which keeps a global view of the system. This view includes the status of each node used in the partition, the type and location of the D tools, the type and location of errors that occurred, and the status of R tools that are being executed. The manager also has the possibility to connect to an operator module, which forms a two-way interface between the operator and the DIR net to perform manual recovery actions.
- The DIR manager is assisted by multiple DIR agents on the different nodes, that do the field work. These agents start up and initialise D tools. They are warned when an error occurs, and forward this information to the manager. These agents take local recovery actions, or for coordinated actions, they start R tools upon request of the DIR net manager. Note that different agents are not interconnected, but can only communicate via the manager.

A hierarchical/centralised structure of the DIR net has been implemented so far. We are also evaluating a distributed (peer to peer) architecture for the DIR net to avoid the single point failure problems of the centralised version. However, performance reasons make a distributed version of DIR net quite unattractive, especially in the area of high performance applications that we are considering. Besides, the DIR net is a lean piece of software that has little probability of failure compared to the data-dependent software faults affecting the applications, which makes failures introduced by the DIR net quite unlikely. Hence, we are leaning towards building support mechanisms for the hierarchical DIR net structure (backups, duplications, message checking, heartbeat mechanisms, etc.).

4 Current status and outlook

A first prototype of this EFTOS framework has been implemented. It comprises basic elements of each of the fault tolerance steps. In a test application, this resulted in successful experiences when software fault injection was applied. While the implementation work continues on different levels, the developed basic elements, structures, and techniques will be tested, integrated into the target applications. These industrial application developers will validate the framework and provide feedback to the developers.

As such, the developed framework will become a flexible, and standardised set of solutions to improve the dependability of a large set of embedded applications at a low cost.

Acknowledgements. This project has partly been supported by ESPRIT project 21012 EFTOS, COF/96/11 and an FWO Krediet aan Navorsers. Geert Deconinck has a grant from IWT. Rudy Lauwereins is research associate of the FWO - Flanders.

5 References

[1] Anon., "Parsytec CC Series", Parsytec GmbH, Aachen (D), Jul. 1995.
[2] Anon., "EFTOS: Embedded Fault-Tolerant Supercomputing", Technical Annex ESPRIT- project 21012, Feb. 1996.
[3] G. Deconinck et al., "Fault Tolerance Requirements in Postal Automation: a Case Study", 4th *IFAC Alg. and Arch. for Real-Time Control*, Vilamoura (P), Apr. 1997.
[4] G. Deconinck et al., "Fault Tolerance in Massively Parallel Systems", *Transputer Communications*, Vol. 2(4), Dec. 1994, pp. 241–257.
[5] H. Kopetz et al., "Distributed Fault-Tolerant Real-Time Systems: The Mars Approach", *IEEE Micro*, Feb. 1989, pp. 25–40.
[6] D. Powell (Ed.), "Delta-4: A Generic Architecture for Dependable Distributed Computing", *Springer-Verlag*, Berlin Heidelberg New York, 1991.
[7] B. Randell, J.-C. Laprie, H. Kopetz, B. Littlewood (Eds.), "ESPRIT Basic Research Series: Predictably Dependable Computing Systems" *Springer-Verlag*, Berlin Heidelberg New York, 1995.

STAMPAR: A Parallel Processing Approach for the Explicit Dynamic Analysis of Sheet Stamping Problems

L. Neamţu[1], F. Zarate[1], E. Oñate[1], G.A. Duffett[2] and J.M. Cela[3]

[1] International Center for Numerical Methods in Engineering (CIMNE), Edificio C-1 UPC, Gran Capitán s/n, 08034, Barcelona, Spain
[2] Quantech ATZ S.A., Edificio NEXUS, Jordi Girona Salgado s/n, 08034 Barcelona, Spain.
[3] Center for Parallel Computing in Barcelona (CEPBA), Edificio D-6 UPC, Gran Capitán s/n, 08034, Barcelona, Spain.

Abstract. Industrial sheet stamping problems generally require enormous amounts of computer resource for their solution within an acceptable time. Explicit dynamic algorithms are commonly used for these analyses since they reduce the memory requirements but they still require large CPU times and parallel processing offers the only realistic alternative. This paper describes the parallel program Stampar used for the solution of industrial sheet stamping problems. The methodologies used for the parallelisation are described as well as an example with timings that show the advantage of this approach.

1 Introduction

The simulation of industrial sheet stamping problems requires the combination of many numerical methods in order to achieve an acceptable result. Industrial problems tend to be geometrically complex and the large deformation, large strain, material plasticity, friction and contact properties of the problem must be considered for even the simplest simulations. Due to the complexity of the analysis the use of explicit dynamic technology has many advantages.

Explicit dynamic programs are more efficient, require less memory and contain simpler algorithms than their implicit counterparts but the analyses require an enormous number of time steps, due in part to the conditional stability of the methods. The natural solution is to turn to parallel processing. The parallelisation of explicit dynamic software is certainly not new [1, 2] but the tools available for parallelisation have improved dramatically and computer networking is ubiquitous; this makes parallel software a viable tool for many industries to utilise.

The work described in this paper relates to the Stampar software program developed by the authors [3] which utilises an explicit dynamic algorithm for the analysis of sheet stamping problems. The program has been developed to run on a parallel processing platform envisaged to be a heterogeneous network of computers including PC's, workstations and/or distributed memory machines.

The methodologies employed in the parallelisation exploit the advantages of explict dynamic techniques and split the finite element mesh into a number of domains based on the mesh nodes rather than the elements. Various domain decomposition methods are available although the strip methods appear to work best for this type of problem.

To obtain maximum performance from Stampar all aspects of the code have been addressed: the computation of the critical time step (computed as a global minimum over all the elements), the triangular shell or BST elements (that utilise a surrounding patch of elements to compute the bending tensor in terms of displacement variables only) and the contact algorithms (which require searching for contact zones between distinct bodies). The final system leads to a simple structure that contains a single data synchronisation and communication point with further communications overlapping the element computations.

The effectiveness of the parallel program Stampar is shown by means of an example from the most recent NUMISHEET conference [4]. This example shows that almost linear speed-up can be achieved with an efficiency of about 80%.

2   Numerical Methodology

The dynamic equilibrium equations are written in terms of the principle of virtual work using an updated Lagrangian formulation. The discretized form of equilibrium is obtained in the standard manner [5]:

$$\mathbf{M}\ddot{\mathbf{r}}^t + \mathbf{C}\dot{\mathbf{r}}^t = \mathbf{R}^t - \mathbf{F}^t - \mathbf{F}_c^t \qquad (1)$$

where $\mathbf{M}$ is the mass matrix, $\mathbf{C}$ the damping matrix, $\ddot{\mathbf{r}}$ the vector of nodal accelerations, $\dot{\mathbf{r}}$ the vector of nodal velocities, $\mathbf{R}$ the vector of external loading, $\mathbf{F}$ the vector of internal forces and $\mathbf{F}_c$ the vector of contact forces, all evaluated at time t. The displacement vector is also determined and denoted by $\mathbf{r}$.

The explicit dynamic numerical procedure used to integrate the equations of motion (1) from time step n to time step $n+1$ is as follows:

1. At time step n the vectors $\mathbf{r}^n$, $\dot{\mathbf{r}}^{n-1/2}$, $\mathbf{R}^n$, $\mathbf{F}^n$, $\mathbf{F}_c^n$ are known at the nodes and the strain and stress tensors, ϵ^n and σ^n respectively, are known for each element.

2. Compute the accelerations at all nodes:
   $\ddot{\mathbf{r}}^n = \mathbf{M}_D^{-1}[\mathbf{R} - \mathbf{F} - \mathbf{F}_c - \mathbf{C}\dot{\mathbf{r}}]^n$ where $\mathbf{M}_D = diag\,\mathbf{M}$ and $\mathbf{C} = 2\alpha\mathbf{M}$

3. Compute velocities and displacements at all nodes:
   $\dot{\mathbf{r}}^{n+1/2} = \dot{\mathbf{r}}^{n-1/2} + \ddot{\mathbf{r}}^n \Delta t^{n+1/2}$, $\mathbf{r}^{n+1} = \mathbf{r}^n + \dot{\mathbf{r}}^{n+1/2} \Delta t^{n+1}$
   where $\Delta t^{n+1/2} = \frac{1}{2}[\Delta t^n + \Delta t^{n+1}]$

4. Compute strains and stresses for all elements: ϵ^{n+1}, σ^{n+1}

5. Compute internal force vector at all nodes of each element and assemble:
   $\mathbf{F}^{n+1} = \int_V \mathbf{B}^T \sigma^{n+1} dV$

6. Check for frictional contact conditions and compute contact force vector $\mathbf{F}_c$

7. Compute external force vector $\mathbf{R}^{n+1}$

8. GOTO step 1 and repeat the process for the next time step

Conditional stability of the explicit time integration requires that the time step size must not exceed a critical time step size related to the lowest period of vibration. Using an approximation to the critical time step size for each element and finding the smallest of these allows a critical time step size to be computed at each step (or at pre-determined intervals) which therefore provides an automatic time-stepping procedure.

3 Parallel Processing

Detailed profiling of the sequential explicit dynamic software [6] on industrial size problems revealed that approximately 90-95% of the CPU time was used in the calculation of the internal nodal force vector F; element strains, stresses and internal forces are computed here.

The advantages of parallel processing for explicit dynamic programs are thus obvious and have been acknowledged for many years [1, 2]: no system of equations needs to be solved and hence almost all the CPU time is spent at the element level. This means that huge speed-ups can be attained by decomposing the mesh into domains and carrying out the computations in parallel.

3.1 The Stampar System

The basic philosophy of the parallel system is that each processor carries out exactly the same function but on a different domain (there are no master-slave relationships). This is effectively a SPMD (Single Program Multiple Data) structure. To do this effectively and with as much transparency to the user as possible it is necessary to carry out the domain decomposition and analysis steps as automatically as possible.

The use of standard coding and PVM communication tools ensures that Stampar can run on a heterogeneous network of computers that may include PC's, workstations and/or shared memory machines.

3.2 Domain Decomposition

Domain decomposition methods have proved to be effective for the solution of explicit dynamics applications since they exploit the coarse grained parallelism and are thus suitable for Distributed Memory Multiprocessors (DMM). Effective domain decomposition ensures load balancing and the minimisation of communications.

The finite element domain is decomposed within Stampar in terms of nodes that are assigned to the various processors [7]. Various domain decomposition methods are utilised: strip partitioning, recursive bisection and multilevel partitioning. However, for most of the industrial sheet stamping applications considered a 1.5D strip partitioning method seems to be most effective. This method consists of 2 steps: in the first step the domain is divided into strips with perfect load balancing. In the second step each strip is divided into boxes using an orthogonal direction to the first one.

3.3 Parallel Computations

With the domain decomposed in this nodal manner all element computations may be carried out independently on each processor. At the end of these computations each processor contains contibutions from all of its nodes to the global vectors. For the internal nodes these vectors are complete and the solution may be readily obtained. However, for the domain interface nodes an exchange of information is necessary so that each processor may assemble the nodal information for the interface nodes within its domain. This is a data communication point when all processors communicate (send and receive) data relating to their external and local interface nodes. Now each processor contains all the necessary information for all the nodes in its domain to carry out the force/mass vector divide and compute the acceleration at each node (see step 2 in section 2).

3.4 Special Parallel Considerations

All parts of a program need to be addressed in order to effectively parallelise any software (see Amdahl's law [8]). Aspects that contain an interdependency between processors result in processor waiting, bad sychronisation and reduced performance. The aspects falling into this category that have been identified and addressed in Stampar are:

(1) The computation of the critical time step that involves the computation of a global minimum over all the elements.

(2) BST elements have been developed that include membrane and bending deformation effects defined only in terms of displacement degrees of freedom [9]; each element requires information from the surrounding patch of elements.

(3) The computation of the contact force vector F_c involves contact search algorithms to locate the zones of contact between distinct bodies. In sheet stamping many surfaces come into contact so this is a very critical part of the Stampar code.

4 Numerical Example: Deep Drawing of a S-rail

The Stampar program has been utilised to analyse the deep drawing of a steel S-rail defined in the NUMISHEET'96 conference [4]. This problem simulates the stamping and elastic springback of the rail (wrinkling and springback cause many manufacturing problems). The complete mesh for the S-rail stamping in Figure 1 shows, from top to bottom, the die, the sheet (12,000 elements using Hill's plasticity model [10]), blankholders and punch (note that the stamping direction is upwards); the total mesh contains 14,654 elements with 27,879 degrees-of-freedom.

The final stamped results agree very favourably with those presented in NUMISHEET'96 (see [4]): the front view of the deformed shape after springback is shown in Figure 2 where the buckling of the surfaces is clearly seen (a comparison with experimental results indicated the good agreement that was obtained).

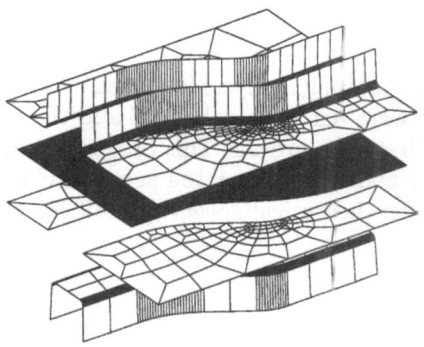

Fig. 1. S-rail initial set-up from top to bottom: dies, sheet, blankholders and punch.

Fig. 2. Deformed shape of S-rail after springback.

The stamping and springback of the S-rail was run using Stampar on a network containing various numbers of SG workstations and using the 1.5D strip domain decomposition method. The speed-up results are shown in Figure 3 where it can be seen that superior speed-ups were achieved when the contact communications were overlapped with the element computations: the speed-up for less than 12 processors is almost linear with about 83% efficiency (with more processors this reduces to about 76% since communications start to affect the overall performance). This deviation point would, in general, be different for different problems. When no overlapping was used the speed-up reduced to about 62% due again to the communications affecting the performance of the system in general. The speed-up curve is smoother however with no obvious deviation point.

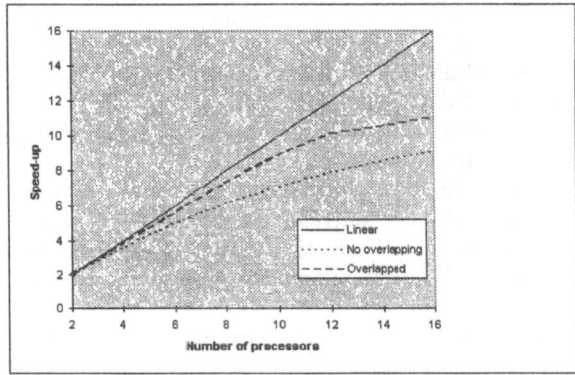

Fig. 3. Speed-up for S-rail stamping problem.

5 Conclusions

Stampar, a parallel software for sheet stamping analyses has been presented. The accuracy and effectiveness of the software has been validated and the efficiency when running on a parallel platform has been proved. This program, running on parallel platforms gives speed ups of about 80% over the sequential version of the software. This is approximately the theoretical speed-up limit for programs of this nature.

Acknowledgements

The authors wish to acknowledge the support of other members of the software development and support teams within CIMNE and CEPBA and the European Community for their support grant through the ESPRIT PCI project STAMPAR.

References

1. Duffett, G.A., Alves Filho J.S.R., Honnor, M.E. and Owen D.R.J. Transient shell analysis using transputer based parallel processing techniques. FEMSA'89 Conference Proceedings, University of Stellenbosch, 1989.
2. Oñate, E. Possibilities of parallel computing in the finite element analysis of industrial forming processes. VECPAR'96 International Meeting on Vector and Parallel Processing, Porto, 1996.
3. Rojek, J., Jovicevic, J. and Oñate, E. Industrial applications of sheet stamping simulation using new finite element models. *J. Materials Processing Technology*, **60**, 243–247, 1996.
4. Lee, J.K., Kinsel, G.L. and Wagoner, R.H. (eds.) NUMISHEET'96 Numerical Simulation of 3D Sheet Metal Forming Processes - Verification of Simulations with Experiments, Dearborn, Michigan, USA, 1996.
5. Zienkiewicz, O.C. and Taylor, R.L. The finite element method, 4th Edition. McGrawHill, Vol. I, 1990, Vol. II, 1991.
6. STAMPACK User Manual: An explicit dynamic finite element package for sheet stamping anlaysis. CIMNE, Barcelona, 1996.
7. CAESAR User Manual: A domain decomposition tool for parallel processing applications. CEPBA, Barcelona, 1996.
8. Amdahl, G.M. Validity of the single-processor approach to achieving large scale computing abilities. AFIPS Conference Proceedings, **30**, 483–485, 1967.
9. Oñate, E. and Cervera, M. Derivation of thin plate bending elements with one degree of freedom per node: a simple three node triangle. *Engng. Comput.* **10**, 543–561, 1993.
10. Hill, R. Theoretical plasticity of textured aggregates. *Math. Proc. Camb. Phil. Soc.* **85**, 179, 1979.

Index of Authors

Lecture Notes in Computer Science

For information about Vols. 1–1211

please contact your bookseller or Springer-Verlag

Vol. 1249: W. McCune (Ed.), Automated Deduction – CADE-14. Proceedings, 1997. XIV, 462 pages. 1997. (Subseries LNAI).

Vol. 1250: A. Olivé, J.A. Pastor (Eds.), Advanced Information Systems Engineering. Proceedings, 1997. XI, 451 pages. 1997.

Vol. 1251: K. Hardy, J. Briggs (Eds.), Reliable Software Technologies – Ada-Europe '97. Proceedings, 1997. VIII, 293 pages. 1997.

Vol. 1252: B. ter Haar Romeny, L. Florack, J. Koenderink, M. Viergever (Eds.), Scale-Space Theory in Computer Vision. Proceedings, 1997. IX, 365 pages. 1997.

Vol. 1253: G. Bilardi, A. Ferreira, R. Lüling, J. Rolim (Eds.), Solving Irregularly Structured Problems in Parallel. Proceedings, 1997. X, 287 pages. 1997.

Vol. 1254: O. Grumberg (Ed.), Computer Aided Verification. Proceedings, 1997. XI, 486 pages. 1997.

Vol. 1255: T. Mora, H. Mattson (Eds.), Applied Algebra, Algebraic Algorithms and Error-Correcting Codes. Proceedings, 1997. X, 353 pages. 1997.

Vol. 1256: P. Degano, R. Gorrieri, A. Marchetti-Spaccamela (Eds.), Automata, Languages and Programming. Proceedings, 1997. XVI, 862 pages. 1997.

Vol. 1258: D. van Dalen, M. Bezem (Eds.), Computer Science Logic. Proceedings, 1996. VIII, 473 pages. 1997.

Vol. 1259: T. Higuchi, M. Iwata, W. Liu (Eds.), Evolvable Systems: From Biology to Hardware. Proceedings, 1996. XI, 484 pages. 1997.

Vol. 1260: D. Raymond, D. Wood, S. Yu (Eds.), Automata Implementation. Proceedings, 1996. VIII, 189 pages. 1997.

Vol. 1261: J. Mycielski, G. Rozenberg, A. Salomaa (Eds.), Structures in Logic and Computer Science. X, 371 pages. 1997.

Vol. 1262: M. Scholl, A. Voisard (Eds.), Advances in Spatial Databases. Proceedings, 1997. XI, 379 pages. 1997.

Vol. 1263: J. Komorowski, J. Zytkow (Eds.), Principles of Data Mining and Knowledge Discovery. Proceedings, 1997. IX, 397 pages. 1997. (Subseries LNAI).

Vol. 1264: A. Apostolico, J. Hein (Eds.), Combinatorial Pattern Matching. Proceedings, 1997. VIII, 277 pages. 1997.

Vol. 1265: J. Dix, U. Furbach, A. Nerode (Eds.), Logic Programming and Nonmonotonic Reasoning. Proceedings, 1997. X, 453 pages. 1997. (Subseries LNAI).

Vol. 1266: D.B. Leake, E. Plaza (Eds.), Case-Based Reasoning Research and Development. Proceedings, 1997. XIII, 648 pages. 1997 (Subseries LNAI).

Vol. 1267: E. Biham (Ed.), Fast Software Encryption. Proceedings, 1997. VIII, 289 pages. 1997.

Vol. 1268: W. Kluge (Ed.), Implementation of Functional Languages. Proceedings, 1996. XI, 284 pages. 1997.

Vol. 1269: J. Rolim (Ed.), Randomization and Approximation Techniques in Computer Science. Proceedings, 1997. VIII, 227 pages. 1997.

Vol. 1270: V. Varadharajan, J. Pieprzyk, Y. Mu (Eds.), Information Security and Privacy. Proceedings, 1997. XI, 337 pages. 1997.

Vol. 1271: C. Small, P. Douglas, R. Johnson, P. King, N. Martin (Eds.), Advances in Databases. Proceedings, 1997. XI, 233 pages. 1997.

Vol. 1272: F. Dehne, A. Rau-Chaplin, J.-R. Sack, R. Tamassia (Eds.), Algorithms and Data Structures. Proceedings, 1997. X, 476 pages. 1997.

Vol. 1273: P. Antsaklis, W. Kohn, A. Nerode, S. Sastry (Eds.), Hybrid Systems IV. X, 405 pages. 1997.

Vol. 1274: T. Masuda, Y. Masunaga, M. Tsukamoto (Eds.), Worldwide Computing and Its Applications. Proceedings, 1997. XVI, 443 pages. 1997.

Vol. 1275: E.L. Gunter, A. Felty (Eds.), Theorem Proving in Higher Order Logics. Proceedings, 1997. VIII, 339 pages. 1997.

Vol. 1276: T. Jiang, D.T. Lee (Eds.), Computing and Combinatorics. Proceedings, 1997. XI, 522 pages. 1997.

Vol. 1277: V. Malyshkin (Ed.), Parallel Computing Technologies. Proceedings, 1997. XII, 455 pages. 1997.

Vol. 1278: R. Hofestädt, T. Lengauer, M. Löffler, D. Schomburg (Eds.), Bioinformatics. Proceedings, 1996. XI, 222 pages. 1997.

Vol. 1279: B. S. Chlebus, L. Czaja (Eds.), Fundamentals of Computation Theory. Proceedings, 1997. XI, 475 pages. 1997.

Vol. 1280: X. Liu, P. Cohen, M. Berthold (Eds.), Advances in Intelligent Data Analysis. Proceedings, 1997. XII, 621 pages. 1997.

Vol. 1281: M. Abadi, T. Ito (Eds.), Theoretical Aspects of Computer Software. Proceedings, 1997. XI, 639 pages. 1997.

Vol. 1282: D. Garlan, D. Le Métayer (Eds.), Coordination Languages and Models. Proceedings, 1997. X, 435 pages. 1997.

Vol. 1283: M. Müller-Olm, Modular Compiler Verification. XV, 250 pages. 1997.

Vol. 1284: R. Burkard, G. Woeginger (Eds.), Algorithms — ESA '97. Proceedings, 1997. XI, 515 pages. 1997.

Vol. 1285: X. Jao, J.-H. Kim, T. Furuhashi (Eds.), Simulated Evolution and Learning. Proceedings, 1996. VIII, 231 pages. 1997. (Subseries LNAI).

Vol. 1286: C. Zhang, D. Lukose (Eds.), Multi-Agent Systems. Proceedings, 1996. VII, 195 pages. 1997. (Subseries LNAI).

Vol. 1289: G. Gottlob, A. Leitsch, D. Mundici (Eds.), Computational Logic and Proof Theory. Proceedings, 1997. VIII, 348 pages. 1997.

Vol. 1292: H. Glaser, P. Hartel, H. Kuchen (Eds.), Programming Languages: Implementations, Logigs, and Programs. Proceedings, 1997. XI, 425 pages. 1997.

Vol. 1294: B.S. Kaliski Jr. (Ed.), Advances in Cryptology — CRYPTO '97. Proceedings, 1997. XII, 539 pages. 1997.

Vol. 1299: M.T. Pazienza (Ed.), Information Extraction. Proceedings, 1997. IX, 213 pages. 1997. (Subseries LNAI).

Vol. 1300: C. Lengauer, M. Griebl, S. Gorlatch (Eds.), Euro-Par'97 Parallel Processing. Proceedings, 1997. XXX, 1379 pages. 1997.